MW00846309

# Cartography.

Kenneth Field

*A compendium of design thinking for mapmakers*

Esri Press, 380 New York Street, Redlands, California 92373-8100


Printed in the United States of America

22 21 20 19 18    2 3 4 5 6 7 8 9 10

Library of Congress Control Number 2017964197



Ask for Esri Press titles at your local bookstore or order by calling 800-447-9778, or shop online at esri.com/esripress. Outside the United States, contact your local Esri distributor or shop online at eurospanbookstore.com/esri.

Esri Press titles are distributed to the trade by the following:

*In North America:*
Ingram Publisher Services
Toll-free telephone: 800-648-3104
Toll-free fax: 800-838-1149
E-mail: customerservice@ingrampublisherservices.com

*In the United Kingdom, Europe, Middle East and Africa, Asia, and Australia:*
Eurospan Group
3 Henrietta Street
London WC2E 8LU
United Kingdom
Telephone: 44(0) 1767 604972
Fax: 44(0) 1767 601640
E-mail: eurospan@turpin-distribution.com

For my dad,
Dr. Dennis Field
1943–1986

# Contents

# Foreword

**Ken walked into the cartography lab at Oxford Polytechnic in late 1989 as a kid with long hair who liked maps.** He sat at the back, quietly, and began listening to what I and my colleagues were teaching: cartography. His first piece of work showed hints of a raw talent, and I recall telling him his cased lines were not quite consistent enough. It was details like this he spent the next few years working on (though he didn't remain quiet for long). I particularly saw how much he helped others in the group lift their game, and that's what marked him out.

**Ken always worked hard and played hard, he was vocal in his thoughts and opinions, and he worked on the student cartography society setting up visits to organisations and suchlike to widen the group's horizons.** He graduated in 1991, and I'm at least partly to blame for his destiny as I sent him the advertisement for the position he eventually got—as a postgraduate research assistant. He likely doesn't know I made a call to his future employers to tell them what I thought of this young man and his ability but it was unnecessary. He interviewed well and got the job, and he was on his way.

**Over the years I've watched his career progress as he was quickly promoted to the position of lecturer and as he started teaching students of his own.** His early career was marked by a huge transition from photomechanical production to desktop mapping and GIS. He had to develop new skills rapidly, as had many cartographers who went before. The pace of change became ever more rapid, yet the core ideas and beliefs (and that appetite for making high-quality maps) were something he continued to impress on his own students, many of whom picked up the baton and went into the industry themselves. Ken moved to Kingston University where he led the GIS programmes and brought cartography back to the core of the geographical sciences. He became involved in many professional bodies and served them in different capacities: for example, as editor of *The Cartographic Journal* for nine years. He was also heavily involved in the International Cartographic Association, with which I have also served extensively. Thus student became teacher.

To Ken –
If you read and understand a
of this you will have no trou
Best of luck!
Roger Ans

Roger's message to Ken inside *Basic Cartography for Students and Technicians*, Volume 2 by Roger Anson

**When Ken left academia to join Esri® in the US, I pondered the move.** In some respects it seemed a waste that his talent was lost to students in the UK but I understood his thinking. The challenge of something new and different and the opportunity to teach beyond academia, albeit in a different way, was enticing. The platform to capture the imagination of new mapmakers who had never gone through an academic route spoke to his sense of trying to help others, just as he had done with his peers at university. I did ask him why he felt he needed to go to America, and his response was simple. 'Have you seen some of their maps?' he said in quick-fire response. 'They clearly need help'. It was said tongue in cheek, and his dry wit has often got him into trouble, but the sentiment came as no surprise to me. He'd seen an opportunity to share what he knows of cartography to a new audience. Rather than sit back and take the easy route, he forced himself into a new life and took on a new challenge—one in which I am delighted to see him thriving.

**It therefore came as little surprise when he told me he was writing a book.** He used my books as a student, and he claims to still have them on his bookshelf and to use them. But changing times need fresh approaches, and although there are many books about the practice of cartography, this book sets itself apart. It's not a traditional textbook. It's a wonderful collection of all that Ken has learnt and the thinking he's developed along his cartographic journey so far. It's an insight into his mind, the search for a better way to make maps, smarter maps, and—more than anything—his drive to help others succeed. It's an honour to have been asked to write the foreword for his book, and typical of Ken for doing things differently. Many would have gone for a big name for the foreword, and I know they would have been equally honoured to write it. But the fact he went back to his mentor speaks volumes. The cycle of being taught, learning, and teaching endures, and is the very essence of how a discipline is shaped. It reflects his ethos of building on what you can learn from the past and using it as a strong foundation.

**It's clear from these pages that Ken continued learning from many other sources as well as me.** I'm glad to see that he did read. He also understood. More than anything, he has carried a torch for cartography across many shifts in its style, technology, and practice. His advice blends a mix of the old and the new, the academic, and the practical, the professional, and the amateur, and should be required reading for students of cartography and, frankly, anyone who wants to make a better version of their map. He promises to buy me a pint of bitter next time he's in Oxford. We've shared a good number of them! I look forward to the next, and I'll hold him to his promise.

**Roger Anson**
Formerly Head of Cartography, Oxford Brookes University
Past president, the British Cartographic Society

Oxford, UK
January 2018

# Preface

## ...or why I wrote this book.

**For most of my career to date, friends and colleagues have often asked why I haven't written a book on cartography.** My reply was always the same—there are perfectly good ones available already. My bookshelf is full of them. But you'll likely not know about them unless you come to cartography via an academic background. So I decided to write a book precisely for those of you who don't have a bookshelf full of them to help develop your knowledge, understanding, and proficiency.

**Making maps has progressed from filling empty spaces with mythical creatures to trying to unravel the complexities in data to present meaning with clarity.** This book sets out to demystify cartography and promote the idea that thinking is key. Maps are rich visual soups of many forms of graphical signage. Approaching mapmaking by thinking about what you want your map to say, how to build something meaningful from visual ingredients, how people read the graphical signage, and what emotions you want to spark is the magic needed to make a better map.

**The process of making maps has up until recently been largely undertaken by professionals (cartographers) but mapmaking has changed.** Data and the tools to make maps have become ubiquitous and so many more people are making maps. The map itself is no longer just a piece of paper designed by a cartographer; but perhaps a slick graphic on a web page built by a coder. Regardless of your background, maps must still be designed with intent. Good, well-thought-out maps do not just emerge as a finished product, and although all maps are designed, some are designed better than others. Cartography as a discipline and profession has much to offer, and learning a little of how a cartographer thinks about mapmaking helps everyone make better maps.

**Technology has made the mapmaking task fast, simple, and reproducible but thinking what the technology is actually doing helps you make a better map.** It is almost incomprehensible to understand how maps were made even 20 years ago. Automation has played a huge role in design and production but in some ways it may have led to a lack of appreciation of what goes into making a good map. Making a map fast does not necessarily lead to a great map.

I have a few basic cartographic tenets:

1. Do it right. There's rarely a short cut.
2. When in doubt, follow what's gone before. Best practice has been developed by many people over many decades and serves you well.
3. Always go beyond the defaults. Defaults are inevitable and can be a good starting point but knowing something of cartography helps you adapt them to your needs.
4. Quality trumps quantity. People like maps but they are drawn to great maps. Quality attracts, inspires, sells, and invites a second look.
5. Study great maps. Learn from those maps held in high esteem so you can appreciate the value of them and why they work so well.
6. Give and take critique. Learning how to assess cartographic quality helps you recognise greatness and do more of it. Learn to accept constructive criticism. It's vital.
7. Break the rules when necessary. If you know the rules, you'll know when you can break them, by how much, and to what effect.
8. Recognise cartographic lineage. Not much is genuinely new. That's OK. Learn from it, use it, and cite it without trying to claim it.
9. Accept that the map is rarely finished. You have to learn when you're done. There will always be tweaks but the old adage of 90% taking 10% of the time and the last 10% taking 90% of the time is often true.
10. Own your work. Put your name on it. Share it and get eyes on it. Build a portfolio.
11. Take it to 11. Making maps can make you incredibly frustrated at times but put your heart into your work and learn to enjoy the ride. The end result is worth it.

As with any undertaking this book is the work of way more than just me. I grew up with maps. My father was a geologist. The maps were bizarre and beautiful. My first geography teacher, Mrs Burston, inspired me. At University I was fortunate enough to learn from the best—Roger Anson, Mike Childs, and many others. I taught for 20 years in universities and learnt from all my students (some of whom hopefully learnt a little from me). I've met some truly great practicing and academic cartographers, many of whom have become close friends.

I am fortunate to work with phenomenal people and am hugely indebted to the friendship, advice, and talents of Wesley Jones and John Nelson, who have helped me shape this book beyond what I envisaged. Two truly talented (c)artists. I'm very grateful to those who contributed words or images, have given permission to reproduce their work, or who have read and commented on draft manuscripts. My sincere thanks go to Esri for supporting this project, particularly Clint Brown and the Esri Press team. I'm eternally grateful for not having my words spun into American English! Many other Esri colleagues have helped practically or by giving me the space to work. Thank you, all.

Finally, to my family and, most of all, my partner, Linda Beale. Without support and encouragement you cannot work on a project of this magnitude, and hers is the most unwavering. Without Linda, there would be no book. Thank you. I love you.

That said, it's all my fault, and if you find any errors they are mine and mine alone!

*'...for there is nothing either good or bad, but thinking makes it so.'*

—Shakespeare

**Cartography doesn't need to be hard and, more than anything, this book is about encouraging thought.** There's plenty of what might be called *rules*, but these are just guidelines for cartography developed from decades of practice and people working out what works and why. Maps should be objective and have scientific rigor but there's plenty of scope for creativity. Any design-led field sits at the intersection of science and art, and learning some of the rules means you'll know when best to break them.

**Put simply, this book lays out what I consider to be manners for the modern mapmaker.** It's a guide to holding your cutlery the right way. No elbows on the table, sit upright, shoulders back, and don't speak until you're spoken to. Without etiquette and discipline in any pursuit, we suffer mayhem, impoliteness, and bad behaviour, and there's plenty of 'bad maps' that evidence this plight. If we know something of the etiquette of cartography, we can more easily identify uncouth maps and try and incorporate better manners in our own work. This is to cartography what the *Haynes Manual* is to car mechanics or *Larousse Gastronomique* is to culinary knowledge. It's a sage, a companion, a guide, a friend, and a compendium of essential information. It's not a lesson in how to make a map 'right'. There is no such thing as a right map or a wrong map. But it will help you make a better map and become a smarter mapmaker by learning how to think a little like a cartographer and give you the confidence to go beyond software defaults or template cartography.

**A lot of thought and experience has gone into making this book.** It encapsulates the wisdom of many people who have taught me and from whom I have learnt. What I have tried to achieve is a translation of cartography from a specialist domain to one that builds a bridge between cartographer and mapmaker. I've tried to make the subject practical and valuable, not only as a reminder to professionals but as a companion to all who need to make a great map. We've all been beginners somewhere along our journey, and we're all amateurs at some things. As a cartographic professional, I hope this supports people in their own cartographic journeys.

# How to use this book

## The purpose and structure of the book.

**This book is structured to mirror the mapmaking process itself—it's non-linear!** Rarely would you make a map in an order that goes from projections, then symbols, then colour, then typography, and then layout—with one practical component applied after the other. Making a map is a process that requires you to think simultaneously about the interplay of the concepts and decisions that underpin your map. Yet most books on cartography are written linearly. Being able to dive into the book to learn a little about a specific aspect is, I feel, more relevant and useful than trying to create an artificial grouping of topics. Sure, some topics are more related than others but I hope this fresh approach to the subject gives you a way of accessing what you need to know, rapidly.

**The book is organised alphabetically, and individual cartographic topics are given their own double-page spreads.** On the left side of each spread you see the title at the top, followed by a short summary line that succinctly describes the scope of the spread. The remainder of the left side of the spread is, in the main, devoted to words. The left, wider column of writing presents the core essence of the topic in about 300 words or less. That was the task in writing—to subdivide cartography into topics of around the same size. Each paragraph leads with an opening phrase, set in bold.

**The main writing is supported by a narrower column to the right.** This column of text is presented in a smaller point size. This styling is deliberate. Maps contain visual structure to enable you ro read them. I wanted to visually structure the words to support the idea that the right column contains detail that is supportive but not core. You'll find interesting historical facts, detail about key people, or simply a little more detail to illuminate what you've read in the left column.

**Every topic has one spread.** No more, no less. But you can find connected topics using the 'see also' links below topics. These links alert you to the spreads that are most closely related.

**See also:** Anatomy of a map | Cartographic process | Defining map design | Defining maps and cartography | Globes | Graphicacy | Types of maps

Colour is used throughout the book to group related concepts. The title of each spread is presented in the colour that, conceptually, it belongs to. These colours are intended to provide a subtle (as in not visually overbearing) way to make connections between topics and also provide a way to navigate the book.

The colour theme is carried through to a small stroke on the left edge of the page. This use of colour provides a way to perhaps peruse the book and immediately see topics that are related. You'll also see colour used for the 'see also' links at the foot of the page. So, for example, this page is coloured as if it belonged to the Foundations category with relevant links to related topics.

Given the book is alphabetically structured I wanted to ensure there were clear breaks between each letter. To achieve this, I invited colleagues to contribute a short description about a classic map. Each of the 25 maps has its own spread with the short description to the left. These words add different voices to the book and provide a way to enjoy other ideas and views. I might have selected different maps but that's the point—this isn't a list of my top 25 maps. It's a collated set from the wider community that act as examples from which to explore and learn.

The righthand page of each spread is reserved for illustrations. I've deliberately avoided a consistent stylistic treatment so each page is different. There are original diagrams and maps made by myself, Wes, and John. There's also external content interspersed. Some classic and familiar examples. Some new and unseen. I've tried to make each spread unique and offer a visually enticing look at the subject through different visual lenses. This is a book I hope is as visually interesting as much as it is useful.

**Opposite (left):** book category information.
**Opposite (right):** some of what cartography is about and what a cartographer does.

Foundations

The background stuff. Fundamentals.

Visual grammar

Using graphics to encode meaning.

Graphical design

Crafting and arranging marks on a map.

Maths for mapmakers

The mathematical frameworks of a map.

Working with data

Understanding data and preparing it for mapping.

Colour

The meaning and use of colour.

Typography

The design and placement of lettering.

Map types

Explanation of different categories of map.

Mapping features

Constructing maps for physical and human features.

Mapping themes

Constructing maps for thematic data.

Composition

Layout and organisation of maps for page and screen.

Exemplar maps

Classic map examples with short descriptions.

# Topics (alphabetical)

# Topics (thematic)

MAP TYPES



MAPPING FEATURES



MAPPING THEMES



COMPOSITION



EXEMPLAR MAPS

# Authors/cartographers

## Kenneth Field (words, maps, illustrations, and design)

A self-confessed 'cartonerd' with a personal and professional passion for mapping. Ken gained his BSc in Cartography at Oxford Polytechnic and PhD in GIS at Leicester University and fell into academia. He spent 20 years in key positions in UK universities before moving to California to join Esri in 2011. He has presented and published widely. He blogs (cartonerd.com), tweets (@kennethfield), is past Editor of *The Cartographic Journal* (2005–2014), and Chair of the ICA Map Design Commission (2011–2019). He co-founded the *Journal of Maps,* is on the advisory board of the *International Journal of Cartography*, is a Fellow of both the British Cartographic Society and Royal Geographic Society, is a Chartered Geographer (GIS), and only the second Honorary Member of the New Zealand Cartographic Society.

Ken is in love with maps, makes maps, collects maps, writes about maps, and is devoted to encouraging others to see the value in quality cartography and helping them make better maps. He attempts to push the boundaries of cartography, and he has received numerous awards for his maps and writing (and kitchen tile designs!). He's in demand as a panellist and keynote speaker and has curated and judged numerous map galleries. Despite evidence to the contrary, life's not all about maps, and Ken can also be found on a snowboard, behind a drum kit, or supporting his hometown football team, Nottingham Forest.

You can view his maps at carto.maps.arcgis.com.

## Wesley Jones (maps and illustrations)

Born in Calgary, Canada, Wes studied geography and history at the University of Calgary. He then trekked across country to specialize in cartography at the Centre of Geographic Sciences in Nova Scotia. In 2009 he moved to California, came to Esri, and has been there ever since.

Wes has a lovely wife and two beautiful children. Besides mapping, he enjoys playing sports, being outside, drawing, and telling stories (whether it be bedtime tales or comic book creations). He can be found on twitter as @wesleytjones.

## John Nelson (maps and illustrations)

Cartographer and user experience designer in Lansing, Michigan. John studied geography and art at Central Michigan University and has a Master's degree in Geographic Science. John works with Esri, and his work has been featured in texts and magazines, and has received the occasional internet blip.

John is a frequent conference speaker on visualization. He blogs at AdventuresInMapping.com and tweets as @John_M_Nelson. When he's not doing these things he's chasing toddlers, wrangling chickens, and generally loving life.

# Contributors

**Linda Beale**
Lead Product Engineer at Esri and Associate Researcher at Imperial College London. Research interests in spatial epidemiology and author of the *Health Atlas of England and Wales*.

**Alberto Cairo**
Knight Chair in Visual Journalism at the University of Miami, and author of the books *The Functional Art* and *The Truthful Art*.

**Sébastien Caquard**
Associate Professor at Concordia University (Montréal). His research lies at the intersection between mapping, technologies, and the humanities.

**William Cartwright, AM**
Professor at RMIT University with research interests in multimedia cartography and geovisualization. Former President of ICA and recognised as Member of the Order of Australia (AM) for services to cartography .

**James Cheshire**
Geographer whose award-winning maps draw from his research as Senior Lecturer at University College London. Coauthor of the books *London: The Information Capital* and *Where the Animals Go*.

**Steve Chilton**
Published extensively in his role as Chair of the Society of Cartographers (coediting *Cartography: A Reader*). He also coedited *OpenStreetMap: Using and Enhancing the Free Map of the World*.

**Jeremy Crampton**
Professor of geography at the University of Kentucky. He has interests in critical cartography and the social history of spatial Big Data.

**Danny Dorling**
Professor of geography at the University of Oxford. He grew up in Oxford. His work concerns issues of housing, health, employment, education, inequality, and poverty—and how all these can be mapped.

**David Fairbairn**
After tasting 'real world' cartographic production with Philips in London, ITC-trained and ICA-inspired, David has spent several decades at Newcastle University promoting cartography to surveyors and geomaticians.

**Martin Gamache**
Runs Art of the Mappable, a custom cartography studio. He has been a cartographer for *National Geographic Magazine* for nearly 10 years and is now Director of Cartography at *National Geographic*.

**John Grimwade**
Graduate of the Canterbury College of Art. John has served as graphics director of *The Times* of London and Condé Nast publications. He has taught widely and runs johngrimwade.com.

**Henrik Hargitai**
Planetary geomorphologist and media historian. He is a Postdoctoral Fellow at NASA Ames Research Center. He studies planetary cartography, fluvial geomorphology, and planetary nomenclature.

**Benjamin Hennig**
Geographer and accidental cartographer passionate about maps. He currently works at the University of Iceland and is also an Honorary Research Fellow at the University of Oxford.

**Nigel Holmes**
Joined *Time* magazine in 1978, eventually becoming graphics director. Despite academic criticism, he's committed to using pictures and humour to help people understand abstract numbers and scientific concepts.

**Daniel Huffman**
Freelance cartographer and an Honorary Fellow at the University of Wisconsin–Madison. You can find more of his thoughts and work at somethingaboutmaps.com.

**Lorenz Hurni**
Professor at the Institute of Cartography and Geoinformation of ETH Zurich. He is the Editor-in-Chief of the *Atlas of Switzerland* (national atlas) and the *Swiss World Atlas* (school atlas).

**Menno-Jan Kraak**
Professor in geovisual analytics and cartography at University of Twente. Author of numerous books on cartography and President of the International Cartographic Association, 2015–2019.

**Mark Monmonier**
Distinguished Professor of Geography at Syracuse University. Author of numerous books, he was elected to the Urban and Regional Information Systems Association's GIS Hall of Fame in 2016.

**Antoni Moore**
Associate Professor in geographical information science at University of Otago. Research interests in cartography and art and Secretary to the New Zealand Cartographic Society.

**Ed Parsons**
Geospatial Technologist of Google®. Ed was the first CTO of Ordnance Survey and has held positions at Autodesk® and Kingston University London where he also holds an honorary doctorate.

**Tom Patterson**
Cartographer with the US National Park Service. He has developed several widely used open-source tools and bases for cartographers, including Natural Earth, and runs shadedrelief.com.

**Robert Roth**
Associate Professor at the University of Wisconsin–Madison and the Faculty Director of the University of Wisconsin Cartography Laboratory. Rob's work focuses on interactive, online, and mobile map design.

**Waldo Tobler**
Professor Emeritus at the University of California in Santa Barbara. Pioneering geographer and computational cartographer, recipient of the Lifetime Achievement in GIS award. Waldo died February 20, 2018, at age 87.

**Corné van Elzakker**
Assistant Professor in the Department of Geo-Information Processing, University of Twente. Corné is past Chair of the ICA Commission Use & User Issues.

# Cover art

## Shouldn't a book on cartography have a map on the cover?

**Angela Andorrer's art adorns the cover of this book.** I first met Angela at the 26th International Cartographic Conference in Dresden, Germany, in 2013. She had set up a small studio and invited cartographers to have their palms painted as maps. Part of the process involved the participants writing a short extract that described the imaginary landscape. Her work is striking, beautiful, and demonstrates some of the pure art of cartography. Here, I share my own handscape, and Angela explains a little of the process and meaning behind her work. I am thankful to Angela for agreeing to share her art for this book's cover. This book represents many people's ideas on the principles and practice of cartography. We are all the product of the work and influence of other experts, and I can think of no better way to demonstrate that than by having cartographers' handscapes leave their imprint on us all.

**My handscape: this is no ordinary landscape.** It's not of this Earth. A small, hitherto unnamed planet lies somewhere distant in our solar system, captured only fleetingly by a distant probe whose signal has travelled light-years to our gaze. The image is blurred yet its surface seems characterized by dramatic, rugged, rocky terrain resplendent in the natural colours of rich mineral deposits yet which has a dark side. Brilliant oranges, yellows, and greens shine out amongst what appear as harsh, unforgiving crevices that litter the landscape. Where the rich bedrock pushes through the surface we see signs of possible life ... linear patterns that perhaps signal that we are not alone; patterns that may only be made by something ... some creature, rather than natural processes ... paths, settlements ... amongst the dry, barren choking plains whose beauty belies the truth. This planet is devoid of water. The atmosphere is a choking stench; a cocktail of noxious gases. Where the dark takes over, the apparent beauty subsides. What exists here? What lives here? These are terrifying places; unmapped and devoid of detail for no mapmaker dares to tread in such a place. No probe can uncover the truth. The surface falls away into the abyss; dark, lonely places where one can only imagine what exists. These places never see sunlight. For here, genuinely, be dragons.

Author Ken Field's handscape.

**Kenneth Field**

**The body is the oldest sign carrier.** Every hand is marked by hills, valleys, paths, and streams, as well as signs, patterns, and types—personal topographies which intensify into landscapes and life stories. I develop new cartographies out of these naturally occuring hand-landscapes: handscapes. I am concerned by how the traditional western landscape painting and cartography embody the trained gaze upon unknown territories. I question the idea of boundaries, hegemony, and ownership, and in doing so I encounter tensions between the politics of the body and painting, and between territories of art.

My artistic media are the colours and the intimate surface of the skin. For a short moment, the process of becoming a stranger to one's own hand is startling. The palm of another becomes a canvas and a map; the starting point of an imaginary journey. The relationship between closeness and distance, between intimate physical history and artistic appropriation, begin to shift.

Many of the cartographers walked around with their painted hand raised and were curious about the visual topographies that would emerge from their colleagues' hands. The core idea was for the cartographers to identify and name typical landform patterns on the canvases of their bodies in a self-reflective process: that they should map their own bodies.

*'We are the real countries, not the boundaries drawn on maps with the names of powerful men.'*

—Michael Ondaatje: *The English Patient*, 1992

My travels on the skin of hands are comforting to me. I find peace and beauty in those landscapes. For a short time there, I can avoid the conflicts that are happening around the world. Instead of foreign countries, we should travel more often to the closeness of the body. However, the existence of the new country and the new painting is brief, as the host body begins to break it down rapidly with corporal sweat. Before the landscape decays it is captured through photography—a snapshot and portrait of the hand owner's individuality. Then the landscape is washed and dissolves.

**Angela Andorrer** (andorrer.de)

# Whose hands?

Friends, colleagues, cartographers.

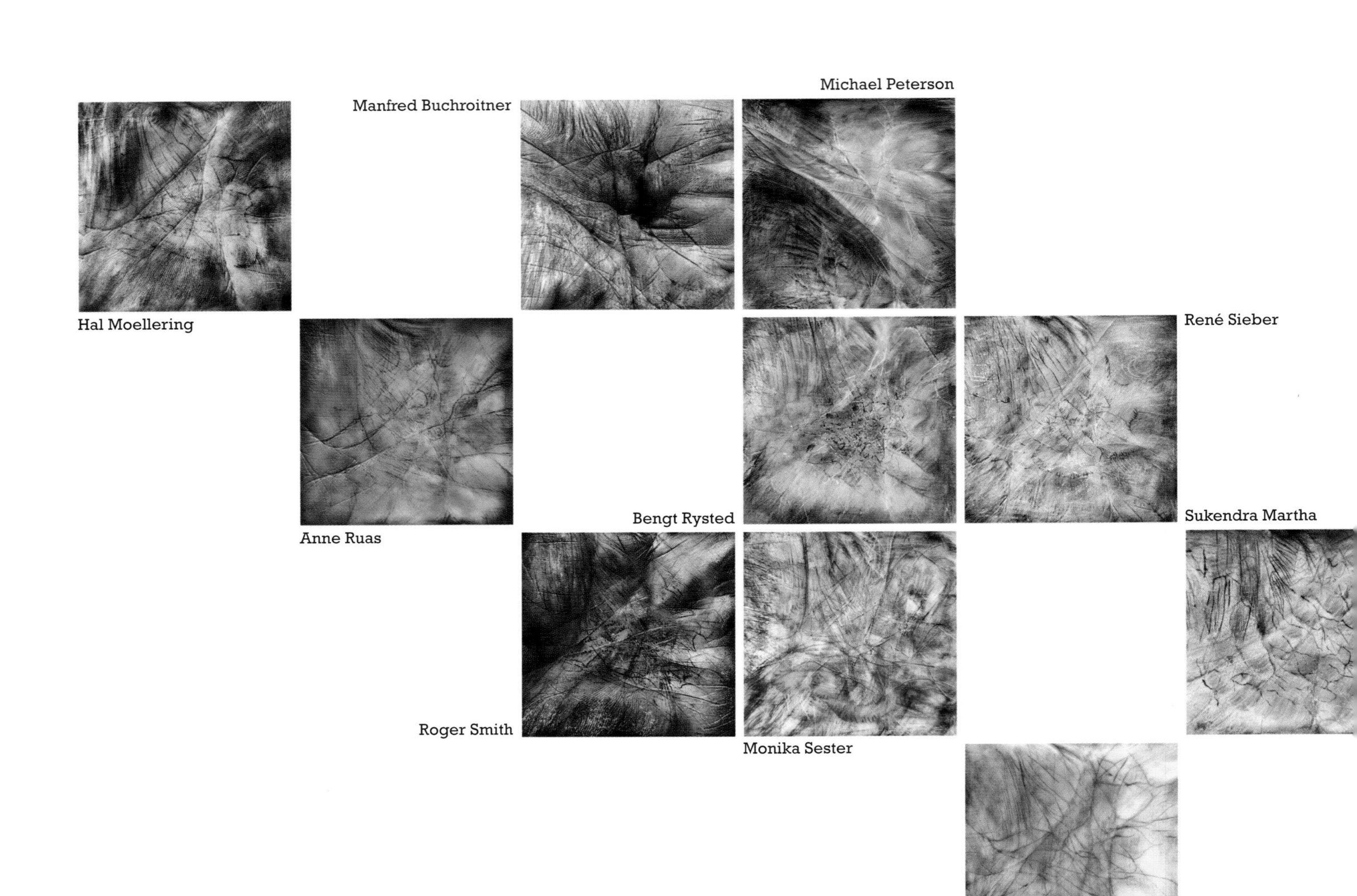

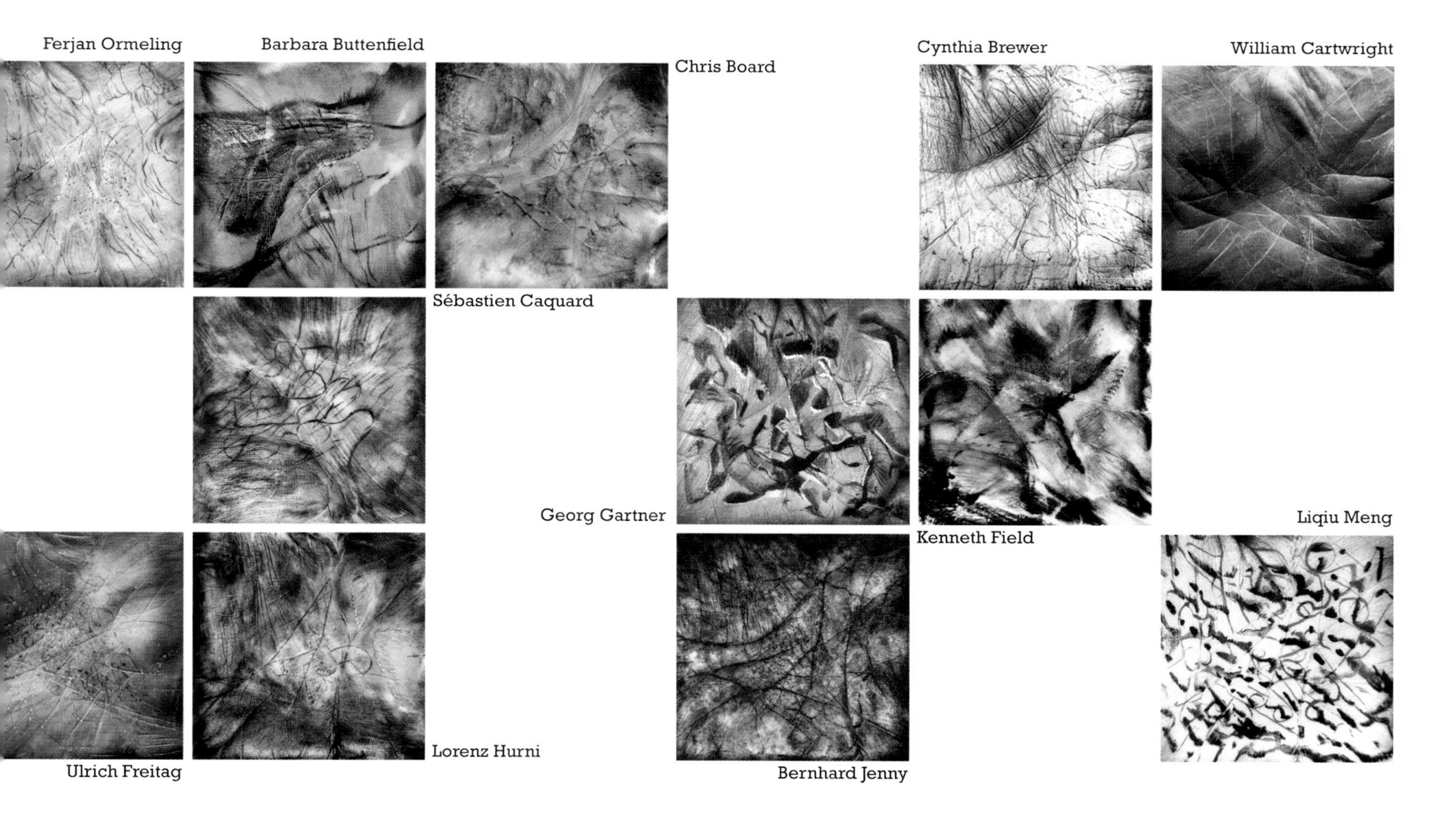
Ferjan Ormeling
Barbara Buttenfield
Chris Board
Cynthia Brewer
William Cartwright
Sébastien Caquard
Georg Gartner
Kenneth Field
Liqiu Meng
Lorenz Hurni
Ulrich Freitag
Bernhard Jenny

# Cartography.

# Abstraction and signage

**All maps are the result of abstraction and the use of signage to represent phenomena.**

**Because the world around us is a complex one, it would be virtually impossible to simply place a small version of it on a map.** There would not be the space to adequately represent all features that exist in the mapped area even in a reduced form. Consequently, all maps are abstractions of reality and are used to display a selection of objects and attributes. All maps are inherently a reduction of reality and so the amount of information you can put on a map will be a reduced form of that reality. This means a map will omit information to a greater or lesser extent depending on scale and purpose. But more than reduction (through selective omission), the features that are mapped are subject to a range of additional processes, such as classification and simplification, that make it easier to understand the true spatial patterns and relationships that exist in reality.

**The way in which we represent features and their attributes is through the design and placement of graphical signs.** These signs do not necessarily take on the appearance of the object in reality but are used to represent the object. The signs should have meaning to enable the map reader to interpret them accurately and appropriately relate them to the real-world object. We refer to the process of encoding meaning into the map as symbolisation. At its simplest level, part of the job of the mapmaker is to design and place symbols that reflect either location or some characteristic of the data. There is considerable scope in the design of symbols and every mark on a map can be considered a symbol in one form or another, from those that represent points, lines, and areas to the typographic components and the marginalia or contextual information.

**Clarity and purpose follow from a careful consideration of abstraction and signage.** Beyond the science of choosing and representing the information, the sophistication of the intended audience, scale, and conditions of use must also be considered so you end up with a clear and concise map that your map reader can easily translate into meaning.

**See also:** Dynamic visual variables | Literal comparisons | Pictograms | Varying symbols

Whereas the principal task of map design is in the decisions taken to select, omit, and symbolise phenomena, there are many processes at work that are difficult to accommodate. For instance, induction occurs when the mapmaker builds signage that depends on some level of inference between mapped features. In this sense, they are applying inductive generalisation to extend the map's content without physically adding more symbols. The use of contour lines is a good example of the process of induction since, depending on the distance between contour lines and whether they get progressively closer or more distant, inference is made about the nature of the landscape. A concave ridge will  be flatter toward the summit and steeper toward its base. Contours will be closer together at the base and further apart at the summit to represent and infer this.

Although certain inferences might be imbued into the map by the mapmaker, the map reader will inevitably play a role in interpretation. Someone with a good sense of map use and reading might find it easy to see and understand such inferences but, for others, the inferences may be harder to see. For this reason it is best to avoid inductive generalisation where possible and, instead, solve your communication dilemmas through good, clean design with more obvious visual cues.

Hudson River
Riverside Park
Central Park
MTA-96th St-1-2 & 3 Lines
MTA-96th St-B & C Lines
97th St
E 97th St
Mount Sinai Medical Center
MTA-103rd St-4 & 6 Lines
MTA-110th St & 6 Lines
MTA-86th St-1 Line
W 86th St
MTA-86th St-B & C Lines
MTA-96th St-6 Line
85th St
MTA-79th St-1 Line
MTA-Museum of Natural History
Riverside Park
E 92nd St
MTA-86th St-4, 5, 6 Lines
79th St
E 79th St
MTA-72nd St-1-2 & 3 Lines
W 72nd St
MTA-72nd St-B & C Lines
The Lake
Henry Hudson Pkwy
Midtown-Edgewater
MTA-66th St-Lincoln Center
MTA-77th St-6 Line
Central Park
Carl Schurz
Columbus Ave
Amsterdam Ave
Central Park W
Park Ave
2nd Ave
1st Ave
3rd Ave
5th Ave
W 57th St
W 56th St
MTA-59th St-Columbus Circle
MTA-57th St-N-R-Q & W Lines
MTA-7th Ave-E Line
MTA-50th St-C & E Lines
MTA-49th St
MTA-47th-50th-Rockefeller Center
9th Ave
8th Ave
10th Ave
11th Ave
Broadway
MTA-42nd St-Port Athty Bus Terminal
W 42nd St
MTA-42nd St-Bryant Park
Metro-North-Gra
Central Termina
MTA-34th St-Penn Stn-1-2 & 3 Lines
NJ TRANSIT-Penn Station-New York
E 34th St
Park Ave Tunnel
MTA-33rd St-6 Line
Queensbridge Park
Roosevelt Island B
Upper Level
MTA-23rd St
MTA-45 Road-Court
9A

# Additive and subtractive colour

## Mapping for screen or print demands a different approach to colour specification.

**Map colours are specified by mixing additive or subtractive colours.** For screens, additive colours red, green, and blue (RGB) are mixed. For print, subtractive colours of cyan, magenta, and yellow (CMY) are mixed. These two systems are not perfectly interchangeable since purity of light is not matched by printing ink. For instance, you cannot mix the very vivid colours made possible on a computer display by using CMY inks on a page. Printing very light colours is also difficult because spacing small ink dots so far apart does not generate smooth colours.

**Additive colour mixing is used for devices that normally have a black background representing no colour transmission.** The opposite is true for printed maps in which you instead perceive reflected light. Light that illuminates paper passes through the printing ink and is reflected back off the paper. The reflected colour is the colour of the ink. If several layers of ink are overprinted on each other, reflected light is absorbed differently by each layer of ink so the result is a mix of the layers. If you use transparent ink in the three primary colours, light will be absorbed by them and, in theory, no light will pass through any combination of the colours resulting in a black image (i.e. no reflection).

**Printing on paper uses subtractive colour mixing and transparent inks in cyan, magenta, and yellow.** Cyan transmits green and blue, magenta transmits red and blue, and yellow transmits red and green. For example, if yellow ink is printed on top of cyan ink on white paper, the yellow ink absorbs blue light but transmits red and green; the cyan layer then absorbs red light so only the green light reaches the paper and is reflected back to the observer. CMY colours are also referred to as *process colours*. Black (K) is usually used as a fourth printing ink to create pure black text and line work and better printed greys. The K means 'key' in four-colour process printing since black is normally printed first and other colours are keyed or registered to it. These pigments are also used in ink and laser printers, which apply percentage coverage of tiny dots to the paper.

**See also:** Elements of colour | Mixing colours | Printing fundamentals | Transparency

Colour is seen through the stimulation of different cones in our eyeballs. For three beams of overlapping red, green, and blue light, all three types of cones are stimulated simultaneously and this creates the perception of white light. Where only two beams overlap, the transmitted light produces a mix of colour that is formed by the stimulation of pairs of cones. Blue and green together form cyan, blue and red form magenta, and red and green produce yellow. By altering the intensity of each of the beams of light, different colours can be created.

On maps, thin linework in colours other than black is often required. For instance, pale blue for rivers and brown contour lines are not easily created. Brown lines can be created by overprinting pale tints of all three process colours. However, registering thin lines on top of each other is often beyond technical limits of the printing process, leading to lines that are blurred. Furthermore, each line is composed of small dots of each of the process colours so this in itself renders a line that can never be sharp. Because of these problems, printers can premix inks for certain colours. Premixed colours are called *spot colours* and, in fact, most national mapping agencies print topographic maps using *spot colours* or a mix of four-colour and a few spot colours. Spot colours can also be printed in percentage tints so open water can be a percentage tint of a blue spot colour.

Printing ink manufacturers produce a range of spot colours. One such system, Pantone, produce spot colours by mixing two or three of a basic set of nine colours plus pure black or white in predetermined proportions. Pantone provide a worldwide standard for colour specification, and if a map is sent for printing with a colour specification that identifies a particular colour as a Pantone colour, the printer will be able to match it.

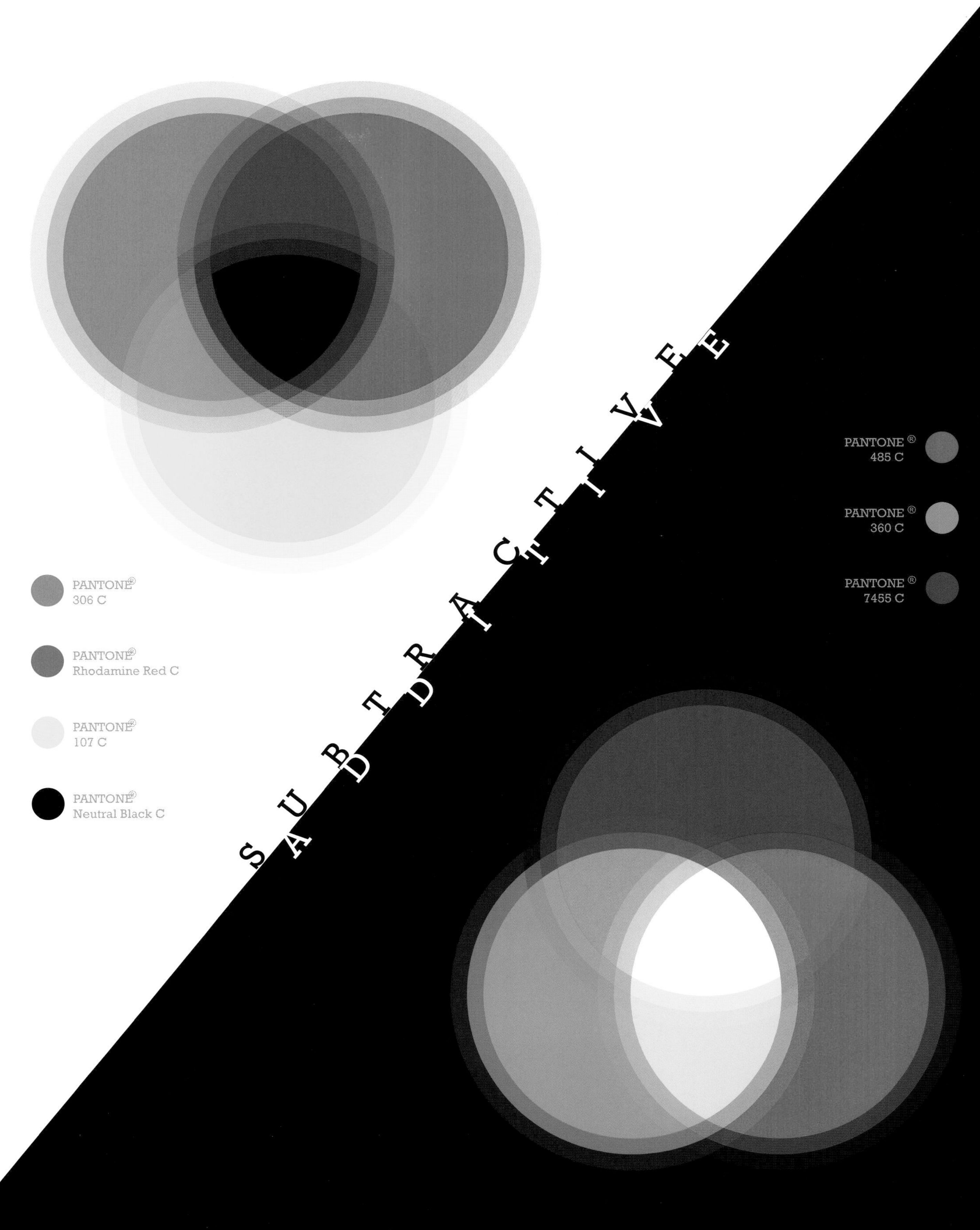
SUBTRACTIVE
PANTONE® 306 C
PANTONE® Rhodamine Red C
PANTONE® 107 C
PANTONE® Neutral Black C
PANTONE® 485 C
PANTONE® 360 C
PANTONE® 7455 C

# Advertising maps

## The use of maps to sell.

**Considering people's general liking for, and trust of, maps, it's no surprise they're used heavily in advertising.** Advertising is designed to create a clear image in a consumer's mind. It does so by being as appealing as possible. Cartographically, the aim is definitely form over function as it attempts to create a favourable comparison to a competitor's product, to emphasise a clear corporate image or to build trust, affinity, and demand for a product. Advertising might be used to show where something exists but omission and exaggeration are often used to build a picture. Graphic clarity is often replaced by the need to convince someone to buy.

**Maps in advertising often play on a theme.** Maps might be used to show convenience or the spatial ubiquity of a service. Conversely they might highlight exclusivity. Maps are used to exploit the consumer, and the map itself is often exploited as a framework for selling some partial version of the truth. For instance, distances are sometimes warped to show places as being nearer or more convenient. Coverage is sometimes shown to be more than it actually is. So the map is used to present a highly generalised version of reality that can communicate the message immediately. Thus, the mapmaker often applies a large dose of artistic licence to exaggerate what they want to show and is creative in masking what they don't.

**Maps used in advertising tend to be some of the most inventive and pictorial.** They often use very familiar shapes such as the outline of the world or of a country which are potent and recognisable symbols in their own right. They play to the fact that the consumer is already familiar with the basic structure of the graphic so the extent to which the image can be modified, yet still retain familiarity, is much greater than for maps that support other purposes.

**See also:** Branding | Copyright | Emotional response | Thematic maps

Perhaps the map that is used to advertise more than most is the one that many people wouldn't even realise is being used in that way. Google Maps™ is the go-to map for millions of people each day. Most will use it for general-purpose tasks such as finding a location or seeking directions but Google did not produce the map as an act of philanthropy thinking that the world needed one consistent map for anyone to use. There is money to be made from maps, and Google itself is neither a map nor a search engine. Although its business has proliferated into a range of markets, at its core it's an advertising company whose revenue streams are predominantly from the money it makes by putting adverts in front of consumers.

Google rapidly realised that the map can act as a proxy for an advertising hoarding. Every time we use the search engine on the map we're looking not only at the map but at the content that is being added. Over time, this content has gradually changed to support more advanced approaches. The map places you at the centre and builds content around you. If you search for a restaurant you are going to see those that Google promotes (because it is paid to do so) or which relate more to you to encourage you to go there. The same is true for a vast range of goods and services that take you to other websites, which may have content on them that Google gets paid to show.

**Opposite:** Mockup of an advert for cellphone and data coverage. The ad wants you to believe in the extent of availability of the product. The map uses perspective, tilt, curvature and occlusion coupled with the symbology to relay that message. Simple to process and suggesting totality, the prism heights are unexplained and even bleed off the page and obscure text to infer it's too much to even show! The words reinforce the idea of coast-to-coast coverage with no gaps (despite the reality). The small legend acts as a disclaimer that is positioned to ignore.

MORE SPEED

MORE POWER

# MORE CAPACITY

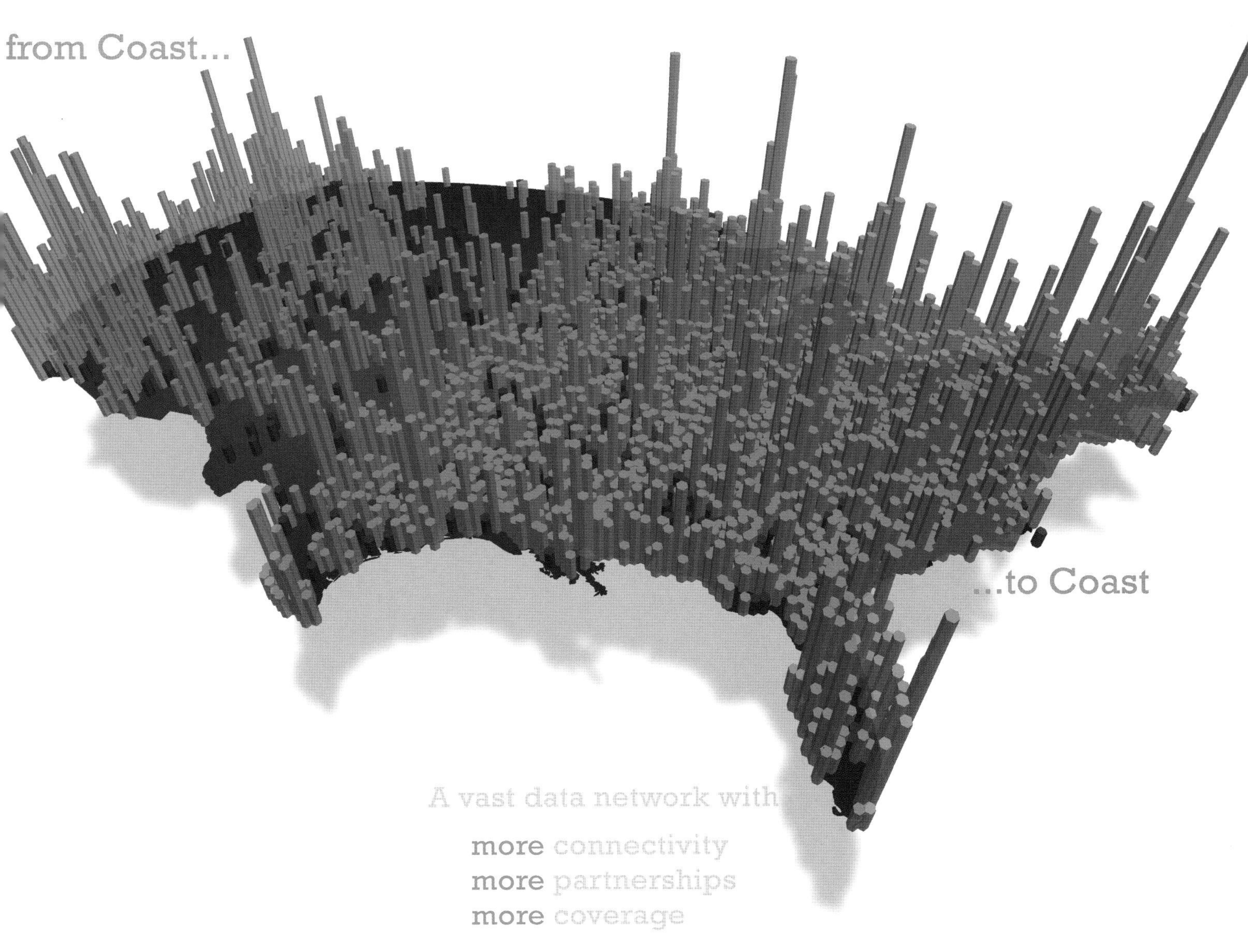


Nation-wide In-Network data usage

Partner Network (roaming coverage) data usage    No Bars = No Coverage

# Aggregation

## Techniques that modify the resolution of feature geometry to alter the cartographic appearance.

**Aggregation transforms multiple clustered point features to represent them as an area.** Aggregation is usually applied when a reduction in scale makes individual points imperceptible or as a result of their coalescence. The main decision is determining at what point the density of features becomes sufficient to warrant merging. Delaunay triangulation is normally used to determine the boundary of the new feature.

**Amalgamation joins together area features of the same feature type where a reduction in scale makes the distance between them indistinct.** Amalgamation can be applied to non-continuous data such as several forest stands that are best depicted as a single forested region at reduced scale. It is also the process used for continuous data such as census geographies where small geographical areas are often amalgamated into larger units at smaller scales.

**Collapse alters the geometric properties of a feature to represent it as another feature type.** For instance a complex areal feature might be collapsed into a single point symbol for representation at reduced scale. A complex building shape might be represented with a single rectangle or a geometric symbol. Symbols may also be mimetic to represent the feature with a pictorial symbol that easily connote its character.

**Merging is a form of the collapse operation that joins linear features such as adjacent railway lines.** Often, original lines are displaced or redrawn in the spatial average location between two input lines and in a space that did not originally contain a line. At a reduced scale, this alteration of position does not create problems but, instead, makes the overall pattern of lines easier to discern.

The various aggregation techniques used in cartographic generalisation convert features that might work well at one spatial resolution or scale to another. They sum or split features into the same or different feature types that, graphically, work better at a different scale.

Thresholds of perception are usually the main guiding principle for determining when these techniques should be considered. When a feature ceases to be recognisable or its detail is no longer distinct, using aggregation to reduce its complexity is useful. Ensuring that the character of the resulting object reflects the original is important. For instance, if amalgamation of polygons is performed on an orthogonal feature such as a set of buildings, the resulting shape should also be orthogonal. For a forest stand that is not orthogonal, the resulting shape should retain the same character.

When complex shapes are collapsed to geometric features, it is important to ensure that they don't conflict with other map content that has a similar geometric appearance. For instance, collapsing city boundaries into point features for smaller scale maps is a common approach but if similar point symbols are already used elsewhere then some distinction would need to be made between the different features.

**See also:** Clutter | Generalisation | Refinement | Simplification

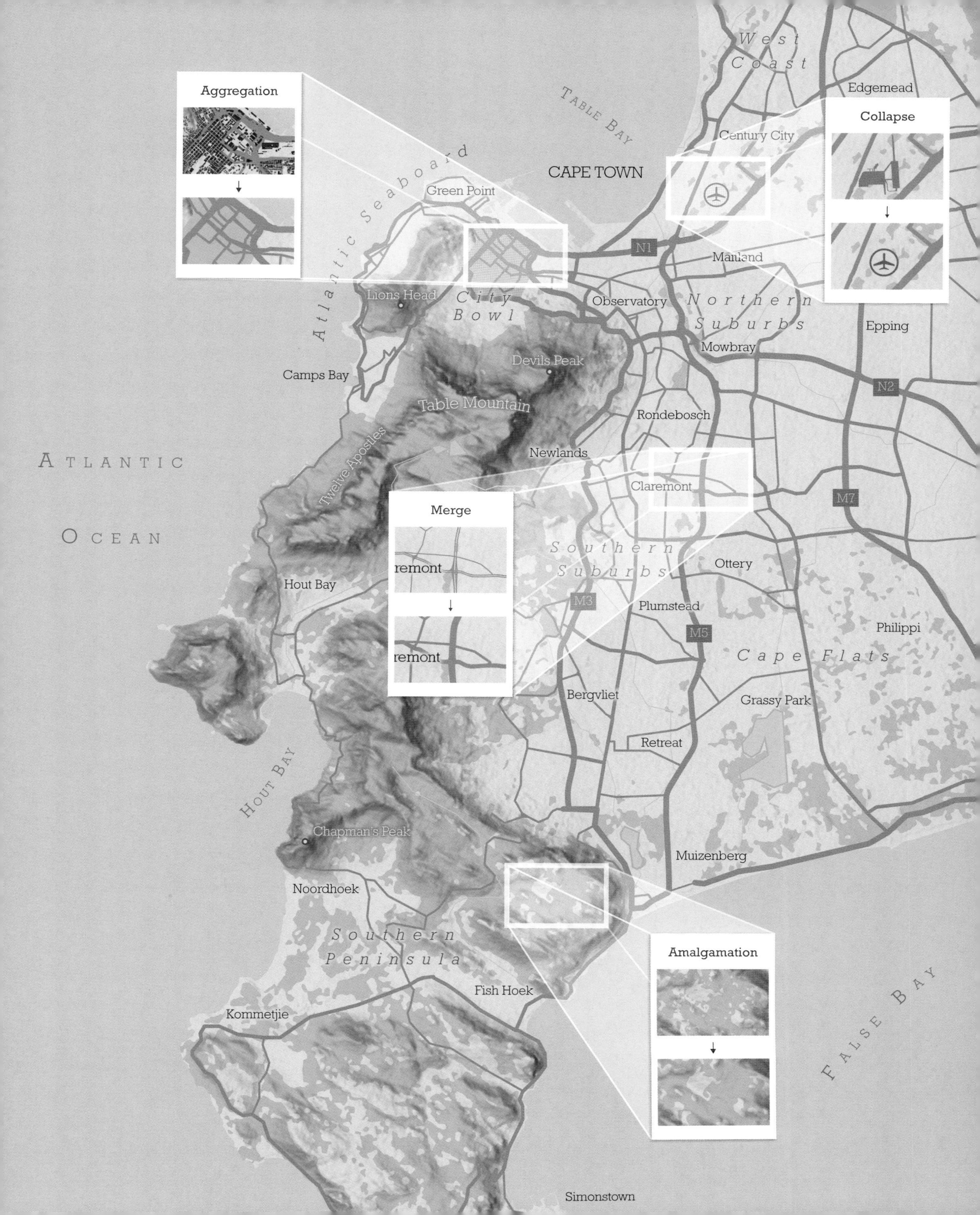

Aggregation
Collapse
Merge
remont
remont
Amalgamation
West Coast
Table Bay
Edgemead
Century City
CAPE TOWN
Green Point
Atlantic Seaboard
N1
Maitland
Lions Head
City Bowl
Observatory
Northern Suburbs
Epping
Mowbray
Camps Bay
Devils Peak
N2
Table Mountain
Rondebosch
Twelve Apostles
Newlands
Atlantic Ocean
Claremont
M7
Southern Suburbs
Ottery
Hout Bay
M3
Plumstead
M5
Philippi
Cape Flats
Bergvliet
Grassy Park
Retreat
Hout Bay
Chapman's Peak
Muizenberg
Noordhoek
Southern Peninsula
Fish Hoek
Kommetjie
False Bay
Simonstown

# Aligning coordinate systems

## Assessing what coordinates your data is in and aligning them.

**All data is created in some kind of coordinate system.** Often data comes simply as a numerical list with coordinates that link back to geography. Before even thinking of the issue of what map projection to use for a map, making sense of the coordinates in data is paramount. Coordinates are usually specified in either a geographic coordinate system (GCS) or a projected coordinate system (PCS). A GCS uses angular units of measurement, usually degrees. The x-coordinate specifies longitude from the prime meridian value of 0 at Greenwich, UK. Values are positive to the east and negative to the west. The y-coordinate specifies latitude from the equator value of 0. Values are positive in the Northern Hemisphere and negative in the Southern Hemisphere. GCS cannot be used to measure ground distance between points, or determine the length of a line or the area of a polygon because angles bear no relationship to surface distance measurements.

**A PCS is used if your map is for measuring distances, lengths, or areas in linear units.** Linear units include the metre, which is defined as the distance that the speed of light travels in a vacuum. Considering that the speed of light travels at 299,792,458 metres per second, distance can be calculated as 1/299,792,458. This is equivalent to 39.37 US inches. The international unit 'foot' is 0.3048 of a metre. The US survey foot equals 1,200/3,937 of a metre.

**Identifying what coordinate system your data is in helps asses what you need to do with it.** If your data is in decimal degrees, it's in a GCS. If it measures position in linear coordinates, it's in a PCS and metadata should assist you in determining what particular PCS is being used. Often the need to deal with projections is because you're using multiple datasets that are in different coordinate systems. You'll need to define the correct projection for each, and then reproject some data into a single consistent projection that lines everything up properly.

Inspecting the data's data, the metadata, can help you understand the coordinate system. Data in GCS will only go ±180 in the x direction and ±90 in the y direction. Larger numbers will be linear units and so the data will already be projected. Identifying the correct GCS involves understanding what the datums are that are typical for the area being mapped. For instance, if your data is global, it's likely WGS84. If the data is based on national mapping, it may use a different GCS. For instance in the United Kingdom the most commonly used GCS is OSGB36. Data may use a local coordinate system based on an arbitrary location as a false origin (0,0). Local coordinate systems are often seen in CAD data but can usually be aligned with other data by transforming it to a known projection.

Geographic transformations may be required to adjust coordinate data to align it with other data. This is usually the case when one dataset uses one GCS and another dataset uses another. The datasets are effectively using two different spheroids. A geographic transformation is a datum transformation to modify one or more datasets to shift to another GCS. Applying transformations will usually require a geographical information system (GIS), which has a library of transformations you can apply.

A common way to see the juxtaposition of data that uses different GCS and PCS is simply looking where it gets plotted. If you add data in a PCS that uses linear units, then overlay another dataset that uses GCS. All the data in the latter dataset will appear around 0,0 on the PCS, extending only ±180 units in the x direction and ±90 units in the y direction. If the PCS is in metres, it assumes the additional data should align—so it will be positioned around a hypothetical spot, Null Island ,which is the intersection between the equator and prime meridian.

**See also:** Datums | Earth coordinate geometry | Map projections | x and y

The line that appears offset references a GCS that is different from the base imagery. Applying a transformation will rectify the two.

Depending on the mismatch, the offset can be up to a few 100 metres.

These elevation points illustrate the vertical difference between the ● ellipsoidal height (a mathematically derived surface) and the ○ gravity-related height (a local calculation).

The difference can be so great that sometimes water modelled with an ellipsoidal baseline can appear to flow uphill on an untransformed surface.

# All the colours

## Computing and graphics engines support a palette of over 16 million colours. This fact has no bearing on cartography.

**Modern computers provide the capability to render 16,777,216 separate colours because they can handle 255 shades each of red, green, and blue.** This vast array gives rise to a number of dilemmas, not least of which is how you make a good choice, but there is a fundamentally bigger problem. The human eye cannot see all those colours, let alone distinguish between them. Furthermore, we all see colour differently so the shades and subset of colours that one person sees when they look at a map is different from the next person and so on.

**In studies of comparisons, humans are able to distinguish and remember only five to seven colours.** This general rule of thumb gets complicated when we add in different contexts such as surrounding detail, brightness, lighting conditions, and animation among many factors. Rather than exploring the idea that we have a palette of 16 million colours, it makes more sense to constrain our choices to a more human-oriented palette. The idea is simply to make better choices that support the final objective of helping the map reader interpret the map easily.

**Whereas applying general constraints in colour choice is appropriate for most uses, any or all of them can be disregarded if they compromise the effectiveness of a specific map solution.** The important dimension here is that you are likely mapping for the greatest number of people in the greatest number of contexts. However, being cognisant of the approximate 10% of males and 1% of females that have some colour deficiency with their sight might also be a consideration. It might not be possible to modify the entire colour scheme to account for this factor, but if this use context is important then it's a key consideration.

Colour often encodes some characteristic of the data so that it remains consistent whenever it is used. As a map becomes more complicated, use coded colours that can be distinguished from the others and which can also be used consistently throughout the map. Colours should be used with simple and familiar associations. This usually means that a colour is universally used for a particular map element though it's worth noting how they work across different background colours. Some colours imply danger or caution and although they might be best avoided they can obviously be used precisely to support that objective.

Bidirectional colour design is where different, distinct, colours are used for only one element of the map and used nowhere else. Although this option contradicts the general idea of reducing the number of colours on a map it elegantly resolves the problem of needing too many colours for easy identification. Geology maps or complex subway maps often take this approach to colour choice.

Unidirectional colour consistency results in the same palette of colours being used consistently across a map or map series. So, for instance, an atlas will ensure the consistent use of colour throughout.

**See also:** Additive and subtractive colour | Colour schemes | Constraints on map colours | Greyscale | Seeing colour | Type colour

**Opposite:** All 16,777,216 colours (well, not quite due to limits of printing technology!)

# Anatomy of a map

## The mapped area and associated map pieces together make the map.

**Maps tend to share common components which each perform a discrete function and, collectively, make the map.** There are two basic components—the mapped area and the marginalia, or the additional map pieces. Not all map pieces will be required on every map. The type of map will lead to different pieces as will the use conditions and the medium of delivery. Maps produced as part of a national map series will have consistent pieces but a one-off map in a book or on a poster may have relatively few.

**The term *map marginalia* can be misleading since the various pieces are often positioned within the mapped area itself.** Certainly, this is more common for digital maps which do not necessarily have formal layouts. Many of the map pieces might be located behind buttons, tabs, or click events instead, which helps to create an uncluttered display.

**Inset or locator maps help orient the map reader unfamiliar with the geography of an area.** Sometimes the location of the map isn't obvious or the map will likely be used by people unfamiliar with the area. Providing a small contextual map that shows the extent of the main mapped area can be extremely helpful. This consideration equally applies to digital maps. The extent of the map may change as a map reader zooms in and out but it remains a useful map piece nevertheless. Alternatively, an inset map may provide a larger scale version of a part of the main map, sometimes used if the data in that area is particularly dense or congested. Insets can also be used to bring disparate parts of the same geography into a convenient view. For instance, thematic maps of the United States often place Alaska and Hawaii in insets because of their detachment and remoteness to the contiguous Lower 48 states.

**Mapped area:** the map itself comprising symbolised points, lines, and areas and typographic components, possibly with the use of imagery.
**Inset/locator map:** showing the larger mapped area in context or detail of a small part of the mapped area.
**Title and subtitle:** the wording that communicates the theme of the map.
**Neatline, frame, or border:** a line that bounds the active map area, frequently constructed from lines of an underlying graticule or grid.
**Graticule:** lines of parallels of latitude and meridians of longitude.
**Grid lines:** a reference grid unrelated to latitude or longitude.
**Graticule or grid line labels:** specifying coordinates in degrees or map units.
**Margin:** an area between the active map and the edge of the display, sometimes literally used to contain marginalia on map sheet series.
**Scale:** a statement of the distance range of the map, graphical or a verbal statement, often not required on small-scale thematic maps or where scale differs across the map because of coordinate system or perspective.
**Legend:** list of symbols and their descriptions used to assist map interpretation.
**Orientation:** a north arrow or other similar symbol to orient the viewer to the map, often unnecessary if a graticule is included or for thematic maps or where the coordinate system means north varies across the map.
**Source:** statement of the origin of data
**Copyright:** statement of permissions granted or of the copyright holder of the map.
**Map series legend:** used if the map is part of a sheet series.
**Text blocks:** description and/or statements of datums and coordinate systems used or of the date of data and revisions.
**Author details:** a statement on who made the map helps give it authority.

**See also:** Form and function | Data arrangement | Functional cartography

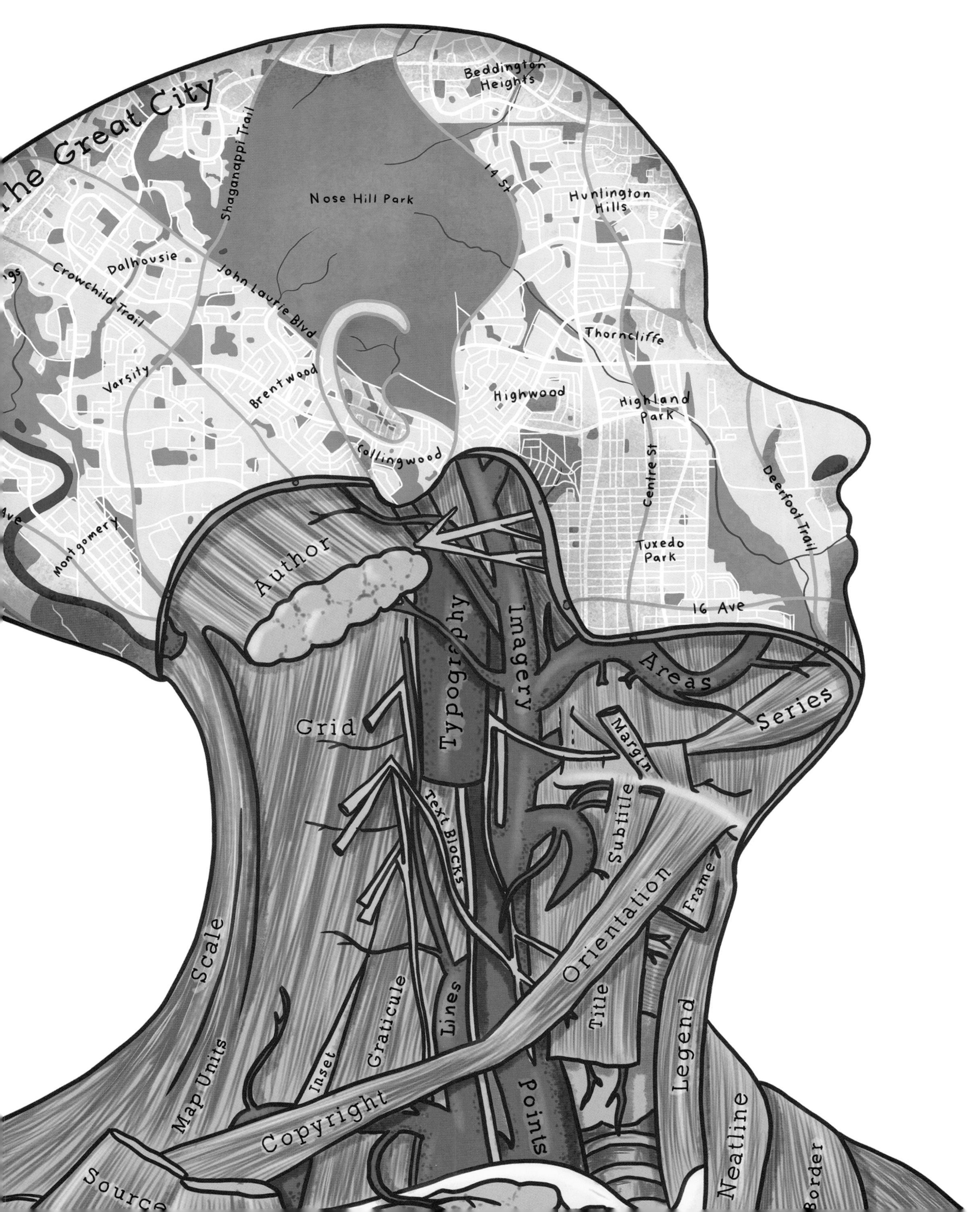

the Great City
Shaganappi Trail
Beddington Heights
14 St
Nose Hill Park
Huntington Hills
Dalhousie
Crowchild Trail
John Laurie Blvd
Thorncliffe
Varsity
Brentwood
Highwood
Highland Park
Collingwood
Centre St
Deerfoot Trail
Montgomery
Tuxedo Park
16 Ave
Author
Typography
Imagery
Areas
Series
Grid
Margin
Text Blocks
Subtitle
Orientation
Frame
Scale
Lines
Graticule
Title
Legend
Map Units
Inset
Copyright
Points
Neatline
Source
Border

# Animation

## A set of sequential images that gives the illusion of movement.

**Animated maps are the cartographic equivalent of making a film.** A sequence of static images are stitched together, each one different from the last and when played at a certain number of frames per second our eyes perceive movement. Animated maps support a range of alternative visualizations of data. They are most often used when data varies over space and/or time or when you want to move the map reader's viewing position relative to the map itself. At their simplest, time slices of different states of a static map can be stitched together. For instance, a series of small multiples placed in a sequence can introduce motion to show how trends of a single variable morph over time.

**Fly-throughs are a common way of creating animated maps for non-temporal data.** They are typically used in 3D cartography, in which the hypothetical camera moves across terrain. As the camera moves, its speed, height, viewing angle, and focus can change to provide different viewpoints.

**Although animations are often simple, the addition of effects and audio can give them a cinematic quality.** Making an animated map can be analogous to making a film in that you go beyond building the map into directing a short sequence. This demands that you consider the storytelling aspect of the film, not only in terms of the sequencing but how the map might be augmented with additional visual and audible effects to facilitate the experience. Many successful map animations veer into the territory of cinematic cartography in which the distinction between a map and a movie blurs.

**Playback is normally handled by standard movie players or browser plug-ins.** The animated map might be published on a streaming service such as YouTube or Vimeo. Users will have scope to start, stop, or fast-forward the animated map and control the speed to an extent but the map will rarely require specialist software for viewing.

**See also:** Dynamic visual variables | Globes | Space-time cubes

Animated maps are normally constructed by creating keyframes of different views of your map. This might be to move the position of the camera or change the map's state. When rendering the movie, the different keyframes act as anchors and the intervening positions will be interpolated to build smooth transitions using tweening. The final map is normally output as a movie file using a common video format. The movie file might then be post-processed to add a range of additional features such as special effects (e.g. modified colours, atmospheric conditions, lens flare), subtitles to add time stamps or key statements, titles, and other inserted static elements (for instance, the legend), or audio such as music or voiceover.

Animated maps differ from static ones in that you are not only creating the content but also determining the way in which the map reader sees the data. You control the order, pace, and focus which differs from how map readers conventionally interact with maps. It's therefore important to consider the extra demands. For instance, pace may need to vary, perhaps faster where simple trends are obvious but slower where more complex ideas are presented. Judgements on the speed of transitions will need to take account of the problem of change blindness in which rapid changes between frames or map state can lead to the map reader missing key information. A fluid animation will require not only a relatively high frame rate but having something interesting that changes on the map. If your data does not vary much, it's going to be a fairly uninteresting film.

**Opposite:** 24 globes, each shown from a different viewpoint. If each globe occupied one frame in a 24 frames per second film we'd see a smoothly rotating globe. Try flicking the left edge of the book.

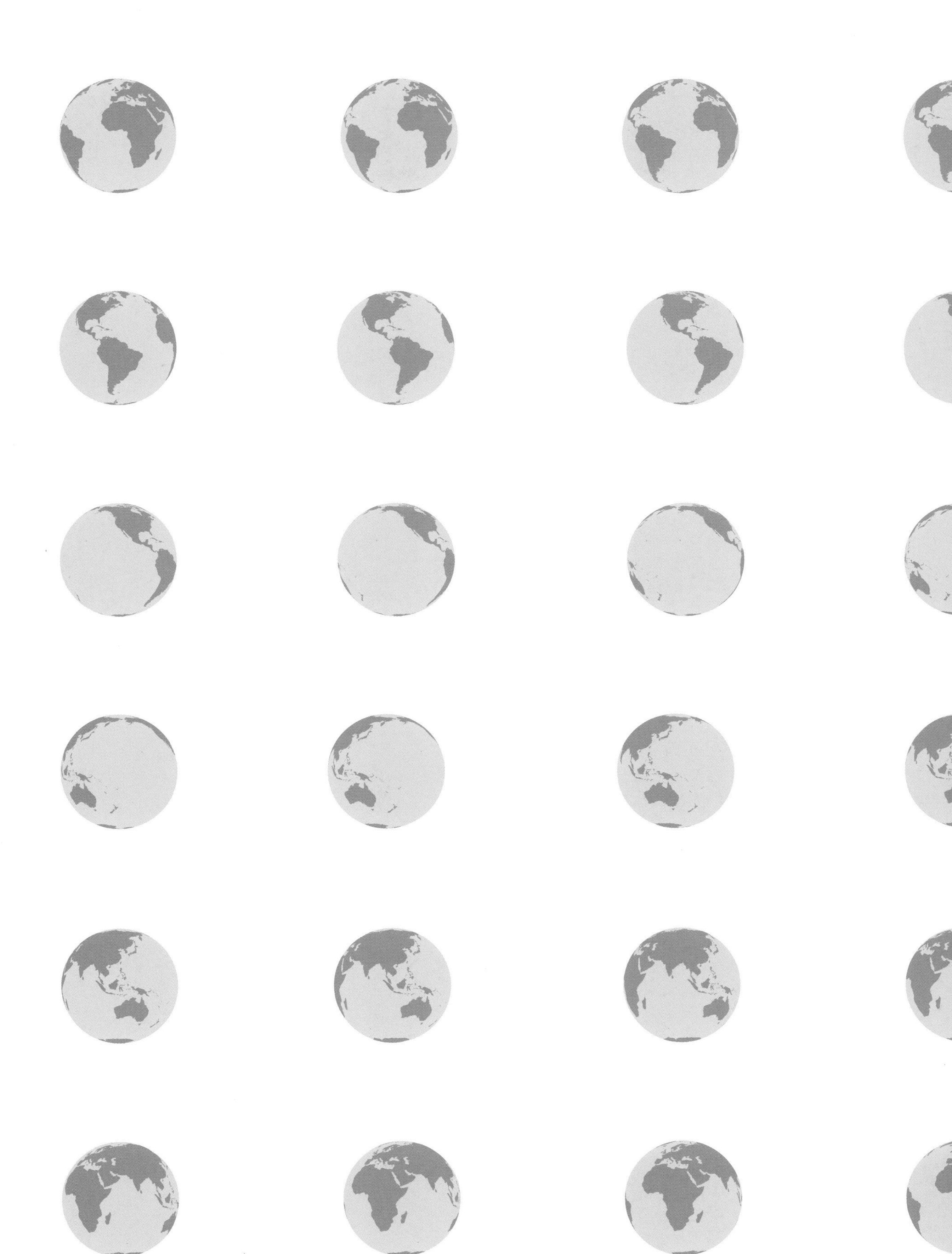

# Arbitrary data classification

## Classification techniques that are easy to understand but don't take account of the data distribution.

**Arbitrary data classifications for thematic mapping are objective.** They do not pay attention to the actual data distribution and apply a classification system on the basis of some unconnected numerical principle. This may be to create an equally divisible set of classes across a range or to classify on the basis of an equal number of ranked data values.

**An equal interval scheme results in each class having the same numerical width.** The class interval (width of each class in units defined by the data) is calculated by dividing the data range by the number of classes. The scheme does not leave any gaps between the classes but it has the potential to leave empty classes if a large number of values fall within only a few of the classes and also display unwieldy numbers.

**A modification of the equal interval technique can be applied to give a more visually appealing class interval.** This simply extends the low and high data values to a rounded number and then uses those values to create an equal interval scheme. Such an arbitrary technique can be created by choosing regular, neat class intervals to encompass the array.

**In a quantiles method of classification, data is ranked, and an equal number of values are placed in each class.** The term *quantiles* refers to the generic form of the classification technique; four- or five-class schemes are referred to as a *quartile* or *quintile* scheme respectively. The number of values in each class is derived by dividing the total number of values by the number of classes. When the number of values per class is calculated as an integer, the task of allocating values to classes is simple. When a real number is calculated, then one should attempt to place approximately the same number of values in each class along the array.

The main advantage of the equal interval scheme is that it is easily understood largely, because of rounded class intervals in the legend. The main disadvantage of this scheme is that the data is not inspected or used to inform the scheme. The class intervals are applied regardless of the data distribution. The more skewed the data set, the more pronounced the maldistribution of the data.

A quintile classification scheme will have the same (or very similar) percentage of values in each class (20% per class). Given this property, it allows the map user to discuss and interpret data in terms of the 'upper 20% of values' or the '2nd quintile' and so on. This also facilitates comparison across maps that share quantile schemes comprising the same number of classes. A further advantage is that the 50th percentile (median) is discernible from the classes. In the case of an odd number of classes, the median will fall in the centre of the middle class. In the case of an even number of classes, it will fall between the two middle classes.

Quantiles are also useful for ordinal data since they are based on ranking of data rather than the values themselves. As with the equal interval scheme, the major disadvantage of quantiles is that they fail to consider the data distribution and might lead to some peculiar class widths or distribution of values between classes. An additional problem is more perceptual in that if you create gaps between classes, the map user might wonder why such gaps occur.

**See also:** Data classification | Eyeball data classification | Statistical data classification

## Equal Interval

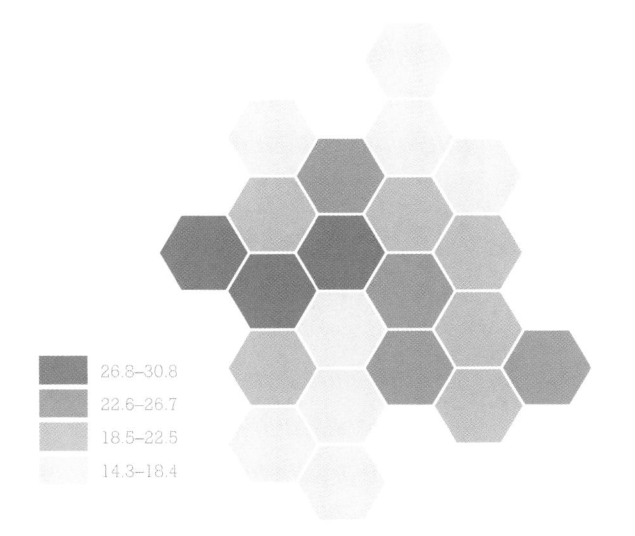


For this data, an Equal Interval scheme distributes the data values across all classes with a slight emphasis on showing low-mid values and isolating the top 3 values.

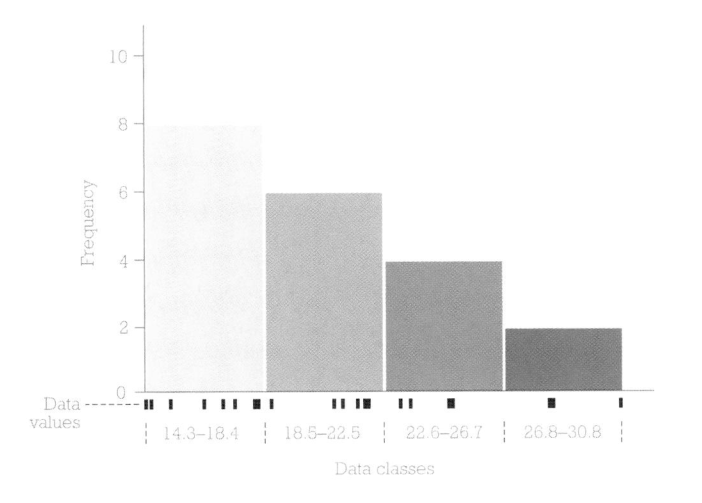


Regular class widths
Irregular number of data items per class

## Quantile

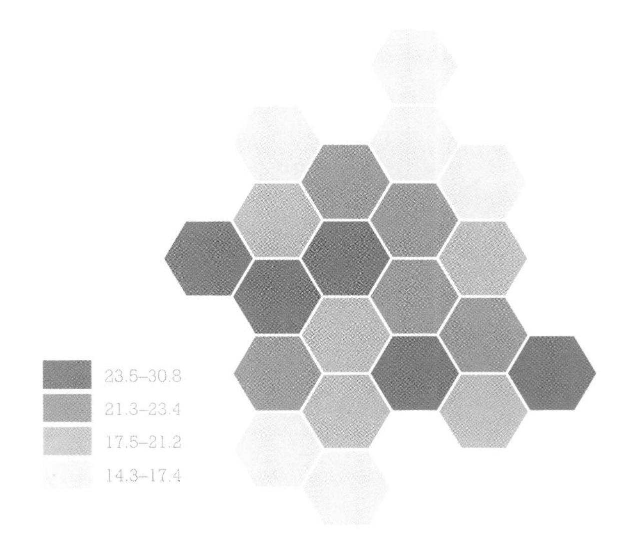


For this data, a Quantile scheme identifies where each area falls in 25% intervals. It shows the top 25% of areas and so on.

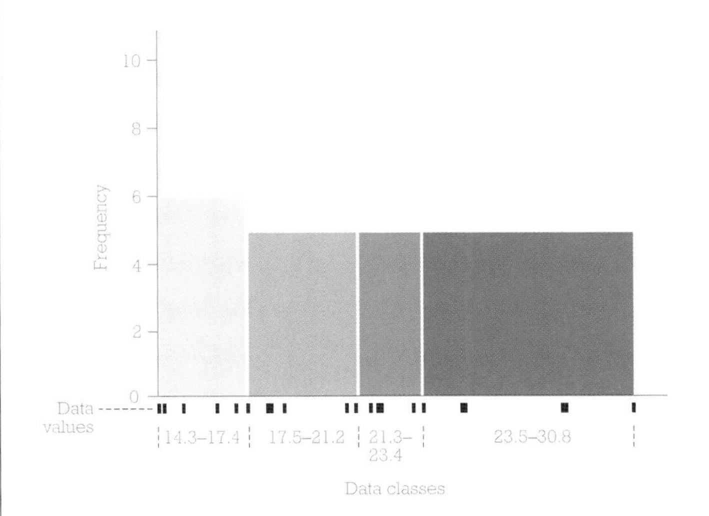


Irregular class widths
Regular number of data items per class

# Areas

## Areas enclose and define the extent of features of a uniform type on maps.

**Area features represent two-dimensional phenomena as two-dimensional symbology.** Real-world features that have width, length, and area will be surveyed as such, and the data will normally consist of a line that encloses a polygon to represent the feature. Topology can assist in defining the feature as an area since it forms either a single or a multi-part enclosed feature that has the same character. The character, or multiple characteristics, can be captured and stored as attributes if the map is database driven. For example, a series of forest stands can be shown as area features all symbolised in the same solid green fill to represent forest, or shown differently to represent different species (e.g. a green fill with an additional pattern that encodes the species information).

**Phenomena can also be referenced as an area.** For instance, a series of weather stations that record data at a point can be manipulated into a map that shows weather conditions, such as temperature, as an area. This makes conceptual sense since temperature exists everywhere, even though the original data wasn't itself an area feature. Statistical surfaces of either sampled data or modelled results can also be shown as areas. These examples are a special case of area data that symbolises volumetric data across an interpolated two-dimensional surface.

**Area features can sometimes be collapsed into point features.** As the scale of a map changes so the representation of areas can be changed. At a large scale, a city might be shown as an area with a clearly delineated boundary. As map scale gets smaller the space for the area lessens as does the need for detail and, eventually, even a large city can be represented as a point with zero dimensionality.

Area data is often captured from aerial photography or satellite remote sensing. Digital photogrammetric techniques or image classification can be used to define the extent of an individual feature. By using the scale of the map, the actual area can be calculated to provide accurate measurements.

Areas commonly act as containers for other data on a map. In fact, many areas do not even physically exist on the ground such as administrative boundaries or areas used for various data collection exercises such as a census. Such areas are artificial constructs yet they perform important functions both in reality and on the map. By varying the symbology of the line work that surrounds an area, you can quickly demarcate adjacent phenomena. The line might not exist in reality but it is used as a powerful map feature.

Many thematic mapping techniques make use of the idea of an area as a container such as the choropleth thematic map. Such maps change the style of the fill to show either qualitative or quantitative information—for instance, land use or population density.

Areas can also be extruded into a three-dimensional feature that encodes an additional variable on the z-axis.

**See also:** Lines | Pattern fills | Points | Texture

**Opposite:** Areas are shown isolated from the complete map. They form the background for this reference map, encoding information through the use of different hues and exhausting space.

# Aspect of a map projection

## The relationship between the point of projection and the orientation of the globe provides scope for a multitude of map projections.

**Projections vary in appearance depending on the conceptual location of the point of projection, the orientation of the globe (its aspect), and its intersection with the developable surface.** For instance, a gnomonic form of projection assumes a point of projection at the centre of the globe. If the point of projection is, conceptually, positioned at the point opposite the point of tangency then we derive a stereographic form of projection. An orthographic form of projection occurs if the hypothetical light source is placed at infinity outside the generating globe.

**In many map projections it is common to position the globe in the normal aspect where north is toward the top and south toward the bottom.** If a reference globe is rotated 90° so that the equator is positioned vertically, then the resulting projection will be in the transverse aspect. Of course, it is possible to rotate the reference globe and achieve any number of alternative orientations. Any orientation of the reference globe other than the normal or transverse aspect is generally referred to as an oblique aspect and tends to give an unfamiliar pattern to the graticule.

**Map projections possess characteristics dependent upon the point of projection, tangency of the developable surface to the reference globe, and orientation of the reference globe with respect to the developable surface.** Although most maps tend to be viewed in a normal aspect with north upward it's often useful to modify the projection to place a specific point or area central to the map's focus of attention. This modified projection will position the map in an oblique aspect.

**Changing the central point or meridian of a projection is of considerable use in cartography to focus attention.** Moving beyond the normal aspect can often bring immediate benefits to a map that otherwise would default to a generic form.

In the normal aspect, Mercator is perhaps the most familiar map projection. It is based on a cylindrical projection in which the pattern of the graticule leaves the meridians parallel to one another. Although this projection preserves shape and is important for navigational plotting, the impact is to massively increase areas with increasing distance from the equator. This is most pronounced at the poles which, in actual fact, can never be shown on a Mercator projection since the projection extends to infinity.

Despite its distortions, Mercator has been successfully adapted for use. The transverse Mercator projection has found many uses in national mapping, particularly when the mapping is for areas with minimal east–west extent. In the transverse aspect, the horizontal axis of the cylindrical projection is aligned with a central meridian.

The ellipsoidal version of the Transverse Mercator was developed by Carl Gauss in 1825 and modified by Johann Krüger in 1912. The Gauss-Krüger projection is the basis for the Universal Transverse Mercator series of projections and also the national mapping in many countries, including Turkey, Germany, and China.

**See also:** Families of map projection | Map projections | Properties of a map projection

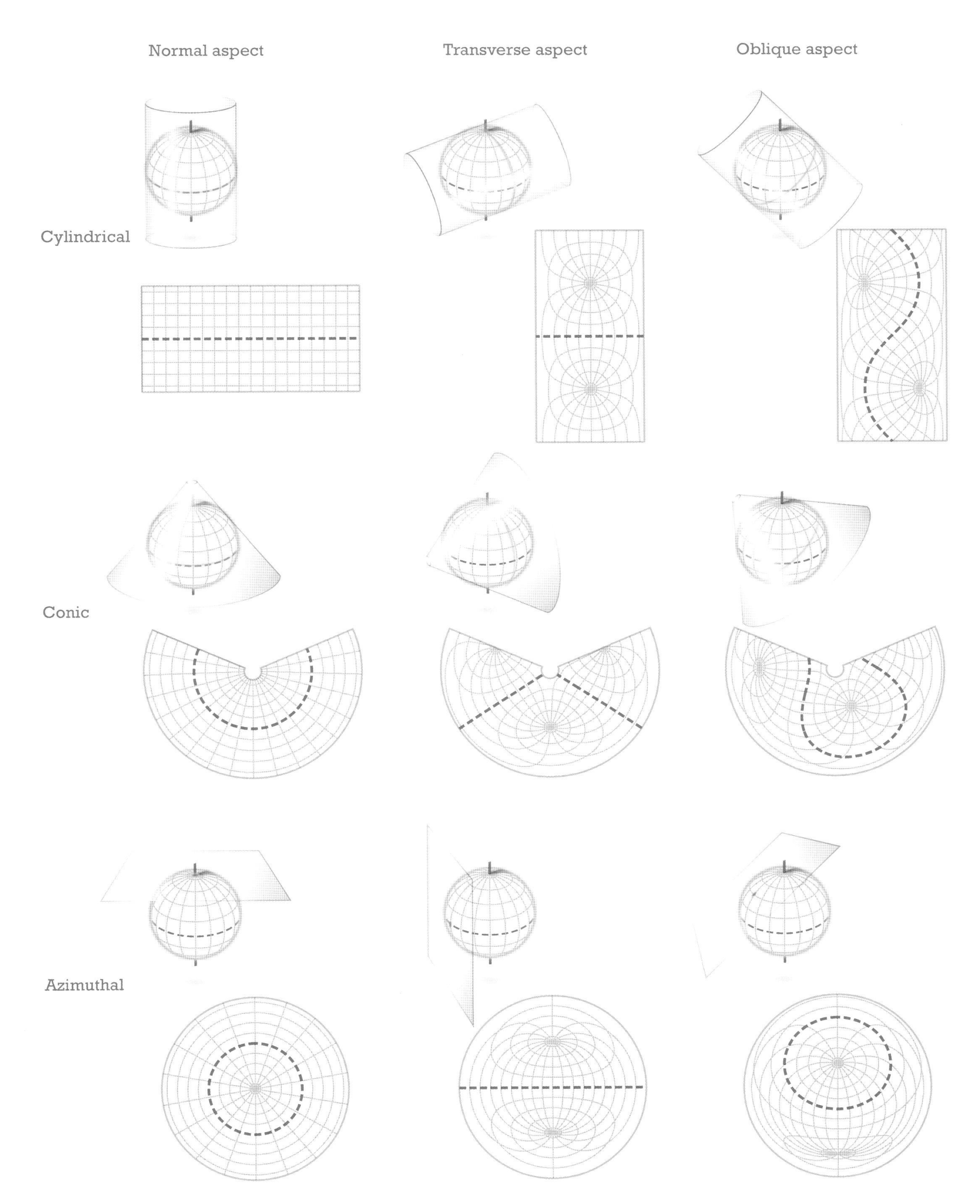
Normal aspect
Transverse aspect
Oblique aspect
Cylindrical
Conic
Azimuthal

# Aspect views

## Viewing the world side-on rather than from-above.

**Aspect views position the viewpoint sideways to the mapped area.** In cartography aspect views are most commonly used in panoramic mapping of a landscape or in virtual globes but can also be used for three-dimensional thematics.

**A perspective view approximates human vision in that objects get smaller as the distance from the viewpoint increases.** This results in foreshortening which describes the scale of features in the line of sight as being shorter than those across the line of sight and can include curvature of Earth at particularly small scales. Although the perspective view is generally a good choice for representing physical landscapes, thematics do not fare as well. Curvature and foreshortening do not support visual comparison of three-dimensional or extruded symbols representing quantitative data.

**An isometric projection is a form of the more general axonometric projection**. In this form of parallel orthographic projection the three coordinate axes are equally foreshortened. Scale across the map and along each of the three axes is the same—all 120°. As with any axonometric projection, the main advantage is to support measurement and visual comparison. However, the distortion of angles of the coordinate axes will appear unfamiliar simply because we view the world in perspective.

**A plan oblique view is a form of parallel orthographic projection which produces two-dimensional images of three-dimensional objects.** The dimensions of the map remain consistent to their planimetric position but elevation (or other objects with height) are modified. Sometimes referred to as forced depth, parallel lines from the source object produce parallel lines on the projected image which intersect the projection plane (map) at an oblique angle rather than the perpendicular as in a proper orthographic projection. It can be used to good effect in cartography, particularly in depicting terrain across a map that otherwise preserves scale in the x- and y-axes. Its main advantage is that we see terrain approximating a side view as we might see it in a landscape.

**See also:** Isometric views | Panoramic maps | Prism maps | Profiles and cross-sections

One-point perspective describes a map which has a single vanishing point situated on the line of the horizon. This gives us the typical image of roads disappearing into the distance as the parallel lines of the road edge converge in the distance. Two-point perspective exists when the map contains two vanishing points on the line of the horizon. These points may be positioned anywhere, and one set of parallel lines might disappear toward one point and another set disappear at the other. Three-point perspective adds a third vanishing point that disappears above or below such that vertical lines also recede in the distance. Four-point perspective is also known as the *curvilinear perspective* and is used to create a 360° panorama.

**Opposite:** *Guide for Visitors to Ise Shrine*
Japan, ca.1950

This intriguing Japanese tourist map demonstrates some interesting design. It's hand painted giving it a certain artistic aesthetic (similar to panoramic maps) and the use of colour presents the landscape in an abstract way but with a golden 'glow'. Although unnatural, this frames the main map. The various towns are illustrated and embedded amongst familiar mountains. The lack of detail presents a pleasant and accessible landscape that invites the reader in.

The most intriguing aspect however is that the oblique representation of the landscape gives way to a diagrammatic and planimetric view of the extended rail link to the right. This allows the map to show the detail of the city network alongside the more compressed and generalised main map. Two very different scales on one map…much like an inset but integrated into the map in a unique way. The painting has less content as it approaches the right edge before it becomes schematic to support the transition, and the upright orientation of the typography works well with other vertical map elements.

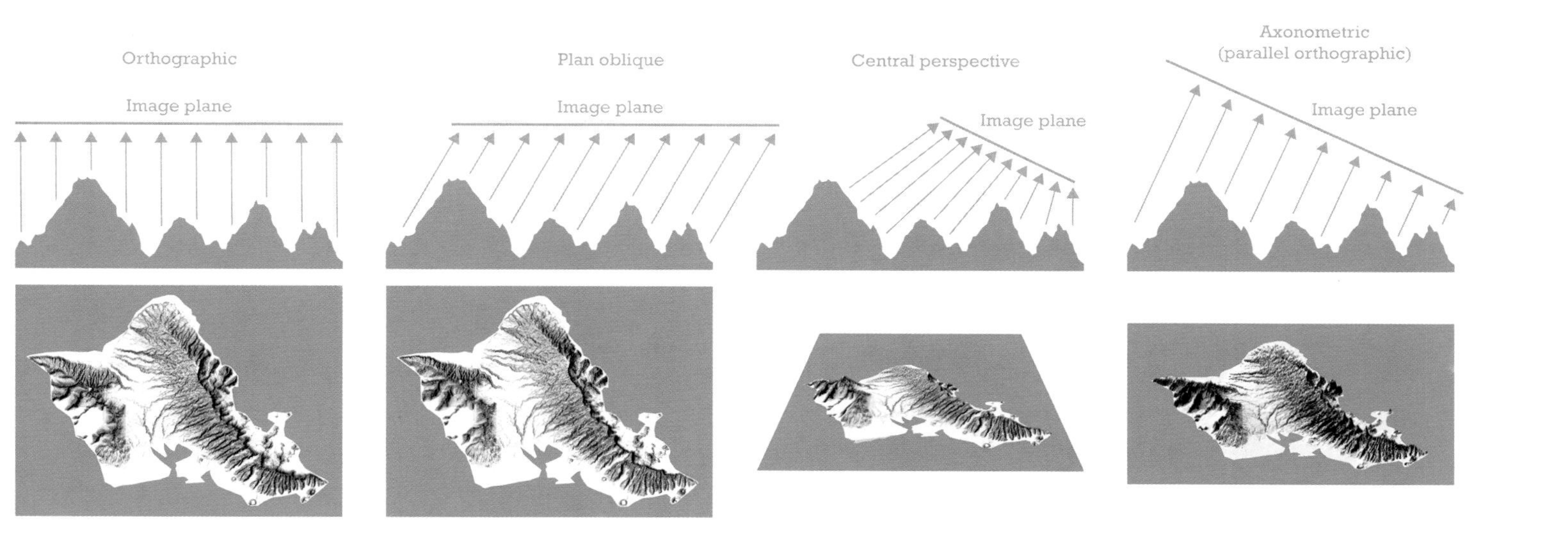

Orthographic
Image plane
Plan oblique
Image plane
Central perspective
Image plane
Axonometric
(parallel orthographic)
Image plane

# Assessing distortion in map projections

**The type and extent of the distortion caused by any map projection can be assessed visually or empirically to aid understanding of its impact on the map.**

**There are several methods for examining the extent and distribution of map projection distortions.** One of the most common is to place a geometric shape, or a familiar object such as the profile of someone's head or an aeroplane, and plot it at several locations on the projection graticule.

**Assessing distortion using graphical means provides a useful method for gaining a quick impression of how a map projection distorts objects at different places on the map.** For instance, plotting an aeroplane on one projection does not imply that the projection illustrates the aeroplane in its correct proportion—merely that if we plot an aeroplane on that projection and then reproject it to the others, it gives us a visual frame of reference for the extent of distortion.

**Alternatively you might calculate a scale factor (SF) to assess distortions.** SF is a numerical assessment of how the scale on the map at a specific location compares to the scale along a standard line(s) or at a standard point. Depending on the way in which the map projection has been applied, there will be points or lines on the map that correspond exactly to the same dimensions as the reference globe. At these points, or along these lines, the map scale is the same as the principal scale. At all other positions, the scale will either be exaggerated or compressed giving a series of local scales that differ from the principal scale.

**The scale factor will vary by about a value of 1.** If SF < 1.0 then objects at the local area are smaller than their true size relative to the principal scale. If SF > 1.0 then this indicates some exaggeration of scale; objects are being represented larger than their real size relative to the particular scale.

**A useful way to understand the distortions produced from converting a curved surface to a planar surface is to peel an orange.** The result is a flattened, torn orange peel where shape, distances, and angles between any two points have been altered from their original relative relationships on the curved surface of the orange. Try it!

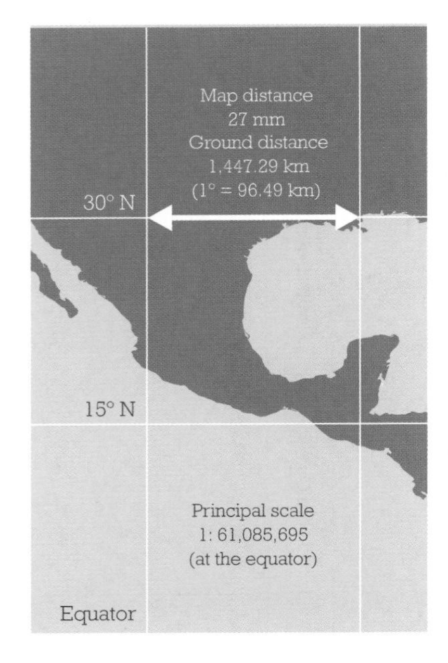


Calculating the local scale at 30° N

$$\text{local scale} = \frac{\text{ground distance}}{\text{map distance}}$$

$$= \frac{1{,}447.29}{0.000027}$$

$$= 1{:}53{,}603{,}333$$

The local scale of 1:53,603,333 is smaller than the principal scale of 61,085,695. The scale factor (SF) at this location is thus calculated as:

$$SF = \frac{53{,}603{,}333}{61{,}085{,}695}$$

$$= 0.877$$

**See also:** Distortions in map projections | Map projections: Decisions, decisions! | Properties of a map projection

**Opposite:** The impact of a map projection on a familiar shape.

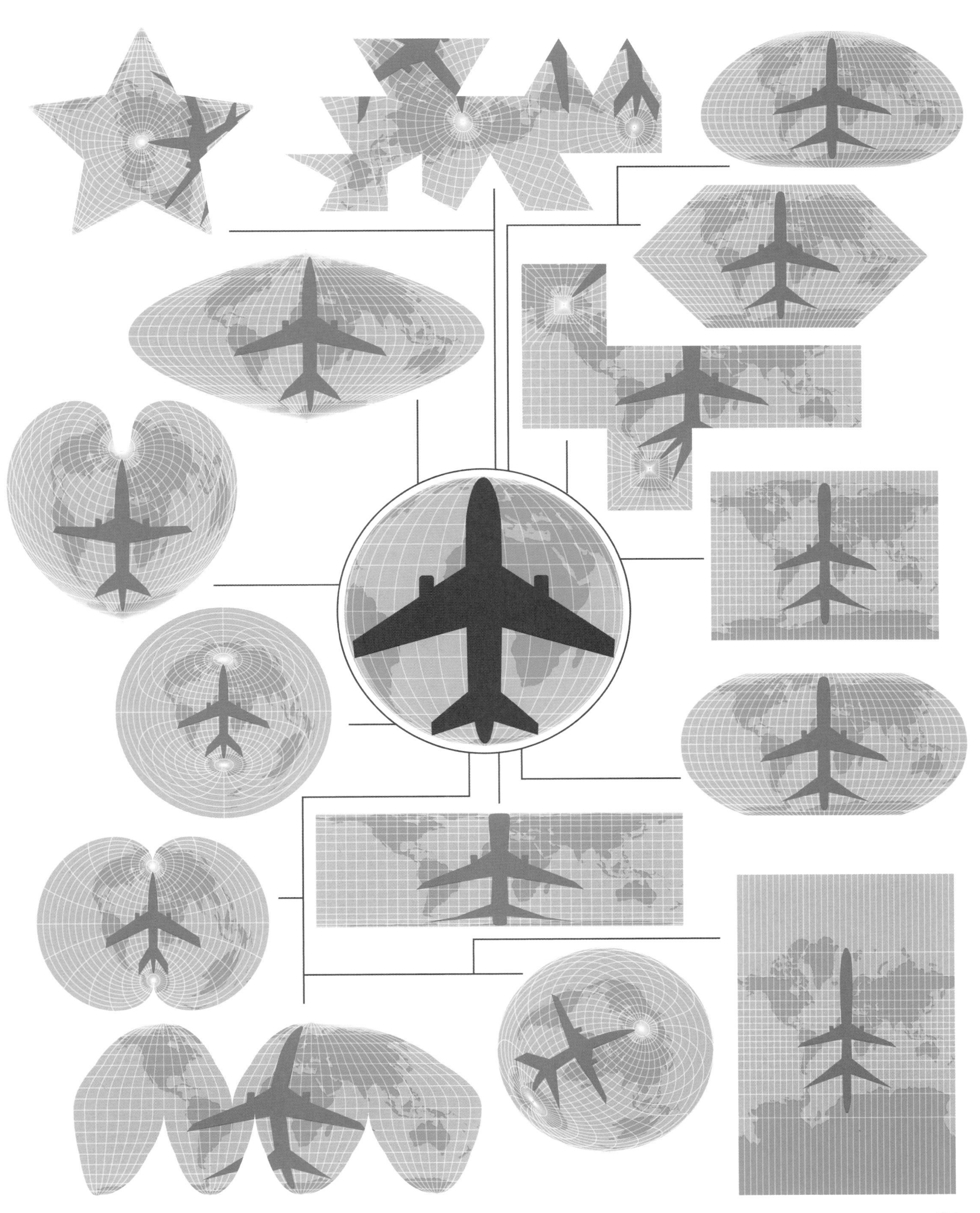

# Atlases

## Quite literally, a book of maps.

**An atlas is a collection of maps usually presented in the form of a book or as a multimedia digital product.** Atlases tend to present information about Earth or specific regions. Typically, maps will present a comprehensive depiction of geographical features. They will also include detailed representations of social, economic, and political features, administrative boundaries, and a range of thematic information. Alternatively, atlases have also been used in other contexts such as to detail the anatomy of the human body.

**Gerardus Mercator is credited as the first to use the term *atlas*.** His *Atlas Sive Cosmographicae Meditationes de Fabrica Mundi et Fabricati Figura* (*Atlas or Cosmographical Meditations upon the Creation of the Universe, and the Universe as Created*) was published in 1595. Mercator used the word to refer to the universe though subsequently, common usage was simply a collection of maps. Mercator wasn't, however, the first to place a collection of maps in book form. Abraham Ortelius is generally regarded as the first to publish a systematic collection of maps in book form in 1570 in his *Theatrum Orbis Terrarum*. And, of course, the term 'atlas' was originally the name of the giant of Greek mythology who held the sky on his shoulders.

**Atlases can be grouped according to a number of aspects.** They can be categorised by format and volume (e.g. pocket, printed), spatial extent (e.g. world, city), thematic content (e.g. political, economic), information level (scientific, school), use (road, city plan), or producer (official, commercial). Digital atlases combine maps with multimedia to support non-linear, dynamic products with far more flexibility and extendibility than traditional printed products. Search, basic analysis, interaction, multiscale zooming, and as an interface to further content all present advantages over the printed counterpart.

Atlases have often been regarded as a status symbol. Dutch cartographer Joan Blaeu published his epic *Atlas Maior* between 1662 and 1672. It comprised multiple volumes (from 9 volumes for the Dutch version to 12 for the French), some 594 maps, and over 3,000 pages of text. It was the largest book published in the 17th century.

The *Atlas of the British Empire*, published by E. Stanford Ltd, is credited as the world's smallest at 5 × 4cm. It was made as a gift to the Queen in 1926 and designed to sit inside the library of Queen Mary's Dolls' House, a toy model 1.5 metres high. The tiny maps even contain inset maps of Great Britain to the same scale.

For over 350 years the *Klencke Atlas* held the record as the world's largest atlas. It measured 1.75 m tall by 1.9 m wide when opened. The 41 maps of European states contained in the atlas were designed as wall maps, to be removed and hung. Many of the maps were created by Blaeu and Hondius. Klencke himself was a Dutch merchant who presented the finished atlas to King Charles II of England in 1660. It is now held in the British Library, London.

In 2012, the Australian publisher Millennium House produced a version of its *Earth* range of atlases designed to challenge the *Klencke Atlas* as the world's largest. *Earth Platinum* measures 1.8 m tall and 2.8 m wide when open, weighs 150 kg, and has 128 pages. It's a comprehensive world atlas featuring a world map for each thematic topic as well as Earth's regions. Gigapixel photography is presented throughout to create stunning double-spread illustrations. The image of the Shanghai skyline is the world's largest image in a book. It is comprised of over 12,000 separate images stitched together into a 272 gigapixel final form. *Earth Platinum* was limited to 31 copies, each retailing at US$100,000.

**See also:** Reference maps | Thematic maps | Topographic maps

Earth Platinum, 2012

EARTH The World's Largest Atlas Millennium House Australia

2.8 metres

Atlas of the British Empire, 1926

5cm

## 100 Aker Wood

E. H. Shepard

1924

Often, the earliest maps we have the privilege of seeing as youngsters are those that accompany our books. Our plastic minds still heavily and wonderfully blur abstract representations of our world with our growing tangible experience. The maps we see, particularly those fanciful illustrations that paint imaginary worlds, might as well be the real thing.

Over time we build our practical shrouds between the imaginary and the possible, dimming our ability to fall fully into a map and live within the brush strokes. The masterful illustrators, however, pull back that curtain and conjure authentic worlds without a sense of condescendence or remote coolness.

Ernest H. Shepard was invited by A. A. Milne to illustrate a world inhabited by Winnie-the-Pooh, in the like-titled collection of stories published in 1926. Shepard's drawn map of the 100 Aker Wood, in the assumed hand of Pooh's young friend Christopher Robin, provides the beautiful and simple context for the adventures undertaken by Pooh and his friends.

If I hold the candle back far enough, even a memory as poor as mine can recall flickers of poring over this sketched landscape, drawing me into the very real world of Pooh and Piglet. Now, as a father of young children, I can appreciate this wonderful map anew and revel again in the vicarious perspective that comes with watching small eyes dart here and there, still possessing that raw gift of imagination.

—John Nelson

CE FOR
PIKNICKS
TO NORTH POLE
BEE TREE
BIG STONES AND ROX
BBITS HOUSE
RABBITS FRENDS AND RALETIONS
MY HOUSE
OWLS HOUSE
100 AKER WOOD
EEYORES GLOOMY PLACE
RATHER BOGGY AND SAD
ME AND MR SHEPARD HELPD

# Balance

## Arranging map components to create visual balance improves composition.

**Achieving balance is the key to so many things—food and drink, political views, the Force, and maps.** But like so many things, actually achieving balance in cartography is not the easiest of tasks. Balance refers to the visual impact of the arrangement of the various components that make up the map. Put simply, a well-balanced map will look much better than one that has poor balance. The objective is to create a visual impact that doesn't appear lopsided or heavy on one side or the other. Balance is a state of equilibrium that, when lacking, leads to visual disturbance.

**Any image has two different centre points—a geometric centre and an optical centre.** The geometric centre is simply the intersection of lines drawn from opposing corners of the page or screen. The optical centre is found by drawing a line vertically from the geometric centre and positioned about one-fifth of the length of that line higher. This is where the eyes naturally look on first inspection of a map or illustration. This becomes your visual fulcrum for balancing map components.

**Map components have visual mass and position.** Mass is modified depending on the location so positioning heavier objects closer to the optical centre and lighter ones further away helps achieve balance. Objects tend to pull less at the centre of a composition than the extremities and so if you position a large map component to the periphery it will tend to pull the visual balance of the entire page in that direction. Map components in the upper part of a map will appear heavier than those in the lower part. Similarly, perceptual studies suggest that objects on the right appear heavier than those on the left. Balance is more than size and position though. It is modified by your symbology, use of colour, and proximity to other components.

One way to approach the task of finding balance in your map is to think of all the different components as separate entities. Imagine them cut out and placed randomly on a table in front of you. The task is to move the various pieces about so that they form a composition that looks right. You might imagine a see-saw positioned around the optical centre of the map with the size and weight of different objects positioned along the balancing beam to create balance.

Visual weight is a function of your symbology. For instance, colour use modifies balance with bright, saturated colours being heavier than lighter, desaturated colours. White can often look heavier than black, and red is heavier than blue. Map components that are of most interest on the map will be implicitly heavier than those that are contextual. Isolated objects will seem heavier than similarly sized and symbolised objects that are surrounded by other features. Regularly shaped symbols and features will appear heavier than irregularly shaped features.

The location of objects in relation to others can also lead them to take on direction, appearing to be pulled across the map by other objects. The shape of an object itself might also create the impression of direction which can modify balance.

It's sometimes difficult to achieve balance simply by virtue of geography presenting you with awkward shapes that are immovable. Rather than battling these constraints, instead work with marginalia and other map components to build balance into the overall composition. For instance, work with page size or page/screen orientation. Fit the map's layout to it rather than the map to a preconfigured layout. Use surrounding space to build balance.

**See also:** Flourish | Layouts and grids | Map aesthetics | Proximity in design

ymmetry
qual weight
n equal
des of a
entral fulcrum
Symmetry
S
symmetry
nequal weight
n equal
des of a
entral fulcrum
ut arranged
create balance
Asymmetry
A
Centre
wo page centres:
eometric centre
nd optical centre
Actual Page Optical Centre
Actual Page Geometric Centre
Centre
C
adial Symmetry
alances based
n a circle
Radial Symmetry
R
Approximate
Symmetry
Near equal
weight on equal
sides of a
central fulcrum
Approximate Symmetry
A

# Basemaps

## Prepublished and configurable basemaps have become a cornerstone of map design and publishing.

**Freely available, authoritative maps have become a disruptive paradigm in modern cartography.** As a consequence, there has been a profound change in consumption of map products. Maps bought once every few years to support a single use (perhaps hiking or a holiday to an unfamiliar place) have largely given way to online, ubiquitous maps that are constantly updated. Initially created to support the delivery of other content, these basemaps have become the de facto pocket reference map. Basemaps go beyond simply a canvas upon which we can overlay other map content. They are reference maps in their own right. They have progressed beyond single-scale slippy maps to multi-scale products that give a seamless overview and rich detail with a click, pinch, or swipe. Different zoom scales work independently as well as in sequence, and you can choose, in seconds, different styles to suit preference or purpose.

**The one-map-fits-all reference product has been replaced by switchable, configurable reference basemaps.** Part of the palette of basemaps offered by different vendors such as Google, Mapbox®, and Esri include those that are designed to be neutral and which support high-quality thematic overlays. Furthermore, you can build your own bespoke basemaps or modify existing ones through accessible APIs or customisation. Raster tilesets can be easily customised and cached and made available through a web server to clients to create canvases that support different cartographies.

**Vector tile basemaps with fast rendering, efficient storage, and seamless zooming have improved the way in which these global maps support our various mapping needs.** The shift from raster to vector also brings the opportunity for cartographers to style their map in different ways which brings back an emphasis on design in relation to basemap use. Prestyled basemaps are undoubtedly useful but the ability to be able to re-imagine vector map data and design it to be specific to the map you're making is powerful and vital to modern map design.

Cartographers have often been accused of doing little more than drawing lines and colouring in. It's a rather negative and stereotyped perception that fails to acknowledge the many different aspects of the job of making a map or indeed, to what purpose the map design is intended to support. However, the ability to style basemap data is important in the context of a coherent map product.

One of the major requirements for a map has always been base data to give context, situate our own data, or simply to indicate the pattern that humans make on the natural landscape. Acquiring base topographic data has become easy through the many open source and paid-for suppliers that exist. But their prestyled nature has perhaps led to less experimentation and, stylistically, many web maps have become similar in appearance or have led to awkward end products with data being overlaid across unsuitable basemaps.

With the increasing availability of basemap styling tools mapmakers are once again being encouraged to experiment and see the design of the basemap data as fundamental to, and part of, the overall map's composition. Moving beyond simply mashing up data across prestyled basemaps is important. The map's design is a context for anything you overlay, and being able to select, omit, and symbolise third party vector data for your own map is empowering.

Sometimes simply restyling builds a new reference map for a specific readership (e.g. a children's atlas or for science fiction fans). Being able to mute the background also gives you a way to build a basemap onto which your own themed data can sit, be seen, and which sits in a well-thought-out hierarchy.

**See also:** Foreground and background | Imagery as background | Mashups | Reference maps | Topographic maps

Friuli-Venezia Giulia
Trento
Ljubljana
Trieste
Veneto
Brescia
Verona
Padua
Venice
Gulf of Venice
Rijeka
CROATIA
Po
Ferrara
Parma
Modena
Bologna
Emilia Romagna
Ravenna
Forlì
Forlì
Zadar
San Marino
Florence
Leghorn
TUSCANY
Ancona
MARCHE
Perugia
Umbria
Italy
Terni
Pescara
L'Aquila
Abruzzo
Rome
Campobasso
Latina

# Binning

## Aggregating point data into logical units provides a visual summary.

**Presenting large quantities of point data on a map is particularly challenging because of the inevitability of overlapping symbology.** This results in what has been referred to as red dot fever; the so-called push-pin approach to mapping points is the simplest way to place such data on a map. Overlapping becomes more acute as the scale of the map becomes smaller and symbols coalesce. Binning provides a way of aggregating and summarising the data for smaller map scales.

**Binning aggregates points such that they fall into logical areal units to create a map of density that exhausts space.** A drawback is the loss of the impression of congested space versus areas with no data points but the obvious advantage is the way to better summarize the core aspect of the spatial pattern of data points, namely how dense is their distribution as a comparison across the map. Put simply, aggregating data can often show its essential characteristics much better than the raw points.

**Binning data can be as simple as aggregating the points into some geographical units such as census areas.** This is generally ill-advised unless the point data has some meaningful relationship with the areas. If it doesn't, the arbitrarily sized and shaped areas will likely propagate the pattern of the areas rather than the points. Instead, binning usually employs a regular grid of tessellated shapes such as squares or hexagons. This is akin to a spatial histogram that uses equal interval areas. The benefit of using a regular grid is the map can show totals, ratios, or percentages. The equal size of the shapes overcomes the problems usually seen with choropleth maps that have unequal sized areas. They are equal in size so equal in a visual sense too.

**It's important that data is projected using an equal area projection prior to binning.** This means any regular bins placed atop will sample an equal area of land. If binning on a sphere then you'll need discrete global grids instead which take into account the spherical or ellipsoidal geometry.

**See also:** Dasymetric maps | Data density | Dot density maps | Simplicity vs. complexity

The spatial summary can simply be the count of points in each binned area or it can be a summary of attributes of the data at each point. For instance, a single point may vary in value which might contribute to a greater extent when combined with other values at other points. The size of the binned areas will be determined by the data you're binning. You do not want a layer of tessellated shapes that overgeneralises the results by being too large. Conversely, the shapes must be small enough to capture the character of the density of the dots but not so small that they are binning a very small number of points and not generalising enough.

With any generalisation of data, finding the balance can be a trial and error process. If the map is to be viewed at multiple scales it's also a good idea to create bins of different sizes and to resample the data at different scales. At smaller scales you'll need larger bins (more generalised) so as the map reader zooms in the bins get smaller, the generalisation is less and greater detail is revealed. This can be very effective when combined with a map of the actual points at a large scale at which they can be seen and distinguished.

Symbolisation for a layer of binned data would follow the same principles as for a choropleth map so darker would suggest more or a diverging scheme could be used if the summary varies around a specific value of interest. Leaving empty bins off the map can be a useful technique to emphasize the lack of points. Multivariate bins can also be developed that use size or hue as symbol characteristics that might encode a different variable

Chicken Joints
&
Burger Joints

Count of **all locations**
when binned into squares

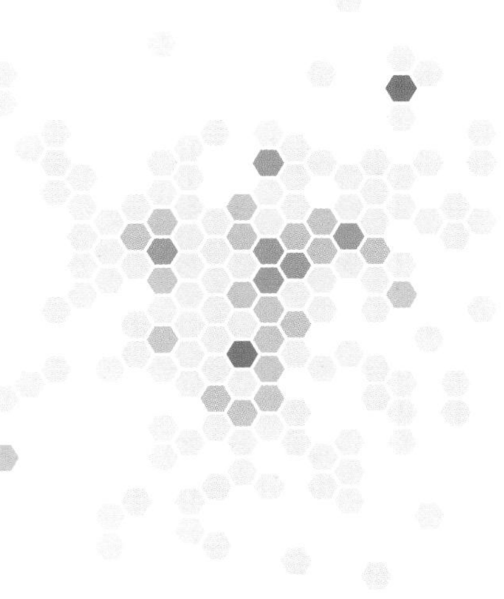

Count of **Burger Joints**
when binned into hexagons

Count of Chicken Joints
when binned into triangles

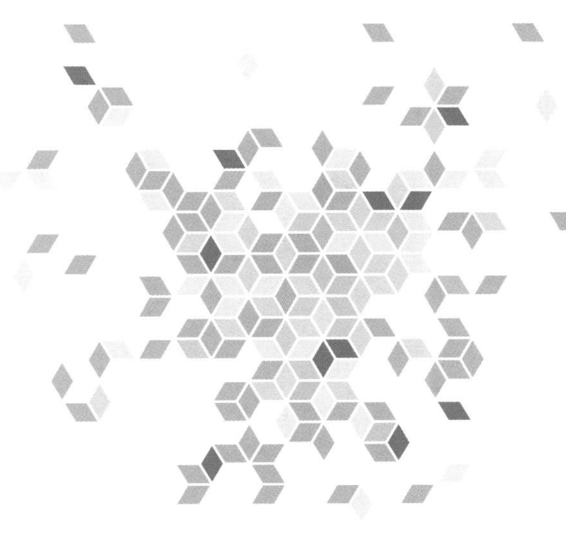

Rate of Chicken Joints vs **Burger Joints**
when binned into rhombuses

# Branding

**Branding provides structure and certainty in a map and related products, usually specific to particular publishers or organisations.**

**The work of many cartographers demands them to apply design and styles according to a set of specifications developed by an organisation.** These specifications give structure and a design framework that clearly situate maps and associated products within the same visual family. One only has to consider the branding used by major cities for their metro rail services to see this at work. Or, think of the different designs that characterise national map series.

**Branding breeds familiarity that builds awareness and loyalty.** People become comfortable with brands, which enables them to immediately associate products with places and the organisation that publishes them. Branding in commerce is designed to sell product, and it's actually a similar concept for cartography. We want to sell a brand, part of which is the cartography that supports the brand. However, consumers have little choice in riding a city's metro because there is only one so why does branding exist? It's done to provide wider exposure and to bring coherence to all the various elements of the design—maps, signage, colours, and typography.

**Communication is the essence of good branding as organisations blend design components to create a whole.** Large branding guidelines often accompany designs, and copyrights exist to protect such branding. National mapping agencies also apply branding to their products and we can easily recognise an Ordnance Survey map from one made by swisstopo, National Geographic, or USGS. They have a look and a feel that generations become familiar with.

**Branding goes beyond look and feel.** Maps often use specific colours, symbology, and fonts that help define them but there will be other clear branding characteristics. The structure of a page, its layout, or the marginalia will all be consistent. Logos are also important and often act as the focal point for a brand. Think of the NASA logo, or the London Underground Roundel or any famous logo. They become iconic and lead the branding in which maps might also be situated.

**See also:** Advertising maps | Copyright | Craft | Emotional response | Purpose of maps

Branding is usually related to an organisation but for the independent cartographer working on one-off projects, branding can also be useful. Developing a style that people identify with can lead to a brand identity. Does the subject matter relate to a certain style or type of map for instance? In fact, do you make maps that are similar in scope which might suit a wider brand identity? Will the reader be expecting a particular look and feel because of who made the map?. Do you need to think about marks, logos, and pictorial components that will allow the reader to inherently and more immediately understand the map and its information as it relates to a wider brand initiative?

Brands have to be self-explanatory. That might not happen overnight but there comes a point where readers will expect a certain look and feel. If they are challenged by the unfamiliar, it becomes an impediment to their ability to retrieve information and can be jarring and uncomfortable.

It's often worth exploring the brand guidelines of many organisations. For instance, the brand guidelines and standards manuals of Transport for London (TfL 2014) and the New York City Transit Authority (Vignelli and Noorda 1970) clearly illustrate the thinking behind their approach to very different subway map designs. They are both synonymous with their cities and help to brand the city itself.

As an independent mapmaker, there's a fine line between persisting with a particular map style and deciding when to change tack. Making all your maps in the same style will ultimately lead to a sense of over familiarity with your work. Your audience may tire of seeing the same design ethos. There's no magic recipe for getting this balance right but being aware of the impact of branding, style, and your audience's capacity for seeing new approaches is useful.

**Opposite:** My own arm getting branded with a likeness of my favourite map.

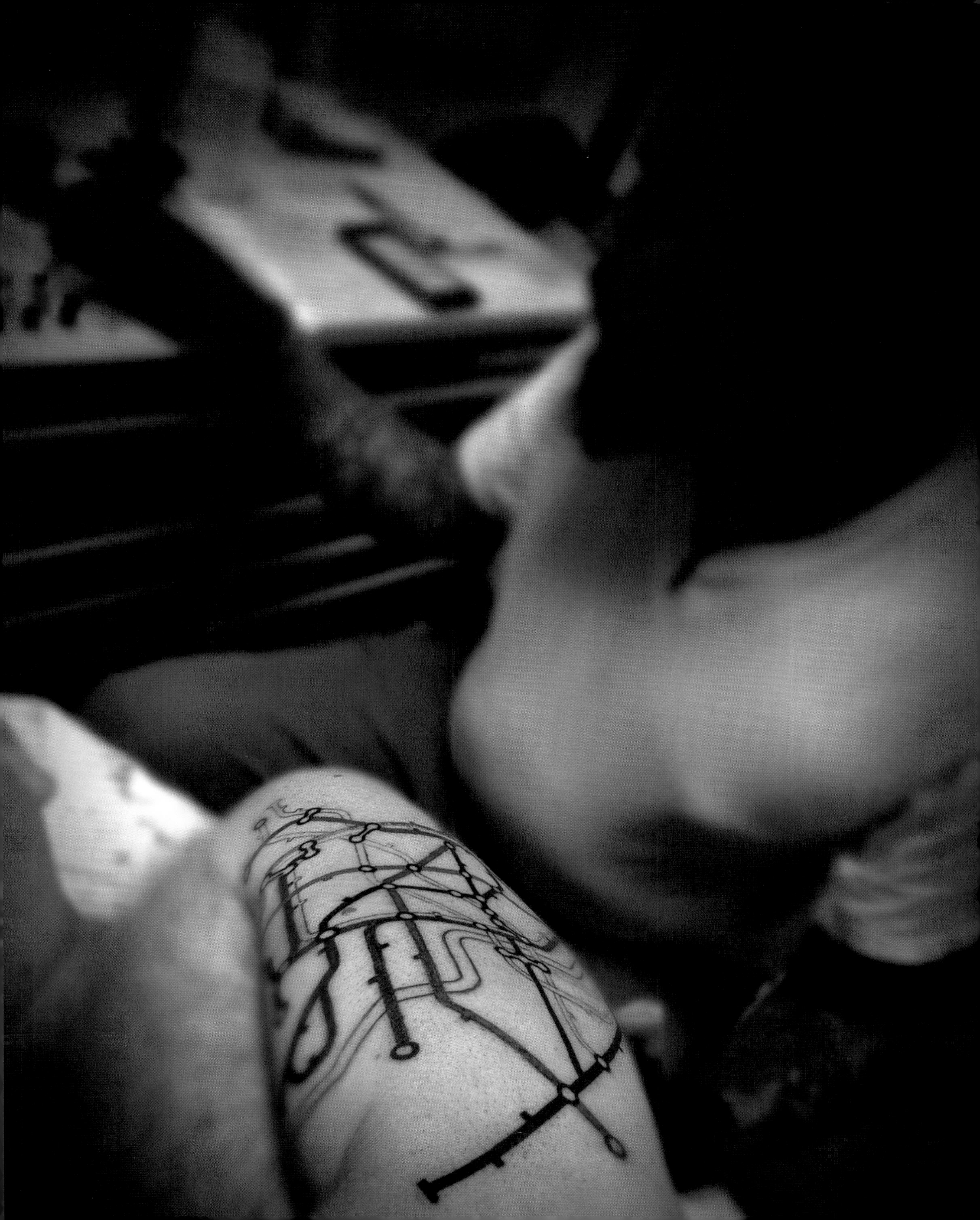

## Airspace: The Invisible Infrastructure

NATS

2014

A brilliant 2.5-minute video animation of the daily air traffic in UK airspace, produced by NATS (National Air Traffic Services). It has a subtle projected DEM underlying dynamic blue flight traces (becoming white in busy areas).

The 'day' starts with the arrival of early morning transatlantic flights and builds as UK airports spring to life, showing 6,000 flights in total. The developing pattern clearly shows both the corridors to the major airports and the stacking patterns over London's airports. It also shows the military training flights over North Wales and the traffic in and out of Aberdeen's heliport.

Produced by Bristol-based company 422 South, using Maya® software, exported to After Effects, and finished in Final Cut X, it makes a complicated picture clear, and nicely complements Joanna Parker's (unrelated) *Britannia Obscura*, whose chapter 'Highways in the Air' explains air route development, restricted airspace, and the origin of differentiated flight paths.

This type of cinematic cartography shows how a short animated video adds to the various mediums which maps inhabit. They can be particularly impactful and eye-catching but they allow animated symbology to be used to express movement in a way that a static map is unable to achieve.

—Steve Chilton

# Cartograms

## Cartograms offer a way of accounting for differences in population distribution by modifying the geography.

**Geography can easily get in the way of making a good thematic map.** The advantage of a geographic map is that it gives us the greatest recognition of shapes we're familiar with but the disadvantage is that the geographic size of the areas has no correlation to the quantitative data shown. The intent of most thematic maps is to provide the reader with a map from which comparisons can be made and so geography is almost always inappropriate. This fact alone creates problems for perception and cognition. Accounting for these problems might be addressed in many ways such as manipulating the data itself. Alternatively, instead of changing the data and maintaining the geography, you can retain the data values but modify the geography to create a cartogram.

**There are four general types of cartogram.** They each distort geographical space and account for the disparities caused by unequal distribution of the population among areas of different sizes. Non-contiguous cartograms are the simplest type where the shapes of the enumeration units are resized according to a variable. Typically this results in an overlapping or non-overlapping version. In a non-contiguous cartogram, topology (adjacency and connectivity) is sacrificed to preserve shape and enable recognition of geographical areas. Contiguous cartograms maintain connectivity between adjacent geographical areas. This often results in dramatically distorted shapes as the data variable is used to warp geography. Graphical cartograms build on the principles of proportional symbol maps but disregard the underlying geography. The Dorling cartogram, perhaps the most well known, uses proportional circles, organised to provide the best adjacency possible. A variant of the Dorling cartogram is the DeMers cartogram, which uses squares instead of circles and thus reduces the gaps between shapes. A final type is the gridded cartogram, which is a compromise between geographic shape and a more uniform topology that enables people to find familiar places through more recognisable adjacencies. Squares and hexagons tend to be the preferred shape for these mosaic-style gridded cartograms.

**See also:** Chernoff faces | Schematic maps | Treemap | Voronoi maps | Waffle grid

Consider the United States map in which states with larger populations will inevitably lead to larger numbers for most population-related variables.

However, the more populous states are not necessarily the largest states in area, and so a map that shows population data in the geographical sense inevitably skews our perception of the distribution of that data because the geography becomes dominant. We end up with a misleading map because densely populated states are relatively small and vice versa. Cartograms will always give the map reader the correct proportion of the mapped data variable precisely because it modifies the geography to account for the problem.

The term *cartogramme* can be traced to the work of Charles Minard in the mid-1800s and various maps since have taken the form of what we now refer to as *cartograms*. Because of the complexity of making a cartogram by hand it was not until the mid-1950s that computerised algorithms began to be developed. There have been many alternatives that to some extent fall into one of the four main categories. Perhaps the most infamous contiguous cartogram is the Gastner-Newman, otherwise known as a population-density equalising cartogram. It does an excellent job of retaining some character of the general shape of individual areas. However, the degree of distortion often renders the map difficult to interpret because of the abandonment of familiarity. Of course, they are attention grabbing and that is often their primary use.

Cartograms do not have to be area based. The distortion of line direction and size can be modified according to some character of its attributes to create a linear cartogram. For instance, subway maps are a type of linear cartogram.

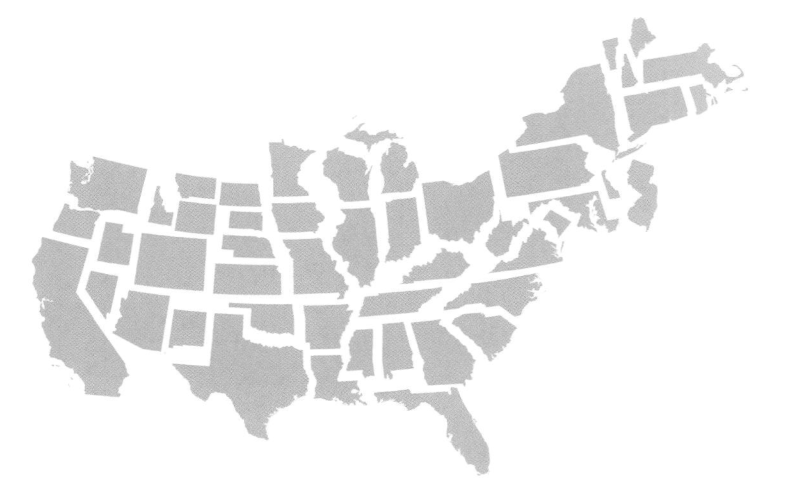

Non contiguous

Shape retained. Area modified. Adjacency not maintained.

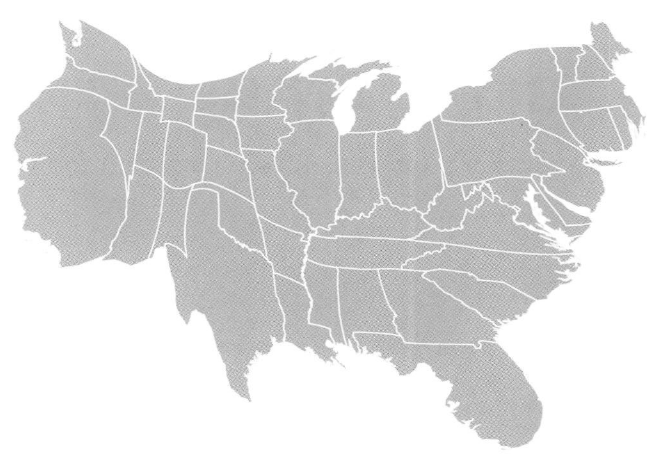

Contiguous (Gastner-Newman)

Shape and area modified. Adjacency maintained.

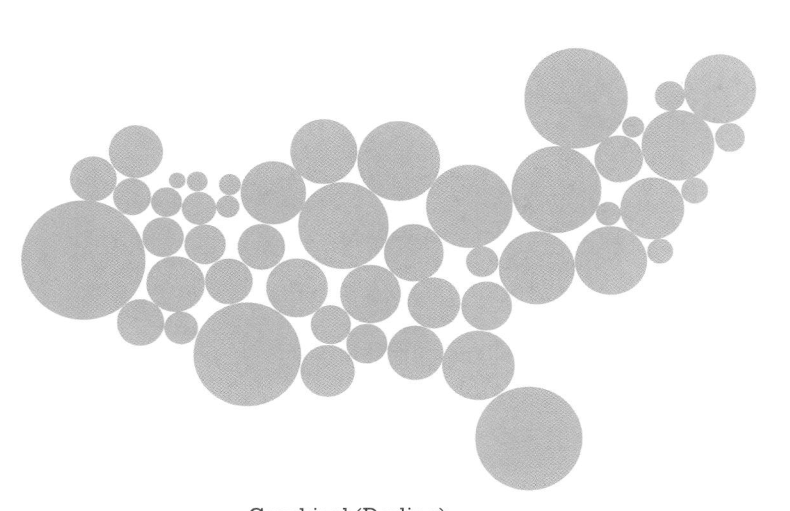

Graphical (Dorling)

Shape modified. Area modified. Adjacency not maintained.

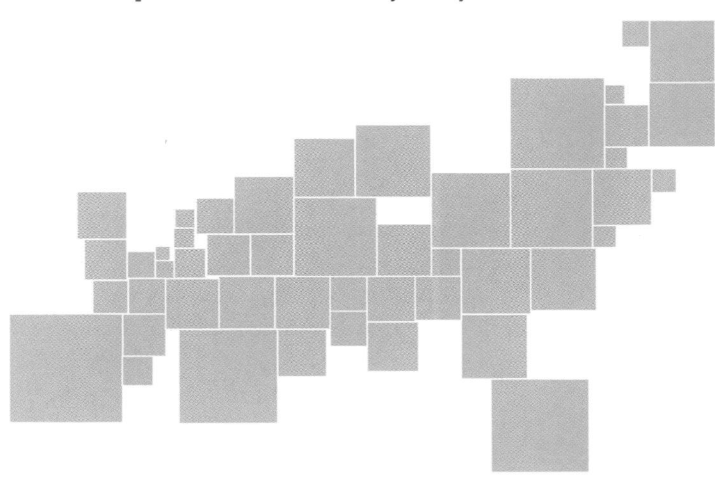

Graphical (DeMers)

Shape modified. Area modified. Adjacency not maintained.

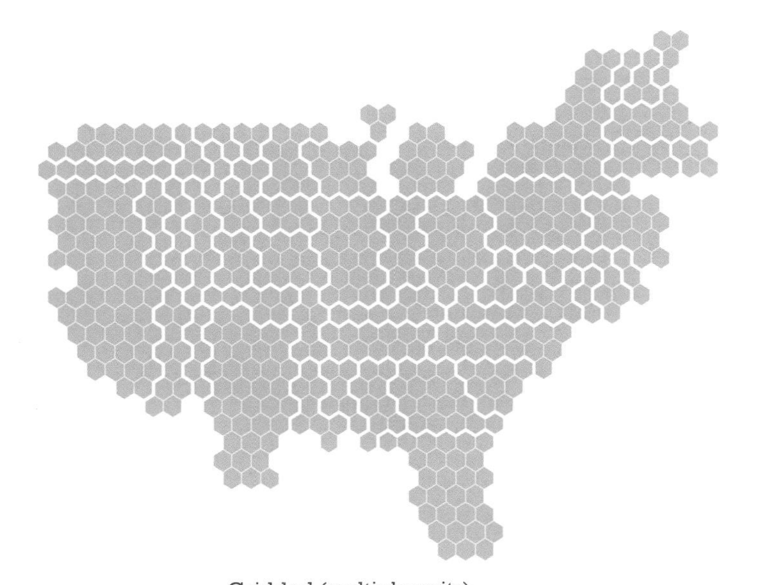

Gridded (multiple units)

Shape modified. Area modified. Adjacency not always maintained.

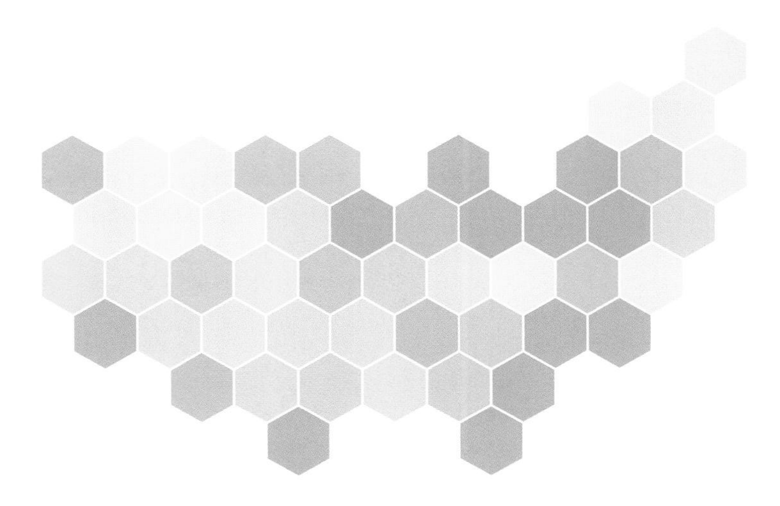

Gridded (single units)

Shape modified. Area uniform. Adjacency not always maintained.

# Cartographic process

## The journey from blank page/screen to map.

**Rarely will a great map simply fall out of your head and onto the page or screen.** Clarity of thought and an appetite for the end result are fundamental to creating an effective outcome. The importance of workflow, or process, is critical. This guides you and acts as a trigger at various stages. Even the simplest of map should establish a workflow because it forces thinking. Given the huge truncation of time that technology has brought to the cartographic process, incorporating time to think has never been more important. It's a good habit to get into because you'll end up making fewer mistakes and develop more nuanced products.

**There are many decisions to make when thinking of making a map.** Medium, format and size, and the story to tell are all large decisions. Whether you're going to use 10% or 20% grey for a line colour is a smaller decision but they are all important. In order to tackle these issues, process is vital. Being flexibly pragmatic rather than dogmatic will help your creativity. Breaking down the larger task into smaller components helps reduce the burden of pressure. Build in time to experiment with alternative approaches to the mapping task rather than ploughing straight into your first idea. Be capable of adapting and iterating. This helps overcome problems and recover rapidly. Being able to focus on different tasks helps compartmentalize the overall task. Sometimes you'll be thinking and formulating your ideas. At other times you'll be sketching and gathering raw data. You make the product by translating ideas into action.

**In truth there is no single cartographic process.** What works for you will be different from what works for someone else. Work out what works for you but try and make it a framework you can stick to. There will be aspects of every job you don't particularly care for but be regimented. They still need doing. Ask for help. Collaborating can be rewarding, and a problem shared is halved. Learn from each project to know what works and what doesn't in terms of both the process and your skills. Finally...if the map isn't working, kill it. Many cartographers have a mass of unpublished work. It's how you learn.

**See also:** Abstraction and signage | Defining map design | Defining maps and cartography | Graphicacy | How maps are made | Types of maps

Remembering that you will never create the perfect map is important. There will always be aspects that could have been different. Someone will likely point something out that irritates them or you'd overlooked. Cartography does not exist in a vacuum and perfection does not exist. Instead of trying to achieve perfection, strive instead for excellence.

1. Planning. Managing your time and resources will keep you on track and prevent scope creep. Time scales are critical. Avoid distractions. Don't overstretch. Build in time for delays.
2. Thinking. Work out when and how you think best. Maybe alone, listening to a favourite piece of music or with a pen.
3. Sketching. Draw out your ideas, literally. Make lists of ideas too, notes that record thoughts, sources, questions, tasks, and ideas. Your memory will not hold all your ideas so they need recording.
4. Listening. Throw ideas around with others. Present and seek feedback. Gain the perspective of experts in the subject you're mapping. Be prepared to be challenged.
5. Researching. Learn about the subject. Ensure sources and content are valid. Ask questions, fill in the blanks, and be interested.
6. Focusing. Be committed. Every aspect of your map deserves meticulous attention. Deal with the minutiae, and the big picture will fall into place more easily.
7. Doing. Find a place that works for you to make the map. Ensure you have the tools at hand and know how to deploy them.
8. Testing. Have others look over your work. Seek a critique and find errors. They do exist, and here's the chance to catch them.
9. Iterating. Change what needs changing. Fix what needs fixing. Don't be so proud that you ignore advice and feedback.
10. Ending. Once you're done, you're done. Don't keep going back to the work as this can become the biggest time sink of all.

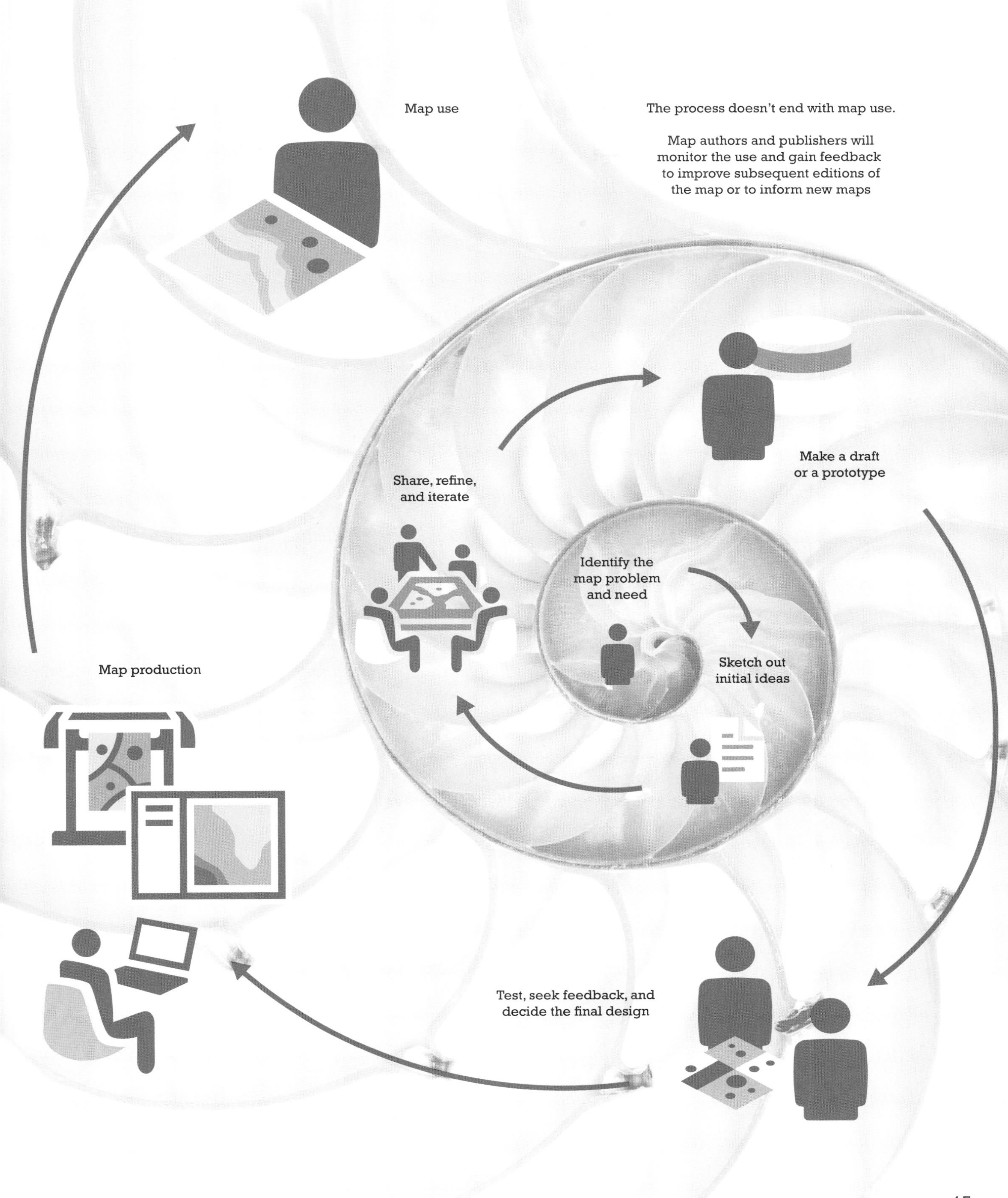
Map use
The process doesn't end with map use.
Map authors and publishers will monitor the use and gain feedback to improve subsequent editions of the map or to inform new maps
Make a draft or a prototype
Share, refine, and iterate
Identify the map problem and need
Sketch out initial ideas
Map production
Test, seek feedback, and decide the final design

# Chernoff faces

**Multivariate data displays can be distinctly geometric. Chernoff faces encode data into an altogether more human representation.**

**Displaying more than one data variable on a map can present complications.** It's difficult to encode more than a few pieces of information into a symbol without risking symbol overload, misinformation, and difficulties for interpretation. Chernoff faces provide a way of encoding multiple data values into a symbol that takes on the form of a human face. Developed by Herman Chernoff in 1973 (Chernoff, 1973), individual facial features such as eyes, ears, nose, and mouth are modified in size, placement, shape, and orientation to represent data. In all, Chernoff suggested up to 18 separate variables could be encoded on a face at one time.

**The broad idea proposed by Chernoff was that human beings easily recognise faces and their expressions.** Psychologically, humans are very good at understanding different emotions from facial expressions so the theory follows that if you design a face in such a way that it emotes the data along the sort of lines of recognition we might interpret, then data might also be better interpreted. Even small changes in the shape of a mouth or an eye can be easily seen, perhaps in a way that a similarly small changes in the size of a segment on a pie chart cannot.

**Chernoff faces handle the different facial features in different ways because they are, in their basic sense, different shapes with different dimensions.** They can also be plotted such that they are centred on the most critical x,y component, and the size of the face can be made proportional to a different variable.

**For some, Chernoff faces are contentious, and the use of a human face can bring unintentional consequences.** Inevitably, insensitive use of colour or modifying facial features that may appear stereotypical can encode unwanted messages. It is perhaps for this reason that they are not often used. Interestingly, the use of facial icons (emoji) are commonly used in SMS (Short Message Service) and instant messaging to relay a simple emotion, and so the basic idea has a wider acceptance in society beyond the cartographic realm.

**See also:** Graphs | Multivariate maps | Pie and coxcomb charts

A variation on Chernoff's original idea would be to create an asymmetrical version such that one half of the face encodes a different set of data from the left. This could be used to show two time periods and, so, a way of showing how a theme changes between two important time periods. It might also deal with the issue that creating symmetrical Chernoff faces inevitably results in wasted space as one side of the face is encoded in the same way as the other. Other variations might be to use actual human faces with variations in expression or, possibly, some development of emoji.

The use of Chernoff faces has provided some contentious examples. In 1977 Eugene Turner made a map called *Life in Los Angeles* where the faces carried emotion. As he said, 'It is probably one of the most interesting maps I've created because the expressions evoke an emotional association with the data. Some people don't like that.' (Turner 2004). His map explored wealth, employment, stress, and the relationship with proportions of white population. Needless to say the map was controversial since affluent white faces gave way to poor black faces. The statement was powerful and proves how the cartographer needs to tread a fine line between statement and overstatement or downright offence.

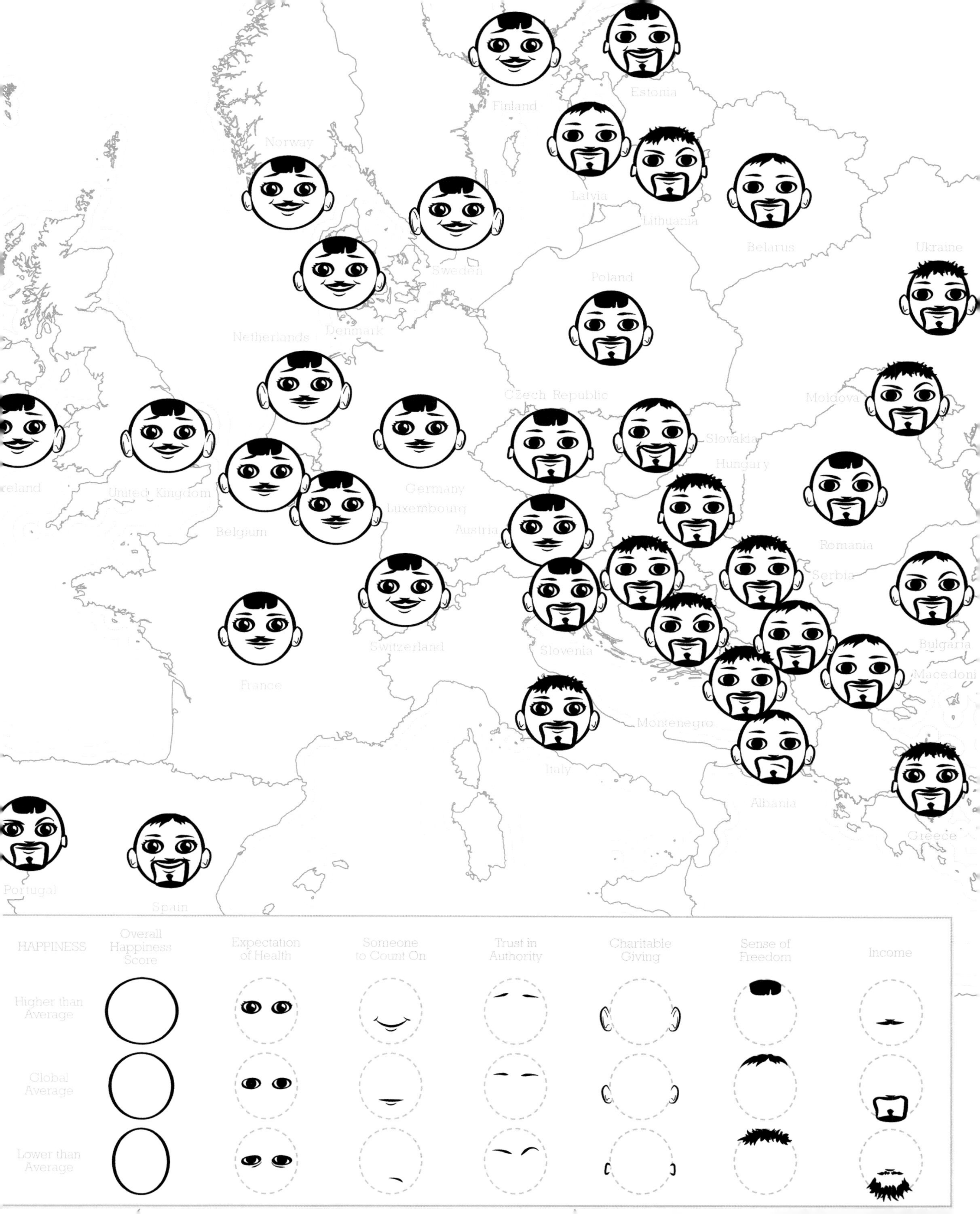
Finland
Estonia
Norway
Latvia
Lithuania
Sweden
Belarus
Ukraine
Poland
Denmark
Netherlands
Czech Republic
Moldova
Slovakia
Hungary
Ireland
United Kingdom
Germany
Luxembourg
Belgium
Austria
Romania
Serbia
Switzerland
Slovenia
Bulgaria
France
Macedonia
Montenegro
Italy
Albania
Greece
Portugal
Spain
HAPPINESS
Overall Happiness Score
Expectation of Health
Someone to Count On
Trust in Authority
Charitable Giving
Sense of Freedom
Income
Higher than Average
Global Average
Lower than Average

# Choosing type

## What typeface should I use? It's a simple question. There's no simple answer.

**Many national mapping agencies have a specific typeface that contributes to their house style and the consistent appearance and recognition of their products.** These typefaces grew out of a mechanical process with a significant cost associated with lettering a map. Type blocks were re-used across different maps so there was very little choice for the mapmaker. They support a consistent style and brand recognition. Occasionally, typefaces are changed as styles, fashions, technologies, or use demand a different approach.

**Modern choices of font are limited only by what's installed on your computer—and if it's not installed, it doesn't take long to find or buy a particular font.** So the cartographer is faced with a different question: What font should I use? For many, they will simply default to Times New Roman or Arial as the obvious fonts because of the default status they hold in word processing software. Others that are commonly used might be Calibri or Myriad. Are they useful in cartography? Possibly–but the possibilities are far richer.

**The answer to the question is as vague as 'the right font for the job'.** But that doesn't necessarily make the choice any easier. The right font must be legible in the context in which it is used. It should connote meaning and not clash with the style of the rest of the map. It must be appropriate, which may result in choosing a sober font or, for a persuasive map, a more eccentric font.

**Fonts can evoke emotions in people.** They might be described as 'too clever' or 'not very interesting', and although we might ponder how a typeface can lead to such emotion, nevertheless it's a fact we have to work with or around. We all have our preferences too. For instance, I spent a considerable amount of time thinking about the typefaces that we would use in this book and, perhaps of more interest, the ones we would not consider under any circumstances. I chose Rockwell, a slab serif font described as sturdy, blunt, and with a no-nonsense character but which harks to a handmade charm. It seemed to speak to the character of the book as a whole.

**See also:** Elements of type | Fonts and type families | Form and function | Guidelines for lettering

Explore how others use typefaces in maps, posters, and graphic design. Fonts are often associated with particular artistic or design movements. Their character almost becomes engrained in a particular reaction so use that expectation to your advantage.

Perhaps the best advice is just don't be boring! Be different and go to the effort of exploring different typefaces and how they might work. If in doubt, revert to classic maps and graphic work to see how others use fonts to develop a particular look and feel. For instance, the designer Massimo Vignelli (whose magnificent 1972 New York Subway map is still held as an icon of information design) limited himself to what he considered the six 'best' typefaces: Garamond, Bodoni, Century Expanded, Futura, Times Roman, and Helvetica.

It's also worth considering how different fonts relate to one another, and rather than selecting a single font, explore how fonts can be paired and which work well together and which don't. Some fonts harmonise well while others create visual clashes. The idea is to find concord and contrast but limit conflict between fonts. They can be very different but still remain complementary which might mean choosing a serif alongside a sans serif font. They will likely conflict if they are too similar.

Old style serifs (e.g. Garamond) generally work well with Humanist sans serifs (e.g. Gills Sans). Transitional or Geometric serifs with their contrast between thick and thin strokes (e.g. Bookman or Bodoni) work well with Geometric sans serifs (e.g. Avenir).

# APPARIEMENT des POLICES de CARACTÈRES

MENU

| | |
|---|---|
| hors d'oeuvre | Rockwell<br>Futura |
| potage | Garamond<br>Helvetica Neue |
| poisson | Caslon<br>Myriad |
| entrée | Noto Sans<br>Noto Serif |
| sorbet | **Rockwell Bold**<br>Bembo |
| salades | Arial<br>Georgia |
| fromage | **Franklin Gothic Demi**<br>Baskerville Old Face |
| desserts* | Comic Sans<br>Papyrus |

*Sometimes you have to really question if a dessert is worth it

# Choropleth maps

## A statistical thematic map that displays quantitative differences in area data using shading or pattern.

**A choropleth map represents quantitative data for areas.** Areas that display relatively less of the mapped phenomena are shown as lighter and areas that display more are shown as darker. The areas might be countries, states, counties, or other enumeration units. You should be able to recognise differences in symbols across the map that reveal areas of high and low with differences and patterns of common characteristics being legible.

**A single data value is represented by each area on the map.** Data should be in the form of ratios, percentages, or rates. Mapping totals is inappropriate because most maps have unequally sized areas which play a misleading part when trying to visually compare symbols across the map.

**Data is most often organised into classes.** A good guide is to use between four and seven classes. Fewer classes tends to over-simplify the map. Too many classes can create difficulties differentiating between symbols on the map. For a single choropleth map, using a natural breaks scheme is a good default as it groups similar data values into the same class. Alternatives such as equal interval and quantile can be useful: for instance, if you are comparing two maps side by side, a quantile scheme creates a uniform scheme for both maps which allows them to share a legend and be comparable.

**Symbols help distinguish different classes from one another.** This is achieved by varying the symbol colour or pattern. There should be sufficient variation between symbols that each can be recognized on the map. If the data contains a critical value (or class), for which the map needs to illustrate differences above and below, then a diverging colour scheme might be used.

**Unclassed choropleth maps display data across an entire shading scheme.** Individual data values are given their own unique shade in sequence. Although this approach is useful for identifying outliers, it is not so useful for identifying areas that share similar characteristics.

**See also:** Data classification | Pointillism | Unclassed maps | Value-by-alpha maps

American geographer John Kirtland Wright coined the term 'choropleth map' in 1938, although the mapping technique itself was first used by Charles Dupin in 1826. Wright cautioned against the use of choropleth maps because of their potential to treat geography as a homogeneous surface (particularly when used to map totals), instead espousing the virtues of the dasymetric map which makes allowances for spatial heterogeneity.

Spectral colour schemes should be avoided because they don't visually convey numerical differences between values of data. Consequently, you cannot decipher the relative differences between classes. Avoid using white or black, or full lightness to full darkness of your chosen colour, at either end of a shading scheme because it suggests absence at the lower end and totality at the upper end. Using white may also clash with your map background.

Mapping totals misleads since areas are different sizes. Emphasis is wrongly given to raw totals in the shading scheme. Mapping derived values, proportional to the area, corrects the map to show data consistently represented across the map and from which proper visual comparisons can be made.

Total (persons)

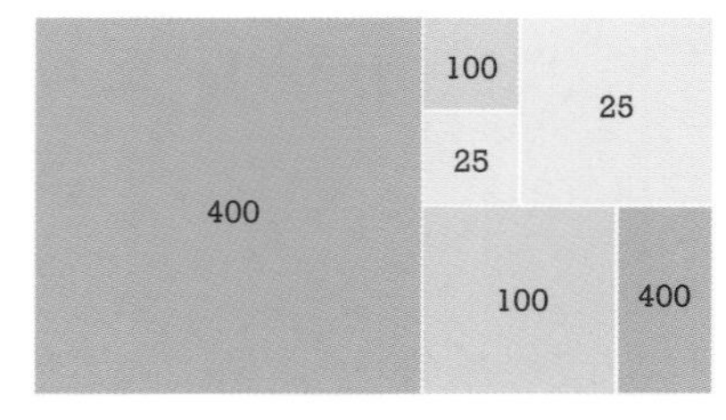


Derived values (persons per km²)

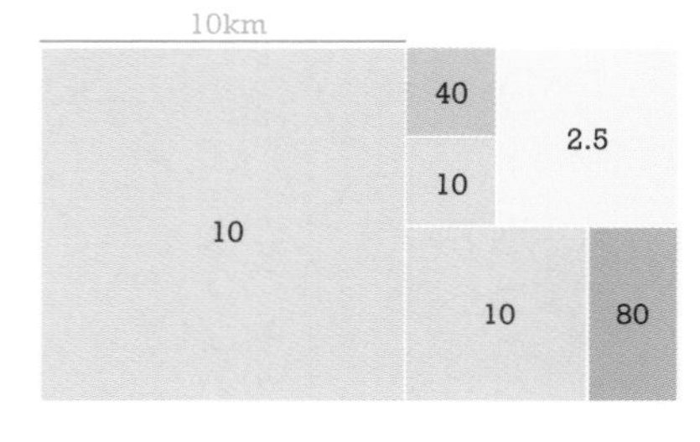

A 5 class choropleth map symbolised using a single hue sequential colour scheme where value changes to connote high to low.

This supports the reader in making comparisons between high and low areas that share similar characteristics.

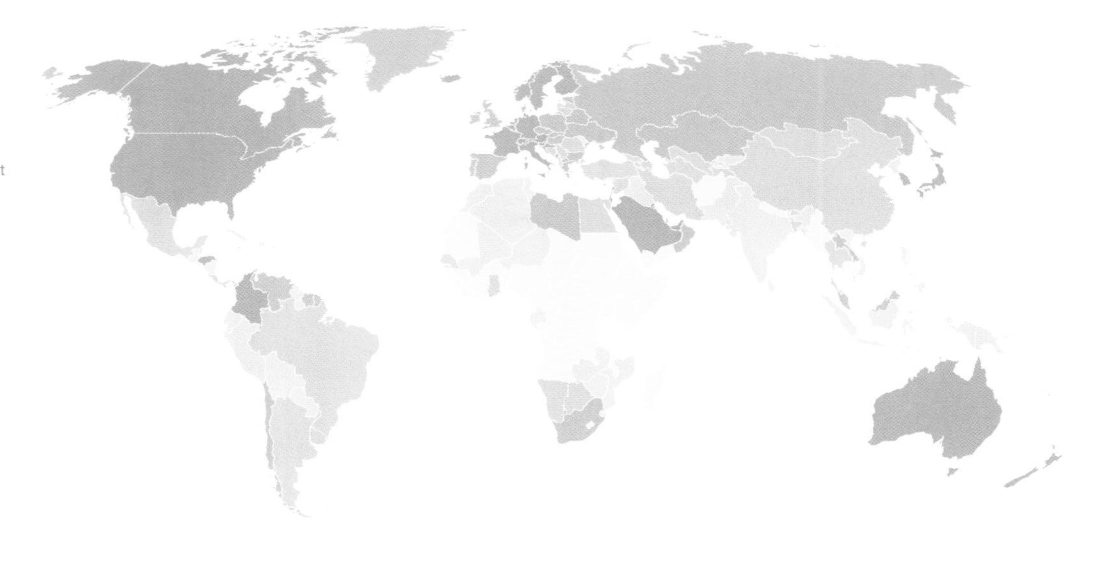

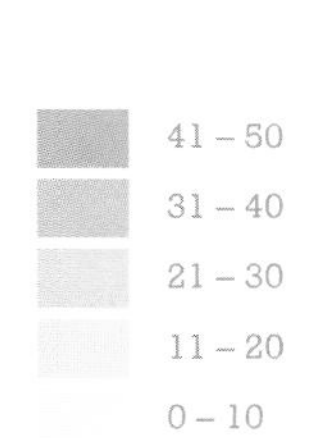


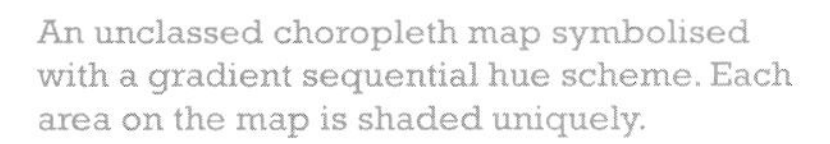

An unclassed choropleth map symbolised with a gradient sequential hue scheme. Each area on the map is shaded uniquely.

This supports the reader in viewing the position of each area in sequence as well as identifying outliers.

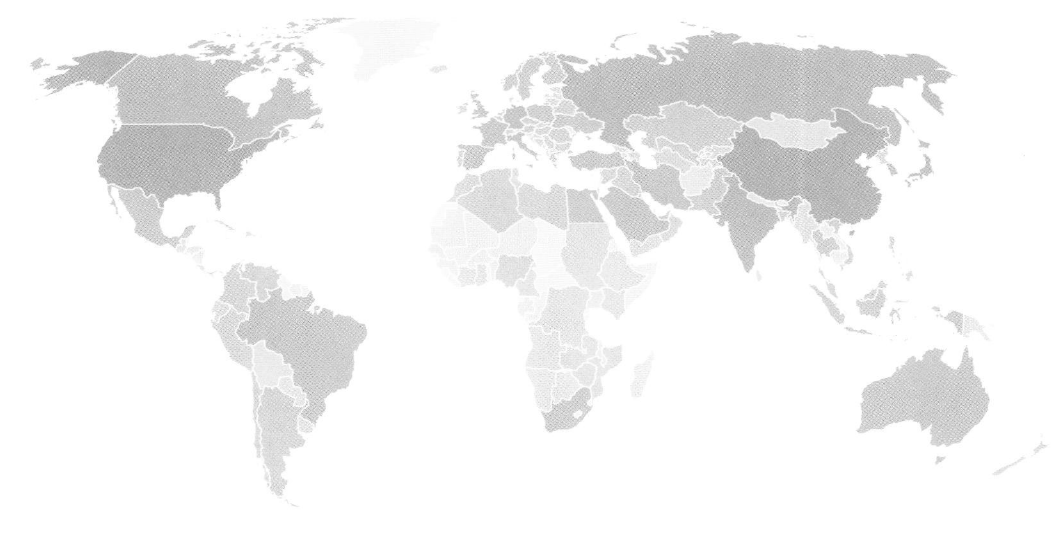

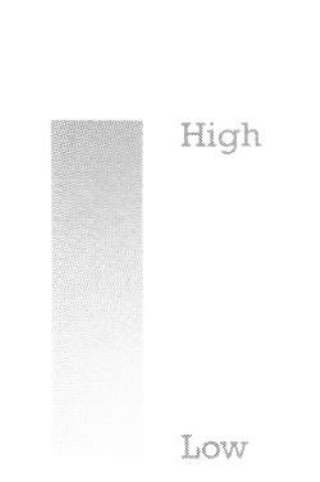


A 6 class choropleth map symbolised using a diverging colour scheme centred around a critical value in the data (in this case zero). This is effectively two sequential colour schemes where value changes to connote high to low.

This supports the reader in making comparisons between areas that share similar characteristics above and below the central value.

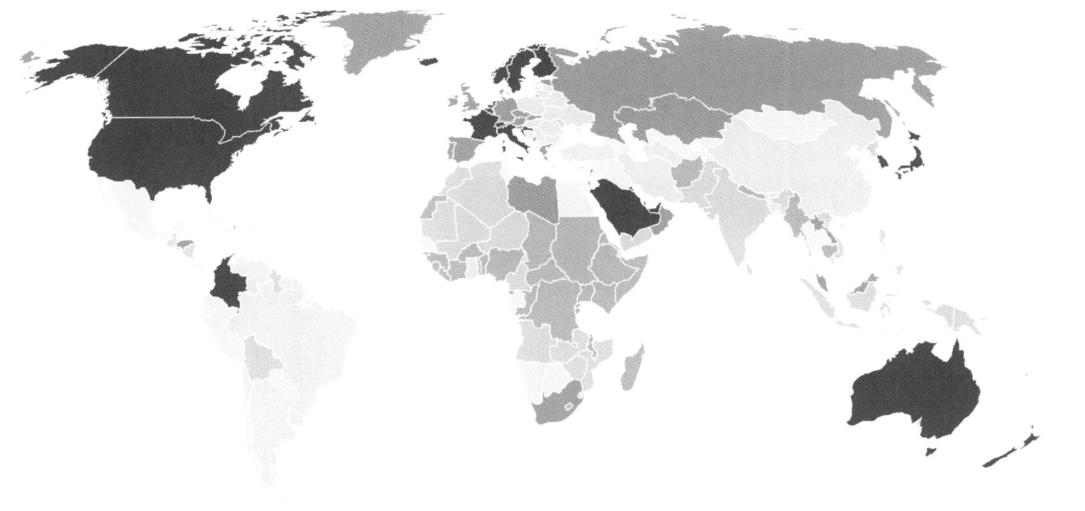

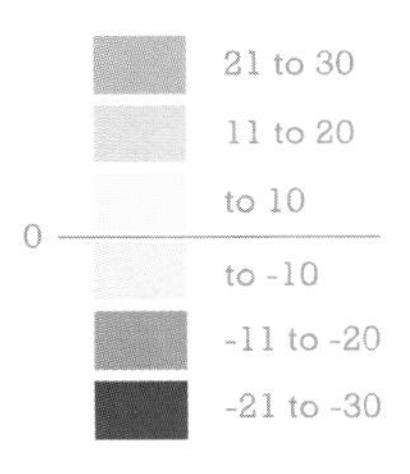

# Clutter

## Circumstance of a map containing more information than is required to get the message across and which affects legibility.

**One of the main tenets of good mapmaking is the removal of any detail and information that isn't required for the map's purpose.** It's a difficult tenet to apply because it requires considerable restraint and resisting temptation to add as much data as you have at your disposal. Clutter is the graphical failure of restraint. The term *chartjunk* (coined by Edward Tufte) is a related term that simply refers to any or all visual elements in charts and graphs that are unnecessary in order to comprehend the information represented, or which are a distraction. In this sense, visual clutter and chartjunk are both important warnings to heed when designing a map.

**In map design, the optimum solution is often the definition of what represents the minimum set of visual elements required to communicate the message.** This usually means significant processing of data and a strong appreciation of generalisation. It also means being careful about marginalia and other map components. Although data generalisation will be key to the map itself, there are many ways that a mapmaker can build visual clutter. For instance, heavy line weights, unnecessary text, complex or overly ornate borders, north points, or scale bars are all easy ways to make a map attractive yet more often have the opposite effect. They can clutter the map and make essential components less legible. Pictures, background patterns, and ornamental flourishes might be appropriate in some contexts such as a map that requires a historical, artistic look and feel. For most maps, they are also unnecessary and could easily clutter the image.

**A more serious form of clutter or chartjunk is the playful (or deliberate) depiction of information that modifies the way in which it might be interpreted.** In essence, it makes the map harder to read and to recover the clear message. It, instead, obfuscates the message. Examples of this type of clutter are depicting detail out of scale in relation to other map elements, backgrounds, or contextual map elements that make it hard to see or compare more important aspects and 3D representations of features which are more easily seen and understood in 2D.

**See also:** Aggregation | Binning | Data (c)art(e) | Data density

Context is everything and there are always exceptions to the general rule. For instance Edward Tufte famously accused Nigel Holmes' 'Diamonds were a girl's best friend' chart as being pure chartjunk.

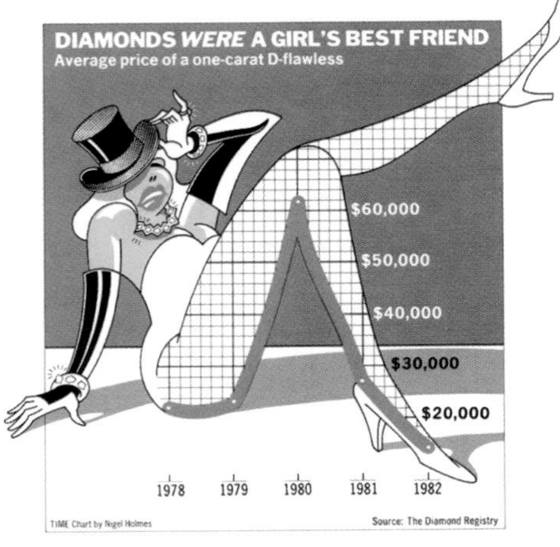


Holmes argues that this form of explanation graphic gives a memorable and arresting way to capture attention that appeals to the reader. He argues there is room for enjoyment, delight, aesthetic appreciation, and wit in graphics. For Tufte the chart is unsavoury, unnecessarily embellished, and counter to his data-ink ratio that emphasises the importance of the amount of ink used for the key message against the ink used for other components.

Knowing when and how to break the basic tenet of avoiding clutter is probably more important than avoiding it altogether. Context is everything.

**Opposite:** *Bomb Sight* by The Bomb Sight Project. Clutter is often equated with simply lots of information yet even 'red dot maps' can reveal structure as this map of WWII bombs in London shows. The map is awash with red symbols showing where bombs were dropped. Ordinarily this would be considered cluttered. Yet here, the result visually demonstrates the extent of the blitz and clutter is revealing and impactful.

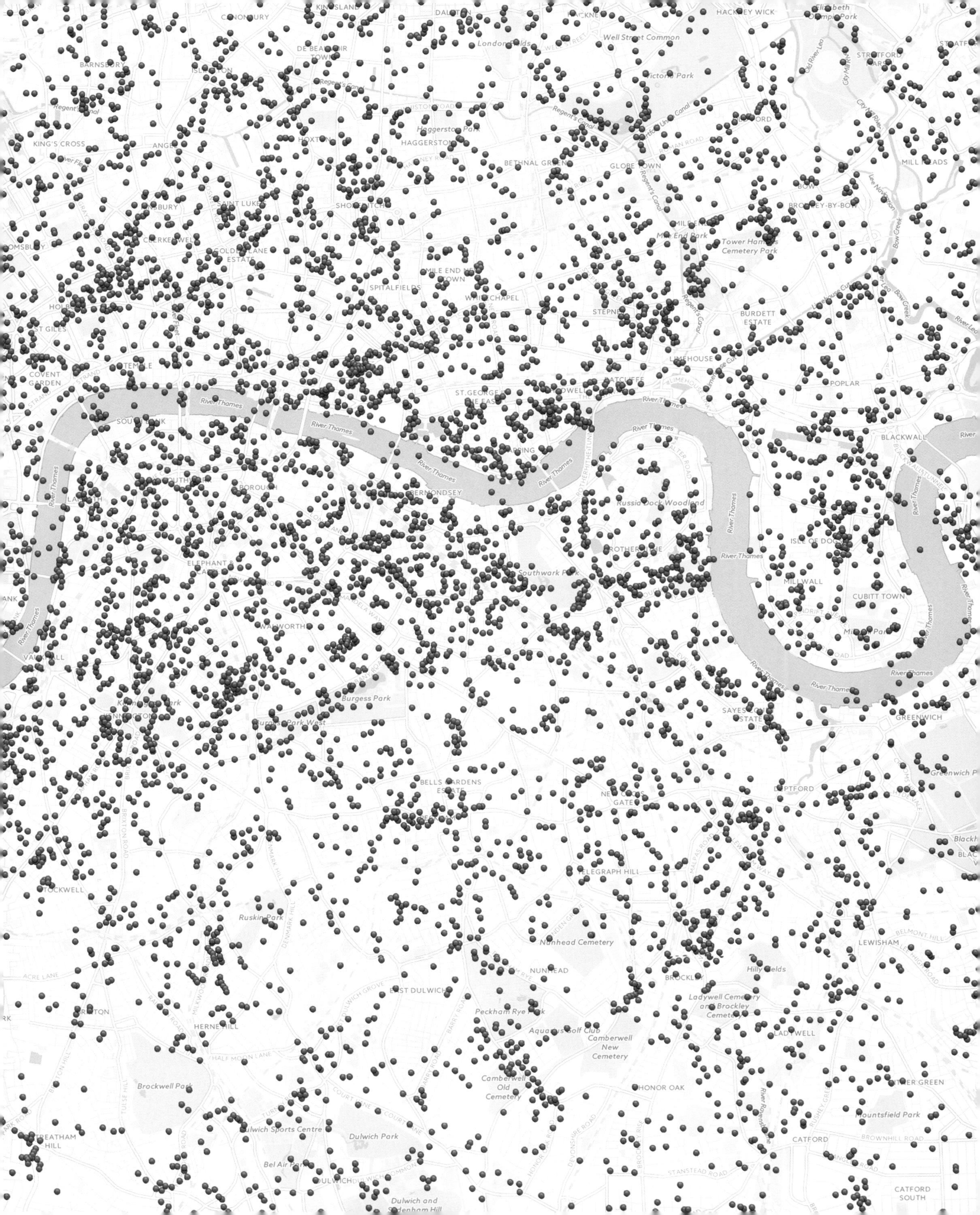

# Cognitive biases

## Systematic deviation from what might be called normal or rational judgement.

**Cartography would be simple if we could assume everyone making or using a map had equal and objective judgement—except we don't.** People are highly illogical and have their own subjectivities that may deliberately or unintentionally mediate the making or reading of a map. These are generally referred to as *cognitive biases,* which can lead to perceptual distortion, poor judgement, or inaccurate interpretation. Such biases grew out of the identification of innumeracy in the 1970s which identified the condition of how humans and decision-making differed. They create stumbling blocks, which can lead to wrong judgements without you even noticing. Although we may not be able to easily control our cognitive biases, let alone those of our map readers, having an awareness of the scope of the issues can help avoid pitfalls in map design.

**Bias arises from a wide range of processes.** We may have difficulty in problem solving, learning, or rapidly processing information. This leads us to make often rapid decisions based on short cuts (heuristics). Storage and retrieval of information can be difficult, which leads to limitations in discerning what is important from visual noise. Our brains may be subject to limited processing capacity through impairment. We may have our own moral or emotional situations and beliefs. We may also be the product of social, economic, or political influence that we may not even be aware of.

**There is practical significance for understanding cognitive bias for cartography.** Appreciating the many fallacies we might make in designing a map should at least be considered. Most of the decisions made in making a map might be thought of as objective or based in sound logic but cartographers are human and they make decisions about every mark on the map. These may very well be influenced by cognitive bias unless we're acutely aware of the pitfalls. Equally we cannot assume rational thought when people read maps. Ultimately people see maps in different ways. Many biases exist that may lead them to read maps in a particular way, forming opinions that the cartographer had not intended.

**See also:** Design and response | Dysfunctional cartography | Emotional response | Error and bias | Ethics | Functional cartography

Biases occur in different ways. Decision-making biases affect how we form judgements and behave. For example, they can be based on a lack of information (ambiguity effect), a tendency to prefer what's popular (bandwagon effect), how biases affect others but not ourselves (bias blind spot), or on the search for information that supports what we already believe (confirmation bias).

Social biases tend to form through interaction with others. They lead to errors of judgement such as giving more weight to the opinions of authority figures (authority bias) or of groups who perhaps have a single spokesperson (group attribution error). We sometimes see single personality traits as reflective of an entire person (halo effect) or that people who belong to the same groups as us as superior (in-group bias). Self-serving bias exerts itself when we are happy to claim responsibility for successes yet do the opposite for failures. Ultimately social biases lead to the familiarity of the status quo which can lead to cognitive dissonance.

Memory biases lead to selective recall. We tend to recall bizarre images rather than the mundane and repetitive. Choice-supportive bias leads to us thinking our decisions were more informed than they were. How we remember things affects future decisions. Similarly, consistency bias leads us to think we're more consistent in decision-making than perhaps we are and we also have a tendency to be self-serving (egocentric bias). We're also good at hindsight bias and overly ascribing positivity to past events or decisions.

The Google effect leads to an inclination to disregard information we know can easily be retrieved online. This is digital amnesia and might partly explain low attention spans and spatial awareness.

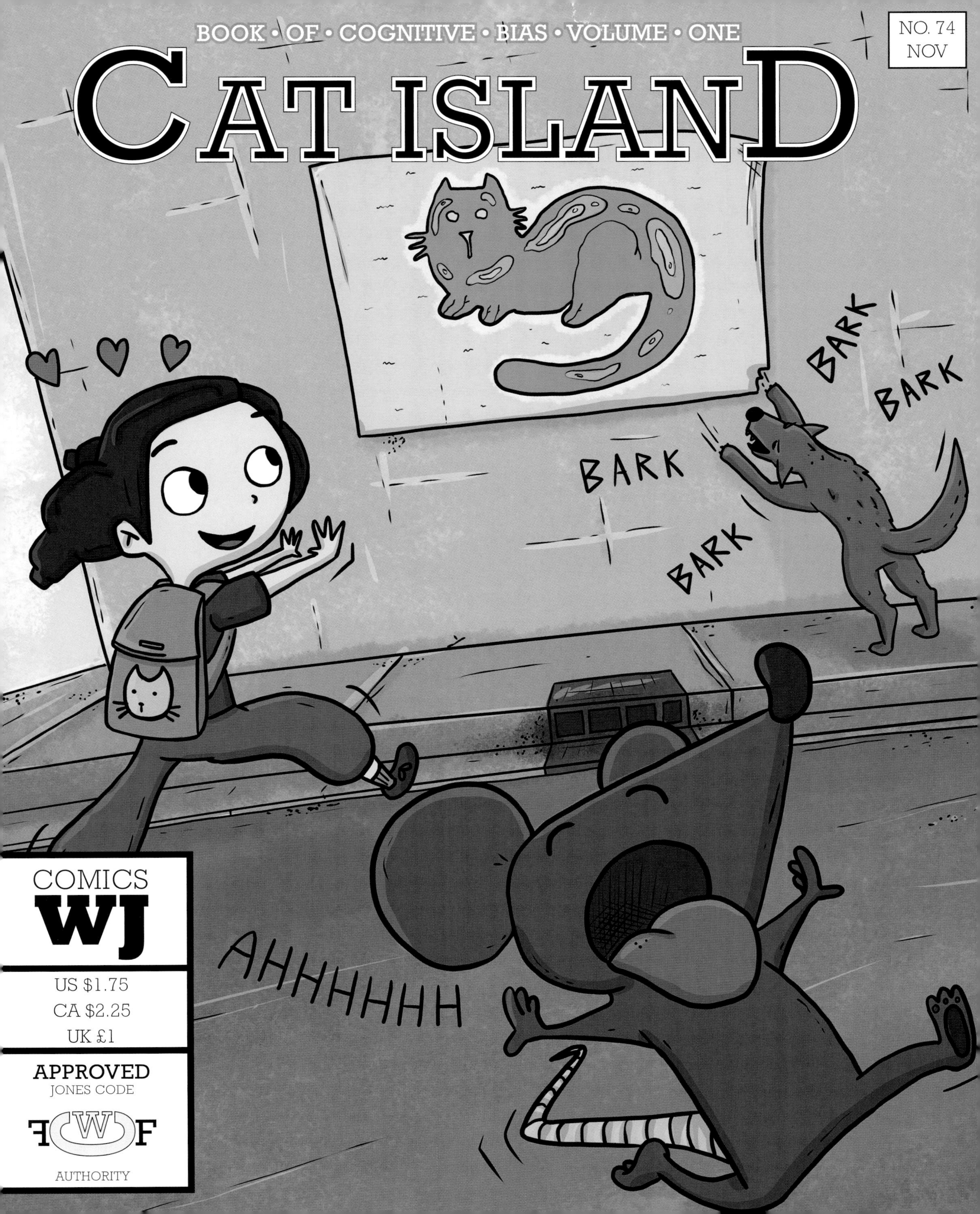
BOOK • OF • COGNITIVE • BIAS • VOLUME • ONE
CAT ISLAND
NO. 74
NOV
BARK
BARK
BARK
BARK
AHHHHHHH
COMICS
WJ
US $1.75
CA $2.25
UK £1
APPROVED
JONES CODE
AUTHORITY

# Colour charts

## A colour chart is an ordered set in some logical or consistent manner which supports colour management and selection.

**Colour charts are intended as guides to aid colour management or selection using known specifications.** Historically, printing ink manufacturers supplied colour charts to illustrate the precise colours they are able to achieve. A colour chart could also be used as a recipe for managing and supplying a map's colour specification to a printer. For instance, PANTONE® colour charts achieve this by informing printers what proportion of basic colours must be mixed together to achieve the desired colour. Colour charts are used less in modern cartography as part of the printing process but they have taken on increased importance as a support for colour choice in the design phase.

**Printed colour charts provide the various values for the parameters of the colour model they represent.** Most commonly, printed charts are based on percentage tints of the process colours CMY, sometimes with the addition of black (K). Many graphics software packages contain colour charts to match RGB to Munsell, PANTONE, and other colours but the result can only be an approximation. When a map is being designed using a computer display for display only on a computer, it is best to use a colour chart or selection system that uses RGB directly, often by mixing various components interactively or by making use of preset colour palettes. For maps that are intended to be printed, specifying colours for the printer to use is optimal.

**For design work, using preconfigured colour charts offers an excellent way to make more informed colour choices.** Many graphics and GIS software packages have a range of predesigned palettes and colour ramps to choose from. Not only do these obviate the need to mix your own colours, they have been designed to suit specific purposes. Third-party colour charts are numerous. Many are developed for general graphics work but some are notable for their intended use for mapmaking. Chief among these is ColorBrewer (colorbrewer.org), which provides a comprehensive set of colour schemes designed to deliver perceptual scaling and for use specifically for cartography.

**See also:** All the colours | Colour cubes | Colour schemes | Mixing colours

Printed colour charts are usually used as a way to reference a particular colour and specify the combinations of inks needed to reproduce a swatch faithfully when printed. PANTONE is widely known for its PANTONE MATCHING SYSTEM®. Others include the NCS (Natural Color System) palette.

For web mapping, web-safe colour palettes have long since been superseded by a range of online resources that provide a rich choice for mapmaking. In addition to ColorBrewer, the Adobe Color Wheel (color.adobe.com) offers an interactive colour design environment alongside curated colour schemes. Although many of these choices help cartographers be precise about colour in relation to their final intended use, many web mapping applications rely on web colour codes specified in hexadecimal format.

Hexadecimal colour coding is a form of specification of the intensity of red, green, and blue components, represented by a hex triplet which is a six-digit, three-byte hexadecimal number. One byte represents a number from 00-FF so a rich red which has an RGB specification of 244, 46, 24 would be specified as #F42E18 in hexadecimal.

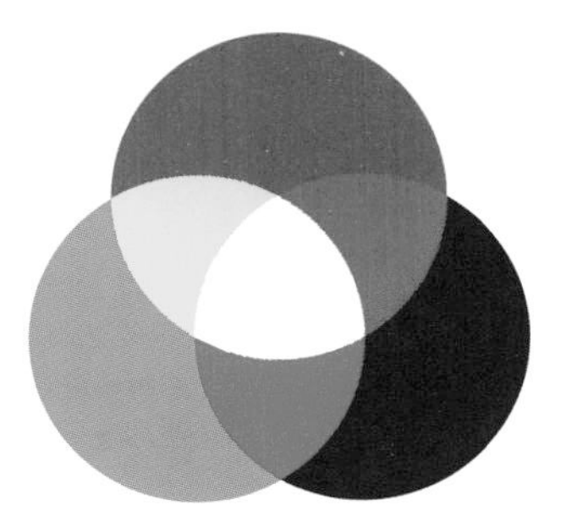

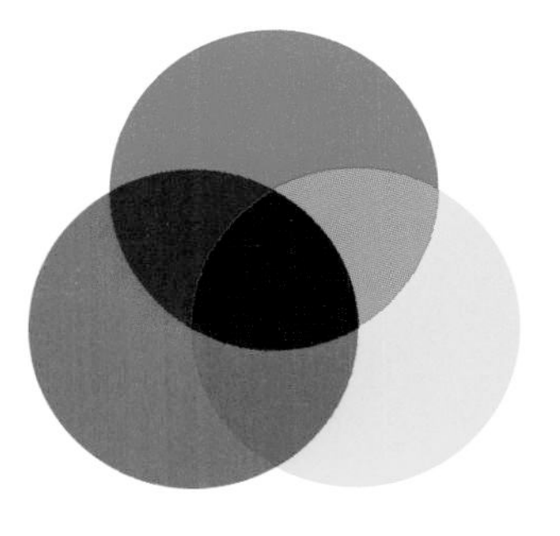

RGB
Digital

CMYK
Print

Spot
Offset Print

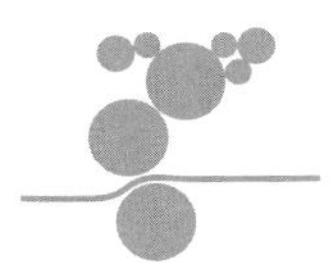

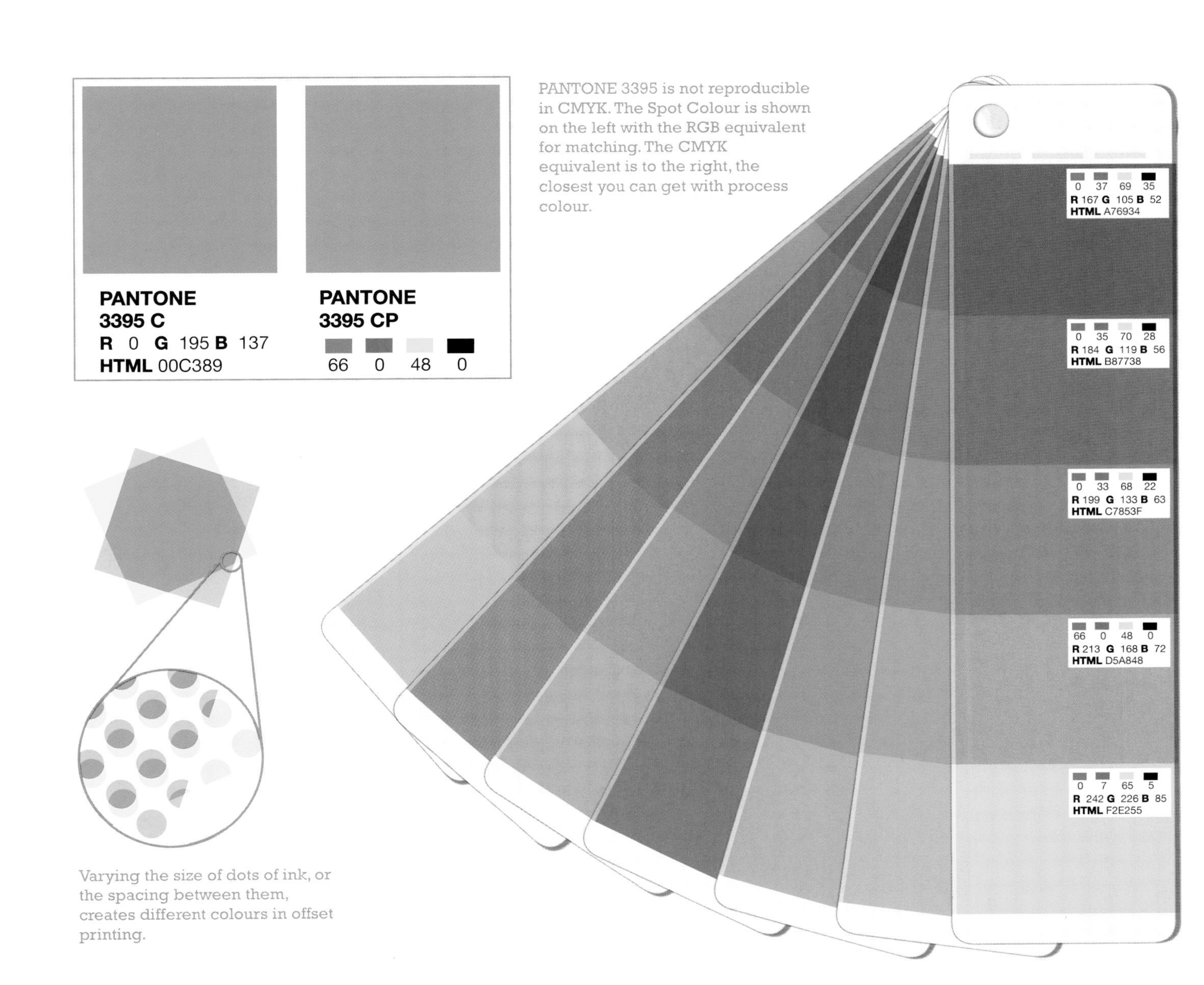


PANTONE 3395 is not reproducible in CMYK. The Spot Colour is shown on the left with the RGB equivalent for matching. The CMYK equivalent is to the right, the closest you can get with process colour.

Varying the size of dots of ink, or the spacing between them, creates different colours in offset printing.

# Colour cubes

## The RGB and CMY colour cubes are used to specify colours.

**The RGB and CMY colour cubes are the most commonly used method for specifying colour.** They are used as a basis for mixing primary colours to specify the amount of pigment (for print) or transmitted light (for screen) of your map colours. There is an important distinction between these colour cubes and other colour spaces (e.g. Munsell) since RGB and CMY are not perceptually scaled. They structure colour in a linear way along three axes that form the overall cube. It's therefore important that creating a ramp of colours from a cube based on regular numeric steps will not necessarily yield a good perceptual colour scheme.

**The RGB cube specifies colour based on the relative intensities of the red, green, and blue additive colours.** Each of the three axes of the primary colours range in value from 0 to 255. Any point within the cube can be given a value for red, green, and blue that specifies the precise colour at that point. Each pair of primary colours mixes to form the secondary colours of CMY. An absence of colour is black and is represented by the values 0, 0, 0. White is generated by full amounts of the three primaries represented as 255, 255, 255. If you're making a map to be viewed digitally, specifying colours using the RGB cube is optimum.

**The CMY cube specifies colour using the three subtractive colours cyan, magenta, and yellow.** The CMY cube is effectively the inverse of the RGB in terms of mixing, and the RGB colours become secondary, occupying the opposite corners in the colour cube. Values are specified in percentages since they relate to the amount of ink coverage in the printing process. White is specified as 0% of the three subtractive colours. Black is specified by 100% of each of the three subtractive colours though in actuality, black is usually added as a fourth ink in the printing process since the three mixed primaries do not produce pure black as a printing ink. Any percentage combination of printed ink of the three colours corresponds to a point in the cube.

**See also:** All the colours | Additive and subtractive colour | Greyscale | HSV colour model | Perceptual colour spaces

Completely desaturated colours (grey tones) occur when values are the same for each of the input colours and are represented as a diagonal line from black to white through both of the colour cubes.

In general, lighter tones for colours are found around the white corner of the cube and darker tones around the black corner of the cube. Points further away from the grey tone line become increasingly saturated.

The RGB cube is routinely used on computer displays to specify colour yet has the disadvantage that conventional components of colour, namely hue, lightness, and saturation, are not used to specify the colours.

The numerical values that specify points in a colour cube are not perceptually linear. They do not match the way our eyes and brains see the visual steps between increments. For instance, a value of 100, 0, 0 does not fall, visually, midway between 0, 0, 0 and 200, 0, 0. Lower incremental RGB values tend to represent smaller visual differences than higher RGB values.

**Opposite:** RGB Colorspace Atlas by Tauba Auerbach, 2011. Digital offset print on paper, case bound book, airbrushed cloth cover and page edges. Three books, 8 × 8 × 8 inches each (20.3 × 20.3 × 20.3 cm). Each book contains 3,712 pages and represents one of the three colour channels.

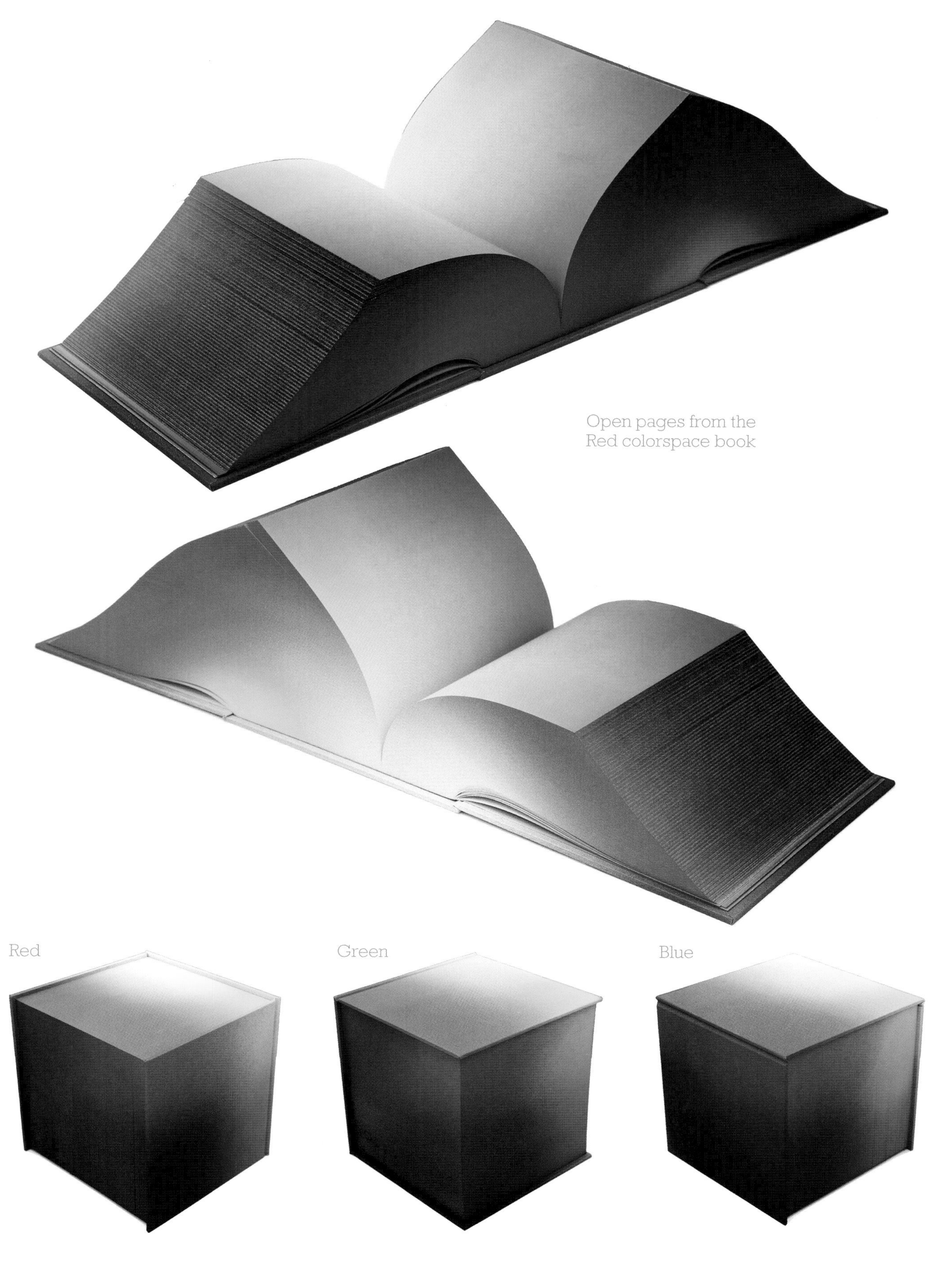

Open pages from the
Red colorspace book

Red

Green

Blue

# Colour deficiency

## Designing for colour-deficient vision leads to the use of particular colour palettes.

**On average, approximately 4 percent of the population has some form of impairment in their colour vision compared with the rest of the population.** This is more pronounced for men, with nearly 10 percent, than women and severity differs between people. Since colour plays such an important role in communicating map detail, there are consequences for ignoring colour deficiencies. Colour deficiencies affect people's abilities to see hues in the same way. The perception of lightness remains unaltered, and this can assist in the selection of colours that accommodate alternative ways of seeing.

**Colour-blindness is either total or partial.** Total colour-blindness (monochromacy) is the rarest form, which results in a monochrome image being seen. Most common is red-green colour-blindness, the main forms of which are deuteranopia and protanopia. Less common is blue-yellow colour-blindness, the main form of which is tritanopia. Each inherited type of colour-blindness results in different cones in the eye having mutated forms of pigment that renders them unable to process specific wavelengths of visible light. Effectively, people with different vision impairments see the electromagnetic spectrum differently, and this leads to an ambiguity in colour perception, slower recognition, and less successful map search tasks.

**Improvements in clarity can be made by using unambiguous colour combinations, supplementary visual variables, and annotation.** Selecting appropriate colours can be achieved by using colour-blind safe palettes, many of which are published as colour-blind safe charts. Avoiding known colour combinations and increasing saturation and value between symbols can also help differentiate, particularly points and lines that are harder to recognise. Ensuring a good level of contrast between a map's figural components and background also helps.

The following pairs of hues are considered a safe set that most people with colour impaired vision are able to see:

- red and blue
- red and purple
- orange and blue
- orange and purple
- brown and blue
- brown and purple
- yellow and blue
- yellow and purple
- yellow and grey
- blue and grey

All sequential, diverging, and qualitative colour schemes can be adjusted to be read by those with colour-deficient vision. A good colour scheme always uses variations in lightness which naturally accommodates colour-deficient vision. Adjusting colours to a safe palette or using the pairs of colours above provides added clarity.

Qualitative colour schemes can be harder to design since many standard conventions will be hard to see. Additionally, it's common practice to design such schemes to be perceptually similar in lightness to avoid overemphasis of one feature type. Choosing hues that cause fewer difficulties and carefully adjusting lightness give a good separation while also remaining suited to normal vision.

There are a number of useful web tools available that can be used to choose colours to accommodate deficiencies. Alternatively, test your map by filtering it to show how people with particular colour deficiencies will see the map. Tools are listed in the back matter.

**See also:** Constraints on map colours | Perceptual colour spaces | Seeing

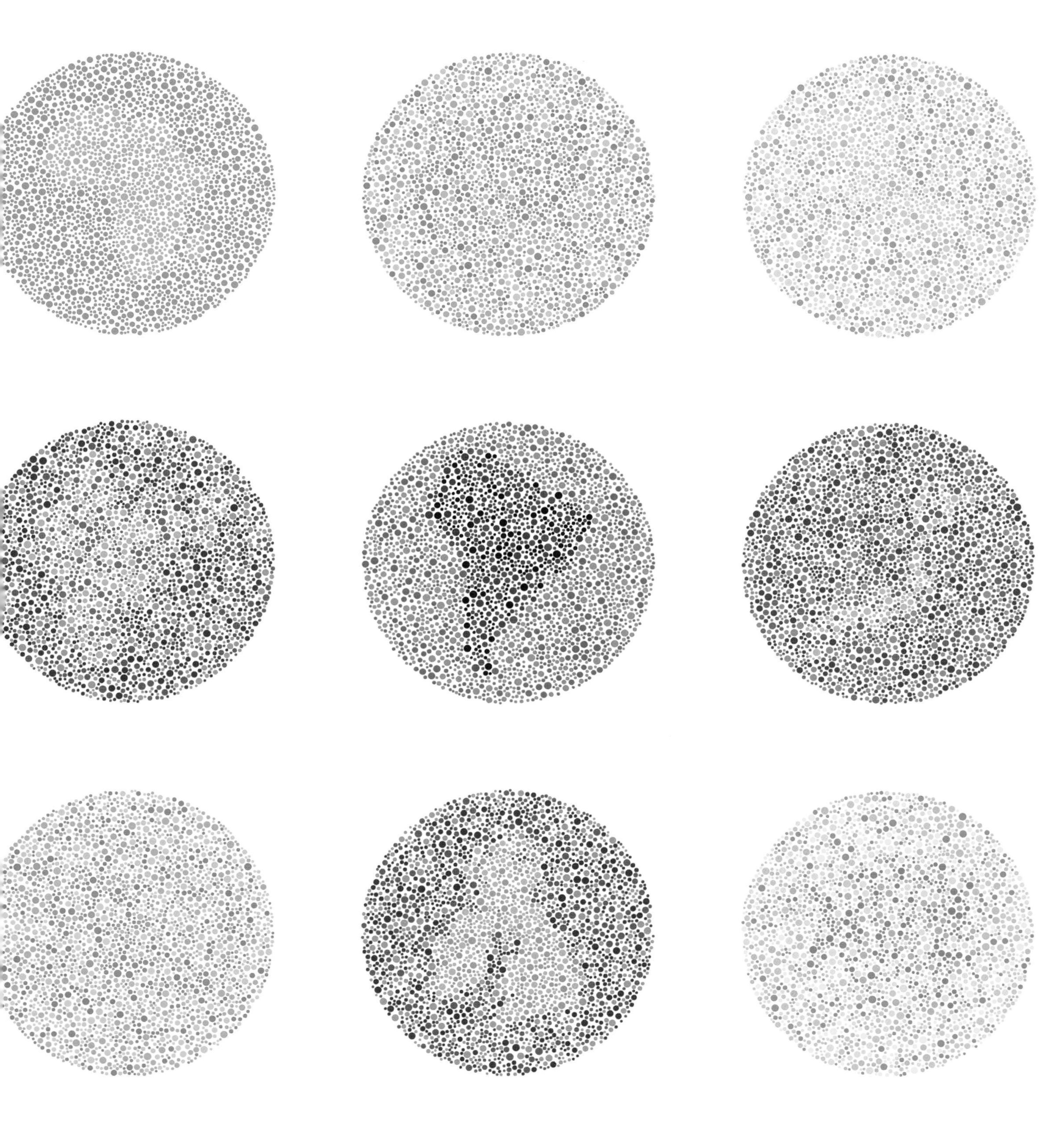

Answers: Africa, Australia, Italy, India, South America, Japan, Mexico, British Isles, Antarctica

# Colour in cartography

## Applying colour to a map is a challenging aspect of cartography.

**One of the major challenges for mapmakers is to employ colour effectively.** Careful use of colour can make a massive difference in mapping the results of your work but the easy access to colour through modern software has its drawbacks. The unlimited freedom to select from graphical systems that generate over 16 million possible colours is daunting. There is always a temptation to overuse colour and end up with a kaleidoscopic result. The problem of choice is compounded by the fact that most maps are designed on a screen that employs the red, green, and blue (RGB) additive system of colour mixing. This translates well to a map published online but not if the map is to be printed using a process that employs the subtractive cyan, magenta, and yellow (CMY) colour mixing system. Problems can arise when matching colours that appear on screen with those that will appear on the printed page.

**Colour takes on a special role in mapmaking.** Map readers tend to have certain expectations based largely on prior experience. This in itself provides a good guide for choosing colours. Departing too far from colour expectation is one way of producing a map that people may struggle to interpret. The goal is to envision your colour use and then set about selecting or mixing colour using the structure afforded by a formal colour model such as RGB, CMYK, or HSV (hue, saturation, and value).

**When using colour as a symbol, you'll need to consider the three perceptual dimensions of hue, lightness, and saturation.** They allow cartographers to encode qualitative and quantitative differences in mapped features. For instance, topographic maps use hue to differentiate between different land uses (e.g. green for forested areas, blue for water bodies), as an aid in the classification of road order, and for representing elevation. Thematic maps often use colour in a different way. For instance, lightness of a fixed hue to indicate class differences, from low to high, in values of the data being mapped. Saturation is the most subtle of colour's perceptual dimensions. If it's controlled poorly, you'll end up with oversaturated 'bright' colours, which are hard on the eye.

**See also:** Additive and subtractive colour | Colour deficiency | Colouring in | Constraints on map colours

Technological change has had a major impact on the use of colour in cartography. Maps produced using offset printing processes were created by using several printing plates, each of which applied a percentage tint of one of the process colours to the printed page. Applying colour in this fashion built up the image and allowed colours to be mixed through a combination of halftone tints. For instance, light blue could be produced as a tint of cyan ink, and orange created by applying a tint of magenta offset with a tint of yellow.

Analog production techniques built printed maps from solid colours, percentage tint screens, and halftone screening. With the advent of trichromatic printing for reproducing photographs, printing negative halftone images in cyan (C), magenta (M), and yellow (Y) allowed a faithful reproduction of an original to be produced, usually with black (K) added to improve the final result, Alternatively, premixing printing inks to obtain the precise colour required (spot colours) was also favoured by the cartographic industry, particularly for producing topographic maps.

Digital imaging processes have replaced a considerable amount of the colour reproduction technology in cartographic workflows. Cartographers are easily able to visualise the appearance of their finished map on full-colour screens before printing or publishing which was not previously possible. However, they vary in their colour requirements. Some require a few colours and are unconcerned if they look different when printed. For others, careful systematic colour design remains a priority to support consistent output and reader expectations.

**Opposite:** PrettyMaps by Aaron Straup Cope, Stamen Design, 2010, pushes the aesthetic boundaries of modern cartography.

# Colouring in

## Cartographic conventions provide guidance on how to apply colours effectively.

**Early hand-coloured maps began to appear toward the latter half of the 19th century which began the development of conventional use of hue.** Some conventions remain from this period, particularly in topographic mapping such as blue being used for water features, green for forested areas, brown for contours, and red for main roads. Geologists also developed standardised uses of colour during this period, and through convention and familiarity they largely remain and are familiar to both mapmakers and map readers. Familiarity with any form of symbolisation assists the mapmaking process since less reference to a legend is required by the reader and information retrieval is optimised.

**Hue is used as a visual variable to identify map features and differentiate between features of a different type.** To show related features, you would choose hues that are related in some fashion such as green hues to illustrate different vegetated landscapes. Hue can also be used to provide perceptual distinction for contrasting features.

**Highly saturated colours provide some of the greatest perceptual differences between colours**. This is perhaps one reason why it's easy to create oversaturated printed versions of maps—because they are designed on screen and with individual data layers that the mapmaker might wish to perceive differently. It's easy to oversaturate map colours. But on a printed map saturated colours are not visually pleasing. Desaturated colours tend to be perceived as more aesthetically pleasing.

**Value (lightness) is conventionally used to represent order or numerical difference on a quantitative scale or for large symbol area fills.** In general, when mapping quantitative differences, larger magnitudes are shaded darker but changing value alone can have unwanted consequences. Inevitably, saturation will also alter as the value of a hue is increased. Modifying colour schemes by applying a subtle change of hue can also help distinguish adjacent colours and improve recognition.

**See also:** All the colours | Colour charts | Constraints on map colours | Elements of colour | Maps for and by children

Some conventions conflict. For instance, the use of green for vegetation can conflict with some conventional representations of relief on small-scale maps which layer varying greens. A standard depiction of elevation in atlas mapping uses hypsometric tints which, if overlaid with forested areas could present problems of visual conflict. Hypsometric tinting was developed to mimic green, cultivated lowlands and barren, brown mountains but it does not necessarily suit all landscapes where, for example, low-lying desert may be fringed by forested mountains.

Using desaturated colours makes colour differentiation more difficult the more you desaturate. Where a large number of different hues are required, distinctiveness cannot be achieved by varying saturation alone.

In colour sequences, lightness and saturation must be controlled together, not alone. For instance, if you choose a red hue and change its value from 10% to 100% across six visual steps, it will appear like that of the left six shades in the bottom diagram below. If you continue to add black then you achieve darker colours. However, the high chroma, 100% red now appears in the centre of the visual scale and it dominates perception. An improvement would simply be to remove the 100% red.

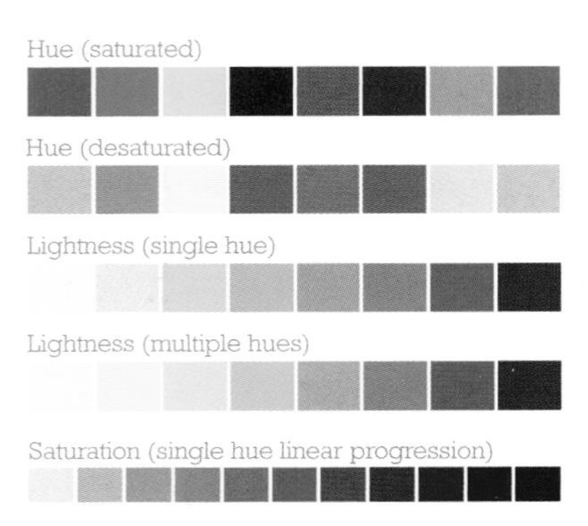


**Opposite:** The adult colouring book craze gave rise to versions with maps. Now you too can simply colour in maps!

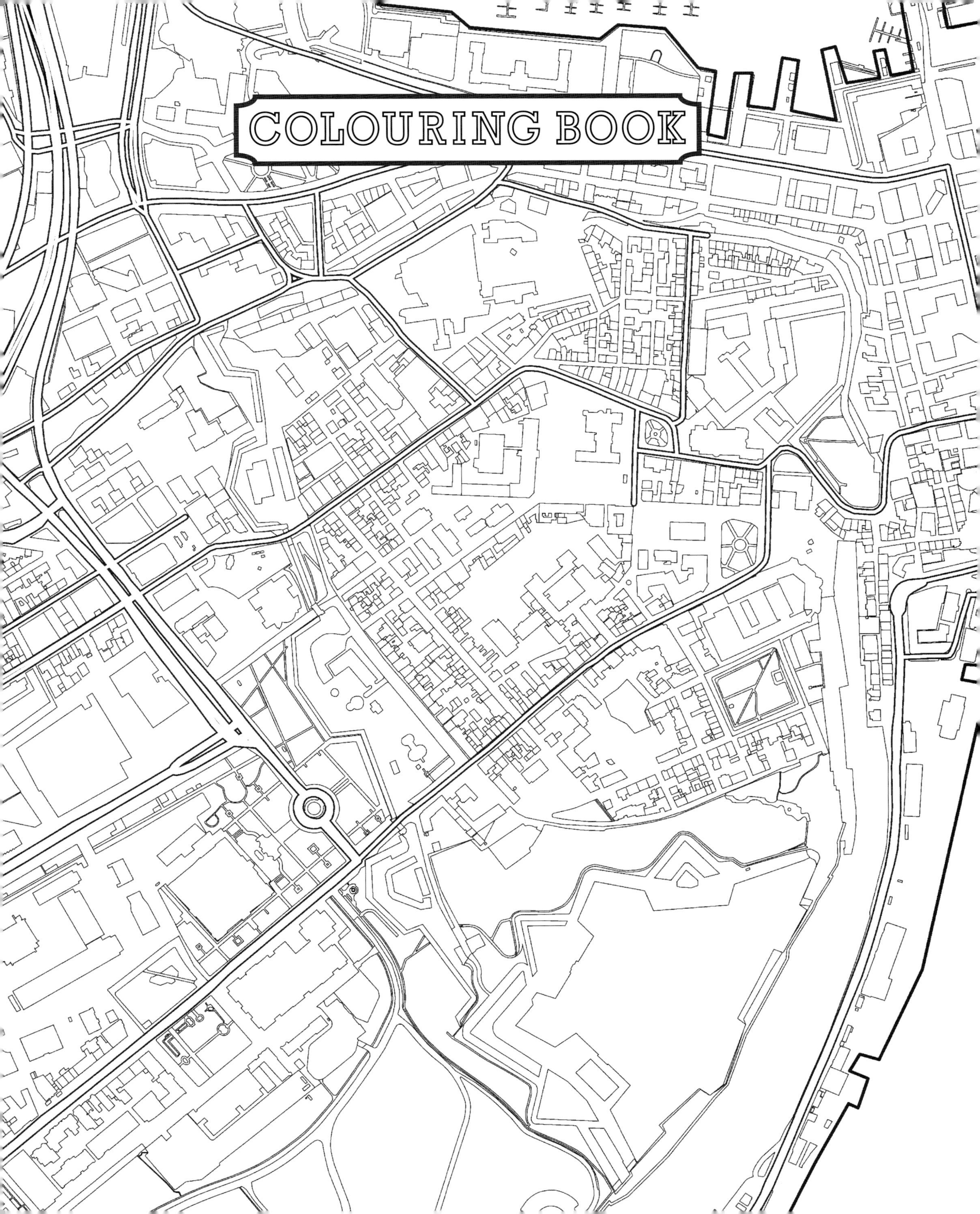
COLOURING BOOK

# Colour schemes

## Qualitative and quantitative datasets are well represented by colour schemes and custom colour ramps.

**Map readers process colour on a map by seeing differences between the perceptual dimensions of hue, saturation, and value.** Designing to map these onto the logical structure of your data is key to supporting interpretation. Colours are structured using sequential, diverging, and qualitative schemes in different ways to reflect the quantitative and qualitative character of your data. More complicated schemes can be built from these basic forms.

**Sequential schemes primarily vary lightness to represent ordered data.** Darker colours denote higher data values and vice versa. A moderate hue shift can also be applied to make different classes distinct though careful control over saturation is required so that no class appears more vivid than the others.

**Diverging schemes are appropriately used where data has different extremes that might be best represented with different hues.** A diverging scheme emphasises the midpoint critical class with a light colour and then the two extremes with two diverging hues. This sort of scheme can also be used to represent 'no difference' or 'no change' as well. Diverging schemes can also use a critical break approach.

**Qualitative colour schemes are used to represent differences between mapped features.** These usually use different hues. For instance, a land-use map might have a range of different categories, and each must be illustrated differently without prominence or suggesting order. The simplest way to achieve this is to ensure the hues maintain a similar contrast with the background colour of the map by controlling the value and saturation of each colour.

**Bivariate (or trivariate) schemes can be used to represent more than one variable at a time.** Each variable is mapped using a sequential or diverging scheme and an overlay constructed by applying transparency to blend layers. Alternatively, creating discrete schemes from pairs of sequential or diverging colours gives greater control over colour mixing.

**See also:** Colour charts | Colour cubes | Hue | Hypsometric tinting | Mixing colours | Saturation

For continuous data in particular, a useful way to define the colour for an array of values is to use an algorithmic colour ramp. This allows you to specify end points (say orange at one end and green at the other) and then for interpolated colours to be defined and applied to the data values. Some software applications plot a linear line through a colour space, selecting a colour for each class according to a linear distance along the line. Considering all the non-linear perceptual problems experienced when seeing colour, this is not the most effective method of colour selection for mapping. Additionally, ramping often cuts though desaturated colours at some point midway through the ramp creating unusual schemes. To overcome this problem, conceptually, an arc should be threaded through colour space. This can be approximated by selecting an appropriate midpoint as part of the colour ramp which the software can pass the line through in ramping colours.

Hypsometric tinting is often used to depict relief where hues are used to show different elevations. For mapping bathymetric depth, it is common to use a blue hue with the principle the deeper the darker.

Spectral schemes are popular in scientific visualization since the full sequence is familiar. However, the scheme is often misused as a sequential scheme that overemphasises values in the data array. Also, different hues do not represent quantities along a linear scale effectively.

Where possible the four-colour theorem is a useful technique to show differences among area features. It simply states that no more than four colours are required to colour all regions on a map so that no adjacent regions have the same colour. It tends to produce a well-balanced overall appearance and is often used on political maps.

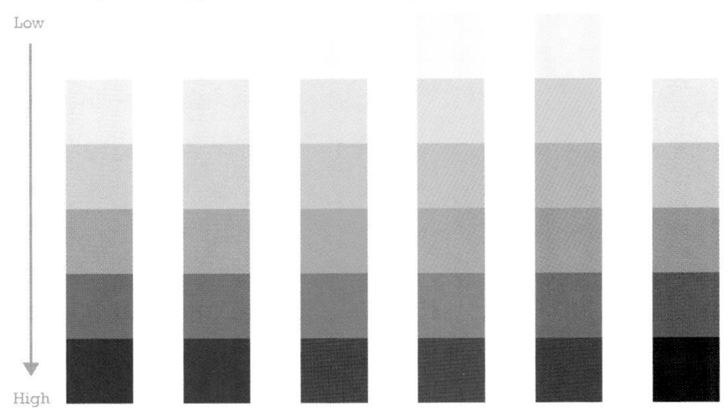

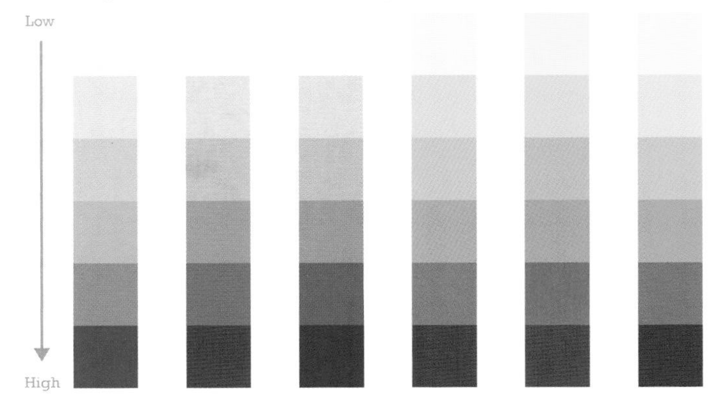


Diverging (quantitative variables)

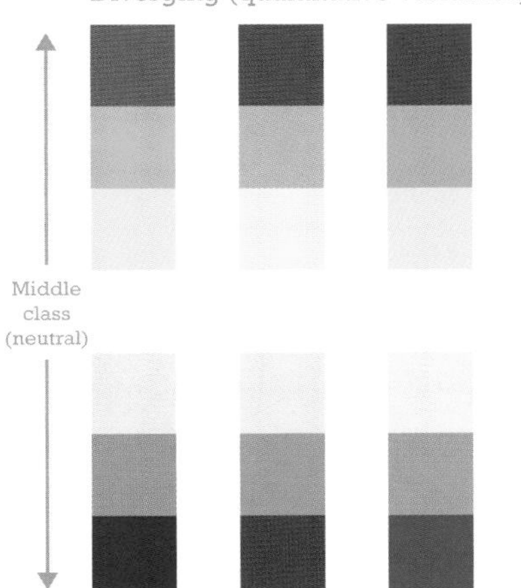


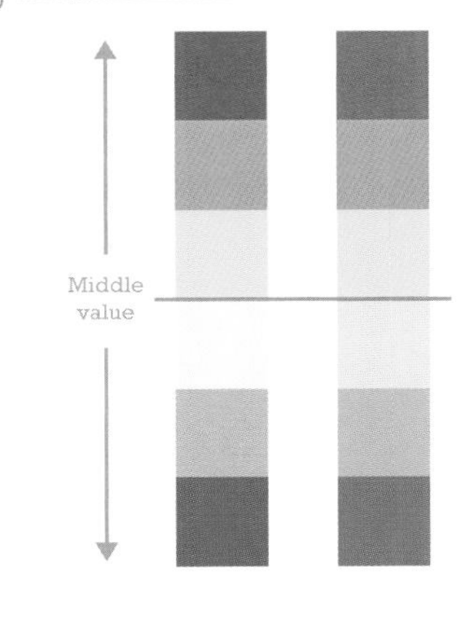


Qualitative

Spectral

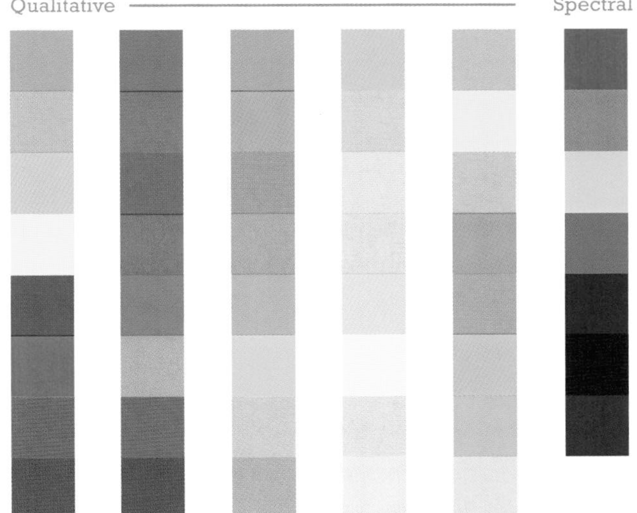

Bivariate Sequential

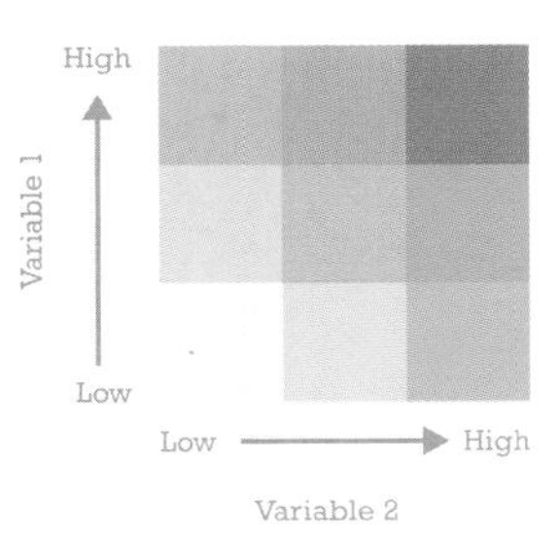


Two sequential schemes creating transitional hue mixtures

Bivariate Diverging

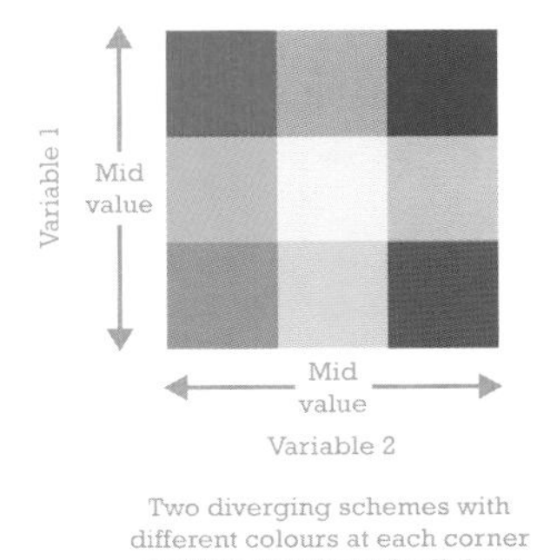


Two diverging schemes with different colours at each corner and hue transitions for lighter midpoints.

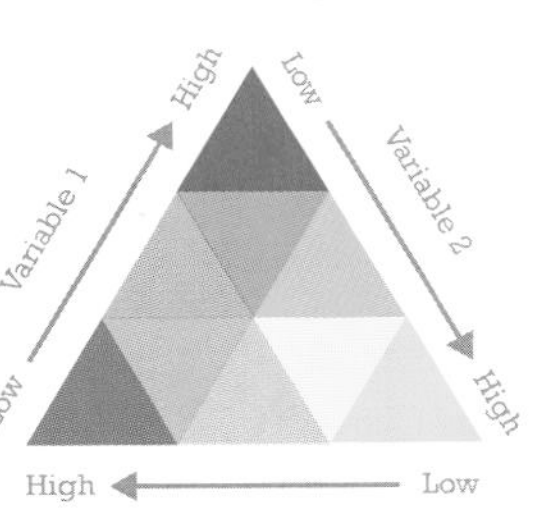


Mixture of different values for three hues on each axis.

Four colour

Four-colour theorem applied to the United States. No two adjacent States have the same colour which creates a well balanced distribution of colours across the map

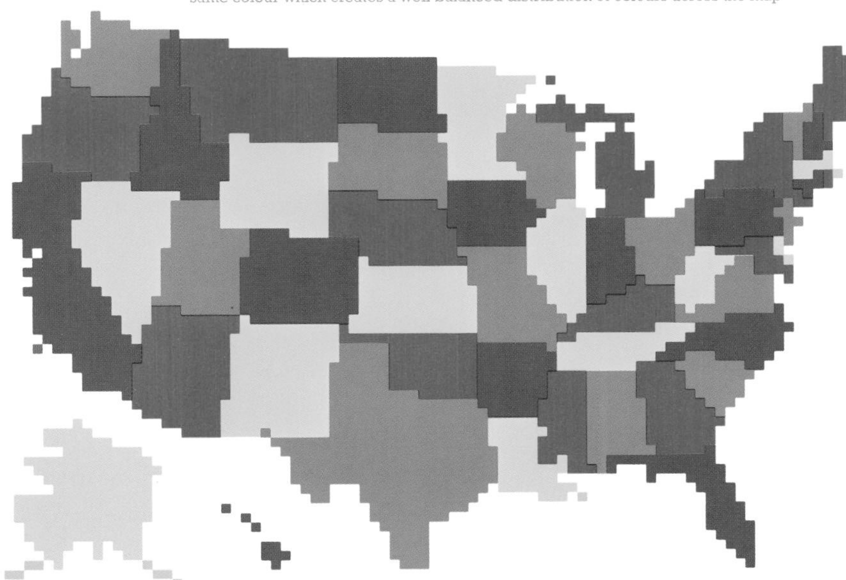

# Combining visual variables

## Knowing what visual variables are is useful. Knowing how to combine them creates meaningful cartography.

**The amount of choice in using visual variables to design effective symbols is almost infinite.** Indeed, one of the main ways to vary symbols is to combine two or more visual variables, conventionally combining a primary visual variable with a secondary visual variable. In this sense, visual variables act as the basic building blocks for more complex symbolisation. Conjunctions of visual variables can strengthen the encoding of information.

**The design and combination of visual variables should be with respect to the specific mapping task undertaken.** One map might be more appropriately designed using value as the primary visual variable to depict quantitative difference, yet on another, size might be more appropriate. Different secondary visual variables used in conjunction will also, therefore, be different. The mapping context (and map type) are thus extremely important considerations that will inform appropriate choice in visual variable selection and how they are combined.

**The choice of which visual variables to use as the building blocks for the design of symbols on a map is a function of many factors.** The effectiveness of information dissemination that each visual variable can provide is neither linear nor absolute. Visual variables are selected in line with the central message of the map and also the level of measurement of the data to be depicted. Other factors such as map user group, map type, scale, and other map content will also impact the final design of symbols, and it would be impossible to provide a more prescriptive set of rules for symbolisation.

**Syntactics provides a reasoned way of assessing the utility of signs relative to one another.** Given the level of measurement of the feature and its attributes, visual variables can be assessed in their ability to support the preattentive seeing of information. Visual variables can be unordered to encode nominal information, ordered (but not quantitative) to encode ordinal information, and ordered (quantitatively) to encode numerical information.

**See also:** Chernoff faces | Information overload

The development of visual variables stems from the work of Jacques Bertin (Bertin, 1983) who framed them as retinal variables since they are seen preattentively. That is, humans see the symbols as immediate rather than having to process them cognitively. Thus we can predict how each of the visual variables are going to be seen and processed which leads to the selection of one over another for a specific purpose.

Perception of visual variables might occur in one of four ways. Firstly, associative perception is where variations are seen with equal weight. For instance, no one colour or shape is seen as more important than another. Dissociative visual variables will, conversely, lead to one variation being dominant over another and so on. Secondly, selective perception leads to the eye being able to focus on one variable among others. This tends to work well for all visual variables except shape. For different shapes of roughly equivalent sizes, it's difficult to see patterns formed by one shape. For this reason shape is useful when individual data points need to be seen—but not for patterns among a lot of the same type of data point. Third, ordered perception leads to seeing ranked variation as variables of the same type are seen as more or less relative to one another. This is the main reason why using hue for numeric data is inappropriate—because in a rainbow colour scheme it's impossible to determine whether blue is more than green or red, etc. In contrast, darker is seen as more than lighter. Finally, quantitative perception allows you to estimate numerical information. Size supports this ability.

**Opposite:** gridded point-based thematic data is symbolised using size, hue, and transparency to encode different characteristics. Balance and legibility is achieved by ensuring the three visual variables work in concert.

# Consistent denotation

**A consistent map, in terms of the denotation of symbols, is one that will likely support the map reading task more effectively.**

**The more consistent we can be in designing map elements that people find familiar, the better we denote meaning.** This also applies to making elements in the map itself perform in a consistent manner across the map, and those elements are, importantly, recognised the same way by everyone. Consistent denotation helps map readers think of elements on the map as categories of information and within those categories are consistencies that support meaning.

**Whereupon once, the cartographer began their task with a blank canvas, we now use computers which give us all manner of defaults.** Defaults for colour, for typography, for line weights, and for patterns are all chosen for us, and even though there is always the option of changing them, this is often the first look at the map we are making. Are those defaults part of a unified set? Unlikely. So it's important to look beyond the defaults to bring visual harmony to the map.

**With regard to colour, we can use connotations to bring some sense of consistency.** Repetition is also important so we use the same colour repetitively to mean the same thing across the map, albeit that the elements might be different—rivers, oceans, lakes all use blue, which supports connotation as well as consistent denotation. This implicitly makes colour-coded map elements easier to understand and interpret.

**In greyscale, the same idea applies, and the use of the same value consistently across a map for the same feature and different shades for allied features can be used.** For instance, a hierarchical road network may use ranked shades ranging from lighter to darker to give some visual consistency to the ordinal ranking of the data.

**Typographic consistency also helps denote meaning.** By limiting the variation of typefaces across the map you assist the map reading process. A good approach is to limit yourself to using no more than two typefaces, often one sans serif and one serif, and simply vary the character to denote different meaning.

**See also:** Abstraction and signage | Cognitive biases | Combining visual variables | Design and response | Emotional response | Legends | Semiotics

Developing consistency in a map is both an intellectual and an intuitive aspect of map design. Often this comes down to experience and a general feel for how little variation you can get away with and still make the meaning among graphical symbols distinct.

A good rule of thumb is to think of your map as not having a legend (or even, ask people to interpret the map without a legend) and see if the logic of your choices can be seen and interpreted independent of looking at the legend. The less people have to resort to interpreting the map through consulting the legend, the closer you are to effectively imbuing meaning in your symbology.

You could also think of consistent denotation being based on a restricted set of rules for making the map. Rather than thinking of having a completely free palette of design to choose from, if you severely limit your own choices it forces you to think more deeply about how best to design and deploy your symbols. If you think of restricting a design to, say, only two colours, can you still make logical graphical forms that convey meaning? The answer is always yes, but it's often a greater challenge in design terms than it is to make a full-colour map.

You will find that, with practice, limiting your choices builds greater confidence in how your selection of symbols works harmoniously to create consistent denotation.

**Opposite:** Those Who Did Not Cross 2005–2015 by Levi Westerweld. On December 3, 2011, a 29-year-old pregnant woman died from drinking seawater after the boat she embarked on in Libya went adrift in the Mediterranean Sea for 16 days. From 2005 to 2015, over 16,000 other individuals were reported dead or missing as they tried to reach European shores fleeing conflict and instability in Africa and the Middle East.

Each red dot on this map shows where a person went missing or died.

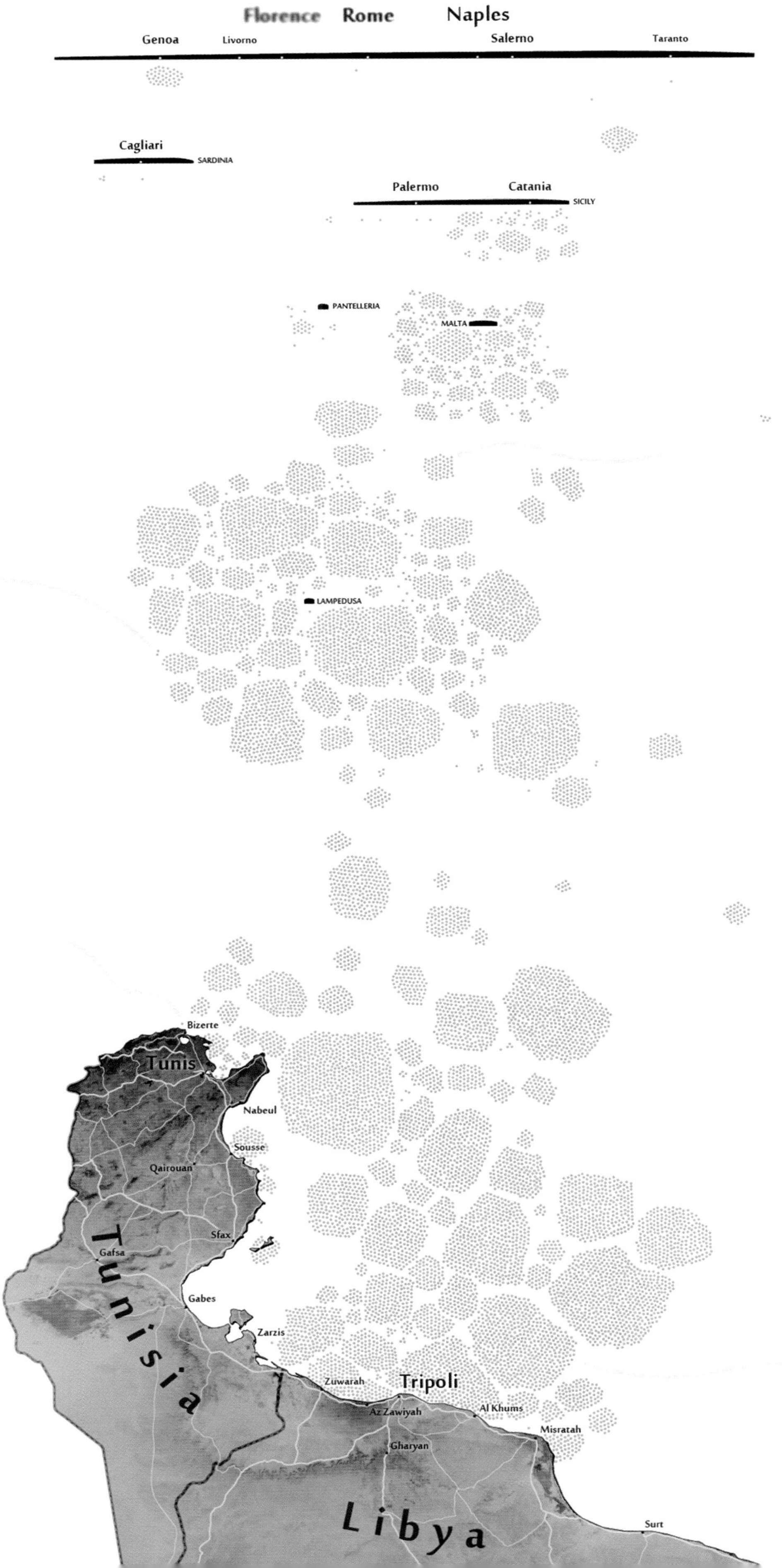

Florence Rome Naples

Genoa Livorno Salerno Taranto

Cagliari SARDINIA

Palermo Catania SICILY

PANTELLERIA

MALTA

LAMPEDUSA

Bizerte

Tunis

Nabeul

Sousse

Qairouan

Sfax

Gafsa

Tunisia

Gabes

Zarzis

Zuwarah

Tripoli

Az Zawiyah

Al Khums

Misratah

Gharyan

Libya

Surt

# Constraints on map colours

## Circumstances of design, production, and use give the mapmaker additional constraints when applying colour.

**The use of colour on a map is a function not only of understanding how to describe and specify but also the way in which it should be applied.** One way of considering this is to acknowledge some of the constraints and how to control for them. If carefully applied, colour can greatly enhance the communication of the intended message and improve the aesthetic quality of the map display. Colour choice on a map, like the use of visual variables more generally, is informed by the way in which map users will interact with the map.

**When designing for interactive maps, colour can be used to allow important information to be as distinct as possible.** Common practice used to be to use a black background when working with vector graphics on the basis that colours stand out more readily against black than a white background. This convention does not apply in modern mapping environments as a white background can just as easily be used.

**There simply isn't a better way of assessing the quality of printed output than printing and checking.** Good colour management systems (that modify colours on a screen to make them comparable to printed versions) are not routine in mapping software so it's always useful to use colour charts to specify colour regardless of screen appearance.

**For all colour schemes, be aware that some colours might cause inappropriate or even offensive connotations.** For instance, political affiliation is not something a cartographer should necessarily imply in a map to ensure objectivity although using colours that are widely acknowledged or associated with certain mapped phenomena is normally a good visual link. For example, using highly saturated red to present a map of death rates in epidemiological studies is not ideal but using the same colour on a political map showing the success of a particular party would work in visual terms. Examples of maps that use skin colour associations to represent different ethnic groups are also considered insensitive.

**See also:** Colour in cartography | Form and function | Printing fundamentals | Seeing

Preparing final maps on a display screen might include maps for broadcast, in-car or smartphone navigation systems, or for the web. A factor to take into account in preparing these maps is that you have little or no control over the final appearance since display resolution and quality can vary dramatically.

A map reader's colour display and resolution may have different settings for brightness, contrast, and colour balance compared with those used in the design phase. A colour can look very different on a range of different display devices (and more so if it is to be projected using a multimedia projector which tends to wash images out). Additionally, retina or other similar high-resolution screens give a sharper result with colours benefiting from greater clarity and luminosity compared with lower resolution devices.

Most colour spaces and models support a colour gamut. The gamut represents the subset of colours that can be accurately represented or reproduced by the colour space or a particular output device. It's important to consider the gamut, particularly when converting between colour spaces and models. Some colours will be out of gamut.

Specifying colour should be made with consideration of the lowest expected use conditions. Testing with poorly adjusted devices, projection equipment, and high ambient light, which all affect the way the map is perceived, is advisable.

Ultimately you have very little control over how your map will be used or by whom so the safe option is to assume people won't read it in the circumstances you envisage. They will almost certainly not have the understanding of colour you have used to make the map so ensure you've encoded colour as unambiguously as possible.

Geologic maps make heavy use of qualitative colours and rely on convention and careful legending.

Diverging colour schemes require distinct bookend hues and a muted mixed centre.

In thematic maps with both positive and negative values, warm colours are generally used for positive values while cooler colours are used for negative values.

The human eye is more sensitive to small changes in lightness than to small changes in hue.

Colours carry with them cultural and visceral associations. Bright red may exaggerate the apparent urgency of a feature.

Rainbow colour schemes invite errors of interpretation and are generally inappropriate for scientific visualizations.

Context has much to do with how we interpret colours. The basemap upon which thematic data is draped should guide choices of hue, saturation, and brightness.

Maps designed for low-light environments use a darker palette and reserve brighter colours for features of primary importance.

Maps used in bright sunlight are generally light in theme and leverage high contrast to ensure visibility.

Contrast drives meaning. When using a dark basemap, higher thematic values ought to increase in brightness…

Cool colours

Warm colours

…and when using a light basemap, higher thematic values ought to increase in darkness or saturation.

Greens are often used for forest and other vegetated areas on topographic maps.

Barren or dry areas are more easily associated with light, pale, earthen colours.

Blue hues are broadly understood to denote water features or bathymetric depth.

Elevation colour schemes tend to follow a similar pattern and are used to convey topography at a large scale.

Urban settlements are typically given hues distinct from that of the mapped natural features.

Employ cartographic conventions where appropriate.

# Continuous surface maps

## Technique for mapping gradually changing phenomena.

**Surfaces represent the geography of values that exist across an entire region.** Although isarithmic maps allow you to create contoured representations of a surface, they classify the data into bands. Instead, a continuously varying surface might be used. Continuous surfaces are interpolated from data that is collected at sample points either by mapping the density of points or the values they represent. Continuous surfaces are good for showing data that is known to vary gradually from place to place rather than having abrupt boundaries.

**Sampling data is usually used for continuous phenomena or for data collected at points.** Spatial samples can be random where sampling points are unevenly distributed across space. Alternatively they might be systematic, or regularly spaced, or stratified, where more sample points are used in some areas than others on the basis of some prior knowledge of the feature's distribution.

**Creating an interpolated raster-based continuous surface requires statistical analysis of the input data.** A variety of methods can be used. Exact interpolators include the actual data values as values in the final map. Smoothing interpolators create a more generalised surface and do not necessarily include actual data point values in the final map. Common methods include Inverse Distance Weighted and Kriging, of which each has multiple variants depending on input parameters to the algorithms. There are many alternatives such as using polynomials, radial basis functions, kernels, or diffusion methods.

**Often different terms are used interchangeably when referring to density analysis for the creation of continuous surfaces.** The use of the term *heat maps* has also become popularised though its use is erroneous. A heat map is a specific form of regular data matrix and not a map. Many so-called heat map techniques used in mapping are actually calculating density or kernel density. The additional erroneous use of spectral colours to imply hot and cold also causes problems and confusion with genuine hot-spot mapping.

Mapping density applies a moving search radius across the input data points to calculate values. Kernel density analysis extends the approach to use values represented at the input points. For both techniques, a smaller search radius creates a flatter overall surface with peaks and troughs of high and low values. A larger search radius creates a smoother surface with lower amplitude in the peaks and troughs.

Inverse distance weighted techniques apply a weighting factor so data points that are closer to one another have a greater influence in determining interpolated values than those farther away. Interpolated values that are closer to input sample data points will have more similar values than those farther away. The resulting map can exhibit peakedness due to this approach.

Kriging takes a geostatistical approach to interpolation. The general method includes information derived from the underlying variogram to generate the surface such as directionality. As it's regionalised and builds trends into the creation of the surface, it can create more appealing final maps. Weighting is included on the basis of the overall spatial arrangement of input points and values rather than just distance. There are numerous types of kriging that allow you to customise the algorithm for different types of data.

The Getis-Ord Gi* algorithm generates a z-score that creates a surface where clusters of high and clusters of low values are spatially aggregated. This is sometimes referred to as 'hot-spot analysis' and represents statistically significant clusters.

Density maps are often used in other aspects of cartography such as the creation of gaze plots as part of eye-tracking studies to explore cognitive responses to map stimuli.

**See also:** Contours | Digital elevation models | Heat maps | Isarithmic maps | Weather maps

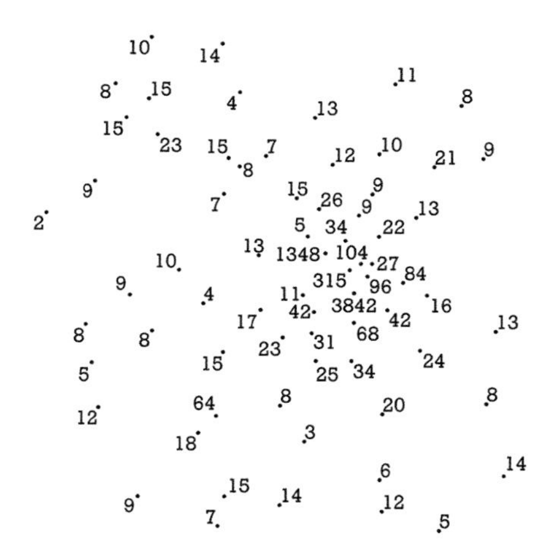


Sample points with data values

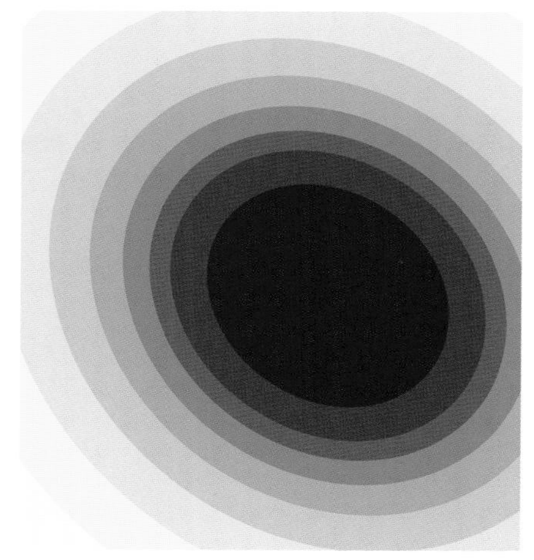

Global polynomial surface

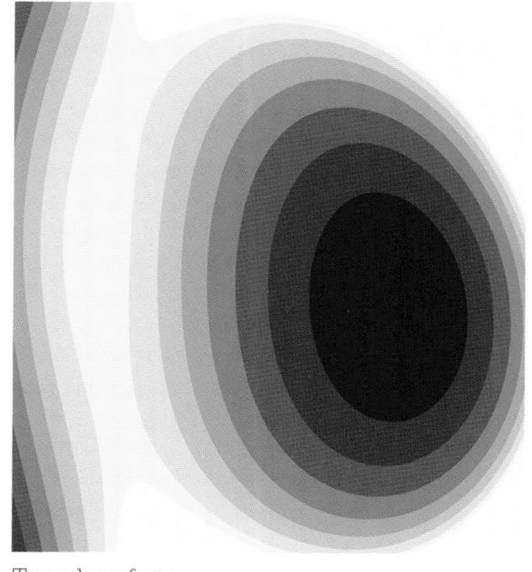

Trend surface

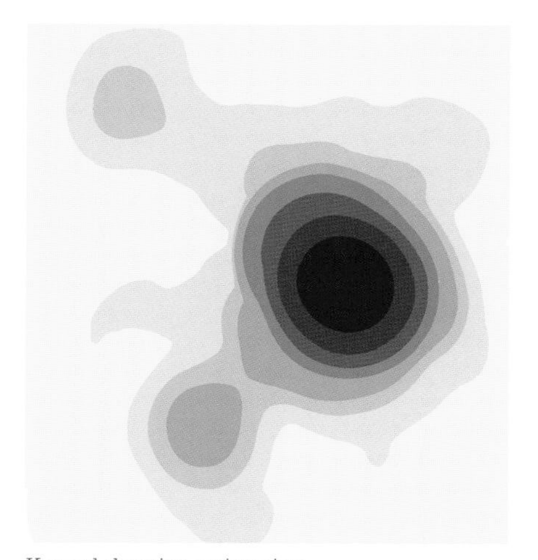

Kernel density estimation

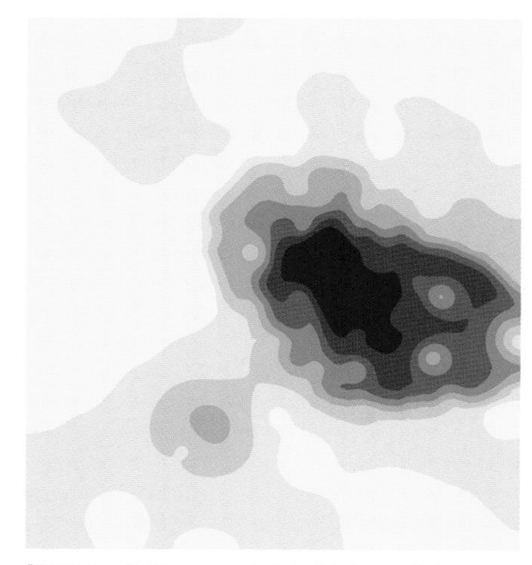

Inverse distance weighted interpolator

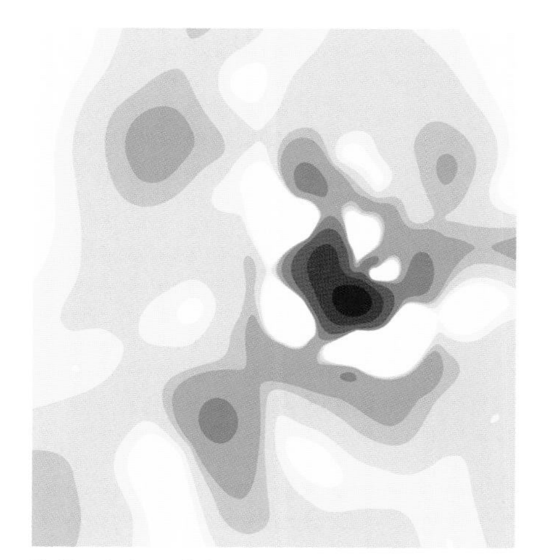

Spline interpolator

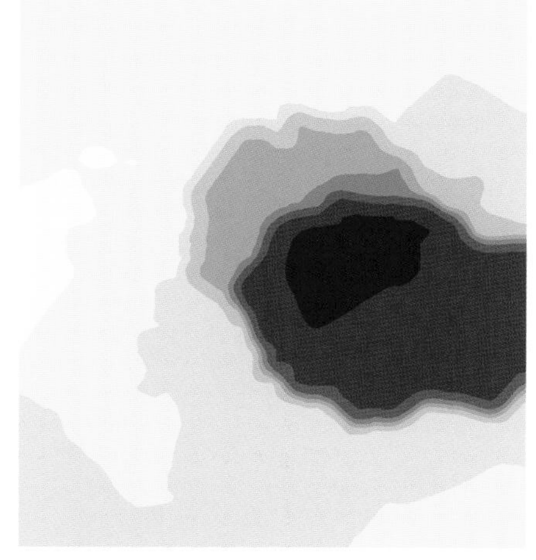

Universal Kriging

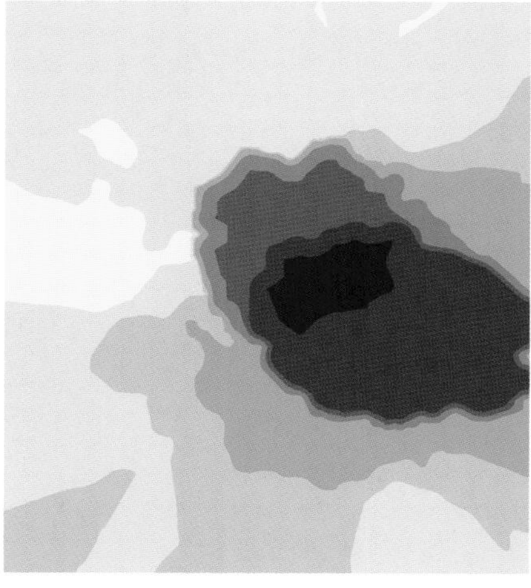

Empirical Bayesian Kriging

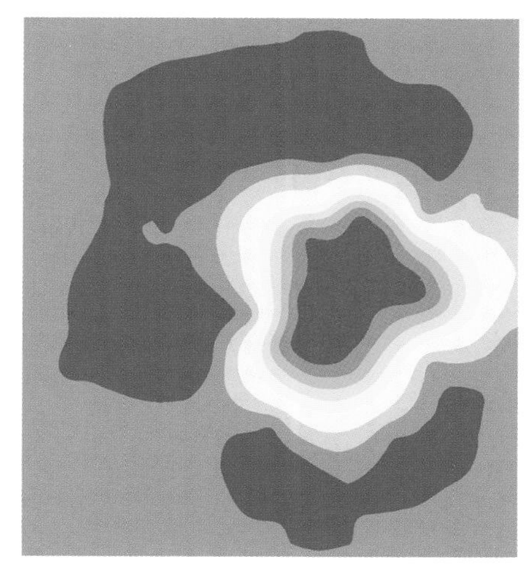

Getis-Ord Gi* statistic (hot-spot) surface

# Contours

## Contours are drawn to join points of equal value so the line has a constant value. Usually used to show elevation but contours also have thematic uses.

**Contours are lines that connect points of equal value, such as elevation, bathymetric depth, temperature, precipitation, pollution, or atmospheric pressure.** In this sense they can be used to map more than just elevation and map various thematic datasets as well. The distribution and separation of the lines shows how values change across a surface. Where there is little change in a value, lines are spaced farther apart. Where values rise or fall rapidly, the lines are closer together.

**Contours are usually constructed from spot height information or a digital elevation model (DEM).** The base contour is the value from which to begin generating contours. Contours are generated above and below this value as needed to cover the entire value range. The contour interval specifies the distance between contour lines. The choice of contour interval and smoothness of the line's curvature is a matter of choice led by the amount of other content on the map and the map scale. You'll likely want to generate a more detailed set of contours for larger scale products.

**Contours can be symbolised in many ways.** Typically these include using a single colour (commonly orange/brown) or a colour ramp that shows lower contours as darker and higher contours as lighter. Value can modify hue when symbolising bathymetric contours. It is also common practice to change the thickness of the lines you might want to use as interval (or index) contours to emphasise them. Interval contours usually represent major values in elevation—for example, every 1,000 m where intermediate contour lines might represent each 200 m in between.

**Not every contour line needs labelling.** After all, these are already abstract representations placed on the map so adding too many labels might clutter the final image. It is usual to label interval contours. Positioning is usually in a stair-step placement pattern around other map detail to assist interpretation of slope direction. Alternatively they can be right reading.

Sometimes, certain physical features need a modified representation to aid interpretation. For instance, closed depressions or basin-like features might be difficult to read among other contour lines. In this case, small right-angled ticks can be positioned to the downslope (inside) of the contour line to focus attention. These tick marks help distinguish between small hills and small depressions.

On some maps, additional types of contour lines may be useful. For instance, supplementary contours (usually thinner lines) might be used where elevation change is minimal and the normal contour interval is spaced too far apart. This might be useful on a floodplain. Dashed contours are often used to represent approximate or indefinite contour lines where the surveyed detail is perhaps unreliable. Contours can sometimes coalesce in areas of very steep terrain in which case they might be merged into a slightly thicker line or feathered out to thin the amount of line work.

The basic contour line is an extremely flexible symbol. It has become well understood despite its abstract character. Because of this general acceptance, there's scope to stretch how the symbol is designed to create new or different effects. Changing the line style, for instance, into dotted lines, and masking parts of contour lines that fall across very gentle slopes, can also provide interesting effects such as an historic dotted contour map. Indeed, although it veers into map art, contours can be used effectively as the only representation of topography on a map.

**See also:** Continuous surface maps | Illuminated contours | Isarithmic maps

250
928
301
609
389
164
991
987
328
223
722
443
794
1099
1011
925
947
47
866
709
423
151
527
378

# Contrast

## Building contrast among map components increases the identification of separate components as well as creates visual interest.

**Contrast establishes differences between map components to ensure they are distinguishable from each other.** When you read a book or listen to a speaker you're immediately turned off if the writing is uninteresting or the delivery is monotone. You use language to create interest in the written word and intonation, emphasis, and tone to add interest to the spoken word. The same is true for maps except you use a graphical language to build visually interesting relationships as a way of emphasising some components over others.

**Contrast creates focus and ensures legibility of unique map objects.** Some techniques can also be used to connote relative importance and modify the visual hierarchy of your map, and so contrast and hierarchy often work hand in hand. Modifications to the symbology of a map, such as size, spacing, shape, arrangement, and the components of colour, can be modified to create contrast. When creating symbols, contrast should be considered a key part of the design process not only to imbue meaning but to support emphasis.

**A lack of contrast can lead to a monotonous design with little or no relative importance between map elements.** This might give misleading messages about the importance of map objects. Maps can easily be seen as monotone in appearance if, for instance, type size is similar throughout; the lightness and size of thematic symbols is too similar; or line weights are the same. If there is little distinction between the mapped area and the background, it can make all the map components appear too similar. The map reader may then have fundamental problems in determining which parts of the map are foreground and which are background.

**Maps that display good contrast tend to be the easiest to read and interpret.** They are also likely to be more visually interesting and have a better balance. Consequently, they will stimulate a map reader to be more interested than a version of the same map that has poor contrast among the map components.

**See also:** Greyscale | Hierarchies | Mixing colours | Proximity in design | Resolution | Shape | Value

Black on white provides the maximum colour contrast. However, white backgrounds can blend into other backgrounds such as paper colour. Black can then appear to be lost among other background elements. In fact, black on a bright-yellow background provides the next best solution which is why you often see yellow halos for black text on a map. This is also why airport signage and many warning signs use black type on a yellow background.

Some of the most visually interesting maps use no more than two colours. Often, black or grey line work is used as the context and a single colour used for the main theme being mapped. This stark contrast is easy for map readers to interpret.

Ensure that your textual map elements have as much contrast as possible so that they are readable across the background map colours.

Typographic contrast can also be achieved by varying size, value, hue, or by using bold, italics or other font styles and spacing.

Reversing line work, imagery, or text, for example white on black, is also a useful technique for creating contrast as well as grabbing attention.

Contrast among map components can be emphasised by making good use of white, or empty, space on the map. Being confident of giving your map components space to breathe often helps the overall impact.

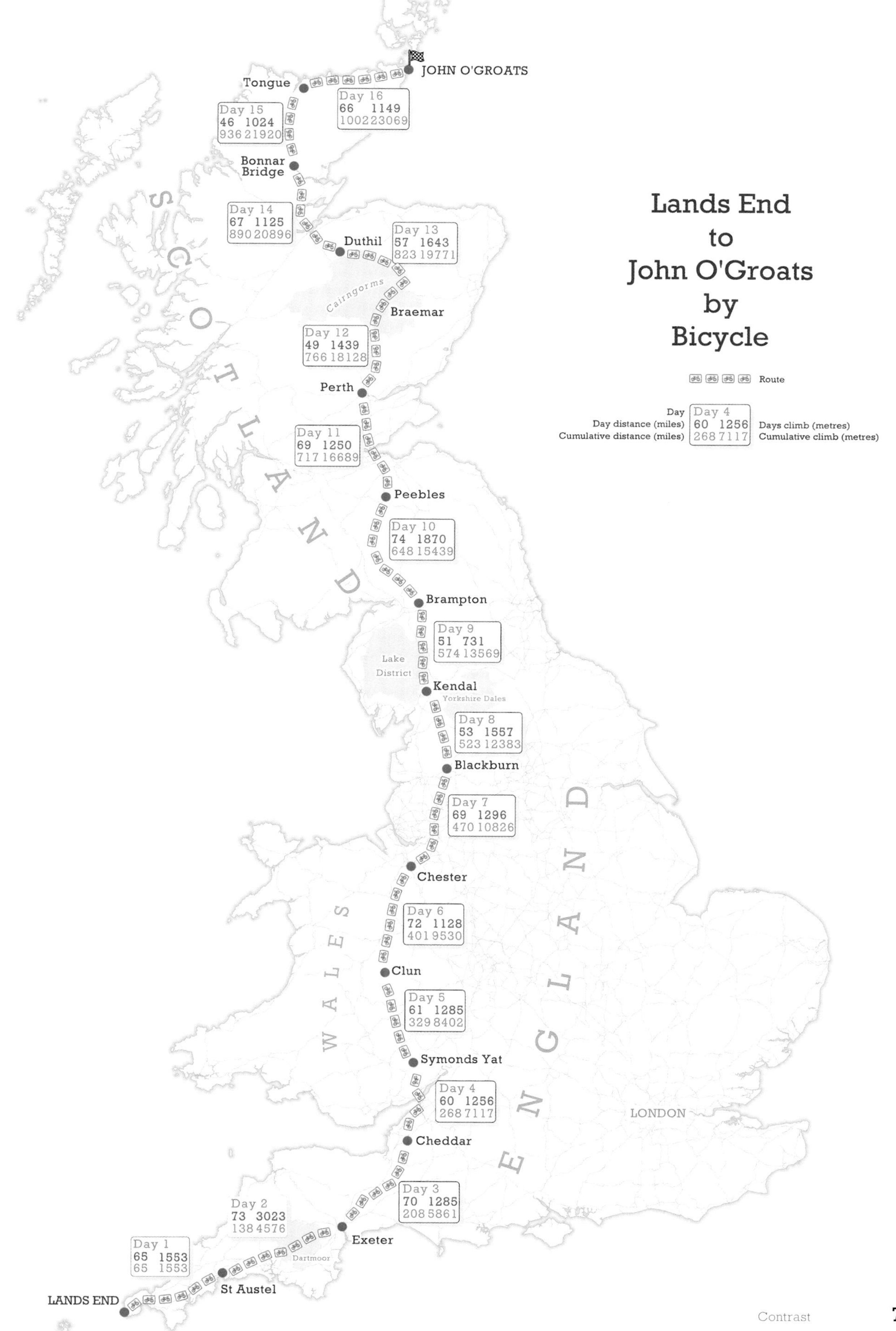

Lands End
to
John O'Groats
by
Bicycle
Route
Day
Day distance (miles)
Cumulative distance (miles)
Days climb (metres)
Cumulative climb (metres)
Day 4
60 1256
268 7117
JOHN O'GROATS
Tongue
Day 16
66 1149
1002 23069
Day 15
46 1024
936 21920
Bonnar Bridge
Day 14
67 1125
890 20896
Duthil
Day 13
57 1643
823 19771
Cairngorms
Braemar
Day 12
49 1439
766 18128
Perth
Day 11
69 1250
717 16689
Peebles
Day 10
74 1870
648 15439
Brampton
Day 9
51 731
574 13569
Lake District
Kendal
Yorkshire Dales
Day 8
53 1557
523 12383
Blackburn
Day 7
69 1296
470 10826
Chester
Day 6
72 1128
401 9530
Clun
Day 5
61 1285
329 8402
Symonds Yat
Day 4
60 1256
268 7117
Cheddar
Day 3
70 1285
208 5861
Exeter
Day 2
73 3023
138 4576
Dartmoor
Day 1
65 1553
65 1553
St Austel
LANDS END
SCOTLAND
WALES
ENGLAND
LONDON

# Copyright

## Restrictions governing use of map data and people's use of your map.

**Copyright laws exist to protect.** It's the legal right that grants you or the producer of the work exclusive rights to its use and distribution, often with some limitations. Different copyright laws exist in different countries and as much as they protect you from having people breach the use of your work, you must also ensure that you are in compliance with the copyright on what goes into your map. Limitations might include fair use or the fact that it covers the expression of an idea and not necessarily the underlying ideas themselves. Put simply, you cannot copyright a 'fact' but you can copyright an expression of that fact which makes it easier to distinguish the more original content you have created.

**Copyright is one form of establishing intellectual property for creative works.** Maps can fall into this realm, which offers protection to the original authors of such works. For instance, in the United States copyright includes placing restrictions on the rights of others to reproduce your work, to prepare derivative works based on your work, or to redistribute, sell, or transfer ownership of your work. Copyright can be officially registered, or you can simply add an appropriate statement on the map asserting your copyright. Copyrights are territorial, and although international copyright agreements exist, it is worth checking the copyright laws of different countries.

**Maps and data you use are also likely to have some copyright associated with them even if you acquire them for free.** It is important to establish what copyright and licensing restrictions exist on the raw map materials you use. Many datasets may be free from copyright but it shouldn't be presumed. Even when you are free to use data there are often statements of compliance that you must include. Always request permission from the copyright holder if you intend to use a part of their work in your own work. Permission can usually be granted in writing or through a contract that specifies quantity, royalty, or licensing fees and any specific terms. Failure to gain permission can result in penalties including payment of profits or damages, court costs, and lawyer's fees. The biggest cost is in terms of credibility.

**See also:** Branding | Ethics | How maps are made | Integrity | Map traps | Who is cartography?

Open, free, or public domain licences are commonplace in cartography. Many datasets are made available to use freely but it's worth acknowledging that free in this context refers not to cost, but to the principles of freedom. Examples of such licensing include the popular Creative Commons (CC) licences. Creative Commons was founded in 2001 as a non-profit organisation to support the legal sharing of creative works. It creates a series of licence options to support people's work which allow copyright holders to specify the conditions under which work can be used or reused.

Different CC licences stipulate whether work can be modified or whether derivative works can be created or whether commercial work can be based on it. If you use data made available under a CC licence, you'd include a specific form of words on your map. Likewise you may make your own work available under a CC licence and require people to include a form of words that you permit.

As an example, prior to September 2012 OpenStreetMap was provided under a Creative Commons Attribution-ShareAlike licence. Subsequently, the licence changed to an Open Database Licence (ODbL) provided by Open Knowledge Foundation. Using OpenStreetMap requires you to include a specifically worded statement on your derived work—© OpenStreetMap contributors—and if you are using the cartography that OpenStreetMap provides through its map tiles, an additional CC-BY-SA statement is required.

"Copyright" is a form of protection provided by the laws of most countries to the authors of 'original works of authorship', including literary, dramatic, musical, artistic, and certain other intellectual works. This protection is available to both published and unpublished works. Copyright protects the original expression of an idea. Generally, copyright law gives the owner of copyright the exclusive rights to do and to authorize others to do certain things with the original work of authorship. The duration of copyright protection may vary by country, but at some point a work ultimately falls into the "public domain" where others may use the work without fear of infringement.

A work of authorship is in the "public domain" if it is no longer under copyright protection or if it failed to meet the statutory requirements for copyright protection. Works in the public domain may be used freely without the permission of the former copyright owner.

"Fair use" is a legal doctrine that promotes freedom of expression by permitting the unlicensed use of copyright-protected works in certain circumstances. For example, Section 107 of the US Copyright Act provides the statutory framework for determining whether something is a fair use and identifies certain types of uses – such as criticism, comment, parody, news reporting, teaching, scholarship, and academic research – as examples of activities that may qualify as fair use. Section 107 calls for consideration and balancing of the four factors in evaluating a question of fair use.

"Copyleft" is defined as an arrangement whereby an author of an original work may allow the original work to be used, modified, and distributed freely by another person on condition that any subsequent, derivative work is bound by the same licence conditions. Basically, the author/owner of an original work is relinquishing or waiving by contract (e.g., an open source-type licence) certain exclusive rights statutorily conveyed by copyright law.

Creative Commons ("CC") is a global nonprofit organization that enables sharing and reuse of creativity and knowledge through the provision of free legal tools. CC is best known for its copyright licences. CC licences are legal tools that creators and other rights holders can use to offer certain usage rights to the public, while reserving other rights. Currently, there are six (6) CC licences comprised of a combination of the following rights:

In contrast to CC's licences, a CC0 dedication to the public domain gives creators an affirmative way to waive all their existing copyright and related rights in their works to the fullest extent allowed by law.

CC's Public Domain Mark is recommended for works that are free of known copyright around the world. These will typically be very old works that have statutorily fallen into the public domain. (It is not recommended for use with works that are in the public domain in some jurisdictions if they are also known to be restricted by copyright in others.)

# Craft

## The act of doing can be as important as design.

**The idea of design suggests it comes before the work but crafting the map can be equally as important.** The majority of this book espouses that a knowledge of design and the fundamentals of the cartographic process feed into the formulation of a piece of work that is then undertaken. The converse can also be useful and richly rewarding. The qualities of craft and doing are not just a practical execution of design. It's not just implementation, it brings additional qualities. Defects in a map can be a function of craft, and no amount of excellence in design can actually make the map. Design can be nullified or exemplified by craft.

**Virtually every map is made by a human to a greater or lesser extent.** Technology may help, but even then, we are invariably at the controls. The quality of the outcome is therefore also human controlled, which no amount of design can ensure. In this respect design is made good through good craft. Perhaps craft is what differentiates the true quality from the masses of maps that we see. A map might be memorable and resonate but only if the craft has elevated it above the average, beyond simply being truthful and functional or even pretty, but to some character that goes beyond mass production. A crafted map has depth, subtlety, diversity, overtones, and a richness that only comes from crafting something. In this sense, the human remains fundamental to the cartographic craft because cartography is a very human craft.

**Craft goes beyond a hobby.** Making mapmaking more of a hobby (that you love doing) than simply a job is a valuable ingredient in good cartography. It helps that you develop an empathy for the craft, the process as much as the outcome. The quality of outcome is not predetermined by design but depends on the judgement, skill, and care that the maker exercises. Quality is therefore at risk throughout the entire cartographic process. The counter is that some level of quality can be a certainty but it tends only to come with a pre-determined process.

David Pye refers to the notion of craft as a distinction between the workmanship of risk and the workmanship of certainty (Pye 2015). In the 1970s he explained the difference as being as simple as writing with a pen as compared to modern printing. With computerised technology some standards are assured. The workmanship of certainty is expressed through automation and repetition. Workmanship of risk is expressed through writing by hand. Of course, the page can be spoiled in many different ways but practice develops skill and dexterity and the risk becomes smaller. You also develop techniques that overcome some of the natural shortcomings of the approach—such as using a ruler to draw a straight line or a French curve to draw a smooth curve.

Speed tends to equate to the workmanship of certainty. Computers no doubt help us make maps and they allow us to replicate but the savings in speed and cost are only borne out by economies of scale. Using a computer for a one-off project isn't necessarily a time-saver per se. But there is more to workmanship than making things faster or being able to replicate them. Craft seeks to make things in a non-standardized way. Individuality is important. This is not to say that maps should all be drawn by ruling pen but that craft remains valuable because it showcases difference and opportunity. It expresses the huge variation that is possible through cartographic practice and in many ways it influences what is then taken up by design and repurposed for the masses. There's an aesthetic richness that often accompanies a well-crafted map. It's a quality that cannot be mass produced but which imbues a very human part of the person who crafted it. It helps push cartography forward and gives those who take the workmanship of certainty route more to aim for.

**See also:** Knowledge and conviction | Design and response | Different strokes | Elegance | Flourish | Graphicacy | Hand-drawn maps | Jokes and satire

**Opposite:** Extract from *The North American Continent* by Anton Thomas.

Portland
NEW HAMPSHIRE
VERMONT
St Lawrence
Kingston
Concord
Manchester
Albany
Utica
Syracuse
BOSTON
MASSACHUSETTS
NEW YORK
Springfield
PROVIDENCE
HARTFORD
RHODE ISLAND
CONNECTICUT
Ithaca
Binghamton
Scranton
NEW YORK
Trenton
PHILADELPHIA
NEW JERSEY
Harrisburg
Atlantic City
Dover
MARYLAND
BALTIMORE
DELAWARE
Annapolis
WASHINGTON
D.C.
Charlottesville
VIRGINIA
RICHMOND
NORFOLK
Lynchburg
Roanoke

# Crispness

## The design character of the edges or fills of map symbols.

**Edges or fills of map symbols can be designed to be fuzzy rather than crisp, which changes their visual prominence.** The boundaries of map symbols are normally sharp, which gives a clear and abrupt change between adjacent symbols or between a foreground symbol and its background. This is for good reason since you often want to show clear differentiation between map symbols. Similarly, fill symbols are usually consistent for the whole symbol. However, by changing the spatial filtering, a softer appearance to these edges or variation to the fill can bring benefits in certain circumstances. Softer edges will help the symbol recede to the background so it can provide a good way to modify the appearance of symbols to shrink the gap between figure and ground.

**Softer symbol edges and fills give the map reader a sense that the data is less certain.** Crisp edges bring a sense of certainty, or assuredness to the symbol. Precise line work and crisp edges speak to the clarity of the message. As edges become increasingly blurred, the effect connotes uncertainty. Perhaps the boundaries are less precise between areas, or the location of a point symbol is only known imprecisely. For fill symbols, some change in crispness gives the impression that while belonging to the same feature there may be variations within. These may not be large enough to warrant a separate symbol but are sufficient to show that the feature is not precisely uniform.

**Modifying crispness provides a way to change the aesthetics of the map.** Blurring the edges of map symbols or of the typographic components will create a distinct look and feel. This might be regarded as the equivalent of applying a soft focus to elements of a photograph. This could be to a particular class of symbols or, consistently, across the whole map.

More often than not, crisp edges are found on reference maps or national map series. Such maps intend to portray a sense of precision and, so, the symbology meets the requirement. Certainly, maps made during the 20th century were more concerned with mapping the precise location of features and the symbology tended to reflect that desire. Print production techniques also limited the ability to reproduce the effect.

With digital technologies the barrier to creating effects such as crispness are no longer a problem. Desktop design, GIS, and illustration software all contain tools that support the creation of blurred edges and fills. This has created a rich and fertile design environment that has led to maps becoming more artistic. Applied subtly, crispness can modify a symbol enough to set it apart but not too much that it changes the message. Softer boundaries and blurred techniques can be used successfully to add to the map's impact and overall design aesthetic.

Visual variation

Increasing crispness becomes figural

For seeing

Distinct Levels

For representing

Nominal Ordinal Numeric

**See also:** Contrast | Foreground and background | Hierarchies | Transparency

**Opposite:** *Health Hangars* by Lateral Office, 2011.

Grise Fiord
Resolute
90 min
60 min
Pond Inlet
Population served = 2,375
10-20 Beds
Diagnostics, Cancer, Tuberculosis
Arctic Bay
Nanisivik
90 min
60 min
Igloolik
Population served = 5,206
17-34 Beds
High Risk Maternity, Infant Care, Disabilities
Clyde River
Taloyoak
Gjoa Haven
Pelly Bay
Hall Beach
90 min
60 min
Qikiqtarjuaq
Pangnirtung
Population served = 4,265
15030 Beds
Mental Health, Addictions
Repulse Bay
60 min
90 min
60 min
Baker Lake
Population served = 6,839
23-46 Beds
Geriatrics, Rehabiltion Services
Coral Harbour
Cape Dorset
Iqaluit
(Pop. 6,200)
Chesterfield Inlet
Rankin Inlet
HUDSON STRAIGHT
Whale Cove
Ivujivik
Salluit
Kangirsujuaq
Arviat
Quaqtaq
Akulivik
HUDSON BAY
Kangirsuk
To Winnipeg
Puvirnituq
To Ottawa
Kangiqsualujjuaq
To Montreal
Aupaluk
0
100km
Proposed Hospitals
Hospital Hangar
Proposed Network
Step3
Step 2
Step 1
Communities
Existing Hospital
Actors Involved
MAP
LEGEND

# Critique

**Everyone's a critic! We all think we know what's best for our own work but seeking, appreciating, and contributing to map critique makes it so much better.**

**We all have favourite maps that we think are great, but rather than just saying a map looks great, how many of us are able to take a critical look, evaluate it, and explain why it looks great?** Having the ability to reflect on and evaluate maps as information products is important to understanding effective cartographic design so you can optimize communication of the intended message in your own maps as well as be able to understand why other maps work.

**Critique and review are not the same as being critical.** The intent is not simply to find fault although it's true that most critique occurs to discover the faults so they can be fixed. Critique is a method. It's a systematic analysis of the map. It should equally focus on what works well and to learn why the map works well. Unfortunately this isn't necessarily an easy process but exposure to quality cartography and assessing a map's merits will begin to shape your own understanding of design. One way to achieve this is through the use of a checklist that allows you to systematically 'grade' a map. By working through a checklist you begin to develop a sense of what one is looking for in performing a critique.

**Understanding the mechanics of making a map doesn't always translate into creating a great map.** What might be intuitive for an experienced professional cartographer is not necessarily so for everyone. But we all start somewhere. Using a checklist approach at least allows you to reflect on your own work and improve your chances of making a great map. Also, it's worthwhile to take a similar approach to any map product whether a large atlas project or a simple new map for a website.

**Critique is supposed to be objective yet many maps include very subjective components.** As in many design-led fields, you'll have a sense of style that you like and others that you don't. Critique tries to go beyond just stylistic likes and dislikes though when a stylistic approach masks a message, you're likely identifying where the cartographer's subjectivity has clouded their judgement.

**See also:** Knowledge and conviction | Different strokes | Who is cartography? | Your map is wrong!

At the outset of performing a critique of your own or other work, take a good look at the map and try to gain a sense of the key, immediate overall message. Is it obvious and central to the composition? Do map components sit in harmony with one another? Is something screaming to be seen or hiding behind something else? Is something missing or is there too much?

Critique and review isn't about trying to find a way to make every map some sort of purist's piece of work. It's about appreciating the importance of getting opinions on a piece of work, understanding why those opinions have been expressed, and what it then means for how you respond. Being able to develop a critical eye will allow you to perform this task as well as understand how to deal with others who comment on your work. Ultimately, exposure to great maps helps build your own repertoire and understanding.

When providing or receiving critique it's important to stick to the map. It's a natural human tendency that people will take criticism to heart, and all too often critique is taken as a personal affront. It's not and should never be about the person—it's always about the map. Equally, when receiving critique about your work, don't take it personally. Make the vital distinction between someone's comments, which are about your work and not you as a person.

**Opposite:** You can use a checklist like this to critique your own work or that of others. It's developed around the concept of assessing the subjective objective and the affective objective. The subjective objective relates to the content itself. The affective objective relates to the look and feel, or the stylistic treatment.

# Map Evaluation Checklist Questions to consider when performing map critique

Name of author

Date of evaluation

Title of map

Map sheet (if in a series)/URL

## Cartographic Requirement

What is the rationale for the map?

What is the purpose of the map (i.e. the substantive objective)?

What is the "look and feel" of the map (i.e. the affective objective)?

Who is the audience for the map?

What is the expected educational level of the audience for the map?

What are the expected conditions of use for the map (medium, distance, light, etc.)?

## Cartographic Compilation and Design

Have all required themes and features been included?

Does the map have appropriate figure-ground organisation?

Is there appropriate visual hierarchy among all themes and within each theme?

Is there appropriate visual emphasis on the important theme(s)?

Is the symbology for qualitative and quantitative data effectively applied?

Do the colours and symbols support the substantive and affective objectives?

Do the font styles, size, and colour support the substantive and affective objectives?

Are the symbols and labels legible?

Are the symbols intuitive and easy to decipher or do they have good explanation?

Is there appropriate use of graphics, images, text blocks, and other supporting information?

Can the map be placed in context both geographically and thematically?

Is the map projection suited to the map's purpose (equal area, conformal, etc)?

Has the map projection been appropriately modified (central meridian, standard parallels, etc.)?

## Map Elements and Page Layout

Does the page look balanced?

Do all the map elements support the substantive and affective objectives?

Are the map elements placed logically on the page?

Are the map and map elements aligned to the page and to each other?

Does the map have appropriate borders?

### Orientation Indicator

Is the grid or graticule appropriately aligned?

Does the grid or graticule have appropriate labeling?

Does the map require a north arrow?

### Scale Indicator

Is the scale appropriate to the map?

Is the scale bar appropriately designed, positioned, and sized?

Are the scale units logical?

### Legend

Have all the necessary symbols and details been included in the legend?

Do the symbols in the legend appear exactly as they do on the map (size, colour, etc.)?

Is there a logical structure related to the function of the legend?

Are the patches, symbols, labels, and descriptions appropriately sized and positioned?

Are the labels logical?

### Titles and Subtitles

Are the titles and/or subtitles relevant?

Are the titles and/or subtitles suitably descriptive (area mapped, subject, date, etc.)?

Are the titles and/or subtitles suitably positioned and sized?

### Production Notes

Are production notes included?

Are the production notes dated correctly?

Are the production notes placed appropriately?

Have copyrighted sources been correctly attributed?

Has the map's assertion to copyright been included?

Have attribution and/or revision details been included?

## Synopsis and Final Check

Given your responses to the above, provide an overall evaluation of the extent to which the design of the map meets its intended aims. You might also include here comments made by independent reviewers of the map as a new perspective that often reveals areas that the map author may have overlooked.

## The Four Point Plan

Comment on four areas that work well and four that might require attention.

**What works well?**

**What doesn't work so well?**

# Curvature of terrain

## The extent to which a slope deviates from a flat, linear surface.

**Digital elevation models (DEMs) are commonly used to assess slope characteristics as a way to represent the terrain.** This normally assumes a linear relationship between one cell and the next but in reality slopes are likely to curve in multiple directions and may be concave or convex rather than straight. We can use assessments of curvature to explore, model, and, subsequently, map nuance in slopes. This is a particularly useful technique when mapping hydrological flow or soil erosion, for instance.

**Hillslopes have five main elements which are defined by their position and slope characteristics—namely, summit, shoulder, backslope, footslope, and toeslope.** Each of these elements has a different profile and, combined in different ways, allows us to model curvature beyond the linear increase or decrease in height between two points.

**Curved slopes differ from linear slopes, which would be represented by evenly spaced contours with more widely spaced contours representing a uniform shallow slope.** A curved convex slope would have contours that are wider at the top of the slope and which progressively decrease in separation as they approach the foot of the slope. A curved concave slope is the opposite with narrowly spaced contours at the top of the slope with increasing separation toward the foot.

**Curvature is a second derivation calculation of a surface which follows two types—profile or plan.** Profile curvatures can be calculated to help assess the overall slope characteristic. Negative values indicate that the overall surface is convex at that cell. Positive values indicate that the surface is concave at that point. Plan curvature occurs perpendicular to the direction of the maximum slope. A positive value indicates the surface is convex perpendicular to the slope at that cell. A negative value indicates the surface is concave perpendicular to the slope at that cell.

Different combinations of profile and plan curvatures allow us to make maps that support the need for understanding particular surface characteristics. Slope affects flow, or movement, downslope and aspect defines the direction. Profile curvature affects acceleration and deceleration at different points down a slope. Plan curvature affects the convergence or divergence of flow.

In reality a real slope is likely to encompass a combination of slope, profile, and plan curvature. All of these can be derived mathematically from a DEM and assessed either separately or be combined to provide a more realistic model of the terrain under investigation.

Techniques to map the curvature of terrain support modelling and analysis but they also provide results that add texture to maps of terrain. By incorporating elements of curvature into other representations of terrain, for example hillshades, a much wider range of landforms can be highlighted than would otherwise exist through linear hillshade calculations, for instance. They often bring a sense of detail to terrain representations.

**See also:** Raised relief | Slope, aspect, and gradient | Small landform representation

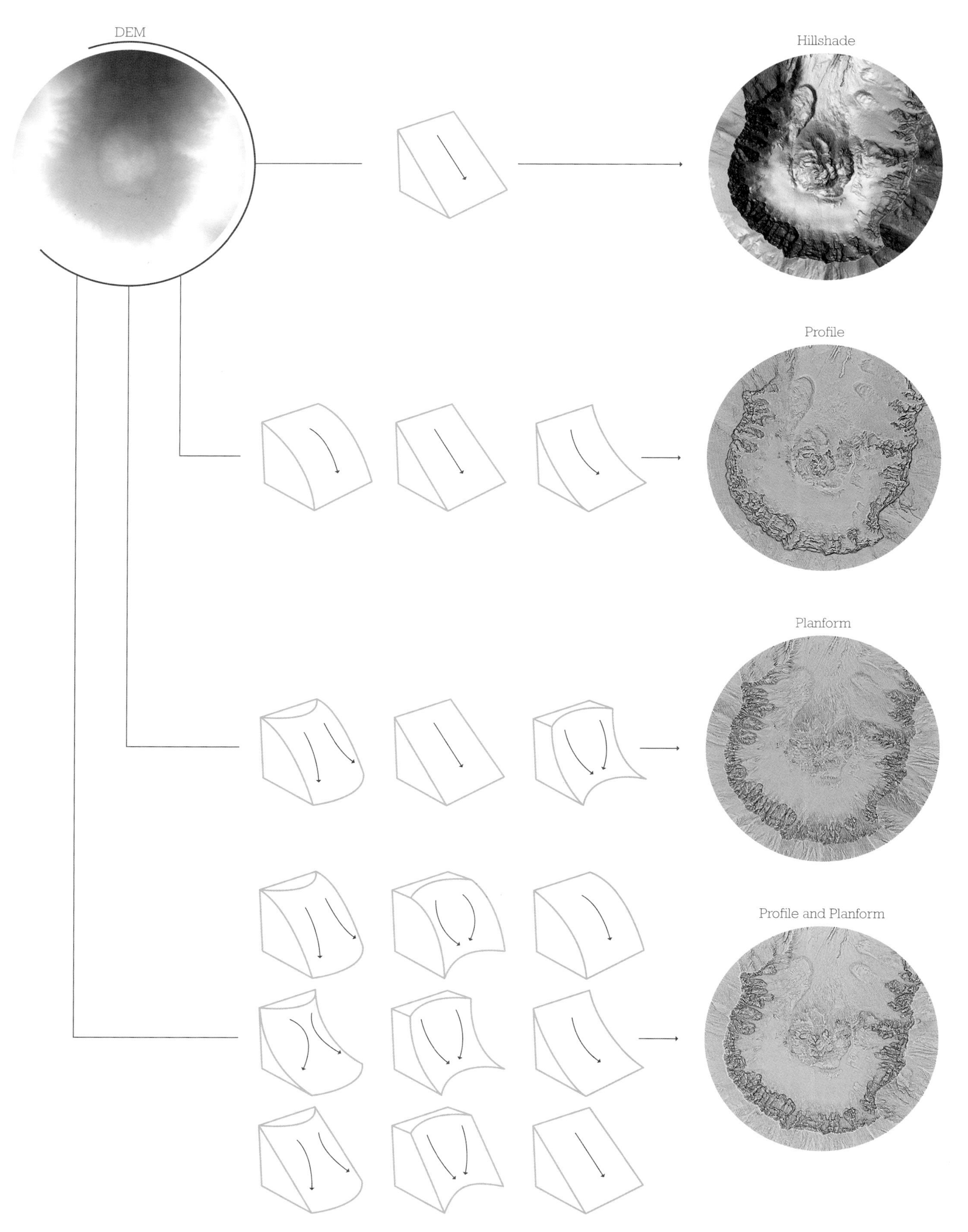
DEM
Hillshade
Profile
Planform
Profile and Planform

## Carte Figurative des Pertes Successives en Hommes de l'Armée Française dans la Campagne de Russie 1812–1813

Charles Joseph Minard

1869

Minard's map drawn in 1869 depicting Napoleon's Russian campaign is without a doubt the most well-known flow map around. Edward Tufte's 1983 quote on the map being 'the best statistical graphic ever drawn' has made it famous, and ever since it is used as the example of its kind.

For some it is even the ultimate map—but such maps do not really exist. Although flows inherently incorporate time, Minard's map is less explicit about time and even direction. It needs deduction of the accompanying text and diagram to get a full sense of time and direction.

It is a story map avant d'lettre informing us about the dramatic faith of the French army via a map and a linked diagram. The map is both schematic (the flow) and exact (the basemap), and shows qualitative (to and from Moskva) and quantitative (number of soldiers) information. The fact that even today its design is not easily generated by current automated (carto)graphic software proves the power of this map.

—Menno-Jan Kraak

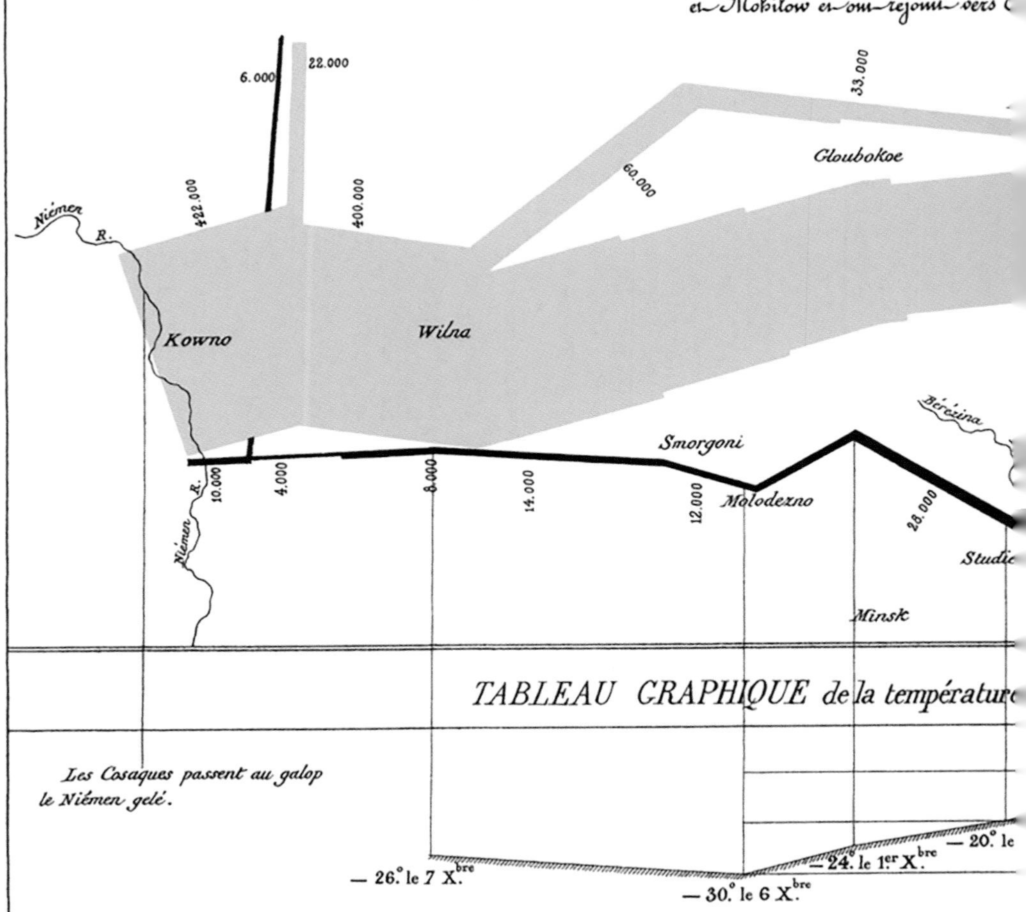

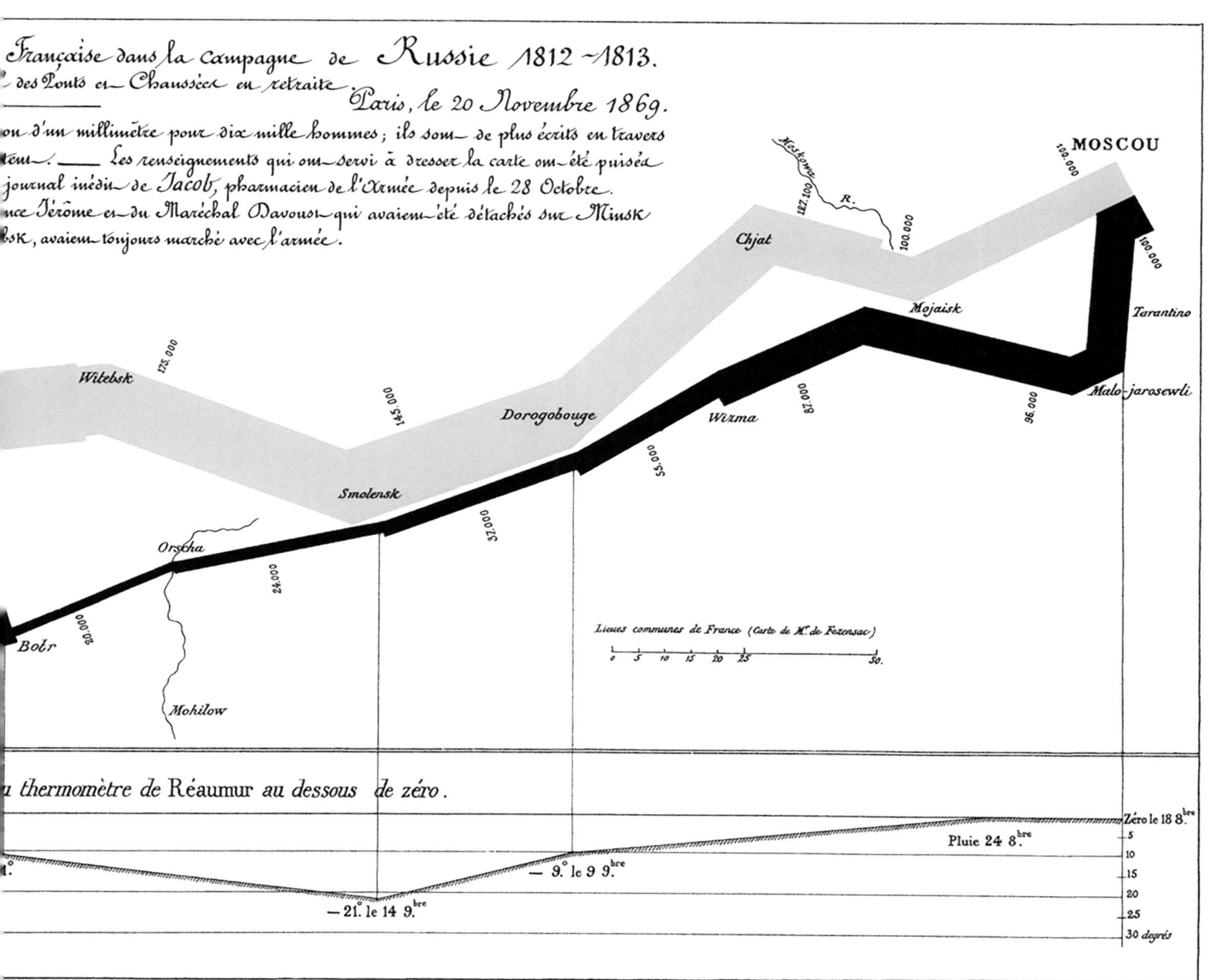

Française dans la campagne de Russie 1812-1813.
des Ponts et Chaussées en retraite.
Paris, le 20 Novembre 1869.
on d'un millimètre pour dix mille hommes; ils sont de plus écrits en travers
tem. Les renseignements qui ont servi à dresser la carte ont été puisés
journal inédit de Jacob, pharmacien de l'Armée depuis le 28 Octobre.
nce Jérome et du Maréchal Davoust qui avaient été détachés sur Minsk
bsk, avaient toujours marché avec l'armée.
MOSCOU
100.000
Moskowa
R.
127.100
100.000
Chjat
100.000
Mojaisk
Tarantino
Witebsk
175.000
145.000
Dorogobouge
Wixma
87.000
96.000
Malo-jarosewli
Smolensk
55.000
37.000
Orscha
24.000
20.000
Bobr
Mohilow
Lieues communes de France (Carte de M. de Fezensac)
0 5 10 15 20 25 50.
u thermomètre de Réaumur au dessous de zéro.
Zéro le 18 8.bre
Pluie 24 8.bre
— 9.° le 9 9.bre
— 21.° le 14 9.bre
5
10
15
20
25
30 degrés
Imp. Lith. Regnier et Dourdet.

# Dasymetric maps

## Dasymetric mapping can be used to modify other map techniques to account for different geographies.

**Dasymetric mapping is a development of other thematic mapping techniques such as the choropleth or dot density map.** It's an attempt to break away from the often arbitrary units used to collect data and then present it on the basis that arbitrary areas fail to recognise real geographies. Whereas standard thematic mapping techniques use prescribed enumeration areas to map the data collected uniformly for the same units, a dasymetric technique incorporates something of the geography of an additional variable. For instance, the real population distribution within an enumeration area will not conform perfectly to the areal extent of the unit or be spread uniformly across its area. A county may have a city occupy a small proportion of the area with a sparse rural population elsewhere. Choropleths or uniform dot density maps do not make this distinction.

**The dasymetric process reapportions data using secondary data that better describes the real geographical distribution of the underlying geography.** Once the new zones are constructed, the data can be reapportioned to them and the standard techniques of choropleth or dot density mapping applied. The result is often a more nuanced map which presents data that reflects some real geographical distribution and not just the abstract geography of a statistical areal unit used to collect the data.

**As with most techniques, the dasymetric technique has advantages and disadvantages.** The clear advantage of a dasymetric approach is that the map will better reflect the actual underlying distribution of the variable being mapped. Many people could be forgiven for assuming that a map of data presented as a choropleth means that data exhausts space uniformly across each area. Not so a dasymetric iteration of the map. One downside, particularly for small-scale country or continental maps, is the inevitability that whatever data is mapped it will take on the appearance of the secondary data structure. If this is based on where people live, any population-based map will tend to simply show where people live!

**See also:** Binning | Data density | Dot density maps

As with many thematic mapping techniques that rely on some form of statistical or transformational process, dasymetric mapping was only popularised with the advent of computerisation and computational cartography in the mid-1900s. This was despite it being proposed in 1911 by Russian geographer Benjamin Semenov-Tian-Shansky who also published many maps using the technique in the 1920s.

Interest in the technique has largely been borne out of a desire for high-resolution population mapping and has been underpinned by the availability of high-resolution data and imagery that can be processed to provide the key secondary data component. Many routine administrative boundaries have become simply too coarse for reporting at large scales or on multiscale web maps where there is an expectation of being able to zoom into small areas to explore patterns. Since a lot of reported data is aggregated into standard reporting units, the dasymetric technique provides a way of building a level of detail into the data that might otherwise not be available.

Of course, transforming data between different boundaries and at a higher resolution than the intended use has consequences. Ultimately, the data is being manipulated in ways it was not designed to support. It's a form of interpolation for which any number of underlying assumptions have to be made when interpreting the result. The most important of these is to realise that it is simply a different way of viewing the data and is no more or less accurate than the original dataset allows. The use of secondary sources can introduce error as an unintended consequence, potentially distorting the new view of the data in ways that are difficult to recognise or quantify.

There's a danger, at large scales, that a dasymetric map can lead to the reader inferring a greater level of detail than the data supports. For instance, a dot density map symbolised as 1 dot per person might be falsely interpreted as showing exactly where that person exists.

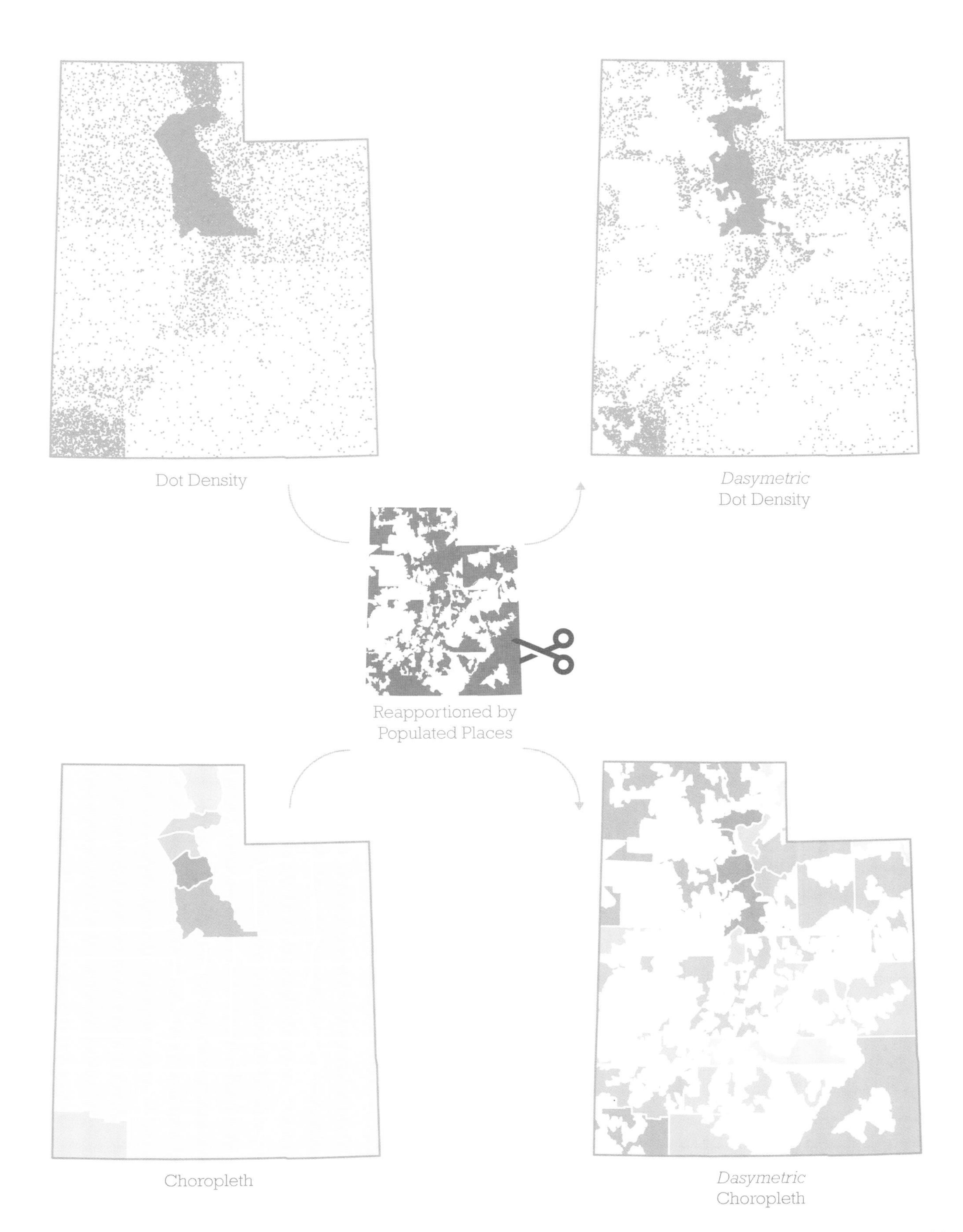
Dot Density
*Dasymetric*
Dot Density
Reapportioned by
Populated Places
Choropleth
*Dasymetric*
Choropleth

# Data (c)art(e)

## From art to carte, data dumps never looked so good!

**With the increased availability of large digital datasets of practically any phenomena comes a natural desire to map.** Although a lot of cartography attempts to present a summary of data through sensible generalisation, classification, and symbolisation, there's also space for alternate approaches. Rather than applying cartographic design principles as understood for decades to reveal a specific aspect or nuance in the data, there's sometimes value in presenting detail. Indeed, many of these maps might be referred to as data dumps because they deliberately aim to show all the data at once. Clarity of expression is found not through the cartographer's lens but in the mapping of millions of points or thousands of lines themselves. Often these data dumps are devoid of any basemap or other contextual content but the pattern they make is at the same time beautiful and informative.

**Analytically these maps provide wonderful visuals and can reasonably be referred to as 'data art'.** Although it may sound counter-intuitive, sometimes adding thousands or millions of small details helps a coherent larger picture to emerge that simply couldn't be seen any other way. There can be value in mapping detail to give clarity to an overall view of data. The challenge for cartography is to understand how these maps can be properly formed and interpreted but the trend will continue as mapmakers move from exploring large data using these approaches to mapping truly big data. Sometimes these data maps are purely aesthetic endeavours but it's pushing the frontiers of cartography to find ways of revealing meaning.

**Opposite:** *Visualising Friendships* by Paul Butler, 2010.

Data art in map form is now a common sight across social media. Aesthetics has always been a part of great cartography and many classic examples of maps were heavily artistic. Perhaps the birth of modern data-map-based data art stems from Paul Butler's 2010 map of 500 million Facebook connections. He created a social graph of the locality of friendships that was an attempt to explore how geography and political borders affected where people lived relative to where their Facebook friends were located.

Butler's approach wasn't just the plotting of data but the careful manipulation of the sum of pairs of friends between each pair of cities. He used a weighted function based on Euclidean distance between the pairs and the number of friends between them to build a picture of relationships. Lines with more friends were plotted on the top and colour was used to show fewer friends as darker and more friends as lighter.

Butler used a variety of data manipulation and representation techniques to achieve the final map which revealed visible continents and some political boundaries. The map wasn't in fact showing these—it was showing human relationships yet the pattern they created resulted in a far more fascinating picture.

In fact, Butler had made the map using cartographic processes but perhaps without really knowing that what he was doing was cartography. Data dumps are rarely going to look good on their own but with an awareness of map design they can be morphed into real works of (c)art(e).

**See also:** Data density | Map aesthetics | Pointillism | Style, fashion, and trends | Typographic maps

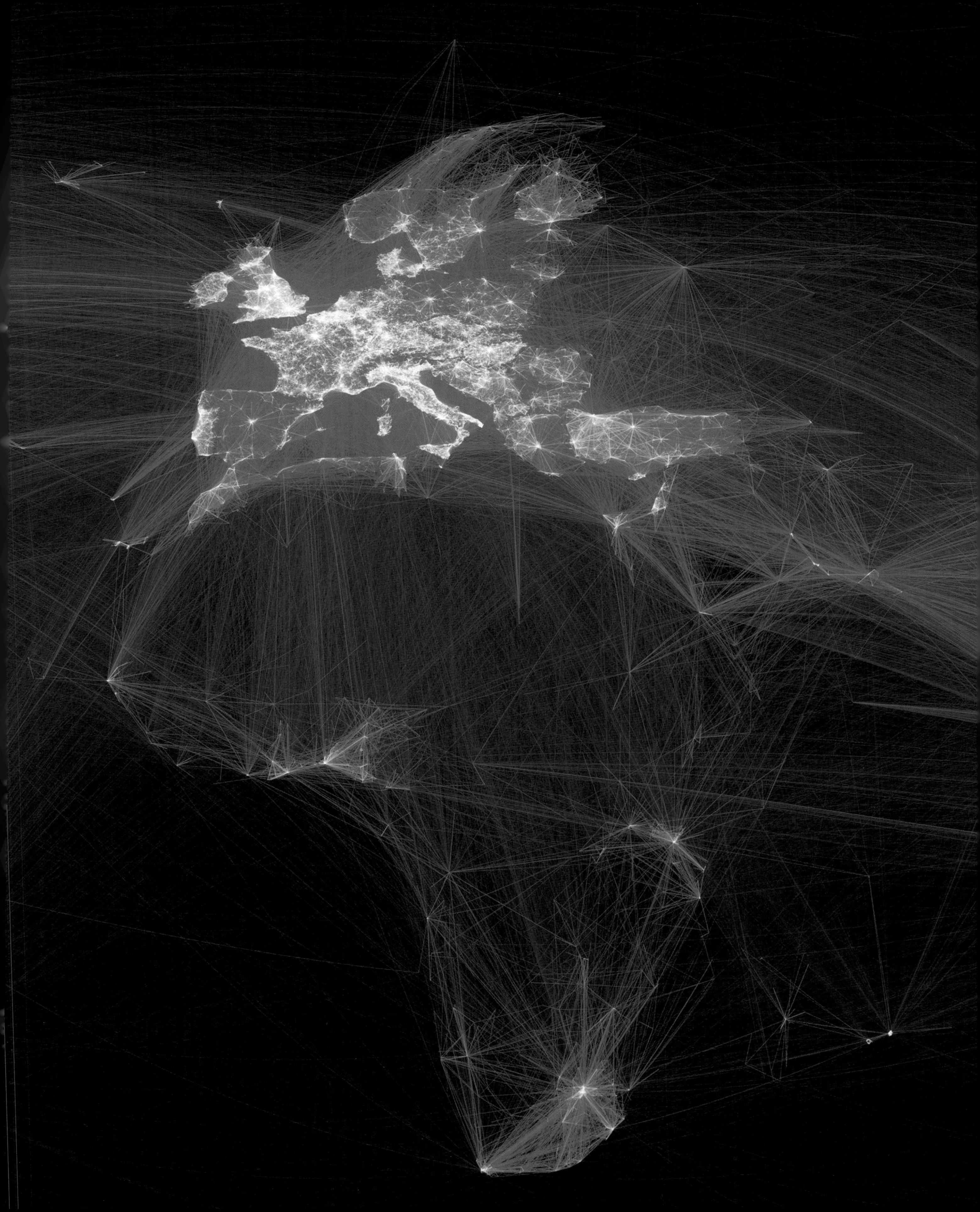

# Data accuracy and precision

*Accuracy* and *precision* have specific meanings in cartography.

**'Make sure your map is accurate' would seem like sound advice but what, exactly, does accuracy mean?** Accuracy refers to the quality of being correct or exact, a description of how close the map is to reality. Accuracy includes the quality of the data used to make the map. Are locations or magnitudes correct or, perhaps, are there errors in the dataset that means the resulting map propagates error, leading to uncertainty?

***Accuracy* shouldn't be confused with precision, which refers to the exactitude of the mapped features.** For instance, the level of measurement of a location or empirical values play a significant role in determining how precise a value is. Precise location may measure an x,y paired coordinate to a fraction of a unit and, so, locating a position to a centimetre is more precise than a metre. Measuring magnitudes to several decimal places gives a greater level of precision to the same values rounded to an integer.

**Data can be accurate and imprecise or, conversely, precise but inaccurate.** They are two separate measurements, both of which can have inherent errors and which build a picture of the overall quality of the data. You may not have much of a say about the precision of the data you use for mapping if you've not been involved in its specification or collection. You may decide to use a less precise form of the data for your map by, perhaps, rounding numbers to create a more generalised form. It's worth remembering that you generally cannot make a map that purports a greater degree of accuracy or precision than the data supports. There is usually a minimum scale at which the data holds up to scrutiny. Purporting to show data at a larger scale than its quality supports should be avoided. For instance, thematic data collected at one level of spatial aggregation cannot be imputed to smaller level boundaries. This is often referred to as the 'ecological fallacy'—the interpretation of statistical data in which inferences about individuals or smaller areas are erroneously deduced from the group to which the individuals or smaller areas belong.

Are the facts you are mapping accurate? What is the veracity of the source and can it be trusted? With reputable sources you can generally be more trusting but it's always worth doing some fact-checking. Accuracy may also be politically motivated. Data from different countries or from suppliers of global data will inevitably show certain geopolitical issues from a particular perspective. Understanding these differences helps you make your own editorial decisions.

How variable is detail across the map? If you are using different sources ensure that the quality of the data is comparable. Including an overview map showing different errors, inaccuracies, or uncertainty at least shows where differences occur.

Are map features where they should be? Maps don't always show features in their correct locations. For instance, many military or sensitive locations are deliberately left off certain maps. Ensuring this sort of policy is often a difficult decision because it's fundamentally a misrepresentation, albeit one made on purpose.

When was data collected? Generally speaking you'll want to make maps with the most current data possible but this isn't always feasible. Updating of topographic data varies. Often, topographic data is based on original surveys many years old with only the largest scale data being maintained as current. Statistical data is often based on censuses that may be 10 years apart so inconsistencies are sometimes unavoidable.

Is data appropriate for the mapping task? Ultimately, the accuracy and precision of data will need to be assessed in relation to the task. Reporting errors and uncertainties at least gives a sense of the possible impact of data quality on the patterns people see.

**See also:** Ethics | Error and bias | Making numbers meaningful | Scale and resolution

# Where is...

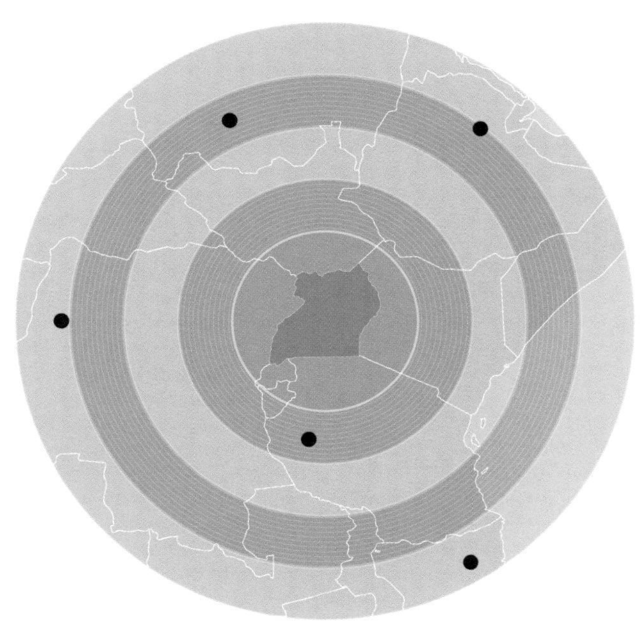

UGANDA
Not Accurate | Not Precise

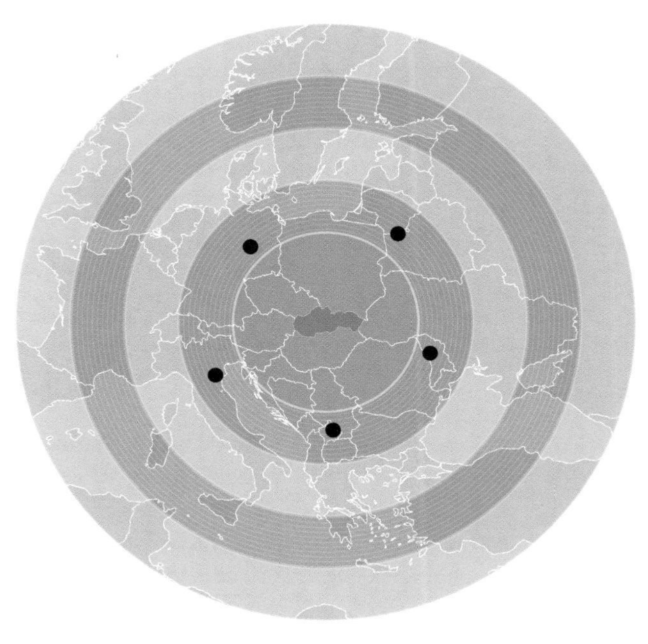

SLOVAKIA
Accurate | Not Precise

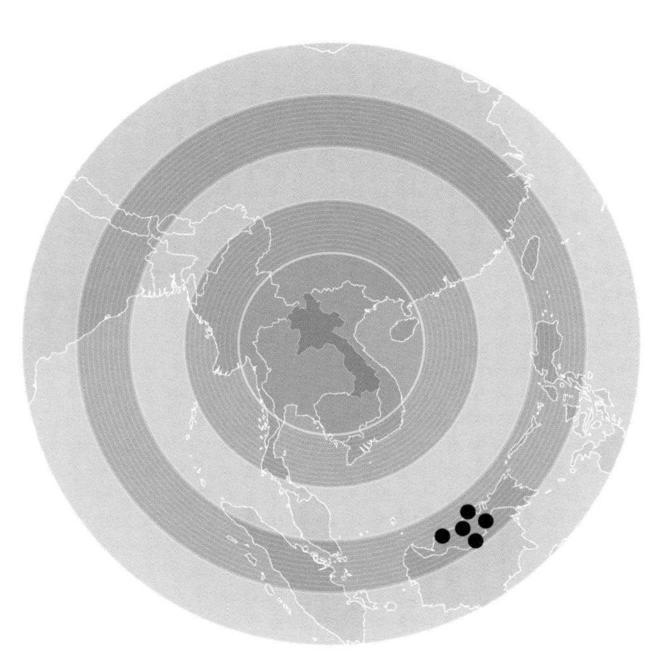

LAOS
Not Accurate | Precise

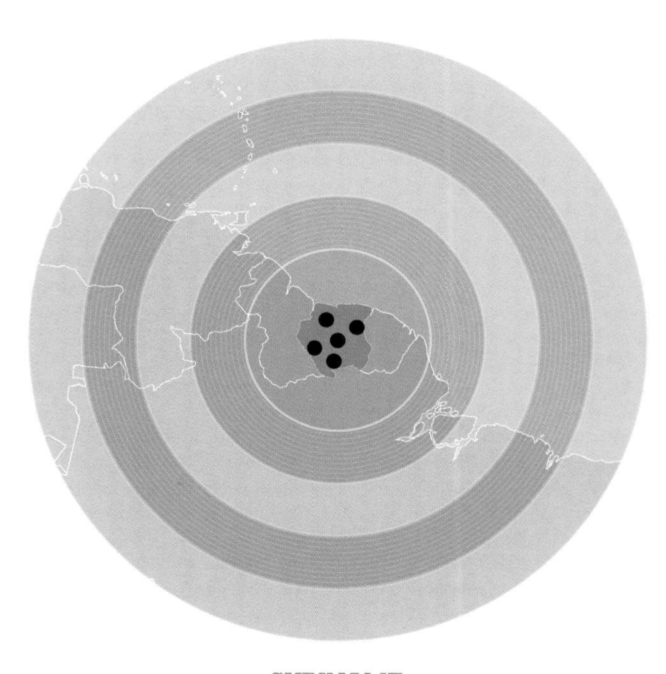

SURINAME
Accurate | Precise

# Data arrangement

## The arrangement of data has implications for choosing map types.

**Geographical phenomena can be arranged along a discrete–continuous arrangement.** The terms 'discrete' and 'continuous' are often used to describe different types of data along a numeric line but they can also be used in cartography in a spatial context. Discrete phenomena occur at distinct locations and are differentiated from other discrete phenomena by the intervening space. For instance, at any moment in time, each individual in a town occupies a discrete location that cannot similarly be occupied by another person. On the other hand, continuous phenomena occur throughout a geographical region. Elevation is an example of continuous phenomena because every latitude and longitude position has an associated value of elevation—there is no empty geographical space between values of elevation for which there is no value of elevation.

**Both discrete and continuous phenomena can also be arranged as abrupt or smooth.** Phenomena that change suddenly are abrupt whereas phenomena that change gradually over geographical space are smooth. For instance, voting patterns for electoral constituencies are likely to change abruptly between neighbouring areas. In fact, this is an example of abrupt continuous phenomena because electoral constituencies exhaust space and there is a value associated with each one, but an abrupt change occurs at boundaries between constituencies. An example of smooth continuous phenomena would be the measure of total precipitation over the course of a year. Abrupt changes in precipitation are not likely across geographical space.

**Different map types are more appropriate than others depending on the arrangement of data.** Selecting a map type that supports the data type and arrangement will help convey the information appropriately although it is possible to convert between data arrangements, for instance by aggregating from continuous to discrete or binning smooth to abrupt.

An example of discrete phenomena that would also change abruptly across space might be the number of employees in a building. The number is discrete in that it occupies a point in geographical space, and it will vary abruptly from neighbouring values because the number of employees might differ dramatically and, of course, not everyone in a region is employed by the same company.

By contrast, the employment status of all people is an example of discrete geographical phenomena (i.e. every person occupies a discrete position in geographical space at any one time) yet the pattern of employment is likely to vary continuously across geographical space. The likelihood that people will be located near others who have similar employment status is high, and there will be no abrupt boundaries across space.

**See also:** Clutter | Data (c)art(e) | Data processing | Digital data | Dispersal vs. layering

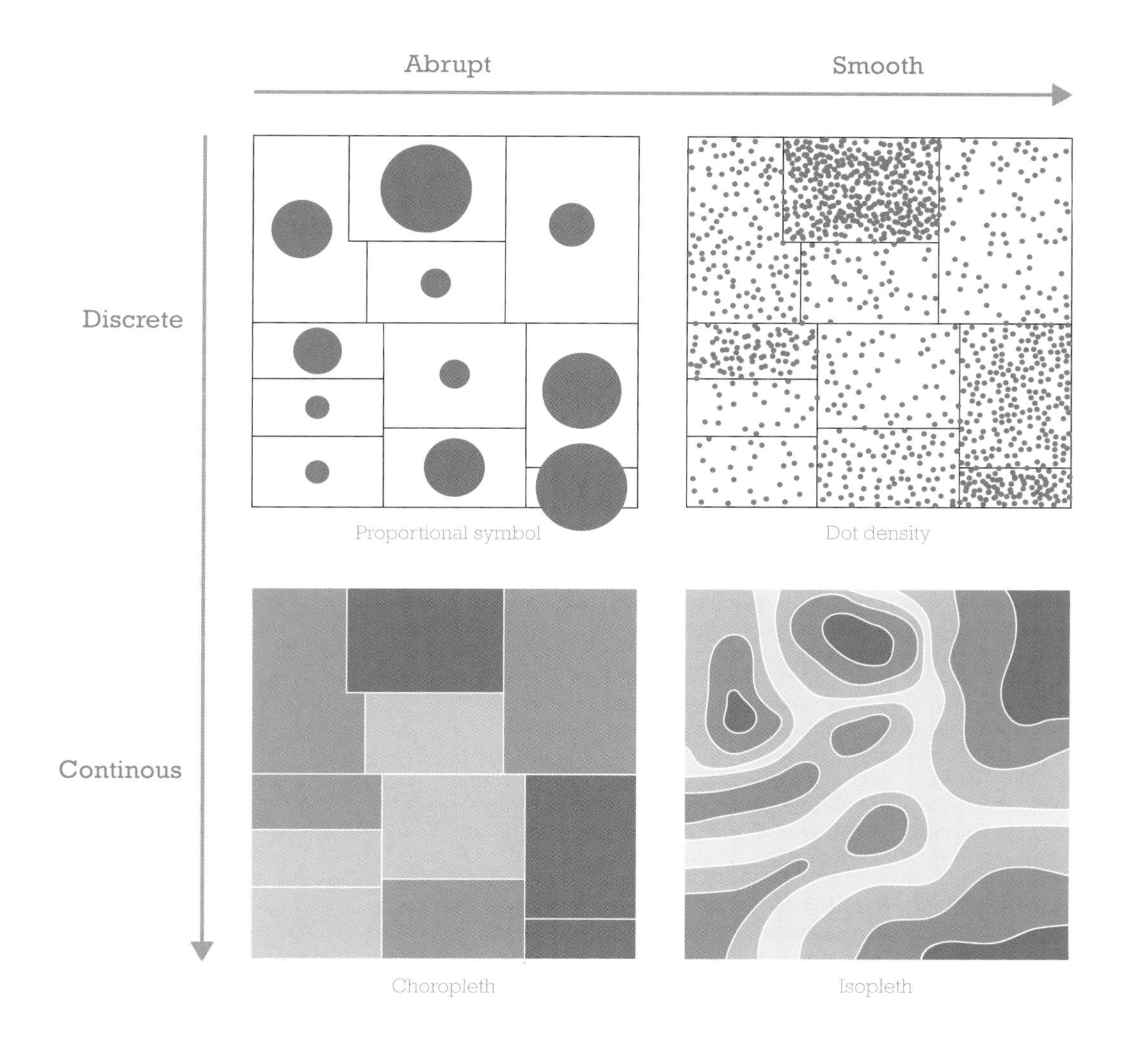
Abrupt
Smooth
Discrete
Continous
Proportional symbol
Dot density
Choropleth
Isopleth

# Data classification

## Data classification processes raw data into something that, when mapped, teases out essential characteristics for display.

**Data classification is the process of transforming raw data into classes that can each be given a unique symbol on a map.** It is important in thematic mapping where the primary focus is to group quantitative values by similar characteristics or qualitative values by type. It is equally important in topographic mapping where, for instance, elevation is classified to derive contour values.

**Possibly the most important decision in classification is choosing how many classes to divide the data into.** For 100 individual values in a dataset, the maximum number of classes would be 100, and the minimum number of classes would be one, in which the class interval encompasses all values. Clearly these extremes are inappropriate but the decision has to be made somewhere along the scale of 1–100. Convention suggests five or six classes are appropriate simply because this yields a number of classes that can be effectively symbolised so map users can discern difference between symbols. It's a good default to guide you. Less than five, and you will lose a lot of detail. More than six, and readers lose the ability to identify perceptually different symbols on the map.

**The antithesis of a classified map is an unclassified map where, in a dataset of 100 unique values, you would require 100 unique symbols to represent each value.** This is graphically possible but cognitively problematic since it requires the reader to do a considerable amount of examining to decipher even basic patterns. Grouping the 100 values into classes gives you fewer unique symbols and aids the map reader in seeing the message. The task is not necessarily easy since by grouping data into classes you are generalising the data and that inevitably means losing detail. You should apply classification schemes that find a balance between effectively summarising the salient characteristics of the data but which can also be depicted cartographically. The retention of detail and ability to see outliers does make unclassed maps useful in that regard.

There are many different methods for classifying data. None are universally perfect but these questions are useful to ask:

- Does the scheme consider the distribution of the array of data values, and is that important to your map?
- Is the scheme easily understood?
- Is it readily applied?
- Is the map legend easy to interpret?
- Is the scheme appropriate for the level of measurement?
- Is the data balanced, skewed, or has a peculiar distribution?
- Is it desirable to use descriptive statistics to inform the classification scheme?
- How many classes are useful?

The following guidelines can also help you create a useful classification scheme:

- classification should be meaningful, revealing, and impartial to the dataset (i.e. not presenting a biased view);
- the class interval should encompass the whole data range;
- class boundaries must not be overlapping;
- where possible, classes should not be empty (i.e. contain no data values);
- there should be enough classes to accurately portray the data, but not too many to dilute the data or produce an overly detailed display; and
- classes should be derived by some logical method understandable by the map reader.

Learning some of these guidelines is a good starting point to know when they can be broken. For instance, if you were making a propaganda map you might want to use a classification scheme that is inherently biased and use your understanding of the data distribution to convey that message.

**See also:** Arbitrary data classification | Eyeball data classification | Levels of measurement | Making numbers meaningful | Statistical data classification

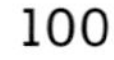

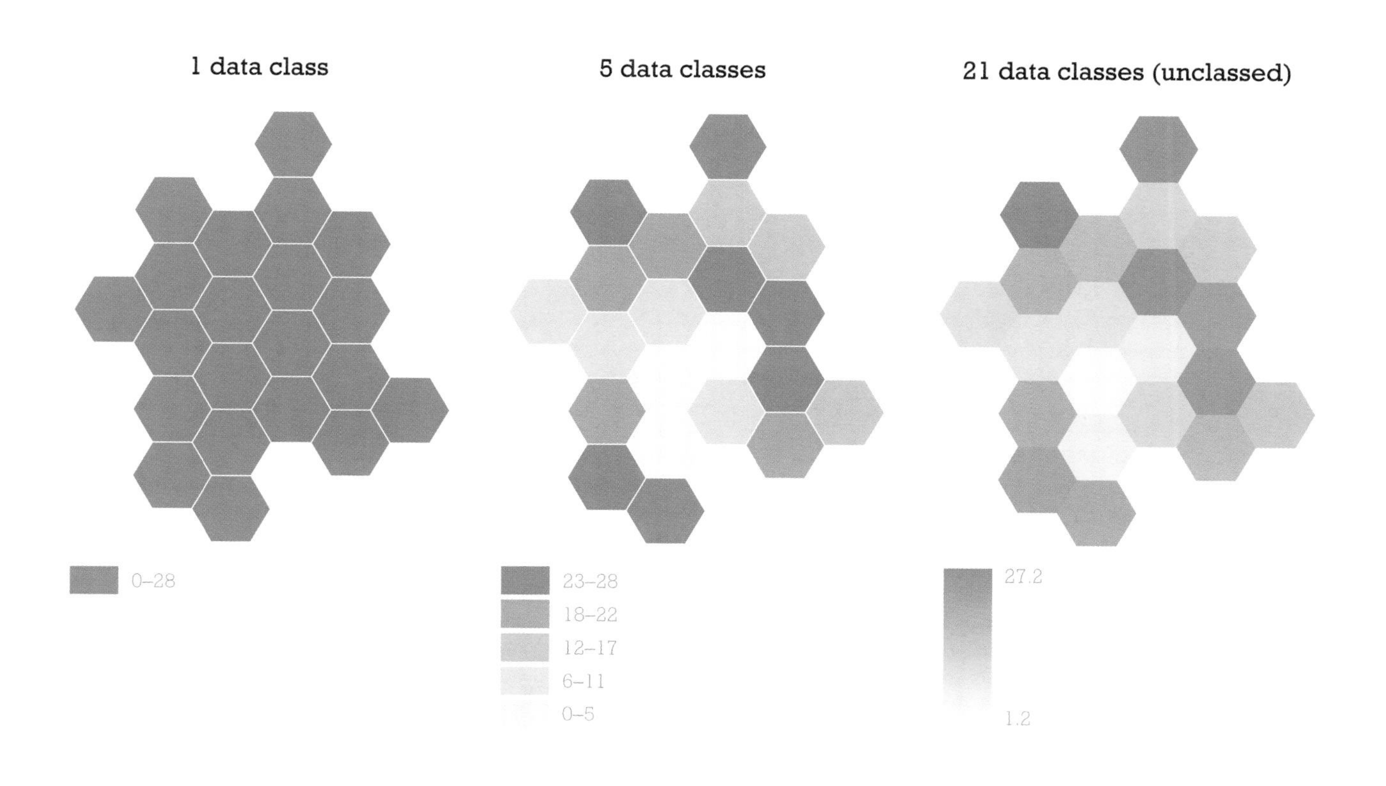

For 21 areas that represent 21 discrete data values, the choice of classes is between one and 21. Using one class simply classifies all data as the same. Using 21 classes (unclassed) gives each area a slightly different symbol from low to high. Using five data classes allows you to see which areas are similar in character. Symbol fills usually vary from light to dark to reflect the data array.

For qualitative data, different hues are normally used to show the different character of areas without implying any order of importance.

# Data density

## Approaches to dealing with a large quantity of data, particularly point-based data.

**Dealing with the representation of dense datasets is a constant cartographic challenge.** Data that overlaps such as coincident points or where the map scale has to be relatively small to encompass the spatial extent of data demands creative solutions. This issue becomes only more pronounced as datasets get ever bigger so you're dealing with not just thousands of pieces of information but, potentially, millions of separate pieces of information, each of which has an equal need to be represented on the map somehow.

**The most common example of a poor cartographic solution to mapping dense data is the classic red-dot fever map.** Point-based marker symbols are all too often used on a one-to-one basis so the map is filled with overlapping point symbols. This is poor information design and results in visual clutter with little chance of any meaning being extracted by the map reader. The one-to-one solution is akin to a data dump but which demands cartographic treatment through generalisation and symbolisation to reveal useful characteristics. A range of techniques deal with this problem by either modifying the symbology or manipulating the points to represent them in summary form.

**Scale is a major factor in assessing how to deal with dense data.** For a single-scale map, a decision on whether the display of individual data points is appropriate can be made once they are plotted at the intended scale. If the overlaps are too cluttered, a range of techniques can be used to represent them differently. At the very simplest level, changing the design of the symbol (specifically the size) helps to reveal more of the original data points. The same idea applies for multiscale maps though at each scale the parameters for each solution can be modified so as you zoom in to larger scales more detail is revealed. At some point the scale will normally be sufficiently large to move from the chosen method of aggregated representation to individual data points.

Applying transparency to each point symbol means that when points are proximal the overlaps will result in a more opaque effect. A drawback is that individual points, which might be important, become less visible.

Binning into a tessellated regular grid provides a visually consistent picture of data density though size and shape of the bins will, to some extent, determine the overall pattern. Once binned, data density can be shown by varying shade or size of the areas.

Converting data points into a density surface creates a continuous, interpolated representation. This can also form the basis for an isarithmic (contoured) representation.

Clustering aggregates individual points based on proximity and symbolises with a single symbol that usually holds the related numeric value. On interactive maps these points decluster as you move to larger scales.

Labels can be used to identify important data points. This might be automated to filter data points the reader is interested in seeing.

Vertically stacked chips can be used to show how coincident points differ by type or value. The result is a spatial histogram though occlusion of chips that sit directly behind others can be a problem.

Separating and offsetting symbols can be used to disperse symbols to a less crowded space, often using a form such as a spiral or leader line that ties back to the original location. This method, however, can result in the pattern of density becoming less clear.

Selecting out only those key data points and changing their symbology can make them distinct from the rest. The addition of labels can reinforce the message.

**See also:** Binning | Heat maps | Simplicity vs. complexity | Simplification

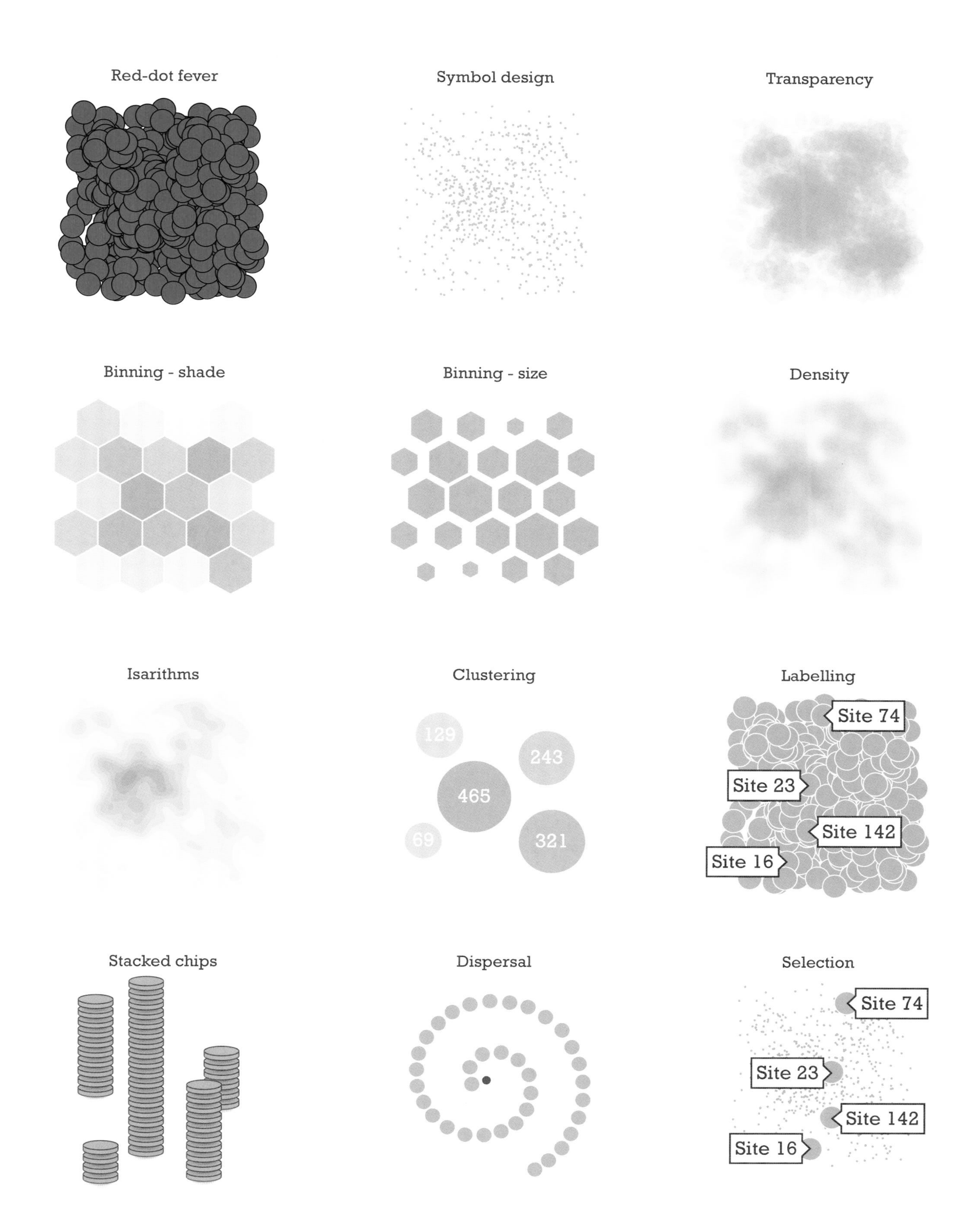
Red-dot fever
Symbol design
Transparency
Binning - shade
Binning - size
Density
Isarithms
Clustering
129
243
465
69
321
Labelling
Site 74
Site 23
Site 142
Site 16
Stacked chips
Dispersal
Selection
Site 74
Site 23
Site 142
Site 16

# Data distribution

## Data distributions can take many different forms, and understanding them supports the cartographic process.

**Many geographical phenomena display a frequency distribution of values that can be represented by a bell-shaped curve.** This is referred to as a 'normal distribution', the principles of which can be used in exploring data for cartographic purposes. If data is distributed normally, percentages of observations occurring in the spaces between $\overline{X}$ and $\sigma$ can be specified. When data is distributed normally, data values that occur at further distances away from the mean are much less likely to occur than those closer to the mean.

**Skewness is exhibited when the peak of a frequency distribution curve is displaced on either side of the mean.** If the majority of frequencies are found on the left of the mean, the distribution is positively skewed. Alternatively, if the tail of low frequencies is on the left and the majority of high frequencies on the right, the distribution is negatively skewed. Skewness is a measure of displacement calculated by

$$skewness = \frac{\sum (x - \overline{X})^2}{n\sigma^3}$$

Negative values indicate negative skewness and vice versa. The higher the value, the greater the difference the distribution's shape is from the normal distribution. Normal distributions have a skewness of 0.

**Kurtosis describes how peaked the data is.** A flat distribution occurs when there are a similar number of data values in each class in the frequency histogram. A distribution that has a high peakiness is one where the majority of data values are in a single frequency class. Kurtosis is calculated by

$$kurtosis = \frac{\sum (x - \overline{X})^4}{n\sigma^4}$$

A normal distribution has a kurtosis value of 3.0. Values above 3.0 indicate increased peakedness. Values below 3.0 indicate flatter distributions.

Understanding the measurement and distribution of numerical data is a key component of handling cartographic data, particularly for thematic mapping.

Classifying and displaying the appropriate aspects of your data objectively depends on understanding the distribution. Bear in mind that software defaults are unable to seek out the nuance and messages in your data so your own exploration and search for patterns is fundamental.

The one characteristic of data is that you'll probably never use a dataset that has a perfectly normal distribution, though many will be close enough to support the statistical summaries that depend upon it.

If your data has a peculiar distribution or odd outliers, then it's probably worth mentioning it in notes on the map to aid in interpretation. Simply burying the facts in your data in a classification scheme makes the resulting map a little untruthful. At the very least it masks the facts from the map reader. Placing a credit to allow people to go to the original data source is good practice to provide readers with a way of verifying the facts of the map themselves.

**See also:** Levels of measurement | Making numbers meaningful | Quantitative statistical maps

Normal Distribution

Frequency

22 Years **Average Age** 63 Years

US Counties

Skewed Distribution

## Equal Interval

Evenly spaced, unevenly filled, ranges.

Suitable for normally distributed data. Legends are efficient and easy to comprehend.

Skewed data, however, will result in a bland map, and clumped data may omit some ranges altogether.

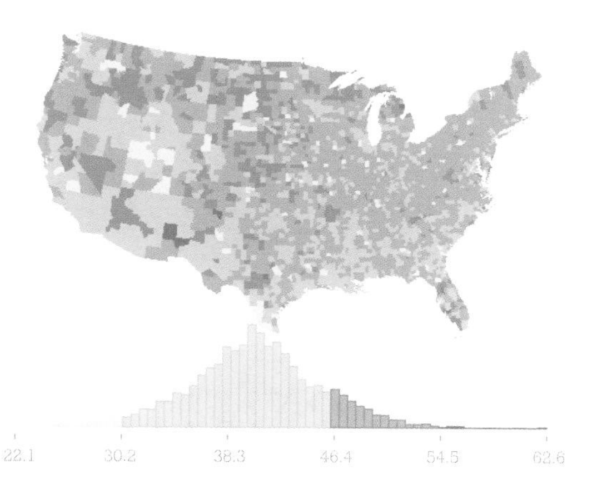

## Quantile

Unevenly spaced, evenly filled, ranges.

Will always depict variability, even if there is little variability in the data.

Results in a reliably lively map but can be mislead-ing, and the legend may seem arbitrary.

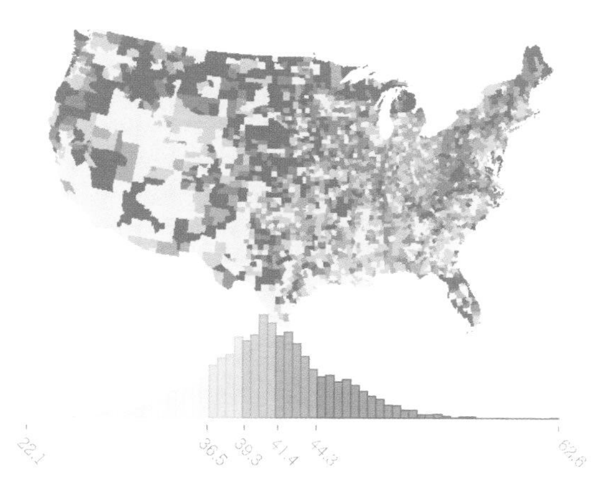


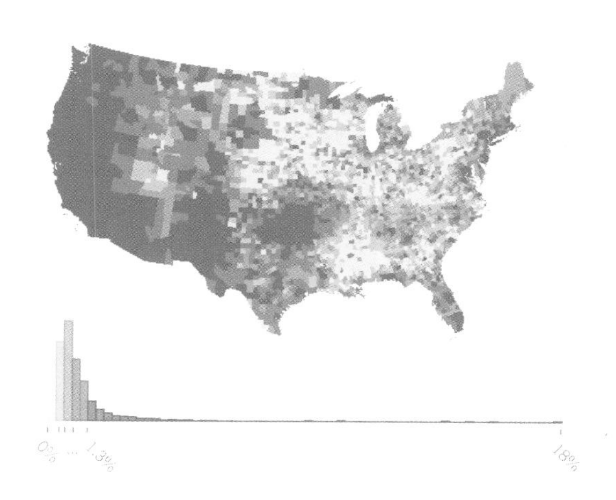


## Standard Deviation

Evenly spaced (in a way), unevenly filled, ranges.

Ranges are created by grouping the "standard deviations" from the average value. Will ensure visual variability.

Good for showing the typical value, with ranges above and below.

The legend should be carefully crafted to not appear intimidating to the reader.

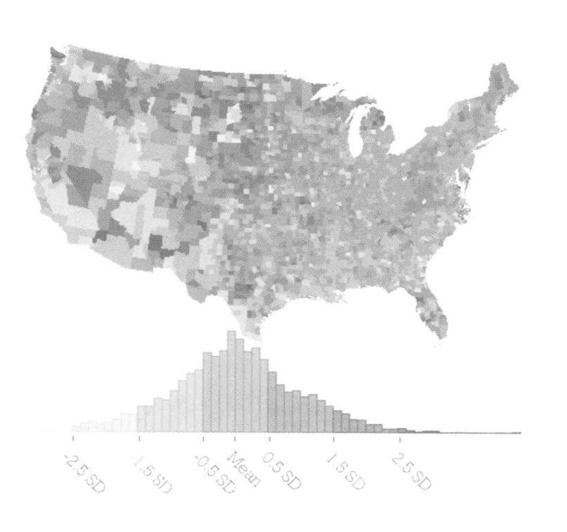


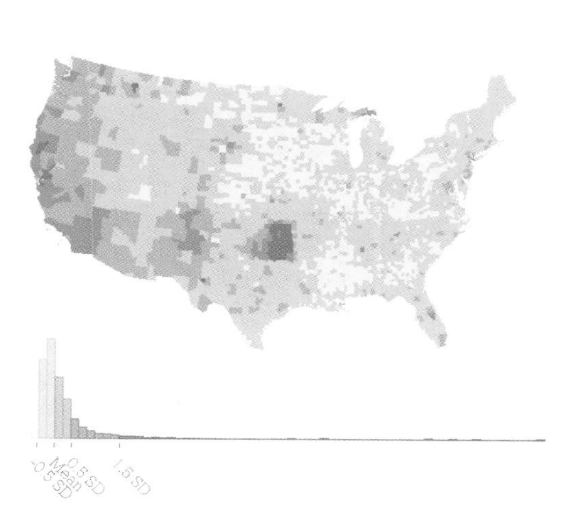

# Data processing

## You must often manipulate the data to be able to map it.

**In most cases, the raw data you have to work with to make a map is rarely in the right form for the map you want to make.** Cartographic design is the process of conveying data in graphic form as simply as possible in order to communicate meaning. The contents of a map may be fairly simple, based on standard data types or routine measures, or they may be more complex based on more sophisticated data manipulation. Data processing is often required to manipulate data into forms that may be mapped more appropriately given the graphical limits imposed by different map types or requirements. In many ways it is one of a suite of processes referred to as 'simplification'—that is, reducing data to retain the essential characteristics and bring them into view for the map user.

**Data processing normally involves mathematical or statistical manipulation.** Mathematical measures are relatively simple and commonly employed to derive ratios, proportions, and percentages. These are commonly applied to deal with data magnitudes and relationships between data values so that they can be effectively expressed using symbols. Statistical analysis is often used to summarise observations, to describe relationships between variables, and to make inferences concerning estimations and tests of significance.

**When making a map, it's often useful to process the data into something more meaningful that clarifies the message you're trying to impart.** Raw data is more than likely not going to work as well and simply positioning raw data on a map might be little more than a visual data dump, though that can often result in some interesting patterns and might lead to a piece of data art.

Cartographic data processing might equally refer to topographic or thematic data.

For topographic (surveyed) data then, the points, lines, and areas that define the location of physical features might be modified to make the map. Scale is usually one of the core reasons for processing data to simplify the complexity of data captured at large scale to represent it effectively at a small scale.

For thematic mapping, the attributes that measure and describe the character of something are the focus of the processing. Data might be aggregated into different geographies or converted from a set of totals to a percentage in order to suit a choropleth map.

**See also:** Error and bias | Point clouds | Statistical literacy | Unclassed maps

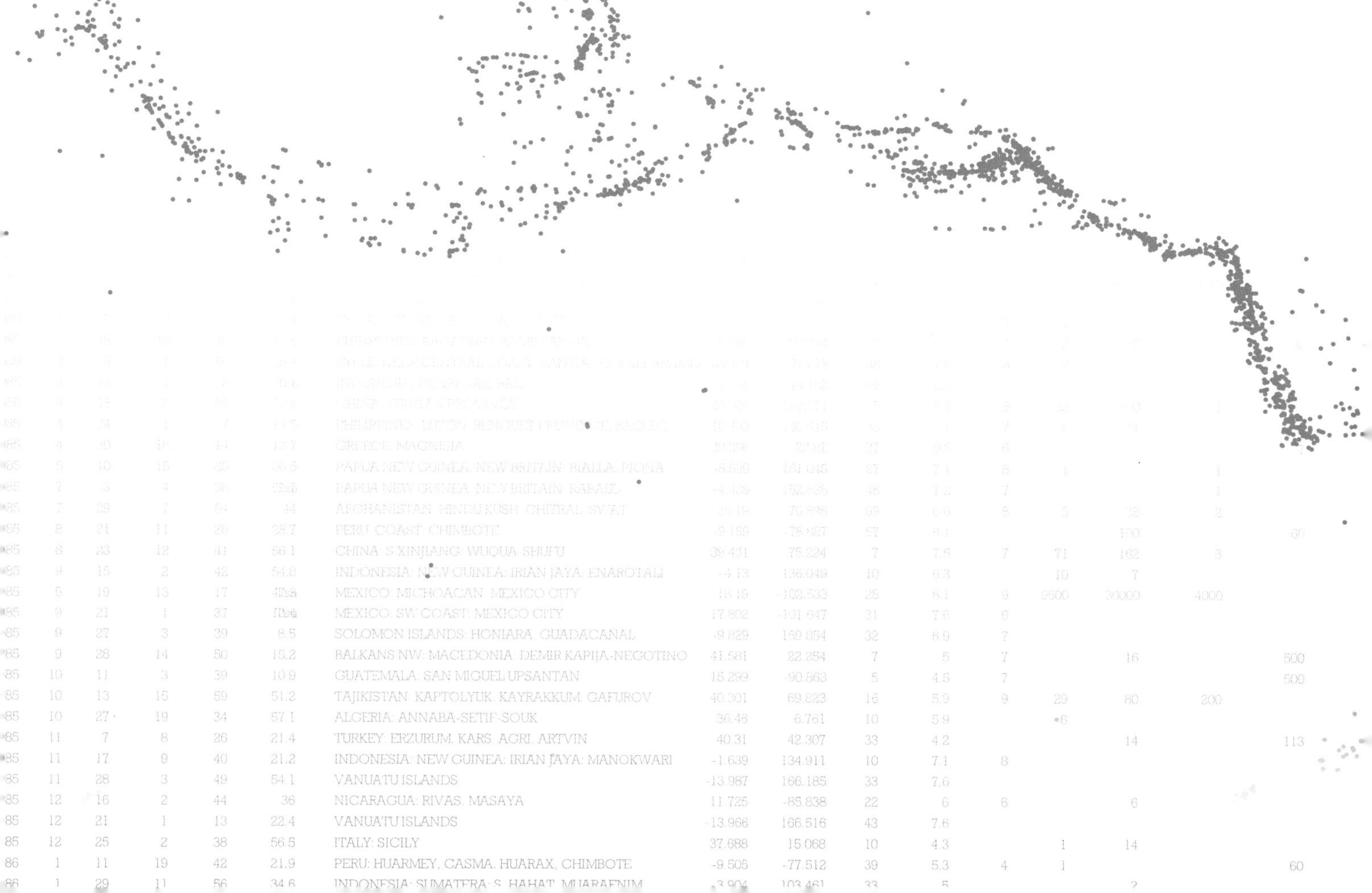

| | | | | | | | | | | | | | | | |
|---|---|---|---|---|---|---|---|---|---|---|---|---|---|---|---|
| 85 | 8 | 21 | 11 | 26 | 28.7 | PERU: COAST: CHIMBOTE | -9.159 | -78.827 | 57 | 6.1 | | | 100 | | 60 |
| 85 | 8 | 23 | 12 | 41 | 56.1 | CHINA: S XINJIANG: WUQUA-SHUFU | 39.431 | 75.224 | 7 | 7.5 | 7 | 71 | 162 | 3 | |
| 85 | 9 | 15 | 2 | 42 | 54.8 | INDONESIA: NEW GUINEA: IRIAN JAYA: ENAROTALI | -4.13 | 136.049 | 10 | 6.3 | | 10 | 7 | | |
| 85 | 9 | 19 | 13 | 17 | [illegible] | MEXICO: MICHOACAN: MEXICO CITY | 18.19 | -102.533 | 28 | 8.1 | 9 | 9500 | 30000 | 4000 | |
| 85 | 9 | 21 | 1 | 37 | [illegible] | MEXICO: SW COAST: MEXICO CITY | 17.802 | -101.647 | 31 | 7.6 | 6 | | | | |
| 85 | 9 | 27 | 3 | 39 | 8.5 | SOLOMON ISLANDS: HONIARA, GUADACANAL | -9.829 | 159.854 | 32 | 6.9 | 7 | | | | |
| 85 | 9 | 28 | 14 | 50 | 15.2 | BALKANS NW: MACEDONIA: DEMIR KAPIJA-NEGOTINO | 41.581 | 22.254 | 7 | 5 | 7 | | 16 | | 500 |
| 85 | 10 | 11 | 3 | 39 | 10.9 | GUATEMALA: SAN MIGUEL UPSANTAN | 15.299 | -90.863 | 5 | 4.5 | 7 | | | | 500 |
| 85 | 10 | 13 | 15 | 59 | 51.2 | TAJIKISTAN: KAPTOLYUK, KAYRAKKUM, GAFUROV | 40.301 | 69.823 | 16 | 5.9 | 9 | 29 | 80 | 200 | |
| 85 | 10 | 27 | 19 | 34 | 57.1 | ALGERIA: ANNABA-SETIF-SOUK | 36.46 | 6.761 | 10 | 5.9 | | 6 | | | |
| 85 | 11 | 7 | 8 | 26 | 21.4 | TURKEY: ERZURUM, KARS, AGRI, ARTVIN | 40.31 | 42.307 | 33 | 4.2 | | | 14 | | 113 |
| 85 | 11 | 17 | 9 | 40 | 21.2 | INDONESIA: NEW GUINEA: IRIAN JAYA: MANOKWARI | -1.639 | 134.911 | 10 | 7.1 | 8 | | | | |
| 85 | 11 | 28 | 3 | 49 | 54.1 | VANUATU ISLANDS | -13.987 | 166.185 | 33 | 7.6 | | | | | |
| 85 | 12 | 16 | 2 | 44 | 36 | NICARAGUA: RIVAS, MASAYA | 11.725 | -85.838 | 22 | 6 | 6 | | 6 | | |
| 85 | 12 | 21 | 1 | 13 | 22.4 | VANUATU ISLANDS | -13.966 | 166.516 | 43 | 7.6 | | | | | |
| 85 | 12 | 25 | 2 | 38 | 56.5 | ITALY: SICILY | 37.688 | 15.068 | 10 | 4.3 | | 1 | 14 | | |
| 86 | 1 | 11 | 19 | 42 | 21.9 | PERU: HUARMEY, CASMA, HUARAX, CHIMBOTE | -9.505 | -77.512 | 39 | 5.3 | 4 | 1 | | | 60 |
| 86 | 1 | 29 | 11 | 56 | 34.6 | INDONESIA: SUMATERA: S. HAHAT, MUARAENIM | -3.904 | 103.461 | 33 | 5 | | | 2 | | |

# Datums

**Datums provide a starting point that gives context to measuring position and to translating positions on a map back to their real position on Earth.**

**A datum is a starting point for defining and measuring location on Earth's surface.** It is the basis of geodesy, the branch of mathematics that deals with measurement of Earth (or any planetary object). It allows us to measure the size and shape of Earth by providing a fixed position from which other measurements can be made. This is usually a known location on the surface of the ellipsoid or the centre of Earth. Horizontal datums describe a point on Earth's surface, in latitude and longitude or another coordinate system. Vertical datums are used as reference points for measuring position of elevation of features such as terrain, bathymetry, water bodies, and man-made structures. WGS84 is a global geocentric datum and represents an approximate definition of sea level for the global ellipsoid. Many different datums exist that have improved accuracy through history or which provide more accurate local mapping on the basis of a local system that is a better fit than a global system.

**For local datums, the ellipsoid is tied to a point on the surface.** For global datums it is fixed to the centre of Earth. Once fitted, latitude, longitude, and elevation coordinates can be derived. Vertical datums can be based on sea levels or gravity and can be based on the geoid, be geodetic, or be based on the same geocentric ellipsoidal models used for defining horizontal datums. Elevations are commonly referred to as being 'above sea level' but that level may differ between different vertical datums and also be subject to differences in sea conditions (wind, waves, and current), atmospheric pressure, topography, and gravitational pressure. Mean sea level (MSL) is a common vertical datum described as the arithmetic mean of hourly water elevation taken over a 19-year cycle. This averages out the variations caused by other short-term changeable conditions although it cannot account for variations in local gravitational differences. That means the height of MSL will vary around the world.

**See also:** Earth coordinate geometry | Earth's framework | Latitude | Longitude | Map projections | Measuring direction

Maps based on different datums will give different coordinates for the same position. This difference is commonly called a 'datum shift' and can be anything from negligible differences to several hundreds of metres. This is the reason that if you stand on the Prime Meridian (0° longitude) at the Royal Observatory in Greenwich, London, and read your position on a hand-held GPS device, you will not see your position as zero. GPS devices use the WGS84 global geocentric datum whereas the original line was drawn on the basis of the Airy transit circle first defined in 1721. The line at Greenwich in fact contains a small error which was immaterial at the time of its creation and which is deflected slightly from a line perpendicular to the modern ellipsoid to the centre of Earth. Modern GPS receivers, based on WGS 84, do not contain this error, which gives a datum shift that positions zero some 102 metres east of the physical line.

Ordnance Survey mapping in Great Britain is based on the OSGB36 geodetic datum, which itself is based on the Airy 1830 ellipsoid. The Airy ellipsoid provides the best fit for Great Britain. Mean sea level for Great Britain is defined at Newlyn in Cornwall based on average sea level between 1915 and 1921. Ordnance Datum Newlyn is therefore the point from which all elevation and bathymetric measurements are made in mapping Great Britain by Ordnance Survey. Ellipsoid-based datums such as WGS84 use a theoretical MSL surface which may vary considerably from the geoid and also from local vertical datums based on a local geodetic datum.

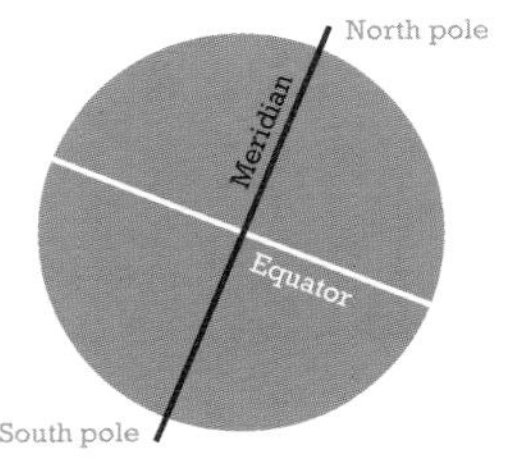


**Opposite:** *Prime Meridian*, Greenwich, London

# Defining map design

## Good design is about thinking...the result of which is the effective communication of information in map form.

**Design as an overarching concept is the optimum use of tools for the creation of better solutions.** That said, it's fair to say that pretty much everything is designed in some way or another. The result of design is to change something with the goal of improvement. Objects such as a car or a vacuum cleaner are designed but so are laws or processes. Maps, as tangible or non-tangible objects, also must be designed. Design also means to create, and it takes on a dynamic character which might be an iterative process or a sequential or repetitive series of actions. Design is something that can make your mapping better through an understanding of what constitutes better design thinking.

**Map design leads to the creation of objects, and it is these objects that create messages.** It is the sum of the cognitive processes that a cartographer applies to the abstraction phase of the cartographic process. It includes the whole range of decision-making related to the choice of map projection, scale, classification, simplification, symbolisation, typography, colour, and much more. What the mapmaker is attempting to achieve through their design is the successful and effective transfer of knowledge between the map author and the map user. To be successful, transfer of data into information should be unambiguous, accurate, and efficient and be brought about by the organisation of all the graphical components that comprise the map. Every mark that is made on the map should be the process of careful and planned decision-making to ensure a structured visual whole that serves the map user's needs.

**Map design is a complex activity that requires both intellectual input and an understanding of visual aspects.** It demands an understanding of the foundations of geography, communication, and psychology as well as the ability to reflect them through a visual medium. That's not to say you need to be expert in all of this—merely to appreciate that these components go a long way to helping shape and build good map products.

It's easy to see how computerised mapping supports the creation of products outside the map design process. Software, of course, must have defaults, and relying on these defaults is arguably the biggest failing that any mapmaker can fall victim to. Software does not know your data or what your intention is in mapping it. Software exists for you to direct and manipulate, and being able to do so through a design lens will yield better maps.

One of the best ways to begin to appreciate the art of good design in cartography is to explore examples that are generally considered to exhibit best practice. Immerse yourself in good work and recognise what works and what doesn't work. There is very little that is new, and many great maps build upon long-established design ideas. There are some examples throughout this book but there are many more.

*'Design in art is a recognition of the relation between various things, various elements in the creative flux. You can't invent a design. You recognize it, in the fourth dimension. That is, with your blood and your bones, as well as with your eyes.'*
*—D. H. Lawrence*

*'Absorb what is useful. Discard what is not. Add what is uniquely your own.'*
*—Bruce Lee*

**See also:** Craft | Defining maps and cartography | Map aesthetics | Prior (c)art(e) | Purpose of maps | Types of maps

## Some principles of map design (as proposed by the British Cartographic Society in 1999)

### 1. Concept before compilation

Without a grasp of concept, the whole of the design process is negated. The parts embarrass the whole. Once concept is understood, no design or content feature will be included which does not fit in. Design the whole before the part. Design comes in two stages, concept and parameters, and detail in execution. Design once, devise, design again. Reader first, reader last. What does the reader want from this map? What can the reader get from this map? Is that what they want? If a map was a building, it shouldn't fall over.

### 2. Hierarchy with harmony

Important things must look important, and the most important thing should look the most important. 'They also serve who only stand and wait.' Lesser things have their place and should serve to complement the important. From the whole to the part, and all the parts, contributing to the whole. Associated items must have associated treatment. Harmony is to do with the whole map being happy with itself. Successful harmony leads to repose. Perfect harmony of elements leads to a neutral bloom. Harmony is subliminal.

### 3. Simplicity from sacrifice

Great design tends towards simplicity. It's not what you put in that makes a great map but what you take out. The map design stage is complete when you can take nothing else out. This is the map designer's skill. Content may determine scale or scale may determine content, and each determines the level of generalisation (sacrifice).

### 4. Clutter to clarity

Maximum information at minimum cost. How much information can be gained from the map, at a glance? Functionality, not utility. Design makes utility functional. Design increases the information transfer process because a well-designed map has clarity. Clarity is achieved by compromise. All designs are a compromise. A compromise between what could be shown and what can be read and understood.

### 5. Emotion expresses, engages, and elucidates

Engage the emotion to engage the understanding. Cartographers acknowledge when creating maps is that it takes something out of them. They have expended some invisible emotional energy in the act of creation. Design with emotion to engage the emotion. The image is the message. Good design is a result of the tension between the environment (the facts) and the designer. Only when the reader engages the emotion, the desire, will they be receptive to the map's message. Design uses aesthetics but the principles of aesthetics are not those of design. You are not just prettying maps up.

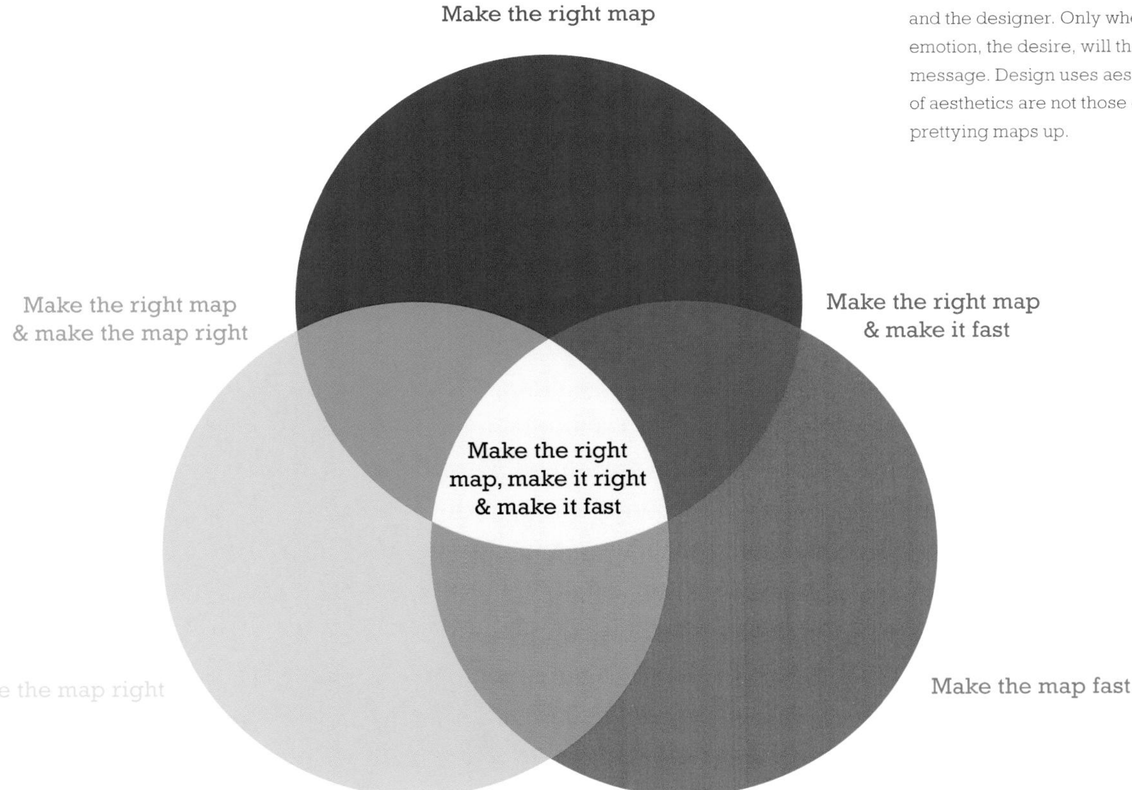

# Defining maps and cartography

**Maps take many forms, perform many functions, and support so many of our daily activities but how is a map and the process that makes it defined?**

**Maps are ubiquitous, many, and varied, and they hold a special place in our world.** We see maps in newspapers or news feeds and learn about the location of world events or local incidents or to decide what to wear based on a weather map. Maps appear on our television screens as well as in books, journals, and leaflets to inform; they appear at bus stops and railway stations to aid our navigation. The internet contains a multitude of maps to serve every possible need. Maps even appear in art, cartoons, and on postcards. Perhaps one of the most common types of maps are those that each of us keeps inside our brains—our mental map of the world around us.

**With such a variety of types, uses, and mediums used for mapping, it is difficult to define precisely what a map is.** A map can be defined as a graphic representation of all aspects of our socio-cultural and natural environment. But it can be defined in other ways too. The International Cartographic Association (the world authoritative body for cartography) has a lengthier definition: 'A map is a symbolised representation of geographical reality, representing selected features or characteristics, resulting from the creative effort of its author's execution of choices, and is designed for use when spatial relationships are of primary relevance same. (International Cartographic Association, 2003)

**Cartography is the study and practice of making maps and globes.** Cartography can also be viewed as being broader than simply mapmaking. It describes a unique discipline; one that requires the study of both philosophical and theoretical aspects of the mapmaking process and which encompasses the study of the map communication process. It is more than simply the study of the technical means by which a map is constructed and requires some understanding of the mechanisms by which 'effective' mapping is achieved. Cartography is multidimensional, and it is perhaps the essence of a deep understanding of the discipline that differentiates cartographers from simply mapmakers.

The term *cartography* is derived from the Greek, chartis = map and graphein = write

Maps can be described as being permanent and exist in 'hard copy' (e.g. on paper) or 'virtually' (in digital or cognitive form); they may be 'visible' and able to be seen or 'invisible' and stored in a database or in our own minds. Furthermore, they may be capable of being handled in a tangible sense or exist only in a non-tangible way. The description of a map using these criteria allows us to recognise how maps that exist in various forms take on different characteristics.

A map existing on paper may be described as being permanent, visible, and tangible whereas a map appearing on a computer screen, while being visible, is neither permanent nor tangible. It is best described as existing 'virtually' and as 'non-tangible'. The map in its stored form on a computer's hard drive is simply a set of files and data that further render the map invisible (although, curiously, the map may be tangible in some sense if it is stored on portable media). A map that is accessible over a network from a database might be described as being virtual, invisible, and intangible.

**See also:** Cartographic process | Defining map design | Information products | Purpose of maps | Types of maps

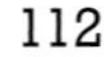

# CARTOGRAPHY

*"Cartography is the discipline dealing with the art, science and technology of making and using maps."*

**Menno-Jan Kraak**

President of the International Cartographic Association

*"Traditionally, map making is graphic rendering of phenomena or relations on the earth. Now the term is expanded to general biobjective representations of spatially distributed relations. For example genetic maps."*

**Waldo Tobler, deceased**

Professor Emeritus at the University of California, Santa Barbara

*"Cartography, to me, is the art and science of simplifying the real world around us to tell a story using pictures."*

**Ed Parsons**

Geospatial Technologist of Google

*"Cartography is the thinking that leads to smarter mapmaking."*

**Kenneth Field**

Professional cartonerd

# Descriptive maps

## Showing explanatory features by combining topographic and thematic information.

**Descriptive maps show where phenomena are located, organizational structure, subdivision, or routing.** They are explanatory and non-quantitative but generally go further than the content of a topographic reference map. The detail on a descriptive map is usually very specific. For example, power lines or sewers, sales territories, or the location of stores. The specific theme is generally mixed with other contextual topographic content so the map becomes neither a pure topographic nor thematic map. The topographic detail included is normally only the bare minimum needed to support the contextualisation of the theme being described.

**Information is encoded using typical map symbology though industry-specific styling and symbology are often used.** The inclusion of quantitative data is generally tangential to the main theme and might be included using marginal elements such as tables or graphs. Quantitative data is not the primary focus of the map. Scale can vary depending on the purpose from small, large-scale maps showing a store location to large, small-scale maps that illustrate the location of all franchises in a company on a global scale.

**Another way of thinking of a descriptive map is that it is a hybrid.** It takes some of the skeletal aspects of a topographic map but incorporates an additional theme. The theme is unlikely to be understandable as a purely thematic map because it's qualitative so it needs some topographic detail. The map is focussed rather than general purpose.

Boundary maps are descriptive. They enable the identification of administrative jurisdictions from a country level down to neighbourhood subdivisions, for example census boundaries or school districts. Jurisdictions may have names or simply an alpha-numeric code. They are illustrated either with line symbology or colour-coded fills with varying hues. Maps might be exploded to highlight a particular area.

Surface or subsurface features are often plotted on a topographic map to describe features. These might be particular physical regions of which land use and geological maps are the main examples, showing soils, bedrock, human use, or mineral deposits. Shapes may be approximate, particularly on small-scale maps with pattern or colour fills used to denote qualitative differences.

Location maps show the existence and position of structures. Icons or mimetic symbols are often used, for instance to denote the location of an airport on a simplified regional map. Distances will often be not to scale, and the map may take on more of a diagrammatic form suitable for use in a brochure or on a poster.

Relationships between themes and the topographic basemap might include the depiction of a service area for a company sales territory, or subdividing space to show the location of the nearest hospital. Relationships may also include the connected paths or routes people and places are linked by. Network maps tend to forgo a lot of topographic detail simply to describe the network itself. They do not attempt to include any further quantifiable information that might redefine them as flow maps.

**See also:** Reference maps | Thematic maps

**Opposite:** *Cloud Cover and the 2017 American Eclipse* by Joshua Stevens, 2017.

Likelihood of clear skies for the Great American Eclipse, 2017

Less likely

More likely

# Design and response

## The layout of a map fundamentally affects the reader's response, and you can use this as part of your design.

**At one time or another, most people will have read a map yet they come to that task with many different and varying experiences.** Unless you are making a map for an extremely niche readership, you're probably going to face the difficulty of trying to author your map to suit a number of different use case scenarios and different map readers. There are ways you can organize the map to feed off people's awareness of other media.

**People generally feel comfortable with the familiar.** In the same way that they are likely to have read a map before, they will also be familiar with the structure of their favoured newspaper, websites they use, and the multitude of different forms we have to complete for banking, health care, taxes, and so on. They all embody very different characters. Some are impersonal and indifferent; others are empathetic. They are designed to elicit a particular response. This characteristic can be mimicked in the design of maps.

**Mapping for neutrality and indifference is actually quite hard.** Perhaps you're mapping simple but pure numerical information, and the map's function is simply to share that pattern or relationship without really caring whether the user needs to be concerned, act upon the results, or deal with the consequences. Such a map can be faceless in the sense that it doesn't need to evoke a response. Fonts and colours can be muted. Layouts can be relatively flat and systematic. You want to give the reader enough to understand the map but without it evoking personal feelings for the subject matter.

**Conversely, you might want to evoke a response.** This might be to lead the reader to action or comment; or support them in doing something as a result of the map. In this sort of map, you might emphasise certain characteristics. You might use colours and fonts in a more dramatic way to evoke certain feelings. Setting out to evoke something particular in a readership has every chance of missing its mark. Fully appreciating the user needs and the persuasive context of your map is vital to success.

Executing a map design is best approached with some appreciation of psychology since an understanding of how people respond to visual stimuli can go a long way to informing your design. Improving the aesthetic quality, comprehension, and usability of your map can be achieved by considering a few key ideas.

Cognitive response is quicker than conscious thought so our brains always begin with a visceral (subconscious) reaction to a map. Visceral designs tend to produce predictable reactions. These include symbology that relates to what we know and expect. Blue for water. Red for danger. Large for more, and so on.

Humans tend to evaluate the cost versus benefit of tasks. When looking at maps if reading and understanding is perceived as difficult, then the map reader will likely not peruse the map. Make the cost low and the reward high! This relates to Hick's law, which states that the more options someone has, the more energy it takes to make a decision and the increased likelihood that they will give up. Keep options to a minimum.

Gestalt psychology describes the way we examine visuals based on proximity and white space. Map elements are perceived as unified when in close proximity even if they are not. Objects that look similar will be perceived as such. Ensure proximity and similarity are sufficiently distinct to avoid these perceptions.

We tend to try matching what we see to known patterns. Recognition is a short cut to understanding so work with what you know people know. Different patterns will feel awkward and compromise understanding. Match expectations to avoid selective disregard and don't overwhelm.

**See also:** Cartograms | Different strokes | Emotional response | Literal comparisons | Wireframing and storyboarding

**Opposite:** *Japan, the Target: A Pictorial Jap-Map* by Ernest Dudley Chase, 1942.

## Visceral Reaction

The focused use of bold red triggers an intrinsic sense of alarm and urgency.

## Recognition

The punctuated use of familiar imperial symbology literally marks territory but also wraps the layout to communicate a distinct and entrenched omnipresence.

## Similarity

A simple but bold palette assigns national affiliation, linking Japan with occupied China via a shared hue.

## Continuity

The radial lines buffering Tokyo invoke directionality and pull the eye to the focal point, and reinforce the theme of target.

## Proximity

Tightly-clustered bombers are perceived as being related in a group, evoking a sense of alliance and shared goals.

## Closure

The centrally-oriented arrangement of the action-line draped bomber clusters face a shared target. The mind paints in an inevitable arrival.

# Different strokes

## Different ways of making a map find favour in different people.

**Ask 10 cartographers how to make a map and you'll get 10 different maps.** That's a phrase I often like to use though I'm unsure if I once read or heard it or just made it up. It simply expresses the fact that if you start with a blank piece of paper or screen, there's so many alternative ways to intersect the art, science, and technology of cartography that you'll inevitably end up with different approaches. This leads to likes, dislikes, and favoured ways to make specific maps for specific audiences. At its worst, it leads to deep dogmatic divides.

**Some approaches to a specific mapmaking task will likely cause consternation in others.** There are always clashes in any design-led field. People like to do things their way, and on occasion they can become blinded to a different way. For some cartographers a rational, scientific approach is favoured and function becomes the predominant aim. Others might consider themselves artists, and they place a much greater emphasis on aesthetics. In effect, this creates two apparently opposing philosophies though the sweet spot in the middle ground is perhaps an optimum position. Where you sit on the continuum will have much to do with your background. Scientific and technical backgrounds are likely to lead to the former approach while graphic design and art the latter.

**Both efficiency principles and decoration have an important place in cartography.** Efficiency in map design demands attention to the substance of the map and communication through clarity and precision. This efficiency principle is described by Edward Tufte (2001) as the data-ink ratio—a measurement of the amount of ink used to represent data in a chart. The principle can equally be applied to maps. Data-ink elements are those that cannot be removed without destroying the integrity of the presentation. Other items (decoration) can be removed because they are redundant or detract. However, maps are decorative. They are often used to decorate and adorn. Decoration comes in many different forms, and being stoically pragmatic in cartography ignores aesthetics which deliver such a rich tapestry of ways of doing. Decoration exists in cartography to enhance presentation.

Tufte's data-ink ratio actually has a formula:

$$\frac{\text{ink that encodes data}}{\text{total amount of ink used to print the graphic}}$$

The nearer to 1.0, the better the graphic is and the less ornamentation there is on a map. Of course,applying this sort of approach to cartography would result in fairly sterile maps. Yes, they might function in a basic sense but their form would be barren. Tufte's dictates are as much an aesthetic choice as a minimalist philosophy.

Sometimes debates between different standpoints spill over. Tufte, professor emeritus of political science and statistics at Yale University, is often lauded for his work and publications on information design and visualization. Nigel Holmes, on the other hand, is a renowned artist and the former art director of *Time* magazine. Their fundamental approaches to information design differ. Tufte prefers minimalist efficiency over ornamentation and believes cosmetic decoration distorts the data. Holmes often incorporates humour and flourish in his work. The Tufte/Holmes debate has become infamous as a way of framing different approaches to visualization.

As with most debates, balance is key. Tufte's demands to respect data and audience should be foundational in any map. Over-decoration is often to the detriment of function. But…decoration can enhance understanding. For a start it can invite an audience who might otherwise be disinterested. It can also support long-term memory recall because we consume a particular visual image of the work and not just a skeleton construction. Beautiful things, of course, can be more than functional.

**See also:** Signal to noise | Stereoscopic views | Style, fashion, and trends | Types of maps

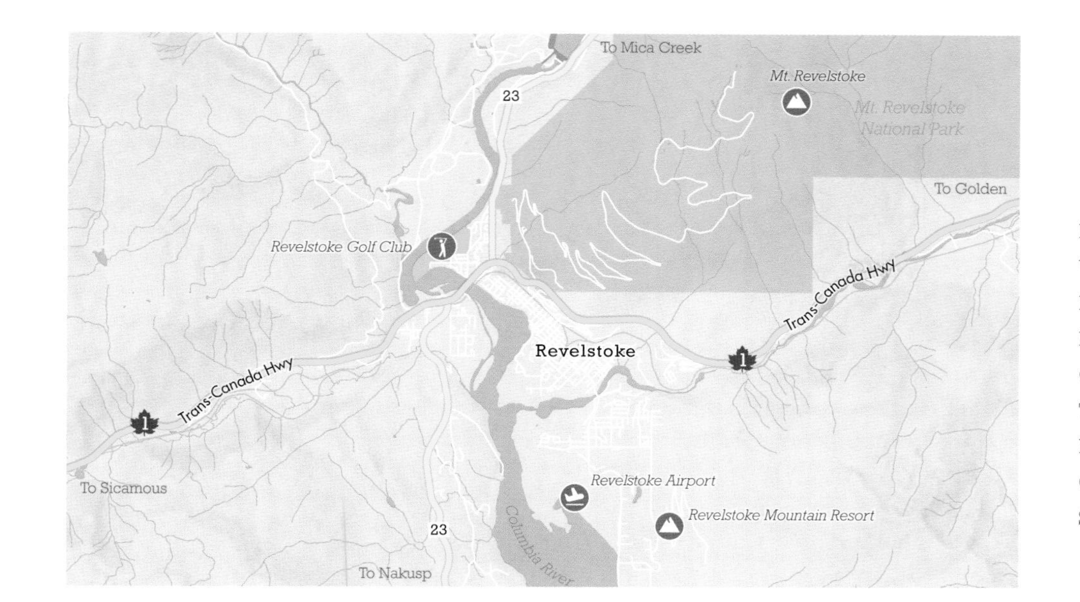


ROAD MAP

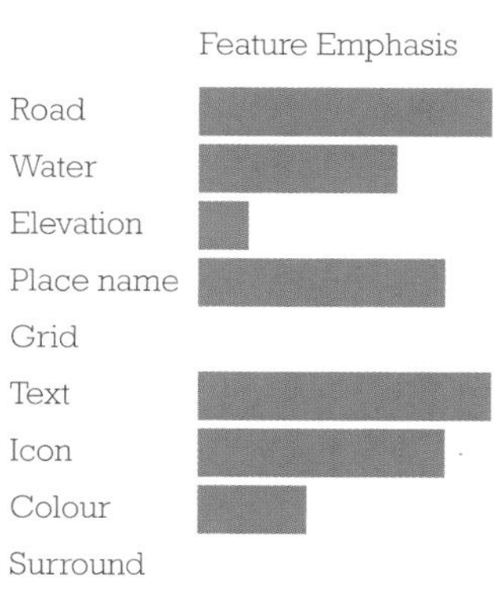


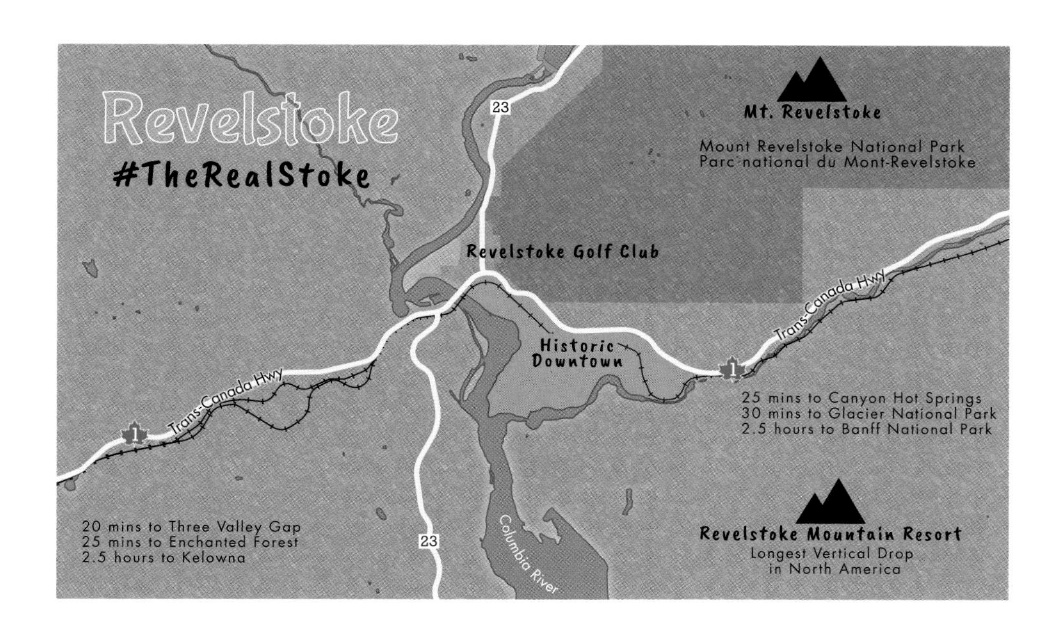


TOURIST MAP

Those 80s tourist placemat sort of maps

Feature Emphasis

Road
Water
Elevation
Place name
Grid
Text
Icon
Colour
Surround

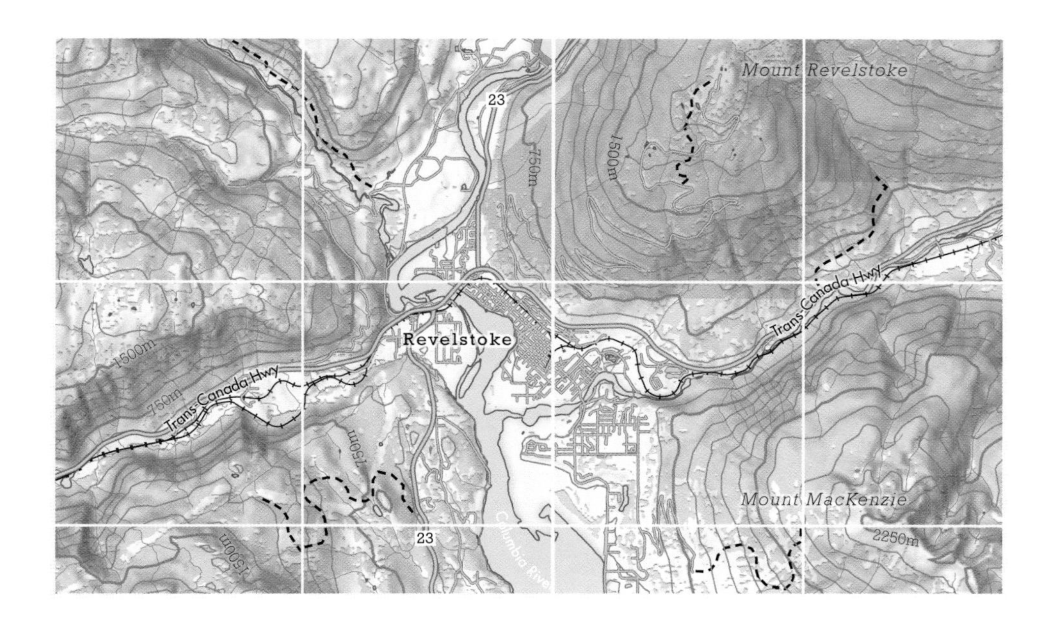


HIKING MAP

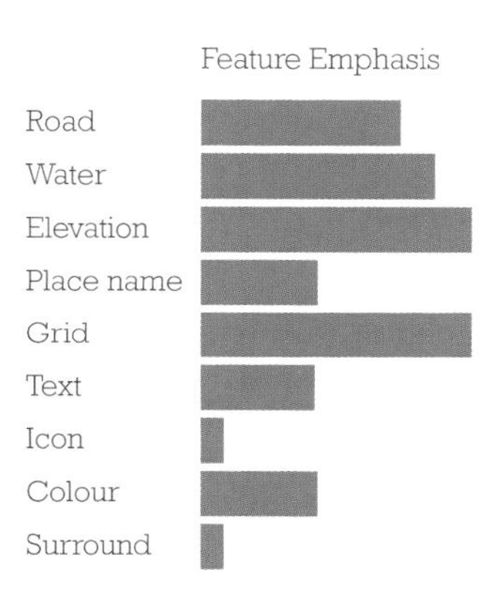

# Digital data

## Maps are nothing without data, and the rise of open data has created vast opportunities, but also challenges, for cartography.

**Maps don't exist without data, and the web has supported many dramatic changes in the availability of data.** Beyond traditional data collected and stored by mapping agencies, much new and interesting data has become emancipated from silos and been made available for people to download and map. The public has also been instrumental in building new datasets and sharing them directly with each other, or augmenting traditional data. For example, our ability to locate ourselves and record information has led to some paradigm shifting datasets such as OpenStreetMap which rivals and often exceeds commercial data as a complete and contemporaneous dataset which supports community mapping as well as many commercial ventures.

**The growth of citizen science and our ability to act as a sensor-network to capture, record, and share volunteered geographic information has supported the explosion of online thematic mapping.** Data is now published rapidly, often with no extra effort (it is passively just shared), and is either already georeferenced or can be geocoded easily and consumed on demand. Cartography used to be based on meticulous surveys that took days, weeks, and months before being wrangled onto a map. It was often out-of-date before it had been published. Data is now streamed live from sensor networks, consumed inside maps automatically, and published in real time to give us a truly on-demand live picture of the world around us.

**The change in data capture, handling, and availability has had profound effects on the cartographic process.** Where once, data was only accessible by a few and cartographers were key to the delivery of maps based on it, now many can access the same data that professional cartographers use. This has increased both experimentation and the sheer volume of maps as well as the number of people making maps. Data will continue to be the bedrock of cartography. Maps will always be required as the principal mechanism to communicate the spatial essence of data where geography is a fundamental component.

OpenStreetMap is perhaps the most widely acknowledged success story of the age of volunteered geographic information. Begun as a collaborative project in 2004 to create a free, editable map of the world it sought to challenge proprietary map data held by national mapping agencies. Data collected on manual, consumer-grade GPS receivers and digitized from a range of other sources has created a crowdsourced digital dataset. This has become indispensable for a multitude of purposes ranging from replacing traditional mapping output to supporting the rapid-response needs of disaster management and support for crisis mapping.

The main output of OpenStreetMap isn't a map, it's the data. It supports many other business ventures that are predicated on its data. Perhaps its biggest contribution to cartography has been in leading the way to the unlocking of datasets from many other sources. Its impact has therefore been profound.

Whatever data sources you use, you have a great responsibility to be faithful to the building blocks from which your map will be created. Data efficacy is paramount and is essentially the processes you use to manage and map data to make an informed map. You should be aware of how data was collected, from where, and based on what criteria. What have you done to the data through processing, and is the approach sound? Are there specific assumptions you've had to make, or has data been removed, cleaned, or summarised, which might have introduced change? What biases might exist that could distort how people interpret the map? Remembering that data and datasets are not entirely objective can be useful. They are human designed and contain errors, bias, risk, and uncertainty.

**See also:** Basemaps | Copyright | Data accuracy and precision | Digital elevation models | Maps kill | Point clouds

**Opposite:** An extract of OpenStreetMap data rendered in ArcGIS Online.

Balls Head
Milsons Point
Kirribilli
McMahons Point Wharf
Blues Point Road
Upper Pitt Street
Kirribilli Avenue
Jefferys Street
Goat Island
Sydney Harbour Bridge
Sydney Harbour Tunnel Southbound
Kirribilli Point
McMahons Point - Balmain East
Ferry: E4, Circular Quay => Darling Harbour
McMahons Point - Darling Harbour
Ives Steps Wharf
Circular Quay - Kirribilli
Circular Quay - Fort Denison
Ferry: E7, Watsons
Circular Quay - Garden Island
M1
Walsh Bay
Hickson Road
Balmain East
Barangaroo Reserve
Dawes Point
Sydney Cove
Nawi Cove
Kent Street
Cahill Walk
Millers Point
The Rocks
Balmain East - Darling Harbour
Pyrmont Bay - Balmain East
Circular Quay, No. 6 Wharf
Farm Cove
Barangaroo
Circular Quay
Macquarie Street
Mrs Macquaries Road
Pitt Street
Royal Botanic Garden
Woolloomooloo Bay
Pirrama Road
Darling Island Road
Barangaroo Avenue
Scotch Row
Western Distributor
Bligh Street
Point Street
Harris Street
Pyrmont Casino
Wynyard
King Street Wharf
Lime Street
Shelley Street
Clarence Street
Hospital Road
The Domain
The Star
Sydney
Martin Place
Slip Lane
Sussex Street
York Street
King Street
Art Gallery Road
Union Street
Miller Street
Pyrmont Bay
Pier 26
Market Street
Forbes Street
Victoria Street
Murray Street
Cockle Bay Wharf
St James
Hyde Park
Domain Car Park
Woolloomooloo
Ada Place
Castlereagh Street
Riley Street
Crown Street
Convention
Brougham Street
McElhone Street
Cross City Tunnel
Fig Street Offramp
Town Hall
Wentworth Park
Darling Drive
Darling Quarter
Kings Cross
Stanley Street
Francis Street
Jones Street
Bulwara Road
Museum
Whitlam Square
Wattle Street
Exhibition Centre
Harbour Street
Dixon Street
Surrey Street
Darlinghurst
Palmer Street
Ultimo
Paddy's Markets
Capitol Square
Campbell Street

# Digital elevation models

## Elevation data supports numerous cartographic workflows for representing terrain.

**A digital elevation model (DEM) represents the terrain surface created from elevation data, usually as a raster grid where each cell value is equal to the height at that position.** A DEM is a fundamental data source used to create a wide range of terrain representations or portrayal of relief such as relief shading and hypsometric tinting. A DEM is a sample of elevations or depths at a specified resolution.

**Although often used interchangeably, DEMs are not to be confused with a digital surface model (DSM) which represents Earth's surface and all objects on it.** Although most commonly a raster grid, some DEMs are in the form of a triangulated irregular network (TIN). DEM data is commonly sourced from remote sensing techniques and in addition to supporting relief portrayal is also important for city modelling and landscape visualization as well as many analytical applications.

**The quality and resolution of a DEM is a function of the data acquisition.** This is referenced as the absolute accuracy of each pixel in the raster grid as well as the relative accuracy of the morphology it represents. Many different factors affect these qualities such as terrain roughness, sampling density, grid resolution (pixel size), vertical resolution, and the various interpolation algorithms used.

**Light detection and ranging (lidar) can also be used to develop surface data.** Lidar measures distance by rapidly illuminating a target with laser light and records distance by measuring the time it takes to return to the measurement instrument. Lidar data is a popular dataset for creating a range of height-based datasets and provides more detailed resolutions of DEM, typically of 1 m or higher. Different return data, usually from airborne lidar sensors, can be used to create DEMs and DSMs from the high-resolution 3D point clouds that are created of the landscape.

Both government and private organizations create DEM data. As examples, GTOPO30 is available for the whole world at a resolution of 1 km but with variable quality. The Advanced Spaceborne Thermal Emission and Reflection Radiometer (ASTER) instrument on the Terra satellite provides 30 m resolution for 99% of the globe. A similar product is available for the United States under the Shuttle Radar Topography Mission (SRTM). Neither GTOPO30 nor SRTM cover the polar regions. The SRTM30Plus dataset combines GTOPO30, SRTM, and bathymetric data to create a global elevation model.

Many national mapping agencies also produce their own DEMs, which can often be at a higher resolution than global products. For instance, in the United States the USGS produces the National Elevation Dataset, which is a seamless DEM of the United States including Hawaii and Puerto Rico using 7.5' arc second resolution.

DEM data alone is not particularly useful for cartographic representation. You can symbolise it to show where is higher and where is lower but that doesn't make for a compelling map of terrain. The real value of a DEM is in using it to derive other representations. For instance, DEMs are the cornerstone to the calculation of contours, or the development of many different forms of hillshade. Because slope, gradient, and aspect can all be calculated between adjacent pixels, it is this post-processing and generation of alternative forms of terrain representation that are of significant value to cartography.

**See also:** Continuous surface maps | Planetary cartography | Point clouds | Slope, aspect, and gradient

**Opposite:** Digital elevation model of Mars and cartographic output.

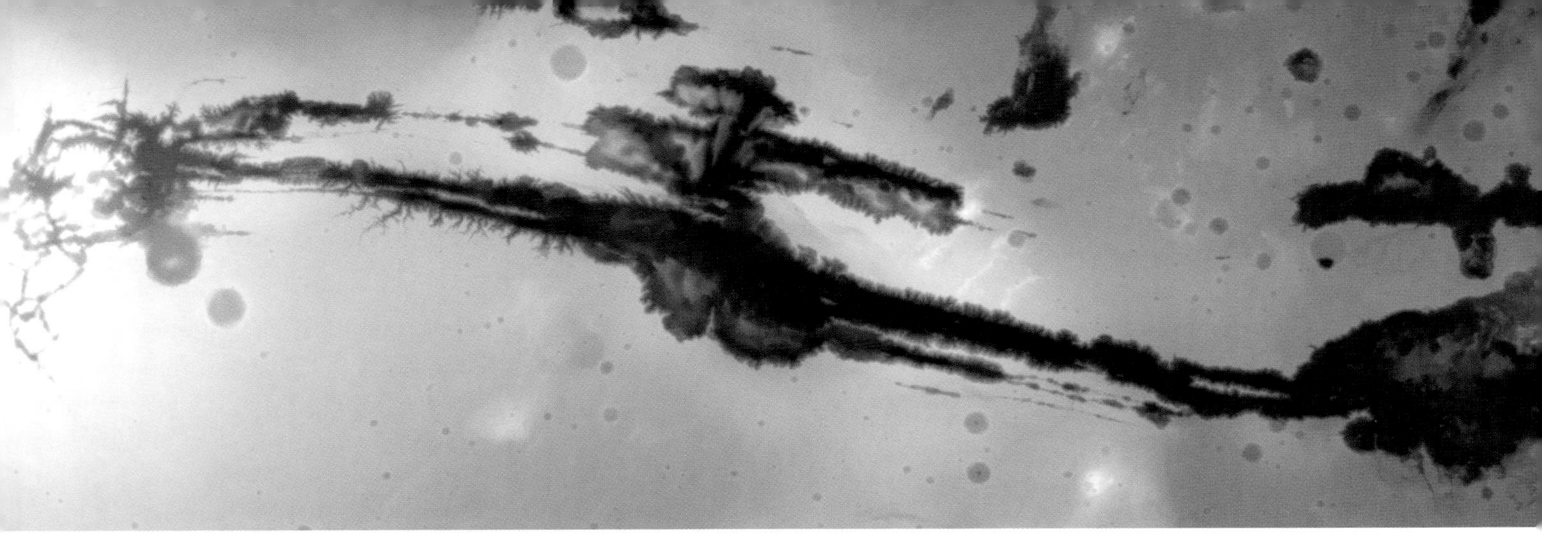

Raw DEM of Valles Marineris, Mars with a black-white stretched colour ramp signifying low to high elevation values.

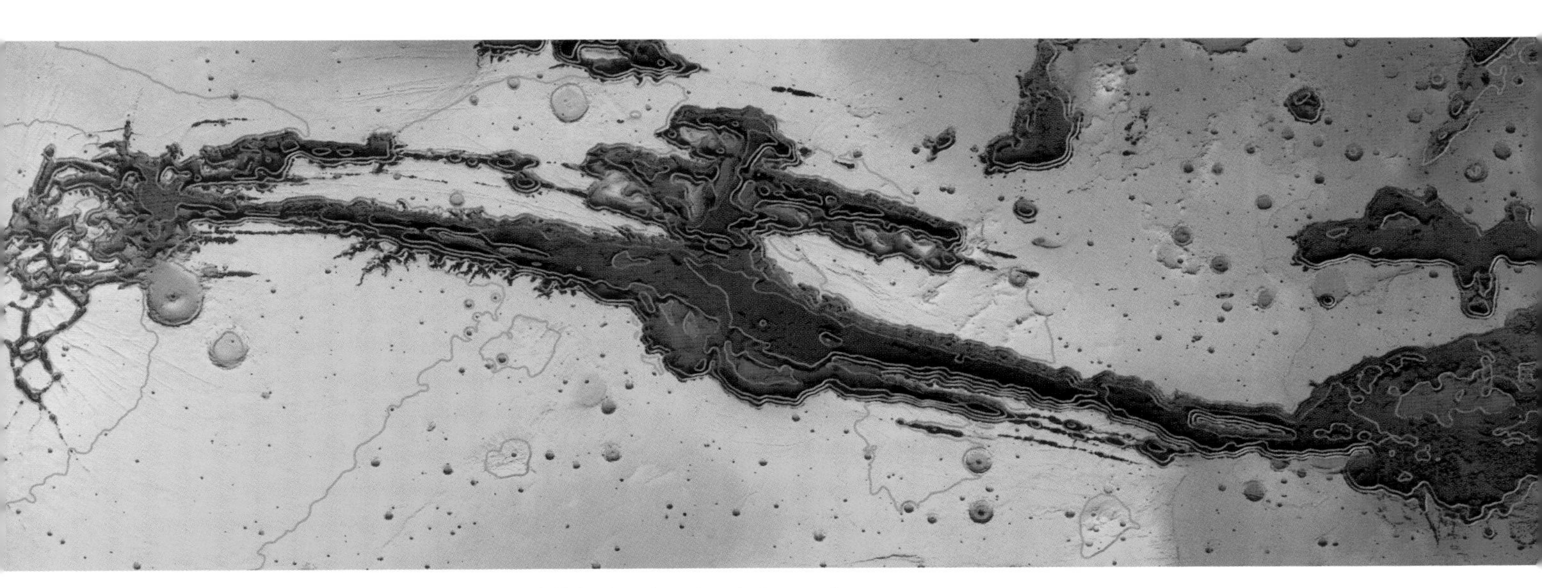

Derived cartographic product incorporating a hillshade, elevation tinting, and contours.

Derived 3D product incorporating a hillshade and elevation tint draped over an exaggerated surface model (x10)

# Dimensional perception

## Increasing dimensionality and magnitudes result in reduced accuracy of perception.

**Symbol complexity is in part a function of its raw dimensionality.** Symbols are one of the forms of point, line, area, or solid which can, in turn, represent location, directionality or flow, quantity, or volume. How the dimensions are symbolised and how they appear among other symbols of similar or different types is the basis of our perception and cognition.

**A point feature essentially has no dimensionality.** Simply showing more points gives a way of showing more of something, and the process of seeing more is fairly easily read.

**A linear feature is one-dimensional, but length and width give a character and a way of symbolising quantity.** The process of assessing the comparison of one line against another is relatively simple. Humans are generally good at assessing relative lengths though the more sinuous a line, the harder this becomes.

**An area feature or symbol is two-dimensional and that allows imbuing a wider range of meaning into the symbol.** Assigning a quantitative measure to an area overcomes the need to over use length of a one-dimensional object. Areas are more efficient containers for symbology and can be used to encode multiple variables through changing size, shape, hue, or value. They also support multiple layering of information to build nested symbology though this also carries with it the danger of information overload.

**Solids (albeit ones drawn on a plane but which mimic solids) are three-dimensional.** They are extremely efficient as containers for information but they are also much harder to understand largely because of the difficulty people have in comparing the relative volumes of solid objects. Pyramids are the hardest dimensional form for perceiving volume (not counting solids with more than six sides). We misjudge volume because of the aspects that we cannot see as they are hidden from view.

The more dimensions a symbol has, the harder the map reader has to work to decode meaning. Sometimes we have little choice as a mapped feature is by definition a point, line, or area. However, it is feasible to represent objects using different feature dimensions. When the mapped task is to reveal information through assessing comparisons, the balance between ease of comprehension and efficient use of space or efficient symbol design is vital to manage. Often three-dimensional symbols are used for graphical effect to capture attention but the question has to be asked whether we want the map reader to make an accurate assessment or simply be made to feel it is accurate.

When we're using familiar measurements the process of understanding is easier. We are all reasonably able to understand certain measures of size, length, and distance but areas and volumes can present problems. For instance, we can all imagine how long a foot or a metre might be but having some visual understanding of the distance of 603 miles is harder. By using visual or descriptive comparisons we might better support people's ability to understand. 603 miles, then, is the straight-line distance between Lands End and John o'Groats in the United Kingdom—from tip to toe.

Several studies have given us clear indications of the over- or underestimation of magnitude of different dimensionality. In fact, humans tend to underestimate magnitude as dimensionality becomes more complex and as magnitudes become larger. One way to adjust for this is to modify symbol design to exaggerate for larger values, thus compensating for the limitations of human perception. This is not without controversy as it could be argued that human perception cannot be deemed to be equal and so compensated adjustments will also be seen unequally.

**See also:** Levels of measurement | Proportional symbol maps | Shape | Size

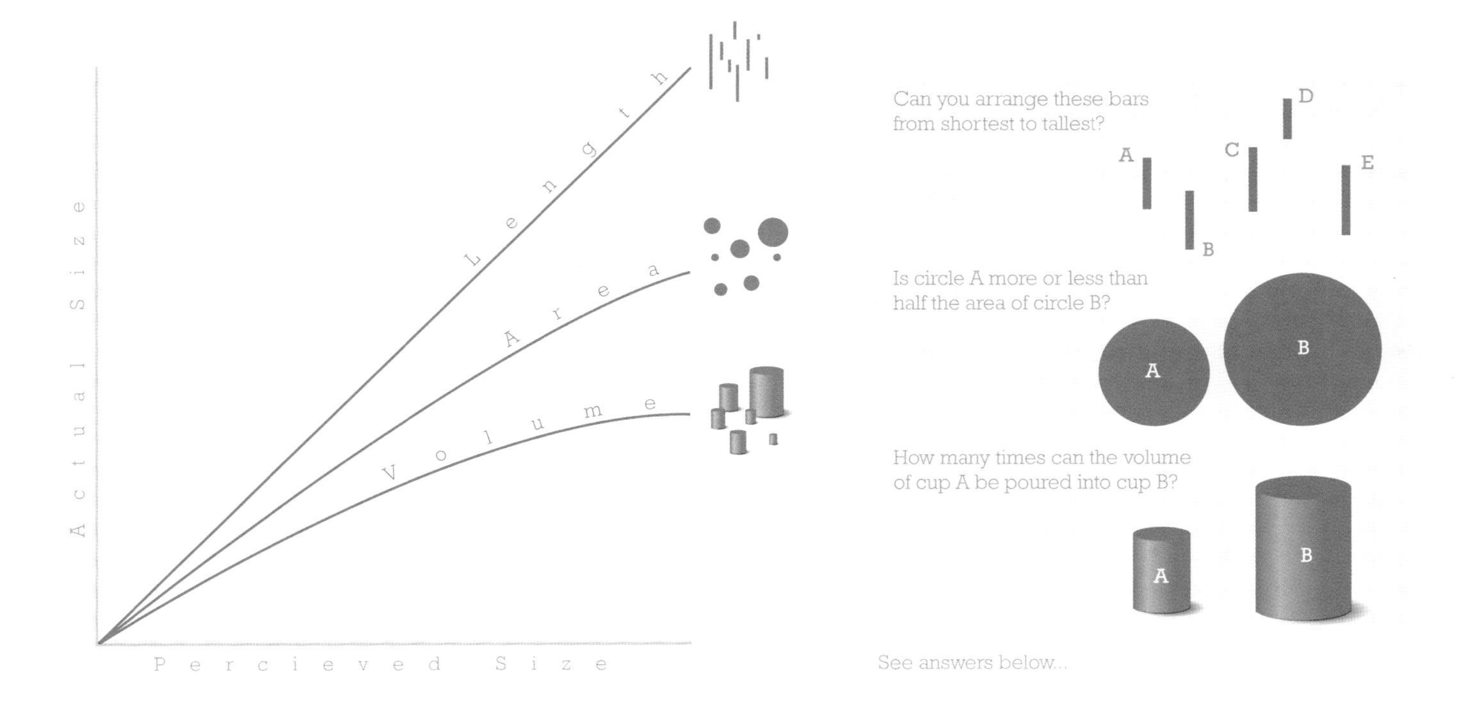


If the smallest symbol represents 3,000 and the largest 8,000,000, about what value does symbol A at left represent? How about B, on the right?

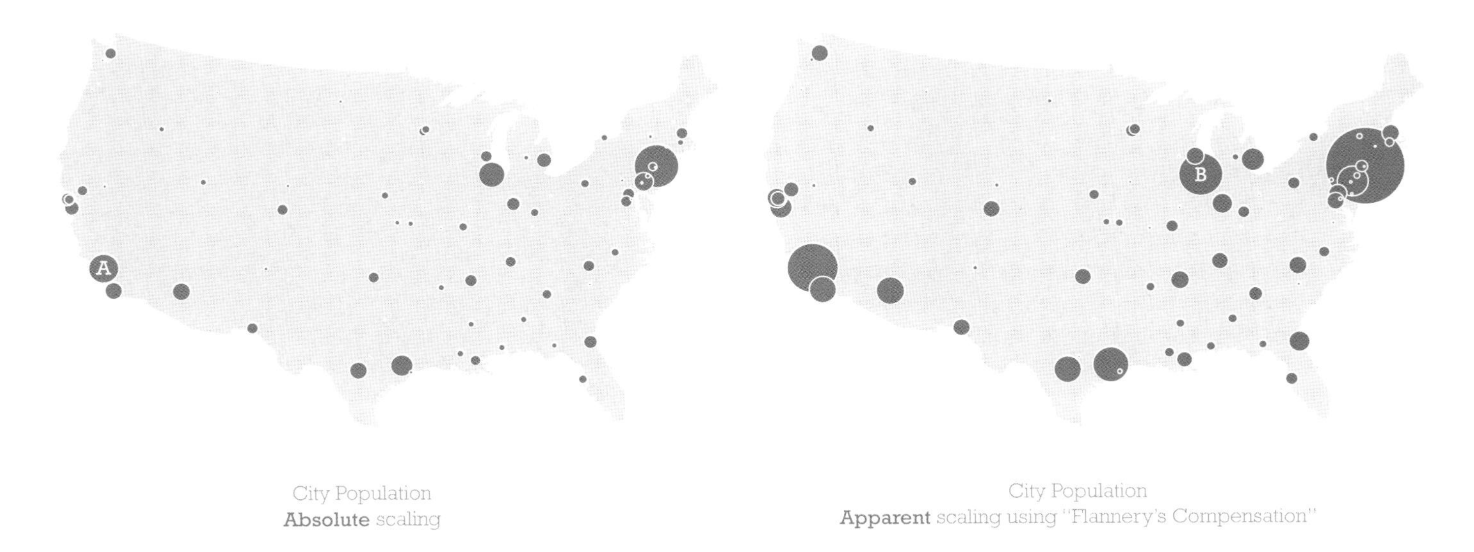


City Population
**Absolute** scaling

City Population
**Apparent** scaling using "Flannery's Compensation"

Answers: D, A, B, C, E | Circle A is less than half the area of Circle B. | More than four and a half times. | Symbol A represents about 3.5 million; symbol B 2.8 million

# Dispersal vs. layering

## Coincident data often benefits from being structured horizontally or vertically.

**Mapped data often contains several depths of information that occur at a coincident place.** One of the cartographic tasks is to develop a structure that brings out the key aspects of the data but we're challenged by overlaid or overlapping data. This challenge is often approached through dispersal of symbols. Alternatively we might layer data to create the illusion of a three-dimensional structure. Both of these approaches can be useful but they also come with compromises.

**It's impossible to simply overlay coincident detail graphically, or through the simple order of layers.** The result of this approach is the decision of which layer is most important and which should be seen first, which puts one on top of the rest and makes the map make sense. Of course, only the piece of information at the top will be seen, and so all other pieces of information are occluded.

**Dispersing information means we apply some offset to part or all the pieces that visually disaggregates them.** This might be in a planar sense or through stacking of data vertically, perhaps like a column of stacked symbols. Dispersal is often accompanied by the use of transparency to allow overlaps, which can all be seen to some extent in two-dimensional space. The Venn diagram is a simple way of thinking about this solution though it can create cognitive compromises since overlapping transparent symbols make a darker sum, which may be an artifact of the symbol used rather than a true overlap in geographical space.

**If we take the layering approach, we generally see how variables interact in combination.** However, it's often difficult to get a sense of the pattern as a whole across the map and of how each component contributes. Map space is used much more efficiently but there is a greater onus on the map reader to disentangle the patterns in the data.

For a dataset with four variables which can be combined in multiple ways, there are two fundamental ways that they can be mapped—firstly to visualise each combination separately through dispersal or to create a multivariate symbol using transparency that supports layering. Of course, with both of these solutions we're encouraging the idea that adding detail is sometimes a way to bring clarity to complex subject matter but thresholds of utility can be rapidly reached.

Although dispersal allows us to immediately recognise symbols that are distinct, we might rapidly exhaust space, and making visual comparisons across a map is only useful between symbols of the same type. Map space includes a lot of repetition and redundancy.

Layering in 3D allows you to encode the z-dimension of a symbol with information. Although it's a useful technique, every time you move into the third dimension you introduce the problem of perspective depth in the map view (causing difficulties for comparing empirical detail across the map that might have foreshortening) and occlusions (as some symbols hide others in the background). These compromises may be acceptable if dispersal itself gives the impression of a different planar distribution in the mapped phenomena.

**See also:** Aggregation | Basemaps | Foreground and background | Hierarchies

**Opposite:** *trees-cabs-crime (in San Francisco)* by Shawn Allen and Stamen Design, 2012, showing a layering approach in 2D through the use of transparency.

TREES
CRIMES
CABS

# Distortions in map projections

## All planar maps contain distortions as a result of projecting 3D to 2D. Distortions vary in type and amount and have consequences for mapmaking.

**The map projection process always involves introducing some form of distortion to the final map.** This is impossible to eliminate and is one reason why it is necessary to understand the impact of any given map projection on the resulting map. Distortions are typically characterised as tearing, shearing, expansion, or compression that modify shapes, areas, distances, and directions on the map.

**In map design terms, appreciating the errors introduced by the map projection process and maintaining control over the distortions introduced is important.** Using a map projection that minimises errors that are most important to the mapping task at hand at the cost of introducing other types of errors is the goal. For instance, if it is important to ensure that areas remain true in size relative to each other, then a map projection should be used that does not distort areal size. Such a projection may, instead, distort angles and shapes but these distortions are, possibly, acceptable given the mapping needs. No one map projection can maintain all properties simultaneously so any map projection is always a trade-off.

**During the 19th century, French mathematician Nicolas Tissot developed a method to illustrate and quantify distortion across a map.** Tissot's indicatrix is constructed by placing small circles on the surface of the reference globe. As the map is projected onto a planar surface, the circles become distorted. Conformal projections transform the circle on the globe into a circle on the planar map. Its size alters across the map but angular properties are maintained. For equivalent projections, the circle transforms into an ellipse on the planar surface. The ratio of the semimajor (longest diameter) and semiminor (shortest diameter) axes of the circle become the reciprocal of one another. Angular properties are not preserved. For some projections, including equidistant projections, circles are transformed into an ellipse that does not preserve conformality or equivalence. The indicatrix gives us a method of illustrating the impact of distortion across a map.

Map distortions vary considerably with scale. The graticule on the globe is effectively a series of quadrilaterals. If a graticule is constructed on the globe to form very small quadrilaterals (for instance if parallels and meridians are drawn at 1° intervals), they are not substantially different from how they would appear on a plane surface (since curvature is minimal across the quadrilateral).

When mapping small earth areas (large-scale mapping), distortion resulting from the map projection process is not a significant issue and can largely be ignored. As the mapped earth area increases in scale (i.e. as map scale becomes smaller), distortions increasingly affect the map. Certainly at a global scale, a world map takes on an extremely different appearance depending on the map projection employed.

Acceptance of distortions is the first phase in understanding them. Knowing what distortion you need to minimise for your own map is the next step. Actively seeking out an appropriate projection and finding a way to make the change from a default in whatever software you use is the final step. It's an all too easy process to ignore but distortions exist and your map will benefit from their consideration.

**See also:** Assessing distortion in map projections | Families of map projection | Properties of a map projection

## Map distortions

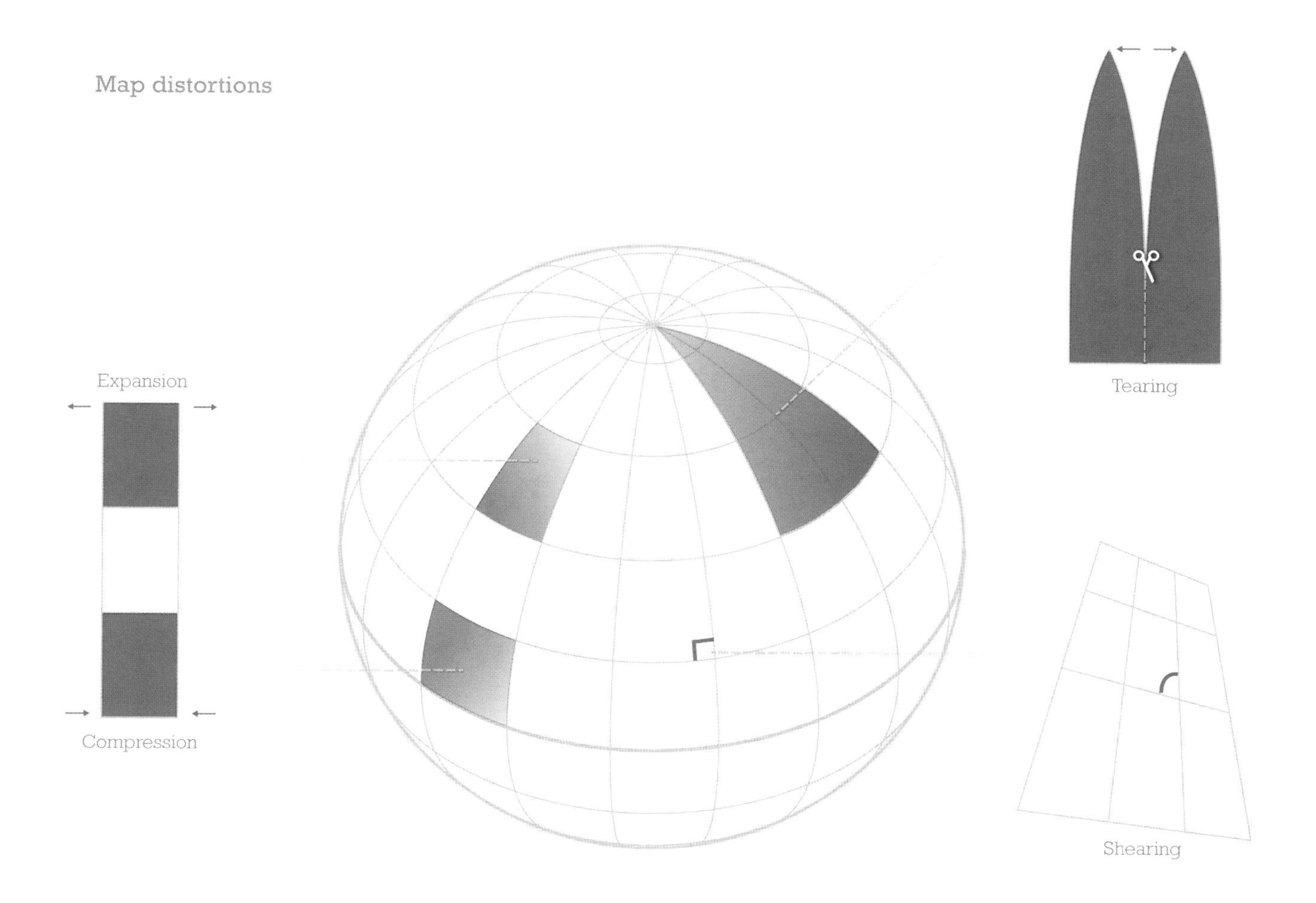


## Tissot's Indicatrix

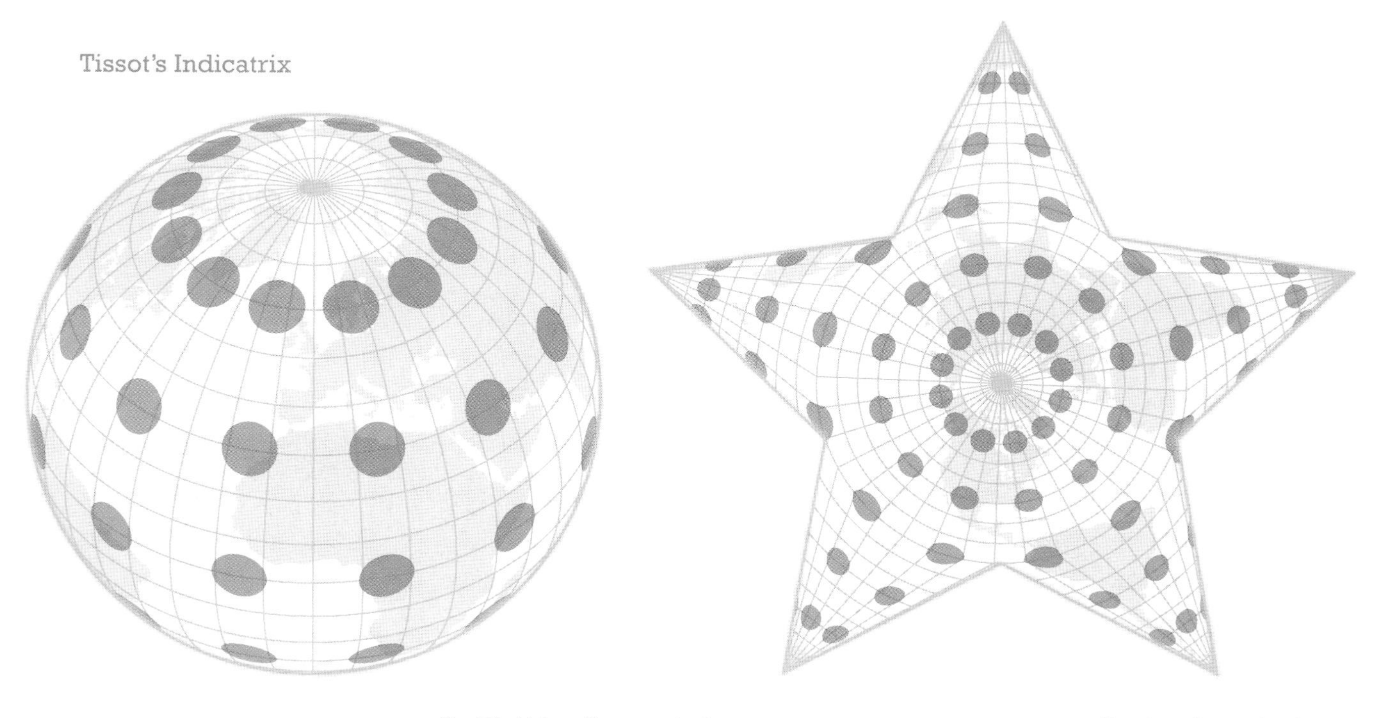

*The World from Space* projection

*Berghaus Star* projection

# Dot density maps

## Dot density maps use dots to denote a quantitative value, which gives an overall pattern of distribution.

**Dot density (or dot distribution) thematic maps use dots as shading symbols, often in a random pattern across a statistical areal unit.** The rationale is easily understood since more dots relates to larger data magnitudes of the mapped phenomena. It is an effective method to represent spatial density and has one advantage over the choropleth in that more than one variable can be mapped at the same time using different coloured dots. At larger scales, dots can be seen to represent their unique colour but with careful application, at smaller scales the mingling of dots can take on the appearance of a mix of colours—for instance, blue and red dots can appear to create a purple mass. This technique can be used very effectively to illustrate areas that have similar values for multiple variables.

**Data is usually in the form of a total number for the areal unit being mapped.** The data should be numerical (quantitative) and represent differences between features on an interval or ratio scale of measurement. When mapping multiple variables, each should be measured on the same scale to allow them to have a functional and meaningful relationship when mapped together.

**Symbols should be designed so that different variables can be easily distinguished from one another.** Differences in hue can be successfully used to show differences in type when multiple variables are mapped with value and lightness being kept relatively constant so that neither appears more important than the other as a result of visual prominence.

**The map reader should be able to gauge relative differences between area densities across the map.** The reader can determine apparent density rather than actual density of the data. Spatial assumptions about distributions should be able to be formed but we would not expect people to be able to recover the data from counting the dots themselves. A legend should enable readers to distinguish the different mapped variables and the value associated with an individual dot.

Designing a dot doesn't sound particularly difficult but the key is to think of the end result and the overall appearance of all dots. The design of the dots requires a balance to be reached between the size and value—the numerical value represented by each dot—so that overall the dot densities make visual sense.

Dot size and value are controlled by the scale of the map since it determines the amount of space available for positioning the dots. There are no definite rules for selection of dot value and size but the map should be neither too sparse nor have huge numbers of overlapping dots. There should be sufficient distinction between the different densities across the map, and the value should be easily understood (e.g. 100, 500 units per dot and so on).

Dot value and size should harmonize with the scale so the total impression is neither too accurate (usually caused by dots that are too small) nor too general (usually where dot values are too high). Finding a balance where the densest area contains dots that are just beginning to coalesce will generally work well.

Overlaying dot density thematic detail across a topographic background gives an impression of a variable surface of dots despite the fact that dot densities are most likely built from values reported in arbitrary administrative areas. For instance, the map opposite is a dot density thematic map but with the administrative boundaries omitted. The dots therefore appear as part of a more organic surface that relates geography to population distribution.

**See also:** Dasymetric maps | Data density | Pointillism | Points

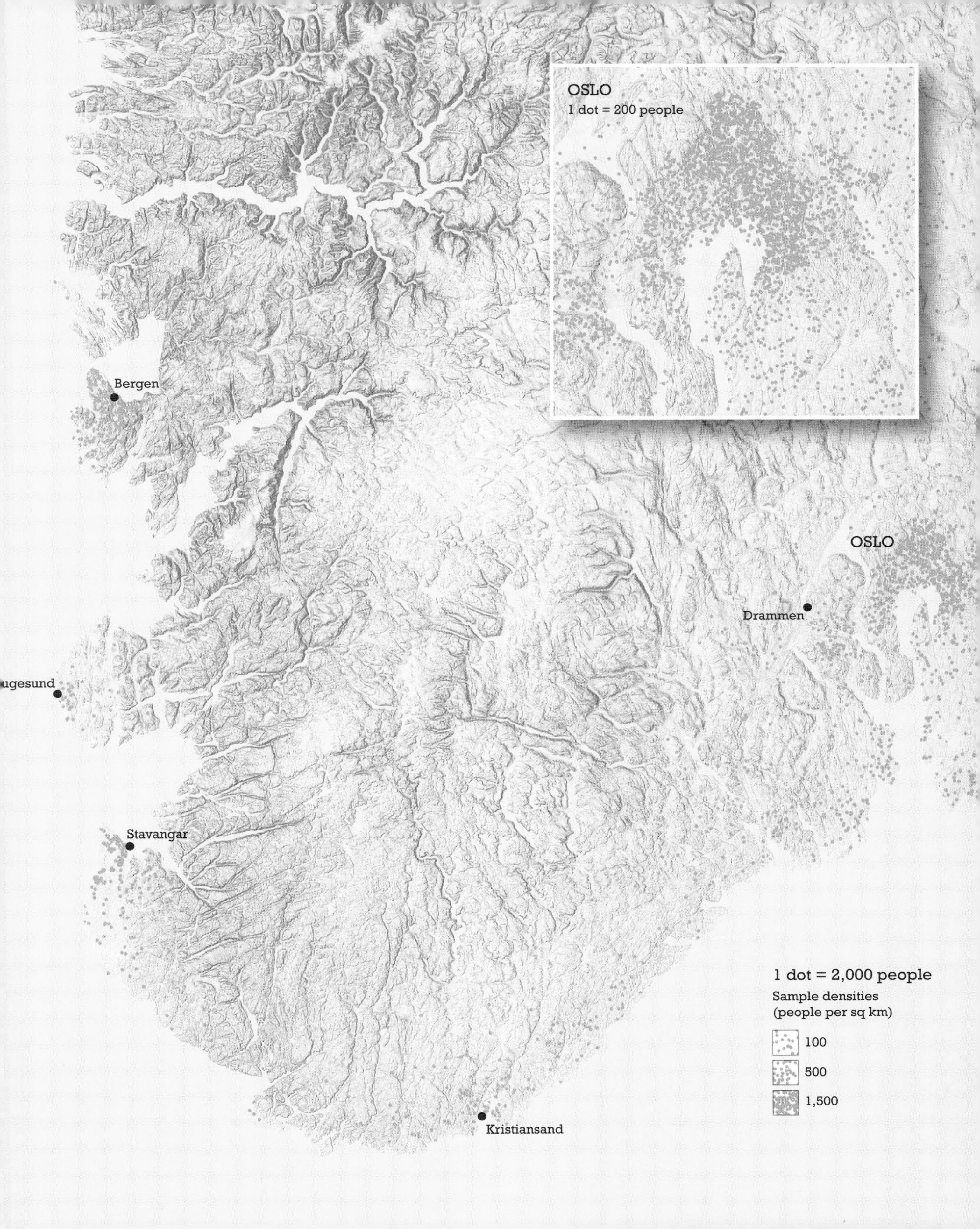
OSLO
1 dot = 200 people
Bergen
OSLO
Drammen
ugesund
Stavangar
Kristiansand
1 dot = 2,000 people
Sample densities
(people per sq km)
100
500
1,500

# Dynamic visual variables

## Ways of representing data designed to support animation.

**With interaction and animation becoming a mature cartographic practice, Jacque Bertin's original static retinal (visual) variables are insufficient to support new displays.** They remain relevant but an extension to the set provides a framework for considering how to represent movement and change effectively. By applying visual variables creatively to individual frames of an animation, it's possible to show dynamic change between different map states. Animations can be a powerful way to communicate and are sometimes preferable over a static map or a map series. An additional six visual variables support dynamic mapping.

**Moment**—a moment can be defined that indicates a state of change between frames of an animation. A single frame shows the moment of change.

**Duration**—a temporal duration is formed by the logical extension of a moment occurring for a specific period of time. Non-temporal durations can be shown by doubling the amount being shown by a doubling of duration.

**Frequency**—rate of occurrence can be shown through a graphical effect such as blinking. Higher frequencies are shown by increased rates of blinking.

**Order**—chronological illustrations of different map states in an animation defines order.

**Rate of change**—rates can be found by dividing the change magnitude by the duration for each map state in an animation. In sequence, the rate of change can be seen. The pace of an animation (in frames per second) can emphasise how rapidly a rate changes.

**Synchronization**—the display of two or more independent variables which support comparisons via the animation

Dynamic visual variables extend the graphical language for cartographic representation from the purely static to support communication through movement. As with the use of static visual variables, they are perceived and decoded in specific ways that support particular requirements such as association, ordering, magnitude, and selection. Dynamic visual variables have particular perceptual properties that define how they should be used.

Moment is only useful when combined with other dynamic visual variables. For instance, showing an animation in which one moment changes results in an unremarkable animated map. Having too many moments results in visual clutter but striking a balance where the animation displays the change of some moments is optimal.

Of more use is duration, which allows the recovery of the change of state in the order of moments.

Long durations should be avoided as waiting for change can in itself lead to problems of data retention and estimation of the temporal component.

Frequency and order are both reasonably effective for illustrating associations and ordering and, to a lesser extent selection.

Rate of change is effective for showing ordering and some changes in magnitude.

Synchronization presents some difficulties simply because there is more than one map at any point in the animation. Focus becomes difficult though associations can be discerned and, to a lesser extent ordering and magnitudes.

**See also:** Animation | Flow maps | Globes | Sensory maps

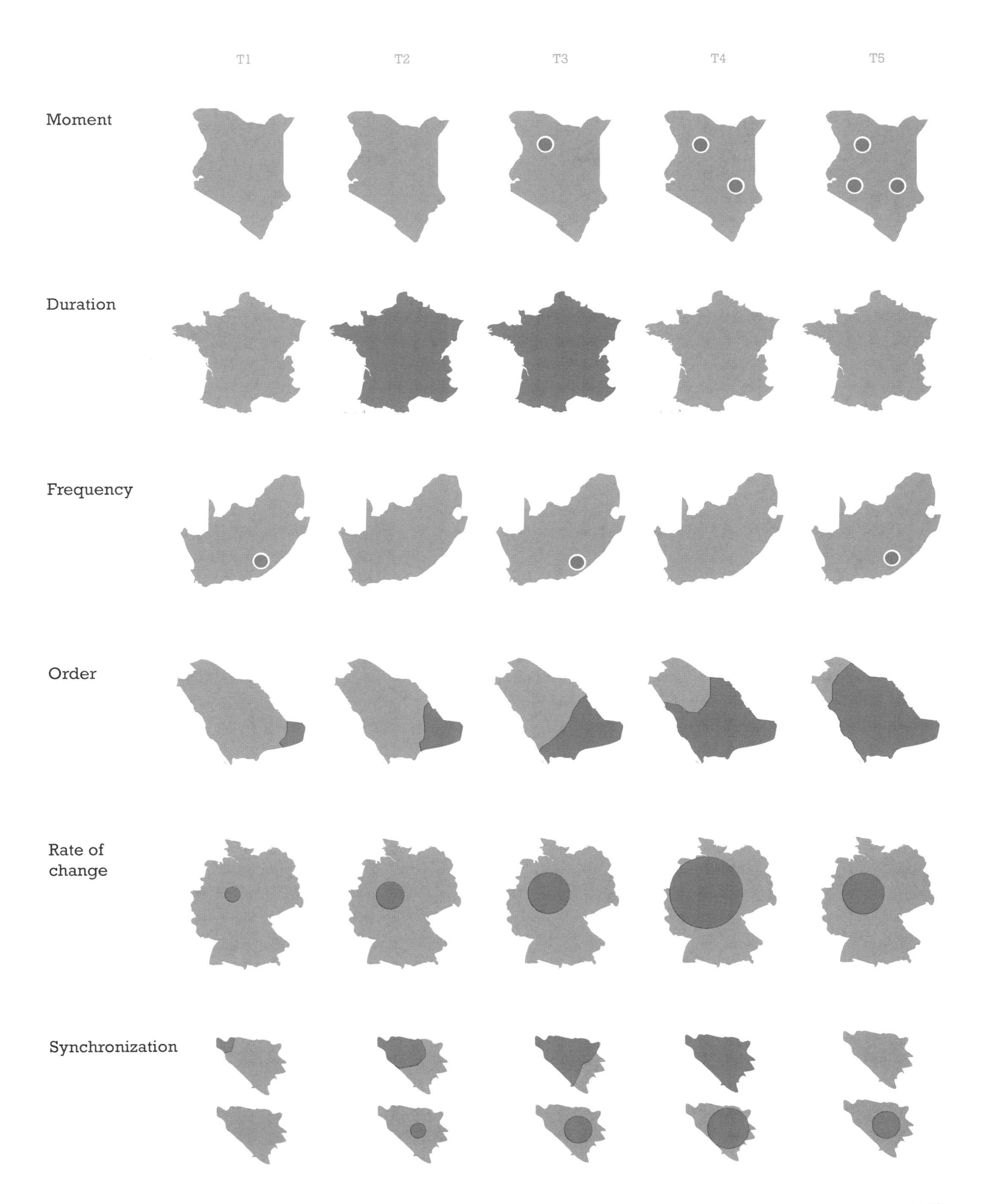
T1
T2
T3
T4
T5
Moment
Duration
Frequency
Order
Rate of change
Synchronization

# Dysfunctional cartography

## Although it's important to focus on how to do things right to make a map, sometimes it's useful to appreciate how you can go wrong.

**A good map embodies a functional balance between the information represented, the skill of the cartographer in revealing that information, and the context in which the information is viewed.** This includes the background, education, and experience of the map reader. Getting these components to balance is where the profession of cartography comes in. The cartographer sits at the fulcrum of the balance. Constraints can lead to design and representation issues that must be handled properly to avoid creating a dysfunctional map; one in which some basic rules might be ignored or where the design fails to fit what we know about how maps are read and used. There are any number of reasons why the best approach may be not to make the map at all.

**Data is often a principal cause for making a dysfunctional map.** Data may be incomplete, skewed, or missing. It may lack meaningful detail or just be too oversimplified to support the idea behind the map you want to make. It may be that the data isn't able to reveal the meaning you'd like or that it is too detailed to be transformed into something understandable.

**Not understanding data can also lead to unintended consequences.** The cartographer has to be aware of when their own limitations impede the design. Being unable to distil data into meaning isn't a crime—but making a map of something you simply don't understand can potentially propagate misinformation. If the objective of making the map is to simply make something (anything) that is compelling, there's often a danger of making a counter-intuitive product or, worse, misleading the map reader through ill-thought-out design solutions.

**Indifference toward the reality in which the map is situated is a failure of the cartographer.** Making an unsuccessful map happens when no one understands what the map is saying or they don't particularly care to spend time with it. The form of the map goes a long way to making it work. Asking basic questions about your work—Is it true? Is it relevant? Is it necessary?—is key. Failure to ask these questions is a good way to end up with a dysfunctional map.

**See also:** Defining maps and cartography | Error and bias | Functional cartography | Information overload | Purpose of maps

It's often useful to ask yourself whether your map meets a range of basic criteria to avoid making some key mistakes. These might be called 'anti-patterns', those which will end up making your map difficult to interpret or which fundamentally lead to misinformation.

Examples include, but are not exclusively:

**Projections**—does the projection support the intended interpretation? For instance, a Web Mercator projection is rarely good for thematic mapping.
**Colours**—does the use of colour support cognitive processing? For instance, sequential colour schemes do not support our ability to determine quantitative structure in data.
**Thematic techniques**—does the technique match the data? For instance, if you have totals, you might use a proportional symbol technique but not a choropleth unless you convert them to a rate.
**3D**—does the use of the z-dimension encode anything useful? For instance, 3D is often used gratuitously, and the perspective view creates cognitive issues for map reading.
**Transparency**—does the use of transparency introduce visual problems? For instance, transparency is often used on one layer of information to make visible a second layer underneath. This often destroys the form of the upper layer.
**Binning data**—does the bin size and shape make sense? For instance, an equal-area projection is fundamental for binning to avoid bins representing unequal areas.
**Symbols**—does the symbol design simplify or overload? For instance, symbols often try to do too much.

**Opposite:** *A Taxonomy of Ideas* by David McCandless (informationisbeautiful.net), 2012 illustrates the relationship of structure and function and how we might describe the quality of graphics (and maps) that fall into particular categories.

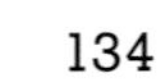

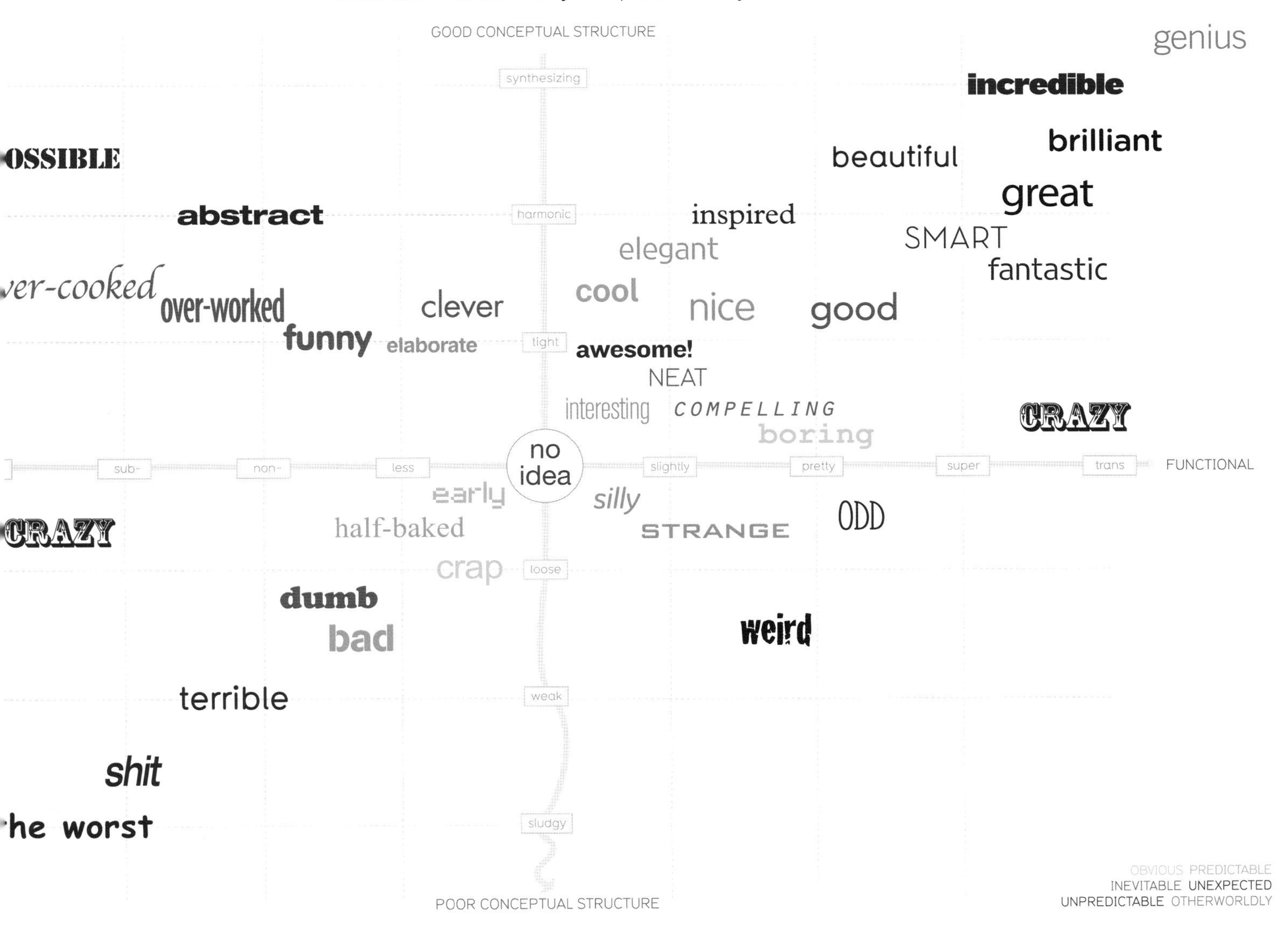
A Taxonomy of Ideas
Structure + Functionality + Unpredictability?
GOOD CONCEPTUAL STRUCTURE
synthesizing
harmonic
tight
loose
weak
sludgy
POOR CONCEPTUAL STRUCTURE
sub-
non-
less
slightly
pretty
super
trans
FUNCTIONAL
no idea
genius
incredible
brilliant
beautiful
great
OSSIBLE
abstract
inspired
elegant
SMART
fantastic
ver-cooked
over-worked
clever
cool
nice
good
funny
elaborate
awesome!
NEAT
interesting
COMPELLING
boring
CRAZY
early
silly
half-baked
STRANGE
ODD
CRAZY
crap
dumb
bad
weird
terrible
shit
he worst
OBVIOUS PREDICTABLE
INEVITABLE UNEXPECTED
UNPREDICTABLE OTHERWORLDLY

## Detail of Area around the Broad Street Pump

John Snow

1854

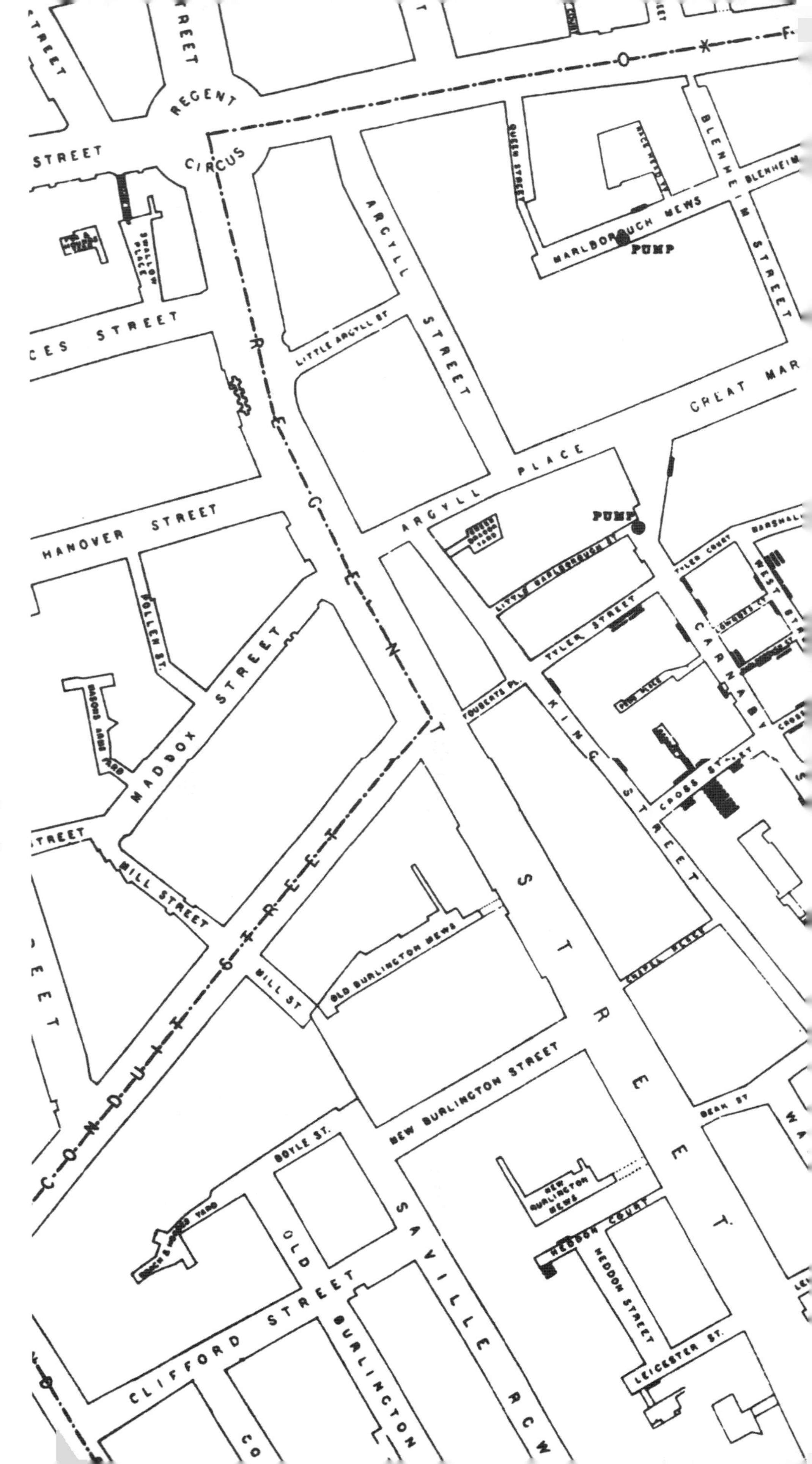


Different stories do the rounds as to whether this map, conceived in 1854, was instrumental in either discovering (exploration) or confirming (analysis / synthesis) John Snow's hypothesized association between deaths from cholera and the location of the contaminated water pump well on Broad Street.

Whatever the truth, this map is one of the earliest and most convincing demonstrations of the power and usefulness of thematic maps. A forerunner to the modern GIS overlay analysis, in which the locations of the 11 community water pump wells in Soho were overlaid by Snow on a map on which he had marked the deaths from cholera, contributed most to the scientific discovery that cholera was carried by water (and not by air, as was the belief up to then). The visual association between the location and density of deaths and the Broad Street pump is clear.

This discovery is seen as one of the founding events of the science of epidemiology or 'geo-health'—as we may refer to it nowadays. The map has attained almost mythical status in cartography though the truth of its construction, by whom, and published some time after the events still causes debate.

—Corné P. J. M. van Elzakker

SOHO SQUARE
NOEL STREET
HOLLEN STREET
POLAND STREET
BERWICK STREET
WARDOUR STREET
DEAN STREET
PORTLAND STREET
LITTLE CHAPEL STREET
CARLISLE STREET
CHAPEL ST.
ST ANNS COURT
PORTLAND MEWS
BENTINCK STREET
WORK HOUSE
DUFOURS PLACE
RICHMOND BUILDINGS
RICHMOND MEWS
BREWERY YARD
EDWARD STREET
DUCK LANE
FRITH STREET
QUEEN STREET
GREEK STREET
BATEMANS BUILDINGS
BROAD STREET
PUMP
CAMBRIDGE ST
BREWERY
NEW STREET
HOPKINS ST
HUSBAND ST
BEARDS COURT
PETER STREET
OLD COMPTON STREET
CHURCH STREET
LITTLE WINDMILL STREET
WILLIAM AND MARY YARD
GREAT PULTENEY STREET
BRIDLE STREET
LITTLE PULTENEY STREET
GT CROWN CT
UP RUPERT ST
KING STREET
GEORGE YD
GERRARD STREET
PRINCES STREET
RICHMOND ST
RUPERT STREET
ARCHER STREET
GREAT WINDMILL STREET
GOLDEN SQUARE
UP JAMES ST
LOW. JAMES ST
LOWER JOHN STR
BREWER STREET
SHERRARD STREET
QUEEN STREET
FRANCIS ST
WELLINGTON MEWS
CATHERINE WHEEL YARD
PARTON SQUARE
ANGEL COURT
MACCLESFIELD
LISLE

# Earth coordinate geometry

## Locating or measuring anything on Earth requires a coordinate system.

**Early exploration was dependent on navigating between locations, taking bearings, and determining the position of one place relative to another.** This requires a mechanism to measure and describe position and led to the development of coordinate systems based on geometry. Although the use of geographical coordinate systems can be traced back to Hipparchus of Rhodes, an astronomer in the second century BC, modern plane coordinate geometry developed from the work of French mathematician Descartes in the 17th century who sought to provide a way of describing the geometry of algebraic relationships. From his work, a graphical system of intersecting perpendicular lines on a plane, containing two principal axes, x and y, was developed and termed *Cartesian coordinate geometry*.

**On both the horizontal x-axis and the vertical y-axis of Cartesian coordinate geometry, a linear measured scale is marked.** Thus, any point within Cartesian space has a measured x- and y-coordinate ($P_{xy}$). Each point has a unique location, and relative location can be easily determined by measuring the difference in x- and y-values between points. Plane (or Cartesian) coordinate geometry is used in mapping in many ways from providing us a way of expressing the intersection of a location as latitude and longitude to the creation of national grids for reference purposes.

**Earth's coordinate geometry is more complex than simple plane coordinate geometry because of its shape.** It becomes even trickier if we want to measure location on an ellipsoid rather than a spheroid. In order to specify a location on the surface of a spheroid or ellipsoid, angular measurement must be used in addition to the principles of plane coordinate geometry. Angles are measured on the sexagesimal scale whereby a circle is divided into 360 degrees (360°), each degree is further subdivided into 60 minutes (60′), and each minute is further subdivided into 60 seconds (60′′).

Natural starting points are required to measure position and which can take a zero value. Because Earth naturally rotates about an axis, the points at which the axis emerges in the north and south provide natural starting points for measuring north–south angular change. These positions are referred to as geographical north (the North Pole) and geographical south (the South Pole). All of Earth's coordinate geometry is based on these two points.

If a plane is passed through the axis of rotation both bisecting it and perpendicular to it, the plane would intersect with the surface of Earth to form a complete circle (notwithstanding the fact that we know Earth has many undulations). This imaginary circle is called the equator.

We therefore have three geometrical properties, namely the North Pole, South Pole, and the equator, that we can use to construct an earth coordinate system.

**See also:** Aligning coordinate systems | Earth's framework | Position | x and y

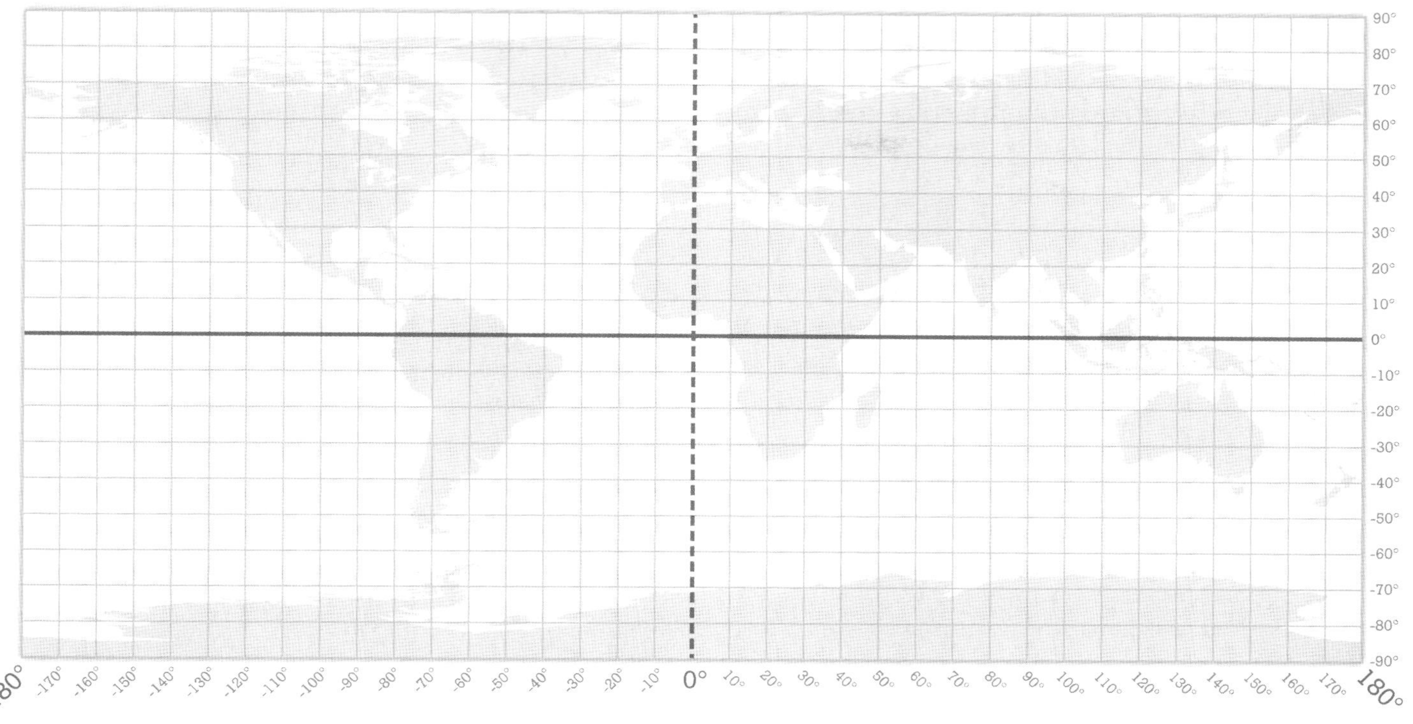

90°
80°
70°
60°
50°
40°
30°
20°
10°
0°
-10°
-20°
-30°
-40°
-50°
-60°
-70°
-80°
-90°
180° -170° -160° -150° -140° -130° -120° -110° -100° -90° -80° -70° -60° -50° -40° -30° -20° -10° 0° 10° 20° 30° 40° 50° 60° 70° 80° 90° 100° 110° 120° 130° 140° 150° 160° 170° 180°


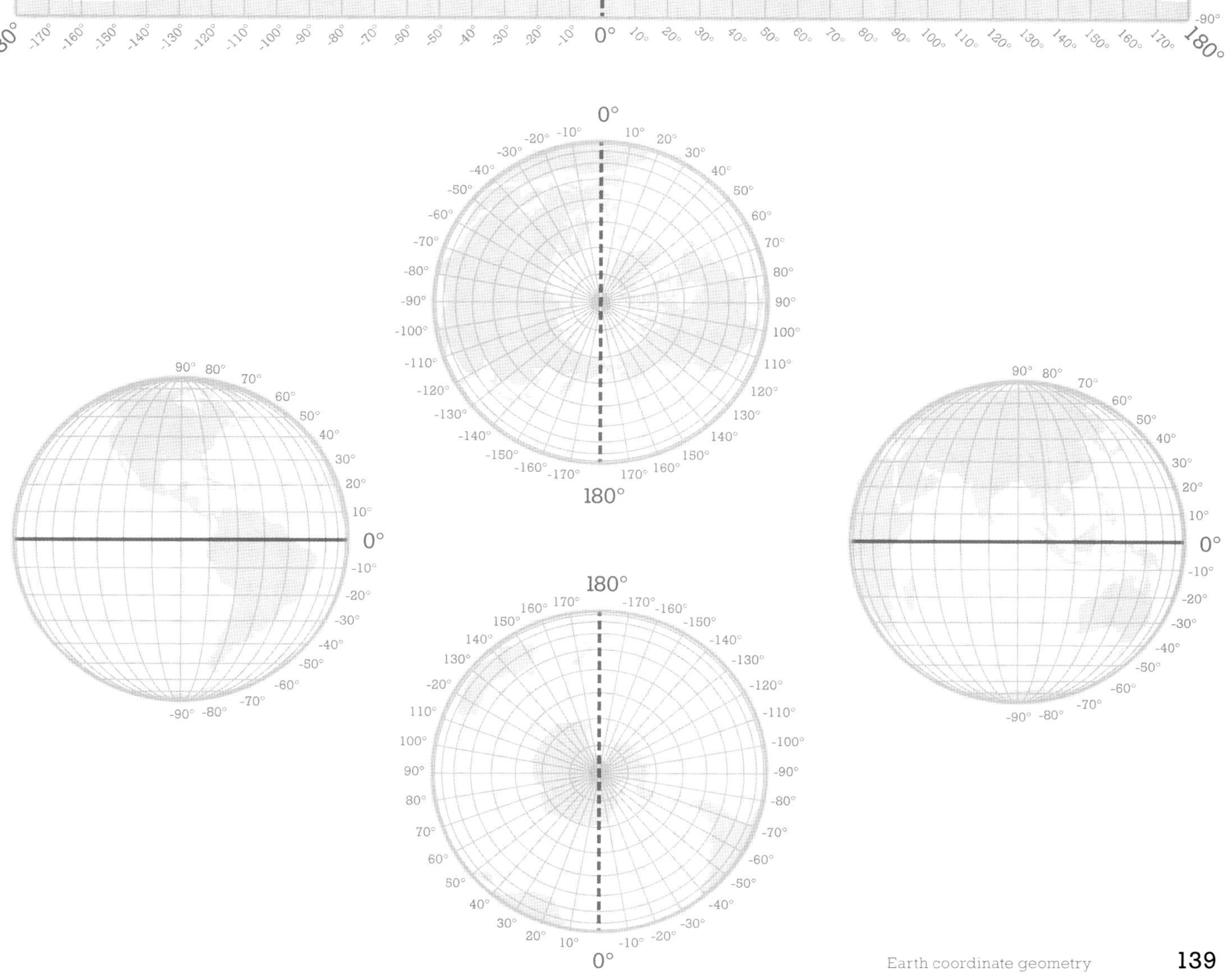

0°
-10° -20° -30° -40° -50° -60° -70° -80° -90° -100° -110° -120° -130° -140° -150° -160° -170°
10° 20° 30° 40° 50° 60° 70° 80° 90° 100° 110° 120° 130° 140° 150° 160° 170°
180°
90° 80° 70° 60° 50° 40° 30° 20° 10°
0°
-10° -20° -30° -40° -50° -60° -70° -80° -90°
90° 80° 70° 60° 50° 40° 30° 20° 10°
0°
-10° -20° -30° -40° -50° -60° -70° -80° -90°
180°
170° 160° 150° 140° 130° -20° 110° 100° 90° 80° 70° 60° 50° 40° 30° 20° 10°
-170° -160° -150° -140° -130° -120° -110° -100° -90° -80° -70° -60° -50° -40° -30° -20° -10°
0°

# Earth's framework

## The fundamental basis of mapping lies in our ability to make measurements of Earth using the framework of a graticule.

**The grid formed by the lines of latitude (parallels) and longitude (meridians) is referred to as the *graticule*, which provides a framework for measuring Earth.** The properties of the graticule relate principally to distances and directions but also have an impact upon area, and knowing how these properties affect measurement is useful before thinking about translating the 3D globe into a 2D surface for mapping.

**The shortest path between two points on the surface of Earth is an arc of a great circle**. A great circle is a plane that intersects the surface of Earth and passes through the centre. The equator and planes formed from traces of meridians form great circles. Great circles divide Earth into two equal hemispheres and can be positioned at any angle that passes through the centre. The equator is the only parallel that forms a great circle. All other parallels form small circles. Other small circles also exist when a plane passes through Earth's curved surface but does not intersect the centre.

**Because of the axis of rotation, it is normal to refer to the direction along a meridian as being north–south and the direction along a parallel as east–west.** One of the properties of the graticule is that, on a spheroid, each parallel and meridian will intersect at 90° since they are perpendicular to one another. The directions formed by the lines of the graticule are referred to as 'geographical' (or 'true') directions. True north is therefore the direction of a meridian of longitude that converges at the geographic North Pole.

**Quadrilaterals formed between pairs of bounding meridians and parallels help define areal measurement.** For identical latitudinal extents between the same two bounding meridians, quadrilateral areas decrease further toward the pole because of the convergence of meridians. On any map that changes the shape of the graticule away from its quadrilateral appearance, a consequential change in shape of any mapped features will also occur. This effect will be more pronounced in higher latitudes because meridians converge more rapidly there.

**See also:** Datums | Earth coordinate geometry | Globes | Measuring direction

Maps may contain a range of alternative orientation information. Magnetic north is determined by taking a compass reading whereby the compass needle aligns itself to magnetic north. This is not the same as true north since magnetic north does not coincide with true north.

Magnetic declination is therefore the angle between magnetic north and true north and can vary over time. Declination is positive when it is east of true north and negative when it is to the west.

Isogonic lines can be constructed along which declination has the same constant value.

The current magnetic north pole (at the time of this book's publication date) is situated at 86.5°N and 172.6°W, northwest of Canada's Sverdrup Island and is moving at a rate of approximately 40 km per year. If the map you are making is for navigation, it is important that it contains up-to-date information on magnetic declination.

*''I'll follow him to the ends of Earth,'' she sobbed. Yes, darling. But Earth doesn't have any ends. Columbus fixed that.'*
—Tom Robbins

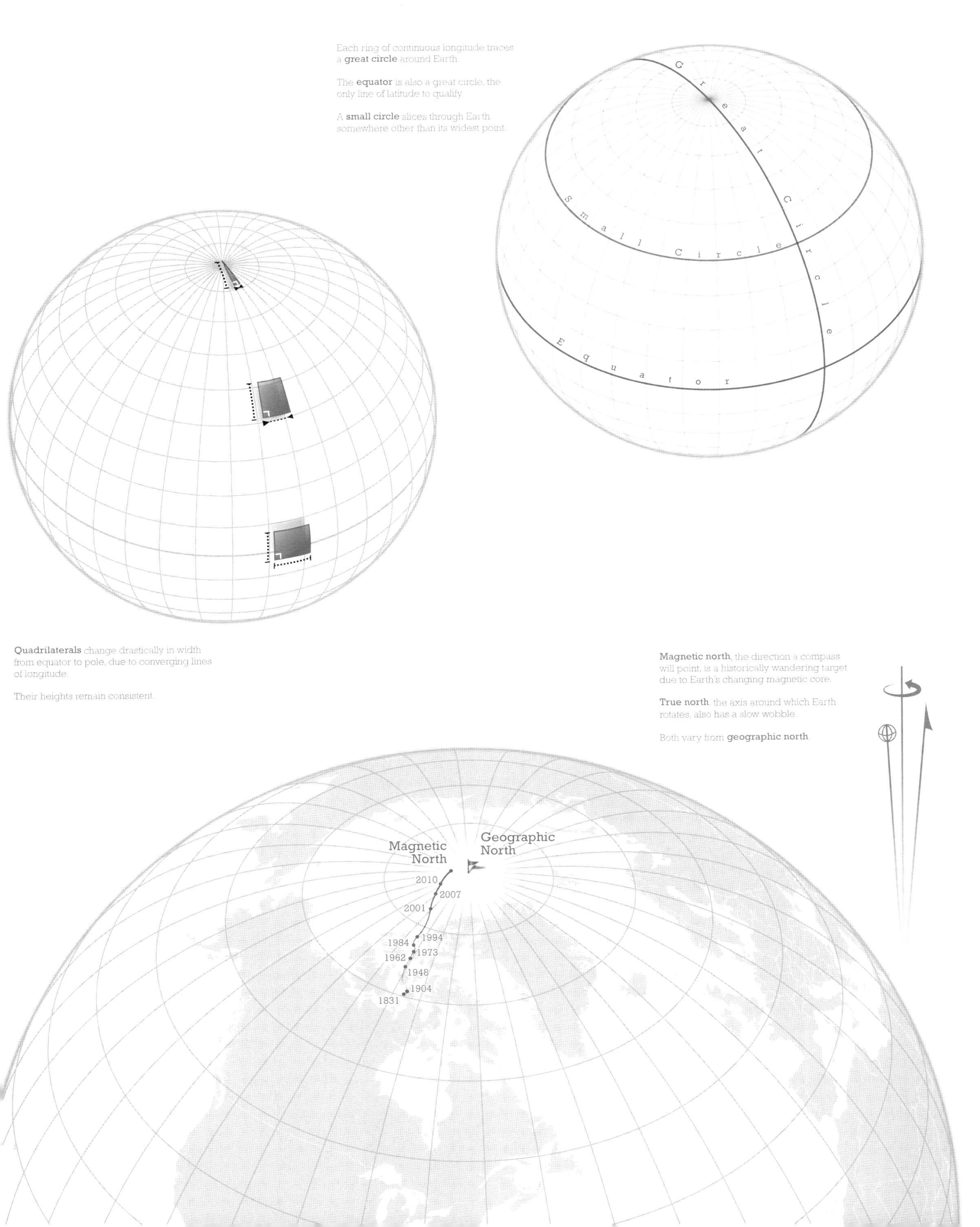
Each ring of continuous longitude traces a **great circle** around Earth.
The **equator** is also a great circle, the only line of latitude to qualify.
A **small circle** slices through Earth somewhere other than its widest point.
Great Circle
Small Circle
Equator
**Quadrilaterals** change drastically in width from equator to pole, due to converging lines of longitude.
Their heights remain consistent.
**Magnetic north**, the direction a compass will point, is a historically wandering target due to Earth's changing magnetic core.
**True north**, the axis around which Earth rotates, also has a slow wobble.
Both vary from **geographic north**
Magnetic North
Geographic North
2010
2007
2001
1994
1984
1973
1962
1948
1904
1831

# Earth's shape

## Earth isn't flat, but neither is it a sphere, and that has consequences for mapmaking.

**The general shape of Earth is ellipsoidal with its minor, or polar, axis being shorter than its major, or equatorial, axis.** The three-dimensional figure is referred to as an *oblate spheroid*. *Oblateness* is the term used to describe the flattening effect and is given by the ratio f = (a-b)/a where a = equatorial radius and b = polar radius. Isaac Newton's first determination of the amount of flattening was 1/300. Modern determinations put flattening at about 1/298.

**Numerous determinations of Earth's radii and extent of flattening have been made on the basis of measuring different locations.** These national surveys create ellipsoids that best fit local areas. Variations between measurements occur because of varying accuracy of the measurements themselves and variations in surface curvature across Earth's surface.

**Earth also has local irregularities.** The figure of Earth is called the *geoid* (meaning earthlike). The geoid does not map neatly onto known highest and lowest points of relief on Earth. In fact, the highest point on the geoid is in New Guinea at 75 m above the ellipsoid while the lowest, at 104 m below, is at the southern tip of India.

**Most reference ellipsoids used for mapping are based on geodetic datums.** They have been derived from an initial starting point somewhere on the surface of Earth which takes account of the geoidal shape. A reference ellipsoid provides a good fit to the surface of the geoid in that particular area. Because Earth is a geoid, the fit of the same reference ellipsoid to other parts of the world may not be as accurate. A large number of reference ellipsoids therefore exist to map different parts of the world accurately.

**Geocentric datums reconcile the centre of a reference ellipsoid to the centre of mass of Earth.** These give a relatively good fit for all parts of the geoidal surface. GRS80 and WGS84 are based on geocentric datums and are effectively a best-fit model that reflects the shape of the geoid.

By the end of the 17th century, the idea of a perfectly spheroidal earth was challenged by the notion that gravitational forces may deform the shape. The English mathematician and astronomer Isaac Newton (1643–1727) developed his theory of gravity based on the principle that Earth rotated around a central axis which created a centrifugal force. Newton postulated that a consequence would be a slight bulging at the equator, offset by a slight flattening at the poles. Newton predicted this flattening effect to be in the order of 1/300th of the equatorial radius. Expeditions to Peru and Finland by the French Academy of Sciences later tested the theory by measuring the ground distance for one degree of angular change north–south between polar and equatorial regions. Polar ground distance was found to be greater—confirming the theory of flattening.

Geodesy is the subset of spatial science concerned with Earth measurement; the determination of precise survey and geodetic measurements. Satellite measurements of Earth from the 1950s to the present show that in addition to the flattening effect at the poles and bulging at the equator, Earth also contains localised depressions and bulges. These irregularities are not visible on the surface; they do not take the form of mountains or valleys but are caused by Earth not being geologically uniform. They occur because of variations in rock density and topographic relief, irregularities of up to 100 m above or below a surface representative of mean sea level (based on the ellipsoidal shape).

**See also:** Earth coordinate geometry | Earth's framework | Globes | x and y

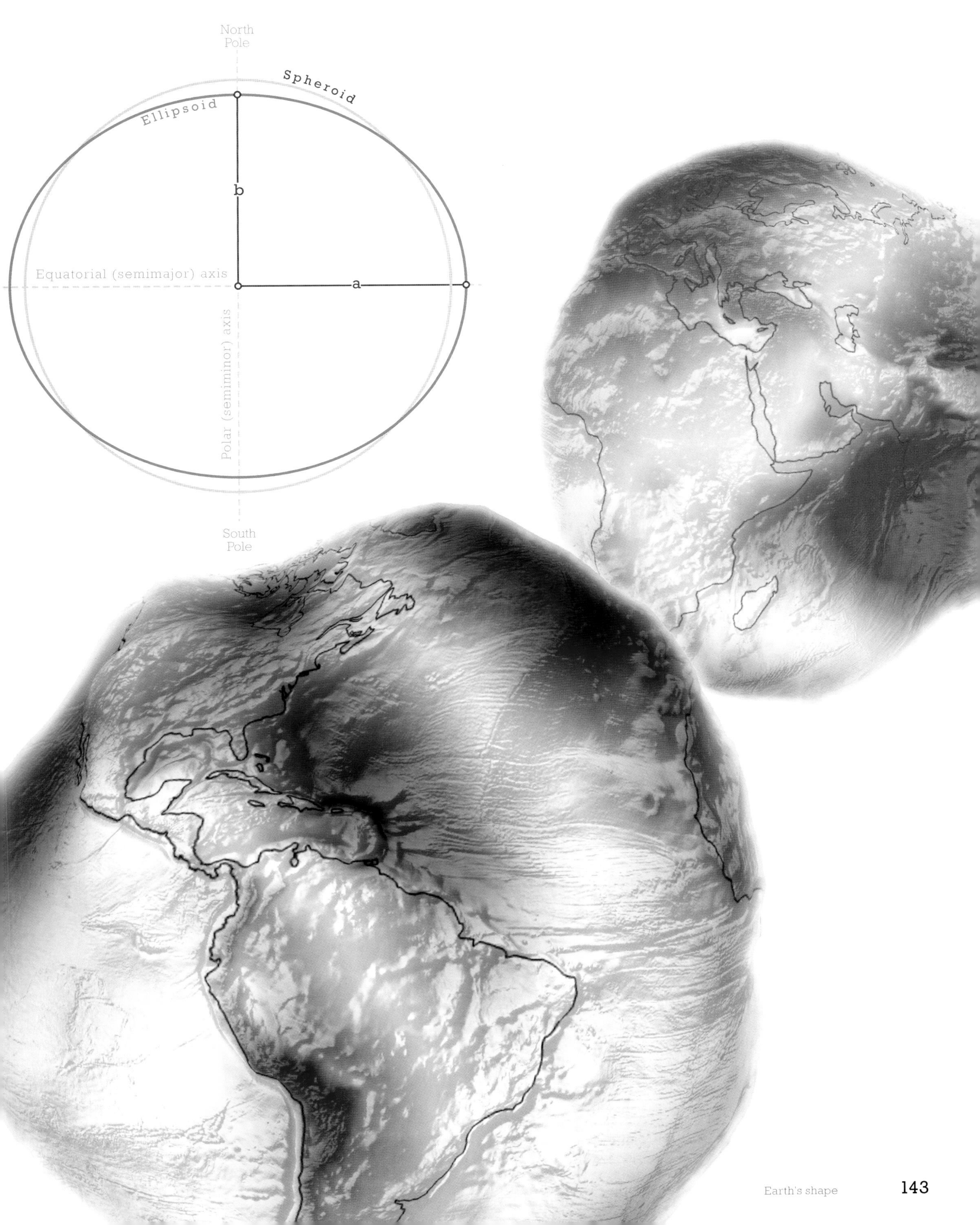
North Pole
Spheroid
Ellipsoid
b
Equatorial (semimajor) axis
a
Polar (semiminor) axis
South Pole

# Earth's vital measurements

## The shape and size of Earth was first measured by Eratosthenes, and was remarkably accurate!

**Eratosthenes (276–194 BC), a Greek scholar living in Alexandria, Egypt, was the first to provide a measurement of the size of Earth.** Since Earth is tilted about its equatorial plane at 23.5° from the sun's equatorial plane, the sun will appear directly overhead in the Northern Hemisphere on the summer solstice. Eratosthenes observed that on this day, the midday sun shone vertically down a well at Syene (near present-day Aswan, southern Egypt). At the same time, the sun cast a shadow at Alexandria, approximately due north of Syene. Eratosthenes measured the length of the shadow from a vertical column of known height and calculated that if the vertical lines at Syene and Alexandria are extended toward the centre of a spherical earth, they will intersect at an angle of 7°12', approximately 1/50th of the circumference of a sphere (360°). To complete his estimation of the size of Earth, Eratosthenes found the distance between the well and column to be 5,000 stades (1 stadia is equal to approximately 192 km). Multiplying the distance between the two locations by 50, he calculated the total circumference of Earth to be 250,000 stadia (46,100 km).

**The modern determination of the circumference of Earth is 40,075 km, which makes Eratosthenes' calculations 15% too large.** There are many errors that, cumulatively, led to over estimation although they will have partially compensated each other. Firstly, the presumption was made that both the well and column were, indeed, perfectly vertical. Eratosthenes also assumed Alexandria was due north of Syene but, in fact, it is not by about 1°. Finally, the actual distance between the two locations is 729 km so the multiplication to determine the final circumference was based on incorrect distance measurement. Given the crude nature of the calculations and, regardless of the errors, Eratosthenes' calculations remain a remarkable mathematical achievement.

The sixth-century BC Greek mathematician and philosopher Pythagoras (582–507 BC) is thought to be the first to propose that humans live on a body of the perfect shape—a perfect sphere.

Aristotle (384–322 BC) provided a more reasoned argument for spherical earth by observing that sailing ships always disappear over a horizon hull first, mast last. If Earth was flat, they would gradually disappear from view as a uniformly decreasing dot.

Eratosthenes might also be considered to be the father of geography. He produced a three-volume work called *Geography*, in which he described and mapped the known world. This rendering included climatic zones and gridlines that were rudimentary lines of latitude and longitude.

The modern calendar might also be attributed to Eratosthenes. During his time at the Library of Alexandria, he devised a calendar on the basis of predictions about the ecliptic of Earth resulting in a calculation that there are 365 days per year. More so, in every fourth year there would be 366 days—the modern leap year.

Eratosthenes also has a deep lunar impact crater named after him which is approximately 59 km in diameter and 3.6 km deep, lying at the western end of the Montes Apenninus mountain range.

**See also:** Datums | Earth coordinate geometry | Earth's shape | Map projections

Column at Alexandria (shadow cast by sun)
7° 12' (1/50th circle)
5,000
stadia
Well at Syene (no shadow)

# Elegance

## Good maps tend to exhibit elegant design.

**Good design is an ethos, an approach, and attainable.** It's not a special sauce that is only known to professional cartographers. Whether you are a practiced cartographer or someone new to making maps, good design can be both hit and miss. We explain good design in nebulous terms, and it's difficult to apply metrics to assess the level of good design a map might exhibit. It deals with the aesthetic quality which is partly subjective. It encompasses the desire for thoroughness and a search for excellence (rather than perfection, which rarely exists). Counterintuitively it also tends to be the case that good design is revealed when there appears to be very little design on show.

**The objective in seeking elegance in design is to build a visual quality to attract an audience to your map.** You also want to sustain their interest and deliver something useful. Imbuing your map with embellishment shouldn't be seen as equating to dressing the map to achieve elegance. It can actually work counter to the very thing you are trying to achieve and undermine the look, trustworthiness, and accessibility of the map. By making your map trustworthy and accessible, you may already have achieved an elegant solution but those aspects are not what people tend to think of when they first look at a map. They are taken in by its appearance, which should ease them into the other qualities you've established in the design.

**Optimising the visual appeal is the hook to get people to look at your map.** The big problem is that what appeals to one person doesn't necessarily appeal to others. Perhaps thinking of your work as being stylish, effortless, or graceful are ways of thinking about elegance. Working to achieve a timeless classic breeds elegance. Conversely, if you are working toward 'trendy' or 'cool', it's likely that the map will be of a time and be more transient, though that's not to say that if the map is designed to live fast and die young then that approach won't work. Ultimately, if the visual appearance appears clumsy, inconsistent, or lacks harmony, then elegance is missing.

**See also:** Clutter | Colouring in | Flourish | Map aesthetics | Style, fashion, and trends

Design thinking can be thought of as a series of conditions that you place on your map. The editorial condition demands that you justify to yourself the value of every mark on the map. There should be nothing on your map that might be superfluous. That's not to say that every mark must contribute to the specific need. It's possible to justify cartographic flourish, but justifying it and why it's on the map establishes the impact of thinking in your map. Being editorially aware is sometimes a challenge. Your investment of time often leads to you wanting to celebrate your work through its visual appeal. This can lead to over design, and being editorially cognisant of when to kill certain ideas actually leads to better work.

Being thorough means you condition yourself to slave over every detail of your map. Everything has to be resolved. Being thorough gives your map authority and trustworthiness. It supports elegant design. Checking and checking again are important processes. Check your data, your sources, the map's spelling, and then check again. It's important background work that pays off.

Elegance is often influenced by the condition of style. Style can be formed through developing or using style guides associated with a client or organisation that places constraints on your design. Or you may have your own style, developed through consistent use of colours, typography, and composition. Even with one-off maps, style should be conscious. You are trying to imbue your map with efficiency and consistency that are supported by a stylish treatment. Embellishments can be important in achieving visual appeal but there's a fine line that can easily be crossed which can lead to distraction. The condition of invisibility of design is worth bearing in mind. The map reader should not see design, they should see content.

**Opposite:** *Pitch Perfect* by Kenneth Field, 2016, which won several map design awards..

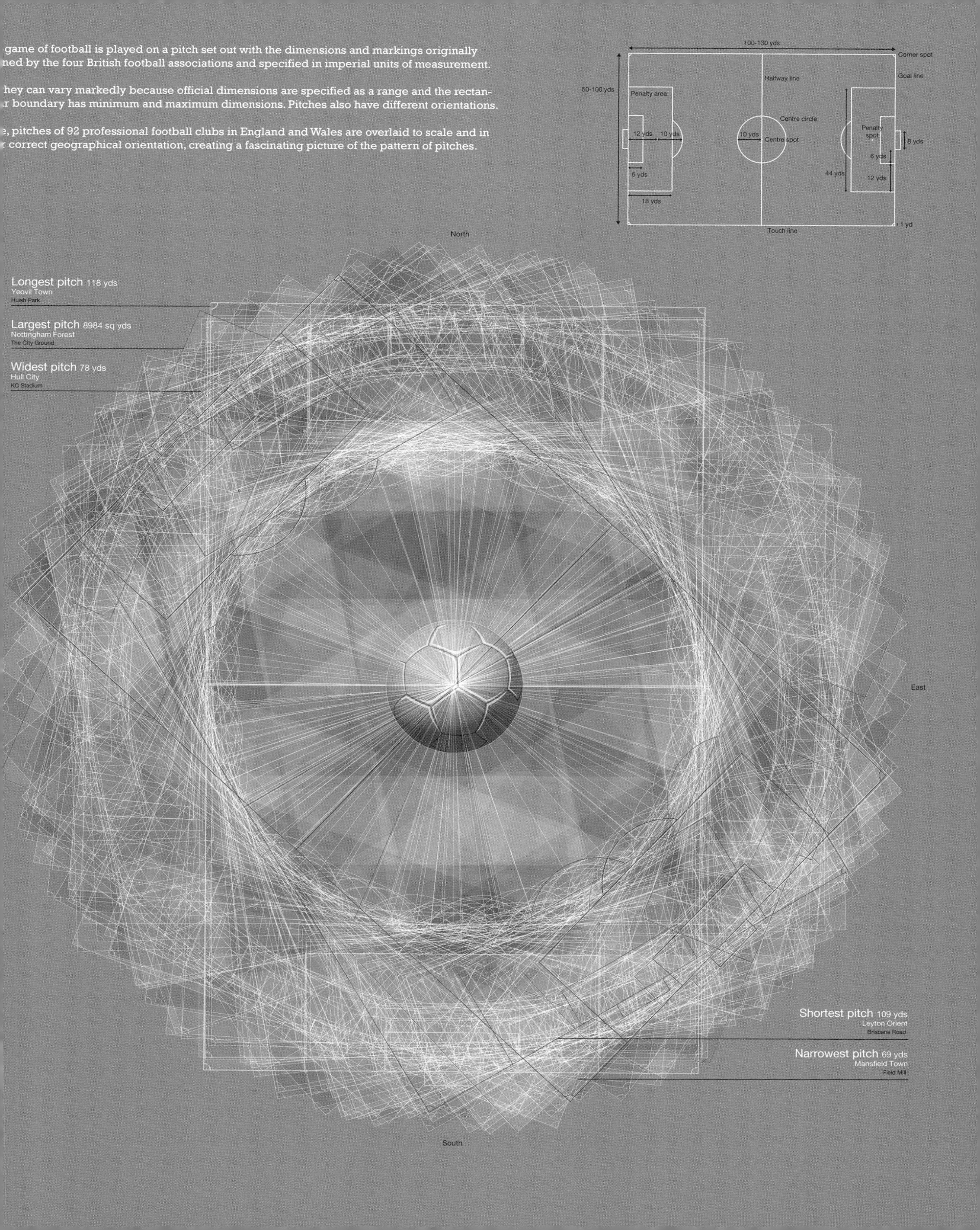

game of football is played on a pitch set out with the dimensions and markings originally
ned by the four British football associations and specified in imperial units of measurement.
hey can vary markedly because official dimensions are specified as a range and the rectan-
r boundary has minimum and maximum dimensions. Pitches also have different orientations.
, pitches of 92 professional football clubs in England and Wales are overlaid to scale and in
correct geographical orientation, creating a fascinating picture of the pattern of pitches.
100-130 yds
Corner spot
Goal line
Halfway line
50-100 yds
Penalty area
Centre circle
12 yds
10 yds
10 yds
Centre spot
Penalty spot
8 yds
6 yds
6 yds
44 yds
12 yds
18 yds
1 yd
Touch line
North
Longest pitch 118 yds
Yeovil Town
Huish Park
Largest pitch 8984 sq yds
Nottingham Forest
The City Ground
Widest pitch 78 yds
Hull City
KC Stadium
East
Shortest pitch 109 yds
Leyton Orient
Brisbane Road
Narrowest pitch 69 yds
Mansfield Town
Field Mill
South

# Elements of colour

## Colour is a combination of three perceptual dimensions that widen the vocabulary.

**Although most people describe colour using the term 'colour', in cartography that description is limited.** Colour can be broken down into three perceptual dimensions that support flexibility in map design. The basic vocabulary used to describe the components of colour is as follows: hue: the dominant wavelength of a perceived colour that we associate with names such as red and blue; lightness: the extent to which a hue varies relative to white (also called 'value'); and saturation: the extent of vividness of a hue (also called 'chroma').

**The three components of colour can be altered to provide a range of ways of applying colour to map symbolisation.** The typical rainbow spectrum places dominant wavelengths of the electromagnetic spectrum into ordered hues and can be laid out in a row or, more conveniently for cartography, a circle which positions RGB and CMY primaries in spectral order. For any hue, lightness can be varied to provide light and dark alternatives (e.g. light blue to dark blue). Holding lightness constant and changing hue provides perceptually equal tints of hues. Lightness can also be used to show ranked data with light hues showing low data values and vice versa. This perception would be reversed if a map employs a dark background when lighter tints are seen in reverse, as higher data values. Changing lightness across a range of colours can include a hue change (e.g. yellow to green) or where hue is constant (e.g. blue). Lightness is relative and is the value that is reflected (or emitted) from the map. In lighter viewing conditions, all hues will appear lighter. Changing the saturation of a hue to make it more or less vivid is not particularly useful in itself but can be used for emphasis. As a hue becomes desaturated, it becomes neutral and grey. Some colours are achromatic such as white, grey, and black since they have no saturation and no hue. Some hues are naturally more saturated in appearance than others and may need modifying when presented as part of a set of colours where one shouldn't be seen as dominant over the others.

**See also:** Additive and subtractive colour | Constraints on map colours | Greyscale | Hue | Mixing colours | Perceptual colour spaces | Saturation

No two people will necessarily perceive or describe colour in the same way. One person's description of blue may be based on the perception of a slightly different colour from the blue that another person recognises. In map design this means you normally prepare maps based on an assumption that the map user has average colour perception.

There are only 11 basic terms that describe colour, namely, black, white, grey, red, green, yellow, blue, brown, purple, orange, and pink. Developed in the 1950s, the American National Bureau of Standards (NBS) built on these 11 terms to develop a schema for describing colour. Descriptors are arranged in two dimensions which correspond to the lightness and saturation of hue. Hues go from grey to vivid in the horizontal direction and from dark to light in the vertical direction. Hues thus form a circle around a central neutral axis of black–white. Although the NBS system doesn't provide cartography with a set of colour specifications that can be replicated consistently, it does provide a language for describing differences between relative tones. For precise colours you still have to go beyond description to specifications.

Being able to define colour with precision is important in order to control colour from input, through graphic display to (potentially) final printed output. The principle is simple—in map design you select a colour (for example, from a colour chart) and then use that colour throughout the entire design process to ensure a consistent appearance in the finished map whether designed for screen or print.

**Opposite:** *Election Pollocks* by Kenneth Field, 2015, uses a particularly colourful palette representing the main political party affiliations. The style mimics a Jackson Pollock aesthetic to connote the messy state of the political circumstances of the time, complete with a paint-splatted border.

Orkney & Shetland
UK General Election, 2015
Paint drip symbols are sized in proportion to the number of votes for each candidate.
Their abstract character gives an overall expression of the relative mix and distribution of votes.
votes
40,000
20,000
10,000
Conservative Party
Labour Party
UK Independence Party
Liberal Democrat Party
Scottish National Party
Green Party
Independent / Other
Sinn Fein
Democratic Unionist Party
Plaid Cymru
Social Democratic & Labour Party
Ulster Unionist Party
Alliance Party

# Elements of type

**Regardless of typeface, each has a set of common elements that help us describe and understand their impact on design.**

**All letters are initiated from a baseline.** The height of the body of lowercase letters is the x-height, which is often referred to when considering readability of different typefaces. Strokes of an individual letter that rise above the x-height are called *ascenders*, with all ascenders in a lowercase alphabet terminating at a common height, the ascender line. Likewise, letters with strokes that fall below the x-height have descenders which terminate at the descender line. Ordinarily, capital letters in a typeface are shorter than ascenders and terminate at a cap line.

**Type families are often referred to as *serif* or *sans serif* (without serifs).** Serifs are small strokes added to the end of main strokes of the letterform. Serifs are not part of the design of every typeface although they are often considered easier to read since the serif allows the eye to flow from one letter to another. Serifs have become less common in modern cartography since sans serif typefaces tend to be considered cleaner and less ornate on a map display. However, pairing serif and sans serif typefaces often works well. The design elements of typefaces can be varied in many ways to create an almost infinite number of different designs.

**Choosing a typeface for a map can be bewildering.** This difficulty is partly because of the array of choice but also because lettering a map is a totally different process from selecting a typeface for a book or passage of text. Type that is set on a uniform background (e.g. white paper in a book) does not have to contend with conflicting backgrounds. Similarly, text is set solid (i.e. contains no letterspacing) and the leading (spacing between lines of type) is often uniform. Conversely, on a map, letters are often spread out across different backgrounds, can be oriented in many ways, might need different leading, and are often interrupted by a range of other marks and symbols. The implications for choosing a typeface are fundamentally that individual letters should be easily identifiable.

Individual letters contain a range of design characteristics. Counters are partially or completely enclosed areas of a letter. Bowls are the rounded elements of letters such as o, b, and g. The lower part of some lowercase letters with descenders are called *loops*. However, the loops of lowercase letters are generally not designed to join yet reproduction of maps at smaller scales often closes this gap which reduces its readability. In map design terms, typefaces that contain loops should be avoided for this reason with preference given to typefaces that are open.

This book is typeset in Rockwell—a slab serif typeface. It's a form of serif that is characterised by thick, angular block like serifs. It was designed by Frank Hinman Pierpont of the Monotype Corporation in 1934. I chose it because I like it, and it works well for relatively short passages of text. I'm under no illusion that the choice of Rockwell will not be to everyone's taste and that in itself demonstrates some of the problems of choosing a typeface for any project. Although you may choose a typeface on the basis of on a rational set of guidelines, ultimately there's a subjective element at play. And that subjectivity might not be matched by your map reader.

Style may also be defined by an organisation so you may have little choice in selecting a typeface. For instance, *National Geographic* is renowned for its distinctive typefaces. In fact, it uses a custom typeface developed by former Society cartographer Charles E. Riddiford in the 1930s on the basis of former hand lettering styles. Digital versions of Riddiford's typefaces are still in use today, providing not only a distinctive house style but an attractive and legible font to reflect and accentuate map features.

**See also:** Choosing type | Fonts and type families | Lettering | Lettering in 3D | Symbols

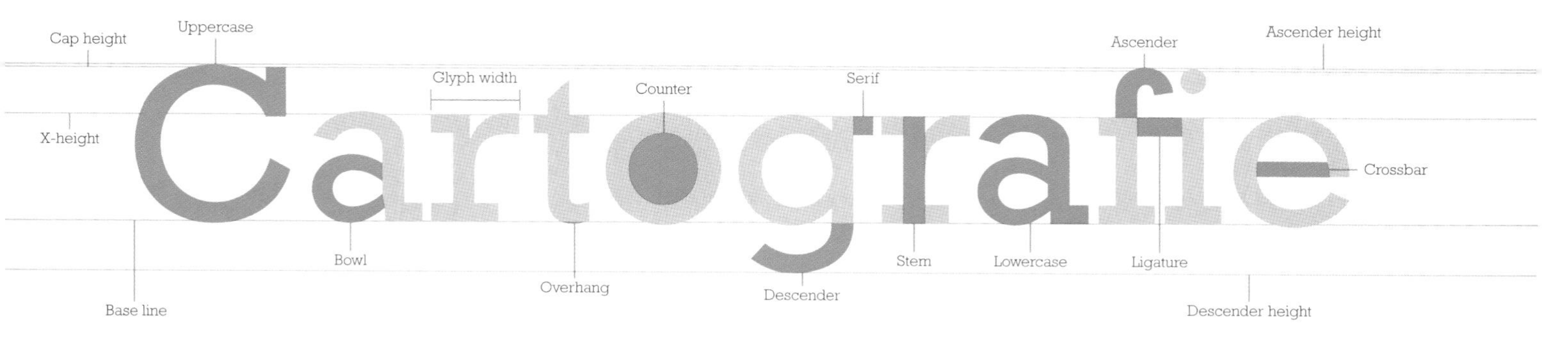
Cap height
Uppercase
Glyph width
Counter
Serif
Ascender
Ascender height
X-height
Cartografie
Crossbar
Bowl
Overhang
Descender
Stem
Lowercase
Ligature
Base line
Descender height

# Emotional response

## Evoking meaning is often implicit in cartography. It can also be explicit. The emotional response of the reader is part of the design process.

**The pattern of components on a map is designed to work as a whole to give meaning to the subject matter and communicate information to the reader.** The relationships among the various marks are complex, and they can be used support each other to provoke an emotional response in the reader. This emotional response may be passive and simply the response of understanding and interpreting the map. On the other hand, it can be active, and the map can provoke a full range of feelings.

**Clear design is always an objective but we can also imbue maps with powerful imagery.** The intent here might be to make a point more forcibly or to give rise to a particular emotional response such as fear, anger, joy, or compassion. In just the same way as a picture is worth 1,000 words, the picture itself can be made to relate to the reader on many different levels.

**At their very obvious, propaganda maps have long been used to bring a sense of right and wrong in warfare.** The graphic elements of propaganda maps are designed to support a very specific response. Certain aspects are exaggerated. Colour is used in a particular way. Perhaps less dramatic, persuasive cartography seeks to be more subtle yet with the same objective. We may see this in certain types of maps used in advertising or journalism where left- or right-leaning political points are being expressed visually.

**Even simple design aspects such as using colour or a type style to bring a particular emotional response can be used to great effect.** Red brings out a particular emotional response. Symbols associated with good or bad can be used effectively alongside or as part of the map design. What humans know, see, or experience are all aspects of reality that we can riff off when we design maps. Simple and direct cartography is often a good basic design tenet but adding embellishments that hook into people's views of the world can raise their effectiveness and make them difficult to forget.

See also: Advertising maps | Cognitive biases | Design and response | Ethics | Flourish | Informing | Sensory maps | Your map is wrong!

There's usually a greater degree of freedom in experimenting with emotional response when making one-off maps. These might be produced individually or perhaps for a newspaper where editorial spin often leads to persuasive maps. The map might be designed to be explicit in its use of imagery and be deliberately suggestive. Many war period maps set about being very deliberate and designed to elicit a specific response.

Designing to garner emotional response can be much more subtle. For instance, a map showing the distribution of Irish surnames from a routine census might be considered a fairly bland topic and of interest perhaps to a relatively small niche group interested in genealogy. But, by using a wide range of Irish themed imagery and stereotypical graphical devices, the map becomes something different. When displayed for a predominantly American audience, it deliberately sets out to work in a particular way. It captures the imagination, speaks to Irish ancestry, and provides a vehicle for discovery. It feeds the questions 'Who am I?' and 'Where do I come from?'

**Opposite:** *Geo-Genealogy of Irish Surnames* by Kenneth Field and Linda Beale, 2009.
**Above:** As displayed in a gallery of maps.

# Geo-Genealogy of Irish Surnames

Gallagher
McLaughlin
Ward
Boyle
Doherty
O'Donnell
McGinley
Sweeney
Campbell
MacGowan
Duffy
Doherty
McLaughlin
Bradley
MacCloskey
Kelly
Mullen
O'Neill
Stewart
Miller
Moore
Johnston
Bell
Kerr
O'Neill
Wilson
Reid
Thompson
Boyd
Hamilton
Graham
Martin
Robinson
Smith
Brown
Hughes
Kelly
Hamilton
Brown
Moore
Smith
Mullan
Wilson
Campbell
Quinn
O'Neill
McKenna
Gallagher
Donnelly
Doherty
Kelly
Murphy
MacCann
Patterson
Murray
Scott
Martin
Graham
Campbell
Thompson
Hamilton
Boyd
Bell
Scott
Wilson
Robinson
Johnston
Brown
Byrne
Joyce
Gibbons
Connor
Burke
O'Malley
Durkan
Lyons
Walsh
Sweeney
Leonard
Reilly
Reynolds
Maguire
Gilmartin
Brennan
Durkan
Scanlon
Connolly
Flynn
Feeny
McGowan
Mulvey
Brady
McLoughlin
Wilson
Dolan
Cassidy
MacManus
MacMahon
Johnston
Duffy
Donnelly
Quinn
Woods
MacKenna
Hughes
O'Hare
Clarke
Sheridan
Reilly
Smith
MacCabe
Connolly
Gallagher
Harte
O'Hara
McLoughlin
Flanagan
McDermott
Farrelly
Lynch
Fitzpatrick
Dolan
Reilly
Murphy
Higgins
Kelly
Moran
Doherty
Gallagher
MacHale
Connor
Regan
McHugh
Conway
Flynn
Quinn
Beirne
Kelly
Farrell
Brady
Smith
Quinn
MacDermott
Murphy
Reilly
Kiernan
Donohoe
Farrell
Reilly
Lynch
Connor
Kelly
Smith
Maguire
Farrell
Reilly
Duffy
McDonagh
Mannion
Forde
Brennan
O'Toole
McDonagh
Joyce
Connor
Murphy
Lyons
Flaherty
Conealy
Connolly
Burke
Walsh
Kenny
Fahy
Daly
Molloy
Murray
Egan
Moore
Doyle
Nolan
Farrell
Cullen
O'Brien
Walsh
Whelan
Murray
Smith
O'Neill
Kavanagh
Keogh
Kelly
Dunne
Byrne
Kennedy
Carroll
Brady
Kelly
Griffin
MacInerney
MacNamara
Clancy
Ryan
O'Halloran
MacMahon
King
Egan
Keane
Kenny
Dunne
Whelan
Lalor
Murphy
Kelly
Connor
Ryan
Mullin
Keane
Moloney
Lynch
O'Brien
Maher
Mooney
Dunne
Fitzpatrick
Delaney
Kennedy
Byrne
Farrell
Doyle
Kavanagh
Brennan
Murphy
Phelan
Lonergan
Hogan
Kelly
Carroll
Shea
Dowling
Delaney
Purcell
Power
Maher
Kehoe
Doyle
O'Brien
O'Connell
McNamara
Dwyer
McGrath
Ryan
Kelly
Nolan
Cullen
Murphy
Ryan
Byrne
Collins
Ahern
Lynch
Murphy
Fitzgerald
Moloney
O'Connor
O'Donnell
Ryan
Hayes
Gleeson
Hayes
O'Brien
Walsh
Kennedy
Morrissey
Walsh
Brennan
Kelly
Whelan
Doyle
Roche
Bolger
Leary
Daly
Foley
Fitzgerald
Moriarity
Walsh
Griffin
Phelan
Sullivan
Murphy
Butler
Brien
Nolan
Walsh
Kavanagh
Connor
Redmond
O'Donovan
O'Crowley
O'Connor
O'Leary
O'Cronin
MacCarthy
O'Callaghan
Power
Whelan
Kelly
Murphy
Walsh
O'Brien
Butler
MacGrath
Flynn
Shea
Brosnan
Lynch
O'Connell
Sullivan
Clifford
Cronin
Mahoney
O'Sullivan
Buckley
Ryan
Foley
O'Hurley
Connell
Sheehan
Barry
O'Riordan
O'Brien
Collins
Connor
MacCarthy
Murphy
Fitzgerald
O'Daly
Walsh
Harrington
Buckley
Murphy
O'Kelleher
Regan
O'Neill
O'Driscoll
O'Keeffe
Walsh
Mahoney

# Error and bias

## All maps are authoritative and accurate, sometimes.

**Maps are regarded as being definitive, accurate portrayals of information.** That's the power they hold as people examine the lines, colours, wording, and symbology. They develop an impression of trust that is error-free, and this breeds confidence, sometimes over confidence in the efficacy of the product. All maps have errors, biases, and uncertainties. Many of these can be related to data accuracy and precision and the way in which these are either dealt with or propagated through the map. Other errors may exist. Biases may be introduced. These may often be unintentional.

**Uncertainty can be a function of the data or the way it is represented.** It is the degree to which the way something is mapped varies from its true value. This might be because of measurement errors, the processing and manipulation of the data, or the way in which the cartographer interprets and then represents the data. Error provides a way of measuring uncertainty in a map. For instance, we can quantify positional errors or determine the extent of attribute errors that might be missing or invalid. Errors may be unintentional, random, and difficult to spot though an understanding of potential error can provide a way to assess the extent to which a map can be trusted.

**Bias is an extension of error and reflects a systematic distortion that is introduced into the map intentionally or otherwise.** It can be introduced through misuse of data that results in a specific error being propagated across the map, affecting all similar features. It might also be added in more nefarious ways to influence the way in which a particular feature might be represented or a message framed.

**Good editing helps to eliminate a large proportion of error.** Checking sources, and checking and proofreading the final map remain vital tools. As with most creative works, it's easy to become so close to your own work that you find it difficult to see problems. Have other people look at your map and pass on comment. Often, an obvious error can be easily spotted by fresh eyes.

**See also:** Cognitive biases | Data accuracy and precision | Ethics | Integrity | Maps kill

Positionally, maps show location yet the precision of location affects the accuracy of the map. Generalised locations cannot be used for maps that are at a larger scale that the data collection scale supports. Positional accuracy also relates to the vertical aspect in the same way. Attribute measurements can also vary in accuracy and precision and similar thresholds of error tolerance defined.

Conceptual accuracy is more of a process than it is because of data. It reflects what is included or omitted, how it is represented, and whether the mediation of the cartographer has been objective. It reflects the level of knowledge the cartographer has about the subject and the mapping techniques used. Factual errors can exist, and sometimes the data may have inaccuracies that compromise the work. Data isn't always necessarily what it seems, and you may unintentionally pass such errors onto the map. Processing errors can often be introduced unintentionally through simply making a mistake. Being aware of the way in which data is processed can mitigate. For instance, simple rounding errors can cause havoc in statistical manipulation of data. Or geographies can become modified through over aggressive generalisation.

**Opposite:** *Locals and Tourists (London)* by Eric Fischer, 2010. The map claims to show 'incredible new detail', 'demographic, cultural, and social patterns down to city level'. The authors suggest it allows you to 'explore stories of space, language, and access to technology'. The data is derived from geotagged tweets. Blue for locals. Red for tourists. Yellow for either (unknown). Geotagged tweets account for around 1% of all tweets. Some 13% are imprecise. Only 67% of all internet users use social media. People who live in cities spend more time on social media. Only 16% of those who use social media use Twitter, and they are most likely to be adults aged 18-29, and male. The map has the appearance of detail and accuracy yet contains all the error, bias and uncertainty of the data.

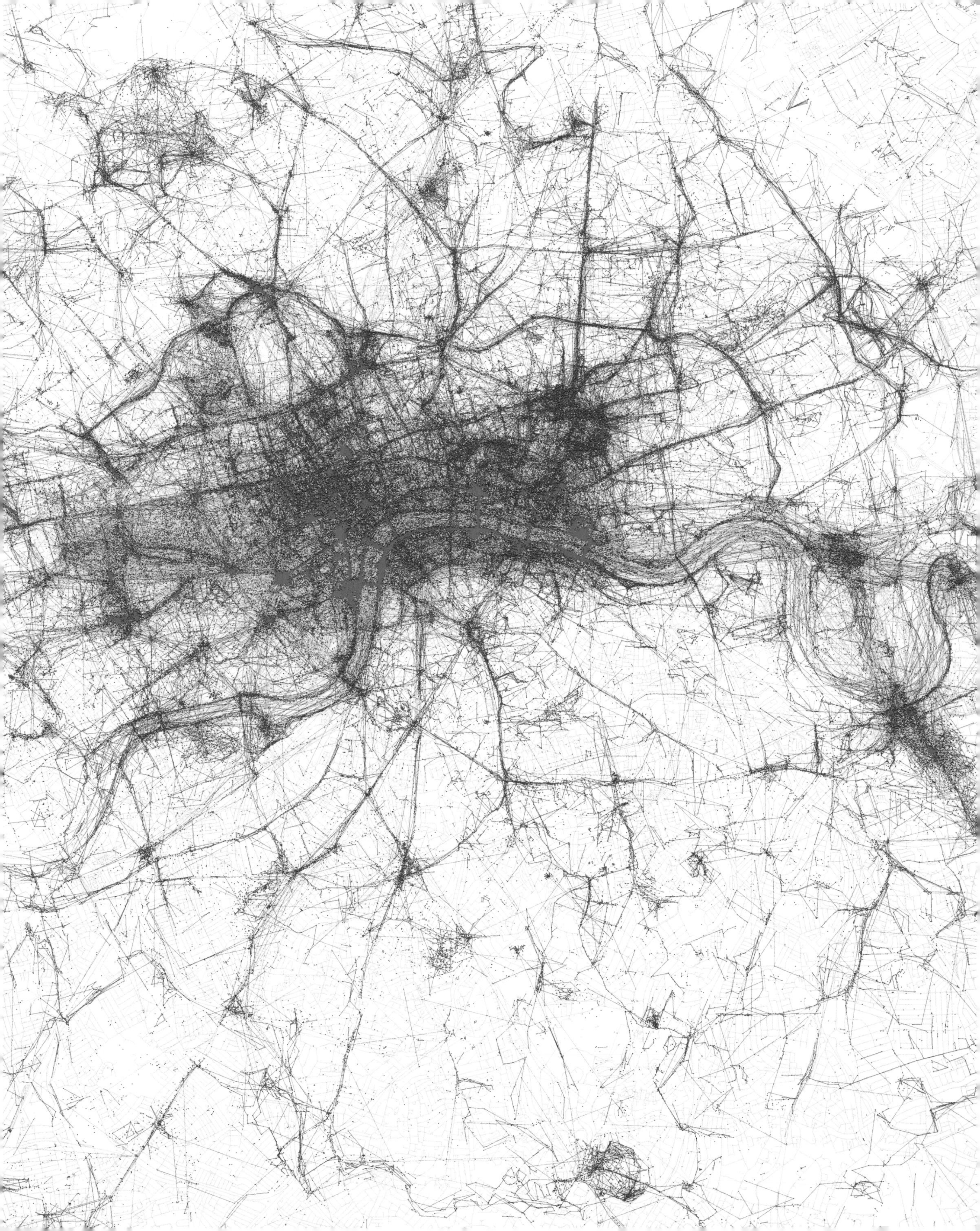

# Ethics

## All maps lie. Understanding how to remain objective is a vital aspect of cartography.

**All too often maps can be regarded as ineffective and poor because they may fail in many ways.** They may be disingenuous about the content; use misguided construction or techniques; or they may even be deliberately designed to be persuasive or propagandist. The ability to accidentally or deliberately make a map in ways that might misconstrue meaning raises questions of ethics or morals in cartography.

**Maps are made by humans and are therefore inevitably going to contain errors.** Errors may result from oversight, poor judgement, or a reliance on software defaults but, sometimes, as a consequence of deliberate action. Mapping is not an exact science; there is no single correct way to map and neither is there a wholly incorrect way (though some techniques can be applied incorrectly). Maps can misrepresent even when the motive is to work toward a map that minimizes the potential for misunderstanding. Here, I've modified Borden Dent's code of ethics for mapmakers:

- have a straightforward agenda, and purpose;
- strive to know your audience (the map reader);
- do not intentionally lie with data;
- show all relevant data whenever possible;
- don't discard data because it might be contrary;
- strive for accurate portrayal of the data;
- avoid plagiarising; report all data sources;
- ensure symbols don't bias the interpretation of the map;
- the map should be able to be repeated by others;
- be attentive to differing cultural values and principles; and
- don't let defaults drive your design. (Dent *et al.* 2008)

**Computers can compound the problems associated with these ethical ideals.** Most mapmaking software easily supports basic mapmaking tasks for anyone to make maps. This is undoubtedly empowering and inclusive but it can also create issues. Software will, on the whole, provide opportunities but are they appropriate to the original data or the analysis that was intended? The purpose and value of the map should drive the use of the software.

**See also:** Knowledge and conviction | Copyright | Integrity | Map traps | Who is cartography?

*'The purpose of this book is to promote a healthy scepticism about maps, not to foster either cynicism or deliberate dishonesty. In showing how to lie with maps, I want to make readers aware that maps, like speeches and paintings, are authored collections of information and also are subject to distortions arising from ignorance, greed, ideological blindness, or malice.'*
—Mark Monmonier, *How to Lie with Maps* (1991)

Consider the impact of a decision to award planning consent for a building that is the subject of massive public objection if part of the decision was based on erroneous mapping. What would happen if a particular road has been omitted from a street map that means an ambulance takes a more circuitous route to the hospital carrying a gravely ill patient? How would people react if financial resources had been misallocated to schools based on the mapping of incorrect information related to population and demographic profiles? These are examples of how mapping can have very real consequences because, ultimately, people make decisions based on what they see on the map. Because maps are so often relied upon and their content rarely questioned, there is considerable onus on the mapmaker to ensure the content and depiction is accurate, objective, and that the message is appropriately communicated.

**Opposite:** A photograph taken by journalist Trey Yingst, 2017, showing the results of the 2016 presidential election in the United States. The map is not wrong but the way in which the results are portrayed is misleading. It maps geography and uses saturated red to show those states that had a large share of the Republican vote. What it fails to show is that many of those same states are sparsely populated and contribute only a small proportion to the Electoral College system that the US election is based upon. It's a map that speaks to the Republican desire to portray the result in a particular way. There are many more truthful alternatives.

Trey Yingst
@TreyYingst
Follow
Spotted: A map to be hung somewhere in the West Wing
2016 Presidential Election Results
3:03 PM - 11 May 2017
4,277 Retweets 8,704 Likes
1.5K
4.3K
8.7K

# Eyeball data classification

## Classification schemes that use the data to inform the class interval.

**Looking at the range and distribution of data helps make a good decision about how to classify it for thematic mapping.** Considering that one of the main aims of a map is to represent data in a way that shows its character, understanding the distribution helps its display. The classification scheme can deliberately group related values and be used to avoid or highlight outliers.

**A natural breaks technique considers the natural groupings of data in the array.** A frequency histogram or dispersion graph can be examined visually to determine logical breaks so that naturally occurring clusters (i.e. values that share similar characteristics) are identified. The purpose of this scheme is to minimise the difference between values in the same class and maximise the differences between classes. This method of classification is subjective and should not be confused with optimal methods that perform a similar task computationally. A modification of natural breaks can be used to remove the gaps between classes.

**A range of optimal computational methods are designed to provide a solution to the limitations posed by other logical or manual methods.** Optimal methods rely on computational techniques to examine the data and place similar data values in the same class by minimising some objective measure of classification error. One of the more commonly used techniques is called *Jenks optimization*, named after George Jenks and Fred Caspall's work in the early 1970s.

**Jenks' technique forms groups within the data array that are internally homogeneous while maintaining heterogeneity among classes.** The classification is based on calculation of a measure called the Goodness of Variance Fit (GVF) such that different options can be calculated and the scheme with the highest GVF, scored between 0 and 1, is determined the optimum. Jenks' technique is commonly used as a computational form of the natural breaks technique.

The main advantage of natural breaks is that it forces visual inspection of the dataset and manual determination of the class intervals. It should follow that it illustrates the data effectively since it is entirely based on identifying clustered values that share similar characteristics. However, because it is entirely subjective, the choice of class intervals might vary between different people. This is inevitable, but if you wish to ensure that your classification scheme is defined by a human rather than an automated process, then natural breaks is the way to go.

The advantage of using an optimal solution for determining classification schemes is that it considers, in detail, how data is distributed. It provides a means of determining the 'best' solution for classifying like values together. It also allows some consideration of the optimum number of classes if we compute the GVF for various schemes with different numbers of classes.

**See also:** Arbitrary data classification | Data classification | Statistical data classification

## Natural Breaks (Jenks)

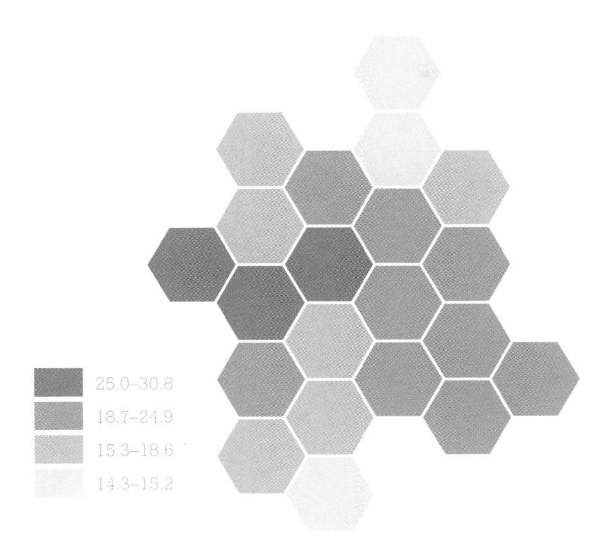


For this data, Natural Breaks classifies the majority of data values to the midvalues and the top and bottom three values are effectively outliers in distinct groups.

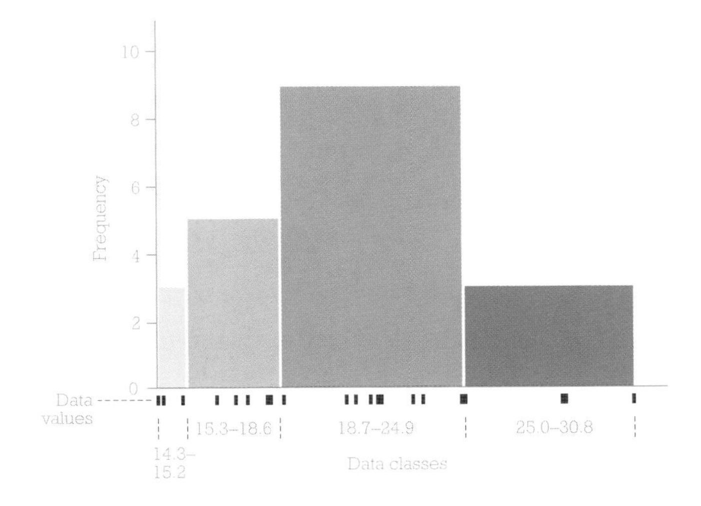


Irregular class widths
Irregular number of data items per class.

## Diagram of the Causes of Mortality in the Army in the East

**Florence Nightingale**

**1858**

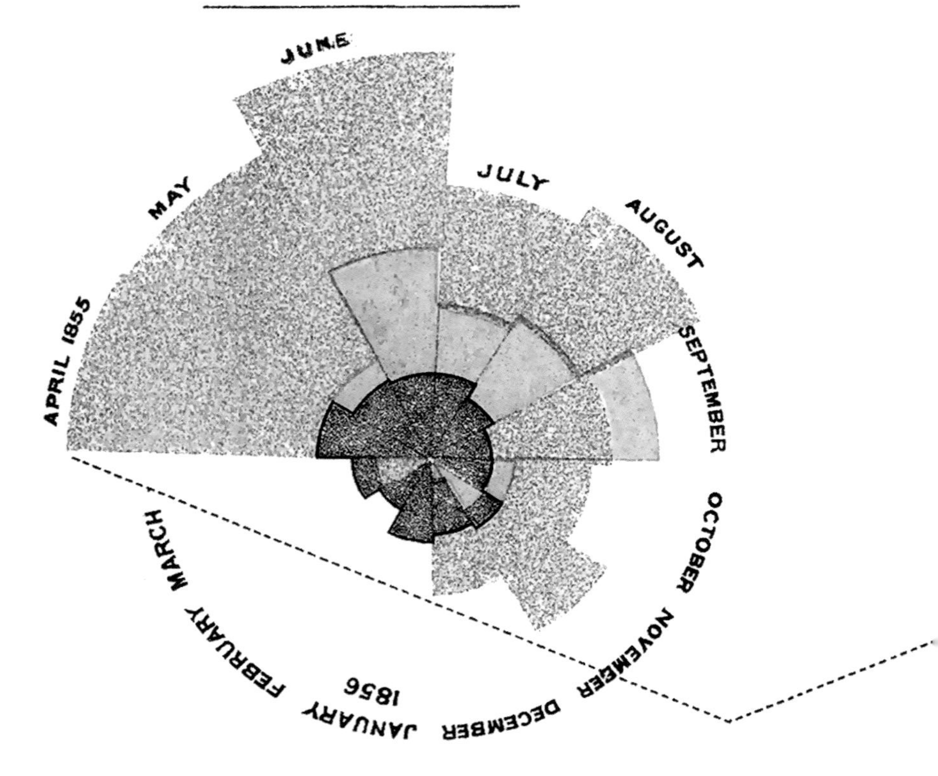


*The Areas of the blue, red, & black wedges are each m*
*the centre as the common vertex.*
*The blue wedges measured from the centre of the circle r*
*for area the deaths from Preventible or Mitigable Zym*
*red wedges measured from the centre the deaths from*
*black wedges measured from the centre the deaths fro*
*The black line across the red triangle in Nov.r 1854 mark*
*of the deaths from all other causes during the month.*
*In October 1854, & April 1855, the black area coincides w*
*in January & February 1856, the blue coincides with*
*The entire areas may be compared by following the blu*
*black lines enclosing them.*

Florence Nightingale was a pioneer in establishing the importance of sanitation in hospitals through her work as a nurse in the Crimean War in the mid-1800s. She gathered data of death tolls and approached the job of communicating this information through applied statistics and graphical communication. She is particularly famous for the development and use of certain graphs, now commonly known as 'coxcomb charts' (otherwise known as 'polar area plots' or 'rose diagrams').

Nightingale realised that soldiers were dying needlessly from malnutrition, poor sanitation, and lack of activity. She strove to improve living conditions for the wounded troops, and kept meticulous records of the death toll in the hospitals as evidence of the importance of patient welfare. In creating the graphs, each sector represents a unique time period and a quantity. Because the sectors all have the same angle, they are visually equivalent when used to represent area. The diagrams overlay three variables for comparison and illustrate the temporal component perfectly by representing each month as a separate segment.

Map or graph? Well, it's a graphical representation of spatiotemporal data. It doesn't need a map backdrop because location is implicit. A great example of innovation as well as clean design to present a clear message.

—Linda Beale

THE CAUSES OF MORTALITY

E ARMY IN THE EAST.

1.

APRIL 1854 TO MARCH 1855.

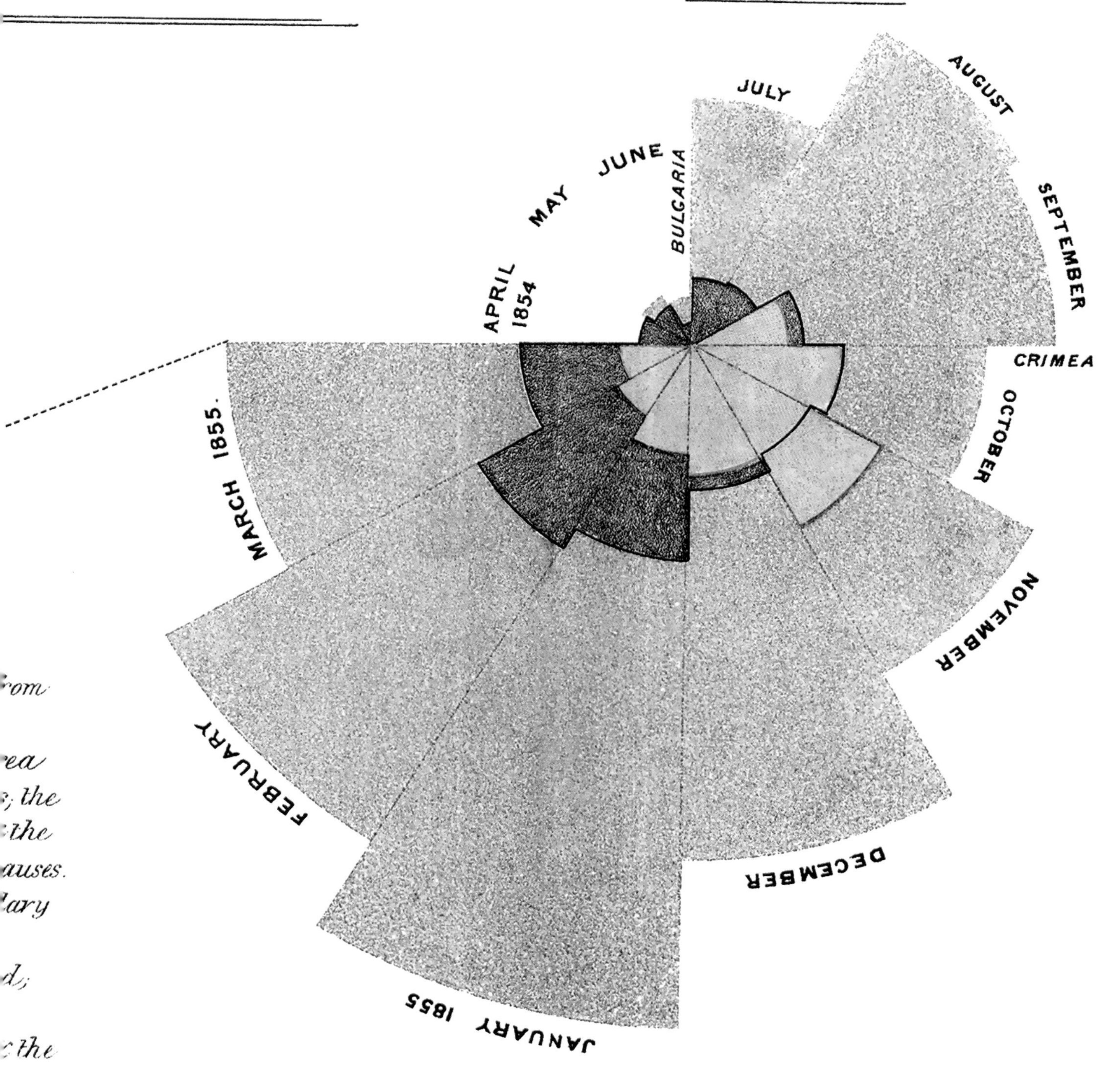

# Families of map projection

## There are three broad families of map projection, which gives rise to maps that have particular appearances and properties.

**There are potentially an infinite number of map projections.** Each is constructed by mathematically transferring features from the sphere to a plane. We can classify map projections into three broad families based on similarities in construction and appearance of the resulting graticule on the map. Conceptually, plotting a map projection can be achieved by forming a projection using a developable surface; namely, a cylinder, a cone, or a plane which intersects with a reference globe This gives rise to the three main families of map projection: cylindrical, conical, and azimuthal.

**The intersection of the developable surface, onto which the map is to be projected, and the reference globe can be in either a simple or secant form.** A simple form is where the developable surface intersects at one point or along a single line—forming either a single point or line of zero distortion (standard point or standard line) which gives the principal scale(s) of the map. A secant form occurs when the developable surface intersects with the globe in more than one place (i.e. it cuts through the globe in some way). This creates two lines of zero distortion (two standard lines and, thus, two particular scales) for the resulting map and creates a situation where combinations of scale exaggeration and compression occur across the map as opposed to the simple form where only exaggeration occurs.

**Some map projections do not conform to the three main families because of their unique mathematical construction.** For instance, pseudocylindrical projections represent the central meridian on a map as a straight line and can often be characterised by an interrupted appearance. Some projections are based on polyhedra and have many alternative shapes and interrupted segments.

Cylindrical projections are, possibly, the most commonly used projection. They are routinely used in atlases and other maps to illustrate the whole world. The normal aspect for a cylindrical projection is referred to as the *equatorial aspect* where the equator forms the standard line. In this case, the standard line is also a great circle. Any oblique aspect cylindrical projection will also have a standard line that represents a great circle

A conical projection in the normal aspect will have an axis of the cone coincident with an axis of the globe. This will give either straight or curved meridians that converge near the poles and parallels that are depicted as arcs or circles. In the simple case of a conical projection, the cone is tangent along a specific parallel, which is the standard line and which has no distortion. Distortion increases away from the curved line also in a concentric, curved fashion parallel to the standard line.

Conical projections are most commonly used to map Earth areas that have a greater east–west extent than north–south.

The aspect of an azimuthal projection is termed *polar* (when in the normal aspect), *equatorial* (when in the transverse aspect), and *oblique* (at mid-latitudes).

Azimuthal projections became important in World War II because of a large amount of circumpolar air navigation. Azimuthal projections are now routinely used in navigation considering the property that directions from the centre of the map remain constant in all directions.

**See also:** Aspect of a map projection | Assessing distortion in map projections | Map projections | Properties of a map projection

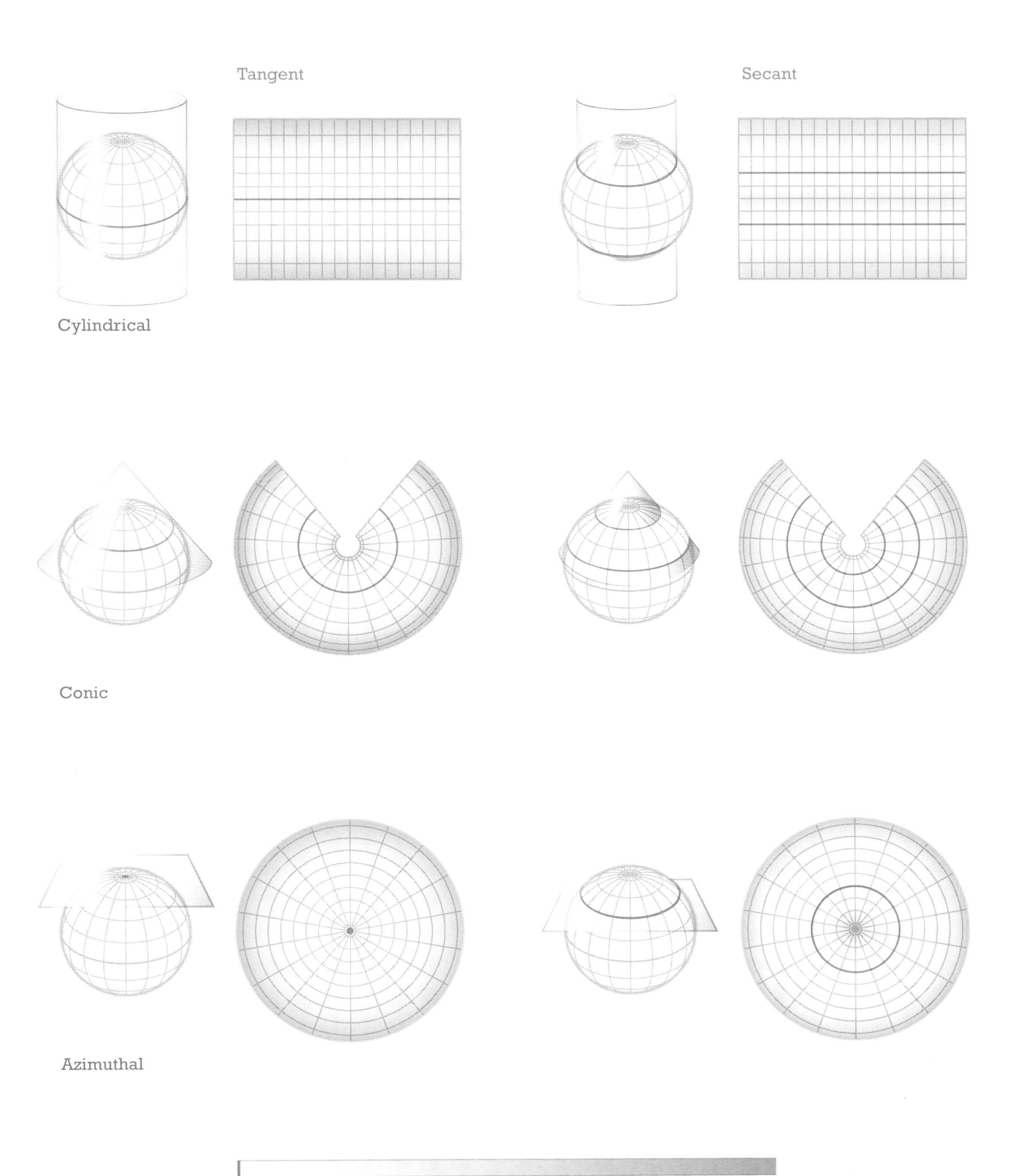
Tangent
Secant
Cylindrical
Conic
Azimuthal
No distortion
Greatest distortion

# Fantasy maps

## Displays of fictional places.

**Maps have often been used to illustrate works of fiction or to give a sense of realism to places that do not exist.** What is striking about many of them is that they take on a sense of realism by using symbology and other traditional map appearances. The development of a sense of place of an imaginary world or landscape is, in part, due to our innate familiarity with real-world cartographies. Fantasy maps therefore contain many of the features we'd expect to see on real maps including land and sea, rivers and mountains, topography and political boundaries, settlements and place-names.

**Fantasy maps in literature often appear as the frontispiece.** They set the scene and plant a firm image in the mind of the reader before they embark on the story. Think of the stories of *Winnie-the-Pooh* or *Lord of the Rings*. And each has a very clear landscape set out. In film, maps also play an important role and act as a character in their own right. The swoop across Westeros in the opening title sequence of *Game of Thrones* sets the scene, placing the viewer in the landscape and is a nod to the literary equivalent. David Lynch famously pitched his vision of *Twin Peaks* to TV executives not with a script but with a map, which immediately gave the story both style and substance.

**Fantasy maps, then, are invented but they do not necessarily show invention.** They repeat the familiar. They also repeat themselves stylistically and in their aesthetic. Many are medieval in appearance since many fantasy maps depict lands far, far away or of a long forgotten time. Robert Louis Stevenson's *Treasure Island* map is in the style of a parchment scroll. Space might even be framed as the final frontier for fantasy maps. Even the *Star Wars* universe has based itself around maps as central characters, principally the holographic plans for the Death Star in *Rogue One* and *Star Wars: A New Hope* or the missing piece of the map that located Luke Skywalker in *The Force Awakens*.

**Opposite:** Robert Louis Stevenson's 1883 tale of action and adventure in *Treasure Island* contained a detailed map on the inside cover that showed the location of a career's worth of looting by the notorious pirate Captain Flint. The map was integral to the story. *X* marked the spot of the buried treasure and so the hunt began.

The map is well drawn, which makes it believable. It contains all the elements of topographic detail you'd expect; depths and rhumb lines finish the nautical chart, and the title cartouche and pictorial flourishes add to the intrigue and resonate with the reader of the book and the characters in the book. It contains annotations which, of course, many maps do as personalised geographies are added over time. It's a vital part of the story, and making the map believable and tangible made the story that much more believable.

The map and the story have become patterns for pirate stories ever since but above everything the parable here is that maps lie. *X* didn't actually mark the spot, so no matter how well designed, aesthetically pleasing, and believable a map may appear to be, they still tell completely fictitious stories.

*'I am told there are people who do not care for maps, and I find it hard to believe.' —Robert Louis Stevenson.*

**See also:** Form and function | Hand-drawn maps | Map aesthetics

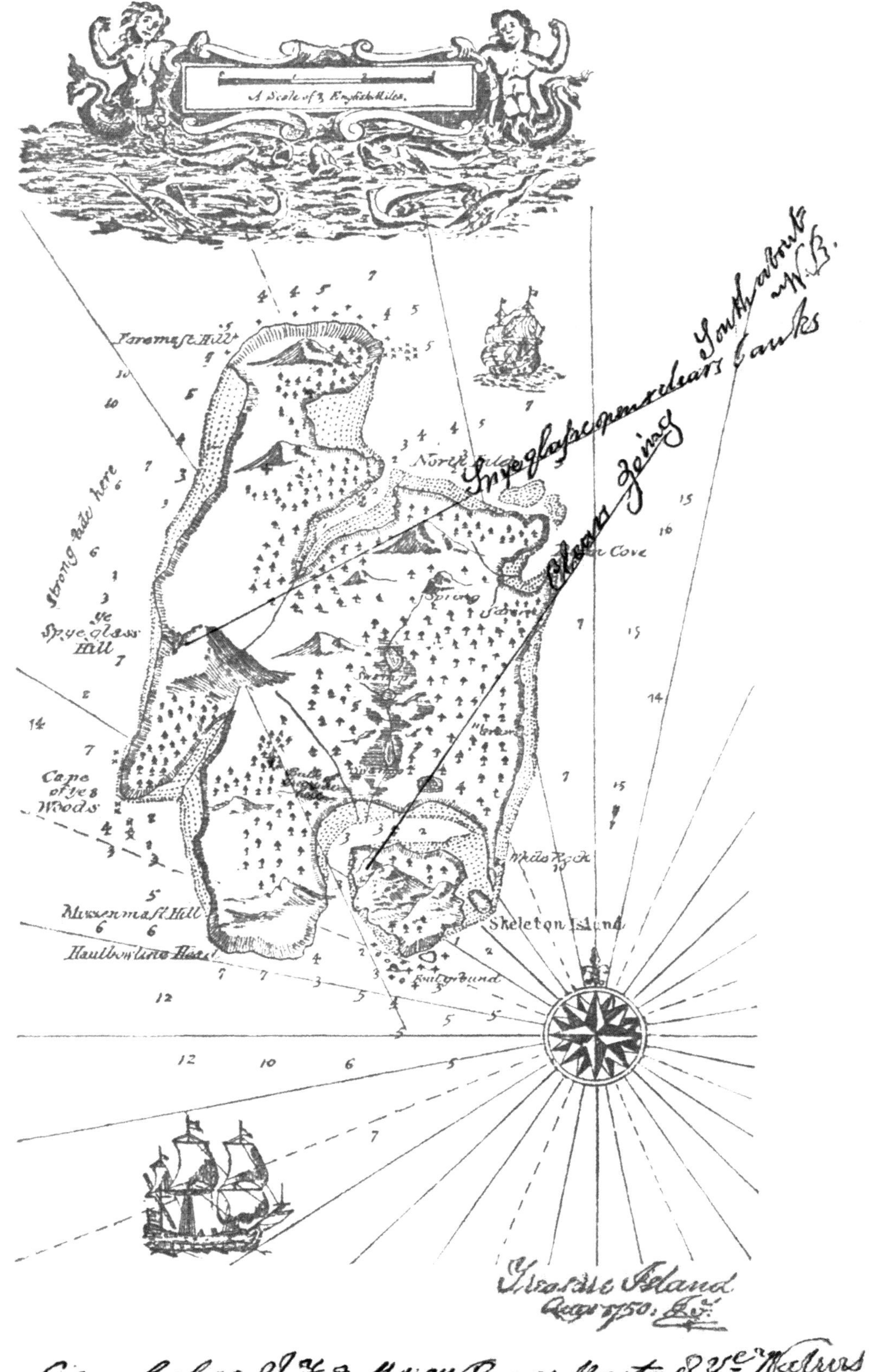


Facsimile of Chart; latitude and longitude struck out by J. Hawkins

# Flourish

## All maps are not the same. Cartographic flourish can be used to create a unique and distinctive look and feel.

**For the most part, cartographers are taught to exercise restraint in their design—to present data simply, clearly, and with unambiguous symbology.** Elsewhere in this book, I talk about keeping the message simple and improving the signal-to-noise ratio. When we look at maps produced by national mapping agencies or news agencies, we see relatively austere approaches to design. There is a consistent look and feel borne out of trying to provide an objective and unemotional map style.

**Styles of cartographic practice change over time, much like any other artistic endeavour shifting as fashions change.** These design treatments and content can be used to modify the entire map's visual appeal. This style encompasses decorative flair as well as the inclusion of components that bring a sense of fun or frivolity. We might consider this aspect of cartography that which deals with flourishes or ornamentation.

**If we consider cartography as we might evaluate many other artistic endeavours, we can see clear parallels.** Architecture, art, and pretty much any commodity today brings with it an array of choice to suit different preferences. A map then can be art deco, it can be minimalist, it can be pop art, or it can be science fiction. Creativity can come in many different forms. It may be that you wish to evoke a certain era or emotion through flourishes. You might apply a consistent palette of colour or form of type that evokes a particular meaning. Modern flourishes might be as simple as applying a decorative border but can also include drawings or the addition of photographs or other content pertinent to the map's key theme. Such work usually functions best as a one-off rather than a repetitive style.

**Humour can also be a powerful aspect of map design.** Making people smile or laugh is a good way to get people to remember the map. They may pass by a less-embellished map whereas humour invites attention. This is a central design philosophy of serio-comic cartography.

**See also:** Different strokes | Elegance | Map aesthetics | Simplicity vs. complexity | Style, fashion, and trends | Vignettes

Maps are works of art. Even our apparently plain-looking topographic maps can be their own works of art. Go back in cartographic history, and typographic flourishes were commonplace when map lettering was produced by hand or engraved. Go back further still when maps were painted and drawn by hand, and because of a lack of content they often included map elements and additions that had little to do with the map itself—they were fillers, such as sea monsters to fill empty space on the map.

Cartouches used to be extremely important and prominent on historical maps. These were usually large map elements that illustrated a key person, perhaps even the cartographer themselves, with emblems, shields, panoramas, or other details. The impression of importance, rigor, and stature lay behind such visual statements. Cartouches can still be useful to provide more context to a map and to bring artistry to a product.

Although much of this book encourages restraint to support clear communication, there is very much a place for artistry, imagination, and creativity in cartography, particularly for one-off maps. Making maps that people want to look at is always key. But applying flourish always straddles a fine line, and it can backfire when you dial your map's glitziness up to 11.

**Opposite:** John Nelson opts for a steampunk aesthetic on this map of voter turnout in the 2012 *US Presidential Election*. Does it work?

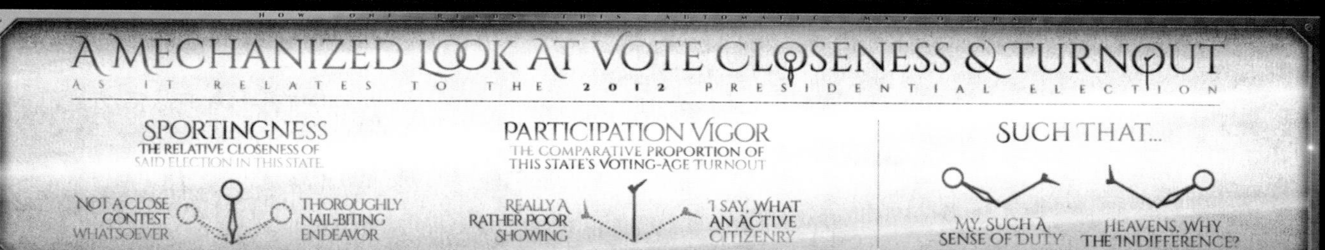
HOW ONE READS THIS AUTOMATIC MAP-O-GRAM
A MECHANIZED LOOK AT VOTE CLOSENESS & TURNOUT
AS IT RELATES TO THE 2012 PRESIDENTIAL ELECTION
SPORTINGNESS
THE RELATIVE CLOSENESS OF SAID ELECTION IN THIS STATE.
NOT A CLOSE CONTEST WHATSOEVER
THOROUGHLY NAIL-BITING ENDEAVOR
PARTICIPATION VIGOR
THE COMPARATIVE PROPORTION OF THIS STATE'S VOTING-AGE TURNOUT
REALLY A RATHER POOR SHOWING
'I SAY, WHAT AN ACTIVE CITIZENRY
SUCH THAT...
MY, SUCH A SENSE OF DUTY
HEAVENS, WHY THE INDIFFERENCE?

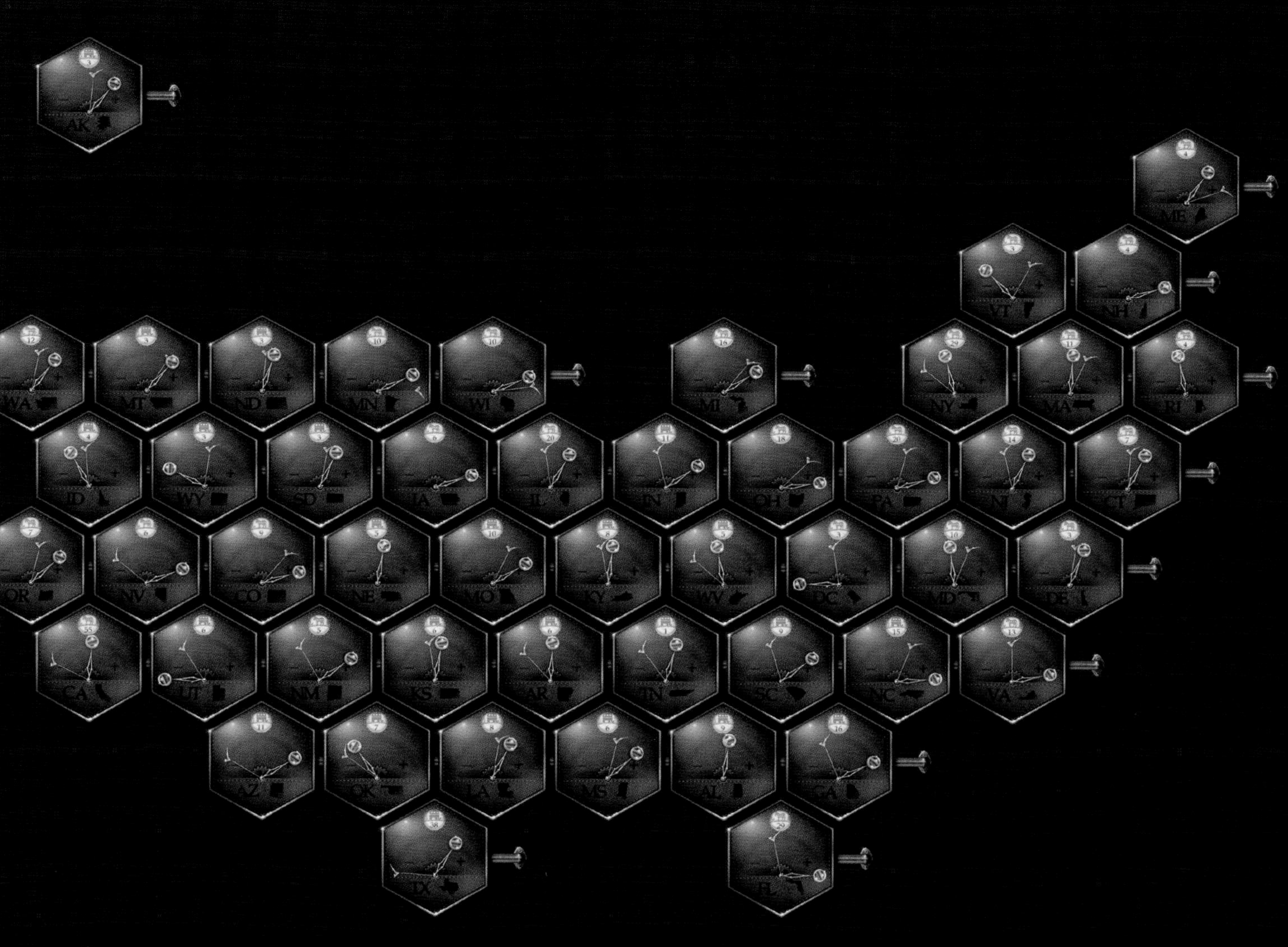
AK
ME
VT
NH
WA
MT
ND
MN
WI
MI
NY
MA
RI
ID
WY
SD
IA
IL
IN
OH
PA
NJ
CT
OR
NV
CO
NE
MO
KY
WV
DC
MD
DE
CA
UT
NM
KS
AR
TN
SC
NC
VA
AZ
OK
LA
MS
AL
GA
TX
FL

HI

# Flow maps

## Representation of linear movement between places either through simple connectivity or the mapping of magnitude.

**Flow maps show linear movement between places and can show qualitative difference between types of flow or quantitative movement through magnitude of flow.** Lines generally change in width proportional to the quantity they are mapping but changes in hue and value, for instance, can encode multiple variables at the same time or for emphasis.

**The technique supports the mapping of totals with widths of flow scaled proportionally to the values for each segment in the network.** Ratios or proportions can also be mapped using this method. If flow is directional, it's usual to incorporate an arrowhead, or some other cap symbol or graphical effect (e.g. tapering or gradient-filled transparency) that signifies origin and destination. When lines branch or merge the widths of the smaller lines should equal the width of the aggregated flow line.

**Flow maps tend to be one of four types: network, radial, distributive, or continuous.** In addition to mapping quantity of flow, a network's organisational structure is also illustrated. Other networks are less defined such as air routes, which tend to show origins and destinations linked by a line, indicative of a route. Radial flow maps (also called 'origin-destination' or 'desire' lines) can show links between paired places. Distributive flow maps are a modification of the radial flow map that illustrates movement from a single origin to multiple destinations, with the width of the line changing at vertices where it splits to a destination.

**Three-dimensional and animated flow maps can be used to emphasise volumetric information or movement.** In 3D, flow can be mapped by using the volume of a cylinder or another 3D shape along lines that represent the three flow types. Another technique that 3D supports is the creation of arcs across a virtual globe such that the linear symbol isn't attached to the ground, but arcs across the space from the origin to destination. Animated flow maps may incorporate a pulsing effect or some other animation that connotes directional flow at speeds that represent velocity.

**See also:** Animation | Dimensional perception | Proportional symbol maps | Size | Threshold of perception

Organisation of lines on a map with multiple flows is always a challenge. An attempt to minimise overlaps or positioning long curved lines around the map should be avoided when possible. Selecting a suitable projection can help, perhaps to show the origin of flow at the centre. Azimuthal projections may be useful for radial flow maps which would render lines as straight from a central origin. Equal-area and conformal map projections are not as important as maintaining other properties such as shape. Curved lines tend to be more aesthetically pleasing than straight.

Lines, then, should be figural and smaller lines positioned on top of larger lines, perhaps using a mask if it's impossible to avoid overlaps. A legend to explain the symbol scaling is also important as are labels to provide a way of interpreting flow. Line scaling should be designed to give a sense of balance between the minimum and maximum widths but datasets with a large range or which are particularly skewed may be more suitably depicted using a classified scheme of graduated widths, or with values above or below a certain threshold depicted in a symbol of consistent width.

Bivariate flow maps can be created by encoding one variable using line width and another using hue (for qualitative difference) or value (for quantitative difference) of another variable.

Transparency can also be useful when your dataset is large and you are trying to depict general patterns of flow rather than supporting the reader in picking out particular routes. Transparency can also be used to denote different time periods with past events becoming increasingly transparent.

**Opposite (main image):** *Flight Paths over Stage 11 of the Tour de France* by ITO World, 2017.

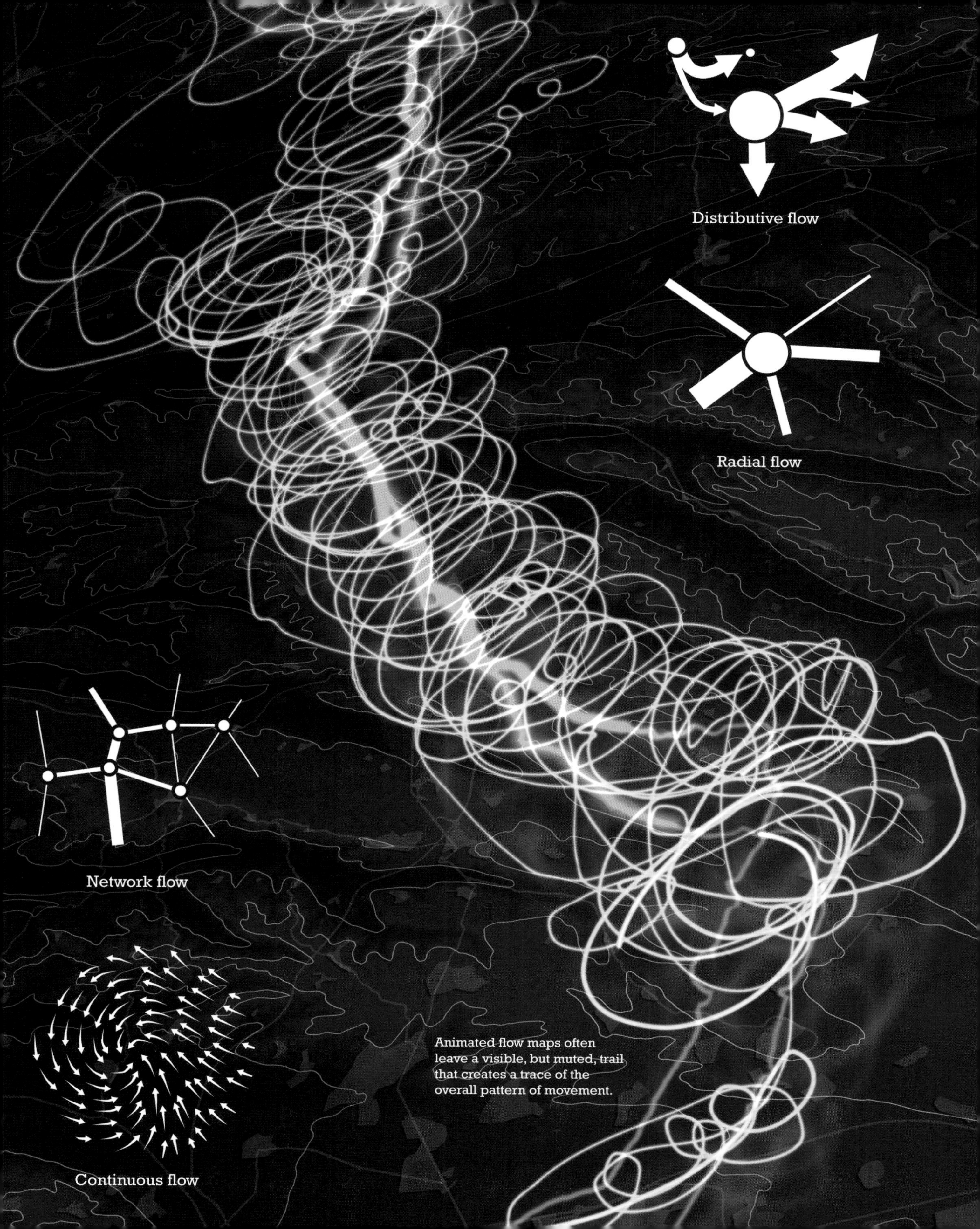


Animated flow maps often leave a visible, but muted, trail that creates a trace of the overall pattern of movement.

# Focussing attention

**Structuring the map to focus attention is more than organisation and visual centering, it's about stripping away the content to reveal meaning.**

**When we read a passage of text we read in serial.** One word comes after another. They form sentences and paragraphs and we build up a sense of what is being said. As human beings we are well drilled in understanding information that is presented to us in a serial, or linear, form. The same cannot usually be said of a map since information is communicated in parallel, and it's up to the reader to disentangle the messages. That's why we use graphical principles like figure-ground and visual hierarchy to give the map structure—to try and enforce a sense of linearity in how people read it.

**What we're essentially asking of our map readers is to undertake a multitasking exercise where we bombard them with a lot of parallel information at once.** For many people, though, multitasking is not something they do particularly well. Creating complex graphics therefore lessens the potential of a map reader to be able to understand the map. It's a fallacy to assume that putting as much as you can on a map provides more information. In many ways a map is a perfect exposition of the idea that less is more.

**Building focus can be achieved implicitly by omitting all but the essential from the map.** In an era of big data when the temptation is to map as much as you can, it's probably more important than ever to reduce the noise on the map. In an explicit way, we can also use graphical devices to give focus. One of the obvious devices is simply ensuring the object and core purpose of your map is placed toward the visual centre of the map and to organise other elements around and in support of that core element. Shining a light, literally through graphical effects, to bring focus on a particular element, or to mute others, can also lead readers to the key aspect.

**There are other ways of ordering the map's message.** For instance, if you make a web map and use a linear storytelling design, it's perfectly possible to convert a map or a series of maps into a sequence that leads the reader. This idea sits at the intersection of a written passage and a film, effectively using storyboarding to deliver a map-based message.

**See also:** Advertising maps | Contrast | Different strokes | Dispersal vs. layering | Hierarchies | Resolution | Seeing | Viral cartography

*'Get rid of half the words on each page, then get rid of half of what's left.'*
—Steve Krug

Steve Krug's quote explores the relationship between design and content. Put simply, omit as much as you can. His first law of usability encourages people to design so that the reader doesn't have to think. Of course, we may want to provoke thinking as a response to the map but the point is to make the process of coming to that conclusion as unhindered as possible. It's up to the cartographer to do the thinking as part of their design to reduce the thinking required on the part of the map reader.

Focussing attention can also be more explicit such as using an arrow to point to a map object or using it to suggest a scanning direction for the eye. The use and position of certain colours, thickened lines, frames, and other graphical stimuli can also lead the eye in a deliberate manner.

Ultimately, as in any mode of communication, the idea is to make the key elements of your work sing and be heard. In writing, we can build a passage cumulatively. In film or music, we build to a climax or a crescendo. In the spoken word, we use intonation, tone, and volume to create interest and ensure the audience gets the key points. In graphics, we use different ways to bring attention to the main aspects and key messages of the map. Reducing the interference by competing graphics and messages will improve the overall effectiveness of the map.

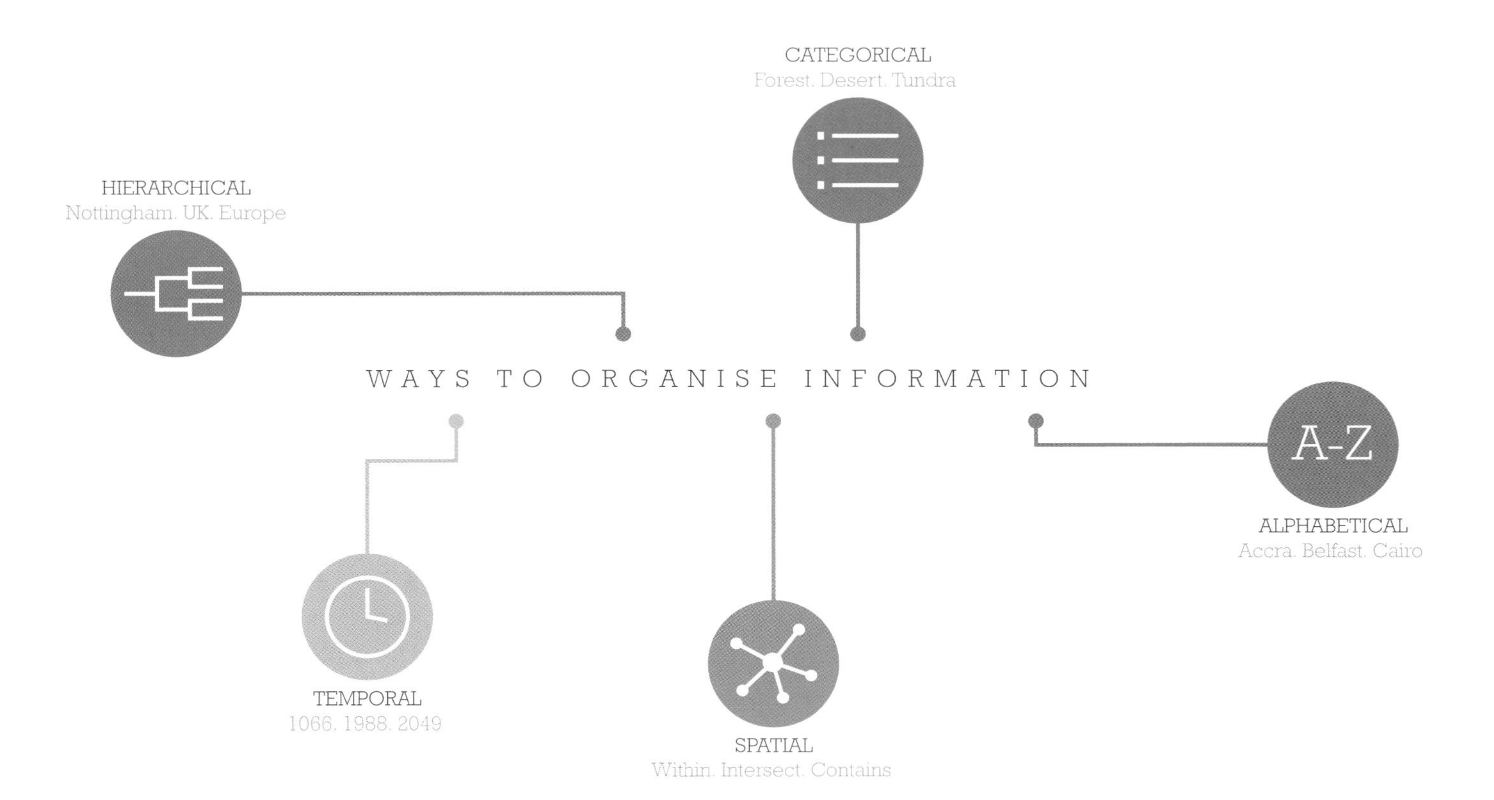
CATEGORICAL
Forest. Desert. Tundra
HIERARCHICAL
Nottingham. UK. Europe
WAYS TO ORGANISE INFORMATION
A-Z
ALPHABETICAL
Accra. Belfast. Cairo
TEMPORAL
1066. 1988. 2049
SPATIAL
Within. Intersect. Contains

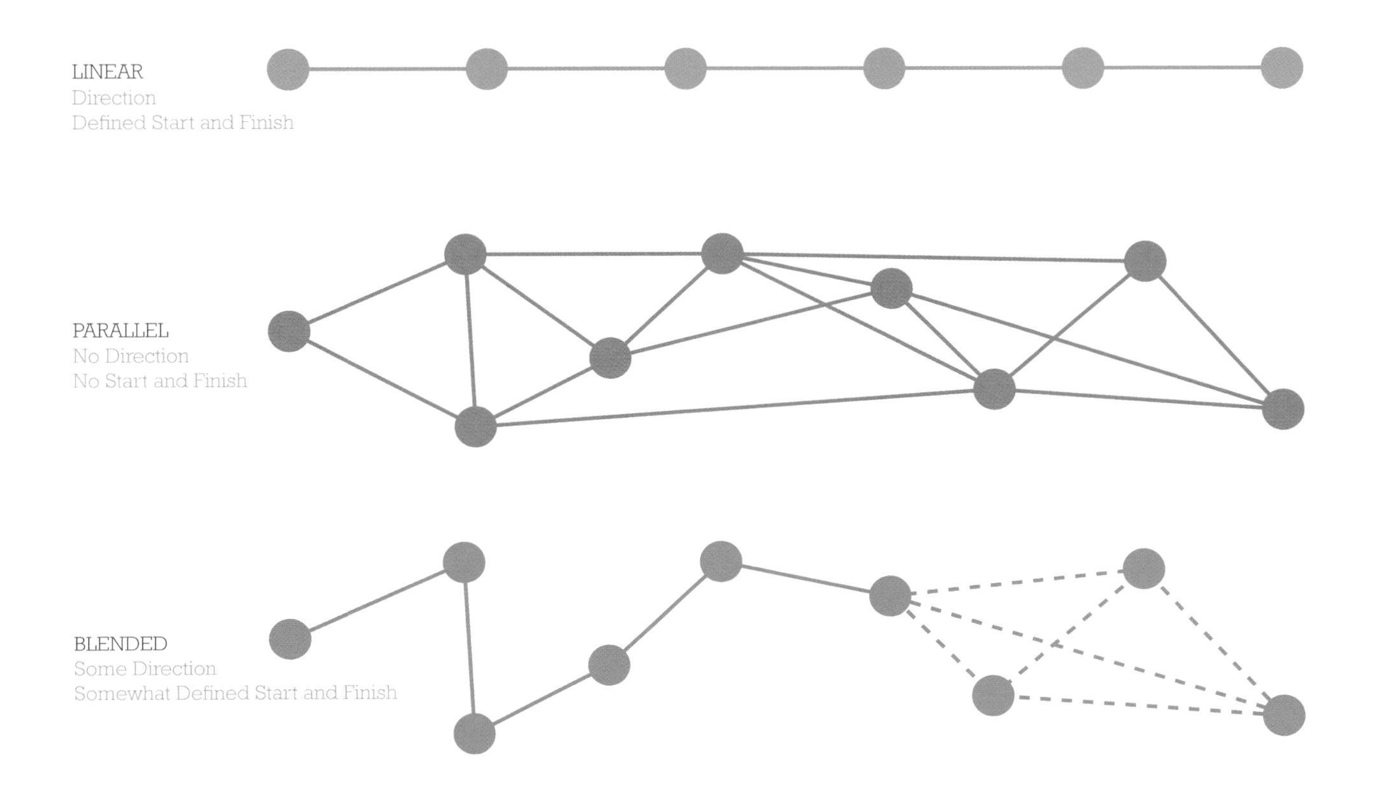
WAYS TO ORDER A MESSAGE
LINEAR
Direction
Defined Start and Finish
PARALLEL
No Direction
No Start and Finish
BLENDED
Some Direction
Somewhat Defined Start and Finish

# Fonts and type families

**A type family refers to a group of type designs that share common design elements and, therefore, the same basic family name.**

**A type family has common design elements among a range of versions of the same basic style.** For each type family, common styles include roman (normal), italic, bold, and bold italic with others such as condensed and expanded also being useful in mapping. Each design version of a particular family and style is referred to as a *typeface*. The word 'font' is used to describe a complete set of characters of type family, type style, and size. A font will include alphanumeric characters, special characters, and punctuation marks.

**In cartography, uppercase and lowercase distinctions provide the map designer with considerable choice.** In written works the use of uppercase and lowercase tends to be applied in a particular way with uppercase characters being used for the initial letter of a sentence and lowercase characters following. Lowercase characters are easier to read because they are less blocky in appearance and are often more detailed than their uppercase counterparts, which makes the overall shape more visible on a map.

**The majority of lettering on a map is normally set in title case where each word is composed of an uppercase initial character and the remainder is lowercase.** Conjunctions and linking words (e.g. 'and', 'on', 'in', 'or', 'with', and so on) are also set in lowercase. Title case is appropriately used in titles, subtitles, legends, point and line labels, and so on. Sentence case would be used where short descriptions or explanatory text is included. The only time uppercase is conventionally used is for short titles, labels for areal features, and sometimes capital cities to differentiate them from other settlements or to show some quantitative difference among labels for features of the same type.

Initially, all typefaces contained a single roman design. During the 16th century, cursive designs began to appear. These became known as italics because they were populairized in Italy. Despite different versions of similar type becoming more common, the idea of a typeface family only dates to the early 20th century. Morris Fuller Benton, as Director of American Type Face Corporation, suggested the idea of a type family which maintained the basic characteristics of the parent design but with individual variance.

The idea of naming variations as bold, semi-bold, and so on was modified in the mid-1950s by Swiss typographer Adrian Frutiger. In creating an ordered series of variations of a typeface, Frutiger introduced a two-digit numerical suffix. The first digit specified the weight between 3 and 8 (light to heavy). The second digit specified the proportion with higher numbers for condensed designs. The second digit was always odd if the typeface was roman and even if it was italic.

Different typefaces can be very different in their use of space. The character width will ultimately determine the amount of space needed, but if you are tight for space, then this may be a key consideration for choice of typeface.

There's sometimes a temptation to use eccentric or artistic typefaces but in general they distract and compromise legibility. Although there is nothing wrong with many of these beautiful typefaces, they do not always lend themselves to cartographic purposes.

Many of the terms used to describe type are used interchangeably. For example, the word 'font' is commonly used to refer to a type family or a typeface, principally because of the erroneous use in common computing applications.

**See also:** Elements of type | Guidelines for lettering | Using words

Rockwell

Rr

ABCDEFGHIJKLMNOPQRS
TUVWXYZabcdefghijklmn
opqrstuvwxyz1234567890.

REGULAR *ITALIC* **BOLD**

Futura

Ff

ABCDEFGHIJKLMNOPQRS
TUVWXYZabcdefghijklmn
opqrstuvwxyz1234567890.

REGULAR *ITALIC* **BOLD**

Garamond

Gg

ABCDEFGHIJKLMNOPQRS
TUVWXYZabcdefghijklmn
opqrstuvwxyz1234567890.

REGULAR *ITALIC* **BOLD**

Lucida Sans

Ll

ABCDEFGHIJKLMNOPQRS
TUVWXYZabcdefghijklmn
opqrstuvwxyz1234567890.

REGULAR *ITALIC* **BOLD**

Times New Roman

Tt

ABCDEFGHIJKLMNOPQRS
TUVWXYZabcdefghijklmn
opqrstuvwxyz1234567890.

REGULAR *ITALIC* **BOLD**

Comic Sans

Cc

ABCDEFGHIJKLMNOPQRS
TUVWXYZabcdefghijklmn
opqrstuvwxyz1234567890.

REGULAR *ITALIC* **BOLD**

Bauhaus 93

Bb

ABCDEFGHIJKLMNOPQRS
TUVWXYZabcdefghijklmn
opqrstuvwxyz1234567890.

REGULAR

News Gothic

Nn

ABCDEFGHIJKLMNOPQRS
TUVWXYZabcdefghijklmn
opqrstuvwxyz1234567890.

REGULAR *ITALIC* **BOLD**

Perpetua

Pp

ABCDEFGHIJKLMNOPQRS
TUVWXYZabcdefghijklmn
opqrstuvwxyz1234567890.

REGULAR *ITALIC* **BOLD**

# Foreground and background

## Designing your map to have foreground detail across a background of contextual information not only grounds the work but builds interest.

**Cartographers often talk of ensuring their map has a good 'figure-ground' relationship.** This simply describes the process of ensuring that some components are placed visually to the fore in the design and some to the background. It's an approach that accentuates one map component over another to structure the map so that an object is seen more prominently.

**More important map components should be promoted to make them stand out perceptually as figures in addition to contrasting them with the map background.** On topographic maps, point and line objects are usually advanced in the map's structure with surrounding area detail receding as ground. A good way to think of this is to consider the difference between mapping objects on an island in an ocean. The ocean is logically the background whereas the land is a little more important. What is more important are the objects on the land. You therefore design the map to have very little detail or colour for the background and increase visibility for the components that sit on the landscape. The reader's eye is then forced to see the important figural components first while, perceptually, the background recedes. Cognitively, the brain processes what the eye sees and builds a structure that emphasises some map components over others. Making the main message of the map the features that are figural is key.

**For thematics, the theme itself should be figural with other detail receding.** Proportional symbols, choropleth shading, and dots all should be seen as the figure with boundary data, labels, and supporting information as the ground.

**Care needs to be taken when establishing figure-ground relationships to not imply meaning unintentionally.** Consider a thematic map (such as a choropleth) where area symbolisation is figural. Applying a shade or colour to the map background can be confusing, as if the background might be part of the mapped area itself. Demarcating foreground and background to avoid confusing the message is an important aspect of getting the visual relationship right.

**See also:** Basemaps | Contrast | Focussing attention | Imagery as background | Layouts and grids | Seeing | Vignettes

A useful analogy to consider when working out how to build a figure-ground relationship into your map is to compare your role in map design to that of a film director.

A film director has a stage upon which they must place the cast. They usually have a set or a location which provides the background, the context. Actors take the stage in different ways. Many are bit-part or support actors whose role is to add to the overall effect of the scene and to complement the set itself. Then there are the lead actors who are prominent in the scene. They not only have the key lines of dialogue but they take centre stage, they are in focus, and they lead the film. But without the set and the supporting cast, the lead actors' work would be sorely compromised.

For instance, Darth Vader is a figural character, but without all the supporting cast and components from views of outer space, hundreds of spaceships and Stormtroopers, he'd just be a guy in a black cape and there's been plenty of those in film.

Set design, dialogue, and positioning of your actors all helps set the scene. Get it right, and the story is compelling and you draw your audience in. Get it wrong, and you're making the map equivalent of *Waterworld*—people living on floating atolls amidst the vastness of the watery background which took centre stage away from the main action!

Background Bathymetry
Bathymetry is more subtle than the hillshade. It becomes lighter nearer the land which helps promote the land.
Channel Wind Direction
O'AHU
Honolulu
Kaiwi Channel
Foreground Hillshade
Hillshade is more intense than the bathymetry. The darkest areas offer the most contrast and jump to the foreground.
MOLOKA'I
Pailolo Channel
Kalohi Channel
Foreground Text
Areas with the most text naturally draw attention.
Kahului
'Au'au Channel
LĀNA'I
MAUI
Kealaikahiki Channel
'Alalākeiki Channel
Foreground Thematic
A drop shadow creates contrast which promotes the thematic arrow.
Alenuihāhā Channel
KAHO'OLAWE
HAWAIIAN ISLANDS
Channel Wind Direction
Foreground Text
The largest text with the most contrast rise to the foreground.
Background Clouds
The cloud layer mutes an area that might compete with the central features.
Hilo
Kailua-Kona
HAWAI'I
Foreground Layout
An interesting layout, like an island chain, promotes the foreground features.
Foreground Vignette
A land vignette helps create contrast which promotes the foreground feature.
Background Colours
Pale blues naturally recede from the foreground.

# Form and function

## Form and function are inextricably linked. When they work in harmony you've hit the cartographic jackpot.

**Cartography is often described as a set of processes that define the form of a map, such that it supports a specific function.** Cartography is more than that but this general idea is an important basic tenet. Form is the general appearance and structure of the map. Function is the purpose which the map serves. Ensuring the form is appropriate and supports the function might seem obvious but it's plagued by pitfalls.

**Defining the function of a map can be problematic.** Often a map is simply made for an ill-defined user or purpose. Without a clear guide or idea, it becomes difficult to craft a form that works. Likewise, with form, there's often a desire for the form to overrule what the data might be capable of showing. We've all had the urge to make a specific-looking map but does the data fit that form? It's generally a good idea for the data to determine the form and not the other way round.

**Mapping, though, is inevitably an exercise in compromise, and there's never a perfect solution to a given mapping task.** Even the basic task of trying to create a choropleth map is riddled with possible pitfalls because geography simply isn't cooperative when it comes to mapping. The cartographer has to negotiate the various pitfalls as they explore design alternatives and select their compromises that are the most optimal or, put another way, least damaging. Do you go for a pleasing and striking form over a standard and, possibly perceived 'boring' form? What about projections—something obvious and easily recognised? Or something less obvious but possibly more appropriate from a technical point of view?

**Cartographic form resides in the internal construction and representation of the data, and function is the external purpose of the map.** There cannot be a priority since they are intertwined. Rather, form follows function as much as function follows form. Ultimately, each needs careful consideration to make a successful map.

The well-worn phrase *form follows function* originates from the world of architecture in the late 1800s and describes how structures should be solid, useful, and beautiful. It represents an aesthetic concern that sees the appearance of a building as being as important as its structure. In fact, the mantra supported the notion that buildings couldn't simply be designed by choosing a standard pattern and that the function had to lead to a consideration of a unique form. It was an attempt to break from the monotony of styles of the past and gave architects freedom of expression. As with many design-led movements, the reaction wasn't always popular and critique suggested that the principle simply led to excessive ornamentation. The idea of form and function has extended to many fields including cartography. Critics of cartography sometimes echo the architectural critique and claim it's not much more than 'making maps look pretty'. This is a misconception of the value and role of cartography but it does require mapmakers to consider form and function rather than assuming form follows function.

The myriad ways in which maps can and are used, the functions they support on their own or as part of other graphics and reports, and the different representations they can take on means there simply cannot be a one-size-fits-all approach to form and function. Each map will require a systematic consideration of the data and the eventual use, the medium and the potential user, as well as the time and technical constraints. There is no magic recipe for making a map where form and function balance and work harmoniously but studying excellent maps and seeing what works and what doesn't is a great way to learn.

**See also:** Consistent denotation | Craft | Design and response | Different strokes | Flourish | Map aesthetics | Style, fashion, and trends

**Opposite:** *Breweries of the World* by Kenneth Field, 2015.

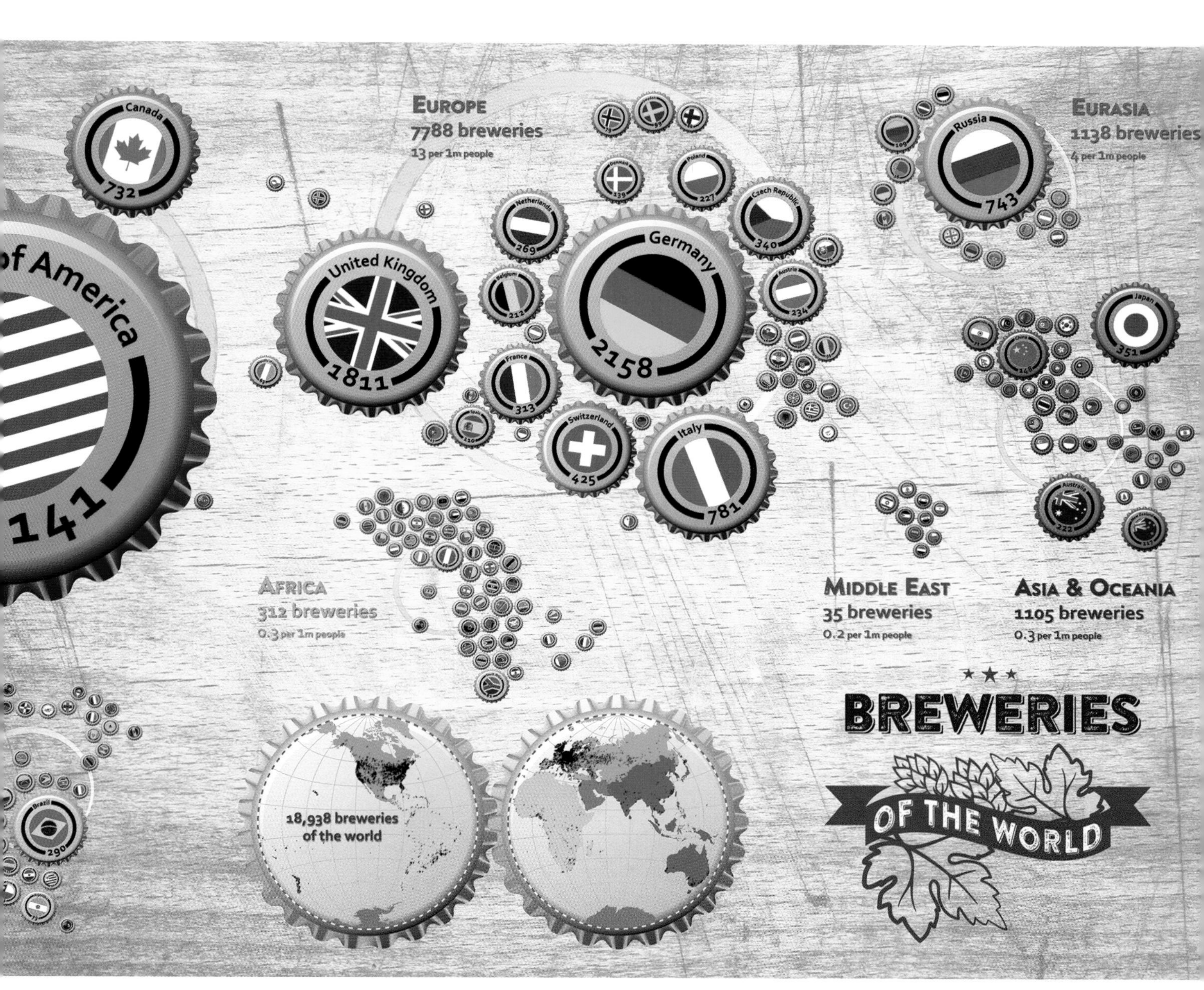


How do you map the locations of 18,938 breweries? Dots on a world map? A choropleth showing breweries per capita for different countries? The answer depends on what the map is for and where it's going to be displayed. This map was designed to function as a large piece of art (20 ft wide) on a brewery wall. It needed to be highly engaging, impactful, and abstract. It needed to be a talking point for visitors to the bar. It therefore needed to stretch the form into something beyond the standard map treatment.

The use of bottletops as a visual metaphor provides a form that speaks to the theme. The Dorling cartogram provides the structure that supports the metaphor as well as the overall function. A tabletop in the background suggests the bottletops strewn randomly across. Adding flags provides some sense of patriotism and allows viewers to relate their own country's 'score'. Beer and maps!

# Frequency distributions and histograms

## Finding patterns among data values.

**Before you even begin to plan a map of empirical data, it's a good idea to do some exploratory analysis.** Understanding the detail in the data and discovering where the interest lies, where the similarities and differences occur, and how the data might be distributed is important in informing the cartographic process. Key to this is understanding the frequency distribution of the data.

**A frequency distribution is an array ordered so that it gives the frequency of occurrence of each value.** This allows data to be summarised for easy inspection and so you can clearly see the pattern of the data distribution across the values. For simple arrays it is easy to illustrate them in a table. For a larger number of observations, tabular inspection of frequencies is not sufficient. Instead, a frequency table is constructed from which a histogram can be drawn. The frequency table is constructed by dividing the total range of the data (the difference between the highest and lowest value in the array) into non-overlapping subgroups called 'classes'. The number of values of data that fall into each class is then counted and referred to as a 'class frequency', which can then be translated into a histogram by plotting the class intervals on the horizontal axis and the frequency on the vertical axis. Columns are then used to depict the frequency for each class interval.

**Creating tabular or charted views of your data also supports choices you may make in creating a map.** Through exploration, you are gaining an insight into the distribution of the data and understanding the patterns it represents. This is a good first step to deciding on a classification method and class breaks for a choropleth map, or the range of values for a graduated symbol map. Certain thematic maps, like the choropleth, are in essence an illustration of the spatial pattern of a frequency distribution.

Most visual techniques for displaying data rely on the basic cognitive process of comparison. For instance, when you look at a bar graph or a histogram you're recognising which bars are taller than others, which fundamentally relies on your ability to compare. You can recognize bars that are double the height of others fairly easily, for instance.

The same process applies when you look at a thematic map since the basic requirement is to see where there is more, compared with where there is less.

Your ability to process, classify, and symbolise data and present it in different ways will always go a considerable way to shaping your message but, ultimately, your map reader is looking for comparisons across the map regardless of how you tell the story.

Graphs are a good way of augmenting your maps. They can be used to support the spatial patterns of the map by placing a series of numbers in close proximity to support the search for similarities and differences. A clean, crisp graph can do an excellent job of adding value to a map display or, even, be used as part of the map display itself.

**See also:** Arbitrary data classification | Data distribution | Graphs | Variables, values, and arrays

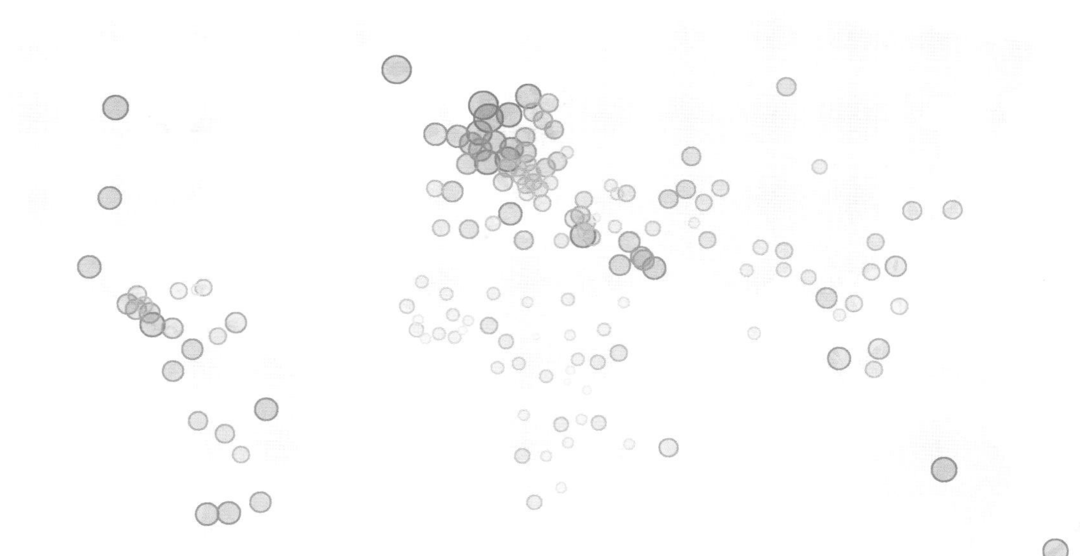

## Happiness Rating

On a scale of 1 – 10, how happy are you?

Least happy
Central African Republic (2.693)

Happiest
Norway (7.537)

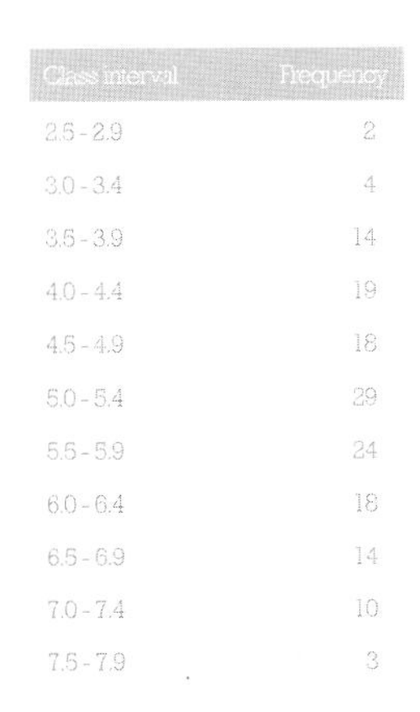

Distribution table

| Class interval | Frequency |
|---|---|
| 2.5 - 2.9 | 2 |
| 3.0 - 3.4 | 4 |
| 3.5 - 3.9 | 14 |
| 4.0 - 4.4 | 19 |
| 4.5 - 4.9 | 18 |
| 5.0 - 5.4 | 29 |
| 5.5 - 5.9 | 24 |
| 6.0 - 6.4 | 18 |
| 6.5 - 6.9 | 14 |
| 7.0 - 7.4 | 10 |
| 7.5 - 7.9 | 3 |

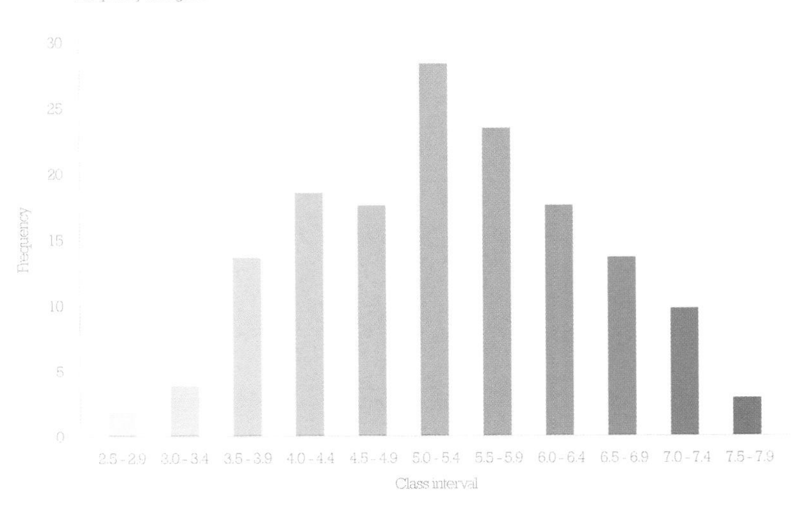

# Functional cartography

## Making a map for someone else is often the hardest part of mapmaking.

**Designing a map with an intended audience or a specific user in mind is a well-worn mapmaking tenet.** Plenty of maps are either self-referential or boutique maps while purporting to be targeted to a wider audience. It's a good idea to avoid making maps that only you, as the mapmaker understand, though that can be quite a challenge. In order to make a well-designed map, you inevitably have to be interested and motivated. It's a hard process to undertake if you are disinterested in the project at hand, even if you're taking on a job from a paying client. Taking a pragmatic approach usually results in a more useful product that conveys information to others. You, as the mapmaker, are rarely the target audience so an ethos that detaches yourself from becoming too involved in the subject matter is a good habit to develop.

**The interplay of data, design, and audience are key to successful cartographic design.** They work hand in hand, and there's a responsibility on the part of the mapmaker to ensure a user-centred design sits at the core focus of the map. Put simply, if the map doesn't work for the person reading it, no amount of explanation or persuasion will make that map any more understood. Despite your hard work and best intentions, the audience will have moved on to something else, and you'll likely not get a second opportunity for their attention. The design therefore has to function and be immediately accessible to the target audience, which might not be the same as the map you'd like to make for yourself.

**It's no longer the case that maps are hindered by a paucity of data.** In fact, the reverse is true, and both geographical and statistical data are more available than ever. The task for cartographers is to wade through this data. Whereupon once their role was to fill in the gaps on the map, now it is to select, refine, and reduce the abundance of data; to omit the visual noise in order to reveal a clear, concise signal via a meaningful map with usable information. Again, this can be challenging but asking questions of the data that you translate into a map that gives your audience answers is a good approach.

**See also:** Defining map design | Dysfunctional cartography | Form and function | Graphicacy | Inquiry and insight | Purpose of maps | Signal to noise

How, then, does one become more pragmatic and objective when designing a map? The key is to go beyond simply collecting and knowing something about the data. You have to understand what the data means to the people who are interested in it. For instance, maps should have intent and support interpretation in a range of many different functions. This may often be underpinned by an agenda that embodies the reasons why a particular dataset is useful to a particular user group so thinking about that requirement is crucial to the cartographic process.

There often exists a motivation behind the map. This may be persuasive, or perhaps borne out of political, social, or economic interests but the map exists to tell a particular story. This is more often the case with thematic than topographic maps, yet even with topography, alternative representations support different objectives. Think, for instance, of how international borders are displayed in different ways and in different places to suit certain users.

Quite often you have to place your own beliefs or prejudices to one side when making a map. By doing so, you'll develop a way of working that removes yourself from the character of the map. This is separate from developing a unique style or look and feel. It supports objectivity in the mapmaking process and helps your maps function for the benefit of someone else.

It's also important when working for a client to ensure they provide a very detailed requirement at the outset. Without one, you'll likely end up presenting your draft only for them to utter the well-worn phrase, 'It' not what I imagined'. Get them to imagine up front, and it will save you time later.

**Opposite:** *Hurricanes since 1851* by John Nelson, 2012. Author prefers the top layout. Client preferred the alternative.

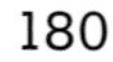

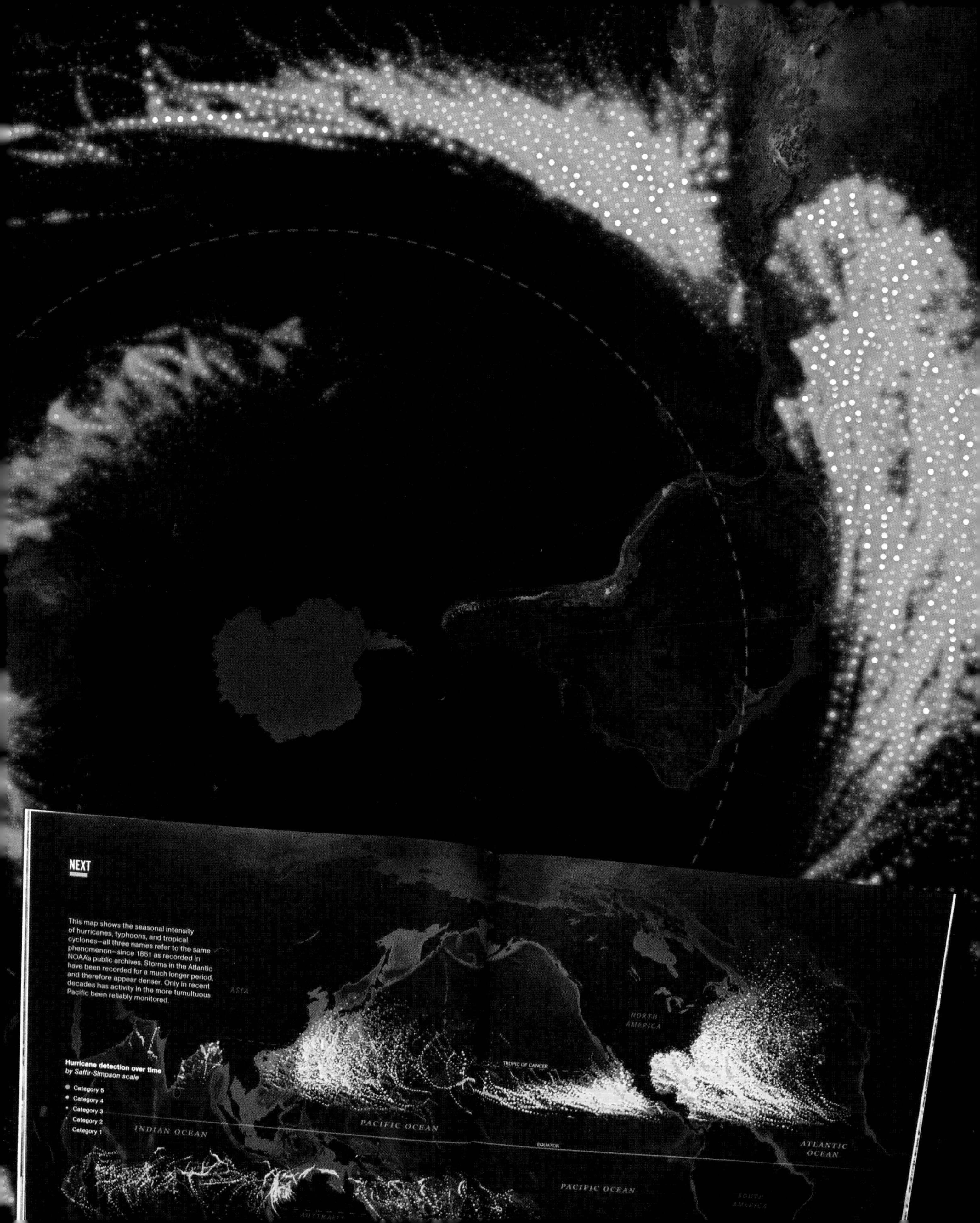
NEXT
This map shows the seasonal intensity of hurricanes, typhoons, and tropical cyclones—all three names refer to the same phenomenon—since 1851 as recorded in NOAA's public archives. Storms in the Atlantic have been recorded for a much longer period, and therefore appear denser. Only in recent decades has activity in the more tumultuous Pacific been reliably monitored.
ASIA
NORTH AMERICA
Hurricane detection over time
by Saffir-Simpson scale
Category 5
Category 4
Category 3
Category 2
Category 1
TROPIC OF CANCER
INDIAN OCEAN
PACIFIC OCEAN
EQUATOR
ATLANTIC OCEAN
PACIFIC OCEAN
SOUTH AMERICA

## The Distribution of Voting, Housing, Employment, and Industrial Compositions in the 1983 General Election

Danny Dorling
1991

More than 25 years after this map was produced, Dorling's cartogram remains a brilliant piece of work in the field of cartograms that is hardly matched.

As one of the earlier efforts at developing algorithms for computer-generated cartograms, it expertly manages to contain seven levels of information within one map. Despite this complexity it effectively communicates the underlying spatial patterns and presents a compelling view of electoral results and socioeconomic circumstances.

In fact, it actually makes use of several mapping techniques. It is a clever use of a cartogram which is still often perceived as being an unusual technique. Secondly, Chernoff faces are used to portray social indicators by mapping facial features to indicators. Finally, it uses the now popular (but by no means new) value-by-alpha concept to modify colours using transparency. By applying these techniques in a unique, combined, and novel way, Dorling is able to display the complex relationship of voting behaviour along with a series of related social indicators, and using the population distribution as the much more logical basemap for this theme.

This cartogram is, simply, a cartographic masterpiece of complexity.

—Benjamin Hennig

# The Distribution of Voting, Housing, Employment and Industrial Compositions in the 1983 General Election.

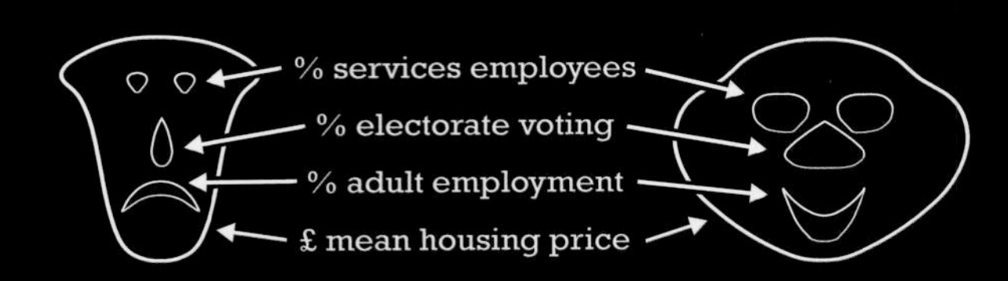


Facial features indicate the social and economic characteristics of the constituencies, colour shows the proportions of the vote for the parties.

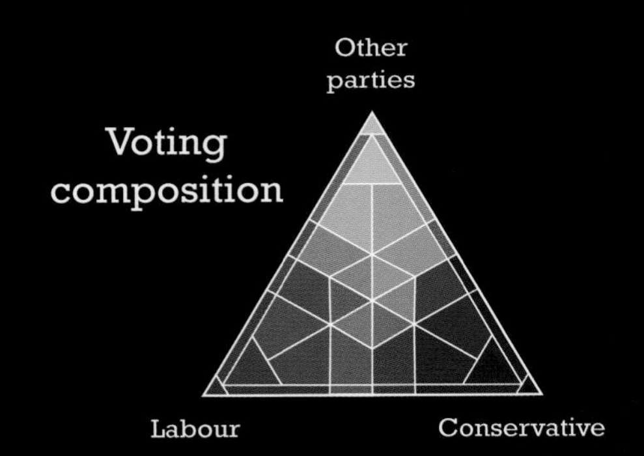

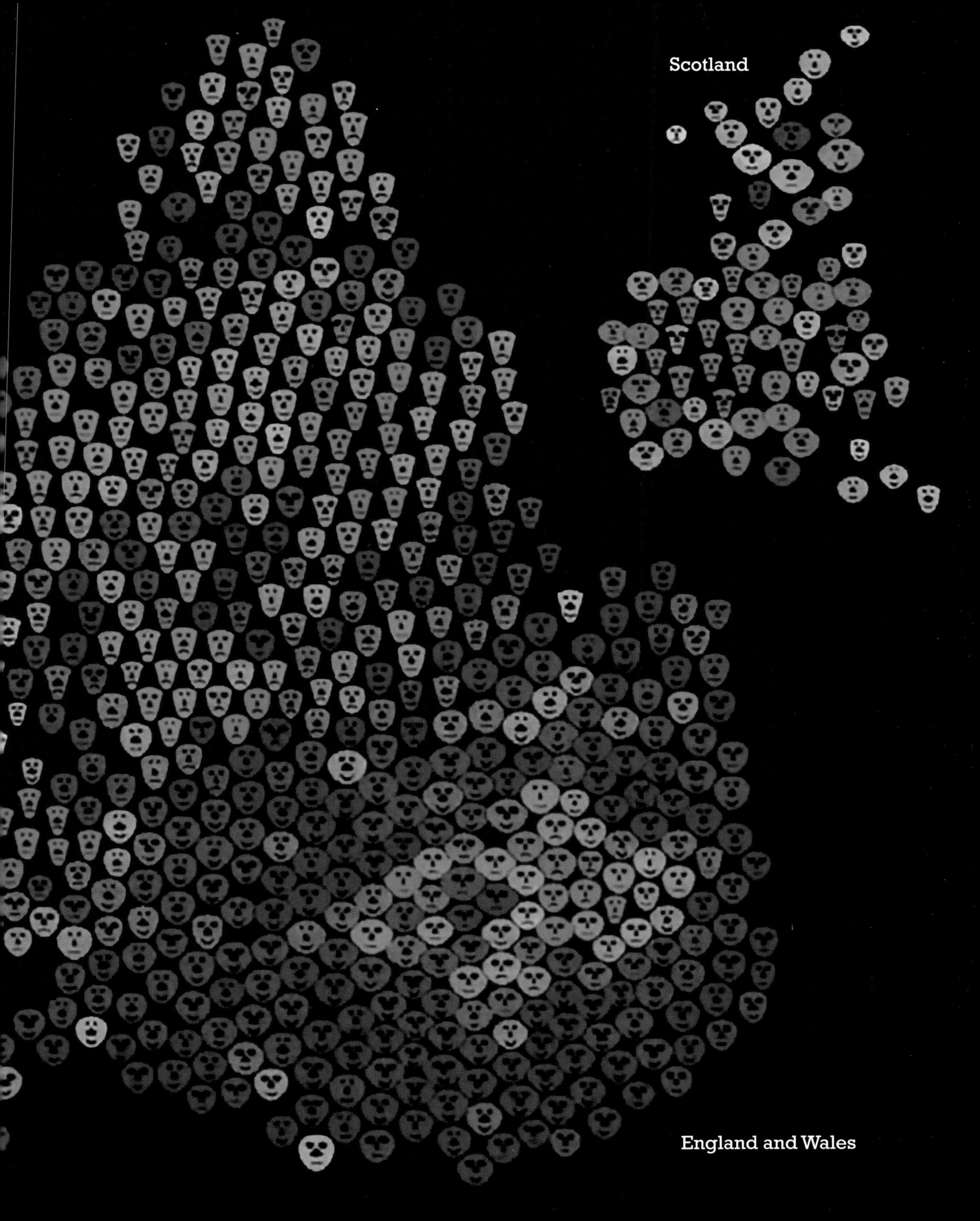
Scotland
England and Wales

# Generalisation

## Generalisation reduces, but clarifies, the information content of a map.

**Generalisation is the overarching process of reducing the information content of a map because of changes in scale, map purpose, audience, or technical constraints.** For instance, when preparing a small-scale map from high-resolution, large-scale data, some of the geographical features will likely need to be omitted or modified to retain clarity. All maps contain some form of generalisation since no map is a 1:1 representation of reality. Whatever your map project or data source, it's almost guaranteed that performing generalisation will improve its appearance and use.

**Generalisation can be performed objectively and subjectively.** Algorithms are often used to simplify the sinuosity of lines from highly detailed versions to less detailed versions for representation at smaller scales. Beyond the use of algorithms, generalisation is partly a subjective process and relies on you to make judgements about how best to render geographical data at a different scale. In truth, generalisation is a mixture of the two processes. Appreciating how it can be applied and what choices work for different data at different scales goes a long way to making a map effective.

**Generalisation is effectively a two-stage process with selection being a preprocessing step prior to the rest of the generalisation process.** Selection involves the identification of features to retain or omit. In compiling a topographic map for a general-purpose audience, detailed base information such as the road layout, water features, urban areas, place-names, and elevation would be retained. For a thematic map at the same scale, many of these features are omitted because it is not deemed important in that context. Once decisions about what to retain or omit are made, processes that modify the graphical character of the mapped objects are usually performed. These processes can include simplification, classification, and symbolisation. Simplification eliminates unnecessary detail in a feature retained; classification categorises geographical data into summary form for map display; and symbolisation assigns graphical coding.

Generalisation used to be a cornerstone of map production. Working from surveyed data, cartographers would pore over every mapped feature and determine how best to represent it at a particular, usually derived, scale.

With the advent of modern mapping databases and the proliferation of open data, maps often display poor (or no) level of generalisation. The reasons for this are two-fold. Firstly, many mapmakers simply won't have the knowledge or the tools to help them in the process. Secondly, generalising data is a time-consuming process, and the tendency in modern mapping is to simply publish the map.

The time it takes to perform generalisation is often overlooked in favour of timely map production. However, performing generalisation on detailed data for display at a smaller scale is always worth the effort. It is one of the key mechanisms to reducing clutter and making the map come to life.

Generalisation shouldn't be performed to make the best use of map space. Geography is inherently heterogeneous and so you'll always face decisions about what to include or omit. The 'Baltimore' phenomenon is a useful lesson in the pitfalls of generalisation. On many small-scale maps, Baltimore, Maryland, USA, is absent, yet smaller surrounding towns and cities with smaller populations are retained simply because of the availability of space for the map label. This has the potential to lead to problems when the map is read and the way in which towns and cities are perceived in relative terms. Making logical decisions is more important than simply using the availability of space as a determinant for generalisation.

**See also:** Scale and resolution | Schematic maps | Simplicity vs. complexity | Simplification | Symbolisation

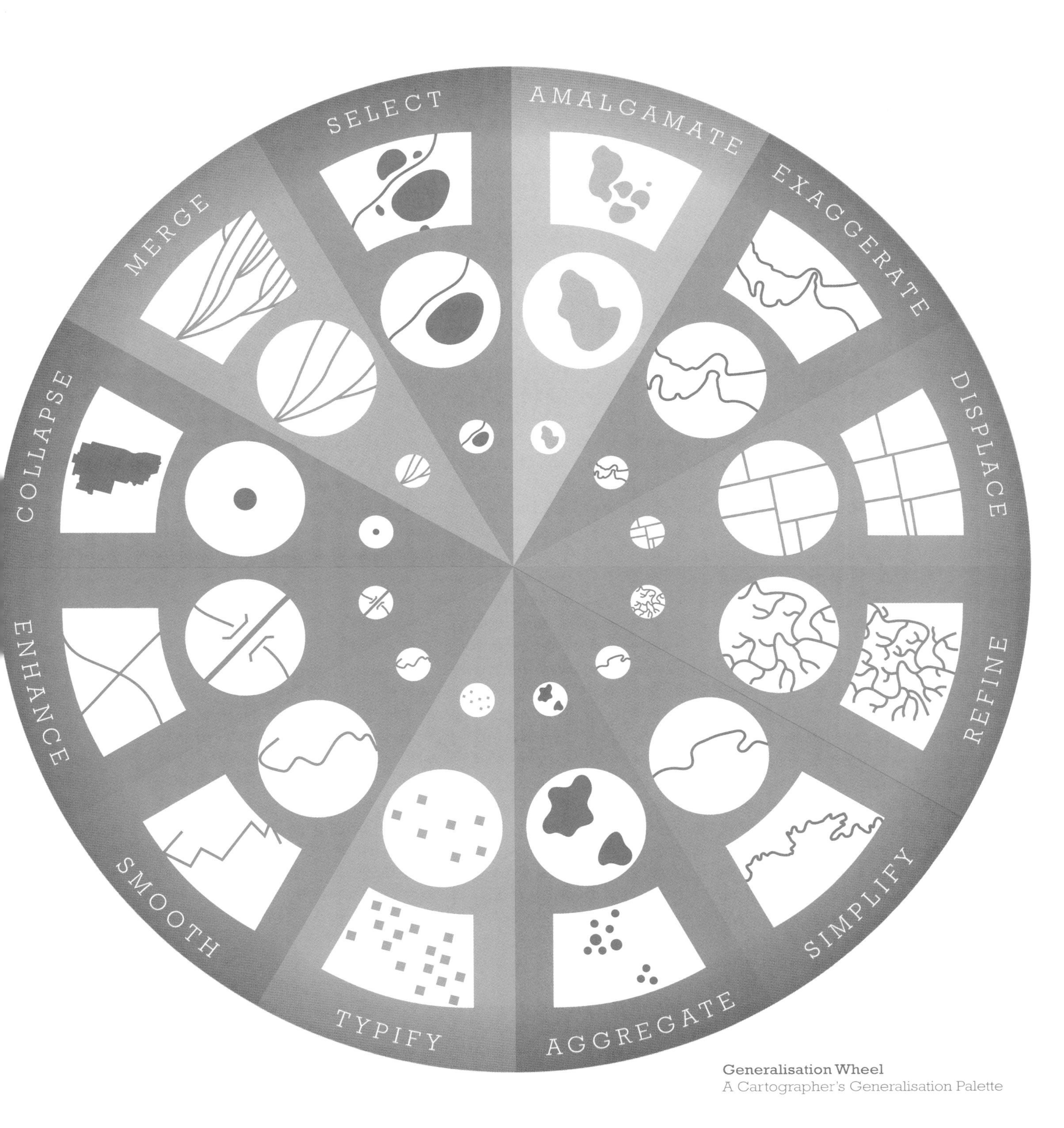


**Generalisation Wheel**
A Cartographer's Generalisation Palette

# Geological maps

## Special-purpose mapping that shows subsurface features.

**Maps are not restricted to the display of surface features, and geological maps show the distribution at Earth's surface of different types of Earth's materials.** The planimetric nature of a topographic map sheet can be used as a framework for displaying information about subsurface features. Rock units and geological strata are normally indicated using different colours, akin to a land-use map in which a qualitative difference between features of different character is shown. Exposure at the surface is often the focus of such maps. Bedding planes and other structural features such as faults, folds, and lineations are often added using symbols that depict dip and strike. These signify the third dimension. Additionally, stratigraphy is often shown using contour lines to indicate the subsurface trends. An isopach map can also be used to illustrate the thickness of strata.

**The patterns on a geological map show relationships between rocks and rock types.** They support the understanding of how these rocks were formed as well as their 3D arrangement beneath the surface. They provide a way of illustrating the prediction of features some way beneath the surface that has both value and importance. Geological maps are crucial for understanding oil reservoirs, aquifers, mineral and ore deposits, and even the likelihood of subsidence, tectonic shift, and earthquakes. Of course, knowledge about what rocks intersect Earth's surface also help in understanding landscapes more generally, for geographers and archeologists.

**Because most geological features are arranged in three dimensions, the geological map is different from most ordinary planimetric maps.** Most of the mapped detail is based on interpretation rather than direct observation. Geology is not only concerned with spatial arrangement of rocks but also conditions under which they were formed as well as geological time. Maps tend therefore to reflect how well understood an area is at a given time.

**Opposite:** English geologist William Smith is credited as the first to have created a nationwide geological map. Smith's legacy is in the first-ever map to collate a full geological record of a whole country into a single map.

Published in 1815, Smith's map is perhaps also an early marker for that most current trend in digital mapmaking—the mashup of third-party basemaps and some sort of overlay. He uses conventional symbols to show urban and rural areas, roads, tramways, and collieries and mines. The geology itself is applied over the top and all hand-drawn using a range of colours to denote the different rock types. One of Smith's approaches was to use the fossil record as a way of establishing the strata as opposed to simply rock composition. His map was more accurate as a result. Indeed, the map Smith created is not far off the modern geological map of England and Wales at this scale illustrating just how accurate he managed to make his map.

The map was made in a range of formats: on sheets, or canvas, or mounted on rollers. In total the map measures approximately 8 feet tall by 6 feet wide. He oversaw the hand-colouring of each of approximately 400 maps, and each is numbered and signed.

It's possible Smith could also claim to have made a map with one of the longest titles. The full title: *A delineation of the strata of England and Wales with part of Scotland: exhibiting the collieries and mines, the marshes and fen lands originally overflowed by the sea, and the varieties of soil according to the variations in the substrata, illustrated by the most descriptive names.* Quite a mouthful and not recommended.

**See also:** Profiles and cross-sections | Rock drawing | Small landform representation | Topographic maps

OF THE
STRATA
OF
ENGLAND AND WALES,
WITH PART OF
SCOTLAND;
EXHIBITING
THE COLLIERIES AND MINES
THE MARSHES AND FEN LANDS ORIGINALLY OVERFLOWED BY THE SEA.
AND THE
VARIETIES OF SOIL
ACCORDING TO THE VARIATIONS IN THE SUBSTRATA,
ILLUSTRATED by the MOST DESCRIPTIVE NAMES
BY W. SMITH.
FIRTH OF FORTH
FIRTH OF CLYDE
THE GERMAN OCEAN
THE IRISH SEA
CAERNARVON BAY
CARDIGAN BAY
St. GEORGE'S CHANNEL
BRISTOL CHANNEL
THE WASH
MOUTH OF THE THAMES
STRAITS OF DOVER

# Globes

## A three-dimensional scale model of a celestial body.

**A three-dimensional model of Earth is referred to as a *terrestrial* or *geographical globe*.** Globes can, of course, be made to represent any other planet or moon and share the basic characteristic that they are spherical. Despite the fact that Earth isn't a perfect sphere, at the size and scale of most globes the oblateness wouldn't be pronounced anyway. Although many globes have a smooth surface, often they have raised relief that corresponds to topography. This is normally exaggerated since even Mount Everest on a 30 cm desk globe would be less than 2 mm in height if rendered to scale.

**Globes have often been referred to as embodying the perfect map projection—because they simply don't need one.** Projections are only used to translate three-dimensional geometries to a planar surface. Globes are therefore the only representation of Earth that contains no inherent distortion caused by a map projection. Parallels of latitude and meridians of longitude are often printed onto the globe to aid location, and they are often mounted at an angle (23.5° to reflect Earth's angle in relation to the sun) to allow them to spin on their axis. They are also highly tangible objects, capable of being spun and reoriented depending on their mount.

**Virtual globes are the digital equivalent and generally support the same functions of being able to be spun and to view Earth from different angles across their curved surfaces.** The ability to zoom into a virtual globe allows detail to be included at different scales and resolutions, effectively forming a digital equivalent of stitched topographic mapping or aerial imagery draped across a digital elevation model at larger scales. Virtual globes have also been used as a basis for thematic mapping though care must be taken because of the distortions that perspective and the curved surface lead to.

**See also:** Animation | Craft | Interaction | Types of map

Globes are some of the oldest cartographic products largely because measurement of the sphericity of Earth is of considerable interest to astronomers, mathematicians, and scholars. The earliest terrestrial globe can be traced back to the third century BC. The oldest surviving terrestrial globe is the *Nürnberg Terrestrial Globe* (or *Erdapfel*) made by German mapmaker Martin Behaim around 1492 though the Americas, as yet undiscovered, are obviously missing.

Globes have long been used as status symbols. Many historical globes have provenance as being made for specific people such as the Coronelli celestial globes made for King Louis XIV of France in the late 1600s. Globes also frequently appear in the frontispieces of historical atlases as symbols of power, domination, and territory. They are also depicted in art, reflecting the tools of various scientific trades such as in paintings by the Dutch artist Johannes Vermeer including *The Geographer* (ca. 1668).

Many modern globes are constructed from thermoplastic, with flat pieces of plastic printed and moulded into hemispheres and joined together. Traditionally, globes were formed from wood or plaster and the map itself printed into gores (strips) which were glued onto the surface. Gores are wider at the equator and narrow to a point at each of the poles. The distortions from the individual flat gores were minimised by increasing the number of gores.

**Opposite:**

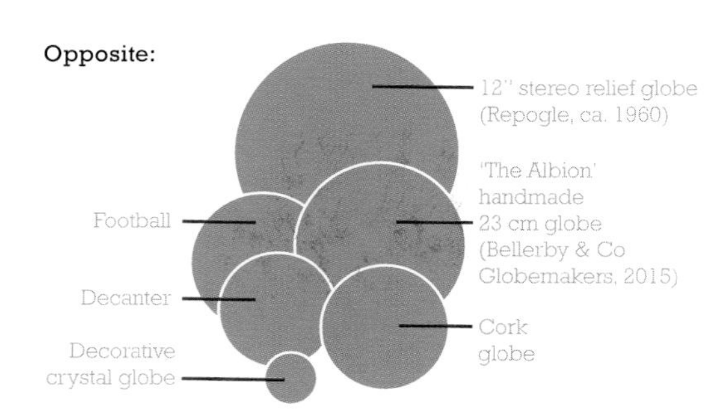

INDIAN
OCEAN

# Graduated symbol maps

## Graduated symbol maps show quantitative differences between features grouped into classes of similar values.

**A graduated symbol map shows how features differ in quantity for the theme being mapped, grouped into classes of similar values.** The data should be numerical (quantitative) and represent differences between features on an interval or ratio scale of measurement. Data is usually absolute (totals). Data is classified using a scheme that reflects the data distribution in a similar way to a choropleth map.

**Symbols should be designed so that different magnitudes of data are easily distinguished.** Symbols should be scaled so that those representing the smallest class are visible and those representing the largest class do not overly smother the map. Some overlaps are inevitable but using graphical techniques such as an outline or cut-outs allow different symbols to be seen. Generally, symbols should be geometric though with only a few differently sized symbols to recognise there is greater flexibility in using non-geometric shapes than a proportional symbol map. You must still support recognition and see comparisons and differences across the map, and there should be enough variation between each class of symbols to make them distinguishable from one another.

**The legend should include a representation of each of the symbol classes to enable readers to understand the classification scheme.** Legends are usually strung out or nested. When viewed, the map reader should be able to efficiently estimate the different quantities mapped in different areas. At the very least, relative differences should be obvious and the reader should be able to determine a pattern across the map.

**When size is used as an ordering visual variable we are ascribing more importance to the larger magnitudes of data.** We visually interpret the symbols as differently sized so we perceive larger symbols as meaning 'more'. When using graduated symbols on a multiscale web map the problems of overlapping symbols are not as acute since zooming in to larger scales reveals more detail and overlapping symbols naturally separate as they are held constant in size.

**See also:** Dimensional perception | Flow maps | Points | Proportional symbol maps | Size | Threshold of perception

Graduated and proportional symbol maps are similar in many respects. They are both designed to represent differences in magnitude but there's a subtlety in the way the data is represented. A proportional symbol map has constantly varying symbols, each one a function of the precise magnitude. Graduated symbol maps use classified symbology. There are many reasons you may prefer one technique over the other. Firstly, using exact values can be problematic when you have very large values, a large range, or a few outliers. These can result in a few awkwardly sized symbols that will dominate the map. Graduated symbols may be a better option because symbol sizes are arbitrary and not directly linked to the data. Of course, the drawback is you lose a lot of symbol variation through classifying the data into only a few broad categories. The pros and cons are similar to the argument between classed and unclassed choropleths. Both are useful techniques but your precise data and map demands will likely determine which is most suitable.

Graduated and proportional symbols had been used on graphs in the mid-1850s by Charles Minard before they first appeared on maps. He used a circle as the shape, sized by one variable and made them into pie charts to symbolize additional variables. Graduated (and proportional) symbol maps can also show multivariate information in this and other ways, but be careful not to overload the symbols too much as, along with overlaps, this can create problems for legibility. Mimetic symbology can also be used but this often causes increased difficulty for people estimating symbol size and value. It is far easier estimating the area of a circle relative to another than it is any other shape. Increasing shape complexity (and three-dimensional symbols) create increased difficulties.

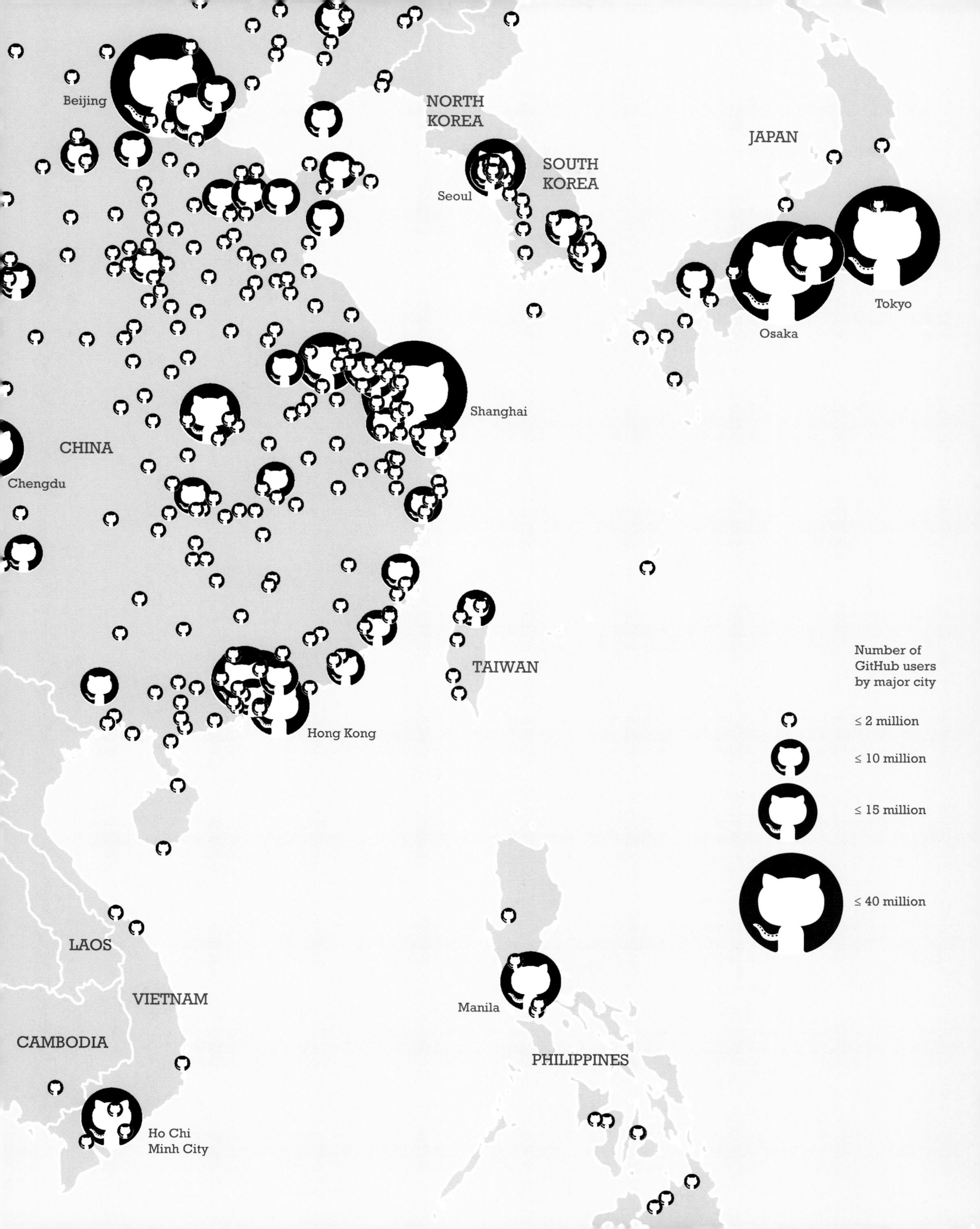

Beijing
NORTH
KOREA
JAPAN
SOUTH
KOREA
Seoul
Tokyo
Osaka
Shanghai
CHINA
Chengdu
TAIWAN
Hong Kong
Number of
GitHub users
by major city
≤ 2 million
≤ 10 million
≤ 15 million
≤ 40 million
LAOS
VIETNAM
CAMBODIA
Manila
PHILIPPINES
Ho Chi
Minh City

# Graphic and dynamic labelling

## Automating the process of map labelling can create consistency.

**Positioning labels by hand has now largely been replaced by automated means.** Many algorithms for automated text placement are built using heuristics or 'rule of thumb' approaches. However, by their very definition they cannot accommodate every conceivable set of circumstances surrounding placement for any given part of a map. It is still important to review placement and modify even where automated techniques are used to place text initially. This step can be crucial to give your map a cartographic polish. Even very small modifications to label colour, size, and position can make a large difference to the overall composition and legibility of the map. It is also worth noting that labelling algorithms only extend to mapped labels and not titles, legends, and other marginalia though templates and map specification sheets can be used for consistency.

**Graphic labels add general information to a map such as the title, subtitle, and source.** They are simple text elements that are not associated with any particular map feature and will stay in the same position and at the same size as they were designed whether on a printed map sheet or the marginalia of a web map. For multiscale web maps, graphic labels tend to remain as they are as well.

**Automated data-driven dynamic labelling supports multiple rules applied in response to how the map is presented.** For instance, you might alter classes of type, set type characteristics, weights, and buffers, allow overlapping labels, control duplicate labels, and constrain placement conventions. This offers a good solution for labelling data as a precursor to creating maps at different scales from the same input data. As scale and extent of the map is altered, labels will be placed and styled dynamically. At smaller scales the number of labels shown may be restricted. At larger scales there is more available space and more labels will be shown, often in more optimal positions. This is a key component of multiscale web maps that supports progressive labelling to give more detail where it's needed and when it's needed in the viewing experience.

**See also:** Elements of type | Guidelines for lettering | Typographic maps

Labelling multiscale web maps introduces particular labelling issues. While putting map detail onto the map is normally the optimum approach so you don't hide map detail, web maps sometimes suffer from graphical overload and using interactivity can be used for the labelling component.

The size of the map seen through the window of a digital device will almost certainly be smaller than a print equivalent and so the space for labels reduced. Using a progressive reveal so that only essential labels are seen at smaller scales with more revealed as you zoom into the map is a useful principle to follow.

Typefaces for web maps will rely on those that are available from either server or clientside (unless a map is cached as raster tiles), and so designing with a restricted choice can limit labels being switched for alternatives that might not fit the design.

As with other content on a web map, labels can be hidden behind actions or in pop-up windows. For instance, on a choropleth map it makes little sense labelling enumeration areas due to the fact the labels will both clutter the map and obscure the map's central detail. Allowing a label to appear as a ToolTip as a reader pans across the map can be helpful. Adding appropriate labels in a click-event such as a pop-up window is also a useful way of adding context to the map.

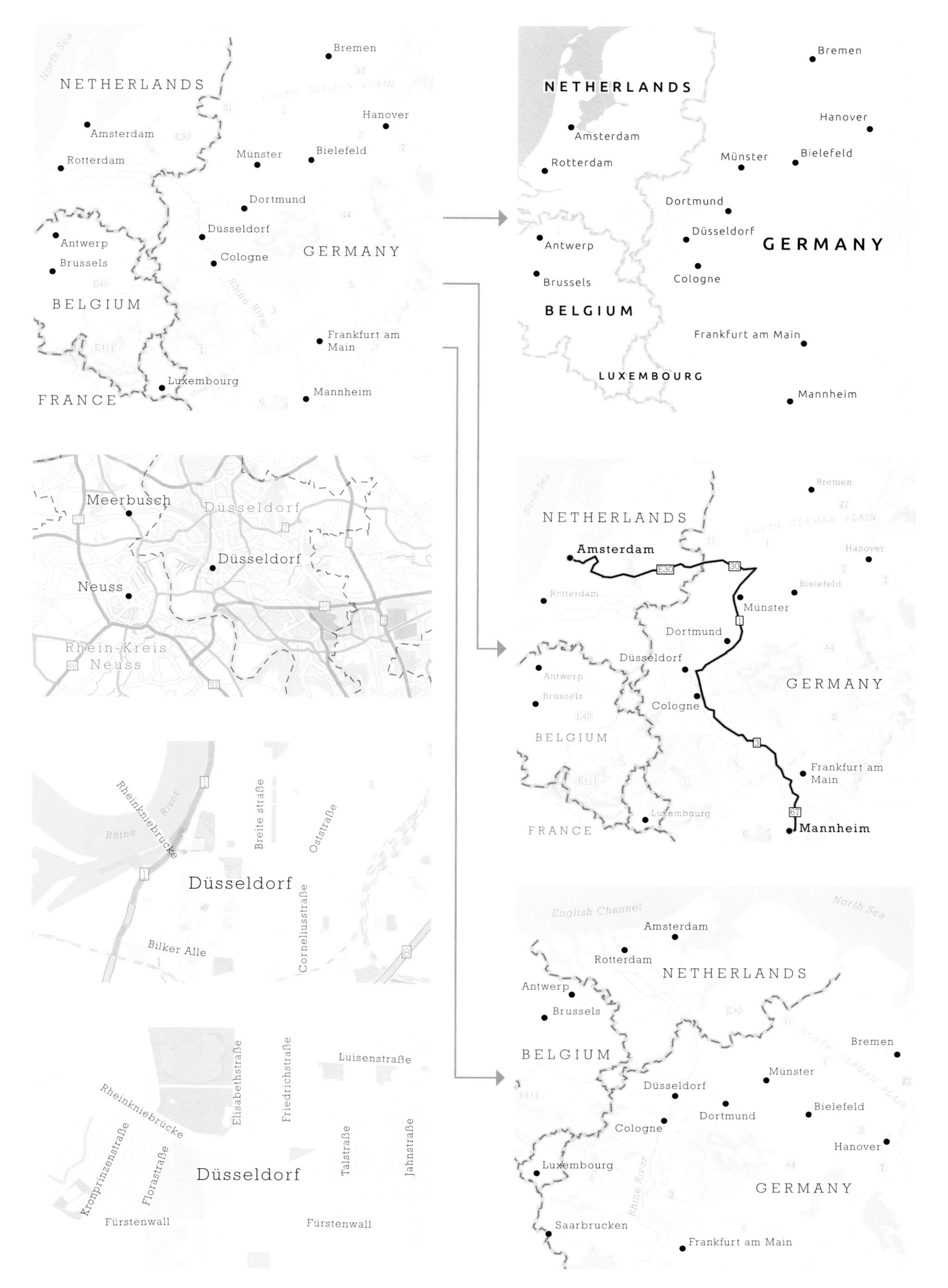
Label adjustment with Scale Change
Label adjustment with Content Change
Label adjustment with User Command Change
Label adjustment with Rotation Change
NETHERLANDS
Bremen
Hanover
Amsterdam
Rotterdam
Münster
Bielefeld
Dortmund
Düsseldorf
Antwerp
Brussels
Cologne
GERMANY
BELGIUM
Frankfurt am Main
Luxembourg
Mannheim
FRANCE
Meerbusch
Düsseldorf
Neuss
Rhein-Kreis Neuss
Rheinkniebrücke
Breite straße
Oststraße
Rhine
River
Düsseldorf
Corneliusstraße
Bilker Alle
Luisenstraße
Elisabethstraße
Friedrichstraße
Kronprinzenstraße
Florastraße
Talstraße
Jahnstraße
Fürstenwall
LUXEMBOURG
English Channel
North Sea
Saarbrucken

# Graphicacy

## Understanding how to communicate visually.

**Schooling is the main way in which we learn how to communicate.** From a very early age we are taught to read, write, and handle numbers. We are taught how to stand before a group of people and present using the spoken word. We call these *literacy*, *numeracy*, and *articulacy* (oracy), and they form the basis of education—the three Rs. We hone these skills as we go through school and higher education. They become the basis of how we work and how we relate to others. But how many of us have any formal training in presenting information using visual forms? This, we might call 'graphicacy'.

**'Graphicacy' was coined in the 1960s by William Balchin and Alice Coleman.** The term was defined as 'the fourth ace in the pack' and a way of conceptualising the geographical importance of skills such as map-reading and drawing (Balchin and Coleman, 1966). Balchin and Coleman had become concerned at the loss of essential skills and proposed the term to promote the idea of teaching it alongside the other three pillars of education. They reasoned that 'in the choice of a word to denote the educated counterpart of visual-spatial ability one must first ask the question what exactly does this form of communication involve. It is fundamentally the communication of spatial information that cannot be conveyed adequately by verbal or numerical means, e.g. the plan of a town, the pattern of a drainage network, or a picture of a distant place—in other words the whole field of the graphic arts and much of geography, cartography, computer-graphics, photography, itself. All these words contain the syllable "graph" which seemed a logical stem for "graphicacy," which was completed by analogy with literacy, numeracy, and articulacy.'

**Being able to communicate visually is at the heart of cartography, and so graphicacy is crucial.** Understanding the language, grammar, and syntax of visual communication is what cartography is about. It also helps us understand it is a wholly different form of communication that in some ways has to be learnt and practiced as much as the three Rs. Graphicacy is not innate though clearly, as with learning languages, some will have a greater predisposition than others.

**See also:** Abstraction and signage | Cognitive biases | Design and response | Form and function | Semiotics | Symbolisation

The written and spoken word have a linear structure and formal rules for organising. Graphics do not conform to the same structural definitions. Maps are unstructured until the cartographer builds structure. Take the following passage of text that describes something of the geography of a region.

Draw, stating your scale, a contoured map to show an island 65 km long from SW to NE which varies in width from 48 km in the SW to 16 km in the NE. The SW coast is much dissected by long, narrow fjord-like inlets, and is fringed by 5 small rocky islands of varying sizes. From this coast the land rises sharply to a plateau some 600 m above sea level and extending through about one-third of the island. The plateau descends to a low undulating plain about 25 km long and 20 km wide. From the plain a range of hills rises to the NE, flanked by a coastal plain about 8 km wide. From these hills, rivers flow to both plains and also from the plateau to the larger plain. The plateau is gritstone. The hills which run down to the coast in the NE, to form cliffs, are chalk. Much of the smaller coastal plain is marsh, but the larger plain, from which two estuaries open, is of well drained alluvial land. In addition to relief and topography, show drainage, possible sites of settlement and lines of communication. Name your island appropriately!

This is the first exercise I was set as a student of cartography at Oxford Polytechnic by Roger Anson. The purpose was to interpret the passage and draw. The extent to which we made a map that reflected the passage was a fascinating challenge. The map, as you see, is a human foot, conceived as Anson got out of the bath one day and saw his soggy footprint on the mat. I've used this exercise regularly, and it demonstrates clearly not only the difference between the written word and the map but also the importance of the processes that drive graphicacy.

**Opposite:** *Anson Island* by Kenneth Field, 1989 (redrawn, 2017).

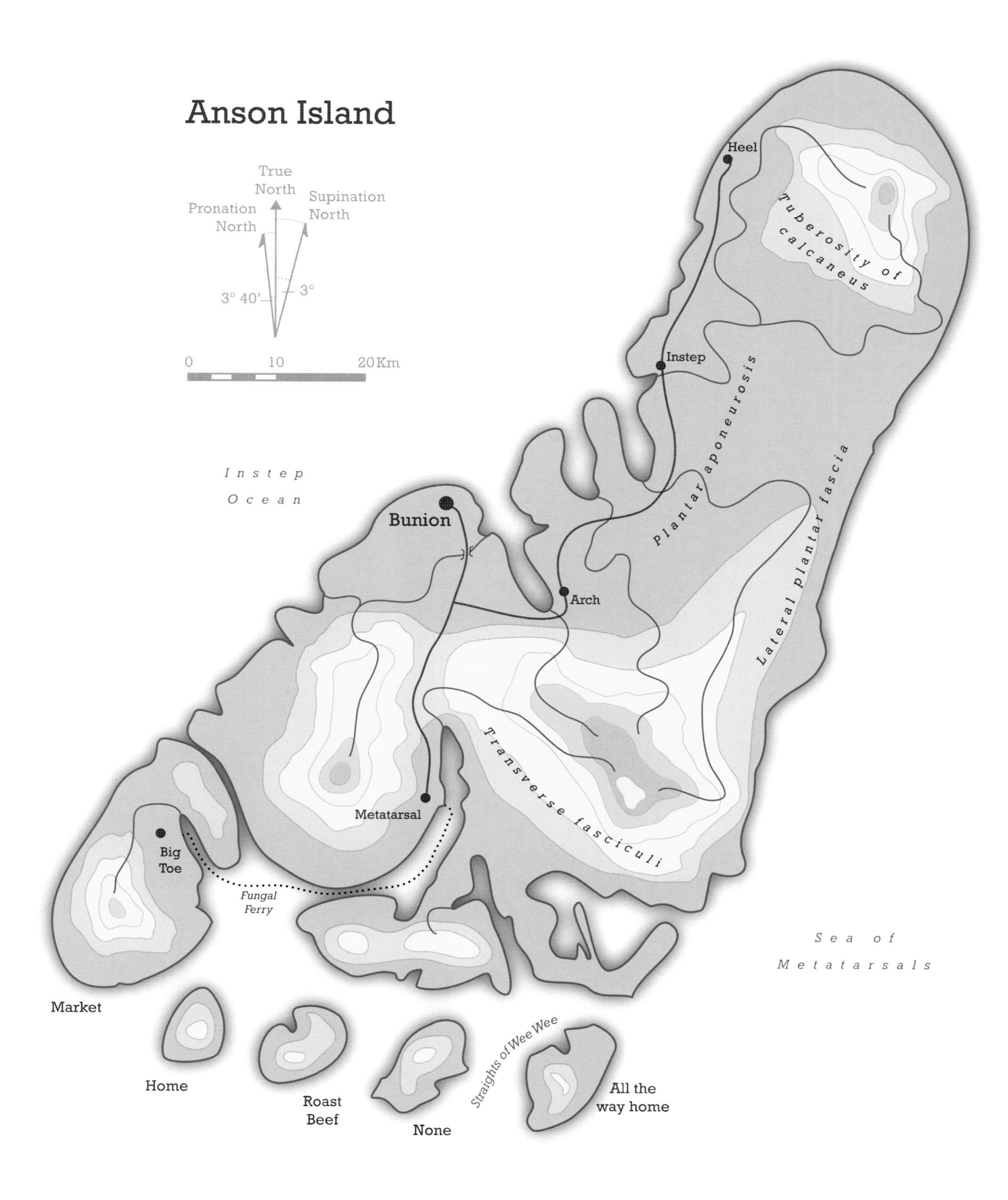
Anson Island
True North
Pronation North
Supination North
3° 40'
3°
0
10
20Km
Instep Ocean
Heel
Tuberosity of calcaneus
Instep
Plantar aponeurosis
Lateral plantar fascia
Bunion
Arch
Transverse fasciculi
Metatarsal
Big Toe
Fungal Ferry
Sea of Metatarsals
Market
Home
Roast Beef
None
Straights of Wee Wee
All the way home

# Graphs

## Graphs used to summarize data can be combined with maps in many ways to show the spatial dimension.

**Graphs play an important role in helping us understand the world around us by summarizing quantitative data into visual forms.** Usually this data is spatial but graphs can be combined with maps in numerous ways. Graphs can be useful in augmenting a map display, perhaps summarizing key details and adding to the representation. They might be positioned around a map as part of the overall composition. When data does have a spatial component you can use the map as a canvas upon which to show a series of minigraphs that display the same variables but at different locations. In this sense, graphs and maps are complementary and can work well together either as symbols in their own right or in pop-up windows on a web map.

**Graphs all have common elements and should be designed to be read independently.** In any graph there is a data region, some form of boundary, and scaled rulers and, in many, this comprises x- and y-axes. Labels, tick marks, lines of best fit, and other elements provide the framework for us to read and interpret the graph. All these elements must be incorporated into graphs used on maps, although if the map is digital, many can be hidden and revealed as rollovers or mouse clicks to avoid visual clutter. Certainly, if map scale is limited, then graphs must be designed in sympathy and use minimal complexity.

**The key challenge of placing graphs on a map is size and positioning.** The size of each graph must remain legible but balanced with an organisation that does not result in overlaps. The map usually becomes the scaffolding upon which the graphs are placed. In a geographic context this can show something of the spatial pattern as different variables may begin to display spatial characteristics. Graphs can also be used in a cartogram form by organising them into a grid that reflects geography in a stylized manner. In this sense, the boundary of the graph itself becomes the equivalent of the spatial boundary of the unit that it represents, and the map presents graphs in a small multiple arrangement.

**See also:** Pie and coxcomb charts | Profiles and cross-sections | Schematic maps | Small multiples | Treemap | Waffle grid

Of the numerous graph types, here are the more common ones to consider.

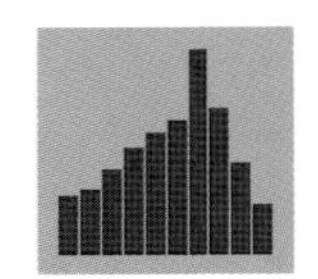

Histograms show the statistical distribution of the frequency of occurrence of data using classes of an equal range of values. Frequency polygons can be created by joining the tops of the bars in a histogram and, if reordered, could be represented using a Lorenz curve.

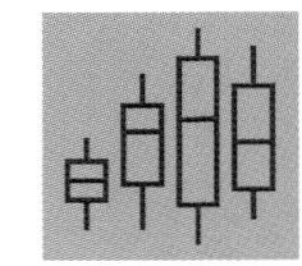

Box-whisper plots illustrate the percentile summary of a dataset. The graph usually displays the data minimum and maximum, third and first quartiles, and median. It can also help visualise outliers.

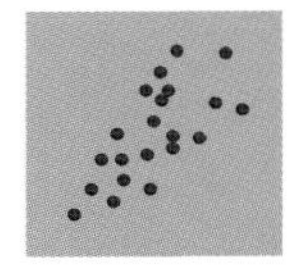

Scatter plots allow you to explore the relationship between two variables that intersect when plotted against different x- and y-axes. A bubble graph extends the scatter plot by encoding a third variable into each data point by changing the symbol size. More complexity can be introduced to show qualitative difference using colour or shape.

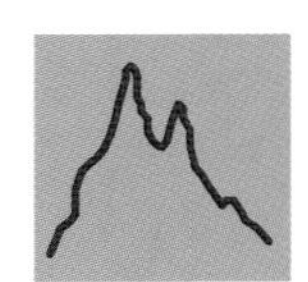

Line and bar graphs can display multiple variables using different line shapes or colours to represent individual variables. Line graphs provide an opportunity to show how patterns of data vary temporally which can be useful on maps that might otherwise not display temporal data effectively. Bar graphs use height to show a quantifiable difference between data categories. The bar widths have no context.

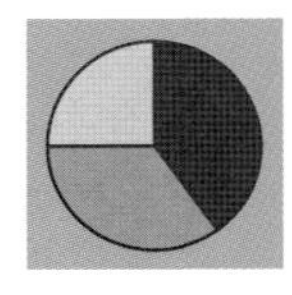

Pie charts show subcategories that add up to 100%, displayed as wedges which vary in size defined by the angle. Different hues are often used to differentiate between individual sectors to show qualitative difference. Pie charts can also form the basis of more complex multivariate charts.

**Opposite:** *London's Tube DNA* by Kenneth Field, 2017, showing the average total weekday passenger use by line and station as gridded small multiples graphs.

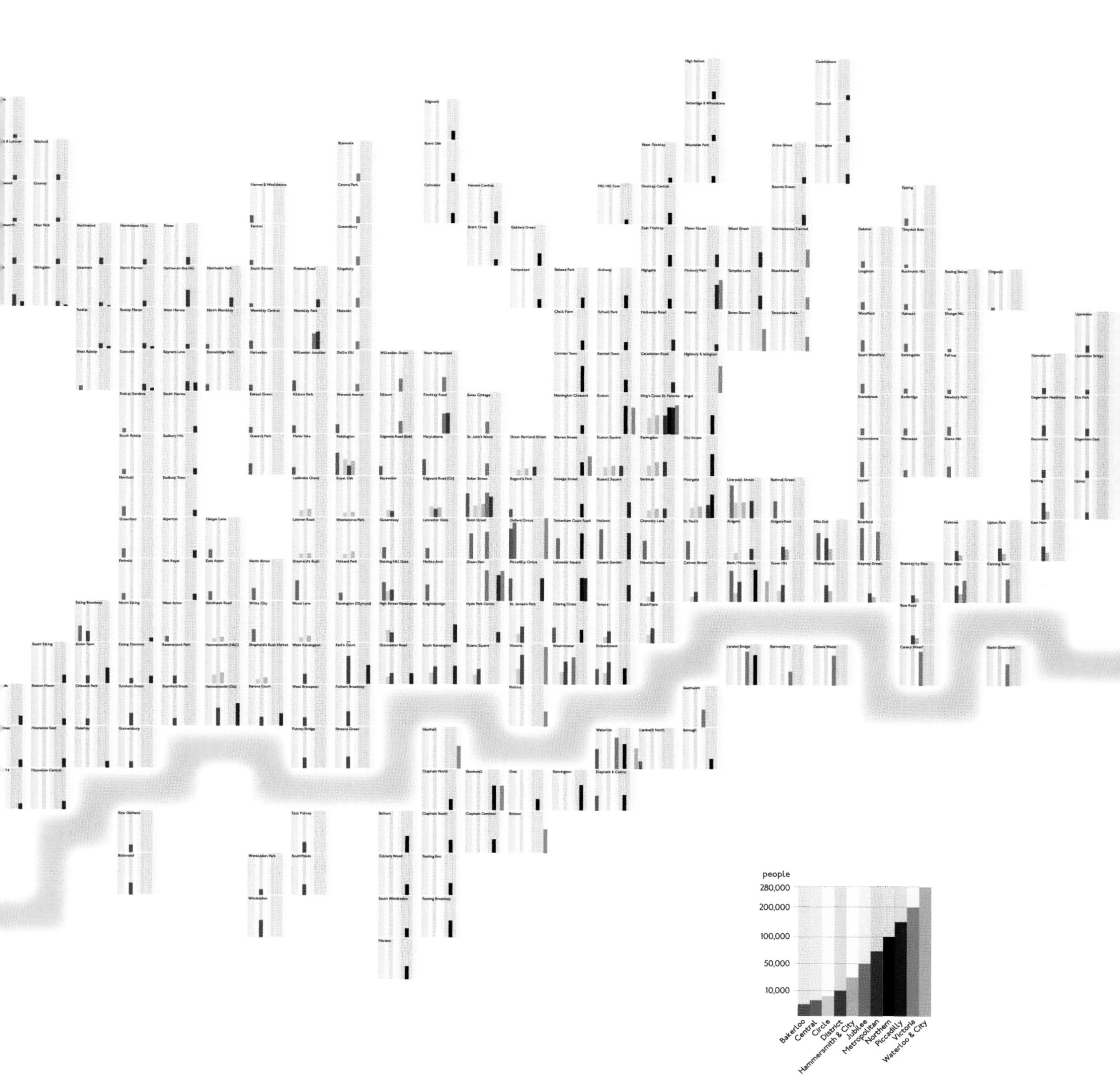
people
280,000
200,000
100,000
50,000
10,000
Bakerloo
Central
Circle
District
Hammersmith & City
Jubilee
Metropolitan
Northern
Piccadilly
Victoria
Waterloo & City

# Graticules, grids, and neatlines

## The skeleton of the map provides a framework for its content.

**Maps are often drawn with respect to coordinate systems, reference grids, and layouts.** Predominantly for topographic maps, these components are important for the overall interpretation of the map and support a range of functions. They give the map context as well as a formal structure.

**The graticule is a network of lines that represent latitude and longitude.** A map's projection will fundamentally determine the appearance of the graticule. The major family of projection—cylindrical, conic, or azimuthal—can easily be recognised from the structure of the graticule. For instance, lines may be straight or curved, and they may be parallel to one another or converging. Separation of lines may be constant across the map or vary from place to place. The angle formed at the intersection of parallels of longitude and meridians of latitude can be regular or vary, being any size from place to place across the map.

**A grid is a set of straight lines that intersect at right angles.** Grids are used to create local reference systems and define position on a map by distance based on a Cartesian coordinate system. Many countries have their own reference grid and nomenclature for specifying position. Grids are often used in conjunction with a graticule though they also sometimes replace a formal graticule as the main locating structure on the map.

**Neatlines specify the frame for a map.** They enclose the map detail and the limits of the mapped area. They are used more in print cartography than online where the concept of map edges rarely exists. That said, frames are useful graphic devices to contain your map and assist in defining context and balance. It's always a useful exercise to frame your map and use it as a container into which your map components are positioned. This can equally apply to a web map even though only a portion of the map may be viewed at any one time.

Although graticules, grids, and neatlines have distinct functions on a map, they are interchangeable in a graphical sense. On large-scale printed maps, neatlines are often formed from the reference grid itself. The shape of the map will therefore always be square or rectangular. On small-and some medium-scale maps, neatlines are constructed from two parallels and two meridians of the graticule. If the projection is cylindrical, the neatlines will enclose a map that is rectangular. For a conic projection, the neatlines will likely be straight when they are based on parallels and curved when based on meridians. The edge of the map nearer one of the poles is likely to be shorter than the one nearer the equator.

Neatlines can also be entirely arbitrary, being straight lines that have no association with either lines of the graticule or the grid. They simply exist to subdivide the map into distinct areas, sometimes demarcating a main map with insets that have been repositioned for the purposes of arrangement and composition.

The separation of the grid or graticule will be a function of the scale of the map. A suitable, evenly spaced separation usually conforms to rounded numbers (in distance or degrees). The symbolisation of the graticule can take different forms such as solid thin lines, tick marks on the bordering neatline, or small crossed marks where lines would intersect across the map. Grids are more usually shown as solid or finely dotted lines. Sometimes lines or crosses are omitted or masked when they intersect with other map detail. Colours are usually neutral, but where a grid and graticule are shown, then different colours should be used to clarify the difference. Although playing an important functional role on the map, the symbols should not be figural.

**See also:** Anatomy of a map | Latitude | Legends | Longitude | Which way is up?

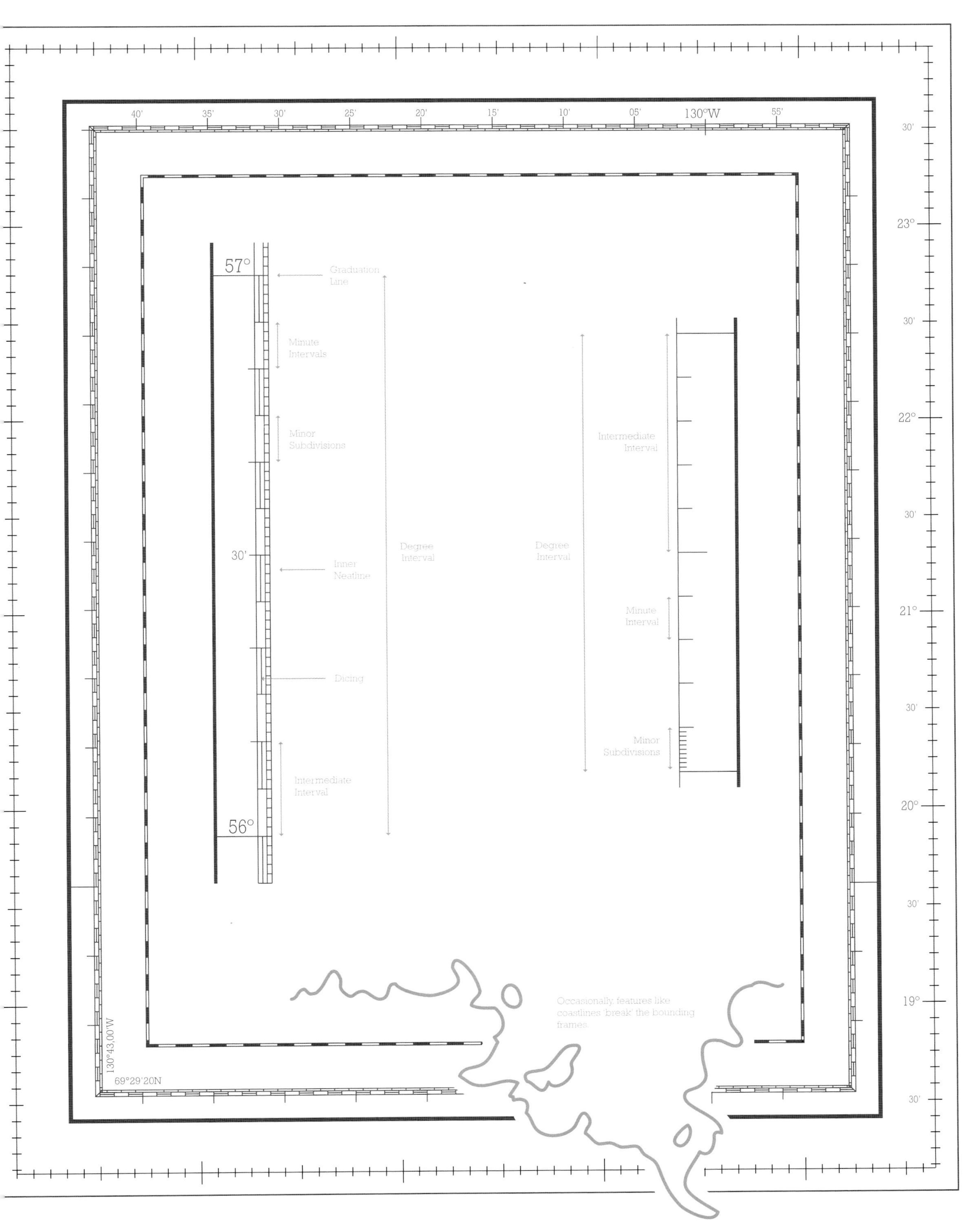

40'
35'
30'
25'
20'
15'
10'
05'
130°W
55'
30'
23°
30'
22°
30'
21°
30'
20°
30'
19°
30'
57°
Graduation Line
Minute Intervals
Minor Subdivisions
30'
Inner Neatline
Dicing
Intermediate Interval
56°
Degree Interval
Degree Interval
Intermediate Interval
Minute Interval
Minor Subdivisions
Occasionally, features like coastlines 'break' the bounding frames.
130°43,00'W
69°29'20N

# Greyscale

## Maps don't have to be in glorious technicolour!

**It's not that long ago that colour was rare in map production.** Colour maps were really only the preserve of topographic maps published by national mapping agencies, or atlases. Most one-off maps or those found in books and periodicals were printed in monochrome—sometimes simple black line work on a white background or sometimes in greyscale often printed using halftone techniques. And web maps? There was a time before they even existed. Colour has become the modus operandi for cartography simply because it's easy to make colour products. Digital maps are consumed on devices with over 16 million colour possibilities. Colour printers and copiers are cheap and ubiquitous. The end result is people's maps default to colour.

**Colour isn't always optimum for map design.** Learning to design in greyscale forces you to be more critical in what you include or omit and how you symbolise. Your palette of possibilities is constrained, and that forces you to think more carefully about your design. You must be creative to build contrast into the map and to make each mark legible and easily distinguished among the other marks. Pattern becomes a useful technique.

**Monochrome isn't simply a desaturation of your map's colours.** There is a difference between creating your map in full colour and having to adapt it to monochrome and starting in monochrome. If you find your map must be reproduced in monochrome, it's likely that simply desaturating the colours to their greyscale equivalent won't work. The nuances of hue and chroma that are characteristic of different colours may lead to similar grey values when desaturated, making symbols all appear the same. Thinking of your map as a greyscale product first and then applying colour later can be a good way to ensure functional difference between symbology.

Apart from the simple fact that colour has a higher overhead of cost associated with the design phase as well as production, monochrome maps can be very effective. It's always worth considering whether, in fact, a greyscale version may even be optimal since black on white provides you with the most contrast available. Using shades of grey as well as shape and pattern also avoids some of the complexities of colour.

It's common for monochrome maps to be referred to as 'minimalist' simply because they eschew colour. Monochrome represents a certain aesthetic. Such styles are commonplace as third-party digital basemaps, designed for a splash of colour to be placed across the top, normally relating to a thematic content. The contrast between monochrome basemaps and colour overlays works well, particularly for digital displays and for supporting the focus on overlaid thematic detail.

The same idea of staying within the greyscale palette can be expanded to maps that stay within other tonal variations. For instance, maps that are created entirely in the blue or the green spectrum of colours. In this case, thematic detail or typographic elements are usually designed in black, white, or greyscale.

Limiting your work to two tone can also be both constraining yet beautiful in design terms. Designing the majority of your map in greyscale but with a key feature isolated in a saturated colour, for example red, allows you to create maps that are visually jarring but which suit the purpose. They are designed to focus attention on a single figural component.

**See also:** Elements of colour | Illuminated contours | Saturation | Value

**Opposite:** Extract from a reference map designed in monochrome.

Gulf of Honduras
Islas de la Bahía
Caribbean Sea
La Ceiba
Tocoa
San Pedro Sula
El Progreso
Río Ulúa
HONDURAS
MOSQUITO COAST
Río Coco
Siguatepeque
Juticalpa
Comayagua
Bosawás
Tegucigalpa
Danli
ensuntepeque
Jalapa
Puerto Cabezas
SALVADOR
Siuna
Río Coco
San Miguel
Choluteca
Somoto
Río Grande de Matagalpa
Estelí
Gulf de Fonseca
Matagalpa
Cerro
Wawashang
Chinandega
Boaco
León
NICARAGUA
Managua
Lake
Río Mico
Managua
Juigalpa
El Rama
Masaya
Bluefields
Lake Nicaragua
Cerro Silva
Rivas
Río San Juan
Indio Maíz
Pacific Ocean
Liberia
Quesada
COSTA RICA
Puntarenas
San Jose

# Guidelines for lettering

**The use of type in cartography and the lettering of a map can be improved by following a range of general typographic selection guidelines.**

**Keep type as simple as possible.** Try to avoid decorative type families that make letters and words harder to distinguish among other map detail. In general, simpler letterforms and sans serif type families work better. Similarly, avoid using bold and italic styles where possible or embellishments such as underlining (visual clutter). Bold type tends to overemphasise map lettering, and sizing of type is usually sufficient to show difference. Italicised text is normally reserved for labelling of hydrographic features, which has become an accepted standard because of the slanted appearance connoting flowing water. Blue is also conventionally used for the type colour for water feature labels.

**Try to avoid using more than two different typefaces on a map.** Sufficient modifications should be possible within a typeface to represent a range of features uniquely. For a consistent appearance, map elements such as the title, subtitle, legend detail, scale, and data source should all be in the same typeface. Where two type families are employed, it is conventional practice to use one family (e.g. a serif family) to label one category of features and another family (e.g. sans serif) for a second category of features. This may be features categorised as cultural or natural on a general-purpose topographic map, for instance.

**Select a suitable lower limit for the smallest type to be placed on the map.** This should take account of the intended audience, reading distance, and lighting conditions among other factors. A lower limit of 7 points is generally accepted for a map to be read at a standard reading distance though smaller sizes might be needed for a label-heavy map. To aid legibility, text is almost always coloured black.

**Where possible, the size of the type should be in proportion to the importance of the feature.** Type for large cities should be noticeably larger than type for smaller cities or towns. Where text is used to assist a connotation of relative importance (for instance, for ordinally measured features), the difference in size between type should be at least 2 points.

**See also:** Choosing type | Fonts and type families | Lettering | Lettering in 3D | Sizing type | Spacing letters and words

Using complementary typefaces that are acknowledged as working well together is often a good way of selecting typefaces for a map. Some font pairings simply work well together because they don't fight for attention. They should contrast but not conflict. One approach is to use a typeface family that has weights and styles that are designed to work well together. Examples such as Lucida/Lucida Sans or Meta/Meta Sans provide both serif and sans serif versions that harmonise.

Alternatively, contrasting different but complementary typefaces can yield harmonious results. For this, selecting similar typefaces won't contrast enough. Old-style serif typefaces combine well with humanist sans serifs. Traditional or Modern serif typefaces pair well with geometric sans serifs.

It's important for the map reader to be able to distinguish between quantitative and/or qualitative labels and features quickly and easily. For this reason, establish a hierarchy among your labels and select typefaces that have sufficient variation to support these needs.

Avoid verbosity. For instance, avoid using the word 'map' in the title as in 'A map to show…'. It's likely obviously a map and should stand on its own without a further description to inform the reader what it is they are looking at. Similarly, the words 'legend' and 'scale' might be omitted along with other words that provide no reasonable additive value.

As with any document, a map should be thoroughly spell-checked. This is particularly important for place-names, which might change over time or where a particular spelling might prove controversial.

# ASIA

*PACIFIC OCEAN* INDIA CHINA Ocean Country

*Kara Sea* *Sea of Okhotsk* Rajasthan Sichuan Irkutsk Punjab Sea Province

*Korea Bay* *Gulf of Thailand* *Ob Bay* Beijing Tokyo Mumbai Karachi Bay Major City

*Lake Baikal* *Lake Balkhash* *North Aral Sea* Morioka Taraz Akola Guilin Vladivostok Keelung Lake City

*Ob* *Lena* *Amur* *Yangtze* *Ganges* *Irrawaddy* *Tigris* Aksay Berdsk Turbat Sircilla Udala Pyu River Town

Testing font size in relationship to importance
and the suitable lower limit of type

## Dymaxion projection

R. Buckminster Fuller

1943

Architect and inventor Richard Buckminster Fuller is best known among cartographers for his *Dymaxion Map*, a polyhedral map projection arguably superior to similar frameworks anchored arbitrarily at the poles. Here, the projection is used as a framework for a map of travel times to Hong Kong.

First published as an article in the March 1, 1943, edition of *Life* magazine, Buckminster Fuller's compromise projection contains far less distortion than other flat maps. The map was printed as a pull-out section designed to allow readers to assemble the map. It divides the globe's surface into a continuous surface without bisecting major land masses, and it is unique in that there is no right way up. It can be read from any orientation and rearranged in many alternative ways. Eight equilateral triangles and six squares, bounded by great circles that share the same scale, were configured to minimize distortion without interrupting land areas.

Fuller had been fascinated with the notion of 'one-world island in a one-world ocean' since the 1930s. He later abandoned the 14-sided cuboctahedron for a 20-sided regular icosahedron, but described the earlier version in his 1944 patent application for an invention titled merely 'Cartography'.

For the latter version, he chose the more durable protection of a copyright and the trademarked name Dymaxion, coined from 'dynamic', 'maximum', and 'tension', and used for several earlier inventions, including structures and an automobile.

—Mark Monmonier

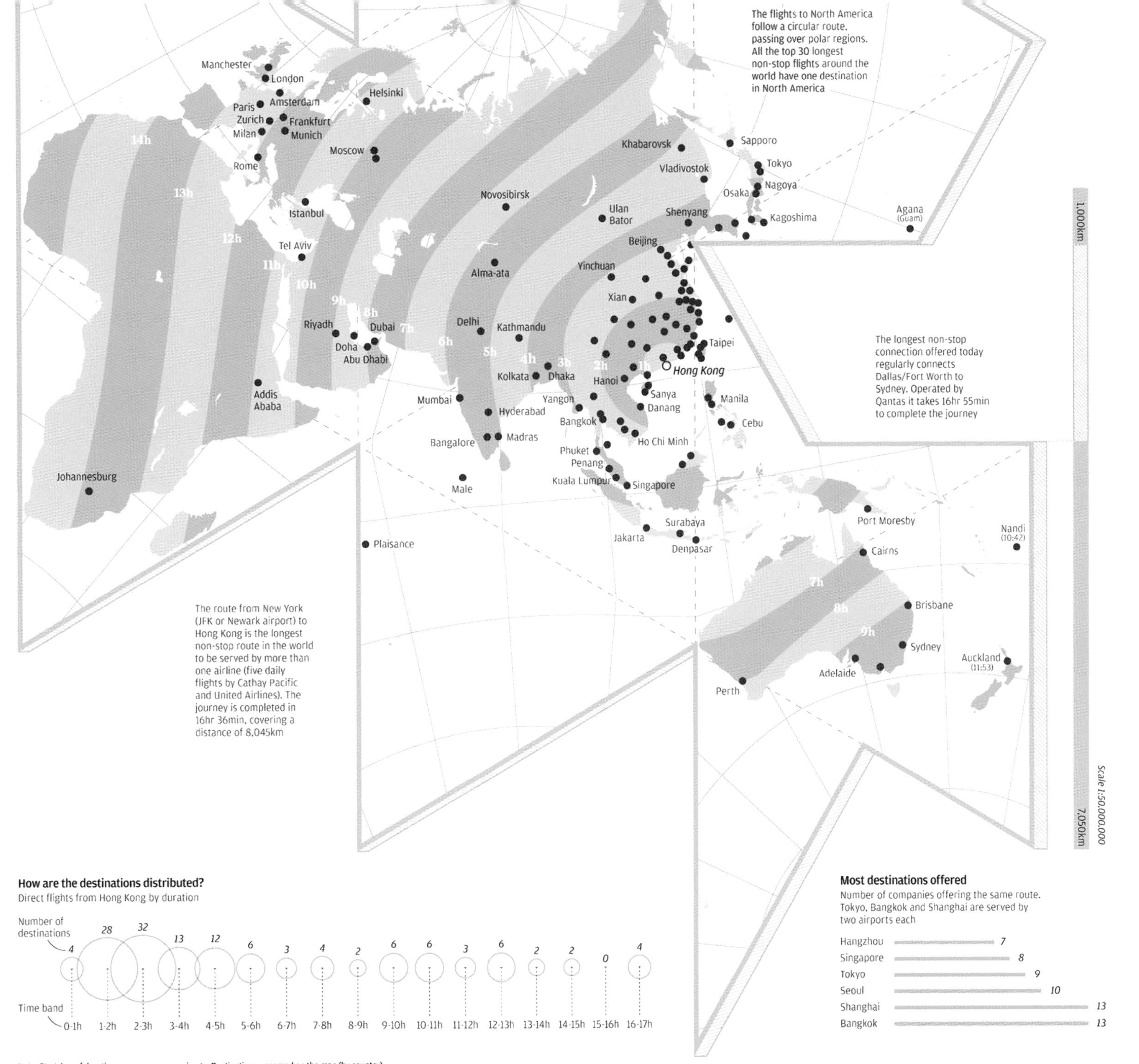


Note: Stretches of duration on map are approximate. Destinations unnamed on the map (by country):
CHINA: *Guangzhou, Meixian, Shantou, Zhanjiang, Haikou, Nanning, Guilin, Changsha, Xiamen, Quanzhou, Fuzhou, Wuyishan, Nanchang, Wuhan, Yichang, Chongqing, Guiyang, Kunming, Lijiang, Chengdu, Fuzhou, Wenzhou, Hangzhou, Ningbo, Nanjing, Shanghai, Zhengzhou, Taiyuan, Xuzhou, Yancheng, Huaian, Wuxi, Lianyungang, Qingdao, Dalian, Jinan, Tianjin.* TAIWAN: *Taichung, Tainan, Kaohsiung.* SOUTH KOREA: *Seoul, Busan, Cheju.* JAPAN: *Fukuoka, Okinawa.* PHILIPPINES: *Angeles, Iloilo.* CAMBODIA: Siam Reap, *Phnom Penh.* THAILAND: *Chiang Mai, Ko Samui.* MALAYSIA: *Kota Kinabalu.* BRUNEI: *Bandar Seri Begawan*

# Hachures

## Method by which terrain is rendered with short lines of varying length and thickness arranged so they point downhill.

**Hachures are a classic technique for representing three-dimensional topography on a two-dimensional map.** A fine line, or *hachure*, is generally drawn in the direction of the steepest topographic gradient. Hachuring across an area creates tonal variations throughout the map. These tonal variations are a form of analytical hillshading, creating a three-dimensional appearance. They cannot be used to determine any metrics such as elevation or slope angle but the density of the resulting lines gives a good impression of slope orientation and steepness.

**Possibly the most renowned hachuring was developed by Johann Lehmann, who introduced the technique to Europe in 1799.** In the Lehmann technique, the thickness of hachures varies so steeper slopes have wider, thicker, and shorter lines, while gentler slopes have longer, thinner lines separated by larger gaps. Hachures are usually drawn in black or another dark tone though applying tonal variation depending on aspect can give rise to illuminated hachures where lighter hachures are on the northwestern slopes and darker on southeastern (assuming a light source in the upper-left similar to a standard hillshade).

**By applying vertical or oblique lighting techniques, different effects can be achieved.** For instance, a technique developed by Swiss cartographer Guillaume Dufour in the mid-19th century virtually eliminated hachures on illuminated northwestern slopes, which accentuated the three-dimensional character of the resulting map.

**Hachures are generally useful for only large- to medium-scale maps.** At small scales there simply isn't enough variation in the graphic marks to provide a detailed depiction of the terrain. The result tends to be a grossly generalised appearance.

Swiss cartographer Eduard Imhof's basic rule for hachure construction:

- Follow the direction of steepest gradient,
- Arrange in horizontal rows perpendicular to direction,
- Length corresponds to the local distance between imaginary contours of equal value,
- Width becomes thicker for steeper slopes, and
- Density remains constant across the map.

In general hachures are more likely to be seen on historical maps for relief representation than modern maps which favour alternative approaches. They have, however, become a technique used to create artistic maps either as a way to replicate a retro style or just as an artistic approach.

Hachures have found additional uses such as symbolising embankments and railway cuttings. On British Ordnance Survey maps, for instance, they have also been adapted from straight lines to ones which are stretched, narrow triangles with a short base at the top of the slope, and the point of the triangle at the bottom of the slope. Such small changes in symbol design allow hachures to be adapted to different uses for representing slope.

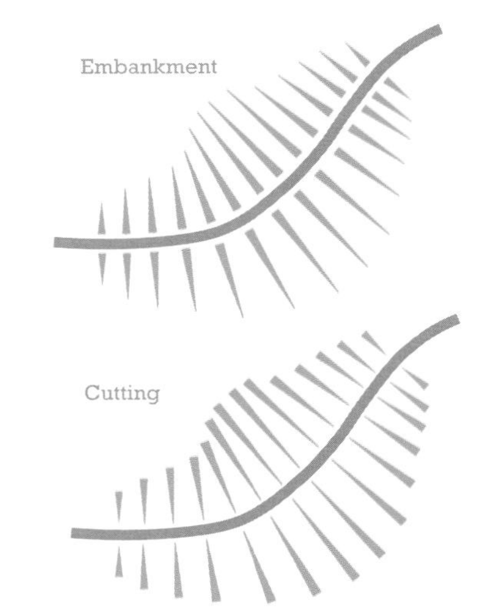

**See also:** Hand-drawn shaded relief | Hypsometric tinting | Styling shaded relief

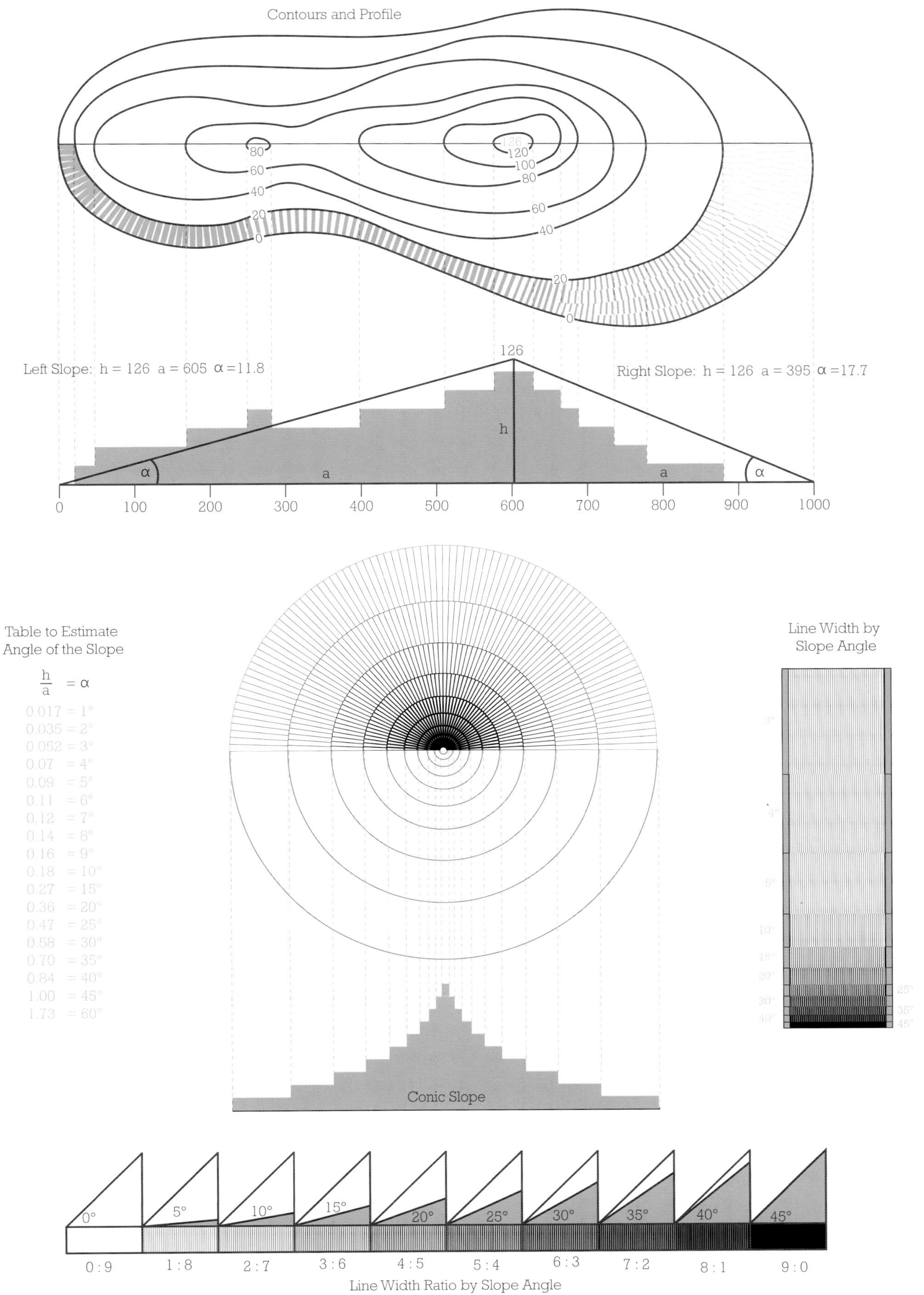
Contours and Profile
126
120
100
80
60
40
20
0
Left Slope: h = 126 a = 605 α =11.8
Right Slope: h = 126 a = 395 α =17.7
126
h
α
a
a
α
0
100
200
300
400
500
600
700
800
900
1000
Table to Estimate
Angle of the Slope
h/a = α
0.017 = 1°
0.035 = 2°
0.052 = 3°
0.07 = 4°
0.09 = 5°
0.11 = 6°
0.12 = 7°
0.14 = 8°
0.16 = 9°
0.18 = 10°
0.27 = 15°
0.36 = 20°
0.47 = 25°
0.58 = 30°
0.70 = 35°
0.84 = 40°
1.00 = 45°
1.73 = 60°
Line Width by
Slope Angle
Conic Slope
0°
5°
10°
15°
20°
25°
30°
35°
40°
45°
0 : 9
1 : 8
2 : 7
3 : 6
4 : 5
5 : 4
6 : 3
7 : 2
8 : 1
9 : 0
Line Width Ratio by Slope Angle

# Hand-drawn maps

Once the mainstay of cartographic production, now artisanal.

**All tangible maps are drawn by hand.** As technology has developed so the reliance on hand-drawn cartography has changed. Historically, maps were both designed and produced entirely by hand. They were drawn and coloured, they were engraved and printed, they included many features that could be drawn only by hand such as hillshading. As photomechanical and, subsequently, computerised techniques have replaced hand-drawn techniques, most design and production is now automated. The pen and scribing tool have been replaced by the mouse or graphics pen. Maps are still made by hand though, albeit mediated through technological wizardry, algorithms, and processors.

**Maps drawn using pen and ink have enduring appeal.** Many historic maps are feted as works of art and have become collectible because there is something very human in the work. The skilled hand of the cartographer can be seen in every stroke of the pen. It's arguable whether modern cartographic production practices have developed sufficiently to match the artistry of many hand-drawn maps. Digital tools continue to develop, and many exist to replicate many of the hand-drawn techniques that can be seen on classic maps.

**Artisanal cartography is a burgeoning landscape for cartographic expression.** Cartography has always been, in part, an artistic pursuit, and in a world where many more maps are now made digitally, we see a lot of bland cartography in design terms. Still, hand-drawn maps inspire and have something compelling about them. The marks of the pen and the shades of the colouring give the map character that is hard to replicate automatically. Often, hand-drawn maps are one-offs or bespoke products. In that sense, the craft of hand-drawn cartography is alive and well in the work of panoramists, illustrators, artists, and the maker community.

**This page is hand-drawn in the style of British fellwalker and author Alfred Wainwright.** In the mid-1900s he produced a range of guidebooks renowned for their exquisite hand-drawn maps and hand-written manuscript. I wanted to try it.

**See also:** Craft | Fantasy maps | Maps for and by children | Panoramic maps | Rock drawing | Jokes and satire

Hand-drawn doesn't always have to refer to pen and ink though some of the very best maps were simply pen and ink. There is a benefit of taking a hand-drawn approach to mapmaking. It takes time to make such a map, and that means thinking about each decision to make a mark. Although analytical techniques bring speed, uniformity, and allow those with less artistic abilities to make great maps, the time it takes to craft something by hand teaches the mapmaker valuable skills as they wrestle with the decision-making process.

The freedom of expression that a hand-drawn map brings often supports the mapmaker's desire to use their imagination, to be creative and, often, to incorporate humour into their work. Daily newspaper cartoons have often used maps as raw material. They are a gift of iconography for illustrators and their shapes can be morphed into a multitude of different characters. Satire and the use of humorous visual motifs also play a large part in hand-drawn mapping.

Ultimately, making a map is an act of passion. You are committing your ideas to paper and with hand-drawn maps there is a definite artistic element. As with any artist you're committing something of yourself to the paper. Noise and clutter is often replaced with honesty and personality. These maps go beyond simply the functional to focus more on form. They are, perhaps, a romantic nod to a long-gone cartographic heritage. Indeed, the number of digital mapmakers who produce maps in historic styles is also noticeable. Harking back to a period of immense cartographic artistry has helped reinvent cartography for the modern mapmaker. This is more than nostalgia. It's about a movement that respects the past, reflects on it, and uses it to make new maps.

**Opposite:** Draw your own map... a nod to the empty map drawn by Henry Holliday to accompany Lewis Carroll's *The Hunting of the Snark.*

Don't forget to sign your work

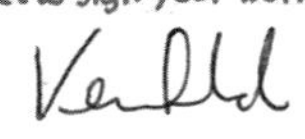

·NORTH·

·WEST·

·EAST·

·SOUTH·

# Hand-drawn shaded relief

## Mimicking hand-drawn techniques is the holy grail of analytical relief shading.

**Manual hillshading relies on the cartographer's knowledge of the local terrain and the ability to figure out the continuously changing lighting conditions and how they affect terrain locally.** Some of the most beautiful hillshading on maps has been generated by hand though it's a very particular form of cartographic artistry. Most hillshading on maps is now controlled by algorithms that process digital elevation models to create the hillshade. A lot of analytical hillshading fails to capture local variation, which causes hillshades to appear flat in some areas, particularly where the light source is parallel to ridge lines. Some techniques exist that attempt to mimic the hand-drawn approach and which at the very least create a different aesthetic appearance for relief shading compared with single-point illumination methods.

**Most relief shading involves one or, possibly, a few different points of light source.** Although there are benefits in deploying a simple, systematic approach to relief shading in that it creates a uniform output, sometimes uniformity is not the end goal. Light doesn't actually illuminate the world in a uniform way. It illuminates from every angle but with different intensities and under different atmospheric conditions. A landscape will look different on a clouded day compared with a bright sunny day, and it will also look different in the early morning or late afternoon compared with midday when the sun is overhead. Generating relief shading from a full sky illumination model allows more accurate depictions of terrain features that trend in all directions. Using multiple sources of light from different angles both in elevation and azimuth creates interesting effects and highlights on terrain, providing an opportunity for an endless array of different atmospheric conditions.

One of the drawbacks of any generic method of relief shading is that they fail to systematically deal with landforms considered unimportant. For instance, valleys are generally less important than mountain ridges and are often represented with a flat grey tone. By modulating for either slope or elevation, you can introduce light perpendicular to the terrain which sets steeper (or higher) slopes as darker while illuminating valley floors and mountain ridges. This has the effect of picking out specific features in the landscape.

Using multiple sources of light from different angles both in elevation and azimuth creates interesting effects and highlights on terrain. Additional weights or intensity of incident light can be assigned to each individual point source allowing even more exaggeration of terrain characteristics. The inclusion of modelled shadows into a resulting relief shading can also provide rich and varying tonal variation.

The holy grail of relief shading is to generate an analytical method that faithfully replicates what a talented human is capable of producing. Combining these artistic representations of terrain with other techniques, such as hypsometric tinting, allows you to create a wide variety of different representations.

**See also:** Contours | Illuminated contours | Pseudo-natural maps | Shaded relief

**Opposite:** Extract from original hand-drawn hillshade of Grand Canyon by E. Mitchell for Rand McNally maps, 1947.

# Heat maps

## A surface representation of data that occurs at discrete points, or not.

**The term 'heat map' has gatecrashed the cartographic lexicon.** It has replaced other, more established, and perfectly useful terms. It's generally used as a catch-all for any map that portrays a density of point-based pieces of information as a surface. The search for a better way of showing point-based data which avoids death by push-pin is a sound cartographic approach. In analytical terms there are many ways to approach the problem. One way is to bin your data into regularly shaped containers such as hexagons, effectively a spatial summary of the point data. Another way is to interpolate lines of equal value across the map to create a surface which then helps you to see areas that display similarly high or low values across the map.

**Interpolation methods collectively result in an isarithmic map.** That is, the planimetric mapping of a real or interpolated three-dimensional surface. Isarithmic maps are used for displaying temperature (isotherm) to atmospheric pressure (isobar); and from height (isohypse) to population distribution (isopleth). Maps resulting from kernel density analysis, the use of K functions, or the Getis-Ord Gi* statistic also create density surfaces that are commonly, and erroneously, referred to as 'heat maps'.

**So, if none of the above are heat maps, what exactly is a heat map?** A heat map is an approach to code a matrix of data values into a graphical representation using colours. It's designed to reveal the hierarchy of row and column structure. Rows in the matrix are ordered so that similar rows are near each other and you see cluster trees on the axes. The matrix is designed to illustrate correlations between variables through linked lines and other axis annotation. The closest spatial representation of data that might reasonably have similarities to a real heat map is a treemap—and that's a cartogram which has destroyed geography for the sake of creating an ordered matrix.

It's important to remember that any interpolated surface is effectively inventing data values for the areas on the map for which you don't have data or sample points. It's therefore of value to consider whether you want data making up for the areas between your data points when you know nothing exists. For instance, make a map of temperature and you'll likely use sample points. It's perfectly reasonable to infer that temperature exists everywhere as a continuous surface so filling in the voids where you have no data is fine. If, on the other hand, you have accident data for road intersections and you interpolate a surface, it makes much less sense. Intersections do not exist across space so filling in the voids with made-up data values is not appropriate.

A map of incident points is isometric data since it shows locations of discrete events that do not necessarily exhaust space. There may be more than one incident at each coincident point. Rather than a simple interpolation of discrete points, one might instead perform several analyses that use the values at each point as a weighting factor. The shape and size of the search area (often called the *kernel*) used to compare nearest neighbours must be carefully considered. A small kernel will create a map that looks much like the original...just discrete points on the map but displayed as splodges. Choosing a kernel that is too large will create an oversmoothed, highly generalised map.

Part of the difficulty with heat maps is that they have developed as part of web mapping software, which means they often allow variation of the kernel using a slider. The kernel is defined by pixel size so when you change scale, the kernel recalculates the density surface. This often results in massively overgeneralised surfaces at small scales.

**See also:** Binning | Data density | Graphs | Thematic maps

**Opposite:** World Cup Final 1966, England 4, West Germany 2.

A density surface of ball possession (not a heat map)

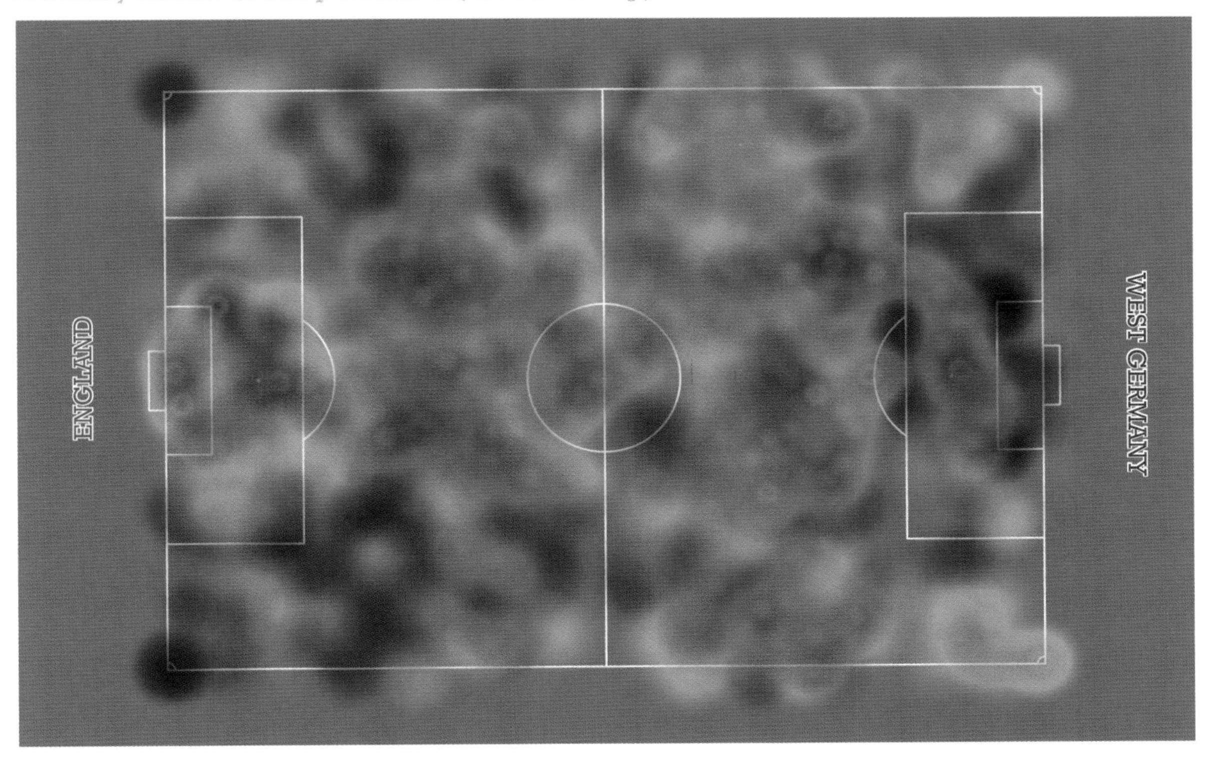


A heat map

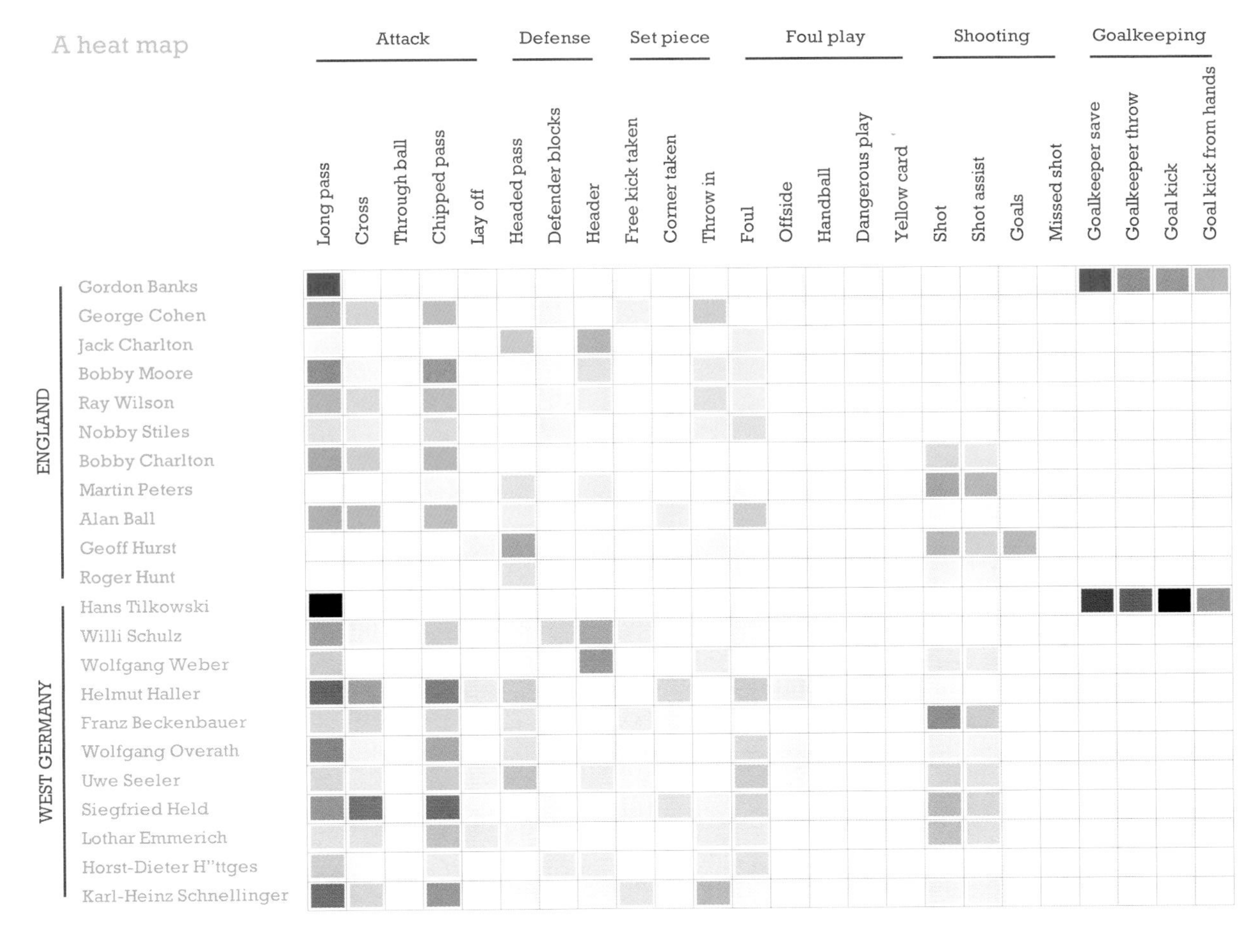

# Height

## Encoding some characteristic of data using the z-dimension of a symbol is a key function of 3D mapping.

**As more map products are created in 3D form, from tiled map surfaces to interactive 3D scenes for instance, encoding the z-dimension of symbols becomes a key visual variable.** 3D cartography is sometimes an attempt to portray the world in a way that is more natural to us, by using perspective views. It is how we see the world so the notion of using not only *x* and *y* but also *z* seems obvious. For natural-looking 3D maps that intend to mimic the view in reality, using photorealistic rendering and then height simply provides an extra surface on which to paint a texture. However, for 3D thematic maps, height can be both a useful additional dimension and a problematic issue.

**For all dimensions of data, height can represent some quantifiable characteristic of the data through extrusion.** For thematic point features, each individual feature is symbolised using a raised stick (or lollipop) where the length of the stick is relative to a quantifiable dimension of the data on the ordinal or numerical scale. Linear features are depicted as an extruded wall-like structure to give the impression of magnitude. For area and 2½ D phenomena, raising the surface of features to reflect a numerical difference can create prisms or a continuously varying elevation surface.

**Height cannot be used for true 3D phenomena simply because the feature is three-dimensional in the first place.** A true elevation surface will provide a 3D scene with different base heights though extruding features that sit atop a real 3D surface is possible. Extrusion of features can also be below the surface. If the function of the map is to support the comparison of the height of extruded objects, then varying the base height by situating it on a real surface, as opposed to a fixed base height, can cause perceptual difficulties.

**Height can also be used to encode time.** This provides a way to show how the character of a mapped object varies spatially and, combined with other visual variables, in character through time.

**See also:** Aspect views | Isometric views | Prism maps | Space-time cubes

The fact that a 3D map is in some sort of perspective or aspect orientation means that features closer to the front of the map are shown at a different scale than those toward the back. This can create problems of comparing like-for-like across the map. A solution is to use an axonometric projection that modifies the viewing angle of the map ensuring that scale is equivalent across the map.

For extruded areal features in particular, orientation of the map becomes important because if a number of high data values, or prisms, appear at the front of the map, their graphical representation will obscure smaller values behind. Using interactive applications to deliver 3D maps overcomes some of these limitations by giving the ability to pan, rotate, and zoom around the map to see parts that might be hidden.

Virtual globes can also cause perceptual difficulties in the interpretation of extruded symbols. The curvature of a virtual globe means symbols will rise away from one another rather than parallel to one another as they would from a flat surface. Comparing the height of objects is difficult in these situations as the curvature and perspective view combined creates a non-comparative viewing experience.

Visual variation

Increased height is seen as figural

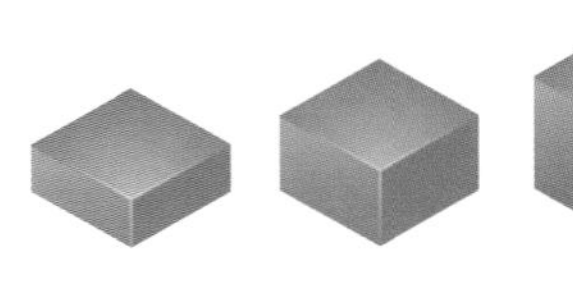

For seeing — Distinct, Levels

For representing — Nominal, Ordinal, Numeric

**Opposite:** *Shan Shui in the World* by Weili Shi, 2016. Hanging scroll. Ink and colours on silk 24" × 55" from automated rendering of Manhattan building heights.

Downtown Manhattan

Uptown Manhattan

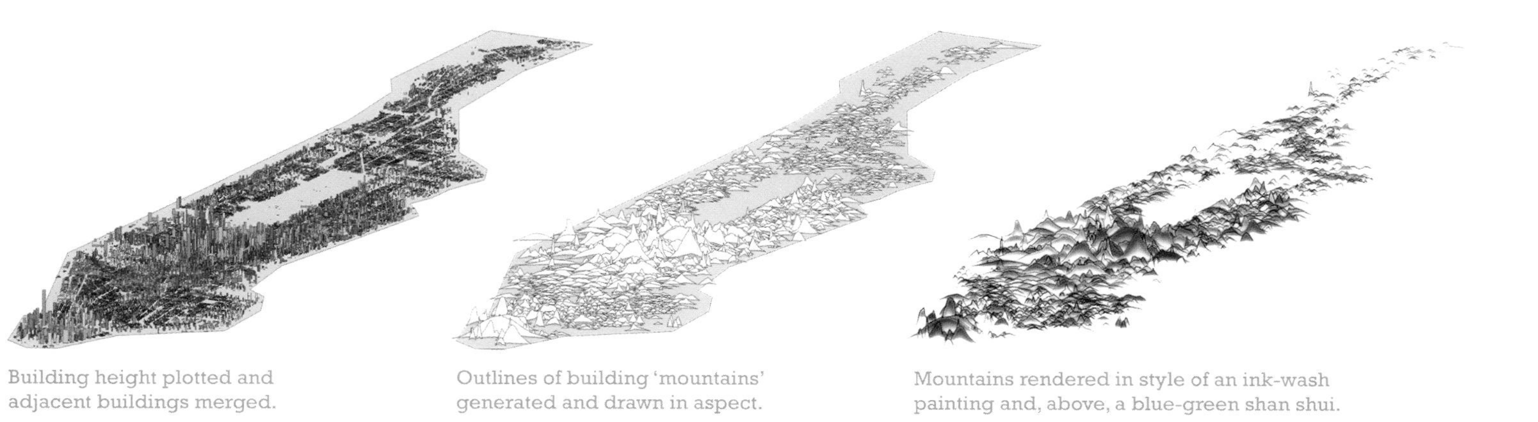

Building height plotted and adjacent buildings merged.

Outlines of building 'mountains' generated and drawn in aspect.

Mountains rendered in style of an ink-wash painting and, above, a blue-green shan shui.

# Hierarchies

Placing map components so that some are seen as a foreground and some as a background builds structure in the map that supports reading.

**Deciding on the relative importance of the various components of your map helps to establish a sense of order in the otherwise chaotic display of graphics.** Thinking about what elements you need people to see and understand in a relative sense to all other map components will begin to build structure into the map. The graphical representation of this process is often referred to as the 'visual hierarchy' of the map.

**Creating a visual hierarchy involves graphically emphasising those objects which are more important and de-emphasising those that are less important.** The overall map composition should be organised to reflect how you want people to read the map. Individual map components should be organised according to a visual hierarchy. For instance, in an overall sense, thematic symbols will probably be more important than the title and the legend so emphasis is applied accordingly. Additionally, if all the text is considered a separate map component, then the title is more important than the subtitle and, in decreasing importance, the legend, data source, scale, and so on.

**The crucial part of creating an effective visual hierarchy is to design the map so the reader's eye is drawn to the most important map component first and the less important ones subsequently.** One way of determining whether you're being successful is to look at the map while squinting and to determine which features stand out more than others. Your design will likely be more successful if the visual hierarchy can be discerned using this simple technique. Two other methods are useful. One is to make a small version of the map and see what components stand out. This forces your eyes and brain to search for the largest and most prominent components among a range of very small components. Finally, generating a greyscale version can also help since darker elements on a light background, or vice versa, will be more prominent. Whichever method you choose, the general idea is to use your natural perceptual and cognitive systems to confirm your choices or provide a mechanism to discover what might need adjusting.

**See also:** Contrast | Dispersal vs. layering | Focussing attention | Seeing | Vignettes

Performing a squint test on your map might sound like an odd way of assessing your visual hierarchy but there is some science behind it. As you close your eyes you're letting less light in, and so your eye's receptors (rods and cones) are working harder to see shapes, colour, and shade. Your brain is also having to work harder to interpret what is being seen, which is the same problem you have when trying to see objects at night without ambient lighting.

What you end up seeing when squinting is the main shapes and the objects that tend to stand out more than others. Your brain attempts to recognise the more obvious things first so you can start to build a picture. It ignores objects it struggles to discern and focuses on what it can understand.

Performing this test on a map is a quick way of literally seeing whether you have built a hierarchical structure sensibly. Do the most important components of your map get seen, or do they blend into the unseen background? If your eye is picking out features that you know should be less prominent, then the map's visual hierarchy probably needs adjusting through changing the visual variables associated with your symbology.

Better hierarchy small

Poorer hierarchy small

**Opposite and above:** Alternative ways to explore hierarchy in map design

Better hierarchy

Poorer hierarchy

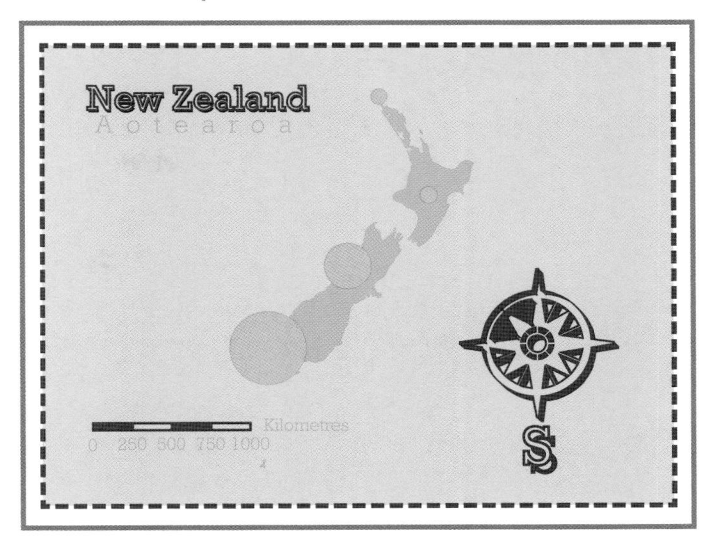


Better hierarchy squint

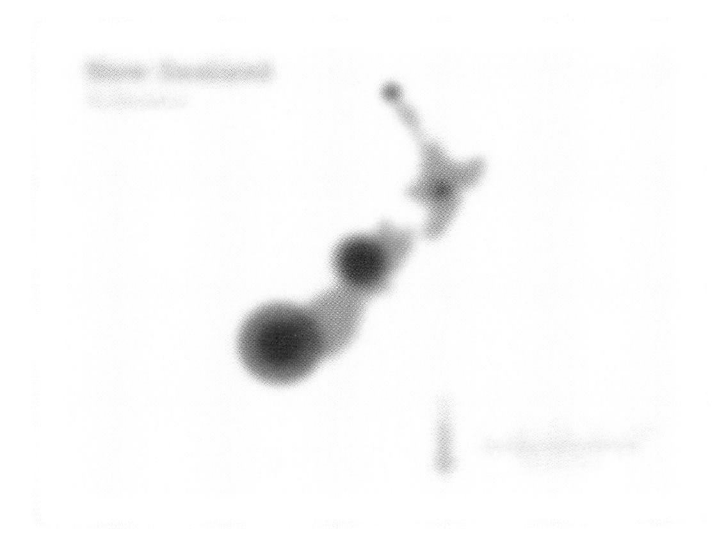

Poorer hierarchy squint

Better hierarchy greyscale

Poorer hierarchy greyscale

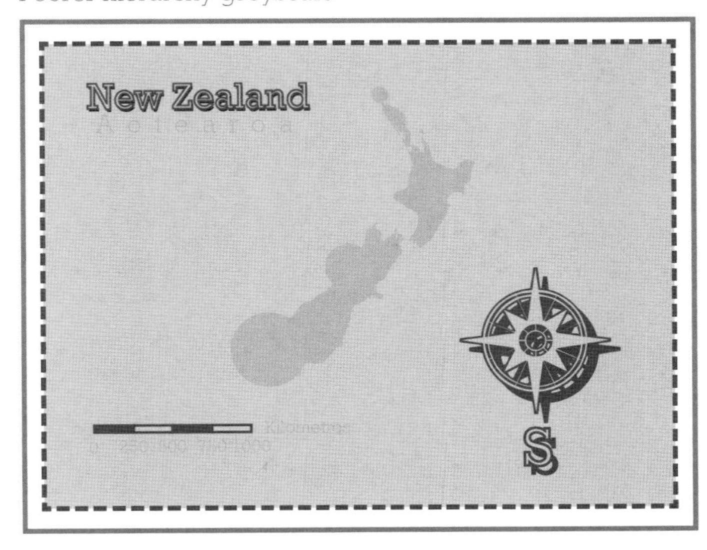

# How maps are made

## Multiple options exist for choosing how to make your map.

**Mapmaking is heavily influenced by technological change.** Throughout history, different production techniques have emerged and themselves been replaced as advances are made or disruptive technologies challenge convention. For much of the 20th century, drawing pens, scribing, and photomechanical production were central to making any map for publication. Those techniques have now been surpassed by computing, which supports a range of opportunities and has vastly decreased the amount of time it takes to make a map. Of course, making a map can be as simple as sketching with pen and paper. In many instances a quick sketch map suits the need perfectly. It may be a map on the back of a restaurant napkin, a beautiful piece of hand-drawn art, or a step in a process to a final published map you intend to design and produce using computer technology.

**For making a high-quality map, your needs, abilities, and finances will strongly influence what you choose.** Desktop cartography is supported by graphic design software such as Adobe® Creative Cloud, Inkscape™, or GIMP. GIS offers another approach through, for instance, ArcGIS or QGIS. The primary difference between the two approaches is graphic design deals simply with graphics, whereas GIS supports data-driven cartography and the full pipeline of data management, processing, and analytics that you'll likely require. Web mapping can be, in part, designed and published via a desktop GIS to web portals or online cloud-based mapping platforms. Both proprietary and open source occupy the cartographic technology space. Online mapping usually requires you to have some ability to use map libraries, APIs, and code (e.g. Javascript) to take full advantage but if you are working in an entirely web workflow you can sometimes bypass the need for desktop software. Choosing the right tools for the right job is more important than being a slave to a particular piece of software. Often, you'll need to mix and match to use the optimum tool for specific jobs.

Interactive mapping applications are often free or low-cost solutions. Increasing sophistication and more options and controls are often found in software you can purchase but there are numerous tools to get you started. The internet is also key for hunting and gathering data, both topographic and statistical, for exploring online map libraries, and using third-party map services. Some web mapping applications extend to include analytical and advanced tools to develop more sophisticated products. Your ability to code and make use of libraries and APIs will likely support or constrain your ability to work entirely in a web-based workflow.

Graphic design software has often been synonymous with cartographic practice. It's a modern equivalent to drawing with pen and paper. Software supports high-end design and production, including extensive control over colour, typography, and symbol design. The ability to support high-quality professional publishing has often been the attraction of such technology though the drawback is that subsequent changes in the data, or otherwise, requires manual updates or a new map to be created.

Geographical Information Systems (GIS) is much more than mapmaking software. It supports the whole sphere of data management, processing, analytical, and display functions you are likely to need. In addition, many systems link directly to the web to support seamless online map publishing. Cartographic techniques are either implemented or easily deployed. Finished cartographic products can be produced through the inherent map design and production functions. GIS is data-driven so updates in data can usually be mirrored in your map rather than creating it from scratch. Many GIS now offer similar functionality to high-end graphic design software.

**See also:** Abstraction and signage | Anatomy of a map | Craft | Graphicacy | Map transformation process | UI/UX in map design

**Opposite:** Made using a GIS but in a hand-drawn style.

ICELAND
SWEDEN
FINLAND
NORWAY
ESTONIA
LATVIA
DENMARK
LITHUANIA
IRELAND
BELARUS
UNITED KINGDOM
NETHERLANDS
POLAND
GERMANY
BELGIUM
UKRAINE
CZECH REPUBLIC
SLOVAKIA
MOLDOVA
FRANCE
AUSTRIA
HUNGARY
SWITZERLAND
ROMANIA
SLOVENIA
CROATIA
ITALY
REPUBLIC OF SERBIA
BULGARIA
MONTENEGRO
ALBANIA
MACEDONIA
SPAIN
TURKEY
GREECE
TUNISIA
ALGERIA
LIBYA
EGYPT

# HSV colour model

## The HSV model allows mapmakers to work directly with hue, saturation, and value when specifying colour.

**The HSV model allows you to work directly with visual variables—hue, saturation, and value—as opposed to proportions of a colour cube.** The colour space is represented as a hexacone-shaped model rather than a cube though the logic of the hexacone relates to a colour cube in one respect. If the RGB colour cube is viewed on an angle with the white corner in the centre, the six additive and subtractive colours would appear as a hexagon surrounding white. This structural organisation of hues in a hexagonal shape is the same in the HSV model as the RGB colour cube viewed in this way. Value changes occur from the apex of the cone to a point central to the hexagon at its base, and saturation changes from the centre of the base to its edges. The HSV model is a non-linear transformation of the RGB colour cube.

**The HSV model can be used to describe colour in a more human way since it directly references the perceptual dimensions.** Humans refer to colour with questions such as, What colour is it? How vibrant is it? How light or dark is it? Hues are assigned an angle 0-360° (sometimes this is normalised to a percentage range) and for each direction have varying levels of saturation outward from the centre. Diametrically opposite colours form complementary hues. Both saturation and value are measured on a percentage scale.

**The HSV model has its disadvantages.** Different hues that have the same numerical value will not necessarily have the same perceived value. For example, green will always appear lighter than red for similar numerical values. Similarly, different hues that have the same percentage of saturation will not have the same perceived saturation. A point midway between two colours does not result in a colour perceived to be midway. Finally, the model's implementation as a method of selecting colour is problematic because lightness and saturation are not normally held as independent variables, and conversions have to be made between HSV and RGB or CMY representations What you specify may therefore not be what precisely results on screen or in print.

The HSV model may seem redundant because it is neither a truly perceptual colour space nor a cube which is used to specify colour in a computerised design environment. It is a symmetrical shaped model but, of course, true perceptual colour is not represented by a symmetrical space. That said, it does provide a good way of helping you explore approximate perceptual colours. Despite one of the major flaws being the difficulty in systematically controlling value or saturation, it is often easier to adjust colours by making these changes rather than the arbitrary values in an RGB or CMY colour cube.

HSV is only one of many alternative colour models. It was developed in the 1970s by computer graphics engineers as a way to enable colour specification in a more intuitive way. The HSL—hue, saturation, and lightness—model is another alternative developed around the same time. HSL was specifically created to relate to the way human vision operates as well as how traditional colour-mixing methods in printing and art were produced. Colour mixing in these senses often occurs through mixing brightly coloured pigments with black or white to develop different tones. Both HSV and HSL offer a way of understanding colour beyond the raw values of colour cubes.

Ultimately, you'll decide what your own preference is in defining and mixing colour whether through the raw values of RGB (light), CMY (pigments), or through tweaking perceptual colour models such as HSV or Munsell.

**See also:** Colour charts | Colour cubes | Colour schemes | Mixing colours

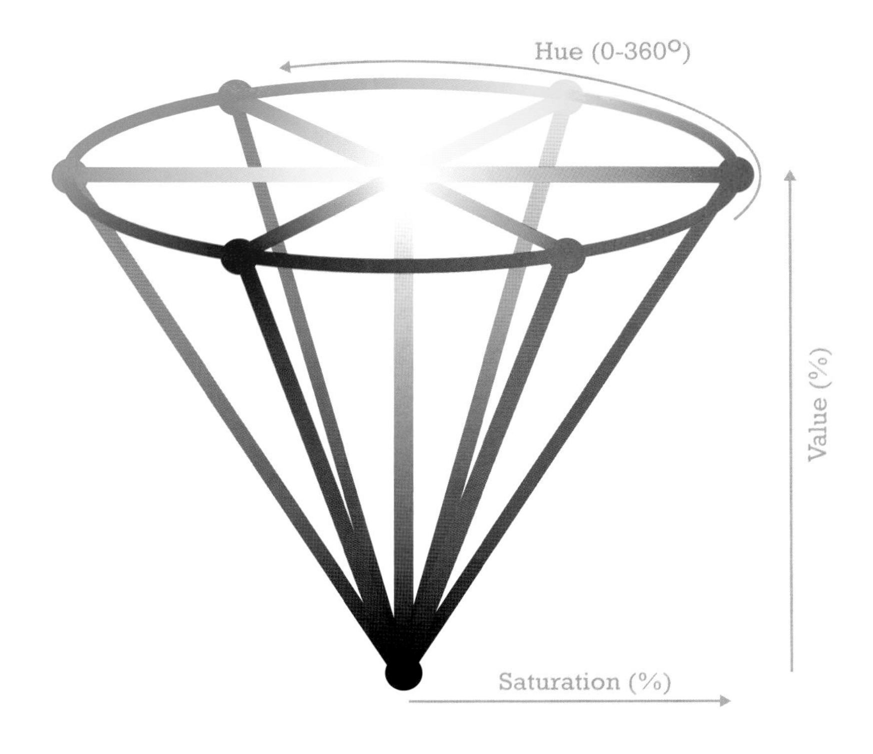
Hue (0-360°)
Value (%)
Saturation (%)

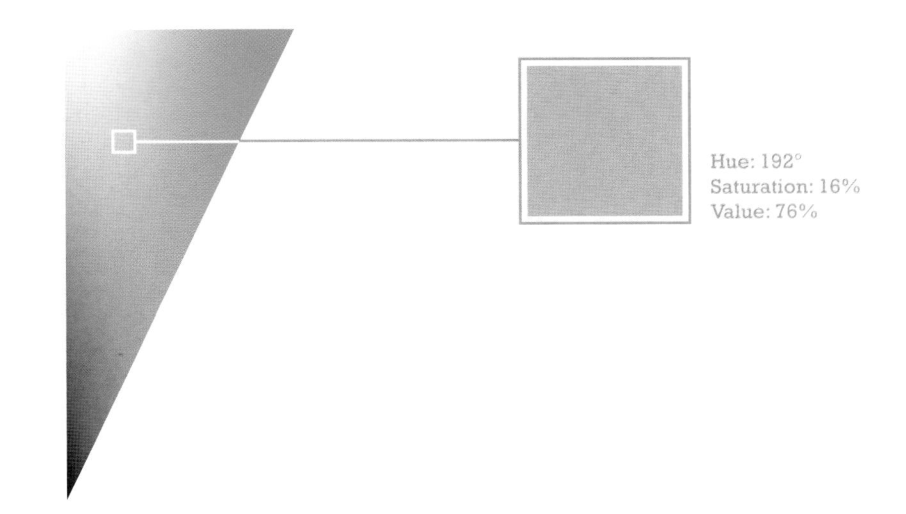
Hue: 192°
Saturation: 16%
Value: 76%

# Hue

## The perceptual dimension of colour that is defined by the dominant wavelength of visible light.

**What we commonly refer to as the colours of yellow, red, blue, or green are actually most accurately referred to as *hues*.** They are defined by the dominant wavelength of visible light ordered from longer wavelengths of reds to shorter wavelengths of blues, which build a spectrum of hues. Hue is independent of the other perceptual dimensions of colour, value, and saturation so it is possible to generate a range of different hues that have the same value.

**Hues are distinct symbols and should not be mixed in a scheme that represents numerical data.** Hue is not particularly effective at displaying quantitative difference in its own right although it is often used in combination with other visual variables to emphasise ordinal differences. For instance, a road classification system might employ different line sizes to indicate order but combine with hue as a secondary visual variable to improve the representation.

**As a widely used exception to the general rule, hue has been used for numerical data representation despite its marginal effectiveness.** For instance, atlases conventionally employ hue to depict relief. This is referred to as *hypsometric tinting* and is a great example of how cartographers break rules to good effect. Hyposometric tinting uses different hues to illustrate predominant vegetation colours at different altitudes so it's a form of mimetic symbol. Its long-standing use has become a generally accepted visual system for depicting elevation, and so the conceptual limitation has become outweighed by general use and understanding.

**In qualitative terms, hue is often used to show nominal difference.** For instance, it is used on a political map to delineate different countries without inferring one is more important than another. More generally, light values of hues will not be seen as much as darker values, which causes difficulties for symbols of smaller sizes, particularly for points and lines. For larger symbols or areas, mid- to dark values, or highly saturated colours should be avoided as they become perceptually dominant.

**See also:** All the colours | Colour in cartography | Greyscale | Mixing colours | Saturation | Value

Hue is often misused to show some sort of quantitative difference. Hues have no inherent visual order so using them to show quantitative differences of data from low to high does not support the map's function. Perceptually, hues cannot be used in their spectral order (or any order) to show changes in magnitude because you're unable to determine which hue, in a sequence, is representing values that are higher or lower than another hue. The eye does not see order in hue so it does not support the cognitive processing required to reconstruct a sense of order.

You can use hue to some limited effect in creating ordered hue progressions by positioning naturally lighter hues at one end and darker ones at the other. By reordering colours in this way, it's possible to create a scheme that appears light–dark for hues of the same value because blues and reds are naturally darker than greens and yellows. The interpretation of light to dark may be perceived but it's not because of the hues themselves but the value they reflect.

Visual variation

Maintain saturation & value to avoid showing levels

For seeing

Distinct Levels

For representing

Nominal Ordinal Numeric

**Opposite:** Extract from *The Alluvial Valley of the Lower Mississippi River* by Harold Fisk, 1944, uses hue (as well as pattern) to delineate historical river channelling.

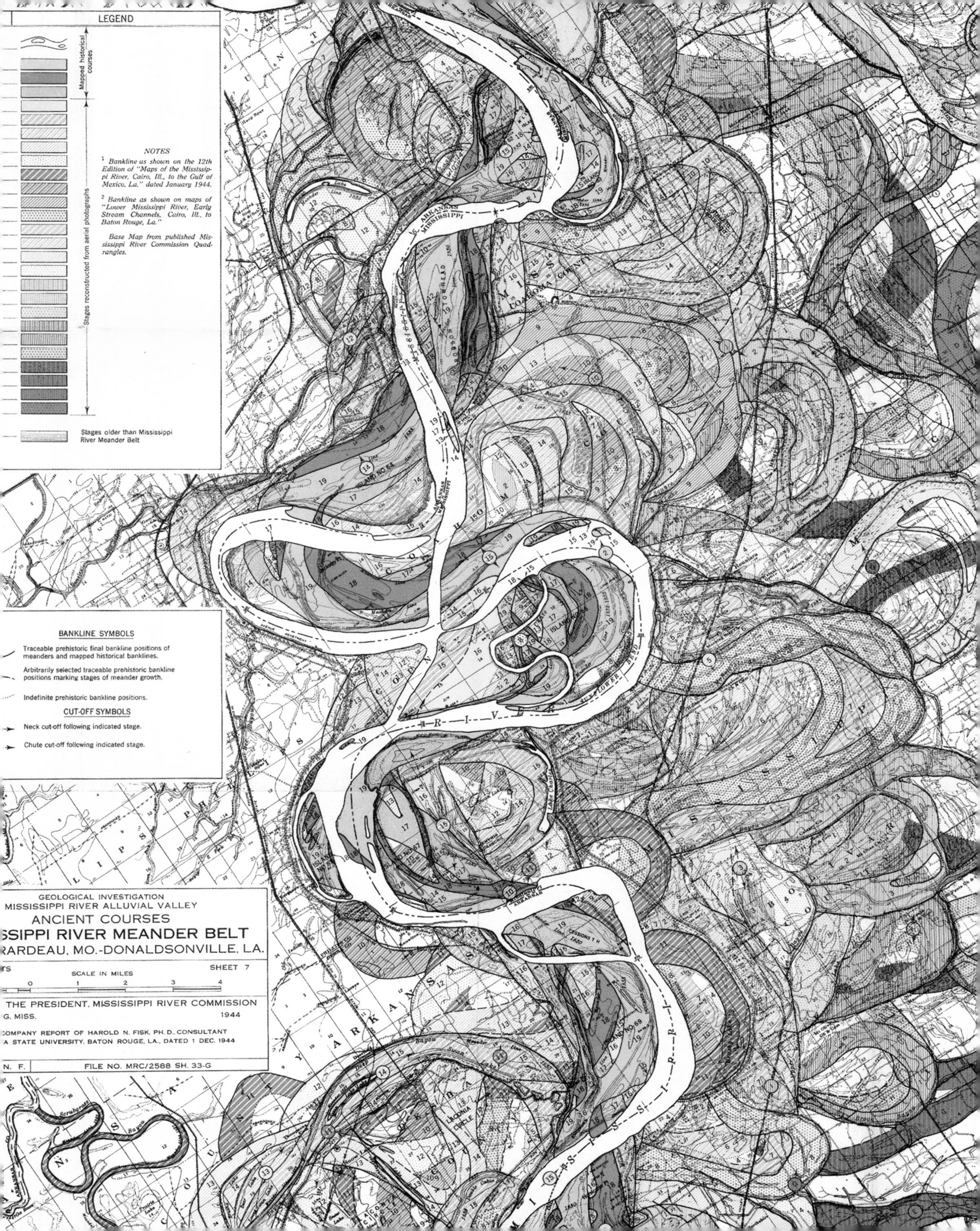

LEGEND
Mapped historical courses
Stages reconstructed from aerial photographs
Stages older than Mississippi River Meander Belt
NOTES
1 Bankline as shown on the 12th Edition of "Maps of the Mississippi River, Cairo, Ill., to the Gulf of Mexico, La." dated January 1944.
2 Bankline as shown on maps of "Lower Mississippi River, Early Stream Channels, Cairo, Ill., to Baton Rouge, La."
Base Map from published Mississippi River Commission Quadrangles.
BANKLINE SYMBOLS
Traceable prehistoric final bankline positions of meanders and mapped historical banklines.
Arbitrarily selected traceable prehistoric bankline positions marking stages of meander growth.
Indefinite prehistoric bankline positions.
CUT-OFF SYMBOLS
Neck cut-off following indicated stage.
Chute cut-off following indicated stage.
GEOLOGICAL INVESTIGATION
MISSISSIPPI RIVER ALLUVIAL VALLEY
ANCIENT COURSES
SSIPPI RIVER MEANDER BELT
RARDEAU, MO.-DONALDSONVILLE, LA.
SHEET 7
SCALE IN MILES
0 1 2 3 4
THE PRESIDENT, MISSISSIPPI RIVER COMMISSION
G, MISS. 1944
COMPANY REPORT OF HAROLD N. FISK, PH. D., CONSULTANT
A STATE UNIVERSITY, BATON ROUGE, LA., DATED 1 DEC. 1944
N. F.
FILE NO. MRC/2588 SH. 33-G
ARKANSAS
MISSISSIPPI
COAHOMA COUNTY
ISLAND NO. 64
ISLAND NO. 67
ISLAND NO. 69
LAGONIA CIRCLE

# Hypsometric tinting

## Method of colouring different elevation values to enhance relief cues.

**Hypsometric tinting (also 'layer tinting' or 'elevation tinting') is a method of applying colour to elevation values to enhance or accentuate the depiction of relief.** Discrete hypsometric tints are formed when a uniform colour is applied between contour lines so that all values of elevation are represented the same way for each contour interval. This is similar in concept to a classified isoline map where values between an upper and lower limit are all symbolised in the same colour. It results in a somewhat abstract appearance with often sharp changes in colour between different contour intervals.

**Alternatively, a continuous hypsometric tint can be applied to a digital elevation model (DEM) that gives a much smoother appearance of changes in elevation.** Rather than selecting discrete colours, the continuous technique uses a colour ramp that gives a continuous succession between a beginning, ending, and intermediate colours. The advantage of this method is each cell in a DEM is assigned a unique colour on the basis of its specific elevation whereas, inevitably, detail is lost in the discrete method.

**With both the discrete and continuous methods, the colours chosen to represent each elevation zone tend to reflect the overarching characteristic of the biome at that place.** So, green is used for low-lying fertile valleys, browns for above treeline areas, and white for snowcapped peaks. These colours also tend to vary for different climatic zones, and so the colour schemes you might find for an atlas of Nepal are likely different from that in the USA. Even within a country, you often find differences as the Pacific Northwest in the USA has very different landscape characteristics from the Pacific Southwest.

**Bathymetric tints tend to revert to a single-hue scheme.** These tints show increasing depth as darker shades of blue. This makes sense in perceptual terms and is a typical illustration of bodies of water.

Hypsometric tinting is an example of how practice has established a symbolisation scheme that contradicts perceptual studies. Spectral colour schemes are not good at displaying quantitative differences because there's little evidence that our eye-brain systems can tell whether green is higher or lower in value than red, or blue or any other colour in a random sequence. Leonardo da Vinci is credited with first using different hues on a map to show changes in elevation on his map of Italy (ca.1503). John Bartholomew Junior (of the Scottish cartographic dynasty John Bartholomew and Son) certainly popularized the technique in the late 1800s in a range of atlas products, later being used in the 1920 second edition of what is now known as *The Times Comprehensive Atlas of the World*.

However, using different hues to represent different elevations has become established and widely used. It's become familiar and so we've become accustomed to seeing elevation presented in this way. What is perhaps more curious is that hypsometric tints and bathymetric tints take different approaches to the way in which elevation or depth is encoded. Different biomes exist at different depths yet most maps retain a spectrum of blues as the way to represent depth. It makes sense. It's the hypsometric tints that are the anomaly.

It's common to use hyposometrically tinted elevation in conjunction with other methods of terrain representation such as labelled contour lines.

**See also:** Hand-drawn shaded relief | Raised relief | Topographic maps

**Opposite:** Alternative hypsometric and bathymetric tinting.

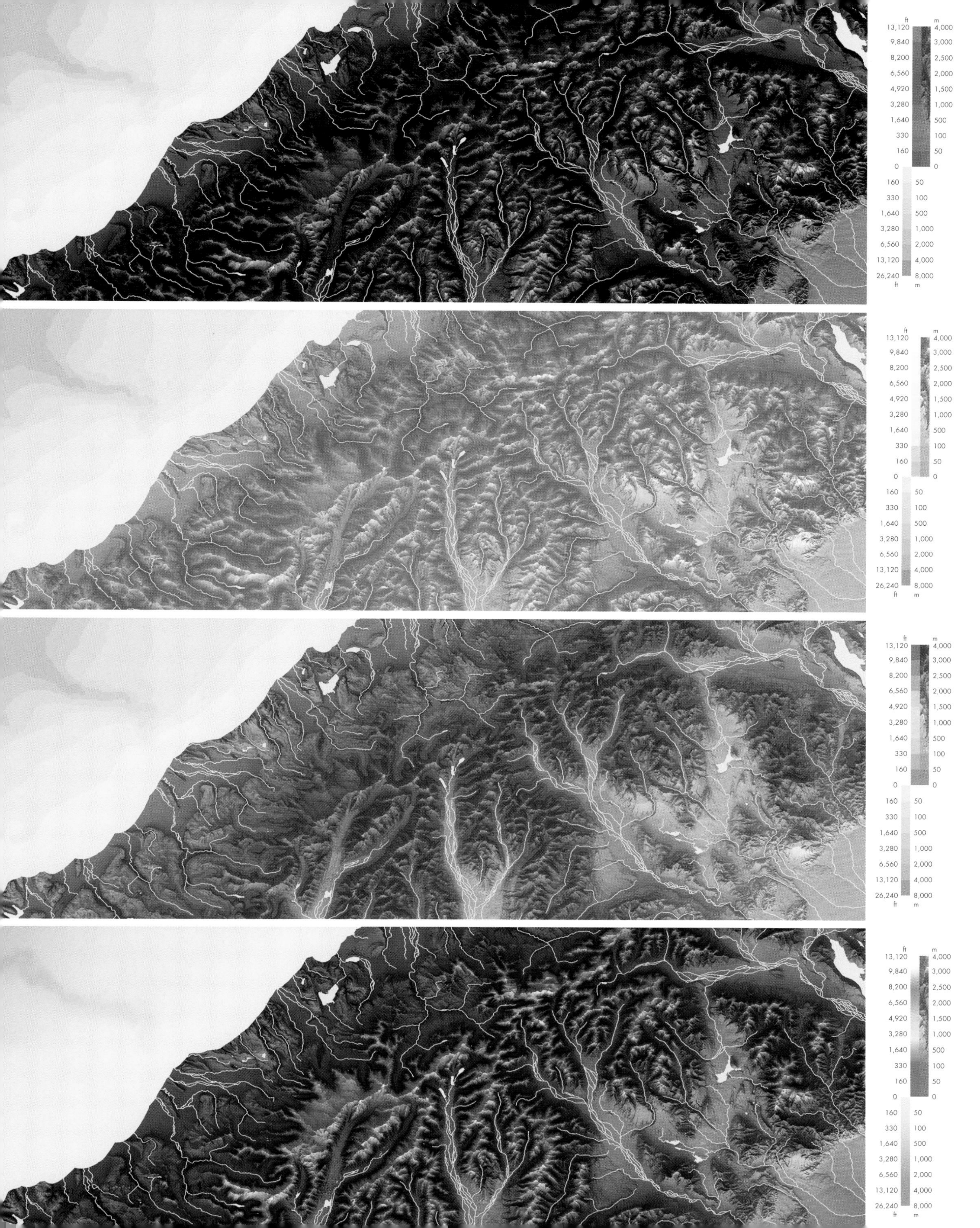

ft
m
13,120
4,000
9,840
3,000
8,200
2,500
6,560
2,000
4,920
1,500
3,280
1,000
1,640
500
330
100
160
50
0
0
160
50
330
100
1,640
500
3,280
1,000
6,560
2,000
13,120
4,000
26,240
8,000
ft
m

# Earth Wind Map

Cameron Beccario

2013–present

The real world is too complex to be fully captured in a single database, so cartographers must decide whether they are treating their data as discrete objects (i.e. things with a clear boundary) or as continuous fields (i.e. things with no clear boundary). Wind speed and temperature fit perfectly into the continuous view of the world, yet for decades weather maps, particularly on TV, have shown them as a discrete object: a single number or symbol placed over key towns.

Increasingly forecasters have developed more continuous interpretations but these still fail to portray the dynamism of the weather. The brilliance of the Earth Wind Map (nullschool.net) of global weather conditions by Cameron Beccario is that it does just that.

Inspired by the hint.fm map of the USA's wind conditions, 'Earth' was the extension of a live wind map for Tokyo developed by Beccario. It has grown to show 10 weather- and pollution-related variables overlaid on moving vectors illustrating wind speed. The map offers one of the best examples of where big data, computer science, and cartography can be combined to have a real impact on how natural phenomena are communicated.

—James Cheshire

## Hurricanes Jose and Maria

## 19th September 2017

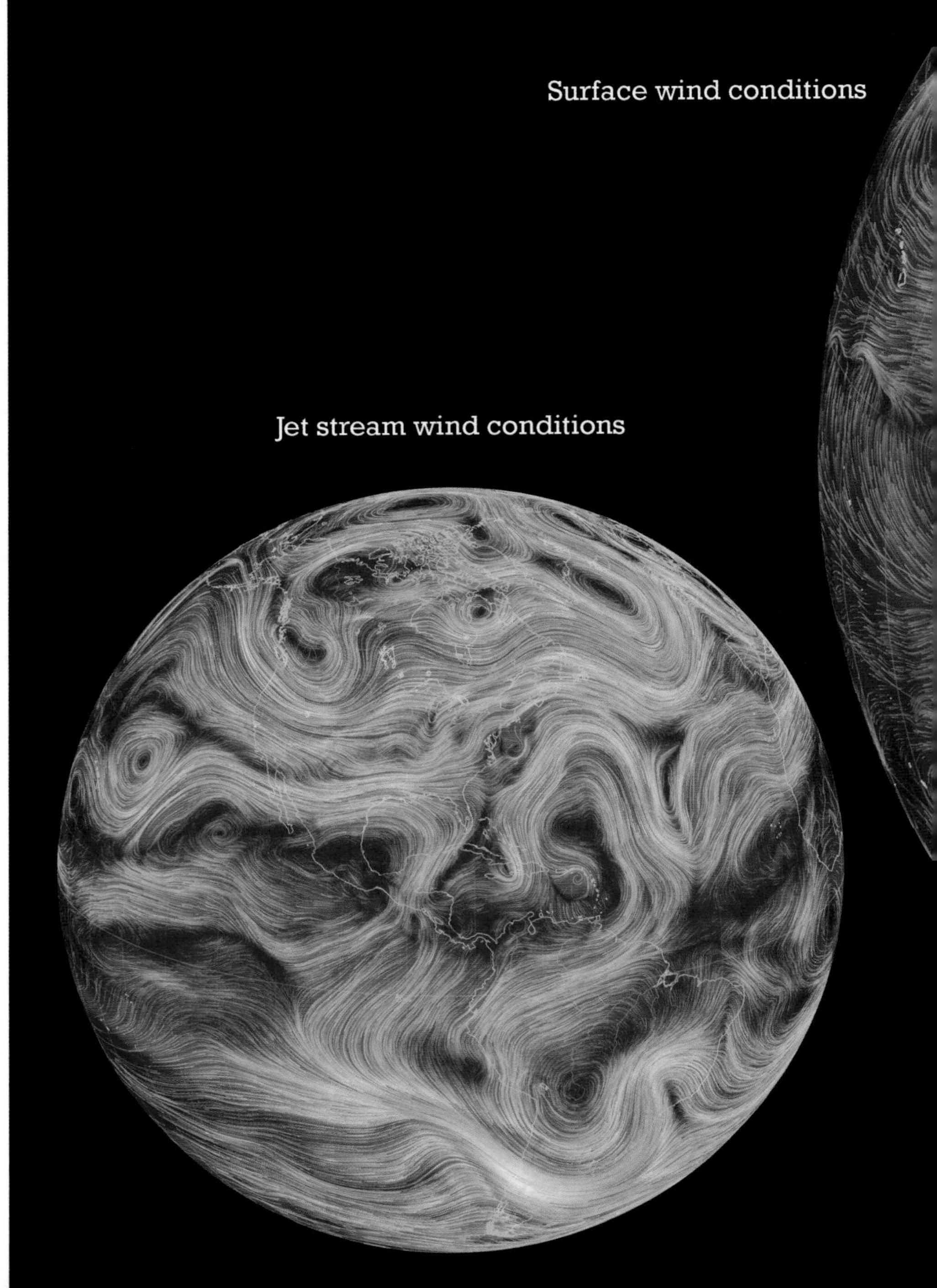

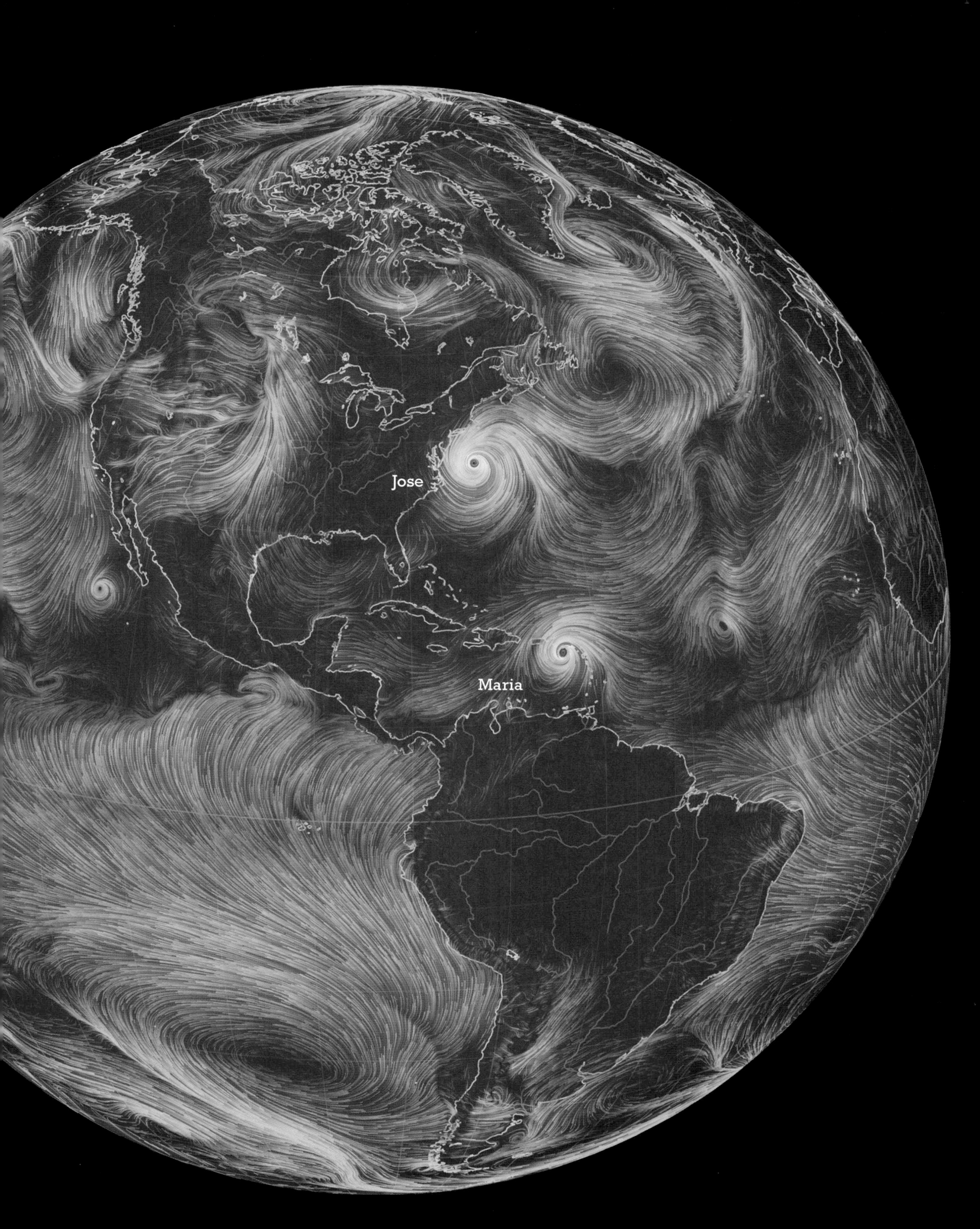
Jose
Maria

# Illuminated contours

**Method by which contours are accentuated to give light and shade and an impression of their illumination.**

**One criticism of contouring as a method to show elevation is that, to the untrained eye, it may not be easily interpreted.** There's no variation in symbology for high and low or to show steepness other than the layout of the contour lines themselves. When you understand how concave and convex slopes appear as contours and what valleys and mountain peaks look like you can begin to infer the shape of the terrain but it may not be intuitive.

**A striking alternative to traditional contours was popularized by a Japanese cartographer, Kitiro Tanaka, in the mid-1900s—illuminated contours.** He employed this technique in a map of a rugged volcanic region of Kyushu, Japan (Tanaka 1950). Tanaka referred to his technique as the 'relief contour method', although it is typically referred to as the 'illuminated contour' or 'Tanaka method'. In his hand-drafted maps, Tanaka began with a background of medium grey and drew white contours on illuminated topography and black contours on non-illuminated or shaded topography, assuming lighting from the northwest (or upper left). In addition, using a calligraphic drawing technique, Tanaka varied contour thickness with the cosine of the angle between the compass direction of illumination (azimuth) and a vector oriented in the aspect direction. He attempted to match tonal variations predicted with the theory of hillshading to tonal variations of black-and-white contours on a grey background with this simple drafting method.

**Illuminated contours gives a planimetric map a three-dimensional quality.** It works particularly well on generalised (highly curved) contours. The technique certainly gives texture and adds visual interest though, as with all the more artistic techniques, care should be taken that the technique doesn't overshadow the map's main focus.

**See also:** Contours | Isarithmic maps | Hypsometric tinting | Value

Illuminated contours can be used simply to create a different look and feel. Beyond that, there are uses which the technique supports.

For instance, often on a topographic map elevation is seen in tandem with bathymetric detail for offshore or for lakes. It's highly likely that the quality and resolution of the data used to create contours for the bathymetry was poorer than that for the elevation. Yet the same symbolisation often exists and, so, the inference is that accuracy and precision would be the same. By varying the symbolisation to use generalised illuminated contours for bathymetry, the difference in representation is sufficient enough to connote a difference.

The technique also works well in areas of sparse or uncertain data simply because the generalised form of the line work tends to support broader curves than tightly nested contour lines. For this reason it can be used on statistical data too, perhaps showing isolines of uncertainty or differences between high and low atmospheric pressure. For this latter example, the subtle addition of depth that illuminated contours brings can enhance the visual message.

Illuminated contours on a dome:

**Opposite:** Lake Tahoe, CA, bathymetry represented as illuminated contours in combination with a hue progression from light to dark showing depth.

# Imagery as background

## Aerial imagery is easy to use as a background to a map, but not easy to use well.

**With the advent of mapping services and basemaps, a rich and diverse set of preconfigured map content is available to use for cartography.** This has afforded fantastic opportunities as it completely eliminates the need for people to source their own data and spend time performing cartographic generalisation and symbolisation. Using third-party basemaps has become de facto, particularly in web cartography. Often the cartography of the basemap and the overlays are in conflict but with careful choices you can find a reasonable choice to suit your purpose.

**Using aerial imagery to provide a background to the overlay of thematic content or focussed detail has become commonplace.** Because aerial imagery is now commonly available and the assumption is that the context provides a realscape for the map, use is prevalent, particularly in web maps. They provide recognition because they are the real map of the world yet they are also incredibly detailed—often far more than the map requires. Topographic maps are abstracted products of aerial imagery precisely because we rarely need the detail of everything to be displayed on the map. At large scales, the detail can be very important and revealing. Less so at small scales.

**The temporal aspect of the aerial image is also important.** For instance, a set of images acquired in the height of summer is not going to support a map whose theme is perhaps set during the winter. Additionally, the vertical component of imagery makes tall buildings appear awkwardly. The taller the building, the larger this visual distortion, and angles can change across an area. Shadows are also likely to be irregular across imagery as individual images were likely acquired at different times. It may also become too pixelated at larger scales, which can be visually disturbing. If using imagery, it's useful to bear these points in mind.

Typically, imagery is placed as a background context. It's usually saturated, and often dark colours become dominant and make it difficult to discern thematic or other topographic content placed on top. For this reason, applying a whitewash using transparency can be a simple but effective way to make the aerial background recede a little (or a lot). If a considerable amount of transparency is applied, it's possible to use imagery to add just a little subtle texture to your normal topographic basemap.

Greyscale aerial imagery, again with some transparency, helps the background to recede a little, and through careful selection of bright or saturated colours on top, it can be an effective way to effectively colorize the parts of the map you are overprinting and which you want people to focus on. This might be to produce a hybrid map where a background of the urban fabric gives way to an overlaid road network. The use of subtle transparencies in the overlaid features can also help such as adding a transparent casing to road lines that de-emphasise the aerial imagery behind the feature. The same principle can be applied to typographic elements overlaid across aerial imagery. Using semi-transparent halos can help the type to be seen across what can often be a hugely varying backdrop.

Ultimately, using aerial imagery is not a panacea for your map's background. It often demands you take a particular approach to your overlaid content to make it visible and legible in the map's visual hierarchy. At the very least, lighten up the background to make it less dominant.

**See also:** Basemaps | Foreground and background | Pseudo-natural maps

**Opposite:** A range of different treatments to use aerial imagery with overlaid content.

White Mine
Dark Lake
Impossibly Bright Flats
Variable Illumination Lighthouse
Heavy Hue Ridge
Verdant Valley
Contrast City
Monochromatic Ferry Line
Mt. Anomaly
Texture Haven
Chroma Dam
White Mine
Dark Lake
Impossibly Bright Flats
Variable Illumination Lighthouse
Heavy Hue Ridge
Verdant Valley
Contrast City
Monochromatic Ferry Line
Mt. Anomaly
Texture Haven
Chroma Dam
White Mine
Dark Lake
Impossibly Bright Flats
Variable Illumination Lighthouse
Heavy Hue Ridge
Verdant Valley
Contrast City
Monochromatic Ferry Line
Mt. Anomaly
Texture Haven
Chroma Dam

# Information overload

## Data rich can become information poor because of the tendency to try and squeeze too much onto the map.

**There's no magic answer to the question of how much information you can or should put on a map.** Notwithstanding the graphical limits of line thickness or font size, a map can act as a container for a vast amount of information. This problem is exacerbated if you're working digitally across the scales of a multiscale web map. The unhelpful answer is you can pack a lot of information onto a map. Actually—that needs modifying: you can pack a lot of graphical marks that represent data but whether it is useful for communicating information is questionable. Simply adding more does not, in and of itself, mean your map is more useful and Edward Tufte's quote *'to clarify, add detail'* should not be confused with simply adding more.

***Information overload*** **refers to the nebulous principle of not putting so much on the map that it complicates, confuses, or makes the meaning illegible.** In practice it's difficult to assess at what point a map becomes overloaded with information, but if a precept of cartography is to reveal meaning among clutter, then the basic task is to remove, not add.

**There is a natural desire, when making a map, to make use of the data at hand.** This often includes a belief that finding a way to incorporate it all makes your map more detailed as well as illustrating your graphical prowess. The key is to practice self-restraint and despite having an abundance of data to simply not use it all—or use it judiciously. Your audience will appreciate not being overwhelmed and being able to read all of less rather than none of more.

**Completeness is hardly ever the goal in cartography.** It is almost always impossible to physically fit all data onto a map. Rather than compromise your map, simply reduce the amount of data you use and simplify the information you communicate. Remember, your reader won't necessarily know or care that you are presenting a selection of data, and as long as the meaning isn't modified you'll be doing them a favour.

**See also:** Elegance | Flourish | Multivariate maps | Seeing | Signal to noise

There are different types of information overload. The obvious example is simply too much information where the resulting map becomes difficult or impossible to use for any purpose. The reason is that the author likely wanted to make a map to cover every potential use!

The converse is the map that has virtually no information at all and serves little purpose. Communicating a lot of information makes the cartographic task harder as you venture into the territory of balancing the need you think the map serves with the larger and more complicated cognitive burden for the reader of locating, recognising, and understanding. Often, the ability to simply fit more onto a map results in a neat cartographic exercise but produces a product of little practical value.

The problem of information overload is a question of usability rather than completeness or accuracy. A subset of your data remains accurate (as long as you're not withholding a vital component that reshapes the message). Work toward being able to share the message in a compelling manner with as little of the data as you need. Whether that is through summaries of data, using aggregated units for display, or down-sampling source information from which you generate generalised topographical detail, selection, omission, and generalisation remain important practices of cartographic self-restraint.

The same is true for symbol design where complicated multivariate, multilayered symbols often hide the interrelationships among data values—obscuring the message. Humans are simply not good at interpreting the nuanced relationship among data presented in overly complicated symbols.

**Opposite:** A page in a student atlas of Belgium illustrating symbol overload and overly complex design that obscures information recovery. By Kenneth Field, 1991. We all started somewhere, and hopefully learn from our mistakes.

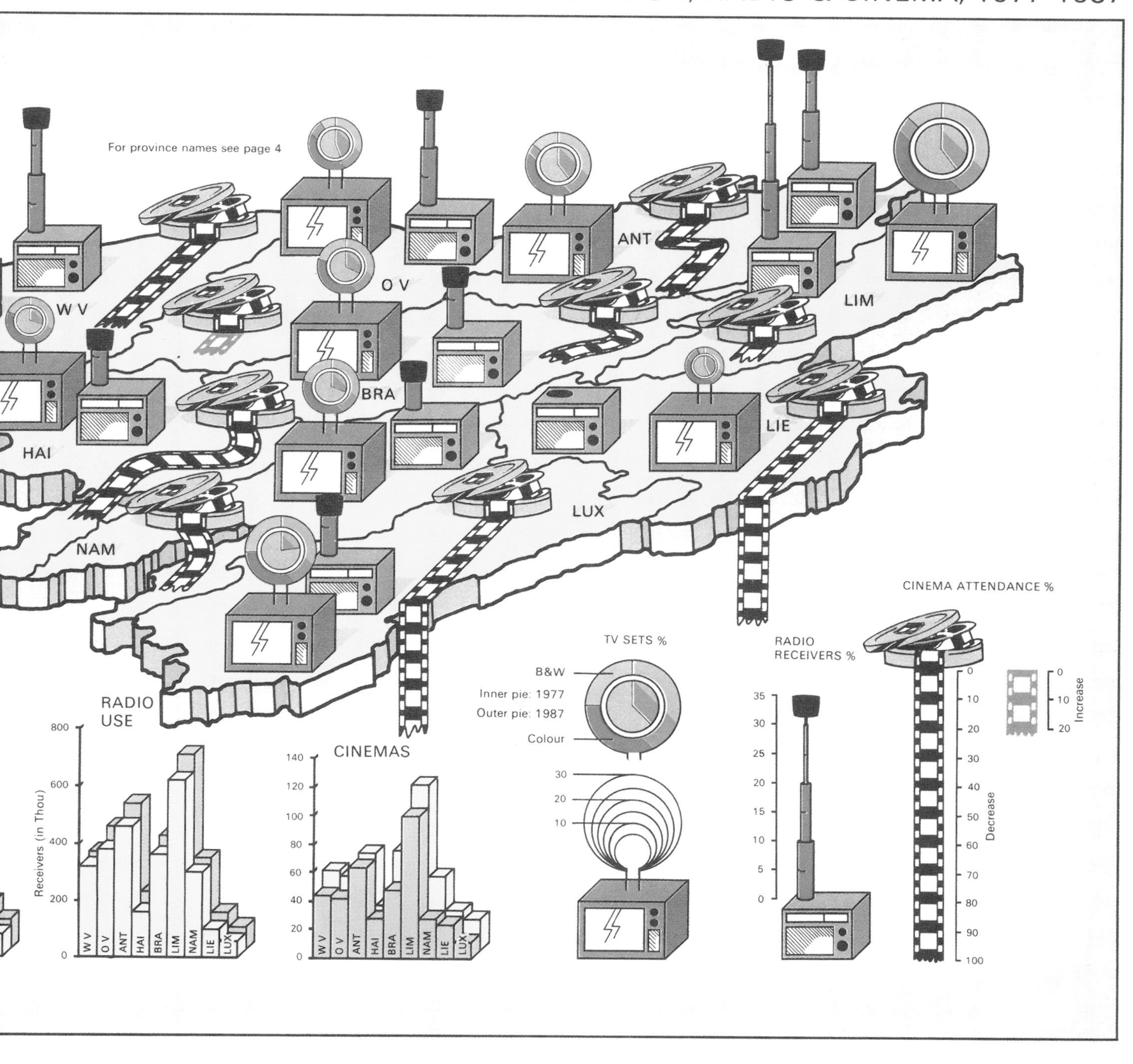
For province names see page 4
W V
O V
ANT
LIM
HAI
BRA
LIE
LUX
NAM
RADIO USE
Receivers (in Thou)
800
600
400
200
0
W V
O V
ANT
HAI
BRA
LIM
NAM
LIE
LUX
CINEMAS
140
120
100
80
60
40
20
0
TV SETS %
B&W
Inner pie: 1977
Outer pie: 1987
Colour
30
20
10
RADIO RECEIVERS %
35
30
25
20
15
10
5
0
CINEMA ATTENDANCE %
0
10
20
30
40
50
60
70
80
90
100
Decrease
Increase

# Information products

## Maps are the quintessential information product, fundamentally designed to communicate spatial information.

**We are all producers as well as consumers of information, which has consequences for how we design maps as information products.** The rise of the internet and the digital age gives us access to, and an unquenchable thirst for, vast amounts of previously unavailable information, uninformation, and misinformation. The nature of information has also changed as it's now streamed to our mobile devices and goes everywhere with us. Where once information was sought after and treasured, it's increasingly a transient commodity.

**Maps have always been one of the fundamental information products of human existence.** Like all information products, a map is a visual system to explain something. Maps should be designed to allow you to find out what you want and to tune out everything else. Information is the message that you are seeking to communicate to your map reader. When expressed in that way it seems uncomplicated—except that the process of achieving that objective in design terms is not necessarily straightforward because we can very easily design uninformation or misinformation instead.

**Maps have always had the potential to lie, either innocently or deliberately.** Uninformation might be the clutter that tends to fog an otherwise clear message. It includes what is probably not that important in the context of your message and which is usually used to embellish, pad out, or (on occasion) mask what you really want to say. Of course, misinformation might be entirely accidental though in all cases there is usually some underlying cause such as naivety, poor data, or a lack of understanding of the data and how to translate it into a map.

**A map, then, is the quintessential information product, graphic, or infographic.** Mapmakers shape information for their readers through the unique form of a product that explains things spatially. You design for information and you create graphics for understanding.

Living in the information age has dramatically altered our lives, and the rules of design have changed to accommodate new ways of thinking about and delivering information. The design of maps must make the journey to the information compelling without distortion.

Maps show where but they can also be used to show why, how, by how much, and many other questions people pose. People have an appetite for information, but instead of being concerned about a lack of information, the challenge for most consumers has changed to being one of sifting out the quality from the quantity. Curiously we want less information but we want that information to be pertinent, objective, accurate, authoritative, and, more than anything, instantly gratifying.

*"Indicium est omne divisa in partes tres "(All information is divided into three parts)* —Paraphrase of Julius Caesar in *Commentaril de Bello Gallico* ca. 40s BCE, which embraces the notion of information, uninformation, and misinformation.

**Opposite and below:** *Stick Chart of the Marshall Islands.* The Marshall Islanders' stick charts were nothing more than sticks to identify water currents and shells to identify islands. They supported navigation to the exclusion of everything else.

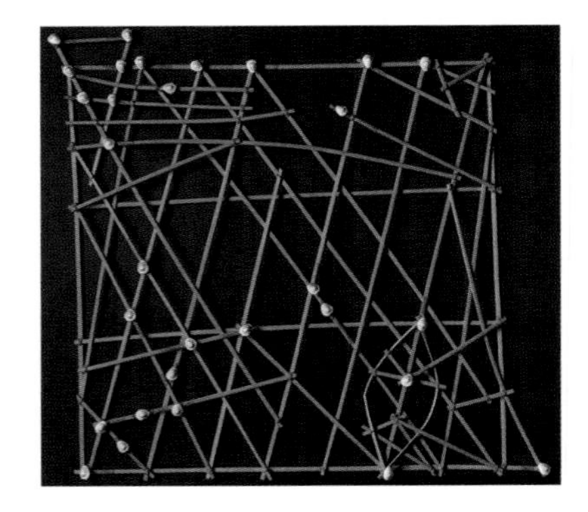

**See also:** Craft | Defining map design | Informing | Inquiry and insight | Map cube | Visualization wheel

ARNO
JALUIT
NAMU
KILI

# Informing

## Maps inform the public understanding of science.

**Science plays a crucial role in our everyday lives.** Almost everything we do depends on science in one form or another be it through the use of objects or simply an appreciation of the weather. Everybody, therefore, needs some understanding of science in order to make decisions that affect what they do, how they do it, and where they do it. Science permeates the media. Feature articles are often science-based even if the science itself is rarely the news. Maps have a crucial role to play in helping inform public understanding of science.

**Maps act as intermediaries between science and the public.** They are visual representations not only of places but of the characteristics of those places. They describe the science of geography at a basic level yet have a much wider role in supporting the dissemination of themed information about so many aspects that underpin life such as epidemiology, economics, humanitarian efforts, and politics. The role, value, and relevance of cartography goes beyond simply the map. It provides a rational, objective way of recording and educating. The same applies to many graphical forms of representation which can be used to show people, in an accessible visual form so they can be better informed.

**Mapmakers must learn to communicate in using maps to inform.** This places demands on those who are experts in one domain to learn something of the expertise required in the domain of cartography to effectively represent their science to the general public. Science and society needs to be willing to do so. It becomes a duty to disseminate. Anyone needing to make a map needs to learn how to take advantage of it to explain their science simply, without jargon and in an accessible fashion. Improving people's general level of understanding of geography but, also, science and society more widely, should be part of the role of maps. Cartographers clearly have a role to counter what might be referred to as a death in expertise in mapmaking.

**See also:** Cognitive biases | Knowledge and conviction | Craft | Critique | Inquiry and insight | Mental maps

A1985 report published by The Royal Society in the United Kingdom on the public understanding of science found that *'scientists must learn to communicate with the public, be willing to do so, and indeed consider it their duty to do so'*. This counters mistrust, scepticism, or a basic lack of understanding and also goes beyond scientific discourse and publications to engage more directly through other avenues. As cartography has become democratised, the need to develop ways of communicating techniques as well as supporting how others communicate has never been more necessary. Communicating cartographic knowledge is part of this process of change.

Maps act as a bridge and a vital media tool. Given the importance of understanding and the need for an informed public, the role of high-quality, objective maps to convey the facts has perhaps never been more critical. The language of cartography is perfectly placed to contribute to the improvement of public knowledge. It requires the scientific community, journalists, and social media aggregators to engage with the theory and practice of cartography, but also, cartographers must provide a mechanism to support the work of others.

**Opposite:** New York's greenhouse gas emissions as one-ton spheres of carbon dioxide gas by CarbonVisuals, 2012. The data is brought to life through a three-minute video that begins by establishing what one ton of $CO_2$ would look like. We are given a comparison to aid interpretation. The animation then develops as spheres are added, pausing to show what the cumulative amount is for a period of time we can easily understand: one hour, one day, one year. The animation is simple, clear, and compelling. It serves to inform a general audience in ways that are easy to understand. The use of familiar buildings, in 3D but lacking detail, offsets the bright-blue spheres, which bring the issue into stark focus giving us a sense of scale for atmospheric pollution. https://youtu.be/DtqSIplGXOA

In 2010 New York City added over 54 million metric tons of carbon dioxide to the atmosphere.

That's nearly 2 tons every second.

75% of the emissions came from buildings.

One ton of carbon dioxide gas would fill a sphere 33 feet across.

If New York City's emissions all merged at the same place, this is the volume of gas that would come out...

One hour's emissions

One day's emissions

One year's emissions

# Inquiry and insight

## Map design relies on a good understanding of geographical processes.

**Map design requires an appreciation of the basic tenets of geographical inquiry and insight.** Geographers ask typical questions such as What? Why? When? But added to these is the crucial Where? question. Where can be quantified by describing position on Earth in terms of coordinates. However, location can be more sophisticated than simply mapping where features exist, and relative location has become key to exploring spatial relationships. For instance, two cities can be spatially described in their absolute sense by defining their latitude and longitude. In a relative sense they may be a certain distance apart. In terms of interconnectedness, the two places may be temporally closer to each other based on different modes of transport. What this means for cartography is that mapping of coordinates; of absolute location, is only one part of the mapping task. Successfully encoding the processes of geographical inquiry allows maps to support insight.

**The sorts of questions we ask of the world, and of data, can be framed around key concepts such as::**
Direction—describes relative location
Distance—describes how far away one feature is from another
Scale—the size of the area being studied (micro-, meso-, and macro-scale)
Location—with respect to coordinates or relative to other phenomena
Areal patterns—describing a distribution over space
Spatial clustering—extent of concentration or spatial proximity
Incidence and prevalence—the existence of a phenomenon and its relationship (coexistence) with other phenomena
Connectivity—describes the linkage between phenomena rather than the phenomenon itself
Spatial interaction—characterises the attraction between places and the spread of movement
Regionality—exploration of similarities and differences of phenomena in comparison with other places
Change—how phenomena differ with the passage of time or under the influence of other phenomena or changing conditions.

**See also:** Informing | Map cube | Purpose of maps | Visualization wheel

Geography is both a social and a physical science that has at its core the study of how phenomena populate and shape space and place. Geography is the science of spatial analysis with many questions based around the attempt to discern spatial pattern and to establish reasons why phenomena are distributed in a particular way. The underlying causal processes that shape spatial patterns are often sought as geographers focus on the spatial dimension of phenomena. Because there is an almost infinite number of subdisciplines or areas of interest, the focus of geographical study can vary immensely. Indeed, any phenomenon that has a spatial location or a spatially defined attribute can be examined geographically. The map is a fundamental tool for anyone interested in geographical patterns.

Waldo Tobler's oft quoted First Law of Geography, *'everything is related to everything else, but near things are more related than distant things'*, was an outcome of the quantitative focus of geography in the 1960s. It's since gained notoriety as a useful principle for the design of geographical inquiry. It sums up the fact that characteristics tend to vary slowly across geographical space. And so, we can recognise and describe patterns relative to this characteristic. Of course, depending on what is being studied, the impact of distance might operate in different ways, and there are also exceptions.

In fact, Tobler didn't stop with just a single law. His second, less well-known, law states, *'The phenomenon external to an area of interest affects what goes on inside'*. This supports the first law since it further defines a relationship between proximal places.

*'Geography is just physics slowed down, with a couple of trees stuck in it'.*
—Terry Pratchett

**Opposite:** An extract from the *Environment and Health Atlas for England and Wales*, 2014.

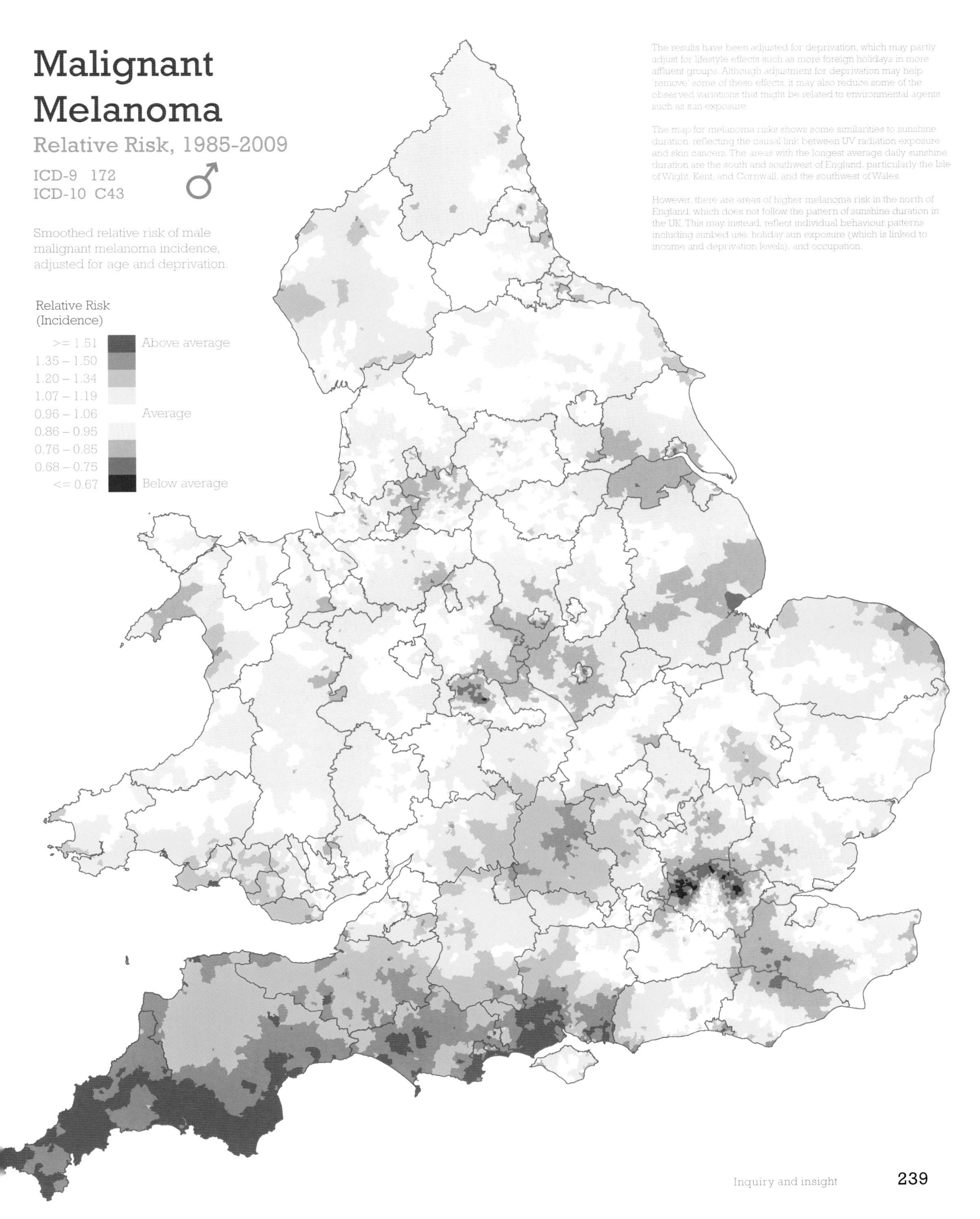
Malignant Melanoma
Relative Risk, 1985-2009
ICD-9 172
ICD-10 C43
♂
Smoothed relative risk of male malignant melanoma incidence, adjusted for age and deprivation.
Relative Risk (Incidence)
>= 1.51
1.35 – 1.50
1.20 – 1.34
1.07 – 1.19
0.96 – 1.06
0.86 – 0.95
0.76 – 0.85
0.68 – 0.75
<= 0.67
Above average
Average
Below average
The results have been adjusted for deprivation, which may partly adjust for lifestyle effects such as more foreign holidays in more affluent groups. Although adjustment for deprivation may help 'remove' some of these effects, it may also reduce some of the observed variations that might be related to environmental agents such as sun exposure.
The map for melanoma risks shows some similarities to sunshine duration, reflecting the causal link between UV radiation exposure and skin cancers. The areas with the longest average daily sunshine duration are the south and southwest of England, particularly the Isle of Wight, Kent, and Cornwall, and the southwest of Wales.
However, there are areas of higher melanoma risk in the north of England, which does not follow the pattern of sunshine duration in the UK. This may, instead, reflect individual behaviour patterns including sunbed use, holiday sun exposure (which is linked to income and deprivation levels), and occupation.

# Integrity

## Maintaining integrity in cartography ought to be considered a matter of fact but there are shades of integrity.

**Truth and integrity in cartography are important and showing something on a map brings with it the veil of truth.** People believe maps and diagrams as purveyors of accuracy and, therefore, of truth. Mapping the facts should be a fundamental objective. Not lying in the way information is represented is, though, not the same as telling the truth and there can be different shades of truth. Historically, topographic maps have left off features that clearly exist in the landscape such as infrastructure deemed to be official state secrets left off Ordnance Survey maps in Great Britain. In thematic mapping the scope to manipulate data or its representation is perhaps easier to accomplish.

**Rational objectives should drive the development of a symbology that fairly represents the data and leaves little margin for misinterpretation.** But even when mapped data is accurate, the way in which it is represented can bring different shades to the truth. For instance, a choropleth map almost always requires the data to be classified. Although the data may all be represented on the final map, the way the classification scheme may have been developed will undoubtedly shape the way the information is represented, read, and interpreted. This might be through a firm choice on the part of the cartographer making the map or, simply, through a lack of understanding of the way in which reality can be shaped.

**Choice of classification scheme is only one way in which truth can be manipulated but sometimes the cartographer has very little control.** Again, using a choropleth map as an example, we usually map into standard geographies—usually enumeration areas that have been used to collect the data. These will almost certainly be predetermined and, consequently, have their own inherent bias as the Modifiable Areal Unit Problem on the opposite page shows. In short, the map is as much a function of the boundaries used as the spatial pattern of the population mapped. If boundaries were drawn differently, data would be collected in different containers and a different visual pattern would likely emerge.

**See also:** Cognitive biases | Knowledge and conviction | Copyright | Map traps | Maps kill | Who is cartography?

Maintaining integrity in how we represent information is not a simple task though it should not be ignored. It is at least a good idea to ensure that data is complete; data is accurate and checked; data is pertinent to the map and that the map is appropriate for the data; that data addresses the issue and focus of the map, to the appropriate level and extent; and that the way you map the data at the very least discourages misinterpretation.

Mark Twain popularised the phrase 'There are three kinds of lies: lies, damned lies, and statistics'. The same might be said of maps and, inspired by the book *How to Lie with Statistics*, Mark Monmonier's classic *How to Lie with Maps* should be recommended reading for any mapmaker. It deals with the use and abuse of maps and teaches how to evaluate maps critically. Monmonier describes how maps must lie as they are scaled-down versions of reality. They flatten a round surface, omit detail, and distort territory. He describes the problems of scale transformations and map editing as well as fact-checking and dealing with mistakes and the apparent information on the map versus the reality. Monmonier explores the idea of seduction in cartography too and how one might modify a map to suit a particular purpose. The point of the book is to express that it's perfectly possible to make your map lie terribly if you are careless or if you do it by choice. Being aware of this should at least make you think even if your ultimate objective is, indeed, to be persuasive! For the map reader, it promotes a healthy scepticism toward maps and a questioning of the reality presented to you through the filter of the cartographer.

*'Everyone is entitled to their own opinion, but not to his own facts'.*
—Daniel Patrick Moynihan

## The Modifiable Areal Unit Problem (MAUP)

'a problem arising from the imposition of artificial units of spatial reporting on continuous geographical phenomena resulting in the generation of artificial spatial patterns' (Openshaw, 1983)

### Scale

tendency for different statistical results to be obtained from the same set of data when that information is grouped at different levels of spatial resolution

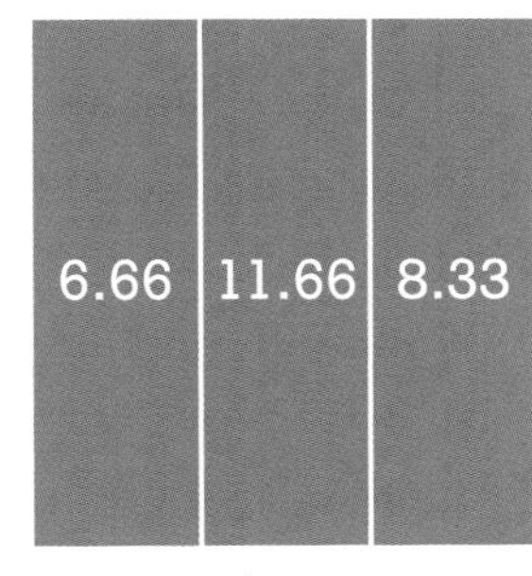


number = 3
mean = 8.88

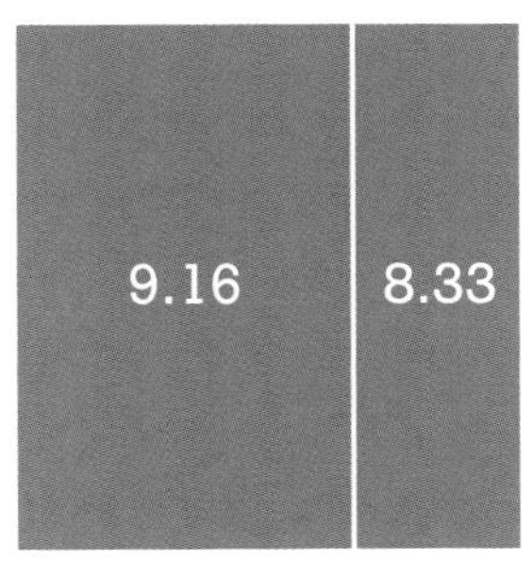


number = 2
mean = 8.75

### Aggregation

tendency for variability in results being obtained through variations in the shape of areas

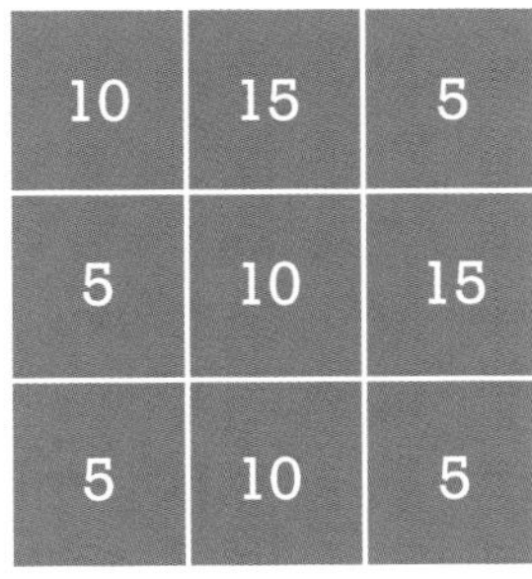


number = 9
mean = 8.88

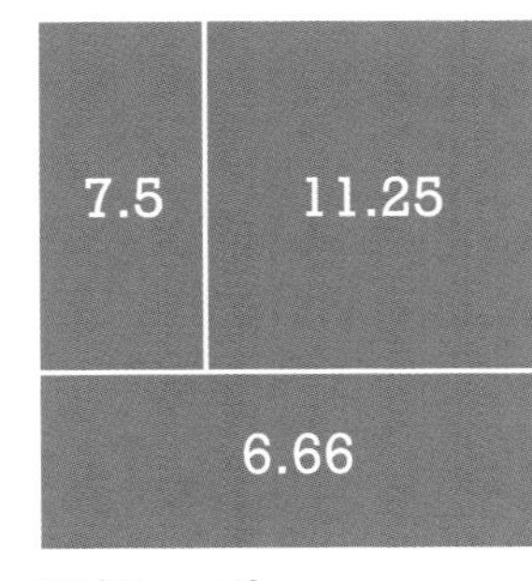


number = 3
mean = 8.47

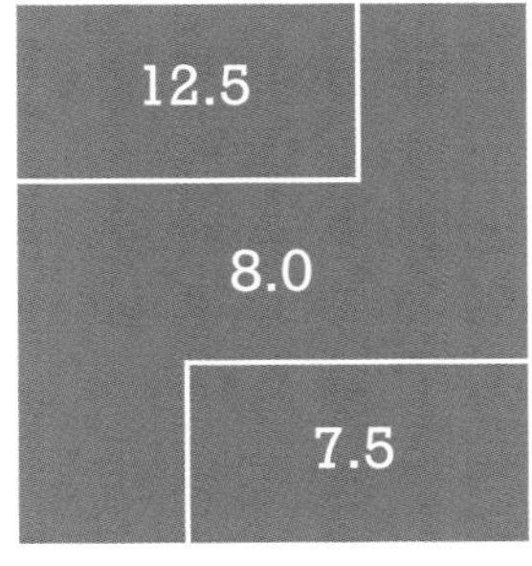


number = 3
mean = 9.33

### Mapped appearance

the impact of MAUP on the map

# Interaction

**Interactivity is the cornerstone of web mapping and adds to the ways in which people interact with maps to retrieve information.**

**Web mapping brings many key opportunities for cartography.** Map design used to have to accommodate a wide variety of potential uses and users and encode every piece of information graphically. Interaction *supports* a wider variety of potential uses. Simply providing the ability to click or hover over map components can reveal labels, text, details of what exists, or a window into further data or links. Panning and zooming allow multiscale maps to deliver information at different levels of detail and resolution. The map has become far more flexible in its ability to support multiple uses because of this increase in interaction between the user and the map itself.

**Beyond simply panning and zooming, you can also build new map interactions into the map use experience.** Identifying and comparing actions can be supported through interactive capabilities. More complex interactions such as brushing linked maps, graphs, and textual elements or focussing (filtering) the map's content based on user choice have also developed the way maps are used.

**Cartographers can take advantage of browser capabilities to finally extend the mapping of certain phenomena into true animation.** Temporal cartography is now much more readily supported, through animating time slices, or using animated symbology to better reflect certain data types (e.g. flow). Animations also give us ways to swipe two versions of the same map for comparative study or to switch layers on and off to see different aspects of the data. The ability to change the navigation, object rotation, and viewpoint provides new ways of positioning the map and supporting different use-case scenarios.

**3D is now also properly supported through interaction.** The static realm has always been the main drawback for high-quality 3D cartography as a single viewpoint inevitably causes occlusions and sometimes a false perspective. The ability to animate, change viewpoint, and interactively rotate, pan, and zoom makes 3D mapping truly usable.

**See also:** Animation | Form and function | Mobile mapping | Simplicity vs. complexity | Your map is wrong!

Cartography used to be about simply encoding information using the processes of generalisation and symbolisation to represent phenomena under the presumption that the map reader would see and interpret the map. This paradigm has been firmly grounded in cartographic research on perception (how maps are seen), cognition (how maps are understood), and semiotics (how maps are imbued with meaning). These are the cornerstones of functional design but interactivity stretches these to suggest accessible design. This notion demands that you add value to the map experience through interactivity. However, just because something can be interactive doesn't mean it necessarily should be.

Maps no longer must rely only on visual encoding to provide information because of relationship between the human and a device that is very much designed for interaction. Here, paper gives way to a computing device where our hands now add to the ways in which we interact via a mouse or through gestures or haptic manipulation. Both the human and the map (via the device) are agents of interaction that have the capability of modifying the behaviour of the other. Maps change according to your requests which, in turn, change the mode of interaction. The once one-way process of reading a map is now non-linear and is much more akin to a conversation or two-way dialogue.

Interaction may go beyond web browsers or touch-based devices and include augmented and virtual reality interfaces. These bring additional opportunities for interaction by placing us in a virtual world where our field of vision and other senses are immersed.

Touch Screen
Visuals
Hand Interaction
Headphones
Audio
Traditional Screen
Visuals
Animations
Touch
Swipe
Drag
Rotate
Double
Pinch
Hologram
Free-form Visuals
Voice Bar
Voice Commands
Keyboard
Type
Modifier Keys
Cursor Keys
Mouse
Click
Drag and Drop
Mouse Chording
Click with Modifier Keys
Gesture

# Isarithmic maps

**Isarithmic maps use interpolated contours to represent data that is usually collected at sample points.**

**An isarithmic map is a two-dimensional representation of a three-dimensional volume.** Two main types exist: an isometric form that is constructed from data at points and an isoplethic form constructed from data that occurs over geographic areas. An isometric map might be constructed from data sampled at weather monitoring stations to create a map of atmospheric pressure. An isopleth map usually represents the distribution of population data, collected using enumeration areas, but without the constraints of the usual boundaries. The purpose of an isopleth thematic map is to show how features differ in quantity as a surface. This can be achieved through representing the volume using contour lines or by using filled contours that are shaded according to the quantitative value being mapped.

**Isarithmic maps are generated from values that represent numerical (quantitative) differences between features on an interval or ratio scale of measurement.** Absolute values can be used for isometric mapping but cannot be illustrated isoplethically because of the inherent problems of using totals for areas that might vary in size or which contain an unequal denominator of the data being mapped. This same issue prevents choropleths from being used to map totals.

**Conceptually, we symbolize linear features using contours representing a line of equal value threaded through the data.** Contours are nested and are labelled with the contour value. Label colours often match the line colour for each value and have a mask specified as the same colour as the background to ensure they knock out the contour line immediately underneath. Labels are normally positioned so that none are on an angle of more than 20–30 degrees to ensure they can be easily read.

Although it's reasonable to assign the same colour and line style to all contours, we can modify the symbology to perhaps vary the lightness (or value) of the symbol's hue to accentuate contours. Lines representing the smallest contour values may be lighter and those representing the largest contour values darker. The areas between contours might also be shaded to create a filled contour version of the isarithmic form, along similar lines to a choropleth map with lighter shades for lower classes of values and darker for higher magnitudes of the classified data. The lines of equal value ought to be rounded and progress in numerical sequence that is readily understood by the reader.

Since contours are normally labelled and the data values are therefore encoded in the map, it isn't always necessary to include a legend. The map reader should be able to efficiently recognise the different contours mapped across the map and see the peaks and troughs in the surface. At the very least, relative differences should be obvious, and the reader should be able to determine a pattern across the map. We visually interpret the symbols to perceive areas of higher values and of lower values.

**See also:** Contours | Data density | Illuminated contours | Isochrone maps

**Opposite (bottom):** Multivariate weather map showing air pressure (Isobars) represented by illuminated contour lines and temperature (Isotherms) represented by filled contours (Hurricane Sandy, 30 October 2012) by James Eynard, 2017.

## Threading an isarithm

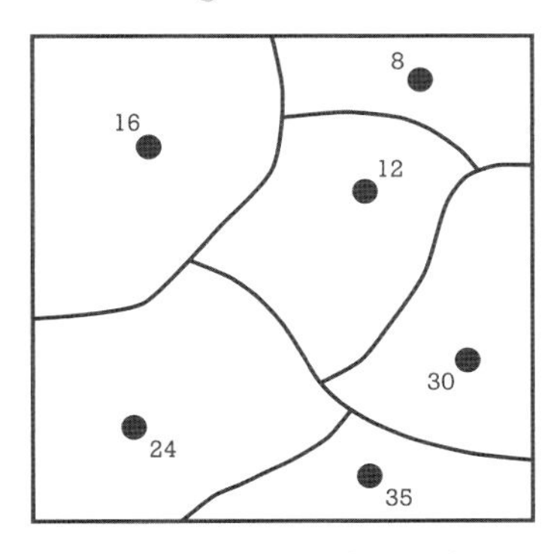


Area data represented as points

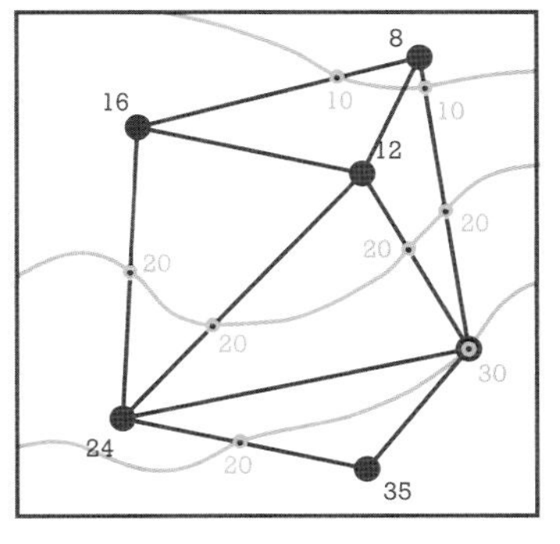


Determine value of isarithms

Thread a smoothed contour through points using Delaunay triangulation

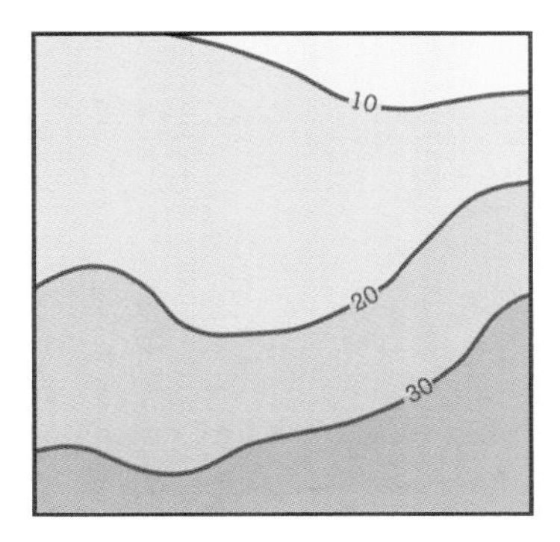


Symbolise using shading for filled contours and label the isarithms

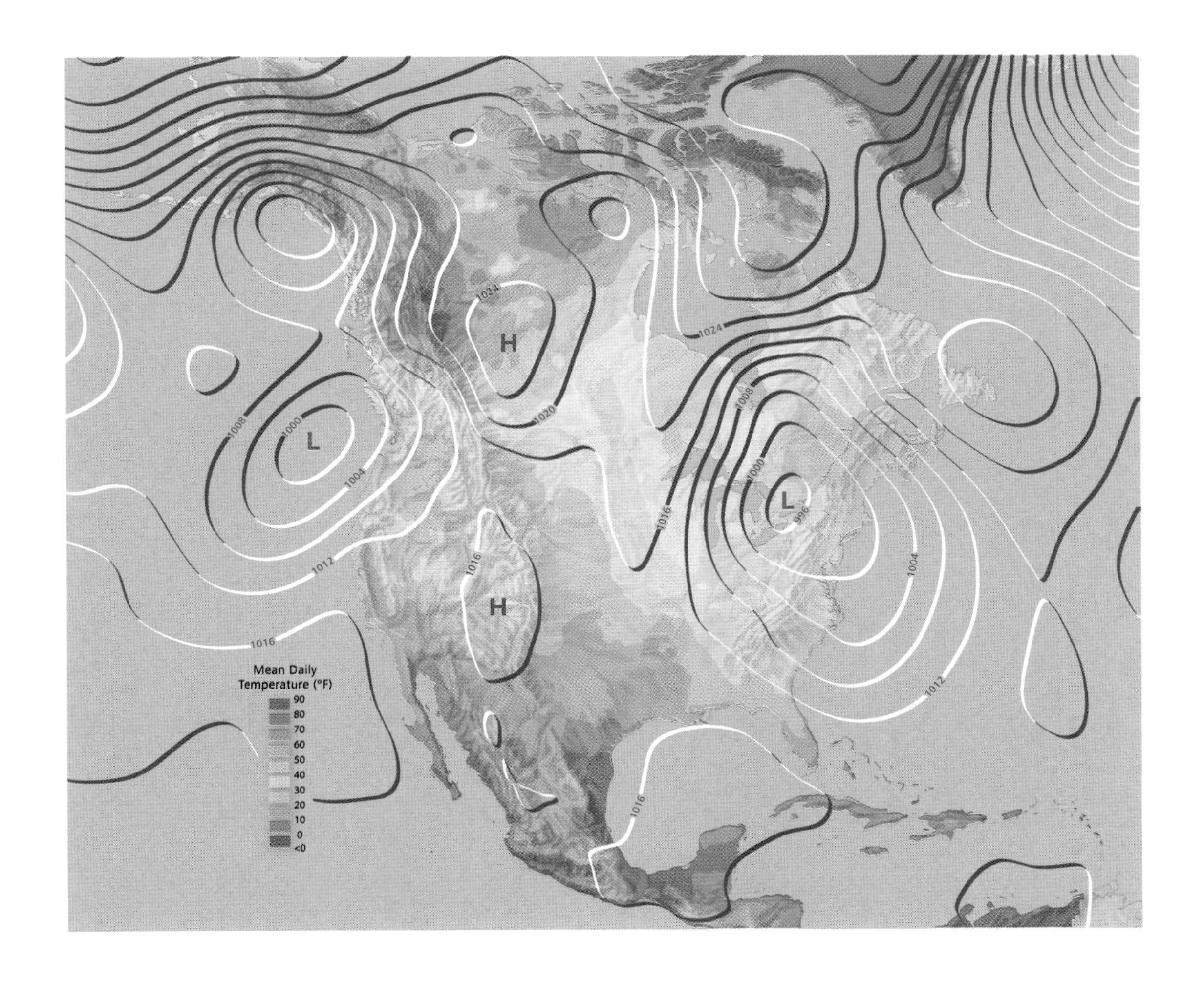

# Isochrone maps

## Where lines of equal value represent the measurement of distance or time away from a feature.

**Conventionally, maps display distance in terms of linear measurement where consistency between features is understood.** However, measuring distance in this way doesn't always make sense. If a map is designed to support an understanding of modes of movement, then fixed geographical scale doesn't necessarily give us a useful picture. People move in different ways on foot, bicycle, car, or public transport. These modes all compress time in different ways, across different networks, and to different extents. Isochrones can be constructed to show distances away from a point or other mapped feature depending on some attribute of the intervening distance such as average speed along a transport network. This simplest form of isochrone is displayed using an overlay showing lines or service areas of equal time.

**Isochrone maps morph the geography itself to create a form of linear cartogram.** Nodes in a network become more important when expressing distance in terms of time. Departing and arriving points and intervening stops are the important characteristics of the map. Stops (and the intervening distance) can be set to be equally spaced to suggest a temporal separation of a particular unit. The faster moving segments in the network become shorter on the map. Although it is the lines of equal value that provide the consistent framework for measurement across the map, the zones in between are usually shaded. This may be through alternating zones of colour (e.g. light grey and white), or by some scheme that varies in value with further distances being represented lighter or darker depending on the emphasis of the map—whether the intent is to concentrate on those areas closest or farthest away.

**See also:** Contours | Illuminated contours | Isarithmic maps | Temporal maps

Geography (other than the network travelled) is largely irrelevant in representing places that are near or far from each other. We also tend to talk in terms of time rather than distance. For instance, it's '10 minutes away if you drive' or 'it'll take you about an hour walking'. The isochrone map is therefore a technique that can provide a thematic layer to visually represent distance in a way that matches how we often refer to it.

Isochrone maps have a long history in transport planning since the late 1800s. Francis Galton created a series of isochrone maps called *Isochronic Passage Charts,* showing travel times from London to different parts of the world. The technique gained popularity and has been used routinely by many subsequent atlas publishers in particular to show the time it takes to travel long distances. Isochrone maps are also used to show travel time by different modes of transport.

In addition to transport planning, isochrone maps can be used in many other contexts to map some aspect of time or other variable through a network. Hydrology makes use of isochrones to show the time it takes for run-off through a drainage basin.

In the majority of this book you'll read warnings against using spectral colours for mapping quantitative data. Yet the maps opposite appear to go against that advice. There are several reasons. First, the colours do not use the entire spectrum. Second, they are generally placed in order of lightness with the exception of the green for the lowest class. This supports the third reason; colours are always used sequentially due to the nature of the mapped phenomenon. Hues are used in this way for hypsometric tinting and this is a variant. Green is slightly darker because the pattern is repeated so it provides a useful break between repetitions. Knowing when and how to break cartographic rules is valuable!

## Sunshine on Leith

The Proclaimers released *Sunshine on Leith* in 1988. The song is renowned for the lyric *But I would walk 500 miles. And I would walk 500 more...*

With a little artistic license, where does 500 miles (and 500 more) from Leith take you and how long would the journey take?

Continental Europe can be reached in 500 miles. The tip of Sweden is 560 miles from Leith in a straight line but 1,000 miles away by road and ferry and would take a 320-hour walk at average pace.

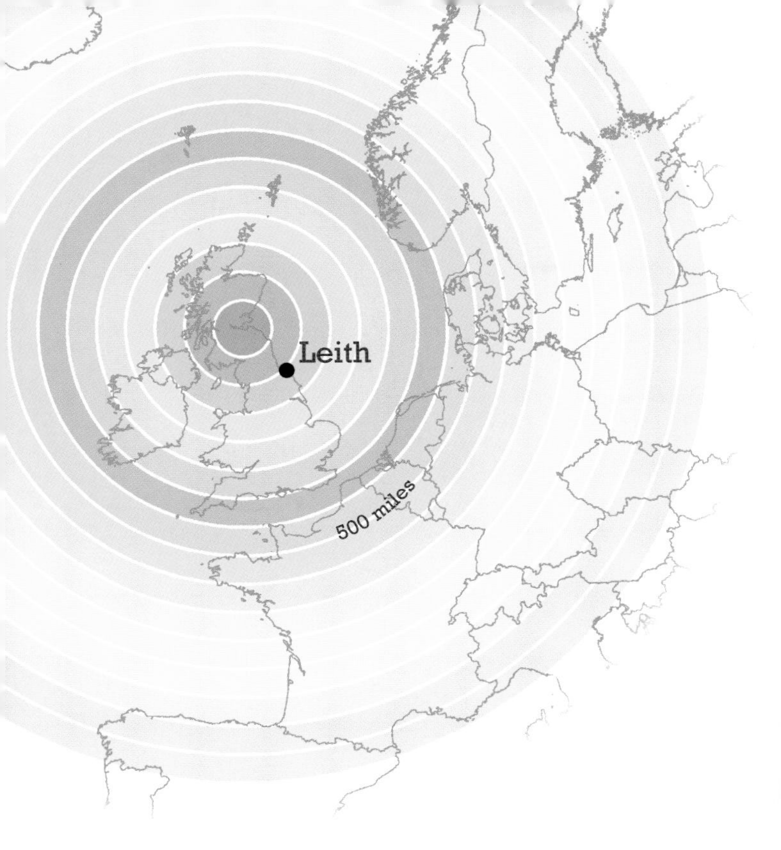


### I would fly 500 miles

Concentric isochrones on an azimuthal equidistant projection at 63-mile intervals.

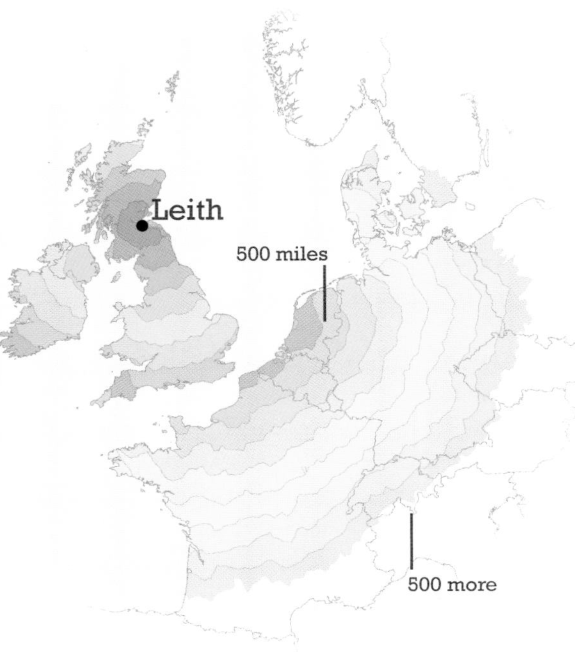


### I would drive 500 miles

Radial isochrones at 63-mile intervals based on travel by the road and ferry network.

| 500 miles | | 500 more | |
|---|---|---|---|
| Hours | Miles | Hours | Miles |
| 0 | 0 | 160 | 500 |
| 20 | 63 | 180 | 563 |
| 40 | 125 | 200 | 625 |
| 60 | 188 | 220 | 689 |
| 80 | 250 | 240 | 750 |
| 100 | 313 | 260 | 813 |
| 120 | 375 | 280 | 875 |
| 140 | 438 | 300 | 938 |
| 160 | 500 | 320 | 1,000 |

### I would walk 500 miles

Linear isochrones at 20-hour intervals showing the walkable routes.

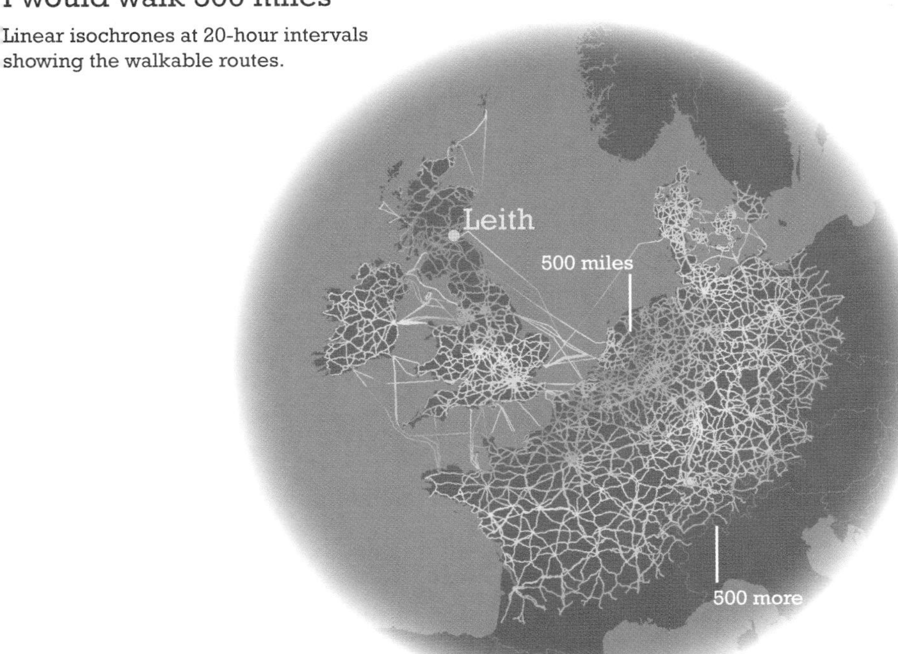

# Isometric views

## A 3D view that supports visual comparison.

**Three-dimensional views often have characteristics that hinder interpretation.** Human sight is stereo, and we see perspective depth. Objects in our near vision appear larger than those that are distant. Three-dimensional maps, graphs, and virtual globes are also often rendered as a perspective view. For topographic maps this is likely to be perfectly acceptable. For thematic maps where extruded features, vertically positioned map symbols, or 3D graphs that sit on a map, this presents problems. The map-reading task demands that we are able to compare like for like across the map yet perspective renders this impossible. The solution is to use an isometric view, which is a form of axonometric projection.

**Axonometric views tilt and rotate so the reader can see the top and two sides of information.** The amount of tilt and rotation impacts readability. When the z-axis scale is relatively small a slight amount of tilt isn't going to reveal much. When there are many high peaks a larger tilt might avoid occlusions. When values of tilt and rotation are close to 0° or multiples of 90° there is little value in the axonometric projection.

**The isometric view specifies that rotation and tilt are constrained so that the angles between each pair of axes are all 120°.** This results in a view in which every object is equally foreshortened, effectively meaning scale is consistent among map features across the whole map. This helps preserve the property that measurements we take or view are comparable across the map. For thematic mapping in particular, this is critical as we use height to encode some quantitative measure.

**In practical terms an isometric view can be achieved manually.** Imagine a cube where the default view is toward one face. First, rotate the cube ±45° about its vertical axis. Next, rotate the cube approximately ±35.264° about the horizontal axis. The cube will now be in an isometric view. You'll notice that the 2D impression of the outline of the cube will form a perfect hexagon, each side of the cube will have the same length of line, and each face on the cube will be of equivalent area.

**See also:** Aspect views | Height | Prism maps

Isometry has a long history in architecture where drawings are required to support accurate measurement. The difficulty with isometry for any drawing or map is that it results in an image that does not seem natural to our eyes. We are so used to seeing in perspective with foreshortening that seeing a view of the same information presented with equal foreshortening can lead to the impression it is distorted. Although measurement is supported, we can sometimes have trouble gauging depth in the map. One consequence of this property is that isometry can also be used to create optical illusions using paradoxical shapes.

Penrose stairs are a classic example of the optical illusion of isometry. The creation of a sequence of stairs that form an impossible staircase gives rise to a continuous loop that appears to both ascend and descend. The implication is that anyone could climb or descend the staircase in an infinite loop yet never get any higher or lower. This is a clear impossibility, yet a visual consequence of the properties of isometry. Although developed by Lionel and his son Roger Penrose in the mid-1950s, perhaps the most famous implementation of Penrose stairs is *Ascending and Descending* by M. C. Escher. Escher also used the technique for his later work *Waterfall*.

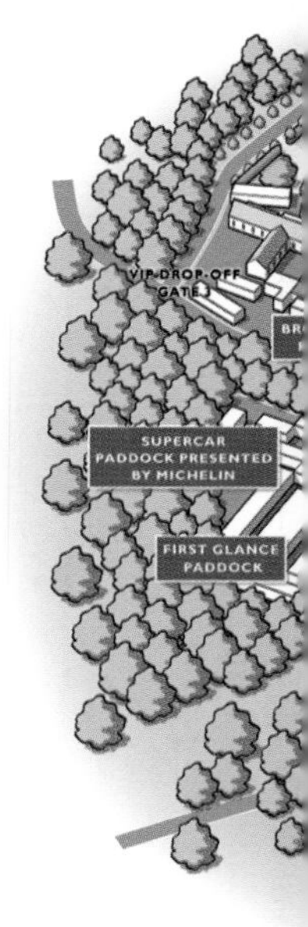


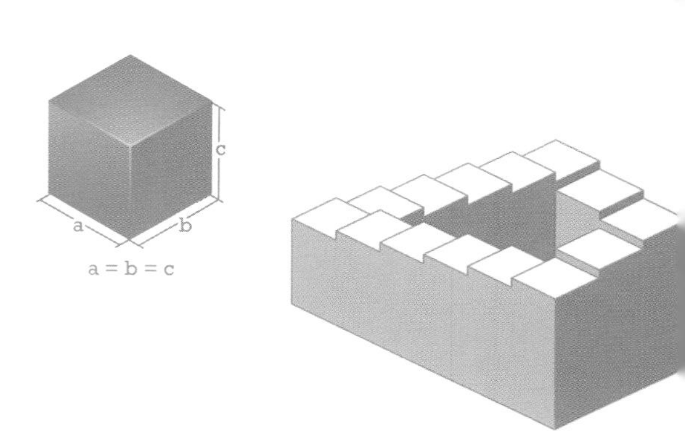


**Opposite:** *Visitor Map for the Goodwood Festival of Speed* by Mike Hall, 2015.

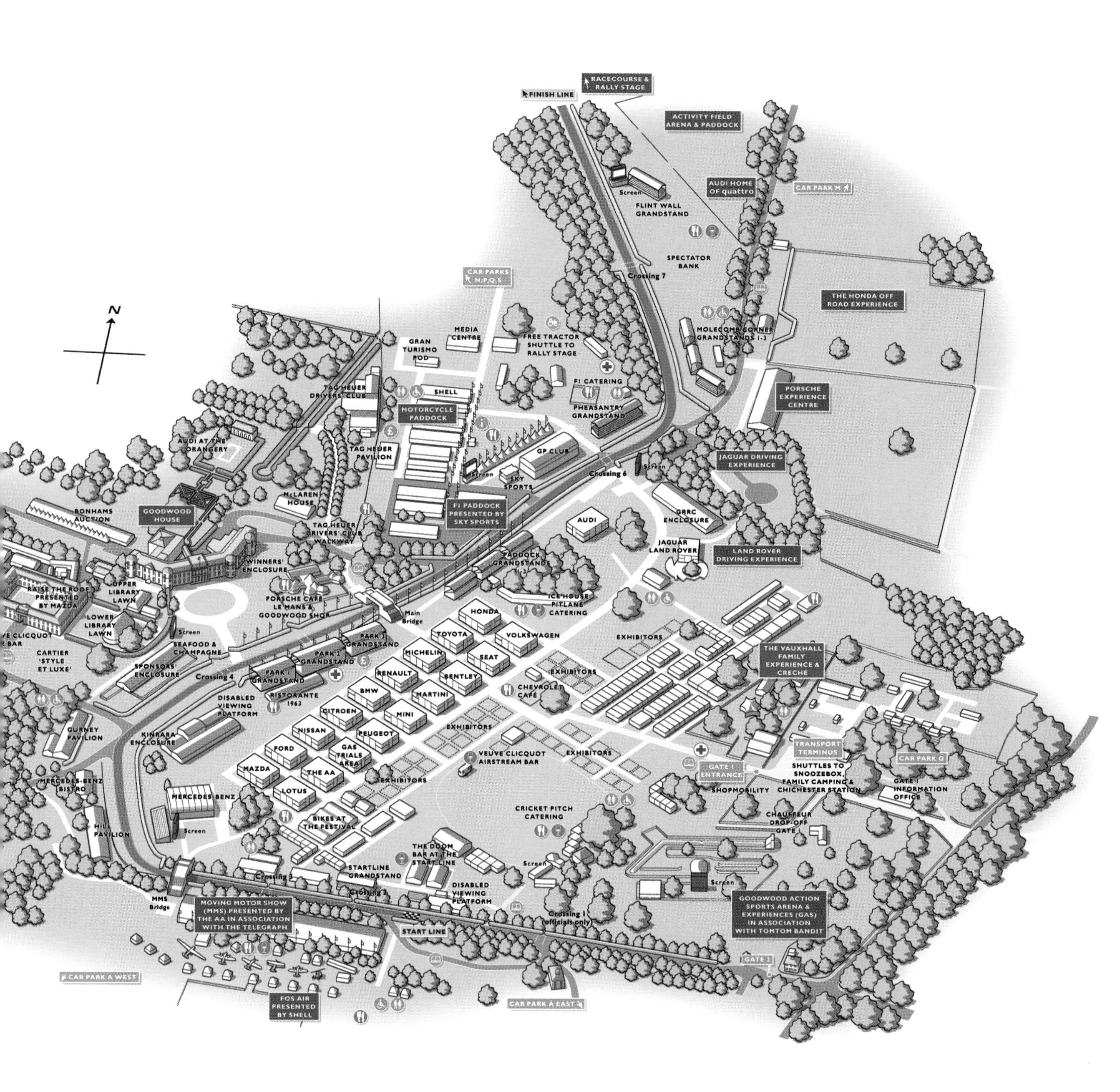

FINISH LINE
RACECOURSE & RALLY STAGE
ACTIVITY FIELD ARENA & PADDOCK
FLINT WALL GRANDSTAND
AUDI HOME OF quattro
CAR PARK M
SPECTATOR BANK
Crossing 7
CAR PARKS N,P,Q,S
THE HONDA OFF ROAD EXPERIENCE
MEDIA CENTRE
GRAN TURISMO POD
FREE TRACTOR SHUTTLE TO RALLY STAGE
MOLECOMB CORNER GRANDSTANDS 1-3
F1 CATERING
SHELL
TAG HEUER DRIVERS' CLUB
MOTORCYCLE PADDOCK
PHEASANTRY GRANDSTAND
PORSCHE EXPERIENCE CENTRE
AUDI AT THE ORANGERY
TAG HEUER PAVILION
GP CLUB
JAGUAR DRIVING EXPERIENCE
SKY SPORTS
Crossing 6
McLAREN HOUSE
F1 PADDOCK PRESENTED BY SKY SPORTS
BONHAMS AUCTION
GOODWOOD HOUSE
TAG HEUER DRIVERS' CLUB WALKWAY
AUDI
GRRC ENCLOSURE
JAGUAR LAND ROVER
LAND ROVER DRIVING EXPERIENCE
WINNERS' ENCLOSURE
PADDOCK GRANDSTANDS 1-3
RAISE THE ROOF PRESENTED BY MAZDA
UPPER LIBRARY LAWN
PORSCHE CAFE LE MANS & GOODWOOD SHOP
ICE HOUSE PITLANE CATERING
Main Bridge
HONDA
LOWER LIBRARY LAWN
Screen
TOYOTA
VOLKSWAGEN
EXHIBITORS
CARTIER 'STYLE ET LUXE'
SEAFOOD & CHAMPAGNE
PARK 3 GRANDSTAND
PARK 2 GRANDSTAND
MICHELIN
SEAT
THE VAUXHALL FAMILY EXPERIENCE & CRECHE
SPONSORS' ENCLOSURE
Crossing 4
PARK 1 GRANDSTAND
RENAULT
BENTLEY
CHEVROLET CAFE
BMW
MARTINI
RISTORANTE 1962
DISABLED VIEWING PLATFORM
CITROEN
MINI
GURNEY PAVILION
NISSAN
PEUGEOT
KINRARA ENCLOSURE
FORD
GAS TRIALS AREA
EXHIBITORS
TRANSPORT TERMINUS
CAR PARK G
VEUVE CLICQUOT AIRSTREAM BAR
MAZDA
THE AA
GATE 1 ENTRANCE
SHUTTLES TO SNOOZEBOX, FAMILY CAMPING & CHICHESTER STATION
MERCEDES-BENZ BISTRO
LOTUS
SHOPMOBILITY
GATE 1 INFORMATION OFFICE
MERCEDES-BENZ
CRICKET PITCH CATERING
CHAUFFEUR DROP-OFF GATE
HILL PAVILION
BIKES AT THE FESTIVAL
THE DOOM BAR AT THE START LINE
STARTLINE GRANDSTAND
Crossing 2
Crossing 3
DISABLED VIEWING PLATFORM
MMS Bridge
MOVING MOTOR SHOW (MMS) PRESENTED BY THE AA IN ASSOCIATION WITH THE TELEGRAPH
START LINE
Crossing 1 (officials only)
GOODWOOD ACTION SPORTS ARENA & EXPERIENCES (GAS) IN ASSOCIATION WITH TOMTOM BANDIT
GATE 2
CAR PARK A WEST
FOS AIR PRESENTED BY SHELL
CAR PARK A EAST
N

# Isotype

## A method for presenting information in pictorial form using repetition to represent quantitative information.

**The International System Of TYpographic Picture Education (Isotype) is a technique for presenting information in pictorial form.** It was originally developed by Otto Neurath in Vienna in the 1920s as the Vienna Method of Pictorial Statistics (Neurath, 1936). Neurath's incentive was to use graphics to bring life to what might be seen as 'dead statistics' and as a way to support recognition and longer-term memory of information using humans' better ability to remember pictures than figures.

**Isotype uses an illustration of the phenomena to make information recognizable.** Icons are generally drawn in silhouette or in profile (side-on) as if they were shadows of the real object. They use simple graphic forms—the simpler the better. Isotype is usually presented as flat graphics rather than in perspective to preserve clarity. Colour is also usually applied consistently to avoid ambiguity. In order to represent quantities of the phenomena, the symbol is repeated in a regularly spaced pattern rather than being enlarged. Repetition is used to present quantity to counter the problems of assessing differences between different proportionally sized symbols. A single-unit isotype can be given a value so that the multiples sum to the overall value mapped. Although mapmakers might understand the unit that varies when we make a proportional symbol (area is relative), it can be argued that some map readers might assume it is the diameter of a symbol that is the unit. Isotype overcomes that since the basic unit remains the same size.

**Isotype provides a strong visual graphical component that can also be used on a map.** Quantities can be graphically represented at a point or for areas. In its most basic form, geometric shapes might be used to create a waffle grid of repeated isotype. For instance, squares can be positioned in a regular pattern. At a distance the appearance is one of a proportional symbol yet the use of isotype supports data recovery.

**See also:** Graphs | Pictograms | Semiotics | Shape | Symbols | Waffle grid

Neurath's basic guidelines for isotype, to make something intelligible and interesting are as follows:

1. Isotype icons should be drawn to the same height, width, and visual weight to ensure visual balance, so they could be counted and so one row didn't seem more important than any other.
2. A greater quantity is represented by more icons, not enlarged icons.
3. For large values, icons could represent a multiple of the value.
4. Use horizontal, rather than vertical, arrangement where possible.
5. Different sized geometric shapes are not good graphics for showing different quantities.
6. Avoid combining isotype with line charts since most contain misleading information. Use bars instead as they give a truer representation of data.
7. Use colour strictly for information and not decoration. Use it sparingly.

The design of each isotype icon is an attempt to create an understandable picture. We routinely do this in map design where we use the vocabulary of abstract marks to bring together an image with meaning and also in point symbol design. We try and evoke meaning through a common, consistent denotation. For instance, wavy horizontal lines are normally interpreted as water.

Placement of isotype would usually be central to the feature being represented but different sized areas may require offset placement of grids of isotype. Alternatively, they are well suited to a more abstract regular placement using a regularly tessellated cartogram as the underlying structure. Isotype also provides a visually compelling addition to a main map through its use as supporting graphical content.

**Opposite:** *Nuclear Nations* by Kenneth Field, 2017. In which I break Neurath's fifth guideline.

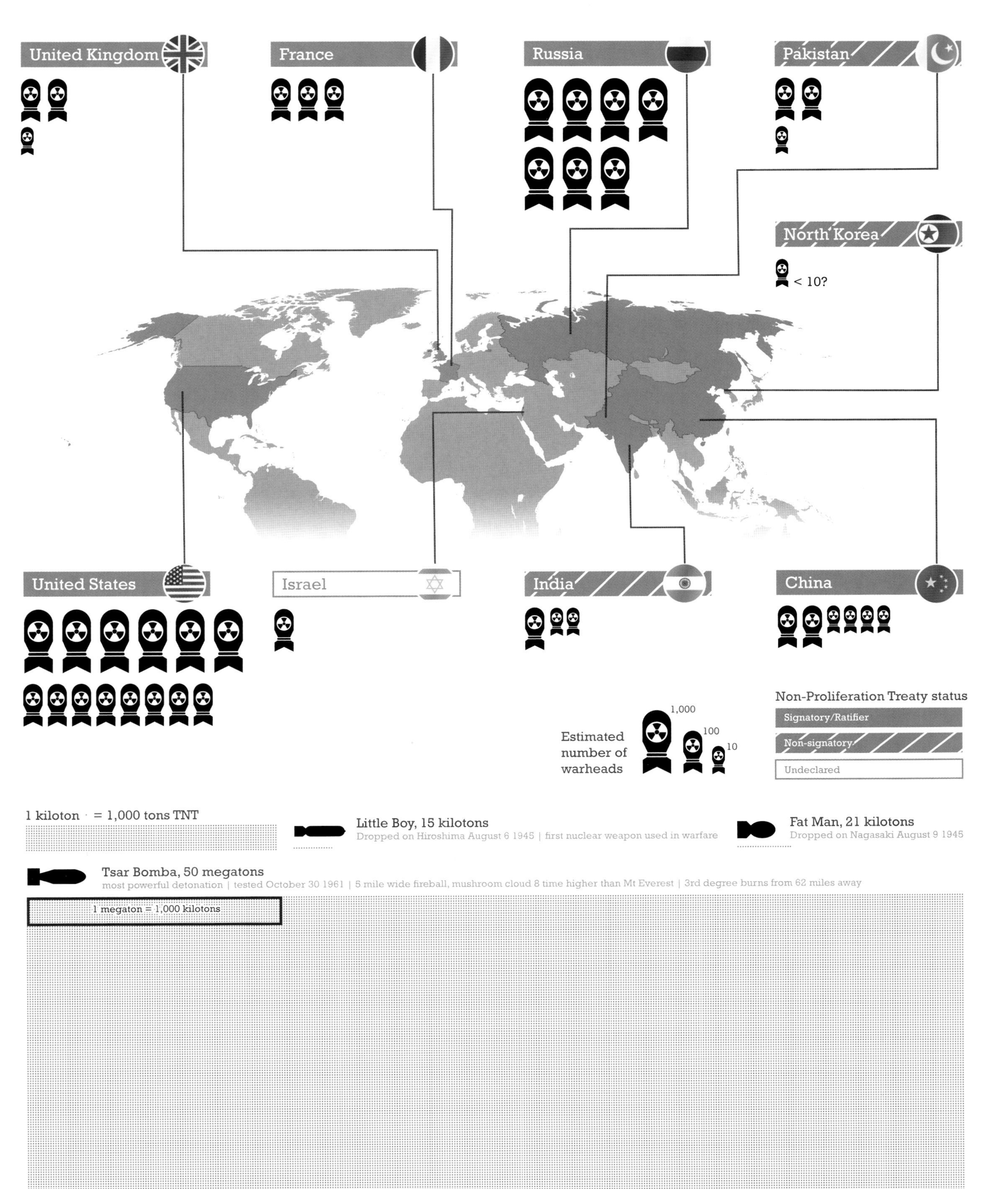

United Kingdom
France
Russia
Pakistan
North Korea
< 10?
United States
Israel
India
China
Non-Proliferation Treaty status
Signatory/Ratifier
Non-signatory
Undeclared
Estimated number of warheads
1,000
100
10
1 kiloton = 1,000 tons TNT
Little Boy, 15 kilotons
Dropped on Hiroshima August 6 1945 | first nuclear weapon used in warfare
Fat Man, 21 kilotons
Dropped on Nagasaki August 9 1945
Tsar Bomba, 50 megatons
most powerful detonation | tested October 30 1961 | 5 mile wide fireball, mushroom cloud 8 time higher than Mt Everest | 3rd degree burns from 62 miles away
1 megaton = 1,000 kilotons

# Geologic Map of the Central Far Side of the Moon

Desiree E. Stuart-Alexander

1978

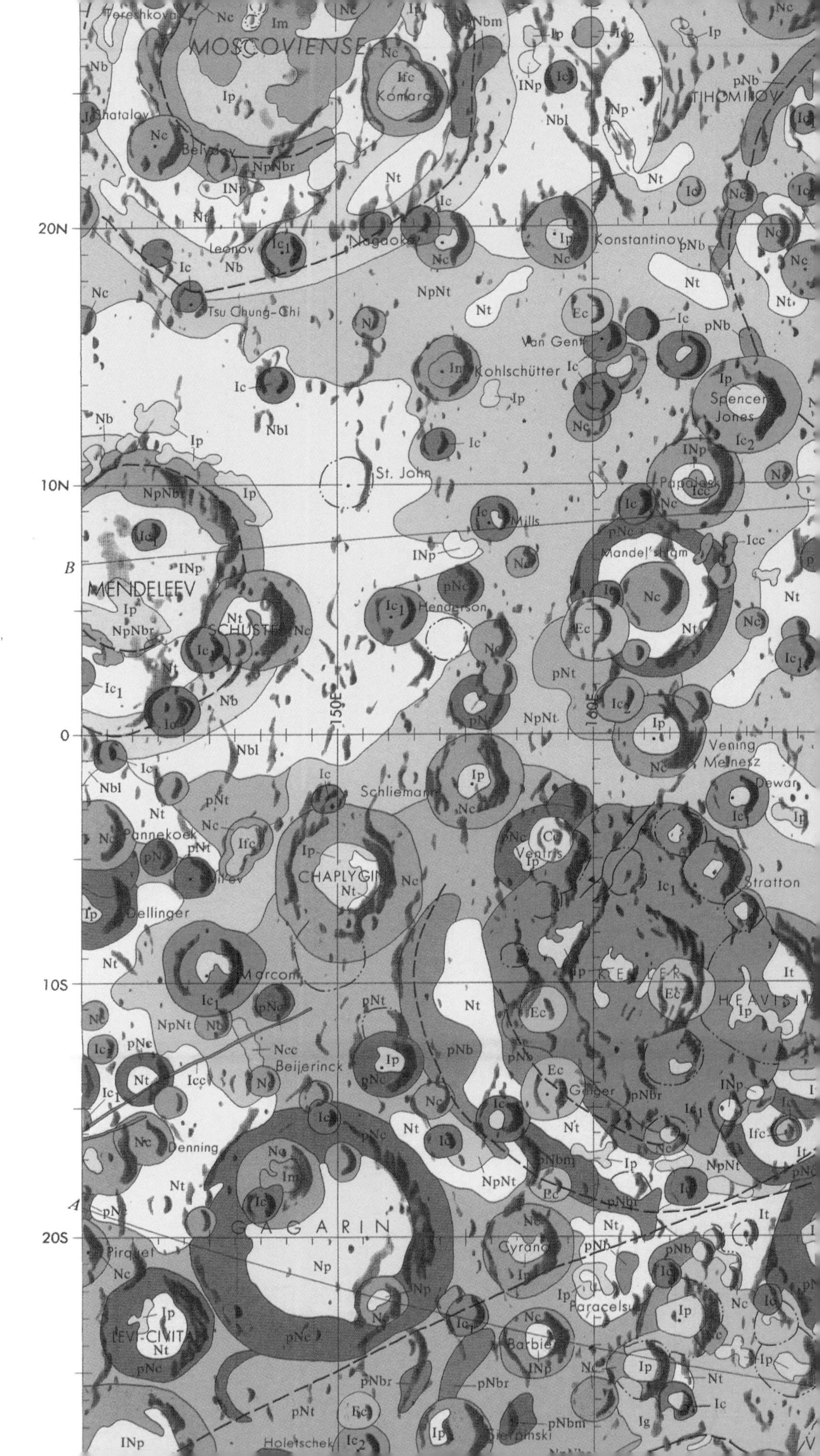


The San Francisco-born Psychedelic Movement, reached all fields of the arts from the 1960s to the early 1970s. The explosion of vivid colours infiltrated into science in the late 1970s when USGS happened to produce its 1:5,000,000 scale lunar geologic map series in its Psychedelia colour scheme.

This first complete reconnaissance mapping of the moon was coordinated by Don Wilhelms. Although the first, 1971 map is greyish pale coloured, those made in 1977–1979 are dazzlingly bright, showing a surface more alien than the extraterrestrial body they represented.

At the same time, paradoxically, these colours made it more familiar, at least for the eye of the 1970s: the moon became part of the standard visual culture.
The interpretation of the lunar terra materials also changed after the first map was published from volcanic to impact. The cornucopia of colours shows the variety of crater ages on the lunar surface.

—Henrik Hargitai

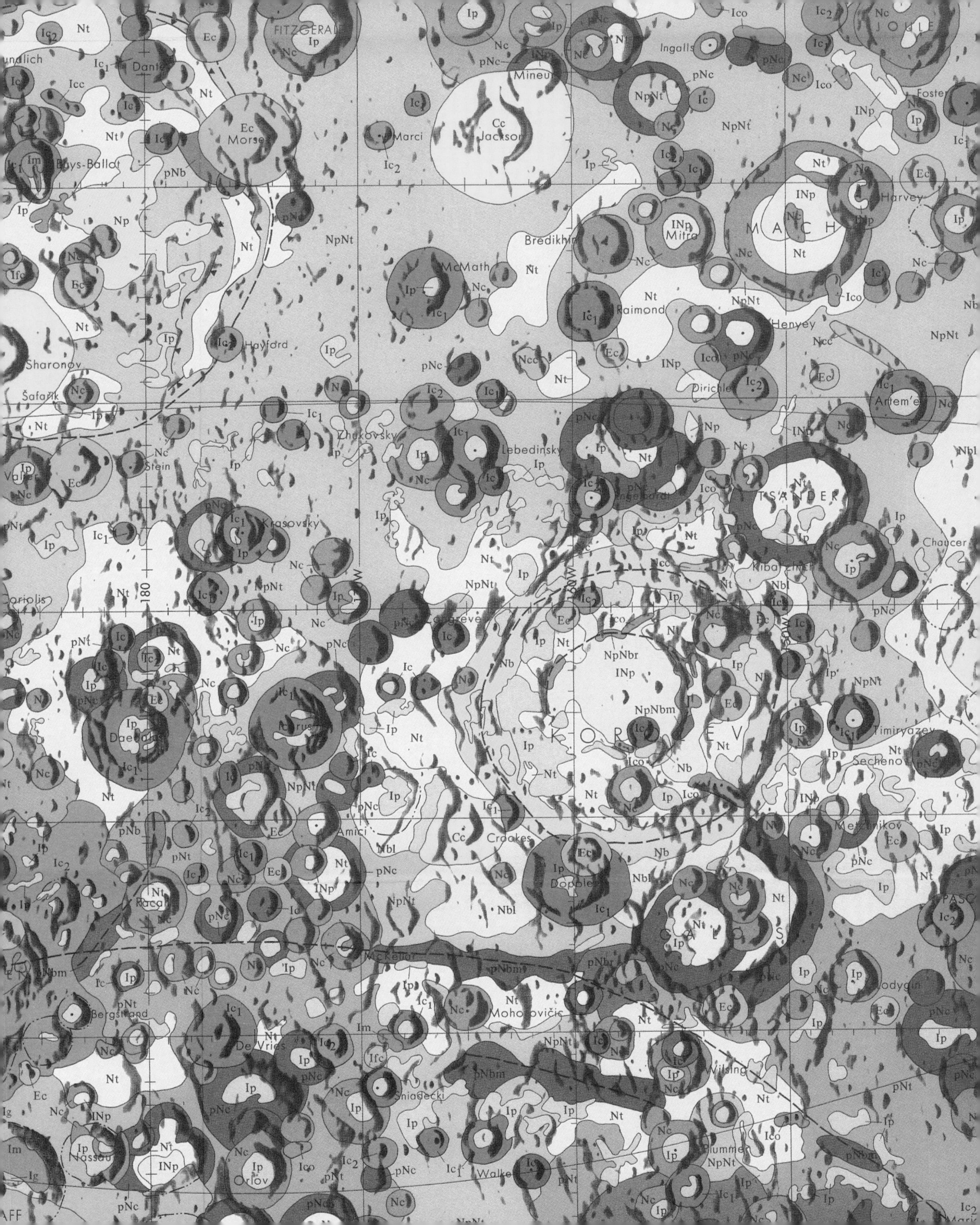

FITZGERALD
JOULE
Dante
Ingalls
Mineur
Morse
Marci
Jackson
Buys-Ballot
Foster
Harvey
MACH
Bredikhin
Mitra
McMath
Raimond
Henyey
Hayford
Sharonov
Šafařík
Dirichlet
Artem'ev
Zhukovsky
Lebedinsky
Stein
Krasovsky
TSANDER
Chaucer
Kibal'chich
Coriolis
Congreve
Daedalus
Icarus
KOROLEV
Timiryazev
Sechenov
Amici
Crookes
Metchnikov
Dopolev
GALOIS
Racah
McKellar
Mohorovičić
Bergstrand
De Vries
Wilsing
Sniadecki
Nassau
Orlov
Walker
Plummer
180

# Jokes and satire

## Cartography as a framework for humour and comment.

**Satire in general is a long-established form of humour that often ridicules, exposes, or criticizes stupidity.** Cartoonists make good use of satire in the context of political, economic, or social states. Satire also has a long history in cartography as a way of painting a picture of a particular region, perhaps identifying restless political neighbours, national stereotypes, or as a way of exerting a biased view of the world. Satire is a particularly opinionated form of cartography that belongs firmly in the minds of the cartographer rather than the objective and scientific qualities we might normally associate with cartography. It criticises government, religion, and has frequently been used to lampoon individuals or trends.

**What of objectivity in cartography?** Ultimately, maps are used to support understanding, and satire can provide an accessible form for many to relate to the map's themes. For instance, to better understand a political situation. Maps are often merely a canvas upon which satire is positioned. The shapes of countries provide excellent forms for caricature or cartoons. They are used as outlines for illustrations and artistry replaces scientific precision. Many satirical maps change the shapes of countries into people or animals that are perhaps symbolic of a country or particular person.

**Maps have always been cloaked in meaning.** Early *mappaemundi* were steeped in symbolic and religious character and were as much about spreading comments on morality as they were in presenting a mathematically surveyed view of the world. Illustrative cartography also has a proud heritage. For instance, the 16th-century metaphorical maps of northern Europe that showed the low countries as a lion (*Leo Belgicus*) became commonplace. It was heraldic and symbolic of the struggle for independence at the time. Whimsical examples such as William Harvey's *Geographical Fun* maps in 1869 took the imagery a stage further and led to the serio-comic genre popularised as ways of commenting on World War I. They provided a visually accessible form of showing warring factions, rivalries, and treaties often through the use of national stereotypes and animal metaphors.

**See also:** Advertising maps | Cognitive biases | Emotional response | Hand-drawn maps | Seeing

Fred Rose's pictorial illustrations or serio-comic maps of the late 1800s depict the threat posed to British interests by Russian territorial ambitions during the Balkan crisis in late Imperial Europe. One of his famous works, *Angling in Troubled Waters*, reflects a particular style of propaganda mapping though he takes a highly artistic approach in an attempt to woo the reader with humour. Rose's concept is simply to take country boundaries and paint a picture of an individual that represents some particular national identity. Rose makes good use of the fishing metaphor to illustrate which countries are fishing and what their catches (colonial possessions) are.

The titles are absolutely critical to the impact of Rose's maps. Without them we may wonder what the imagery represents yet the carefully crafted words give us not only a clear key to the metaphor but also lead us to the same impression that the mapmaker intends. Certain countries are fishing and being antagonistic or even threatening. Other countries are shown as innocent and under attack. Even the fact that the map has a subtitle explaining it is a serio-comic map is important—it's the clear message that this is meant to be taken as satire. Of course, such a statement may be construed as an attempt to avoid accusations from those who may not necessarily appreciate the humour.

As a means of stirring debate and controversy, these types of maps are a particularly provocative way of capturing the imagination.

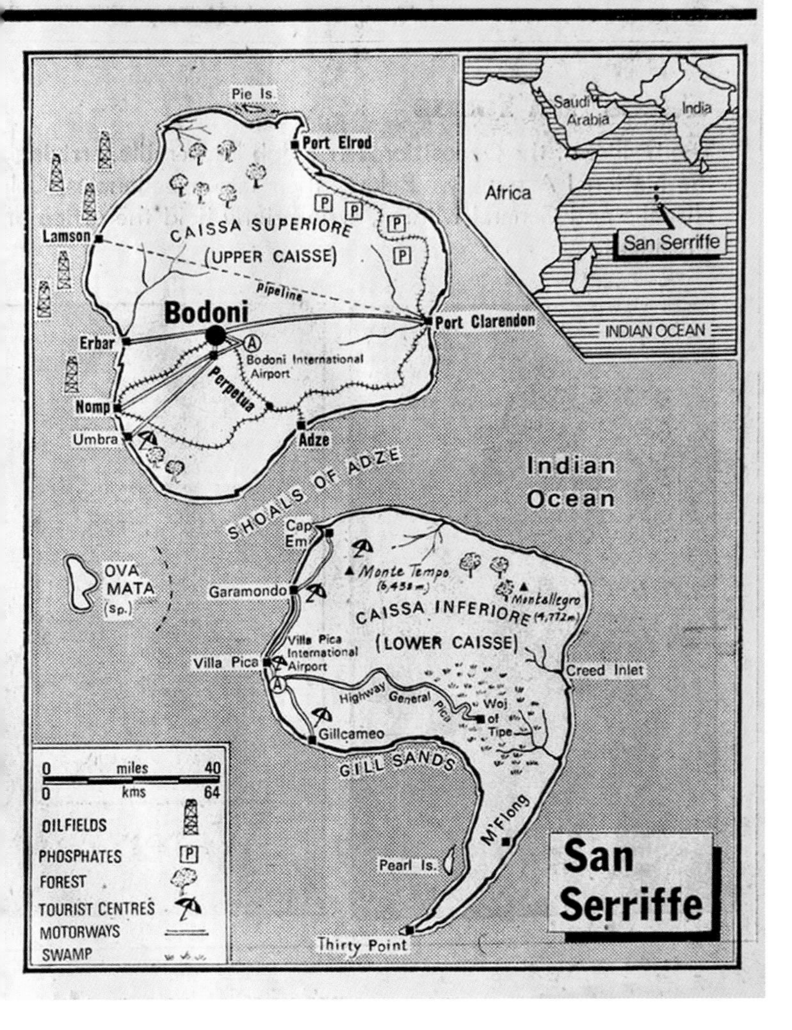


On April 1st 1977, The Guardian produced a 7 page travel supplement in their daily newspaper reviewing a small tropical island named San Serriffe – 'a small archipeligo, its main islands grouped roughly in the shape of a semicolon, in the Indian Ocean'. The supplement celebrated San Serriffe's ten years of independence from Great Britain and contained a small map.

San Serriffe's history was explained as having been discovered in 1421 by the English, colonised by Spain, and Portugal and gaining independence in 1967. The island itself is near the Seychelles and had about 1.8 million inhabitants as of the 1973 census.

The map and the supplement were beautifully crafted with adverts and photographs of the environment and people. There was even a request for people to send in their very own Kodak moments! There are two main islands, Caissa Superiore (Upper Caisse) and Caissa Inferiore (Lower Caisse), the latter having a promontory at Thirty Point. The capital city is Bodoni.

The design, of course, was perfectly suited to the main purpose…as an expertly crafted April Fool's joke. The island is not real and the map a pure figment of the imagination, led by Philip Davies of The Guardian's Special Reports department. The effort remains one of The Guardian's most elaborate and successful April Fool's jokes which they have revisited many times since to catch up on how the island is developing. Of course, virtually every element of the island related to typography, a perfect 'printing' related joke for the newspaper.

April 1st 2014 saw the latest addition to the San Serriffe legacy with a beautifully observed new 'tribute' map, Comic San Serriffe by Craig Williams.

Marrying the fantasy of the original with the typeface that designer's love to hate is deserving of its own place in the San Serriffe pantheon. This isn't just a copy that mimics the style of the original newspaper production (although it does, right down to the pattern fills, drop shadows and knock-out masks for the labels)…the new shape of the islands requires adjustments to the page layout to give it a suitable balance.

There's a keen eye on ensuring the cartography works; that each element sits well in the new landscape; and that each element, including the point symbols and the north arrow has a Comic Sans tweak.

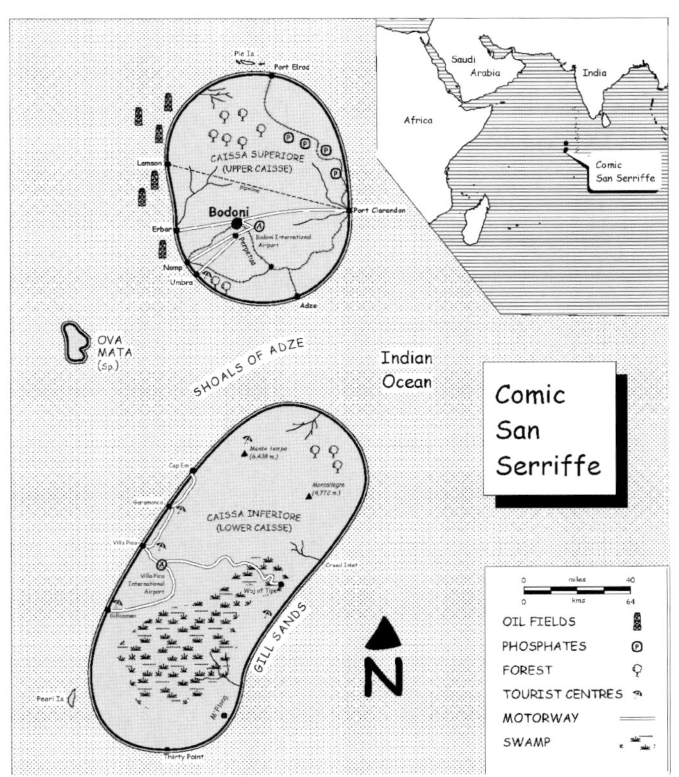

# Atlas of Global Geography

Erwin Raisz

1944

Two popular atlases from the early 1940s changed the way many people saw and understood the world. One was Richard Edes Harrison's *Look at the World*, and the other was Raisz's *Atlas*. Both helped readers escape the flatland of school maps (usually with their Mercator projection distortions) by showing the continents as they would appear on a sphere.

In Raisz's atlas, the page on Africa is a great example: we see the whole continent dramatically curving round the globe, emphasising (but not distorting) its size. At the top of the page, Europe and the Middle East disappear in perspective into the far distance. As a kind of forerunner to Herbert Bayer's 1953 *Geo-Graphic Atlas*, all Raisz's pages include cultural and political facts, illustrations, and small maps.

The final pages of the atlas present some world problems of that time, charted and drawn in a variety of different and interesting ways.

—Nigel Holmes

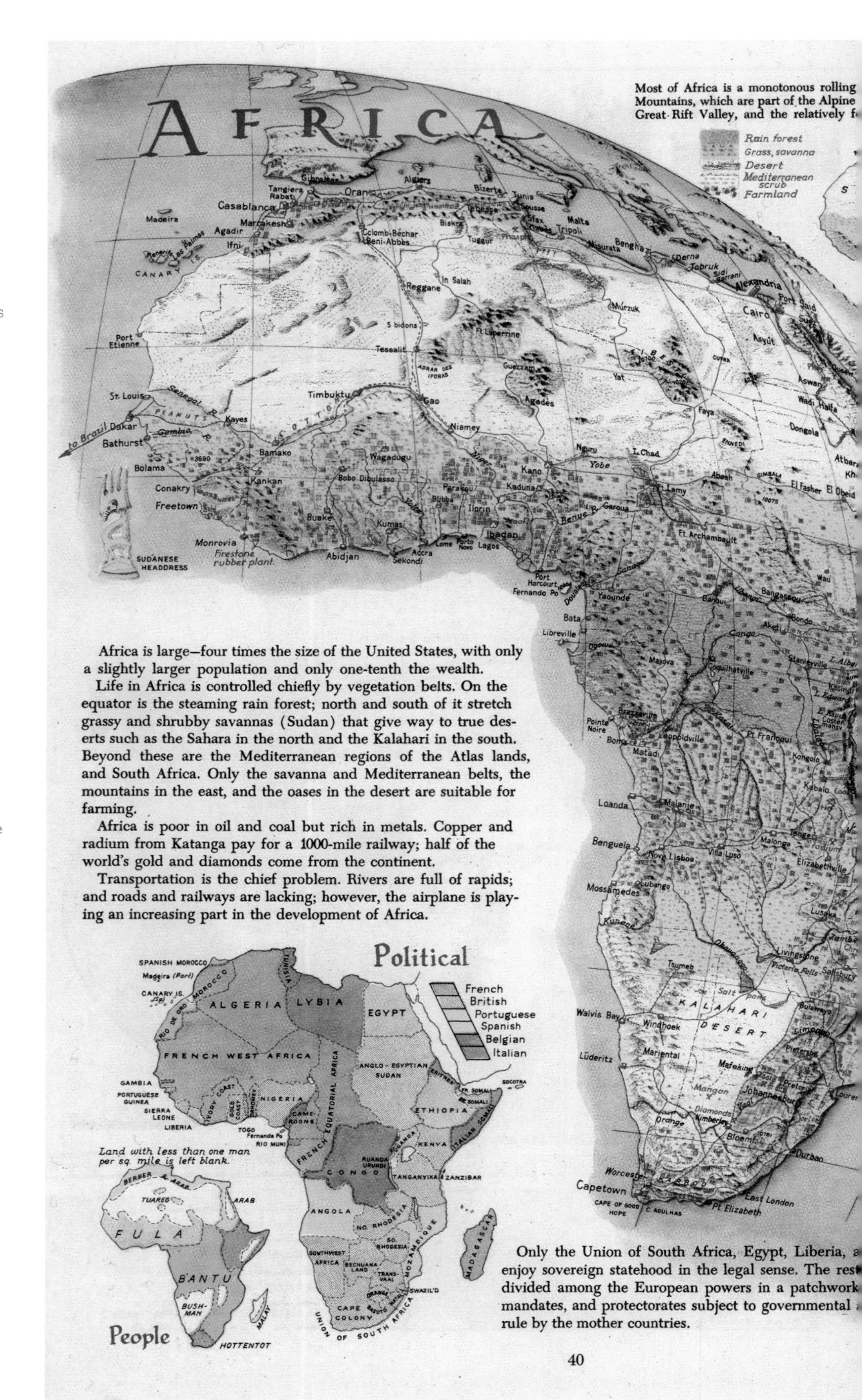

AFRICA

Most of Africa is a monotonous rolling
Mountains, which are part of the Alpine
Great Rift Valley, and the relatively f

Africa is large—four times the size of the United States, with only a slightly larger population and only one-tenth the wealth.

Life in Africa is controlled chiefly by vegetation belts. On the equator is the steaming rain forest; north and south of it stretch grassy and shrubby savannas (Sudan) that give way to true deserts such as the Sahara in the north and the Kalahari in the south. Beyond these are the Mediterranean regions of the Atlas lands, and South Africa. Only the savanna and Mediterranean belts, the mountains in the east, and the oases in the desert are suitable for farming.

Africa is poor in oil and coal but rich in metals. Copper and radium from Katanga pay for a 1000-mile railway; half of the world's gold and diamonds come from the continent.

Transportation is the chief problem. Rivers are full of rapids; and roads and railways are lacking; however, the airplane is playing an increasing part in the development of Africa.

Only the Union of South Africa, Egypt, Liberia, a
enjoy sovereign statehood in the legal sense. The res
divided among the European powers in a patchwork
mandates, and protectorates subject to governmental
rule by the mother countries.

40

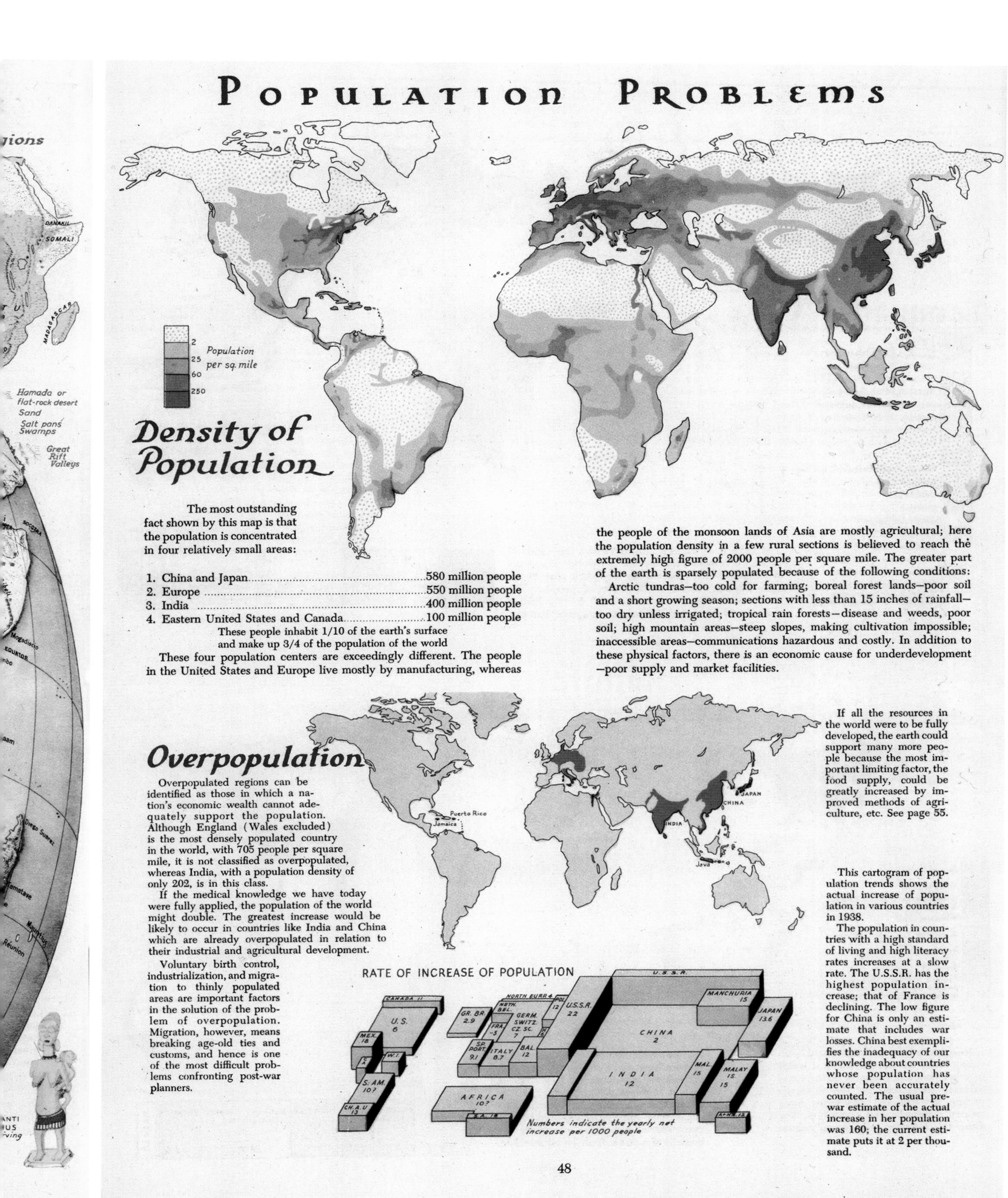

# Population Problems

## Density of Population



The most outstanding fact shown by this map is that the population is concentrated in four relatively small areas:

1. China and Japan........580 million people
2. Europe........550 million people
3. India........400 million people
4. Eastern United States and Canada........100 million people

These people inhabit 1/10 of the earth's surface and make up 3/4 of the population of the world

These four population centers are exceedingly different. The people in the United States and Europe live mostly by manufacturing, whereas the people of the monsoon lands of Asia are mostly agricultural; here the population density in a few rural sections is believed to reach the extremely high figure of 2000 people per square mile. The greater part of the earth is sparsely populated because of the following conditions:

Arctic tundras—too cold for farming; boreal forest lands—poor soil and a short growing season; sections with less than 15 inches of rainfall—too dry unless irrigated; tropical rain forests—disease and weeds, poor soil; high mountain areas—steep slopes, making cultivation impossible; inaccessible areas—communications hazardous and costly. In addition to these physical factors, there is an economic cause for underdevelopment—poor supply and market facilities.

## Overpopulation



Overpopulated regions can be identified as those in which a nation's economic wealth cannot adequately support the population. Although England (Wales excluded) is the most densely populated country in the world, with 705 people per square mile, it is not classified as overpopulated, whereas India, with a population density of only 202, is in this class.

If the medical knowledge we have today were fully applied, the population of the world might double. The greatest increase would be likely to occur in countries like India and China which are already overpopulated in relation to their industrial and agricultural development.

Voluntary birth control, industrialization, and migration to thinly populated areas are important factors in the solution of the problem of overpopulation. Migration, however, means breaking age-old ties and customs, and hence is one of the most difficult problems confronting post-war planners.

If all the resources in the world were to be fully developed, the earth could support many more people because the most important limiting factor, the food supply, could be greatly increased by improved methods of agriculture, etc. See page 55.



This cartogram of population trends shows the actual increase of population in various countries in 1938.

The population in countries with a high standard of living and high literacy rates increases at a slow rate. The U.S.S.R. has the highest population increase; that of France is declining. The low figure for China is only an estimate that includes war losses. China best exemplifies the inadequacy of our knowledge about countries whose population has never been accurately counted. The usual prewar estimate of the actual increase in her population was 160; the current estimate puts it at 2 per thousand.

48

# Knowledge and conviction

## Clear thinking leads to conviction in map design.

**Having firmly held beliefs can be both positive and negative.** The focus of this entire book is on thinking about the design process. Part of this consideration requires you to think about some of the basic tenets, the fundamentals, and the vast body of research that underpins cartographic theory and practice. Having an awareness of what has gone before clarifies your own thinking. Knowing how to go beyond and how to modify beliefs to develop your map builds upon this strong legacy. By following established thinking you'll make a good map. By being able to extend your thinking you might make a great map. Developing your own conviction should be based on confidence rather than arrogance.

**Design choices can be clear cut.** Quite often there are clear paths to make your map. Certain techniques will be more or less obvious, and for the bits that are left, often common sense is a strong guiding light. Sometimes, though, you'll have choice and different viable options, and that's when you'll need guidance and inspiration. That usually means looking at maps, reading what others think about map design, and seeking discussions of what works and what doesn't. You'll likely be influenced by your own tastes, the industry you work in or are most familiar with, and the voices you initially hear. Some of these voices are loud and extol their own strong convictions (hard to believe but I'm often accused of this). With experience you can learn to determine what you hear or see that chimes with your beliefs and what you are able to filter out in developing your own cartographic convictions. There's nothing wrong with challenging long-held beliefs but knowing how to do so from a position of understanding and an appreciation of when facts become opinions and which can be challenged is important. As you gain experience so confidence will grow. Knowing the rules allows you to also know how to break them.

**Cartography shares many of its core ideas with other design-led fields.** By widening your appreciation of other disciplines and seeing their value, you can better establish a philosophy that drives the search for your own convictions.

**See also:** Cognitive biases | Critique | Integrity | Style, fashion, and trends | Who is cartography? | Your map is wrong!

There appears to be very little that links cartography to the design of consumer electronic products. Yet designer Dieter Rams, who is known for the timeless qualities of his industrial designs for the German company Braun, has influenced the world of design more generally. He approached the process of design by questioning his own convictions and sought to answer a simple question when considering the quality of his work. The question was 'is my design good design' and the answer formed is his 10 guiding principles:

Good design:

1. is innovative,
2. makes a product useful,
3. is aesthetic,
4. makes a product understandable,
5. is unobtrusive,
6. is honest,
7. is long-lasting,
8. is thorough down to the last detail,
9. is environmentally friendly, and
10. is as little design as possible.

Although these principles are often celebrated in terms of design, more generally they map very well onto the way in which we might think of making a map. They help us determine whether the techniques we use will work to make the map trustworthy, accessible, and elegant. Of course, not every one of Rams' principles need be followed for every project. Being innovative doesn't have to be the aim for every map. Not every map must be long-lasting. Making maps environmentally friendly may be altruistic considering that the medium of paper is clearly not, nor are hosting servers that might drive your web map. Perhaps more useful in equating this last point to cartography is to have awareness of the impacts of your choices not only on the process and resources but on the people that are involved, not least those with a vested interest as well as the map reader.

**Opposite:** *Interpretation*, by Wane, 2015.

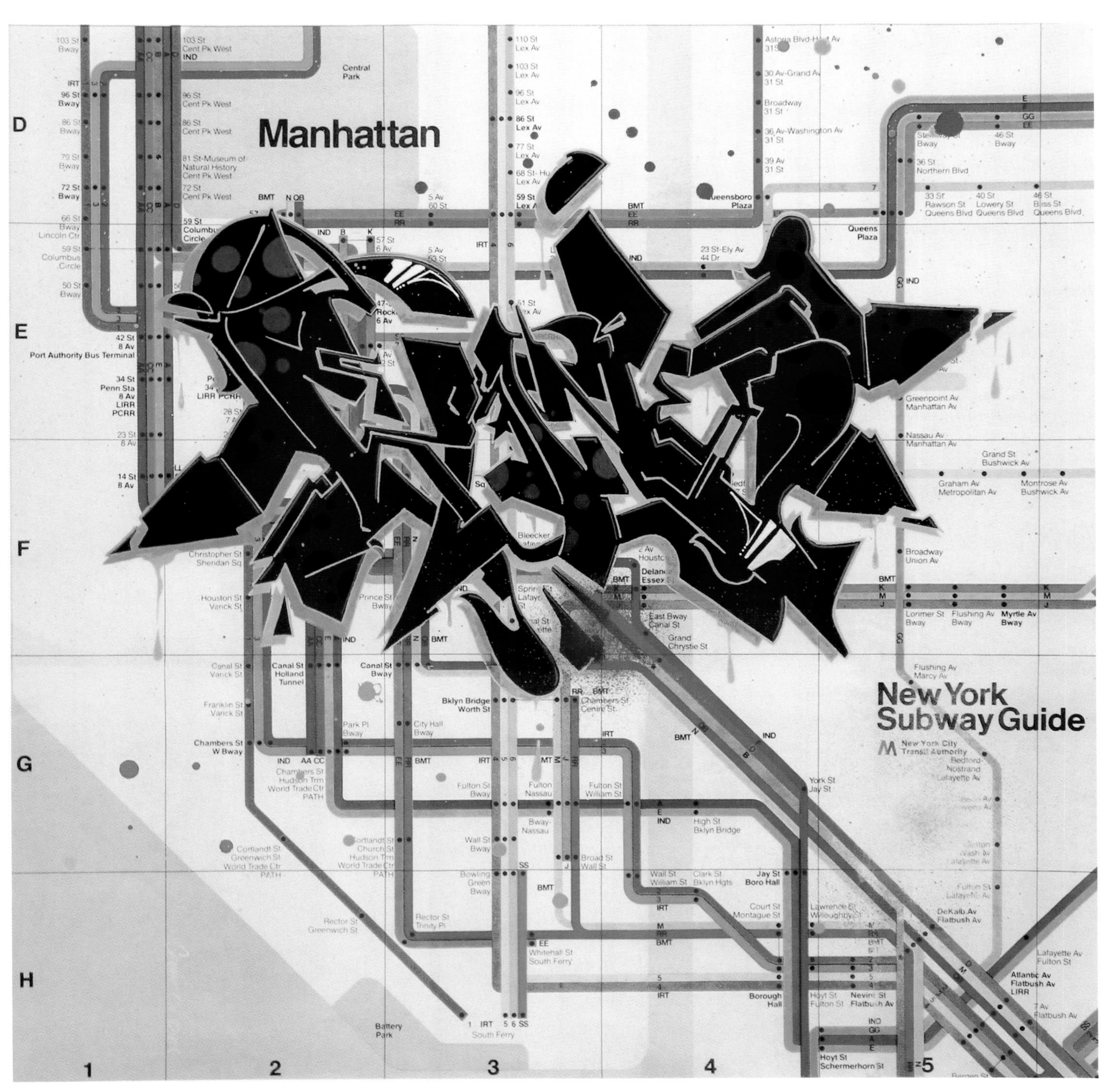


The 1972 map of New York City's subway system by Massimo Vignelli has often been seen as a polarising map in design terms. For every one person who characterises it as a striking piece of design or commends its beautiful aesthetics, another will point to the geographical liberties it took.

Vignelli's map only lasted seven years and was replaced in 1979 with a more topographically correct version that included aboveground detail, intermodal transport links, and dramatically thinned line work. This replacement was designed by a committee of 12 people. While the London public grew to understand that Harry Beck's subway map was grossly distorted yet a perfect solution to navigate from A to B, New Yorkers simply didn't appreciate the same approach.

Vignelli himself explained the issue at a conference in 2010: 'I think the real reason is space. But not because Manhattan is too small, it's because they want to put too much information that doesn't belong in the diagram. That's why. All of a sudden there is …and there is no reason. I mean, all you want to know is [how] to go from A to B'.

Vignelli had been tasked with streamlining the wayfinding task and bringing New York into the future. He succeeded in meeting the design brief yet people eventually reacted by demanding a 'map' as they perceived it in the more traditional sense. His knowledge of design and conviction of approach were outweighed by public opinion.

To emphasise the juxtaposition between art, utility, and public acceptance, here, street artist Wane has added a graffiti tag across an original 1972 map, in the style of subway car graffiti and with respect to the original map and designer.

## Google Maps

Google

2005—present

(extract from November 2017)

You think making one map is difficult? Try mapping everywhere. Oh, and at all (er...20) cartographic scales. Google Maps were not the first 'slippy' web maps, but they were the first that made the user experience visually seamless, transparently usable, and geographically immersive. How? The mapmakers paid equal attention to the representation, the interface, and the underlying data (plus capital, engineers, server warehouses ...).

But design!? As old guard critics called Google Maps austere, banal, and [blah blah blah], we saw consistency across scales, coherence between map and interface, and clarity of the information 'we wanted'. And, the cartographic design of Google Maps improved with each new iteration. Today, designers regularly use ideas behind Google Maps version 1.0 without attribution: APIs, tilesets, markers, mashups, and so on.

But! what followed was truly remarkable: maps no longer are a static capture of somewhere, out there. They are here and now, always with us, contextualising and enriching a place as we experience it. You no longer unfold a map, the map unfolds your mind. #thanksgoogle

—Robert E. Roth

Café Restaurant De Reiger
4.4 99 reviews
Café Restaurant De Reiger
Anne Frank Huis
Westerkerk
Homomonument
Noorderkerk
Winkel 43
De Zagerij
Ko Chang
La Trattoria Di Donna Sofia
Eatery "Harlem Soul Food"
Thai & Co
Amsterdam Centraal
Sexmuseum
Sint-Nicolaaskerk
DoubleTree by Hilton Amsterdam...
Parking Centrum Oosterdok
OBA Amsterdam (Openbare Bibliotheek...
Muziekgebouw aan 't IJ
Mövenpick Hotel Amsterdam City Centre
Club Panama Amsterdam
NEMO Science Museum
Museum Ons' Lieve Heer op Solder
BODY WORLDS
De Oude Kerk
De Nieuwe Kerk
Museum of Prostitution Red Light Secrets
Madame Tussauds Amsterdam
Nieuwmarkt
Abraxas
DE WALLEN
CENTRUM
Green House Centrum
Rainarai
Houseboat Museum
Rakang
The Amsterdam Dungeon
Amsterdam Museum
Amsterdam
Zuiderkerk
Museum Het Rembrandthuis
Allard Pierson Museum
Waterlooplein markt
Cafe de Jaren
De Krijtberg
Bloemenmarkt
Pathé Tuschinski
Pathé De Munt
Museum Willet-Holthuysen
GRACHTENGORDEL
Tassenmuseum Hendrikje
Hermitage Amsterdam
Melkweg
DeLaMar Theater
Apple Store
Holland Casino Amsterdam Centrum
Foam
Museum Van Loon
Soup en Zo - Spiegel
Magere Brug
Koninklijk Theater Carré
Rijksmuseum
I Amsterdam Sign
Van Gogh Museum
Stedelijk Museum Amsterdam
Heineken Experience
Albert Cuypmarkt
OUDE PIJP
Sarphatipark
Red Light District Deluxe
DE PIJP
DUIVELSEILAND
NIEUWE PIJP
KATTENBURG
Het Scheepvaartmuseum
WITTENBURG
Amsterdam Roest
NIEUWMARKT EN LASTAGE
Verzetsmuseum
Micropia
Hortus Botanicus Amsterdam
ARTIS
De Gooyer
WEESPERBUURT EN PLANTAGE
Muiderpoort
CREA
Tropenmuseum
Dappermarkt
Oosterpark
DOOSTERPARKBUURT
OLVG, locatie Oost
OOSTPOORT
Volkshotel
WEESPERZIJDE
TRANSVAALBUURT
HOUTHAVENS
Westerpark
HAARLEMMERBUURT
OVERHOEKS
Oeverpark
EYE Filmmuseum
A'DAM Lookout
Cultural Center Tolhuistuin
IJdok
DISTELDORP
Noord
VAN DER PEKBUURT
Café de Ceuvel
VOLEWIJCK
Noorderpark
TUINDORP BUIKSLOOT
IJPLEIN EN VOGELBUURT
Volkstuinpark Buitenzorg
Java-eiland

# Latitude

## Latitude allows the north–south measurement of location.

**The north–south angular change above and below the equator is referred to as *latitude*.** All points that share the same latitude form a small circle around Earth called a *parallel*. Parallels of latitude are, conceptually, formed in the same way as the equator with an imaginary plane intersecting Earth's surface at points of equal latitude. These parallels of latitude are, conceptually, equivalent to the y-axis in plane coordinate geometry.

**Latitude on a spheroid is defined as the angle formed by a pair of lines that extend from the equator to the centre of Earth, and then from the centre to the location.** This is more correctly termed 'authalic latitude' because it relates to latitude calculated on a spheroid. Values of authalic latitude range from 90°N to 90°S about the equator, which has a value of zero (0°).

**Latitude can also take on alternative values depending on the shape of Earth's surface used for mapmaking.** If a reference ellipsoid is used, then geodetic and geocentric latitude can be measured. Geodetic latitude is the angle formed by a line from the equator toward the centre of Earth and a second line perpendicular to the ellipsoidal surface at the location measured. Geodetic latitude will only take the same value as authalic latitude at 0° and 90° where the line extending from the location will intersect with the line toward the equator at the centre of Earth. At all other latitudes, the perpendicular line extending from the location measured will pass through the equatorial plane at some point other than the equator. Geocentric latitude is the angle formed by a line from the equator toward the centre of Earth and a line drawn from the location to the centre of Earth. This line will not be perpendicular to the surface on an ellipsoid.

North and south designations are commonly replaced by plus and minus signs in handling digital data for mapping (+90° to −90°). Additionally, the measurement of latitude is normally given using the sexagesimal system but this can be translated into the decimal degree system (which has a base unit of 10 rather than a base unit of 60). For example, 51°24'13"N can be expressed as 51.4036.

For the same location, values of authalic, geodetic, and geocentric latitude will vary slightly. In most instances where small-scale mapping is the objective, variations in the shape of Earth are inconsequential, and Earth is considered to be best represented by a reference spheroid and authalic latitude is used. However, latitudinal measurements become important when large-scale, detailed mapping of local areas is the objective. Surveyors will commonly use geodetic latitude to measure their precise position since ground surveys are normally fixed to a particular reference ellipsoid used for surveying purposes in that country. Geocentric latitude is commonly used for satellite-derived locational positioning. This is due to most measurements being based on geocentric datums such as WGS84.

**See also:** Datums | Earth coordinate geometry | Earth's framework | Longitude | Position | Spatial dimensions of data

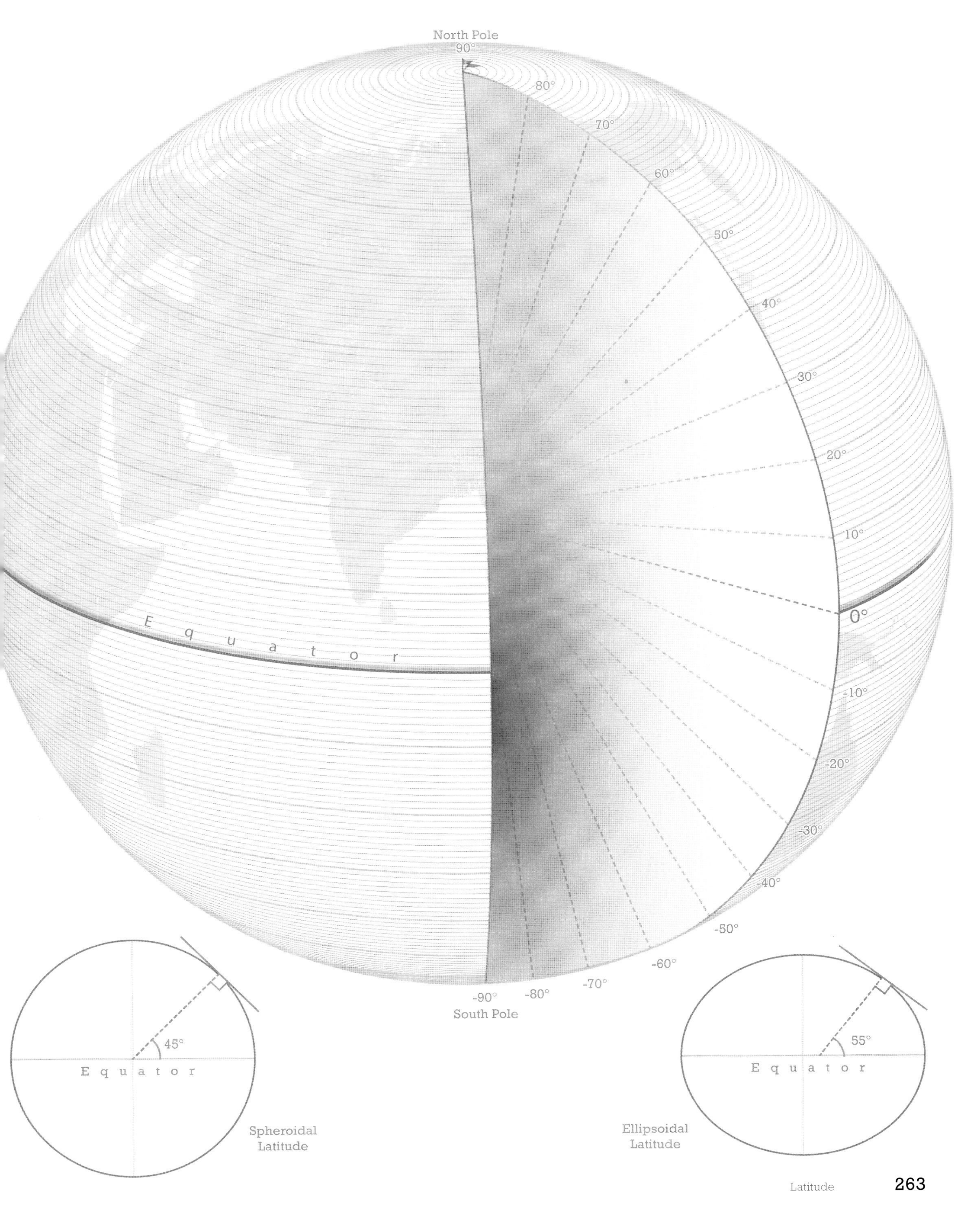

North Pole
90°
80°
70°
60°
50°
40°
30°
20°
10°
0°
-10°
-20°
-30°
-40°
-50°
-60°
-70°
-80°
-90°
South Pole
Equator
45°
Equator
Spheroidal Latitude
55°
Equator
Ellipsoidal Latitude

# Layouts and grids

## Designing layouts is about visual balance and organising information in a way that supports the task of reading. Grids can help.

**Behind every map layout lies an invisible framework that guides the position of elements.** It comprises the structure that the cartographer or designer used to organise the various visual components on the page. Grids are a core component of this process because they provide the structure of the map onto which the content is moulded. Using grids to position and align goes beyond map marginalia but applies the same for a large-format infographic, the page of a book, a poster, or any other form of visual communication.

**There are two main types of grid—skeletal and interval.** A skeletal grid acts as a fixed structure with uniformly spaced sections and spacings. An interval grid might have similar spacings but allows sections of varying size separated by a fixed measurement. For example, in a two-column grid with six sections, they would be equivalent in height on a skeletal grid whereas one might have fewer sections in an interval grid separated by column and section spacings of a fixed size. The point here is that some elements of the grid are fixed, which gives it a visual structure.

**Alternative layouts might work on the principle of the golden ratio (1:1,618, φ).** This can be mathematically defined as a + b is to a as a is to b. Many artists, architects and others in design fields have used this golden ratio in their work because the proportions are deemed to be aesthetically pleasing. They can also be used effectively in cartography to arrange maps and other elements in a way that map readers are more likely to instinctively find visually pleasing.

**Grids and proportions help us build balance for a map page and make it inviting and easier to read.** They underpin the various processes of distribution and alignment that we might subsequently use to ensure map elements are properly arranged and spaced. Although the application may differ, the same principle is used for map-based web pages and applications. The organisation of elements to create visual balance and harmony is important to create a unified product.

Generally speaking it's a good idea to make the layout (perhaps even the size and shape of the printed product) fit the map rather than making the map fit the layout. Map scale, format, and production will all impact on the final choice of layout. This might even be a function of the commercial availability of paper of certain dimensions. Most standard paper sizes are rectangular so a basic decision of viewing the map in landscape or portrait must be made. Either way, the visual balance, or optimal viewing, position is slightly above the middle centre of the map—the so-called 'visual centre'.

Preparing preliminary layouts and rough sketches is always recommended to ensure you give the map and surrounding elements sufficient space. Experimentation and exploring alternatives always leads to better decisions. The use of white, or negative, space is also important so the overall layout isn't overly busy or cluttered. Positioning map elements should be logically developed, unambiguous, and graphically-balanced. In addition to the rules of perception which govern the perceptual order of map elements, there are many ways to position elements. Dominant positions are upper- or lower-left corners, in the middle of the top or bottom, or at the top of a separate column.

Attempt to align and distribute map elements with clear and logical reason. Anything positioned irregularly will look awkward in the composition. The map itself should balance around the visual centre—the fulcrum of the layout. Try and build continuity among the position of related elements in structure and relative importance. Where possible, avoid broad white borders which tend to make the map look smaller.

**See also:** Balance | Graticules, grids, and neatlines | Legends | Mashups | Small multiples

## Skeletal grid column

Title and subtitle at top

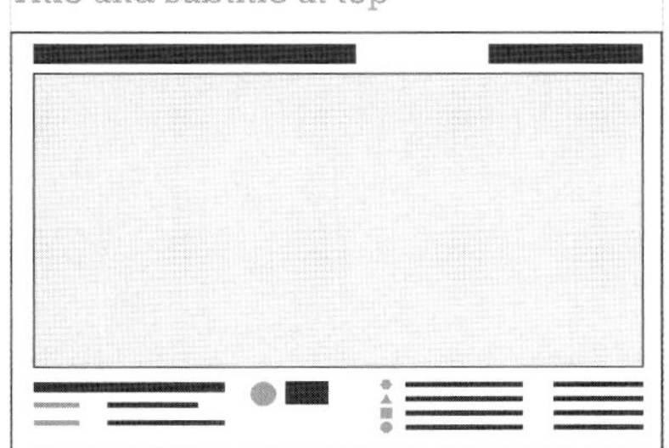

All marginalia beneath the map

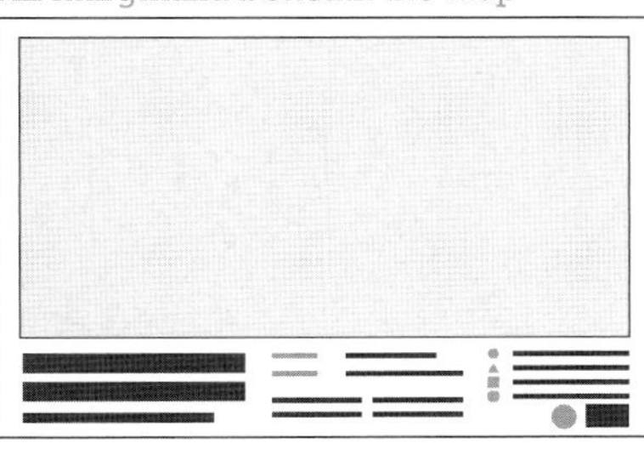

Map bleeds off page, detail in a column

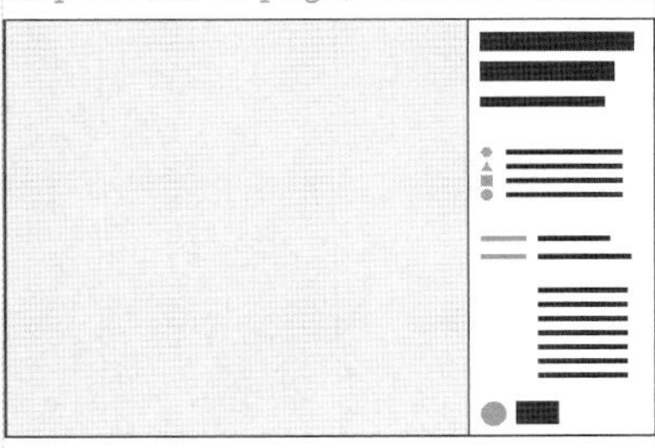

Map shape, detail in a column

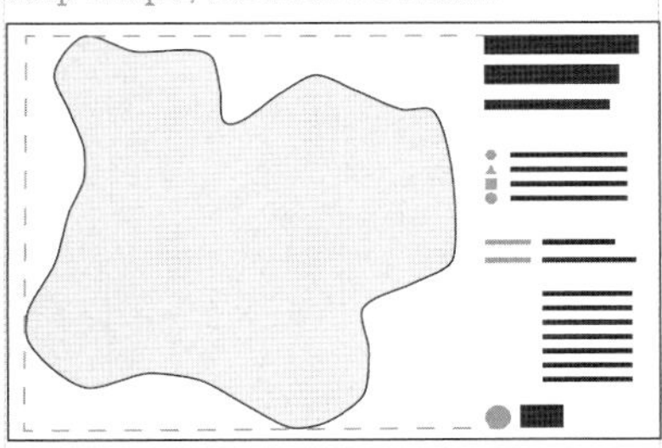

## Interval grid column

Map shape, detail around

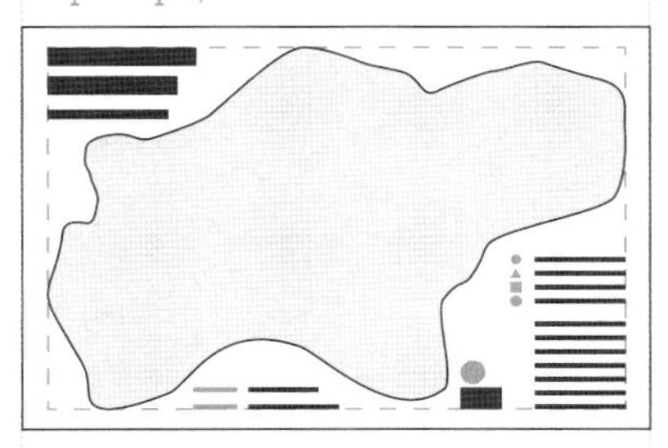

Web map, marginalia in drop-down

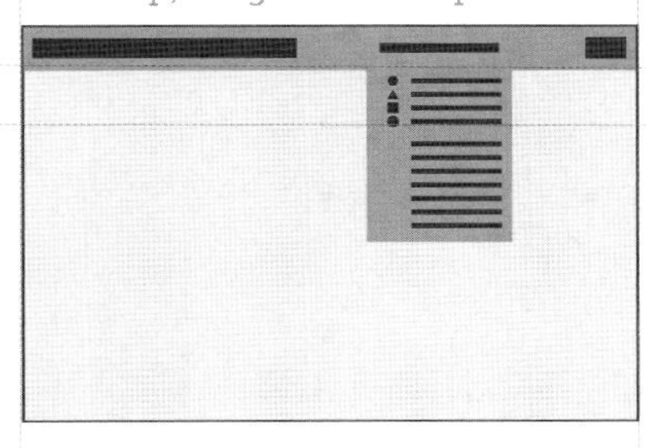

Web map, marginalia in pop-outs

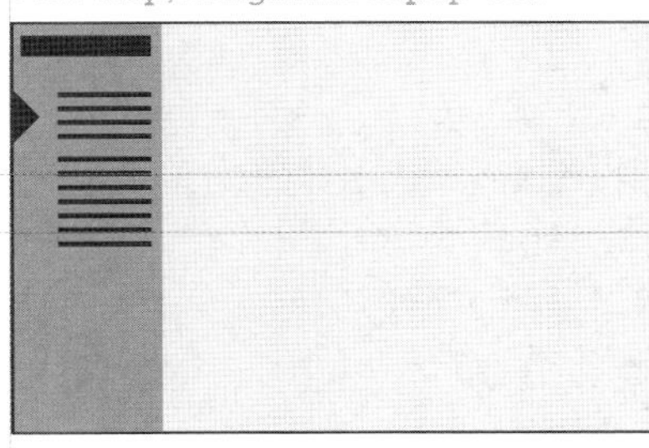

Web map, click, hover, floating pop-ups

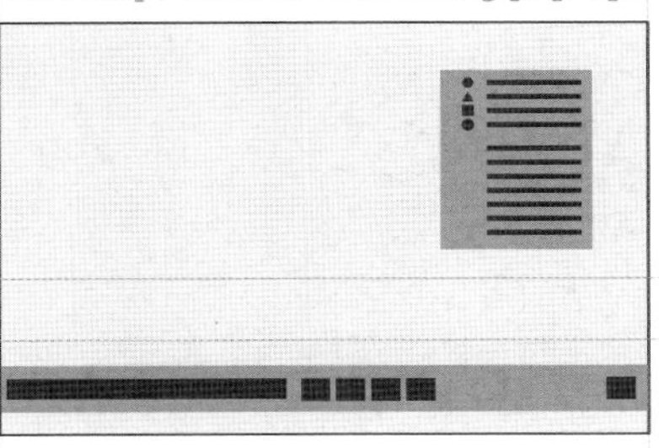

Web map, storytelling template

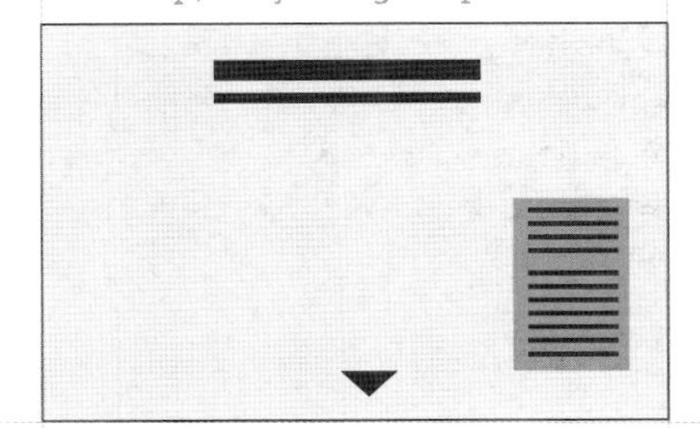

## Golden ratio

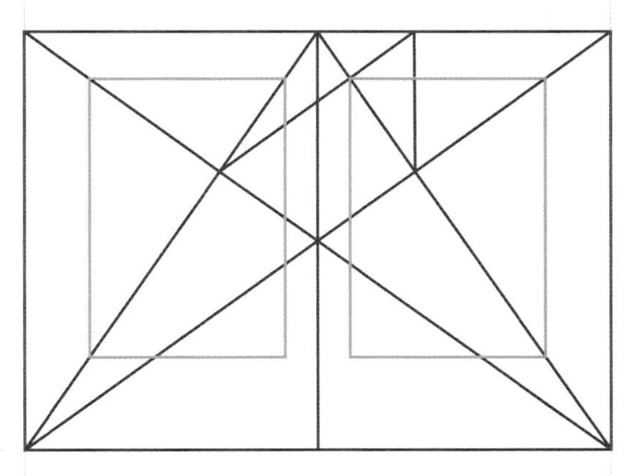

## Golden spiral

a = 42.39mm b = 26.2mm

$$\frac{a}{b} = \frac{a+b}{a} = 1.618$$

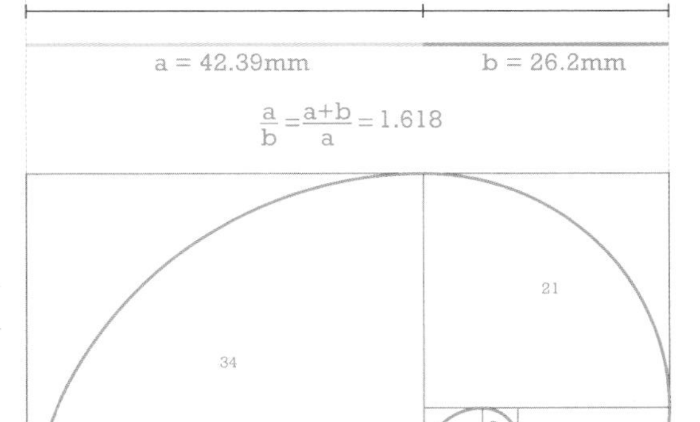


In mathematics a Fibonacci number sequence is that where each number is the sum of the preceding two numbers:

1, 1, 2, 3, 5, 8, 13, 21, 34, 55, 89, 144, ...

The golden spiral is based on an approximation of the sequence of Fibonacci numbers which equate to length of the edges of successive squares. This creates identically proportioned rectangles, each of which has a ratio of 1.618 of the length of the long side to that of the short side.

# Legends

## The design of the map's legend supports interpretation of the map's content.

**The legend is a map component that helps decode the classification and symbolisation of the map's data.** They identify what the graphics represent, indicate any particular characteristics, and when quantitative information is encoded they should enable the map reader to estimate values from the mapped symbology. Some map symbols might be omitted from a legend if their form and function are obvious but be included if they enable the map user to interpret the map effectively. The style of the legend should be simple and efficient. The depiction of symbols should be exactly the same as those found on the map. As with the title on a map, the legend is obvious and does not need to be embellished with the words 'legend' or 'key' (in fact, the word 'key' isn't particularly useful in describing a legend!).

**Legend headings are often included to further explain the map's theme or to subdivide categories of symbol.** It is common to state the unit of measurement and the enumeration area in a legend. Legend headings are normally placed above the legend detail, horizontally centred or horizontally aligned to the left.

**The legend should be of sufficient size to be useful.** It should show the symbols effectively but it should not occupy an inordinate amount of space or visually challenge the major map components on the page. Often, the legend can be hidden behind a drop-down or swiped control for a digital map, which releases map space for content rather than marginalia. The legend heading should be smaller than both the title and the subtitle of the map. Any further definitions or explanatory text should be smaller than other type in the legend. Position is usually dictated by other map components and available space but it should be visually centred within its available space.

Symbols are normally placed on the left and their definitions on the right in keeping with normal reading direction. The map reader can thus read the symbol first and then the definition. Symbols are normally vertically distributed and horizontally centred.

Textual definitions are normally horizontally aligned to the left. Where definitions consist of ranges of numbers, they are normally horizontally aligned on the basis of the separator.

Separators are usually a hyphen, en dash, or the word 'to'. Where negative numbers are included in the range, the use of *to* is preferred to avoid confusion with the negative sign. Spaces are usually included on either side of the separator.

Definitions that have numbers greater than 1,000 should include commas, and decimal numbers smaller than one should include a leading zero.

Where space for a vertically oriented legend is not available, it is acceptable to orient the legend in a horizontal style with definitions horizontally centred below the symbols.

When depicting areal symbols, legend boxes should be connected for a range of values. Connected boxes assist in emphasising the continuous nature of measurement of a single variable. Where a qualitative distinction is made between areal symbols, they should be separated, again to emphasise the level of measurement. Irregularly shaped areal symbols are used appropriately when the feature is itself irregular such as a water body, marshland, or forested area.

**See also:** Graticules, grids, and neatlines | Informing | Navigating a map | Your map is wrong!

## Features

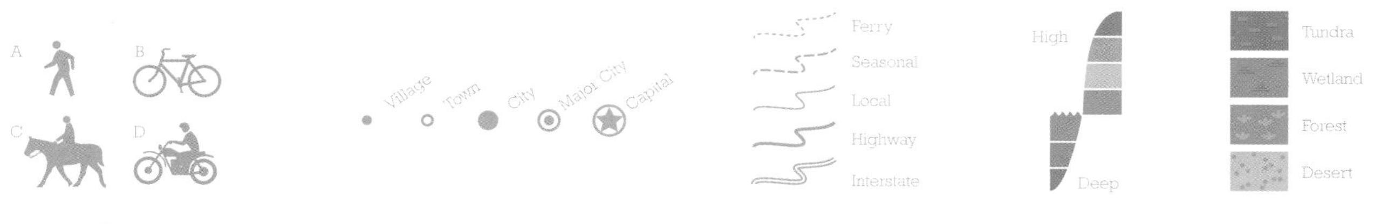


Category Class Physiographic

## Thematic

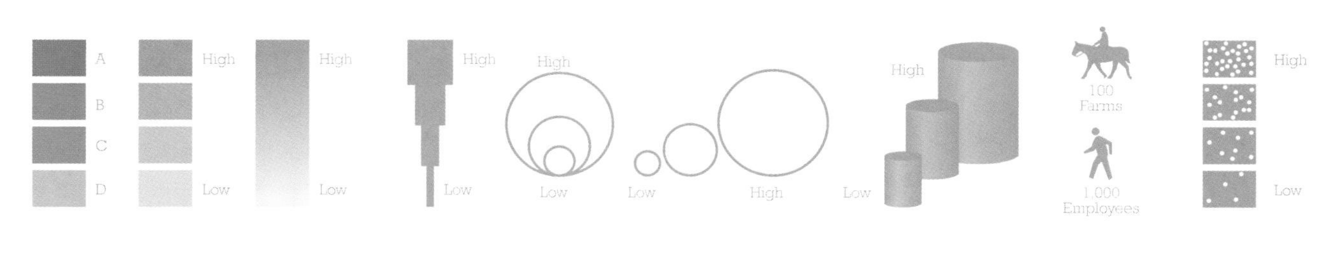


Categorical Discrete Continuous Graduated Representative Density

## Multivariate

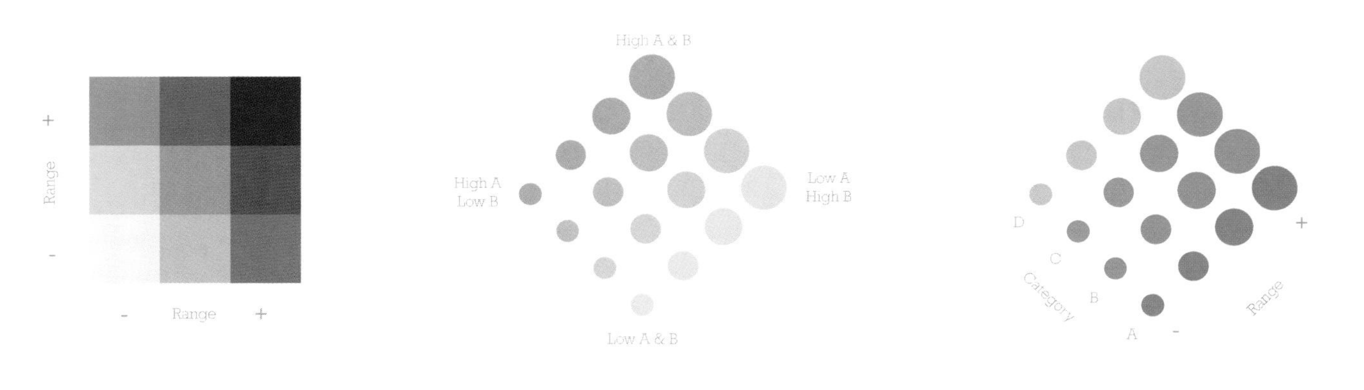


## Multipurpose

Title as Legend Chart as Legend

# Lettering

## The overall function of map lettering is to provide a bridge between the map symbols and the reader.

**Map lettering is most commonly used as a way to augment symbol encoding and to locate places we reference using language.** On general reference or topographic maps, lettering is used to name places (cities, rivers) and label components (scale, graticule, title). Lettering on thematic maps tends to be restricted to marginal elements with little or no lettering included on the map itself although text often appears in pop-up windows and through other click and hover events in online maps to support the description of thematic content.

**All lettering is, ultimately, written language, which is a symbol of meaning in its own right.** In this sense it has a functional purpose and is a functional symbol. In its most straightforward form, map lettering is a literal symbol since the meaning is encoded in the arrangement of letters into the names of features. As with most symbols, map lettering can not only improve communication through good design but poorly designed lettering can hinder communication.

**The design of lettering changes its appearance as a symbol and is used to denote different meaning.** For instance, increasing the size of lettering might indicate some level of importance; location might be signalled by placement; and the extent of a feature can be shown through spacing individual letters. Style, size, and placement are important variables that support different ways to reference features within the framework of the map. If the design character is changed, it can become a nominal symbol to distinguish between features of different types. For instance, all hydrographic features might be labelled with blue type with upright letters being used for open water and italicised letters for running water (rivers). Varying the size and other design characteristics also provides lettering with the ability to be used as an ordinal symbol. For instance, hierarchy of urban settlement size (based on population size) can be indicated by varying size and other characteristics of the lettering.

The functions of type on a map should not be underestimated and neither should the difficulty in making it effective. Although the vast majority of the mapmaking process is to abstract reality and use the visual language of symbols to create a cartographic representation, bluntly, lettering does not exist in the real world to describe many of the features that require map lettering. Real-world features have names but not labels and type on a map has colloquially been referred to as a *necessary evil* as it both takes up valuable space but is fundamental to our ability to describe the map and locate features by name.

Physical features will almost always have a name, and that name is more than likely needed on a topographic map so the place can be described relative to other places, or to support navigation. Keeping type on a map to a minimum allows more space for mapped content but, often, the labels are crucial. Getting a balance and avoiding repetition (e.g. multiple labels for the same feature) is a reasonable aim.

Historically, thematic maps rarely had any map lettering since the map of the thematic data was its own basemap, replacing topography. For interpretation, labels are useful and are also useful when added atop a thematic base. A trend in web mapping has been to vary an overlaid thematic map's transparency to reveal labels on an underlying topographic basemap. This is ill-advised since it dulls the map itself, introducing a mix of the thematic detail and underlying colours. The advantage of seeing labels through the map is outweighed by the impact on other aspects of the map's design.

**See also:** Elements of type | Guidelines for lettering | Lettering in 3D | Type colour | Typographic maps | Using words

| Use | Description and Use |
|---|---|
| **Descriptive text** | Reflects features that are symbolised on map by point, line, area |
| Narrative | Names of objects |
| Descriptive | Additional property of feature (e.g. scenic route) |
| Warning | Dangerous nature of feature (e.g. sunken wreck) |
| Functional | Locatable ground feature (e.g. rescue post) |
| Regulatory | Legal information (e.g. area of land) |
| **Analytical text** | Links user with attribute of features |
| Confirmative | Spatial relations (e.g. distance between features or bearings) |
| Determinative | Tables placed alongside map |
| Interpretive | Difficult to get information from a map so provided (e.g. quickest route is…) |
| Reference | Text alongside map |
| Categorisation | Categorisation of a theme in codes (e.g. geological codes) |
| **Positional Text** | Text to describe or confirm location, in space or time |
| Geocoding | Grid reference positions |
| Measurement | Relative position (e.g. at edge of map, 20 miles to…) |
| Temporal Position | Text to give time of events (e.g. historic battles) |
| **Metadata** | Refer to the nature of source data to map as a whole (e.g. reference ellipsoid) |

# Lettering in 3D

## Making maps in 3D brings additional considerations for the design and placement of typographic components.

**Typographic components normally lie flat on a planimetric map but they become distorted when draped across a 3D surface.** It's a common practice to drape maps across a surface such as a virtual globe, but any that contain typographic components burnt in (such as a pre-rendered basemap) will lead to distortions when the surface is anything other than perfectly flat. In particular, steep slopes will stretch text across a short horizontal distance but a long vertical distance. Where possible, ensure that any draped surface is devoid of type and treat labelling the map as a separate part of the design process.

**On maps that take advantage of the third dimension, labels can be positioned upright across the flat, or perspective, map to give the impression of 3D.** The use of shadows can help to accentuate their location. This might also be the case for isometric maps which exist on a plane but which represent human and physical features in pseudo-3D. This sort of billboarded text might have the addition of some depth to the characters to give them a volumetric appearance.

**Typographic components toward the front of a 3D display will appear larger than those in the distance.** This may be desirable as the map reader rotates and interacts with a virtual globe. On the other hand it might modify the hierarchy of meaning that you have encoded into the map. Making labels appear/disappear and size according to their relative position in the depth field of the display can adjust for such perceptual issues.

**In 3D space, typographic components will usually move with the viewing angle and rotation of the map to remain right reading.** In this sense, labels might be centred on a feature so they remain centred and front-facing as the map moves. For instance, a point feature may be represented with an upright pin and the associated label centred above it. As the map moves, the label will remain centred above the pin. The same might be true for area features similarly labelled.

One of the benefits of a static map display is that you're working with known parameters. Once a label is placed, it remains in position relative to all other map content. In 3D this is no longer the case. As the zenith, or viewing angle, reduces, labels will begin to appear on the same plane and overlap. As a map is zoomed in and out, different space is made available. As a map is rotated, label extents may begin to conflict with other map components that come into view. These circumstances make labelling 3D maps challenging.

If you're dealing with 3D typographic elements, you're also likely to be working in an interactive environment or possibly one which results in a short movie and that means progressive omission, depth priority, and vertical space become available to offer different ways to deal with label placement.

Progressive omission can be used to weed out some labels, so as the map reader zooms into a larger scale, more labels appear. Coupled with the use of depth priority, you can speficy what labels at what depths in the view are seen. Vertical space can be used to position labels higher or lower. This might be to position them above other competing map objects (e.g. buildings) or to highlight one label or class of labels among others.

For linear features, labels can still be positioned to appear along the path of the feature, but as an interactive 3D map is rotated they can flip so they always remain right reading.

**See also:** Elements of type | Globes | Lettering | Type colour | Using words

**Opposite:** Some alternative options for labelling a 3D map.

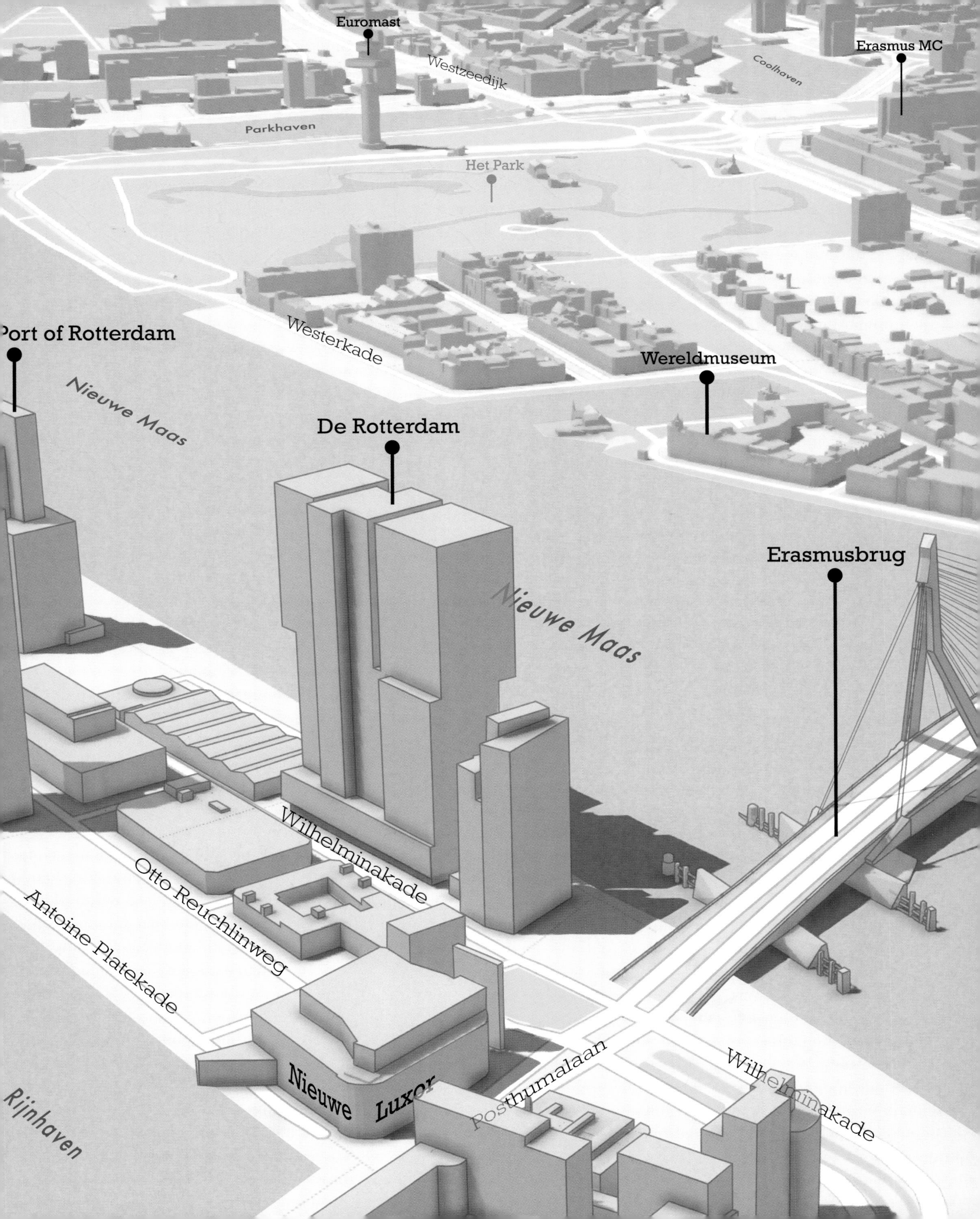

Euromast
Westzeedijk
Coolhaven
Erasmus MC
Parkhaven
Het Park
Port of Rotterdam
Westerkade
Nieuwe Maas
Wereldmuseum
De Rotterdam
Erasmusbrug
Nieuwe Maas
Wilhelminakade
Otto Reuchlinweg
Antoine Platekade
Nieuwe Luxor
Posthumalaan
Wilhelminakade
Rijnhaven

# Levels of measurement

## The way that data is measured has consequences for how it can be mapped.

**Data can be measured on nominal, ordinal, interval, and ratio levels, which increase in detail.** Measurement assigns numerical values to phenomena to represent certain facts about them, and it is this translation of phenomena into data that ultimately allows it to be mapped.

**Nominal scaling is the simplest level of measurement and is akin to a binary form of measurement where everything can be measured as 1 or 0 (or yes or no).** Phenomena are classified into groups that are 'similar in kind' and labelled with identifiers to indicate their difference. Nominal data is not based on any numerical measurement and better described as measuring qualitative, rather than quantitative, difference. Nominal data allows the identification of commonality or difference between different phenomena.

**Ordinal scaling measures a rank or hierarchy between phenomena.** Phenomena are arranged from least to most or vice versa to discern relative position. There is no attempt to identify the distance between each data value. Some statistical analysis is possible on ranked data, for instance to identify relative positions of ranks between different datasets.

**Interval scaling positions individual events in rank order and identifies the distance between the ranks.** Individual data values are numerically scored with equal units used across the scale being measured. In this sense, the interval scale allows an analysis of the magnitude of data values at points along a measured scale. There is no natural starting point (or zero) for an interval scale, and the data is said to have no absolute values; instead data is relative.

**Ratio scaling is similar to interval scaling in that it involves ordering individual events with known distances separating each event.** However, the difference on a ratio scale is that magnitudes are absolute because they have a known starting point. The ratio scale of measurement has zero as its starting point.

A land use map is an example of a nominally scaled dataset in which each different land use type is distinct. On a nominal scale, urban is different from rural, and with finer levels of nominal scaling, we might characterise rural populated and rural unpopulated.

Ordinal ranking is often used to create an overview of the distribution of thematic data. For instance, at a small scale a map of atmospheric pressure may show difference as low, medium, or high. Ranking data into deciles (or some other statistical arrangement) also provides a way of summarising, say, median household income thematically.

The classic example of a phenomenon measured on the interval scale is a map that shows temperature difference above and below zero (i.e. it doesn't have an absolute zero). Values of a degree centigrade (and the interval between them) are the same at 22 degrees as they are at −6 degrees.

Again, taking temperature as an example, if you measure and then map on the Kelvin scale, you begin at absolute zero and all temperatures vary positively away from this value. Other examples include any mass or magnitude of data that cannot have negative values and which also therefore begin at zero. For instance, 20 kg is precisely double 10 kg; hence the term 'ratio' since there is a ratio of two to one in this example. Elevation above sea level is measured and mapped on the ratio scale with a mountain 24,000 m being double the height of one measured at 12,000 m.

**See also:** Frequency distributions and histograms | Making numbers meaningful | Nominal data | Ordinal data | Ratio and interval data

Using a race between our four animal friends as an example, different aspects can be measured.

On a nominal scale, entrants might be entered into the male or female race. Certainly their species makes each one different on a nominal scale. They might each have a unique race number which identifies that they differ from another runner. They may represent different nationalities.

On an ordinal scale, the British bulldog wins the race and is conferred 1st place, the Canadian beaver comes in 2nd place, and so on. It does not matter whether the beaver finishes 2 seconds or 2 minutes in front of the American eagle in 3rd place. They would still be measured 2nd and 3rd. The Australian kangaroo is 4th or last depending on how you want to describe that position.

If we measure the finishing time of each animal by the time of day this gives us an interval scale of measurement where we can examine the finishing times relative to one another and also the times between people ranked in different places. Units of time are not measured against an absolute zero — they simply give an indication of how the times differ between runners and one runner can be identified as being x minutes different from another in the time it took to finish the race. Additionally someone finishing their race at 4 p.m. has not necessarily taken double the time than someone who finished at 2 p.m.

Finishing times can be measured on a **ratio scale** as defined from the starting gun being timed at zero. Thus, if the winner finished in precisely 2 hours, they would have finished twice as quick as someone who finished the race in 4 hours.

# Lines

**Lines are a core mark on a map. Understanding and defining their function is key to making them count.**

**Lines on a map have three main functions.** Firstly, they define edges and boundaries between adjacent areas that might be demarcated through, perhaps, political determination. They contain a space that shares a particular characteristic. Secondly, they connect one place to another whether that is two features on a map or whether it's a feature and an associated label (perhaps through the use of a leader line). Finally, the line can represent a linear component such as a road, railway, or movement of some phenomena. This final function differs from the first two in that the line itself might represent some two-dimensional character such as direction of travel or magnitude. The line acts as a container for some aspect of the phenomena that it represents.

**With so many functions that a line can perform, the task for the cartographer is to establish symbology that delivers information.** Importantly, this should connote the meaning of each of the functions which can be easily understood while not overcomplicating the symbol structure. For instance, it's fairly common on road atlases to demarcate higher class roads by drawing them with two or sometimes three parallel lines. The use of parallel lines is often referred to as *casing*, as it's effectively a single, often thicker, inner line cased by two outer lines. By definition, this increases the complexity of the symbol and gives the map reader additional cognitive load in unravelling the complexity, but it does make the symbol, and therefore, the feature, seem more important than other lines on the map. A single, thicker line can often perform the same job but with a much cleaner look though the drawback is the line may become too figural and out of visual balance with other map elements.

**It's impossible to get away from the fact that most maps are made up of multiple lines and types of line.** Attempting to reduce notational complexity will result in improved efficiency in communication of information. Making lines less sinuous, making them simpler in form, and minimising the type and variation of line work reduces complexity.

**See also:** Areas | Flow maps | Orientation | Points | Schematic maps | Shape | Strip maps

Symbols that represent lines of different sorts on a topographic map must be carefully distinguished. Roads, rivers, coastlines, and boundaries are all very different and will sometimes merge with each other unless they are well designed and sufficiently different. Certain styles are fairly well understood such as a blue line for a river, a black line with small marks at regular intervals for a railway, and a dashed line for boundaries. These qualitative differences between type can be fairly easily reconciled through symbology, often where there is a well-understood precedent to follow.

Many linear phenomena demand a graphical structure that shows some quantitative difference between features of similar types. For instance, a map where all boundaries of different importance or all roads of different type are shown in the same symbol will fail to communicate relative importance. Homogeneity of line symbols should be avoided if the features have a meaningful difference in the real world. Ranks can be classified or real quantifiable differences can be shown by varying the width or character of line. The simple idea is that more important or higher ranked features get heavier line weights (thicker lines) or a combination of dashed and solid lines. Keeping a general consistent character is useful so avoid varying colour or line style too much for ranked features of the same general type. Try to ensure you exaggerate differences between line symbols to ensure they are noticeable.

For quantitative features, the empirical value itself can be used to vary the actual width of a line of a consistent style. This is often used for thematic maps where the width of a line indicates some characteristic of the flow of movement of phenomena along the line.

**Opposite:** Lines are shown isolated from the complete map. They sit atop the area features.

# Literal comparisons

**Maps are usually best understood when the theme is portrayed relative to something the map reader already understands. But you can take it too far!**

**It might seem like stating the obvious but maps rarely look anything like the thing they are representing.** For example, on a topographic map we show physical landscapes, features, and their relative position horizontally and vertically. For this, we use a series of graphical marks that look nothing like our experience of the real world—contours being the prime example. Maps, then, are illustrations and geographic fidelity can get in the way of communicating detail so we distil it to some graphic primitives as a way of summarizing the feature as a visual comparison to the real object.

**The graphic primitives need not necessarily look anything like reality.** Think of a typical transit map—it's actually a diagram with links between nodes. The complexity of a subterranean rail network with all the various systems and characteristics are not needed so, instead, we distil to an illustration—an interface metaphor that people instantaneously know how to interact with. People do not need to know what the underground rail system looks like to enable them to know how it works or what line they need to take between two stations. Simplifying to an interface metaphor is sufficient.

**Metaphors, in a visual sense, are excellent graphical mechanisms we can use to help map readers better understand something they might be unfamiliar with.** Metaphors afford the opportunity to represent something using a symbol or set of symbols that are abstract and which are often used to represent something else. We use graphical metaphors to clarify or identify ideas—to enable people to reflect on something they visually understand in order to understand something less familiar.

**A simile goes one step further by making a direct comparison between one graphical component and another of a different kind.** For instance, we may think of mapping the flow of ideas by using a design that is more often used to show the flow of water. By using a depiction that is explicit in one context, we can take advantage of it to show something less familiar.

**See also:** Design and response | Dispersal vs. layering | Isotype | Pictograms

The general idea of using a map reader's experience in other contexts is not restricted to maps of course. In language more generally, we use literal comparisons, metaphors, similes, and all manner of techniques to describe. The idea here is to mimic what we do in spoken and written language to take advantage of the same ideas in a visual sense.

Literal comparisons, metaphors, and similes can also be used successfully to personify a visual representation, sometimes for humorous purposes, sometimes to shock, and sometimes for comparison. By using strong visual comparisons, we can easily build meaning into a map beyond the data itself. Some of the classic serio-comic maps use literal comparisons and visual metaphors to great effect, often veering into the territory of using stereotypes to depict mapped information relating to different nationalistic characters.

**Opposite:** Making very strong visual statements can sometimes become controversial. The map on the facing page was drawn by Nigel Holmes as a deliberate attempt to give meaning to the mapped data. The literal comparison of a foetus taking the form of a map about the pro-life/pro-choice debate was never published because it was deemed too risqué. The map and the data, of course, is not literally a foetus but by asserting the data in this way a comparison is set up between the map's data and the imagery that helps people understand the context and meaning. It's certainly a strong image and one might argue it reinforces the map's message, particularly if the editorial context in which the map is placed takes a particular view. The image was withdrawn from publication in *Time* magazine one hour before publication.

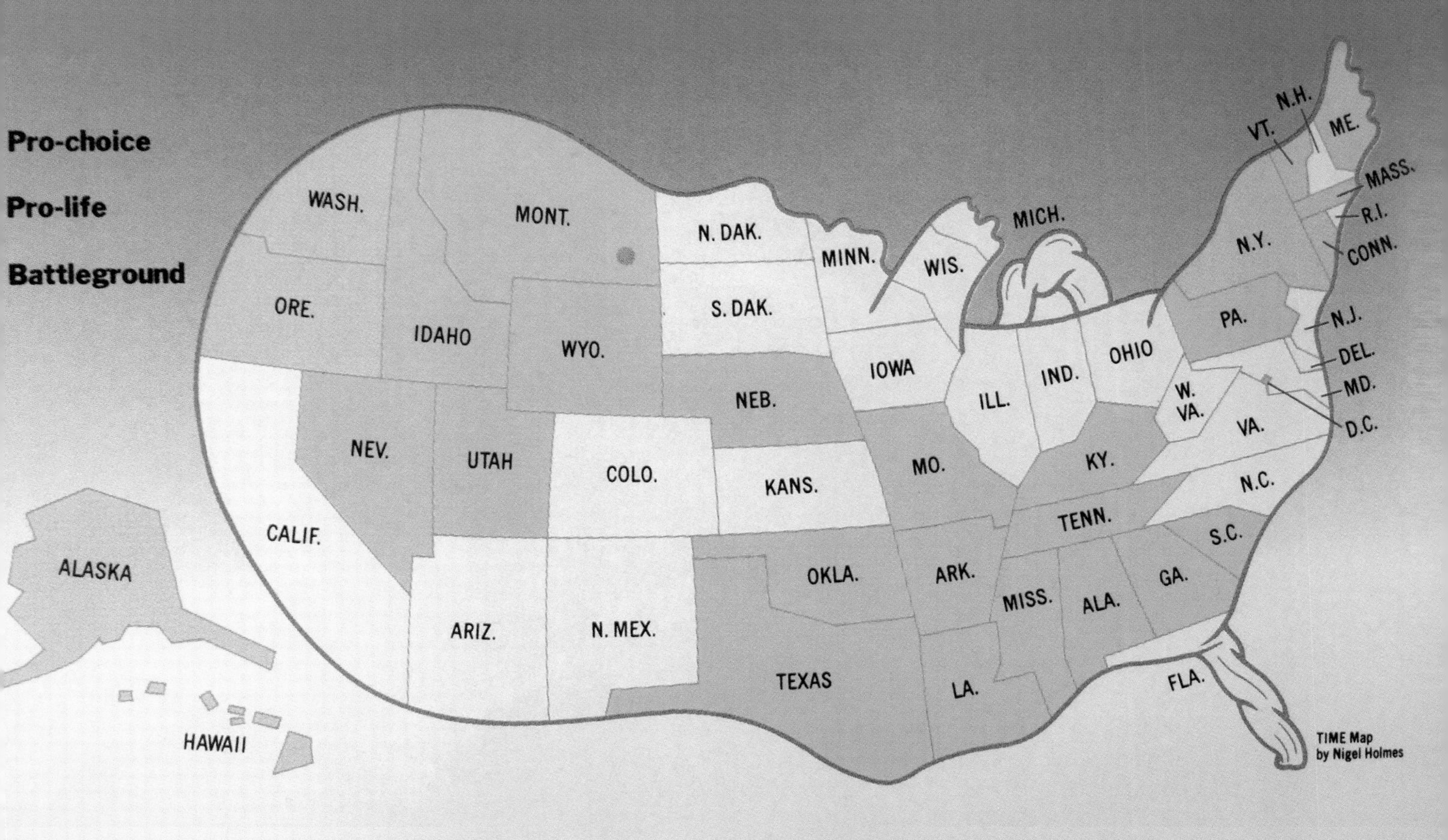

Pro-choice
Pro-life
Battleground
WASH.
MONT.
N. DAK.
MINN.
WIS.
MICH.
VT.
N.H.
ME.
MASS.
R.I.
CONN.
N.Y.
ORE.
IDAHO
WYO.
S. DAK.
IOWA
NEB.
ILL.
IND.
OHIO
PA.
N.J.
DEL.
MD.
D.C.
W. VA.
VA.
NEV.
UTAH
COLO.
KANS.
MO.
KY.
N.C.
TENN.
S.C.
CALIF.
ALASKA
OKLA.
ARK.
MISS.
ALA.
GA.
ARIZ.
N. MEX.
TEXAS
LA.
FLA.
HAWAII
TIME Map
by Nigel Holmes

# Location

## The most fundamental of visual variables, describing the position of a feature relative to its frame and other positions.

**Location is fundamental to cartography as it is the visual representation of where features are positioned.** Location describes the position of a map symbol relative to the frame, the coordinate system, and to other features. It takes visual prominence over other visual variables because it is an inherent part of the structure and narrative of a map. Regardless of whatever additional visual variables are applied to the feature, its position is typified by where the symbol is placed.

**Location defines the spatial aspect of the map's design but additional details can also be encoded.** A feature's position reflects where the phenomenon exists, and this might be represented by a point, a line, or an area symbol. Maps would be fairly meaningless unless the characteristics of the feature aren't, in some way, also represented. Other visual variables are used to show differences in type, magnitude, or importance as a way to differentiate among symbols on the map. For instance, changing the size of a point symbol can be used to show different ordering of towns on a small-scale map.

**Optically, locations that fall toward the centre of a map will appear more figural than those positioned elsewhere.** In planimetric maps the symbols that are more centrally located on a map will always appear more figural because your eyes tend toward the visual centre of a map on first investigation. Map locations toward the periphery will recede as ground. Central locations will therefore be seen as more important in the overall visual hierarchy of the map. This has always been a factor in making maps for print with the most important phenomena given centre stage. It is also important for digital cartography since the centre of a screen is, in perceptual terms, no different from a piece of paper. Interaction allows you to refocus and position alternative locations centrally in digital maps.

Location doesn't simply relate to where a symbol is positioned in planimetric space. If your map is in 3D, there are additional aspects of the impact of location as a visual variable. In an oblique representation, symbols that are nearer to the camera (or viewing) position will also be seen as figural. This is inherent in viewing objects in perspective. If height is also applied to the map symbol, it will inevitably accentuate the figure-ground difference perceived between the map symbol and those peripheral to it.

Ultimately, you can make only a minimal amount of modification to the location of a map symbol since you fundamentally need it to express the position of a phenomenon. Offsets can be used effectively, particularly in areas of crowded symbology, but leader lines may need to be used to link the symbol back to the more precise location. At smaller scales, aggregated symbol representations may be used, which then dissolve into individual symbols as scales become larger. For 3D, the use of an isometric projection can overcome the problem of nearer objects being seen as much larger though, of course, the centrality of symbols still makes them figural compared with peripheral symbols.

Visual variation

Increasing focus & centrality becomes figural

For seeing: Distinct, Levels

For representing: Nominal, Ordinal, Numeric

**See also:** Abstraction and signage | Latitude | Longitude | Position | Symbolisation | Symbols

*Cartography Corridor*

# Longitude

## Longitude allows the east–west measurement of location.

***Longitude* is the term used to define east–west angular change.** It is more difficult to measure than latitude because there are no natural points or lines on Earth that can be used as a starting point.

**The lack of a geometrically determined fixed reference line of zero degrees, called a *prime meridian*, from which to measure longitude, presented early navigators and cartographers with difficulties.** Rudimentary measures of position could be determined from measurement of time since Earth rotates about its axis 360° in each 24-hour period which is 15° per hour. This gave a method of calculating local time relative to other, known fixed points of reference, which could be converted into angular degrees. If a record of time at an agreed fixed point could be kept; and the difference in time between the local time and the reference point calculated; then this time could be converted into angular degrees, which meant it was possible to determine how many degrees of longitude had been travelled east or west.

**Developments in accurate timekeeping still only provided a means of measuring relative longitudinal position.** During the 19th century the Royal Observatory at Greenwich (London, UK) was chosen as being zero degrees largely because two-thirds of ships used the Greenwich meridian as their known fixed position. A line passing north–south through Greenwich therefore takes the longitudinal value of 0°0′0′′ and is referred to as the 'Greenwich meridian' as well as the 'prime meridian.'

**Longitude is therefore defined as the angle formed by a line that projects from the intersection of the prime meridian and the equator to the centre of Earth, and then back to the intersection of the equator and the 'local' meridian that passes through the position.** It ranges from 180°W to 180°E of the prime meridian and can also be represented decimally. For example, the position measured in the sexagesimal system at 0°18′13′′W is measured as −0.3036 in decimal degrees.

In 1714 the British Parliament's newly formed Board of Longitude announced a competition (with a prize of £20,000) to construct an accurate timepiece to solve the problem of calculating longitude at sea. The marine chronometer was invented by John Harrison in the 1700s and allowed navigation accurate to within 1.25 nautical miles and provided a mechanism for the accurate determination of longitude. Countries were able to specify a fixed location within their own boundaries for calculating time, and hence longitude.

The first experimental marine timekeeper, H1, made by John Harrison between 1730 and 1735:

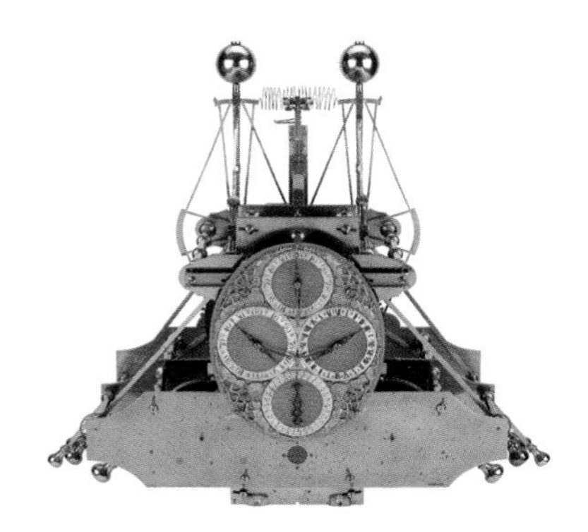

The Greenwich prime meridian also became a convenient starting point for the International Date Line and, along with the line at 180°, forms a circle north–south around Earth that defines the East and West Hemispheres in the same way as the equator separates the North and South Hemispheres.

The distance from the equator to the North or South Pole is 6,222 miles (10,013 km). That gives 69 miles (111 km) for each degree of latitude except the distance per degree of latitude reduces slightly the further from the equator because of Earth's flattening (oblateness).

**See also:** Datums | Earth coordinate geometry | Earth's framework | Latitude | Position | Spatial dimensions of data

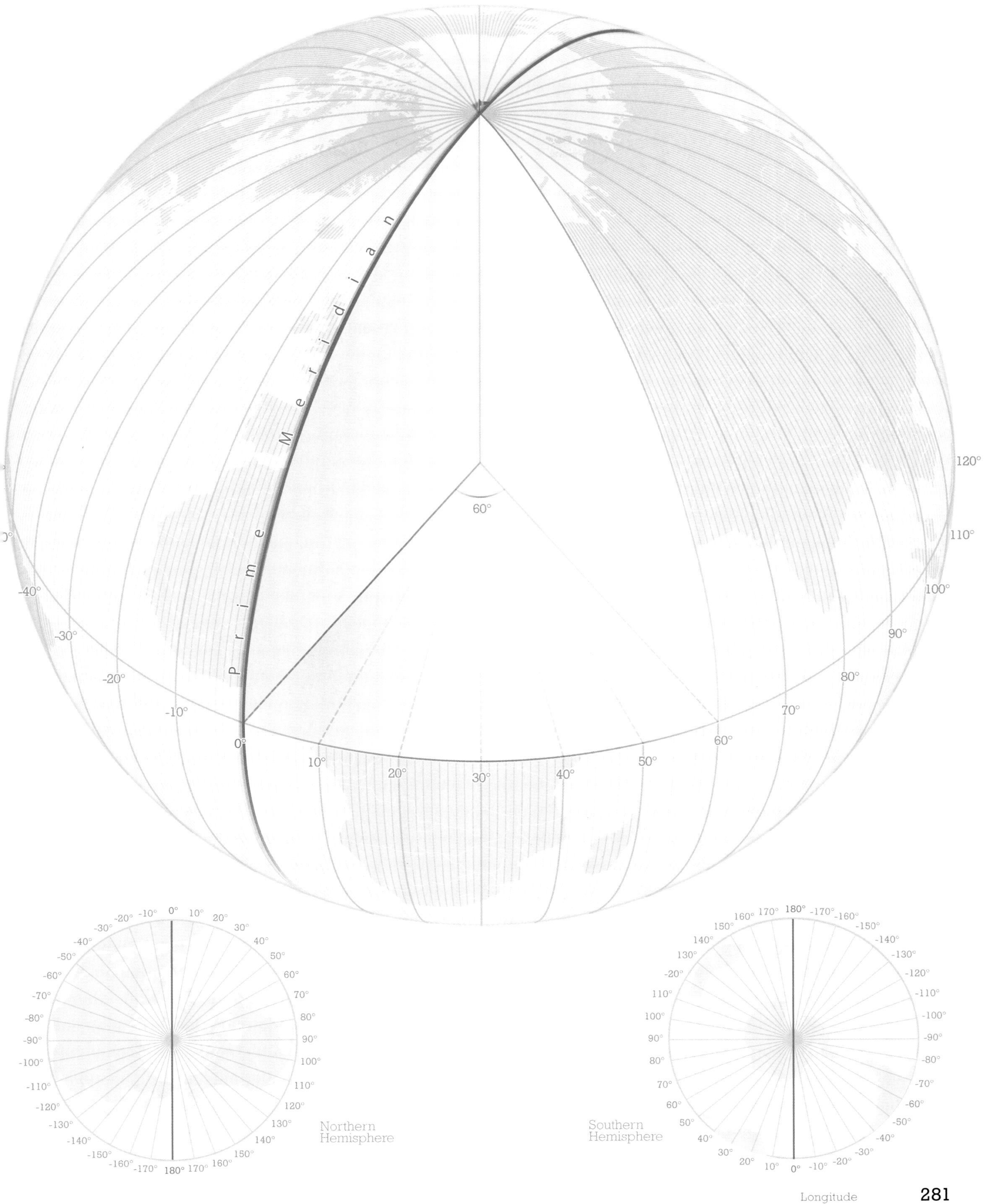
Prime Meridian
60°
120°
110°
100°
90°
80°
70°
60°
50°
40°
30°
20°
10°
0°
-10°
-20°
-30°
-40°
Northern Hemisphere
Southern Hemisphere

# The Heart of the Grand Canyon

National Geographic Society
1978

It was Bradford Washburn's Everest map that set me on my way toward mountain cartography, but when I first saw his Canyon map it blew my mind, opened my eyes, and challenged my notions of what a topographic map should look like and how it should be made much more than his Everest map.

The first thing I noticed and loved about this piece were the rich, saturated colours. An entire sheet filled with lovely textures and contrasting values. It feels more like a painting of the Grand Canyon than the hyperaccurate map that it is.

There is no way to mistake this for a USGS quad. At another level of detail, the vegetation textures on the plateau, the Swiss cliff hachuring, and Tibor Toth's shaded relief work are all masterfully combined in a feast for the eyes.

As a young cartographer, the work that went into this map was not only an inspiration but it set the bar for what cartographic excellence was. It still does.

—Martin Gamache

OTTOMAN
AMPHITHEATER
NORTH KAIBAB TRAIL
Sumner Butte
Zoroaster Temple
Bradley Point
Demaray Point
CLEAR CREEK TRAIL
TONTO PLATFORM
U.S.G.S. Gaging Stations
KAIBAB BRIDGE
Tunnel
COLORADO RIVER
GRANITE GORGE
ZOROASTER CANYON
Clear Creek
Howlan Butte
The Tipoff
Emergency Telephone
Window
Eighty-five Mile Rapids
Zoroaster Rapids
TONTO TRAIL
Cremation Creek
Pattie Butte
LONETREE CANYON
Eighty-three Mile Rapids
O'Neill Butte
Newton Butte
Boulder Creek
Yaki Point
KAIBAB TRAIL HEAD
YAKI POINT ROAD
MULE TRAIL
Picnic Area
EAST RIM DRIVE
COCONINO
Shoshone Point
Lyell Butte
TRUE NORTH

# Making numbers meaningful

## Maps are made of numbers. And all hold different meanings.

**All maps are ultimately made from numbers.** Topographic maps are made from paired coordinates. Sometimes these exist singularly, and sometimes they exist in a sequence to form linear or enclosed features. Thematic maps are generally maps which display empirical and statistical data. Encoding meaning into these numbers is the essence of cartography. Numbers are powerful primitives in cartography, and each has different circumstances and purpose. Whether you are making a map to navigate by or a map that shows patterns of disease, the task is to take the numbers you are working with and fashion a visual mechanism that brings them to life.

**There are many examples of great early information design that might not necessarily be great examples of cartography but did their job of making the numbers meaningful.** Charles Minard's map of Napoleon's march is a simplified diagram with little that marks it out cartographically. It does a superb job of summarising the key statistical detail of the disastrous march. It's an anti-war statement, and Minard gave the numbers meaning. Florence Nightingale's coxcomb charts of the mortality of the army of the east is another. The information is certainly spatial but the diagram might struggle to be viewed as cartographic. The development of a new type of diagram is its true contribution—a way to see how categories of data changed in magnitude over time. John Snow's map of cholera in Soho, London, can certainly be improved upon in cartographic terms but the principal objective was met—he made the numbers meaningful.

**We often look at ornate historical maps with a sense of nostalgia, imagining that they are somehow superior to their modern counterparts.** These examples teach us, however, that simplicity is often far more useful than flourish. Giving the numbers space to breathe and to establish clear patterns is paramount. Often, the value of maps is that they show the importance of the ideas they communicate, often for the first time. Elegance on its own does not necessarily contribute to clear cartographic or information design.

**See also:** Data processing | Design and response | Ratios, proportions, and percentages | Statistical literacy

Snow, Nightingale, and Minard are a few who successfully made numbers meaningful and popularised their use in cartography. However, William Playfair was possibly the first to use numbers as a way to inform, persuade, and campaign. He realised that turning numbers into something visual made those numbers more digestible and impactful.

Born in 1759, Playfair was a Scottish engineer and economist and is often considered the father of graphical methods for statistics. He is credited as the inventor of many types of diagram, including the line, area, and bar chart as well as the pie chart and time-series chart. That's quite a curriculum vitae!

The clarity of much of Playfair's work provides a model for modern cartography not just in terms of the graphical devices you can employ but of the search to encode meaning into the visuals of the map. The imagination to turn numbers into a visual mechanism that communicates more readily lies at the heart of the meaningful design and use of charts. It's a cornerstone of successful information design.

**Opposite:** *Exports and Imports to and from Denmark & Norway from 1700 to 1780* by William Playfair, 1786. This is arguably the first ever graph of data ever made, expressing numbers in a graphical sense to make them meaningful. It appeared in Playfair's *Commercial and Political Atlas.* Its basis is a standard line chart but he encodes other information in subtle ways. The area between the two lines quantifies the trade balance between the two countries and immediately illustrates the balance in favour of and against England. Subtle design ideas thus translate into graphical excellence.

Exports and Imports to and from DENMARK & NORWAY from 1700 to 1780.

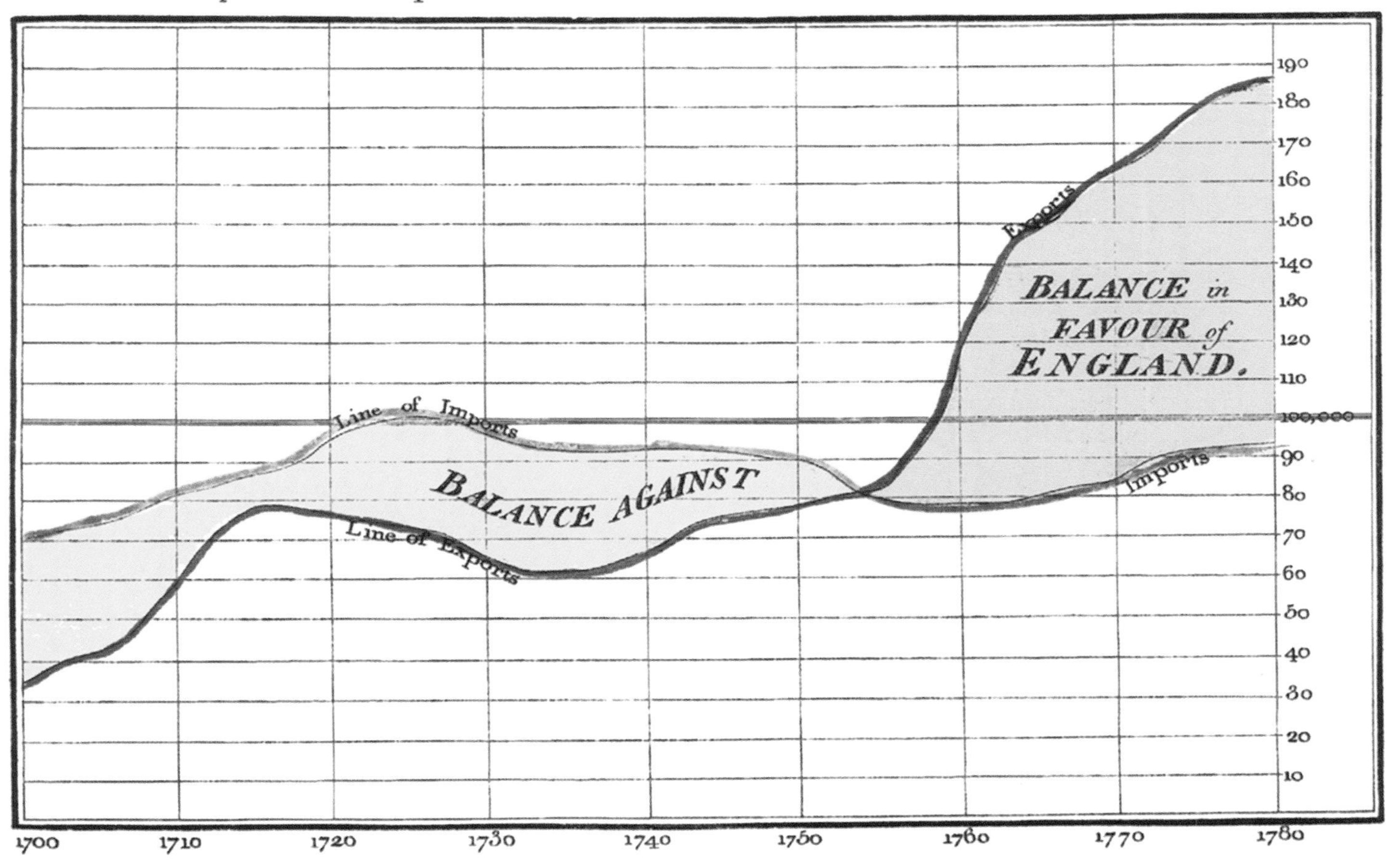


*The Bottom line is divided into Years, the Right hand line into L10,000 each.*

*Published as the Act directs, 14th May 1786, by Wm. Playfair*

*Neele sculpt 352, Strand, London.*

# Map aesthetics

**The aesthetic quality of a map is often hard to define but considering maps as art helps us appreciate their visual value.**

**The maps we love to look at are often also some of the most beautiful but beauty in any artistic form is in the eye of the beholder.** Some people prefer a beautiful topographic map of a mountain range. Others, a well-constructed statistical map. Design encompasses not only a consideration of the relationship between form and function but also an element of artistic expression.

**For many, the aesthetic quality of their maps is simply a by-product and not something to frame their thought processes on during the design phase.** However, aesthetics in map design are important since a map with poor colours, careless placement of typographic elements, over-generalised line work, and crudely classified data may be just as intrinsically accurate as one where care has been taken. The latter is more likely to be taken seriously, enjoyed, or revisited because of its aesthetic quality—its visual appeal. It might be said that one of the key abilities of a cartographer is to intuitively know what 'looks right' and this often holds true yet everyone who makes a map holds a view of what, visually, pleases them or their audience. Having a sense of what people enjoy looking at helps build popular, visually appealing maps.

**There is undoubtedly an element of subjectivity in map design because it is, in part, a creative process.** This has many parallels to art more generally. For instance, the aesthetic of Andy Warhol is very different from that of Rembrandt or Vincent van Gogh. Appreciation of different artistic genres has much to do with intuition and judgement, which is conditioned by training and experience, but the more you immerse yourself in looking at maps of all types and styles, the more you will see hints of what you like and techniques you can draw upon.

**With practice, intuition develops and style begins to form.** Maps are as much the subjective, aesthetic product of a map designer as they are a scientific, objective document. Exploring alternative styles yourself helps develop your own sense of cartographic taste and, perhaps, your own style.

**See also:** Data (c)art(e) | Design and response | Elegance | Emotional response | Style, fashion, and trends | Your map is wrong!

*'The "art" of cartography…is not simply an anachronism surviving from some pre-scientific era; it is an integral part of the cartographic process'.*—John Keats, 1982

In other words, cartography is not just about making maps pretty!

Three elements can be explored when discussing map aesthetics: harmony is the relationship between different map elements; composition is the arrangement of the elements and the inherent balance achieved; and clarity deals with the ease with which a map user can recognise the map's elements.

A map is likely to be seen as aesthetically poor if it fails to deliver on any of these counts. The fact that cartographers express themselves, through design, in different ways means that you can achieve many different designs of ostensibly the same map content. Indeed, if you were to peruse the shelves of a bookstore stocked with a range of different atlases, you would see a plethora of different house styles reflecting different approaches to design. An Ordnance Survey atlas looks different from a Times Bartholomew atlas which, in turn, looks different from a National Geographic atlas. They might all have much the same function but they are an expression of the design style developed by the individual organisations over time.

What pleases you, in aesthetic terms, is also partly a function of what you're familiar with, and that has much to do with where you're from, your education, experience, and likes/dislikes. What map styles are fashionable also changes. Some maps might be considered timeless classics. Some are more fleeting in their appeal. Change may also take place over time as preferences are modified and the search for a 'modern, updated' look and feel become desirable.

Sapporo
Sea of Japan
Japan
Osaka
Tokyo
Nagasaki
Pacific Ocean

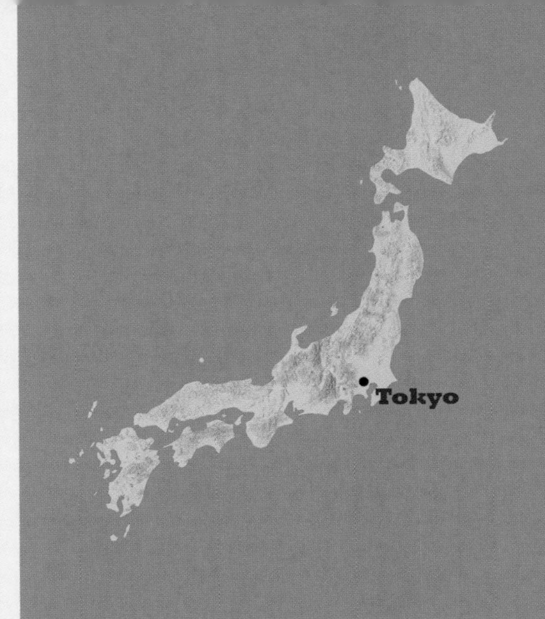
Tokyo

Sea of Japan
Sapporo
Japan
Tokyo
Osaka
Nagasaki
Pacific Ocean

日本

Sapporo
Sea of Japan
JAPAN
Osaka
Tokyo
Nagasaki
Pacific Ocean

OF JAPAN
SAPPORO
TOKYO
OSAKA
NAGASAKI
PACIFIC OCEAN

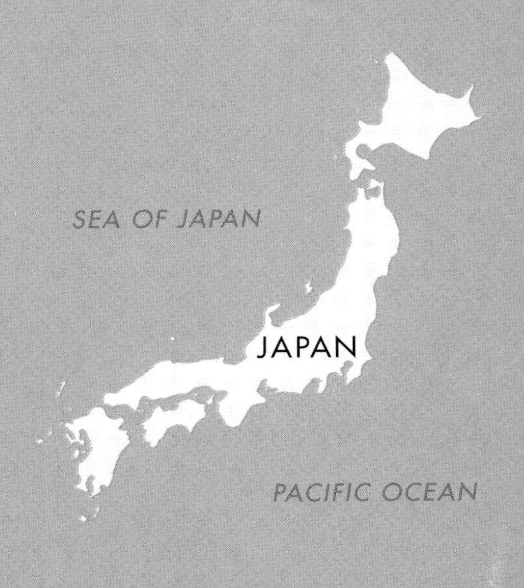
SEA OF JAPAN
JAPAN
PACIFIC OCEAN

Sapporo
Sea of Japan
Japan
Osaka
Tokyo
Nagasaki

# Map cube

## How maps work and the functions they support can be represented as a map cube.

**The map cube, originally developed as (cartography)[3] by Alan MacEachren and D. R. Fraser Taylor in the early Nineties, shows how maps work.** It's an extremely useful way of considering how the purpose of your map is driven by other factors such as the amount of interaction required to read it, whether the map is for private study or public consumption, and whether you're showing phenomena that are largely known and understood or whether you're attempting to reveal new patterns and insight.

**The presentation of geospatial data to the public using low-level interactive products (e.g. a static map) to display known patterns is represented inside the cube in the lower corner.** Synthesis, analysis, and exploration can be plotted on a line that extends from the communication corner to the 'private use, presenting unknowns, high interaction' corner: visualization. They aren't necessarily linear, but there's a conceptual progression and at other points in the cube the balance between the three axes may differ markedly.

**Visualization goes beyond what you might see as the traditional function of cartography to communicate.** It is stimulated by growth in not just computing and geographical information systems but multimedia, virtual reality, and exploratory data analysis. Developments in each of these areas of computer science have influenced mapmaking such that maps are key components at all points in the map cube. The role that maps and cartography play in the changing technological world is itself changing. The power of cartography in harnessing geospatial data and structuring it by visualising it as maps is crucial to not just effective communication of data but also effective visualization.

Thinking about your map in relation to the components of the map cube is a useful way to focus on the functions you need to support, and the form that the design needs to take.

Whether you are designing a printed map or a web map, considering who the map is for, the level of interaction you need, and whether the data is simply for presentation or query is key. These three aspects will go a long way toward shaping how you design the map itself.

For many web mapping applications, the map is now simply one component of a dashboard experience. Other components give the reader different opportunities. These might include tabular displays and rich interactive graphs as well as the functionality of interaction across the components. As a user selects from one component, information may update elsewhere. This sort of application might fall somewhere in the exploration or analysis realm in the map cube. Many data relations are previously unknown as the user searches for patterns. They might process their data and perform statistical analysis to discover new insights. The interaction between user and components and also among components is high, and this sort of work is likely done by individuals. Sharing results might require a different product, perhaps which falls more in the synthesis area.

Thinking inside the box of the map cube to conceptualise your product requirements is a useful way to ensure your maps work outside the box for their intended purpose.

**See also:** Defining map design | Form and function | Simplicity vs. complexity | Space-time cubes | Types of maps | Visualization wheel

**Opposite:** *The Map Cube*, after Alan MacEachren and D. R. Fraser Taylor, 1994

**Visualization**
Highly interactive graphics to reveal unknown spatial patterns to an individual

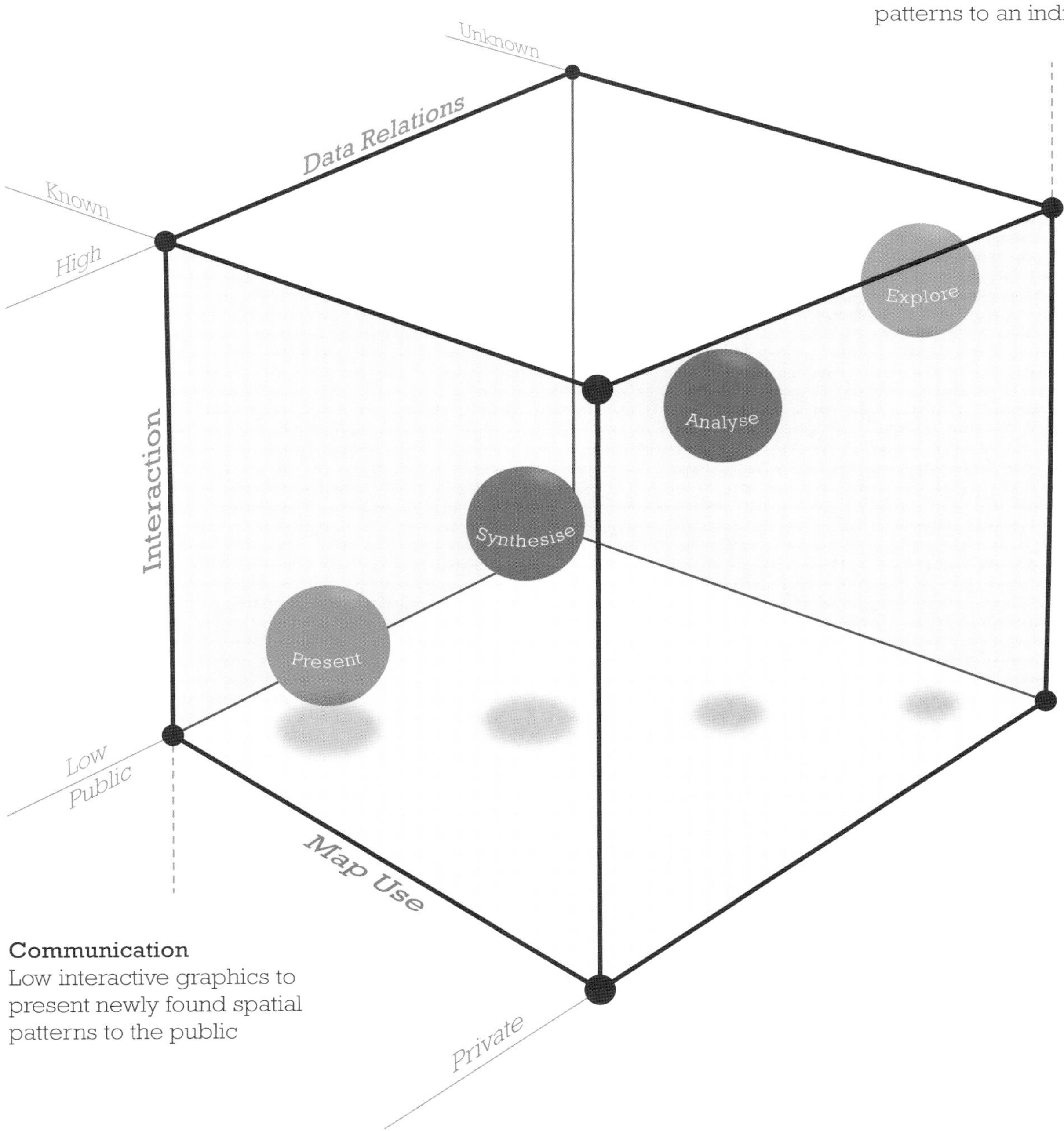


**Communication**
Low interactive graphics to present newly found spatial patterns to the public

# Map projections

## Getting locations of features from the curved surface of Earth to a flat map involves projections. It's maths … but we can think of it in a simpler way.

**A map projection is the systematic arrangement of Earth's (or a reference globe's) meridians and parallels on a plane surface.** Earth's real shape is a geoid that for mapping is represented by an ellipsoid (e.g. WGS84). For many mapping purposes (particularly small-scale mapping where large-scale positional accuracy is not required), the ellipsoid is then further reduced to a reference globe (also referred to as a *nominal* or *generating globe*). Finally, this reference globe is used to transform the graticule and locations onto a map via a map projection. A map projection is therefore the translation of the spherical surface to a planar surface. It is the last step in the process of measuring positions on Earth in order to represent them on a map.

**Conceptually, a map projection is analogous to placing a light bulb in the centre of a translucent reference globe and projecting the graticule onto a flat surface.** This is also a suitable analogy for explaining how distortions occur. If, for instance, we wrap a piece of paper around a globe in a cylindrical fashion such that the paper touches the globe along a single line, the paper's surface is tangent to the reference globe and the line will be projected at exactly the same scale with no distortion on the paper. This line is referred to as the *principal scale*. It is also known as a *line of zero distortion* or a *standard line*. Because there is a conceptual gap between the illuminated globe and the paper at all other locations, the gap causes a distortion effect as the line is projected. The larger the gap, the larger the potential distortion.

In the map projection process, the reference globe is defined by a nominal scale. This normally describes the radius of the reference globe such that for a reference globe with a 20 cm radius, the nominal scale is calculated as:

map distance / Earth distance

= globe radius / Earth radius

= 20 cm / (6378 km × 100000 cm)

= 1 / 31890000.

= 1:31,890,000.

The importance of the nominal scale in mapping is that every feature on the reference globe will have the same scale. All meridians, parallels, great circle arcs, and other geometrical properties will share the same, consistent scale. This is also referred to as the map's *principal scale*.

**See also:** Aspect of a map projection | Families of map projection | Map projections: Decisions, decisions! | Web Mercator

Simple cylindrical projection

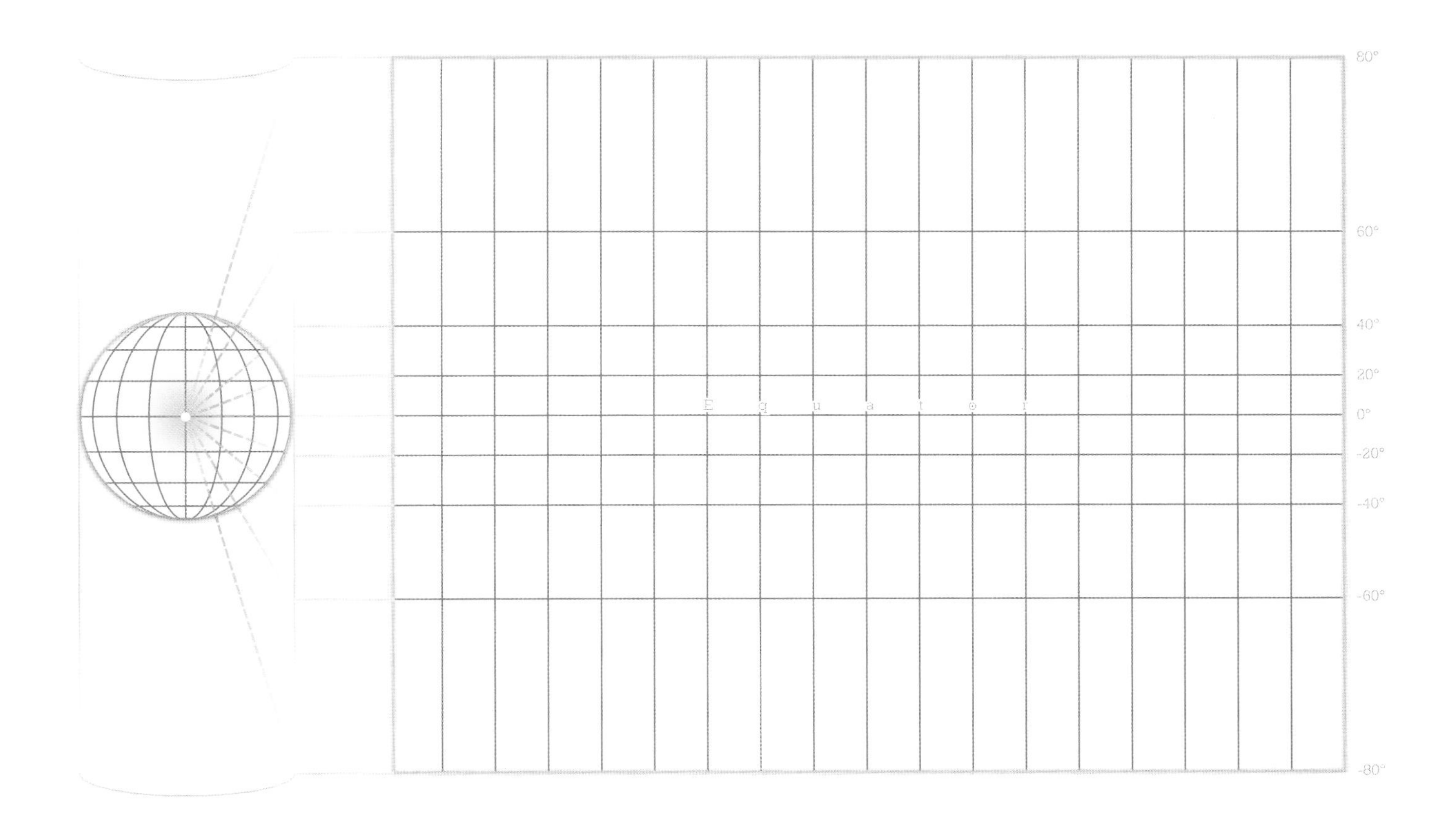


Approximating Earth's surface

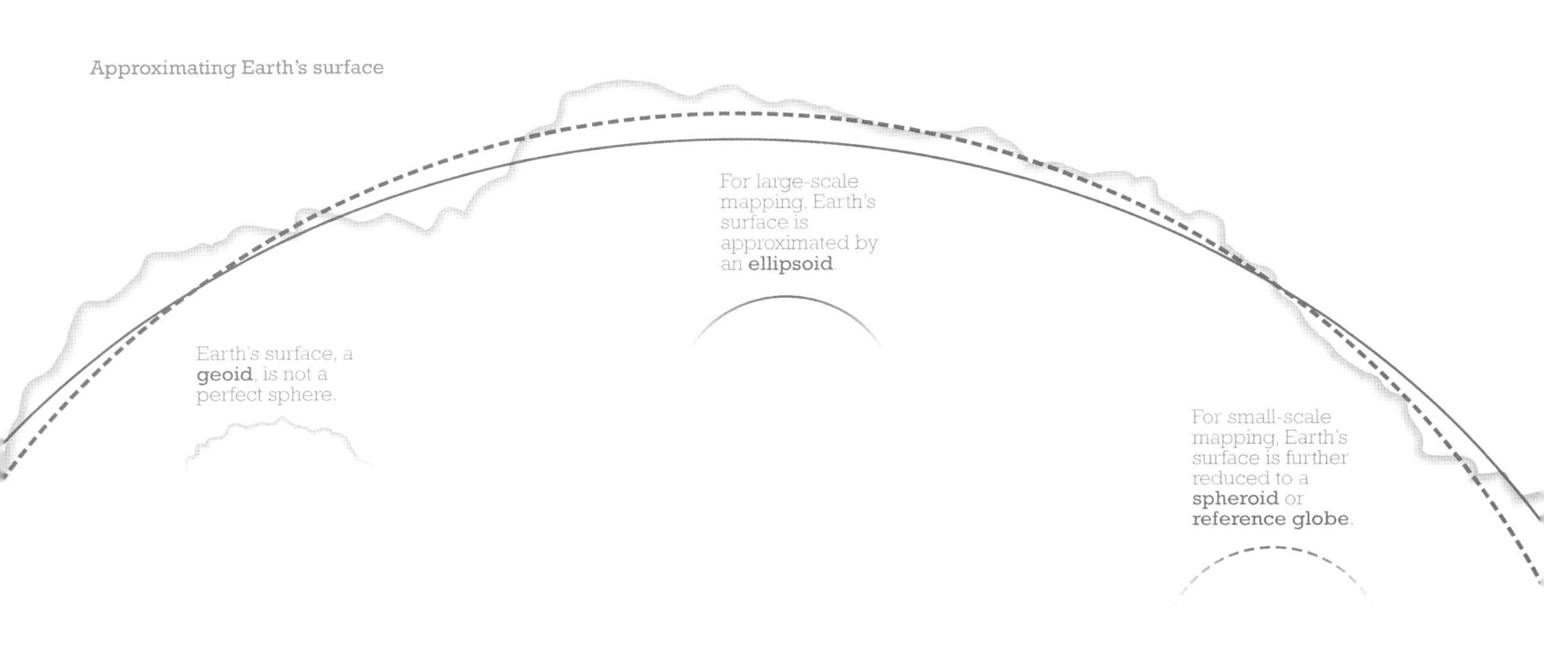

# Map projections: Decisions, decisions!

## Making a decision on which map projection to use is based on function, scale, and appearance.

**With all the possible choices of map projection to choose from, selecting appropriate projections can be bewildering.** A simple rule of thumb is to match the projection properties to the map purpose. So if a map is to be used for analysing or measuring angular relationships in navigation or surveying, choose a conformal projection. For comparing one mapped area with another, opt for an equivalent projection. If distance measurements are important, then go for an equidistant projection. Since the primary focus of many thematic maps is to compare the value or pattern in one area with another, area ratios must be maintained so an equal-area projection is most suited. Flow mapping and other types of map that depict or track movement are most appropriately depicted using an orthographic projection. Two alternatives would be the azimuthal equidistant, to show the correct directions and distances across the map from a central point, or the gnomonic, which depicts all great circles as straight lines.

**John Snyder's hierarchical list of projections organised around the Earth area to be mapped, the projection property (equivalence, conformality, equidistance), and characteristic (class, tangency, aspect) remain a valuable reference (Snyder 1987).** His approach is well suited to the process of map projection decision-making since you will most commonly begin with a basic idea of the part of Earth and extent of the area that is being mapped. Adaptive Composite Map Projections are an operationalised version of Snyder's selection guidelines for choosing map projections. They combine several map projections which have been stitched together to create a seamless, morphing map projection that varies depending on the region displayed and the map scale. The composite projection changes the map's geometry to scale, the map's height to width ratio (size), and the central latitude of the area being displayed by replacing projections as appropriate and modifying parameters to adjust to a more suitable approach. Both provide excellent guides for deciding on an appropriate projection.

The rapid rise of Web Mercator as the default projection for web mapping has been problematic. As a form of cylindrical projection with one line of tangency, it inevitably contains some serious distortions as you move north and south away from the equator. This is a difficulty for many maps, particularly thematic maps which depend on being able to visually compare areas point by point. That said, most GIS and web mapping software allows the creation of web maps that can be published using non-Web Mercator, and the extra effort will be repaid by a map that performs much more closely to the function required.

In practical terms, which map projection to choose relates to the following questions:

1. What projection properties must be preserved, as far as possible, for the mapping task, and which are less important?
2. Are the deformational characteristics acceptable, considering where the mapped area lies in relation to the projection?
3. Can the projection be manipulated to improve certain characteristics such as recentering it?
4. Will the shape of the resulting map be familiar to map readers, or does the pattern of the graticule create an overtly awkward appearance?
5. Is the data to be used in the mapping task in a suitable format for projection or reprojection to the chosen map projection?
6. Is the projection supported in the software to be used, or will it need to be created?

**See also:** Distortions in map projections | Properties of a map projection | Web Mercator

| Global maps | Property | Characteristic | Named projection |
|---|---|---|---|
| World | Conformality | Constant scale along equator | Mercator |
| | | Constant scale along a meridian | Transverse Mercator |
| | | Constant scale along an oblique great circle | Oblique Mercator |
| | | No constant scale anywhere on map | Lagrange |
| | | | Eisenlohr |
| | | | Mollweide |
| | Equivalence | Noninterrupted | Eckert IV & VI |
| | | | Boggs Eumorphic |
| | | | Sinusoidal |
| | | | Hammer |
| | | | Other miscellaneous pseudocylindricals |
| | | | Any of the above except Hammer |
| | | Interrupted | Goode's Homolosine |
| | | | Briesemeister |
| | | Oblique aspect | Oblique Mollweide |
| | | | Polar azimuthal equidistant |
| | Equidistance | Centred on a pole | Oblique azimuthal equidistant |
| | | Centred on a city | Mercator |
| | | | Miller cylindrical |
| | Straight rhumb lines | | Robinson pseudocylindrical |
| | Compromise distortion | | Stereographic conformal |
| Hemisphere | Conformality | | Lambert azimuthal equivalent |
| | Equivalence | | Azimuthal equidistant |
| | Equidistance | | Orthographic |
| | Global appearance | | |

| Regional maps | Directional extent | Location | Property | Named projection |
|---|---|---|---|---|
| Continent, ocean, or smaller region | East–west | Along the equator | Conformality | Mercator |
| | | | Equivalence | Cylindrical equivalent |
| | | Away from the equator | Conformality | Lambert conformal conic |
| | | | Equivalence | Albers equivalent conic |
| | North–south | Aligned anywhere along a meridian | Conformality | Transverse Mercator |
| | | | Equivalence | Transverse cylindrical equivalent |
| | Oblique | Anywhere | Conformality | Oblique Mercator |
| | | | Equivalence | Oblique cylindrical equivalent |
| | Equal extent | Polar, equatorial, or oblique | Conformality | Stereographic |
| | | | Equivalence | Lambert azimuthal equivalent |

| Mapping purpose | Map content | Property to preserve |
|---|---|---|
| Reference map | Physical and geographical characteristics, administrative units; topographic features such as shorelines, rivers, roads and relief | Equidistance |
| Physical map | Relief and hydrography; outlines and directions of river flow; valleys, mountain ranges and orographic features | Equidistance or equivalence depending on specific content |
| Climatic/meteorological | Depiction of isolines to study patterns | Equivalence |
| | Maps with considerable interpolation of isolines to determine gradients | Conformality |
| | Direction, velocity and strength of wind; shapes of isobars and isolines | Conformality |
| Geological | Bands of geological structure | Equivalence |
| | Fault lines | Equidistance |
| Soils, Earth's surface, ocean floors | Areas of soil zones etc | Equivalence |
| Population | Population density | Equivalence |
| | Migration | Equivalence (& conformality) |
| Navigational & aeronavigational | Path between origin and destination | Conformality (centred on route) |

Choices for map projections (after Bugayevskiy and Snyder 1995)

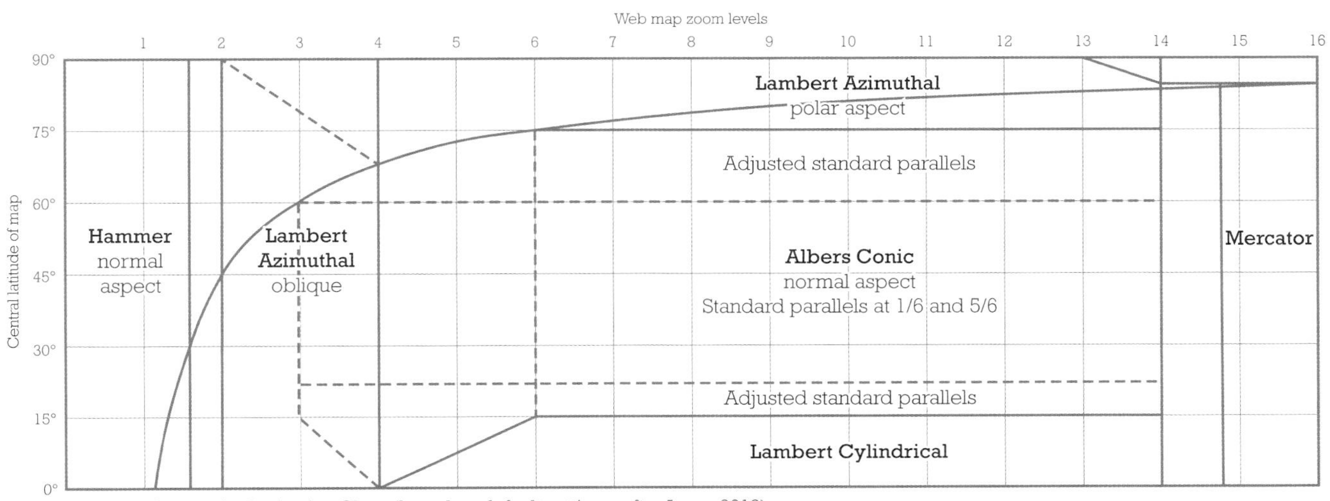


Adaptive Composite Projection Chart (based on default settings; after Jenny 2012)

# Map transformation process

## Functional process that results in a map product.

**Maps are designed and produced as part of a wider cartographic process.** The process helps define and guide mapmaking from initial idea to use. The creation of a map is the result of three major transformations which help characterise different stages. Cartography encompasses all the stages though an individual cartographer may only be involved in a small part of the process that contributes to the whole.

**The first transformation involves data collection.** This involves the capture and reporting of data used to make the map from ground survey, remote sensing, and compilation of topographic detail to the multitude of additional datasets mapped as themes such as censuses. This transformation takes the complex geographical world and effectively samples components that represent it as points, lines, areas, and attributes.

**The second transformation involves the processes of converting the raw data into a map.** These processes are broadly defined by the principles of generalisation. Data is selected (or omitted), simplified, classified, and symbolized to transform it into marks that represent the essence of the captured data. This transformation is often the primary role of the cartographer, and the result is a map.

**The final transformation occurs through map use.** A map becomes used when it is read, analysed and interpreted, and used for a specific purpose. The map reader is an integral part of the overall map transformation process since they should be able to recognise the original geographical environment, mediated by the cartographer, in the final product.

Each of the map transformations has the ability to introduce error, bias, and uncertainty and modify the final map. The capture of original data can be distorted through errors, poorly calibrated equipment, misinterpretation, and poor compilation. The mapmaking transformation can propagate these errors or introduce further errors. The processes of abstraction, although necessary, lead to considerable data loss to bring features into view. At this stage you may even make an informed choice to represent data differently, potentially for persuasive or propagandist purposes. The map reading process can add complexities since every map is seen through the particular lens of an individual. Although the map transformation process includes the fallibility of human beings at all stages, this final transformation is perhaps the one that can be controlled the least. Everybody is different. An individual's skill, experience, and needs will shape the way the map is seen.

A key role of the cartographer is to make effective decisions to reflect the first transformation properly while accounting for as much of the final transformation as possible. There are always ramifications of the decision-making process for the second transformation stage. This is particularly important when the people involved in the second transformation stage are not cartographers. Anyone has the ability to make a map but the second transformation's requirements are consistent regardless of personality or skill set. For instance, coders, journalists, and marketeers all make maps. The fact that they are not necessarily trained or practiced cartographers does not obviate the need for the second transformation process. It behoves anyone involved in making maps to learn some of what defines the second transformation to ensure they are capable of properly transforming the recognised environment into an effective map.

**See also:** Cartographic process | Data processing | Digital data | How maps are made | Mental maps

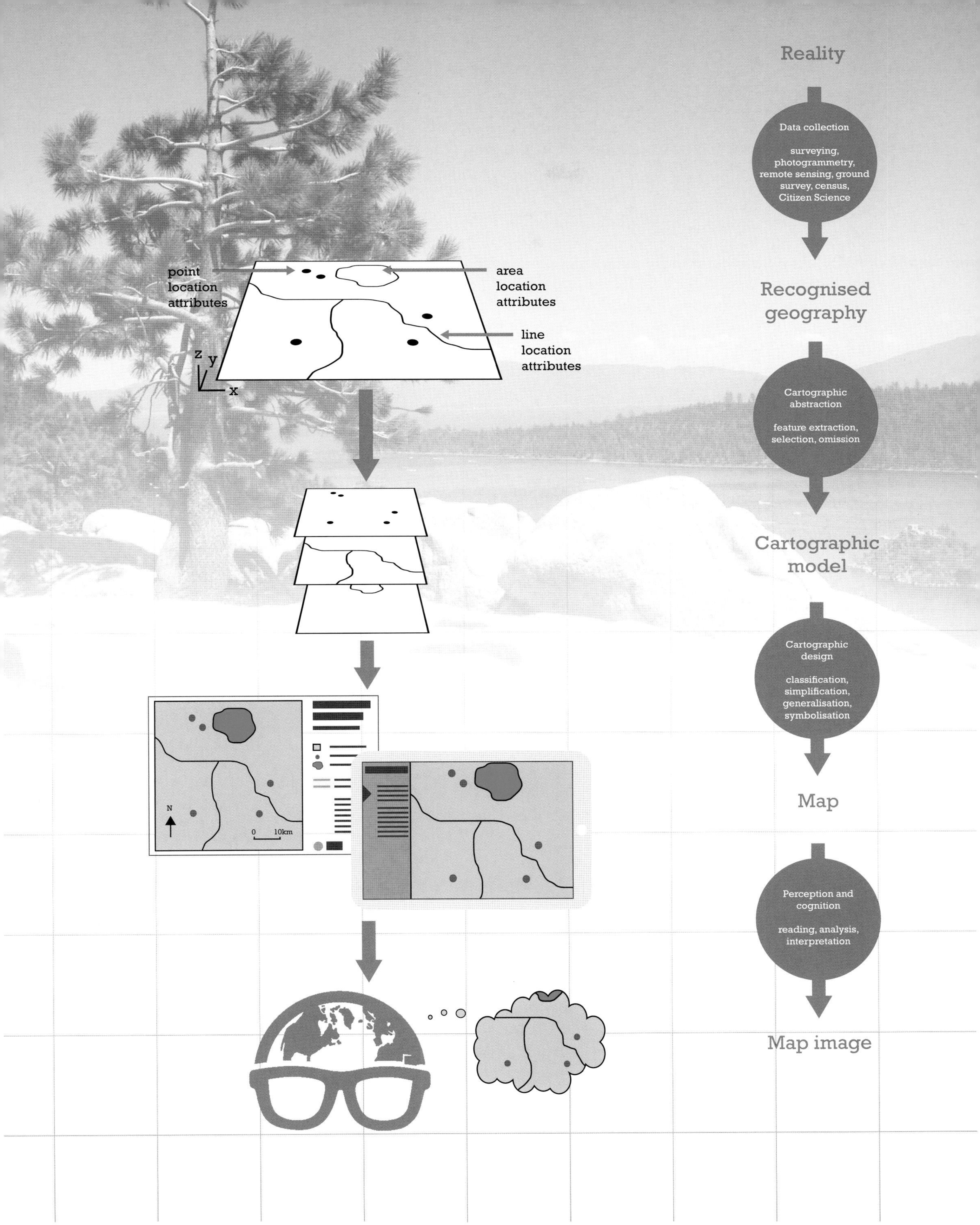
Reality
Data collection
surveying, photogrammetry, remote sensing, ground survey, census, Citizen Science
Recognised geography
point location attributes
area location attributes
line location attributes
z y x
Cartographic abstraction
feature extraction, selection, omission
Cartographic model
Cartographic design
classification, simplification, generalisation, symbolisation
Map
N
0 10km
Perception and cognition
reading, analysis, interpretation
Map image

# Map traps

## Techniques used to combat cartographic plagiarism.

**Plagiarism of cartographic works is as much a problem as is the misuse of many types of creative work.** The effort of collecting the data, and designing and producing a map is a resource-intensive process. Map companies and national mapping agencies in particular are keen to ensure that their work is properly copyrighted and isn't used without their knowledge or proper remuneration. Copyright traps—map traps—have become well practised as ways of embedding visual motifs into a map which, if copied, could be easily traced back to a particular product.

**Creating fake places might be considered a contentious practice.** Map accuracy is often taken as a hallmark of cartographic practice so the inclusion of deliberate falsehoods and fake information might be deemed unsound. Fake features are normally small, relatively insignificant, and do not impact the map's utility. Perhaps a small street is added in a residential neighbourhood on a large-scale map, or a few additional curves to a road or river on a medium-scale map, or the addition of a fake town on a small-scale map.

**Sometimes map traps appear in the form of how a feature is represented.** The initials of the original cartographer have been found in rock drawing on Ordnance Survey maps and small animals inserted on Swiss maps. These sorts of map traps are nothing more than cartographic expressions deliberately inserted for humour by the map's makers. They can work as map traps.

**The practice of using map traps has even been celebrated in literature.** One of the world's most famous literary cartographers, Slartibartfast (from Douglas Adams' *Hitchhiker's Guide to the Galaxy*), was famous not only for his award-winning coastline of Norway but for signing his name in glaciers—a map trap for any other designers of planets who might have been attempting to copy his designs.

Despite claims that it was because of human error, up until 2009 Google Maps included a fake town in England called Argleton, which had previously only appeared on printed maps from Tele Atlas. Perhaps the most famous case of cartographic plagiarism became a legal battle in 2001 when the Automobile Association (AA) of Britain paid a £20 million settlement to Ordnance Survey after it was caught plagiarizing. Proof was in the form of map traps that were embedded in the original Ordnance Survey maps that were subsequently found in the AA maps. The traps were fingerprints and based on stylistic features and proportions.

The earliest known whimsical map animal found on Swiss maps was the white spider found at the top of the Eiger from maps of the 1980s. Drawn by cartographer Othmar Wyss, it was simply an audacious attempt to inject a visual joke into the map he was creating. In a 1981 map of Interlaken, cartographer Friedrich Siegfried incorporated the picture of a face in his rock drawing, and a drawing of a climber was inserted to mask missing data on the Italian side of another of his Swiss maps. The animal theme continued when Werner Leuenberger included a fish in a lake in a 1989 map sheet, drawn as a braided stream network, and in 2016 a marmot was found on 1:25,000 and 1:10,000 maps of Aletschgletscher. Cartographer Paul Ehrlich created it before retirement in 2001 as a way of including some whimsy but without changing the map's meaning. The marmot seemed to him a perfect animal to include in this alpine environment.

The Geographers' A–Z Map Company openly claims there are about 100 map traps in the London A–Z street atlas. For instance, Bartlett Place appears, and although it names a real geographical feature, the street is named after one of the A–Z cartographers. The real place-name is Broadway Walk.

**See also:** Branding | Integrity | Old is new again | Who is cartography?

**Opposite:** A rogue's gallery of map traps.

Fake settlement of Agloe,
Delaware County NY, ca. 1930

Haggerston Park's fake ski slope
London A-Z, ca. 2005

Marmot
National Maps of Switzerland
Sheet 1269 Aletschgletscher, 2011
1:25,000

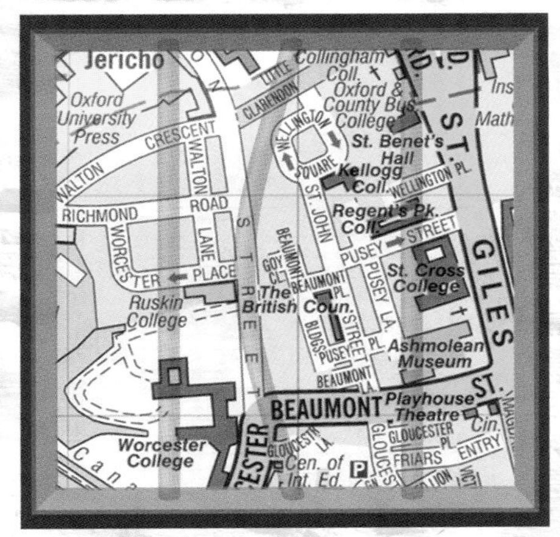


Fake street Goy Close,
London A-Z, ca. 2002

An elephant drawn into the contours,
Gold Coast Survey Department 1923
1:62,500

Fake Town of Argleton,
Google Maps, 2008

"The white spider", fearsome snow on
the north face of the Eiger
National Maps of Switzerland
Sheet 254 Interlaken, 1981
1:50,000

Hiker
National Maps of Switzerland
Sheet 39 Flüelapass, 1997
1:50,000

# Maps for and by children

## Designing to accommodate different needs.

**One of the principal ways we learn about our world is through the use of maps during childhood.** Many geographers and cartographers often point to a beloved atlas or map they recall as sparking their interest in the world around them. Rarely are these maps that we read and use during adult life. They have a different content as well as an altogether different aesthetic. Many are actually toys rather than maps first and foremost because learning through play is one of our fundamental modes of learning.

**Maps designed for children simplify and symbolise in specific ways.** They are often brightly coloured and tend to eschew much of the colour theory we might apply to adult maps. They are often highly pictorial. They use pictures of people, often standing on the maps and engaged in activities such as farming, travelling, or playing sport, for instance. The pictures might represent a particular theme or something that happens at a particular place. The people may be in national costume. Perhaps the map is filled with animals typical of that part of the world. The pictures themselves are often cartoonish in style. Typography is limited, often to brief labels and, sometimes, a little descriptive text. Ultimately, maps designed for children simplify but their design is anything but simple, and they must still support learning and understanding through the medium of the map.

**Maps made by children are also often pictorial.** Children tend to use a map as simply a component of a drawing rather than as a framework for the display of topographic or other detail. A child's rudimentary drawing skills often generate fascinating maps that spurn cartographic convention. Imagination tends to be the driving force, and although it is common to see colour used in conventional ways (green for land, blue for water), the overall designs are more likely to be illustrative and devoid of true scale and other more adult approaches to cartography.

Maps for children come in many different forms. Wall maps packed with illustrations and atlases that can be read are common mediums but map toys also bring interest and excitement to children.

Hugg-a-Planet, by Peacetoys, is a soft toy in the shape of a planet with a central message that we can wrap our arms around the world and give it a hug. It's an object that supports all sorts of play activities in young children, and the map itself is colourful and attractive. This doesn't need perfect cartography, just the sense of the shape of the world that children can come to grips with, literally.

Of course, it has a secondary purpose as an educational aid and allows children to begin to learn about geography by using the Hugg-a-Planet to show where countries and places are. Beyond its appeal to children, Hugg-a-Planet has become somewhat iconic as a metaphor for global peace. It resonates with those seeking to explore environmentalism and a pleasing object to have lying around as a cushion with meaning.

Eleanor Leia Field, age 1 (and a bit), navigating Dublin Zoo

**See also:** Design and response | Form and function | How maps are made | Hand-drawn maps | Who is cartography?

**Opposite:** *Antarctica* in *Collins Children's Picture Atlas*, by HarperCollins Publishers, 2015.

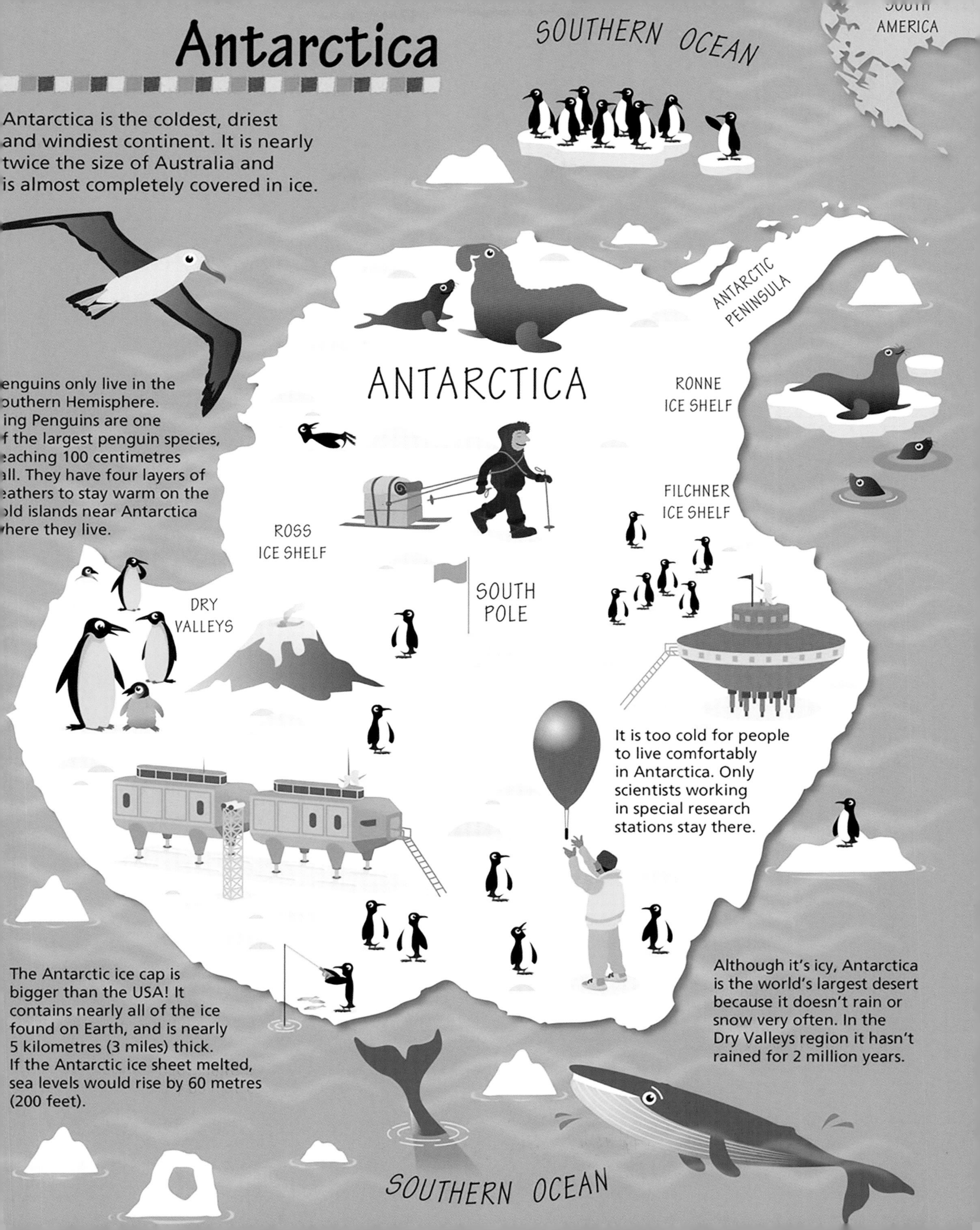
Antarctica
Antarctica is the coldest, driest and windiest continent. It is nearly twice the size of Australia and is almost completely covered in ice.
SOUTHERN OCEAN
SOUTH AMERICA
ANTARCTIC PENINSULA
ANTARCTICA
RONNE ICE SHELF
FILCHNER ICE SHELF
ROSS ICE SHELF
SOUTH POLE
DRY VALLEYS
enguins only live in the outhern Hemisphere. ing Penguins are one f the largest penguin species, eaching 100 centimetres all. They have four layers of eathers to stay warm on the old islands near Antarctica here they live.
It is too cold for people to live comfortably in Antarctica. Only scientists working in special research stations stay there.
The Antarctic ice cap is bigger than the USA! It contains nearly all of the ice found on Earth, and is nearly 5 kilometres (3 miles) thick. If the Antarctic ice sheet melted, sea levels would rise by 60 metres (200 feet).
Although it's icy, Antarctica is the world's largest desert because it doesn't rain or snow very often. In the Dry Valleys region it hasn't rained for 2 million years.
SOUTHERN OCEAN

# Maps kill

## Cartography can, literally, be a matter of life and death.

**Maps are powerful objects and can sometimes become a major factor in life and death.** For many maps, the cartographer has little to be concerned about other than trying to be truthful, impartial, and communicate some information. Sometimes a map might offend if it takes a particular editorial view. Rarer, a map may be crucial in decision-making that places human beings in danger. Conversely, they may support the preservation of human life, for instance through humanitarian mapping of natural disasters that support emergency planning, evacuation, and resettlement.

**Maps are important tools for military use.** Their use in planning as well as ground operations actively supports the fundamental needs of warfare. Historically, maps have been used to plot troop movement and plan out battles. Troops on the battleground must have a basic knowledge of orienteering with maps. Trench maps were crucial in identifying snipers and observation posts. Topographic maps showing gun ranges and compass bearings assisted the identification of targets. In modern drone warfare, digital cartography and accurate positioning are crucial to delivering a drone strike on target. So-called 'friendly fire' and mistaken identities are often a result of poor cartography and targeting as well as misidentification.

**Satellite navigation and digital mapping has also been implicated in deaths.** Although it can be amusing to see reports of satellite navigation errors that send trucks down narrow paths, people following turn-by-turn directions and ending up in a river, or people aimlessly walking along major highways, there are more severe consequences. We implicitly trust the databases and digital maps that drive satellite navigation to be accurate but, of course, they contain errors which often place people in danger. Humans are also fallible and can enter the wrong address or location. There have been numerous reported cases of ambulances going to the wrong address, sometimes many miles away from the intended victim who has later died.

**See also:** Critique | Dysfunctional cartography | Error and bias | Integrity

**Opposite:** The Cavalese cable car disaster.

In 1998, 20 people died in the Cavalese cable car disaster when a United States Marine Corps aircraft cut through a gondola cable in the ski resort while on a training mission. A gondola fell some 80 m to the ground, killing all 20 passengers. The plane itself suffered light damage to its wing but was able to return to base. The cause of the accident stemmed from the maps the crew of the plane were using not showing the gondola or the aerial cables.

The accident wasn't simply a case of cartographic error. Italian military maps, as well as commercial road maps of the area, did mark the ski lift and the cables. However, the pilots were not in possession of the Italian maps since the United States had a policy of not using maps made by foreign countries. Instead, the crew had been given an American military chart that did not show the ski lift or the cable. The ski lift had been in existence for 31 years at the time of the accident, linking the valley floor to the peak of Mount Cermis.

The inquiry subsequently found the plane was flying too fast and too low, but if the map had shown the existence of the cables, then the pilot would at least have had accurate information upon which to base decisions. The events surrounding the accident cannot wholly be blamed on the map but it played a crucial part. The accident was a result of human failings. Decisions taken by the pilot to fly too fast and low were contributory. Decisions by the Pentagon to prohibit the use of local maps places greater responsibility on derived mapping to support specific needs. Ultimately, both mapmakers and users must be sceptical and critical of the map they are using since maps are, ultimately, the product of human creation and can contain errors. Some can kill.

# Mashups

## Combining digital layers of mapped information to create a new product.

**Perhaps the biggest impact of web technologies on cartography has been the rise of the map mashup.** A mashup is constructed by people taking available data and marrying it with other data or data they provide themselves and then repackaging it. Data often comes in the form of map services. A typical mashup might use a third-party basemap and an operational layer of content. Additional layers such as text or switchable layers of content might also be incorporated.

**Map mashups are a democratised form of mapping.** They allow anyone with the basic technical skills to engage in the activity of mapping and this, in turn, leads to a proliferation of maps on all kinds of subjects. Some support citizen science, while some merely overlay the location of houses for sale or some other spatial dataset that can be scraped, or used. These sorts of mashups are often characterised by a weak appreciation of cartography since they rely heavily on the way in which a map service has been pre-rendered.

**Mashups are aggregated.** They bring together content to be visualised to reveal new insights based on the precepts of Web 2.0. Web standards introduced the notion of shared common data formats. APIs further support mechanisms that you can use to mix and match different organisational services to develop new services, products, and information.

**Mashups have developed over time.** The 'map sandwich' metaphor was born out of frustration that many mashups simply don't work because too much information obscures the basemap. Mashups can quickly become unreadable. Rather than having two layers, a map sandwich is built from multiple layers, often placing the operational layer between terrain and an overlaying reference layer (boundaries and names). This is more akin to the compilation of a traditional map but extends the mashup concept. With the advent of customisable vector map tiles, the ability to tailor different layers of a mashup supports more nuanced cartography and supports the ability of mashups to become fully formed cartographic products.

**See also:** Basemaps | Information overload | Page vs. screen | Web mapping

**Opposite:** *Housing Maps*, the very first Google Maps mashup, by Paul Rademacher, 2005.

Google released Google Maps on February 8, 2005, after it acquired Where 2 Technologies, a company formed by Jens and Lars Rasmussen. On its own, Google Maps was a slippy map, which allowed users to find places on the map. A critical moment in the evolution of the map mashup came when Google Maps was reverse-engineered to position other content on top of the map service. Paul Rademacher integrated Google Maps onto his own website and used it as a basemap to overlay data from Craigslist®. This created a map that showed homes and rentals from Craigslist on top of Google's map data. The map mashup was born.

The notion of mashing up one dataset with another was almost a secondary process. The simple idea was that real estate listings were best viewed and browsed on a map rather than being read in list form. As with many datasets that have a spatial component, this notion holds true, and the product becomes a descriptive map of a theme of information seen in concert with contextual topographic reference detail.

Rademacher had used data without asking permission, identifying the web as being open and supporting the development of new products based on existing ones. Although this is a possible matter of interpretation, Google realised the opportunity for having their map be used to support the development of map mashups, and they subsequently released an API in June 2005. The API specifically supported this approach, allowing developers to integrate Google Maps into their own websites and overlay their own data. The release of the Google Maps API effectively emancipated digital basemap data available for others to use.

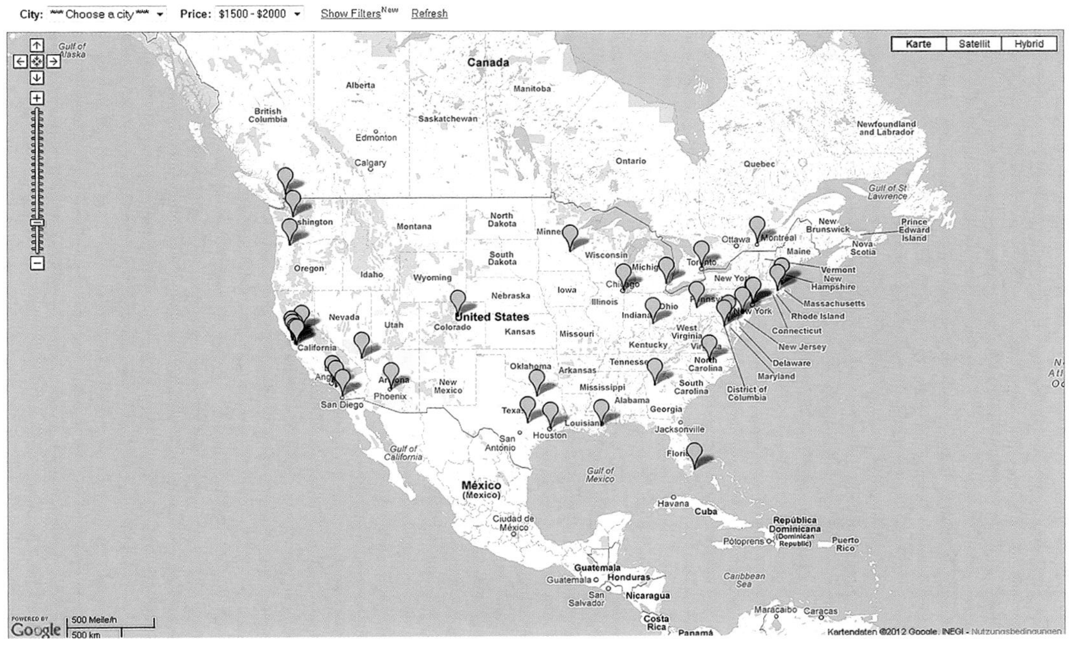

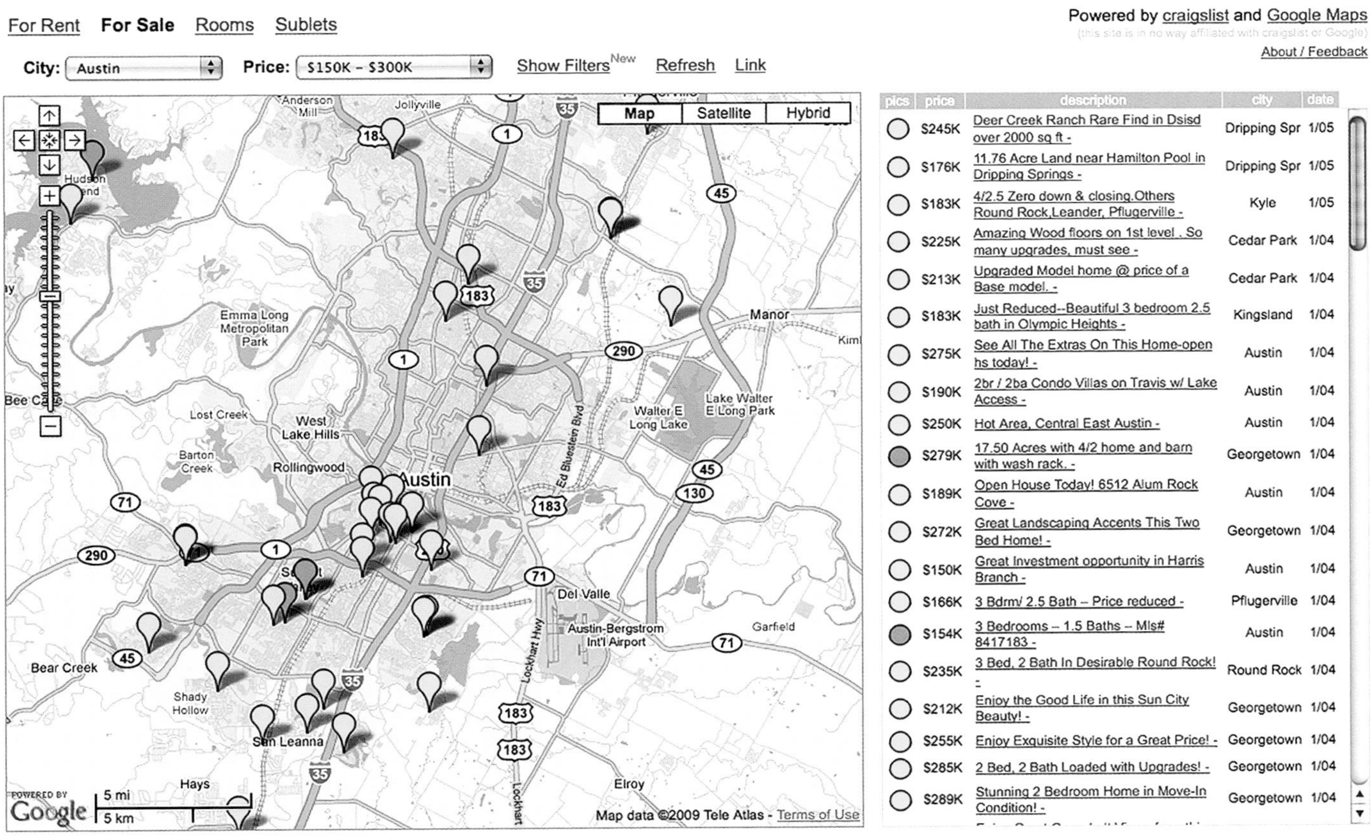

For Rent
For Sale
Rooms
Sublets
City:
Price: $1500 - $2000
Show Filters New
Refresh
Powered by craigslist and Google Maps
About / Feedback
Karte
Satellit
Hybrid
Canada
United States
México (Mexico)
Gulf of Mexico
Drag the map with your mouse, or double-click to center.
Click on an icon to select a city.
In each city, click on the and icons to see listings. You can also click the and icons next to each listing.
Yellow icons have pictures.
Set the price range using the drop-down menu above.
500 Meile/n
500 km
Kartendaten ©2012 Google, INEGI

For Rent
For Sale
Rooms
Sublets
City: Austin
Price: $150K - $300K
Show Filters New
Refresh
Link
Powered by craigslist and Google Maps
(this site is in no way affiliated with craigslist or Google)
About / Feedback
Map
Satellite
Hybrid
Austin
Manor
Emma Long Metropolitan Park
West Lake Hills
Rollingwood
Lost Creek
Barton Creek
Bee Cave
Walter E Long Lake
Lake Walter E Long Park
Del Valle
Austin-Bergstrom Int'l Airport
Garfield
Bear Creek
Shady Hollow
San Leanna
Hays
Elroy
Jollyville
5 mi
5 km
Map data ©2009 Tele Atlas - Terms of Use
pics
price
description
city
date
$245K Deer Creek Ranch Rare Find in Dsisd over 2000 sq ft - Dripping Spr 1/05
$176K 11.76 Acre Land near Hamilton Pool in Dripping Springs - Dripping Spr 1/05
$183K 4/2.5 Zero down & closing.Others Round Rock,Leander, Pflugerville - Kyle 1/05
$225K Amazing Wood floors on 1st level . So many upgrades, must see - Cedar Park 1/04
$213K Upgraded Model home @ price of a Base model. - Cedar Park 1/04
$183K Just Reduced--Beautiful 3 bedroom 2.5 bath in Olympic Heights - Kingsland 1/04
$275K See All The Extras On This Home-open hs today! - Austin 1/04
$190K 2br / 2ba Condo Villas on Travis w/ Lake Access - Austin 1/04
$250K Hot Area, Central East Austin - Austin 1/04
$279K 17.50 Acres with 4/2 home and barn with wash rack. - Georgetown 1/04
$189K Open House Today! 6512 Alum Rock Cove - Austin 1/04
$272K Great Landscaping Accents This Two Bed Home! - Georgetown 1/04
$150K Great Investment opportunity in Harris Branch - Austin 1/04
$166K 3 Bdrm/ 2.5 Bath -- Price reduced - Pflugerville 1/04
$154K 3 Bedrooms -- 1.5 Baths -- Mls# 8417183 - Austin 1/04
$235K 3 Bed, 2 Bath In Desirable Round Rock! - Round Rock 1/04
$212K Enjoy the Good Life in this Sun City Beauty! - Georgetown 1/04
$255K Enjoy Exquisite Style for a Great Price! - Georgetown 1/04
$285K 2 Bed, 2 Bath Loaded with Upgrades! - Georgetown 1/04
$289K Stunning 2 Bedroom Home in Move-In Condition! - Georgetown 1/04

# Measuring direction

**Direction, or azimuth, can be measured using angles from a known starting point such as geographical or magnetic north.**

**As well as forming the shortest path between two points, great circles are important because they also contribute to the measurement of direction on Earth.** As the arc of a great circle crosses Earth, it intersects with meridians in turn. The angle made between each meridian and the arc is termed an *azimuth*. Azimuths are normally measured in sexagesimal units clockwise starting with geographical north as the origin although it should be noted that magnetic north may also be used to determine an azimuth.

**The angle made between the arc of a great circle and each meridian changes.** This presents difficulties for navigational purposes. For example, mariners or airlines will ideally want to travel along arcs of great circles since this is the shortest and most efficient route between origin and destination. However, their route must be constantly adjusted as the azimuths between their position on the arc and geographical north change constantly. This creates an impractical navigational system.

**Alternatively, lines of constant direction, *loxodromes* or *rhumb lines*, can be defined.** These lines intersect all meridians between origin and destination at the same angle. A small circle that intersects parallels as well as the equator will always intersect meridians at right angles. A loxodrome is formed where the small circle intersects Earth's surface. Azimuths remain constant because meridians converge at the poles.

As with any map, the projection is important to support the function. For measuring direction, the key to navigation, a conformal projection that maintains angles is necessary. The Mercator projection is a good choice in this context because its special property is conformality.

On a Mercator projection, a rhumb line will be drawn as a straight line whereas the arc of a great circle (the shortest path) will be drawn curved.

On a planar surface, the rhumb line will give the shortest path between two points. At low latitudes or over short distances, this is useful for plotting the course of a vehicle. However, over longer distances and/or at higher latitudes, a great circle route is much shorter than the rhumb line between the same two points.

'Is there a cartographer on this flight?' isn't often heard. Being on an airplane and hearing the dreaded tannoy announcement 'Is there a doctor on board?' usually fills passengers with dread. For those with a PhD in the mapping sciences, it's embarrassing to have cabin crew ask you specifically, but many flights might make use of those of us who know a bit about cartography. I've lost count of how many times a fellow passenger has studied the in-flight map and asked why we're flying in a curve to our destination. Seeing the aha moment when it's explained is gratifying.

**See also:** Earth's framework | Earth's vital measurements | Spatial dimensions of data | Which way is up?

Great Circle
Rhumb Line
Seattle
London
Great Circle
Rhumb Line
Seattle
London
Seattle SEA

# Mental maps

## Intangible maps that we form in our minds.

**Some of our earliest maps form in our minds as children.** They place us at the centre, and our world gradually forms around them as we gain experience and understanding of the world around us. We continue forming images of space and place into adult life and often fall back on them to help navigate an unfamiliar place through recall. We use landmarks to act as breadcrumb trails.

**Mental maps are a function of our experience.** They are constructed through what we know and where we go. But more than that, they are a function of our education, the books we read, the films we watch, and the places we visit. Mental maps can transcend our physical surroundings, and we can imagine far-off places, even ones we've never experienced firsthand. Self-centred views tend to become replaced by geocentric views as we position ourselves in the world rather than framing the world around us. Understanding relative position becomes easier as we see places in relation to one another and not necessarily connected by a discrete route with landmarks.

**Mental maps are used in everyday tasks.** We route ourselves in towns and inside buildings with reference to our mental maps. We also share them with others through describing or even sketching, and we can assign measurements such as distance, time, or direction to the detail. Most mental maps will form planimetrically. We're not particularly good at thinking in 3D and imagining the world as a sphere. This has much to do with the fact that on a daily basis our locale is essentially planar.

**Mental maps are often amusing.** Maps that purport a humorous view of one part of a country or region to another are borne out of stereotypical views of others and other places. Cartoonists have often used this form of mapping to support satirical commentary.

Drawing mental maps reveals the difficulty people often have in translating a picture in their mind to a piece of paper. Geographically and geometrically, the drawn representation will often contain errors, even if relative detail is accurate. Most will draw the greatest amount of detail for places they know best. Accuracy is also likely to be higher for places with which you are most familiar. Those places will likely be drawn larger, because they are more important to you. The remainder of the map will be less accurate, be drawn smaller, and have greater distortions and errors. Put simply, you'll likely draw and emphasise places you know and care about more than those you don't.

Many examples exist of drawn mental maps that evidence how people see parts of the world. In their classic text 'Mental Maps', Peter Gould and Rodney White explore some of the characteristics of differences. Asking people to draw maps of where they would like to live differed between people who lived in different areas. Regional differences became apparent. Home was normally seen as a place to live but other areas that had cultural significance became important even if the person had never visited there. Division and environmental segregation were also evident, particularly in studies of schoolchildren whose mental maps revealed deep fear, anxiety, and racial boundaries ascribed to physical features in their real environment (Gould and White 1986).

**See also:** Emotional response | Map transformation process | Seeing

**Opposite:** *Ye Newe Map of Britain*, by the Doncaster and District Development Council, ca.1975.

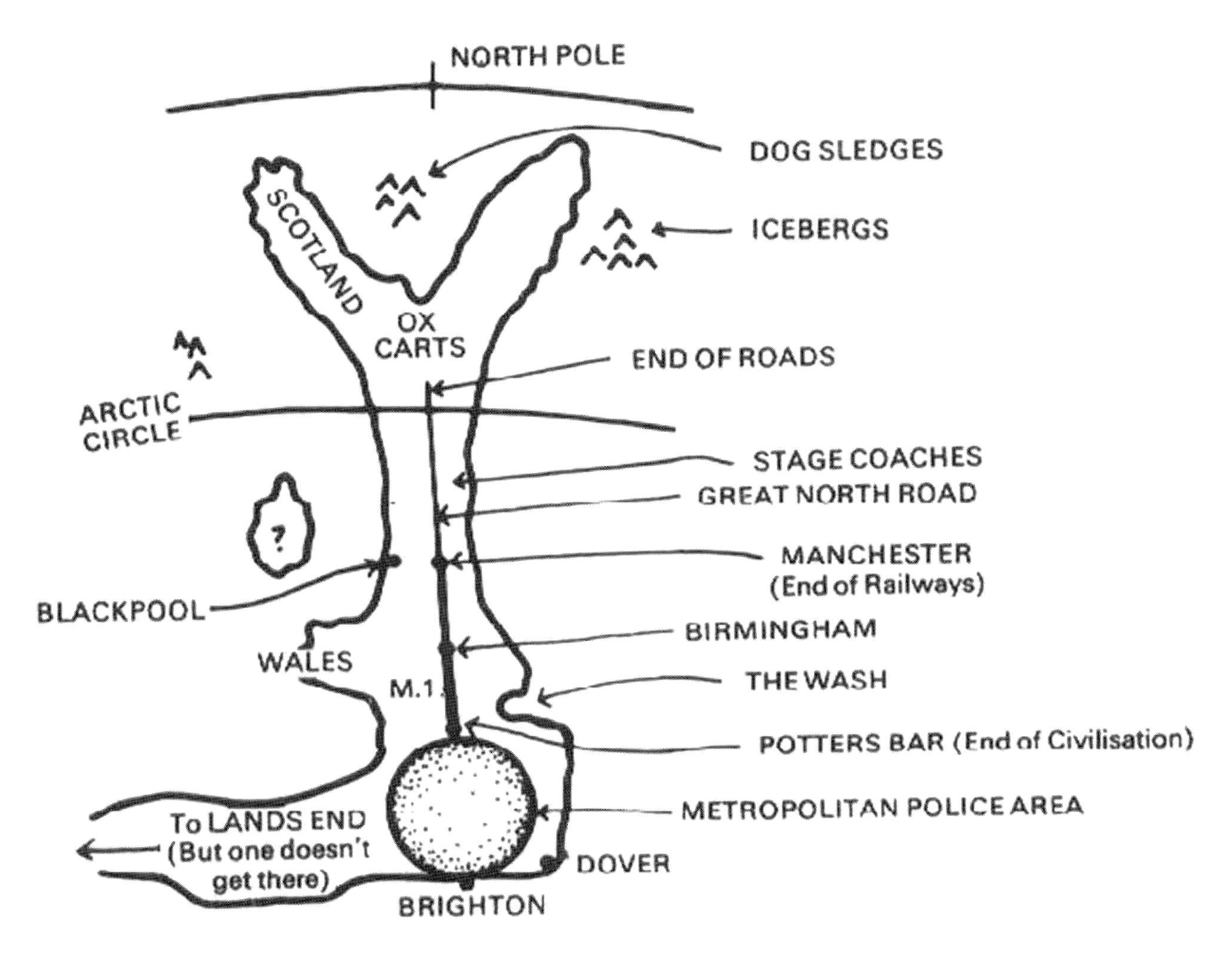


Stereotypical views shape mental maps. Here, drawn from the perspective of a Londoner and showing their view of The North, this 1970s *Ye Newe Map of Britain*, by the Doncaster and District Development Council, was designed to challenge spatial biases and encourage southerners to make more informed decisions.

The shape of the country is massively distorted with the southeast exaggerated and everywhere else drawn smaller and less important. London is defined by an arbitrary boundary reflecting the police force administrative area, which amusingly touches Brighton (often known as London on Sea due to the large influx of London residents to the seaside on a hot summer's day). Wales is but a pimple, and Scotland's shape bears no resemblance to reality. The entire geography of the UK hangs off a single road — the M1, the main northbound road but by no means the only one. Indeed, the M1 finishes in Leeds and doesn't even go through Birmingham or Manchester where it seems to magically turn into the Great North Road. Scotland is so far away it's within the Arctic Circle and not at all far from the North Pole.

# Mixing colours

Mixing from the primaries allows you to specify your own colours.

**Mixing colours using the RGB (red, green, blue) or CMY (cyan, magenta, yellow) primaries allows you to generate your own colours and colour palettes.** Selecting colours from preconfigured charts sometimes leads to them taking on a familiar appearance as certain schemes are often associated with specific software or colour charts. To add to your colour palette or develop your own colours, mixing provides the ultimate flexibility. In a computing environment, colour is mixed using the RGB colour cube. When printed, these colours are translated to a CMY equivalent, but for better reproduction, mixing directly in CMY is more useful if the intended output is print.

**When mixing hues using RGB or CMY specifications, it is useful to view the colour cube with the white point in the centre.** This creates a hexagon with the RGB and CMY primaries at the points of the hexagon. New hues are created by mixing proportions of adjacent hues. For instance, in RGB a yellow is constructed from equal amounts of red and green (on the scale from 0 to 255). Equal amounts of blue and red create magenta, and equal amounts of blue and green create cyan. The same hexagon can be used to mix CMY if percentages are used. For instance, using the same percentage of yellow and cyan will create a green. Using more cyan than yellow will create a bluish green.

**Once a hue has been mixed, you can change the value and saturation.** Higher percentages of CMY in subtractive mixing produces darker colours. Conversely, higher values of RGB in additive mixing yields lighter colours. Saturation is controlled by adding the third primary to the mix and controlling relevant quantities. For instance, large differences in primary colour amounts produce more saturated colours whereas nearly equal amounts of the three primaries will give desaturated colours. Desaturated colours can also be mixed by adding percentages of black (K) to the CMY palette to give a four-colour process printing specification.

General tips for RGB mixing:

- Set hue using one or two primaries.
- Use equal amounts for a purer hue.
- Higher RGB values produce lighter colours.
- Control saturation by adjusting the lowest RGB number.
- Make systematic changes to RGB colours in a sequence to improve perceptual appearance of visual steps between adjacent colours.
- Use larger differences between RGB values for darker colours (lower values) to emphasise visual steps.

General tips for CMY mixing:

- Set hue using one or two primary inks.
- Higher percentages of the primaries produce darker colours.
- Changing the lightness of yellow produces very little difference.
- Add black to desaturate colours or adjust the saturation by varying the primary with the smallest value.
- Make systematic changes to CMY colours in a sequence to improve visual appearance of visual steps between adjacent colours.
- Use larger differences between CMY values for darker colours (higher percentages) to emphasise visual steps.

**See also:** All the colours | Colour charts | Colour cubes | HSV colour model | Perceptual colour spaces

CMYK colour mixing

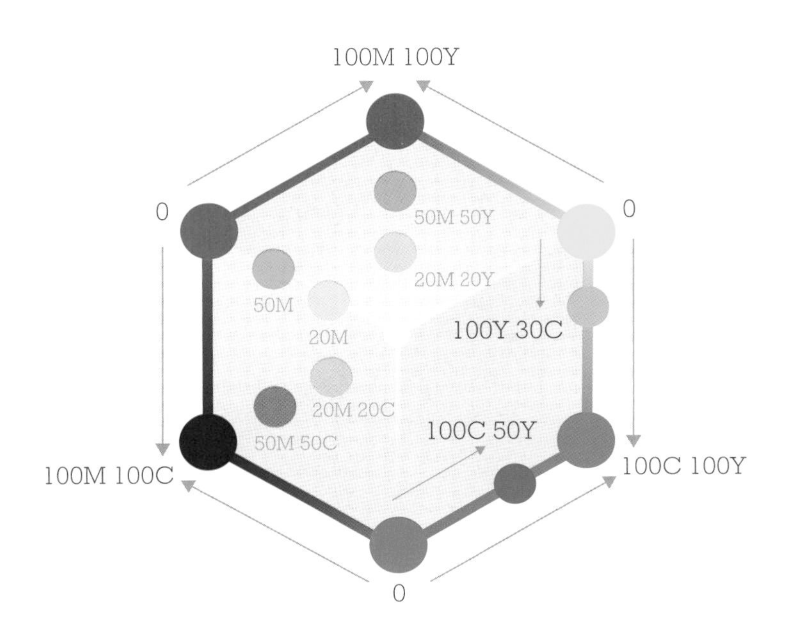


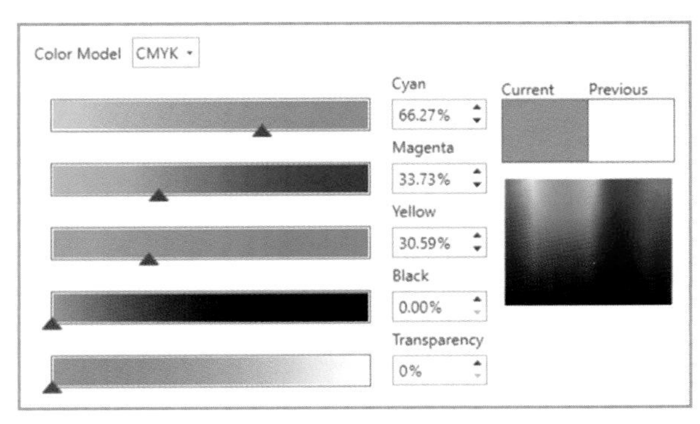


RGB colour mixing

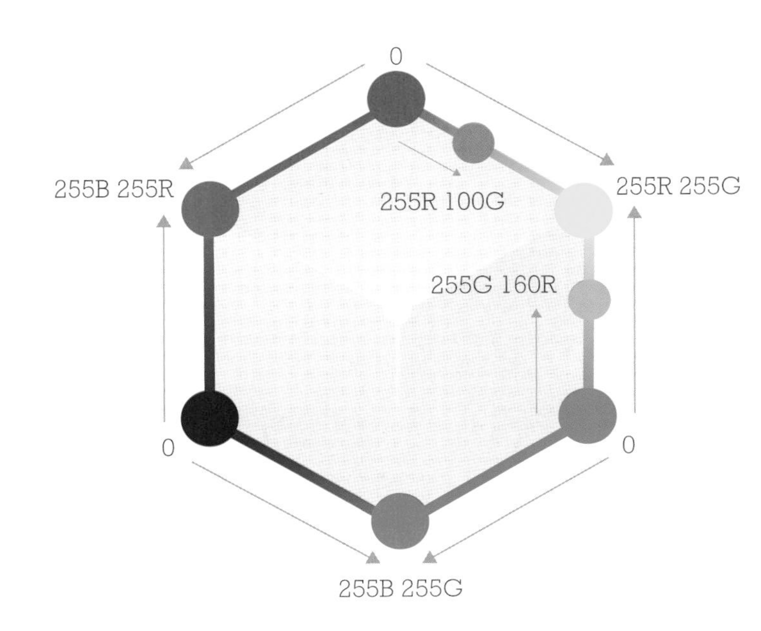


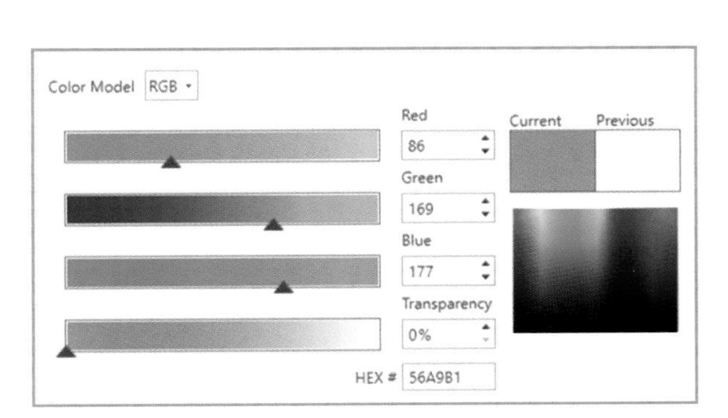

# Mobile mapping

## Mobile maps put you at the centre of the map. They move around you and give context to your location.

**Maps have become location aware as you move around, and they update to represent features using real-time content that is specific to where you are.** It's a paradigm shift from the paper map when you were not at the centre of the sheet. In fact, all too often you found yourself on the edge of a sheet and very often onto a separate sheet. So maps are now edgeless and in some respects scaleless too as you can move swiftly through scales from small to large and back again. Maps now have movement that is tailored to you as an individual user. More than that, cartographers used to map things that didn't change very much. Now, maps show rapid change. Basemaps are updated overnight so when you switch on your maps in the morning they contain the very latest topographic detail.

**The content you see on maps is now heavily derived from sensor networks, live data collection streams, and through wireless transfer.** This rapid data collection and update has largely been supported through automating many of the collection and delivery mechanisms for map data and eliminating human input in the sifting and sorting process. This has meant maps can move rapidly and has largely solved the age-old problem of map printing, production, and storage. This collapse of time has meant mapmaking has gone from years to months to weeks to seconds in only a very short space of time.

**Mapping movement itself has also changed as technologies now support interaction, animation, and immersible 3D.** Static cartography was always challenged to support people's need to recover information. Pop-ups and other gesture-based map interactions support data discovery and truly exploratory mapping needs. Maps now go beyond showing—they allow people to investigate more deeply and interact with the world as they travel through it. Maps also now include new interface elements, and this has meant that the map itself no longer has to struggle to represent everything or, indeed, to omit detail to preserve clarity. The map might now be seen as the interface to data discovery on the go.

Mobile mapping might also be used to describe the process of collecting data from a mobile vehicle. This sort of data collection has become routine with vehicles being equipped with photographic, radar, laser, lidar, and other remote-sensing equipment. These image sensors rapidly collect data as the vehicle moves, and that data can be streamed back to an organisation and used to rapidly make or update maps, datasets, or imagery.

GPS has facilitated these mobile mapping systems and has led to direct mapping without the need for detailed and time-consuming post-processing.

The consumption of mapping applications in the real world has also brought about development in the areas of augmented reality. This provides a combined view of the real world with additional geoenabled content. For instance, a mobile phone's camera can be used as a viewfinder, and the device GPS, gyroscope, and compass can be used to calculate position with content appropriate to that location supplied as an overlay for the image. This might be maps, retail information, adverts, or any number of other location-based services.

**See also:** Imagery as background | Page vs. screen | UI/UX in map design | Web mapping

**Opposite:** Navigation with contextualised content on a mobile map.

me: 7:09 pm (JST)
Temperature: 26°C
Wind: 2 m/s
N
Matsubara Dori
143
Temple 100m
Hotel 20m
Restaurant RSVP 7:30 25m
Hotel 15m
Restaurant 5m
Restaurant 1m
つけもの処
伊勢屋
京料理・仕出し
旅館
ちとせ
民宿 ちとせ

# Multivariate maps

## The display of two or more variables on the same map, usually with symbology that visually encodes data combinations.

**Multivariate maps display two or more variables at once on the same map display.** This is usually through a symbol design that shows the combination of variables and not simply the display of multiple layers on the map. In order to manage complexity, the number of variables shown is normally two to three. Bivariate choropleth maps are perhaps the most common form, in which two variables are mapped using different sequential colour schemes, with each individual enumeration area shown in the colour resulting from a mix of the two colours that represent the two variables. Trivariate choropleths can also be produced in a similar way through the mixing of three sequential colour schemes.

**Legends for multivariate maps can become unwieldy, because with each additional class for each variable, the number of possible combinations increases exponentially.** For a bivariate choropleth map with each variable classified into three classes, the map, and legend, results in nine possible combinations. For a four-class map, there are 16 combinations and so on. The legend is vital for this sort of map though as the complexity of the map would be impossible to decipher otherwise. Because of the complexity and problems of interpretation, it is rare to see bivariate choropleths of more than the three- or four- class variety although technically it's possible to combine two unclassed choropleths, resulting in a cornucopia of colour combinations.

**Other map types can also display multivariate relationships.** Bivariate or multivariate dot maps are also possible where different variables are encoded with different hues or hue/shape combinations. The resulting combinations should at the least remain discernible on the map but the overall effect provides a visual mix of the symbols and colours to create a unique area fill for the enumeration areas. Bivariate flow maps display relationships with the width of a line representing one variable and the hue, value, or saturation of the fill of the line representing another. Similarly, proportional symbol maps can encode different variables through size and fill.

Trivariate maps, where three variables are mapped, are best organised by using a triangular graph. Three connected variables are plotted on each side of an equilateral triangle in even gradations and in the same direction. It is only possible to map percentages so, effectively, if you know two of the values, then a third could be imputed. Symbol shades are usually formed by three hues varying light to dark.

Proportional (or graduated) symbol maps are often used in a bivariate or multivariate form. With size of the symbol encoding one variable, its fill or outline (or both) can be modified with different hues or values to represent a second variable. Bivariate proportional symbol maps can often be more successful at displaying the functional relationship between two variable than a bivariate choropleth. More information can be added through the creation of nested bivariate proportional symbols (to perhaps reflect two time periods) or even the use of the circular shape of a proportional symbol to contain a pie chart or doughnut chart.

Symbol overload can be a difficulty in designing multivariate maps. It is simple to design symbols where every dimension—colour, value, saturation, shape, size, rotation, transparency—can denote a single variable. The ability to tease out complex interrelationships between variables is not well supported by simply building complex symbology. Analysing data, and then mapping the key relationships using a bivariate technique will enable the map reader to decipher the relationships more easily.

Caution is advised when using graphs as symbols. For instance, pie charts are popular but often there's little about the spatial pattern that a mass of pie charts can reveal.

**See also:** Descriptive maps | Graphs | Isotype | Simplicity vs. complexity | Small multiples | Space-time cubes

**Opposite:** *Five Years of Drought* by John Nelson, 2016.

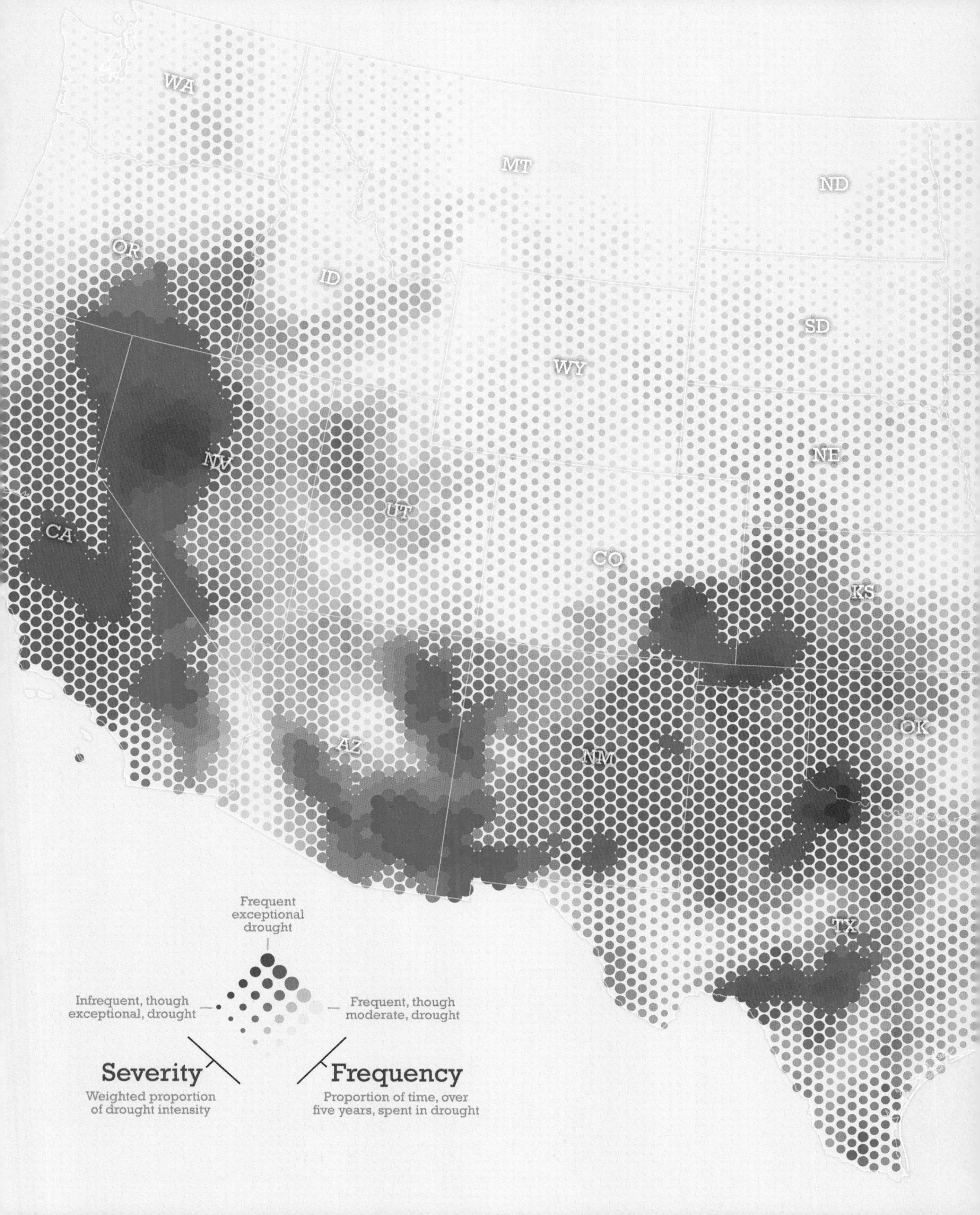
WA
MT
ND
OR
ID
SD
WY
NE
NV
UT
CA
CO
KS
OK
AZ
NM
TX
Frequent exceptional drought
Infrequent, though exceptional, drought
Frequent, though moderate, drought
Severity
Weighted proportion of drought intensity
Frequency
Proportion of time, over five years, spent in drought

## Islandia

Abraham Ortelius

1603

This is a wonderful example of a map that seamlessly combines an amazing level of cartographic detail considering when it was made, with sublime artistic content.

The land mass of Iceland features finely crafted shaded mountain and glacier topography, vignetting of the coast, with similar attention to comprehensiveness and accuracy for the numerous fiords, offshore islands, and human settlements.

The overall impression is of an inaccessible land barrier, a safe place away from the fantastic beasts that festoon the surrounding waters. These range from the plausible 'sea-calf' to hybrid creatures that could descriptively be called 'whale-hogs', 'lizard-rays', or 'mer-horses'. In fact, the entire island, perhaps by design—the shape is still some way off the modern map representation of Iceland—takes on the image of a monster, a dragon that has fire in its belly. This is Mount Hekla erupting, a reminder of the unstable tectonic foundation of this supposed haven.

—Antoni Moore

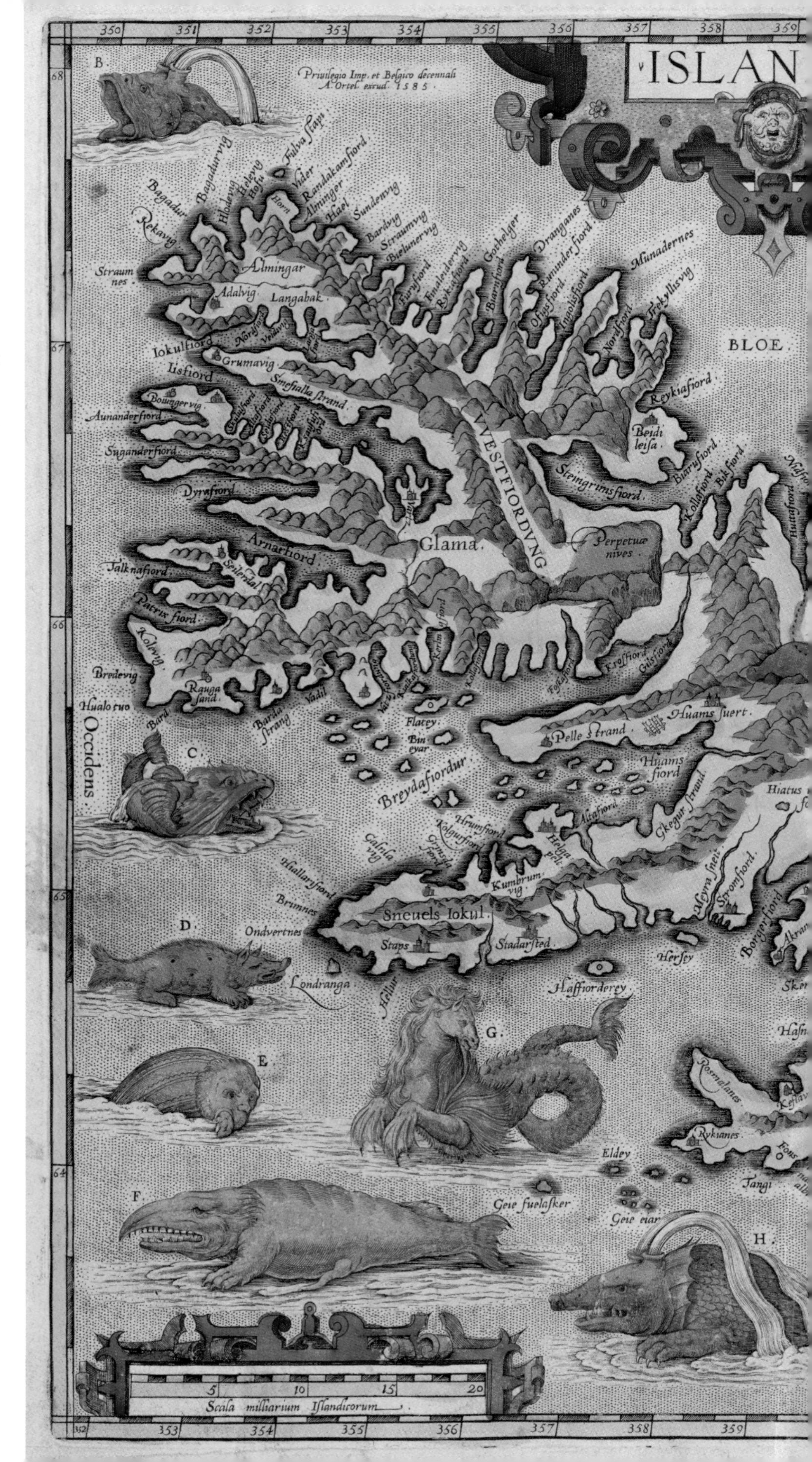

Septemtrio
A
Grims ey
Rauda gnupur
Rafsn haffn
Fulmungavig
Langanes prom.
Sumingavig
Kolla fiord
Skautuvig
Fagranes
Grimelsfiord
Midfiord
Finnafiord
Sandvig
Digranes
Q.
Strand
Hof
Vopnafiorder
Kurbar
Iokus a
Balanes
Reydar fiord
Skirdu closter
Fliotzdallr
Garavig
Beru fiord
Pap ey
Oriens
Aradal
AVSTLENDIN GAFIORDVNG
Hierskeyd
Iokuls a
Breid
Skin eyer
Brolangs eyer
Horna fiord
Horn
P.
Ingols hofdi
N
O.
Astuta vulpe cularum vena tio, in nidis volu crum inuestigandis atque diripiendis
Medalland
Kirke bar closter
Pyrkiebar closter
Skabt a.
Fiske notn
Hekla perpetuis damnata estib. et niuib. horrendo boatu lapides euomit
Mydals Iokul
Eyafialla Iokul
Solheima Iokul
Oddi
Vaccæ marinæ
Corui, et falcones albi
Iokul a
K.
Eldor
WESTMANNA EIAR
M.
I.
L.
Moridies
Equorum tanta hic velocitas, vt continuo cursu 20. milliaria conficiant.
Eyrarbach
Porlarks haffn
AVLVOS
Seluopa
Fontes ferui dissimi
SKALHOLT sedes episcopalis, cui adiuncta est schola
Blaskoger heyd
Langedal
SVNDLEN DINGAFIOR DVNG.
Kluka
Skialbred
Reykia Dallur
Getlands Iokul
Arnafelds Iokul
His notis distinguitur limes inter vtramq; dioecesim
Sand Iokul
Fodinæ sulphureæ præstantissimæ
Suartar notn
Mokrufeld
Munke tuere closter
Bardur dalur
Grenested
Muli
Mynotn
Huseuig
Rykia heydur
Lundey
Flat ey
Rolsker
Huallatursfiord
Huallsfiord
Skialfandafiord
Eyafiordur
Huisey
Olafsfiord
Heidisfiord
Siglunes
Siglesfiord
Skagafiord
Kolbeins dalur
Surbak dalur
HOLA, altera sedes episcopalis, cum schola
Modur valler closter
Holgur dalur
Hialta Dalur
NORDLEN DINGAFIOR DVNG.
Sub eodem tecto homines, canes, sues, et oues
Reme sted closter
Malm ey
Drang ey
Hofs os
ILLVSTRISS. AC POTENTISS. REGI FREDERICO II DANIAE, NORVEGIAE, SLAVORVM, GOTHORVMQVE REGI, ETC. PRINCIPI SVO CLEMENTISSIMO, ANDREAS VELLEIVS DESCRIBEB. ET DEDICABAT.

# Navigating a map

## The way we read and recover information changes with technology.

**Communication evolves with different technological advancements.** Books themselves have not changed dramatically over time but the mechanisms we use to read them certainly have. The move from a traditionally bound paper product to e-readers and web-based content has implications for how we read. The same is true for maps as developments in web technologies and hardware devices now deliver maps digitally in addition to paper maps and atlases.

**Paper maps and atlases might be considered a mature technology.** They are well known and understood, and designing for the medium is similarly developed. It's a reasonable assumption that most people would recognise and be able to find their way around a paper map (this doesn't imply an ability to read a map though). However, the shift to the screen as a medium may be more recognisable for younger people who are less used to handling paper products. The counter may be true for older people who may be less familiar with screens. In some sense, such changes in technology create a phase of transition where different users are more comfortable with the products they are most familiar with.

**Page and screen are no better than one another but they are different.** The printed product has specific physical limitations that constrain design but also give the reader a minimalist and familiar experience. Using a screen to view a map creates additional demands on the reader including hand movements, gestures, and the ability to search using keywords. The map is therefore navigated in a different way and so its design must include an assessment of how map readers will move around the map rationally.

Navigating a paper map is predominantly achieved through seeing and visually untangling the messages so order and structure must be carefully encoded. It's also easy to order pages of an atlas, and a reader can flick back and forth. Screen navigating changes the paradigm since the reader sees only one screen and a segment of the map at a time. The whole is reduced to a partial view and often demands panning and zooming. By not showing the big picture, the medium hinders the view of the big picture. Even allowing for a site map or tabbed mechanisms, the process of flicking through a sequence of maps in a digital atlas is not as easy as with a printed product. It's certainly harder to jump from one part of a digital product to another, which in part explains the revival of classical terms such as *scroll* and *tablet* to connote how a reader should move around a digital product.

Although many have hailed the death of the printed map, its demise has, so far, been greatly exaggerated. The elegance, clarity, and comfort of a printed map has, arguably, yet to be matched by electronic displays. Printed maps require no batteries or internet connection though it's true they get rather soggy on a hike in torrential rain!

Possibly the biggest difference between page and screen is that we associate digital map reading and navigation with non-linearity. The map itself becomes an object that demands inquisition and use though people tend to approach reading a digital product as a surface reader—constantly moving from place to place in search of an answer as they would surf any web product. A paper map tends to support deep reading whereas the map is inspected as a rich document. Paper is optimised for map design, delivery, and use in a way that screen isn't because the codex for screen navigation has yet to be fully imagined.

**See also:** Focussing attention | Globes | Graticules, grids, and neatlines | Interaction | Legends | Using words | Which way is up?

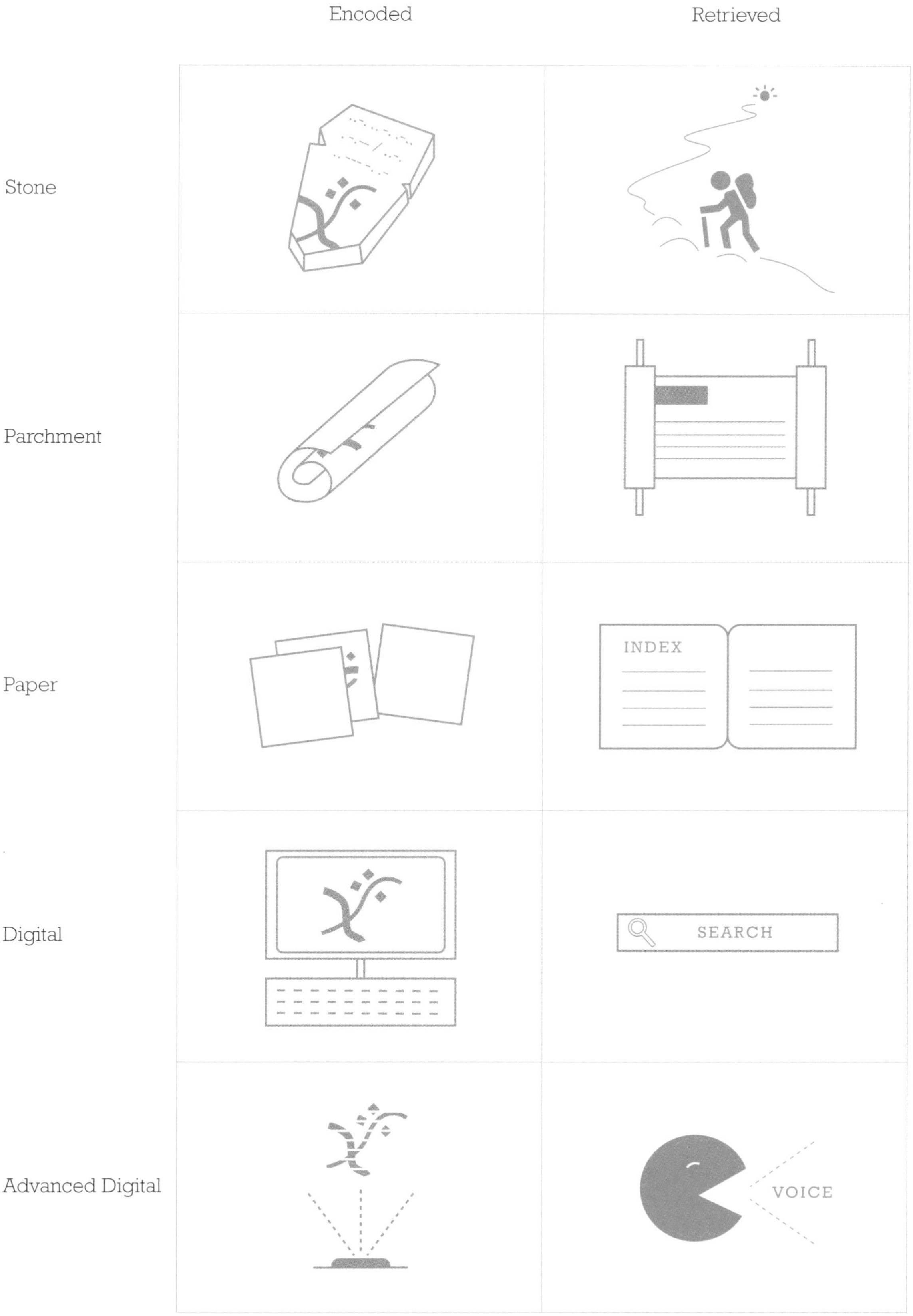

| | Encoded | Retrieved |
|---|---|---|
| Stone | | |
| Parchment | | |
| Paper | | INDEX |
| Digital | | SEARCH |
| Advanced Digital | | VOICE |

# Nominal data

## Mode is the most useful way of summarising nominal data.

**In a cartographic context, many events or observations may occur discretely but be recognised using a different geographical unit.** For instance, census data is aggregated from individual data that has unique spatial locations into a summary measure that is then mapped to best describe some aspect of the values for the area.

**For nominally scaled data, the best descriptor is mode.** The variation ratio gives an index that describes how values are arrayed in the distribution; it defines the dominant class of value and provides an assessment of how well the mode describes the distribution. The mode is simply the name or number of the class in the distribution that has the greatest frequency. For instance, in a survey, each participant might be asked to respond A, B, C, D, or E to a question. The following table shows the modal value as B since it has the highest frequency of response.

| Question response | Frequency |
|---|---|
| A | 28 |
| B | 123 |
| C | 42 |
| D | 5 |
| E | 2 |

The variation ratio is calculated as

$$v = 1 - \frac{f_{\text{modal}}}{N}$$

$$v = 1 - \frac{123}{200}$$

$$v = 0.39$$

where $f_{\text{modal}}$ is the frequency of the modal class, and $N$ equals the total number of occurrences in all classes. The smaller the value of $v$, the better the mode describes the distribution. The nearer $v$ is to 1, the poorer the mode is at describing the distribution. The mode and variation ratio are the most useful indices of central tendency and dispersion for nominally scaled data and can be used to provide additional information for a mapped dataset.

**See also:** Levels of measurement | Making numbers meaningful | Ordinal data | Ratio and interval data | Unique values maps

Isolating the value of data that is most typical in a dataset is often a great way to focus on a particular aspect of the data. For mapping, this has obvious benefits because it's actually a process of generalisation that helps you omit a considerable amount of data to show a specific component clearly.

It's worth remembering, though, that often the most common data value may not be the most interesting or useful aspect of the data. Outliers are sometimes more interesting and more important, and so always reporting a summary doesn't always tell the most interesting story in your data.

|  | Clover Loop | Sparrow Sprint | Juneau Roll | The Bear Crawl | Willow Run |
|---|---|---|---|---|---|
| Day | Mon | Tues | Wed | Thurs | Fri |
| Race | Endurance | Sprint | Endurance | Cross-Country | Downhill |
| Method | All-Ride | Heats | All-Ride | All-Ride | Heats |
| Surface | Hardwood | Cinder | Paved | Mixed | Dirt |
| Venue | Indoor | Indoor | Outdoor | Outdoor | Outdoor |

## Jack-o-lanterns

**Denis Wood**

**2010**

A map is an abstraction of reality, and those who decide to push their work even further into that realm create the maps I enjoy most.

I am especially attracted to maps that are artistic, original, have a clear intent and, most importantly for me, are fun. Denis Wood's *Jack-o-lanterns* is simply enjoyable to look at.

At first glance, it appears to be merely an interesting image—a seemingly random display of pumpkin faces, but as you continue to inspect the map, a whole new level of intricacy and intimacy is revealed, and that is what makes this map shine.

The faces aren't random, but in fact show the distribution of homes that have pumpkins in the area that Wood's atlas *Everything Sings* is situated: Boylan Heights, his neighbourhood in North Carolina. Each symbol indicates a home displaying one or more lanterns.

Perhaps a silly topic, but what an interesting way to show it, and, of course, it lends a very human character to the geography of the area, which would otherwise go unmapped and be lost. When I see a map such as this, I want to emulate it. To me, that is a sign of a good map.

—Wes Jones

# Old is new again

## Looking to the past can provide us with strong cartographic influences and direction.

**Ask people what their favourite map is, and they will likely pick a historical map.** For many, John Snow's map of Cholera in Soho is the perfect embodiment of early thematic cartography and a map that was purported to have saved lives. In fact, it didn't, though it helped establish the mode of transmission of the disease. But for everyone who chooses Snow's map, there are many others. It bears witness to the breadth of cartography that exists.

**Cartographic classics are often regarded as being the first of their kind.** They perhaps brought together a unique idea, an innovative or modified graphical approach, and over time they've been consistently cited as great examples. We become nostalgic when we cite our favourite maps because they hark to a time when the maps became powerful mechanisms to communicate meaning. Time has also inevitably elevated some to the canon of cartographic classics. Perhaps modern equivalents resonate less, simply because they struggle to be heard above many similar maps, there is relatively little true innovation anymore, or that they have yet to benefit from the passage of time during which they become better appreciated.

**There's something more useful we can glean from older maps.** They often give us a clarity of purpose more than their modern counterparts. Maps such as Snow's were simple but conceptually brilliant. They embody an intelligence in information design and were visually strong both at the time and, as has proven, through time. We see the same in William Playfair's graphical approaches or Charles Minard's exquisite flow maps.

**Channelling these classics is an excellent way to frame current practice and learn from the masters.** Many of these examples of historical cartography continue to be relevant because they found a way to harmonise form and function. They developed a graphical notation that has stood the test of time and which we should continue to explore as a way of supporting current mapping.

**See also:** Critique | Prior (c)art(e) | Jokes and satire | Style, fashion, and trends | Who is cartography? | Viral cartography

Nostalgia is a positive way to think about cartography. Yes, we can look at and admire old maps as beautiful images but they can also teach us best practices in representing data simply, clearly, and with a style that transcends cartographic fashion.

Maps are increasingly ephemera. Whereupon once, they took months or even years to survey, design, and produce, they are now made or updated rapidly, sometimes beyond the speed of thought. Many maps burn brightly for a short while yet also fade rapidly. The one aspect of the cartographic process that modern practice has truncated is time, and that is possibly one of the most valuable commodities when making a map. Because maps used to take so long to produce, it was imperative they were clearly planned and carefully executed. Errors could be costly, often requiring a map to be started from scratch again. Modern cartography has given people tools to design and produce rapidly. But this has arguably been at the cost of giving the map the time needed to think everything through properly. Exploring old classic maps teaches us what works and what doesn't.

Dwindling attention spans and increased expectations also impact our modern tastes in maps but we recognise styles and fashions in maps too. Style used to be thought of as belonging to a particular look and feel, perhaps that of a national mapping agency or a news organisation. These days it has more to do with an individual and the search for a cartographic approach that riffs off well-known and loved imagery as a means of rapidly piquing interest. Looking to the past to reimagine techniques or make a map in a particular style ensures a continuum of great cartography even though old is often new again.

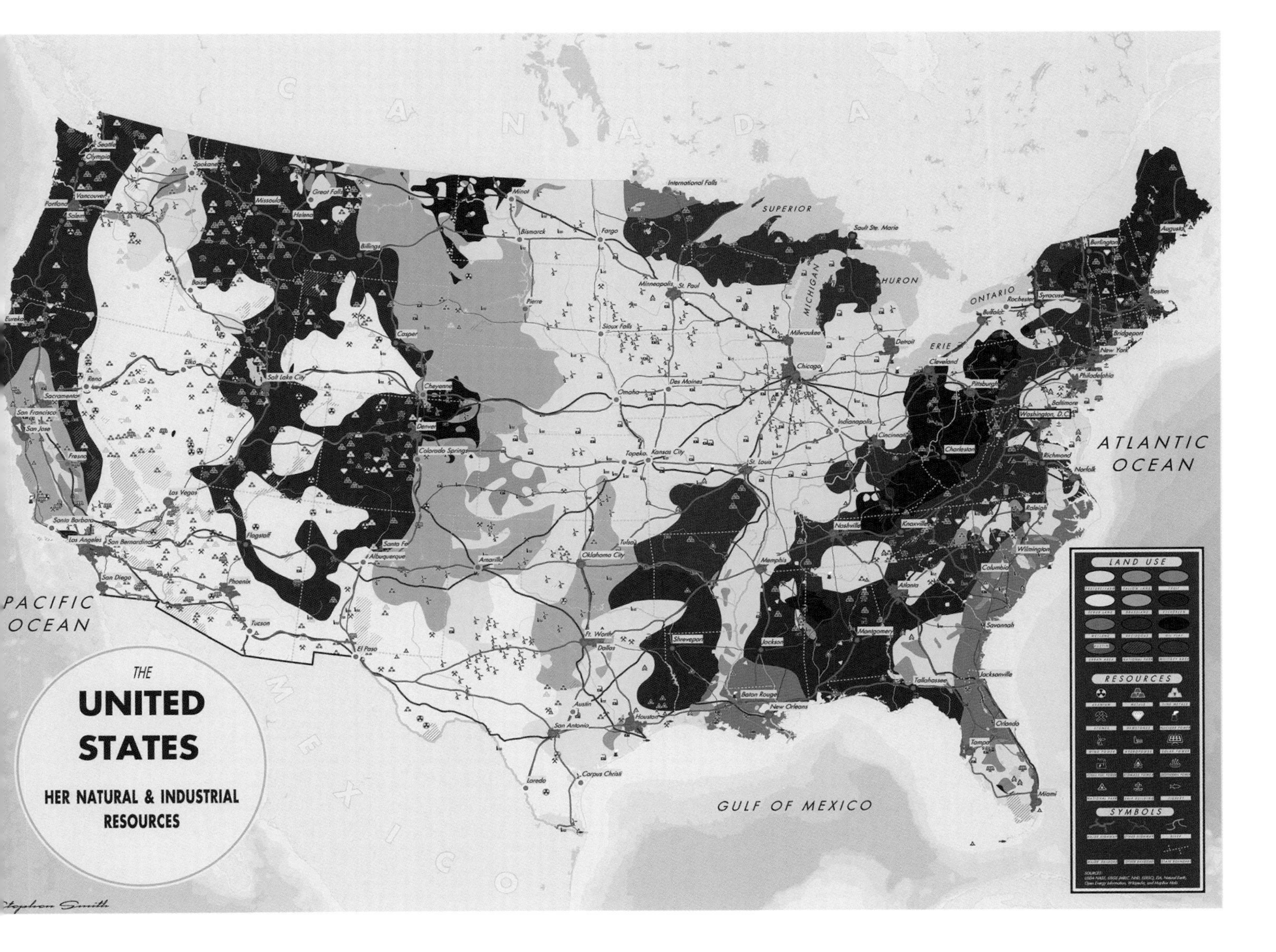

The above map by Stephen Smith, published in 2014, garnered many plaudits and won numerous awards including some for which this book's author was a judge. Despite some 100 years+ of collective cartographic wisdom, judges hadn't spotted that the map wasn't as original as first presumed.

While noting its retro appearance and style, we were unaware of the existence of a strikingly similar map produced by the British Information Services sometime between 1939-1945 (shown left).

This leads to a number of interesting points. First, even experienced cartographers are not aware of everything and we learn something new every day. This is particularly true of the vast canon of historical cartography. It's part of what makes cartography fascinating. Second, taking design cues from what has gone before is important because a lot of design is already out there somewhere and informs our work. There's very little that is truly original and replicating an historical map with up-to-date information is an entirely valid approach. Finally, the US map is a beautiful map and the fact it's an homage to an earlier work doesn't downgrade its artistry.

Would the US map still have won awards if the judges had been aware of the original? Undoubtedly yes. It stands on its own. It's also important to note that Smith cites his inspiration which evidences best practice in recognising what has gone before.

# OpenStreetMap

## Not a map. A database and a movement.

**OpenStreetMap isn't a map as such because the main output is data.** The idea of making a free, open, and editable mappable dataset and the movement it has inspired has revolutionised mapping. In 2004, the United Kingdom in particular was both one of the most detailed and accurately mapped places on Earth yet, paradoxically, most of the data was proprietary and the preserve of national mapping agencies. Inspired, Steve Coast set about creating the phenomenon that is OpenStreetMap. Volunteers using GPS receivers undertook ground surveys and collected data and notes, which were uploaded to the OpenStreetMap website (www.openstreetmap.org). They could categorise and attribute the data into a consistent schema and edit data submitted by others directly in the browser to improve accuracy and detail. The community-driven map was born.

**Early adopters were mainly those in countries that had relatively poor access to open map data.** US contributors were initially conspicuous by their absence mainly because of the relatively rich and available data they already enjoyed. Contributors are varied culturally, geographically, and technically. This, coupled with uneven data coverage, has led to some criticism of the accuracy and consistency of the dataset yet its value, particularly for rapid disaster management and relief and recovery efforts, has been invaluable.

**Maps can be created directly in the browser or the data can be processed using a number of methods and used in other applications.** It is this flexibility and ability to use the data under, formally, a Creative Commons (CC) licence and now an Open Database Licence (ODbL) that has led to its wide uptake.

OpenStreetMap is used in many commercial applications but perhaps most importantly it has been critical in the disaster management and response for major natural disasters.

An impressive momentum has built among the community resulting in more people making and editing the data, making maps from the data, producing new and better software tools for the data, and generating new and exciting products. Much of the use has been from non-geographers and non-cartographers, which represents a massive departure from those more usually interested in maps. This burgeoning growth has led to massive and often disruptive innovation, some of which is truly innovative yet you're more likely to see a terribly designed map made with OpenStreetMap data than you are a good one. This is in many ways inevitable as far more people who have non-geo backgrounds become users of OpenStreetMap data. That balance will shift as more people come to recognise not only the value of such a resource, but how to see it as more than just a technical challenge and use it effectively in tandem with knowledge and understanding of cartography. As for affecting the way map data is made available, OpenStreetMap has changed the landscape in profound ways. The freeing up of even the most tightly controlled geographical datasets is a work in progress but one from which there is no return.

**Opposite:** Port-au-Prince, Haiti, as mapped by the OpenStreetMap community in response to the 2010 earthquake. Volunteers responding to the 2010 Haiti earthquake mapped the roads, buildings, and refugee camps of Port-au-Prince in just two days. As a way to generate rapid data for crisis mapping and to support disaster response and recovery efforts, it's hard to see a better approach than that which OpenStreetMap supports.

**See also:** Basemaps | How maps are made | Reference maps | Topographic maps | Web mapping

Port-au-Prince, Haiti, January 3, 2010

Port-au-Prince, Haiti, January 29, 2010

# Ordinal data

## Median is the most useful way of summarising ordinal data.

**For ordinally scaled data, the median best describes central tendency, and one of several decile ranges can be used to describe the dispersion of values in the dataset around the value.** The median is the point in an array that neither exceeds nor is exceeded in rank by more than 50% of the values in the distribution. In a dataset with an odd number of values, the median is simple to determine. For datasets where there is an even number of values, it is only possible to say that the median position is half way between the two middle ranks—this is not the same as saying it is half way in terms of value (since the data is on an ordinal scale and only relative ranks apply).

**Quartile values are found in much the same way as the median.** The lower (first) quartile separates the lower 25% of values from the upper 75% of values and is referred to as the 25th percentile. The second quartile is the median. The upper (third) quartile is the division between the lower 75% of values from the upper 25% and is called the 75th percentile. Quartile deviation is calculated as

$$\frac{upperquartile - lowerquartile}{2}$$

where upper quartile – lower quartile is referred to as the interquartile range, which identifies the 50% of all values that are closest to the median value. Identifying the quartile deviation allows a comparison of distributions.

**The interquartile range can be useful in cartography to establish which values in the data are most similar to the median value.** Maps are often constructed to illustrate the upper quartile, lower quartile, and interquartile range classifications.

The median value for the data shown below (which has 21 values) is the value ranked in 11th position. Its value of 21.2% is the middle value for the array and might be referred to as the typical value for this dataset. Other values are described as being higher or lower relative to the median value.

| Value (%) | Rank | | |
|---|---|---|---|
| 30.8 | 1 | | |
| 28.4 | 2 | | |
| 28.3 | 3 | | |
| 24.9 | 4 | | |
| 24.8 | 5 | | |
| 23.4 | 6 | upper quartile | Interquartile range |
| 23.2 | 7 | | |
| 22.0 | 8 | | |
| 21.9 | 9 | | |
| 21.7 | 10 | | |
| 21.2 | 11 | median | |
| 20.8 | 12 | | |
| 18.6 | 13 | | |
| 18.1 | 14 | | |
| 17.6 | 15 | | |
| 17.4 | 16 | lower quartile | |
| 17.0 | 17 | | |
| 16.3 | 18 | | |
| 15.2 | 19 | | |
| 14.6 | 20 | | |
| 14.3 | 21 | | |

**See also:** Levels of measurement | Making numbers meaningful | Nominal data | Ratio and interval data

| | Clover Loop | Sparrow Sprint | Juneau Roll | The Bear Crawl | Willow Run |
|---|---|---|---|---|---|
| Hazard Rating | 5th | 4th | 3rd | 2nd | 1st |
| Distance | Long | Shortest | Longest | Mid | Short |
| Spectator Capacity | Large | Largest | Median | Less | Least |
| Terrain Variability | Low | Least | Median | High | Most |

# Orientation

## Describes the direction or rotation of a symbol.

**Many symbols are right-reading with respect to a graticule or neatline but they can be changed through rotation.** Differentiation between map symbols can often be achieved by varying the rotation of a symbol. Changes in orientation connote qualitative difference as opposed to quantitative difference. Rotation might be performed simply to show difference among a range of other variables or, instead, as a way to explicitly show direction. Rotation of symbols tends to work better for geometric shapes. Mimetic symbols will likely not be suited to rotation, and, of course, adding arrowheads will support the idea of directionality if that is the function of the rotation.

**For point symbols, there is an infinite number of orientations while keeping the location of the symbol constant.** If point symbols are aligned in a grid, a wide range of patterns can be achieved if the symbols are all rotated in the same way. By rotating symbols differently, other patterns can also be produced, for instance circles or fans. If point symbols are not organised in a regular grid, then such patterning cannot be performed.

**Generally, it is best to restrict orientation of non-aligned point symbols to four directions.** This restriction allows distinguishing of symbols among a range of symbols. The more rotations that you employ, the harder it is for the map reader to discern difference between them.

**Orientation of marks that make up a linear symbol should be aligned only along the axis of the linear segment or perpendicular to it.** Bizarre patterns will result from any other orientation of graphical marks for linear symbols. The exception is when lines have innate orientation and that is what defines them. The pattern of adjacent or organised lines gives a structure to the overall appearance.

**See also:** Pattern fills | Shape | Size | Texture

The rotation of a symbol can lead to a perception of a different shape entirely depending on the angle at which the map is viewed. For instance, if a symbol is made up of small linear marks and is rotated, the rotation, when viewed planimetrically, will be seen as intended. However, if the angle at which the map is viewed is not perpendicular to the plane of the map, then symbols will be seen differently. As the viewing angle tends more toward being parallel to the plane of the map, then the difference becomes more pronounced.

It is less common to see orientation used to differentiate between mapped variables when colour is used. Changing hue is seen more as a differentiating visual variable than orientation.

Orienting symbols to show direction is fundamental to many symbology schemes such as the marks used on weather maps. Rotated lines with arrowheads are routinely used to show wind direction, and intensification, often in a cyclic motion, is shown by rotating lines of symbols aligned to a grid. A weather plot also has direction built into the symbol's representation. The orientation of the symbol indicates wind direction while sky cover and wind speed are shown using other marks. When seen in a pattern, they generate a view of the overall movement of the air mass.

Visual variation

Ensure angular difference is distinct

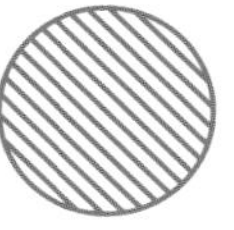
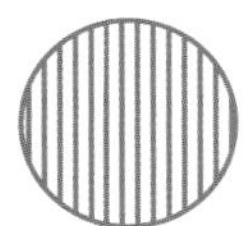

For seeing: Distinct, Levels

For representing: Nominal, Ordinal, Numeric

**Opposite:** *Beyond the Sea* by Andy Woodruff, 2017.

# Beyond the Sea

When you stare out into the ocean, what's on the other side? Finding the answer isn't as simple as drawing a straight line on a map, because the Earth is not flat and coastlines are not straight.

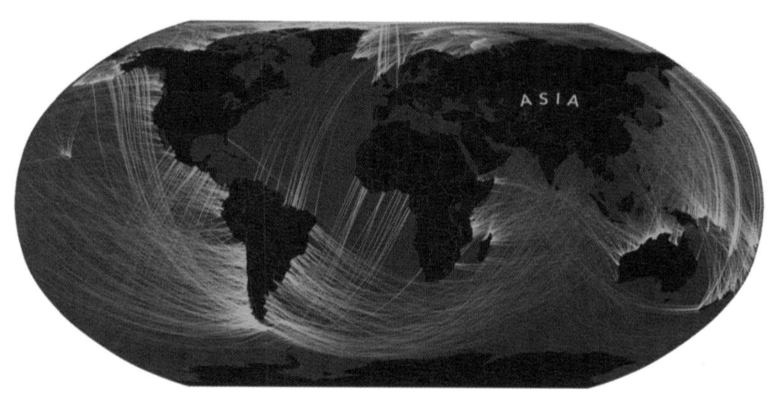


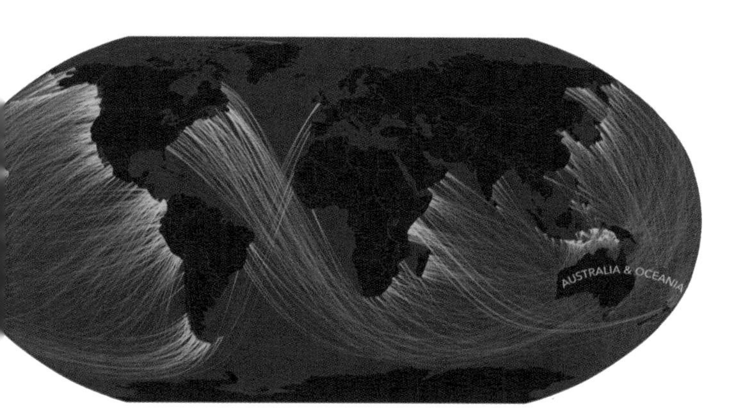

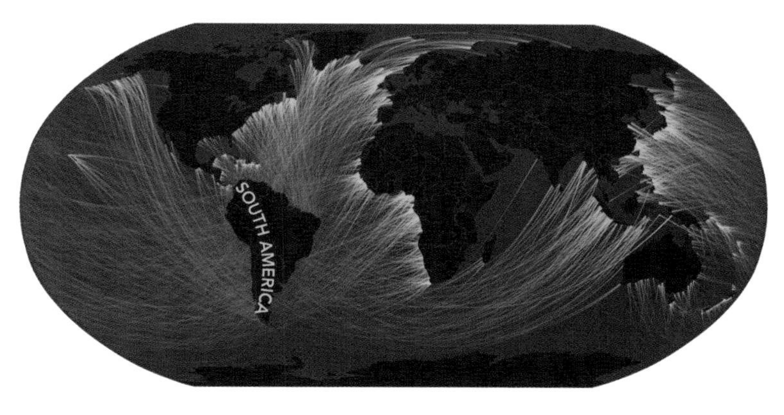


The maps here show coastal points around the world (except from Antarctica) that directly face each continent, taking into account the orientation of the coastline at those points, and a round Earth. The brighter end of each line is the view origin, that is, a point from which you can "see" the continent that is labelled on the map.

## Karte der Gegend um den Walensee

**Eduard Imhof**

**1938**

The Swiss Eduard Imhof (1895–1986) was the founder of the Institute of Cartography at ETH Zurich and a cartography professor between 1925 and 1965.

He authored numerous maps and atlases, such as the *Atlas of Switzerland*. In the 1930s, he also had a major influence on the design and choice of scales of the new official National Map Series for Switzerland.

Imhof was very much up to the clarity of cartographic map representations. He advocated an elaborate and balanced mixture of map elements. A classical topographic map should consist of 'immediate', image like elements such as shaded relief and fictitious, 'indirect' elements such as line art.

In his 2 × 4.8 m gouache wall painting *A Map of the Area around the Walensee* (1:10,000) of 1938, he tried to achieve a naturalistic representation of the landscape with more immediate elements. It may be compared to today's texture-based landscape renderings.

—Lorenz Hurni

# Page vs. screen

**The way we use and interact with maps has changed. Paper and screens require different cartographic considerations.**

**There's no doubt that the beginning of the 21st century has brought fundamental changes in the way we use maps.** Paper has been the dominant medium of production for maps for most of the 20th century. Interactive devices that provide an electronic view of the world are perhaps now more prevalent though paper certainly isn't dead yet! The web is now a source and a delivery mechanism. Screens are now a common medium of interaction, which changes the way people use maps. As a result, what has also changed, profoundly, are the ramifications for cartographic practice to support these new modes of use.

**Our understanding of print and how to design for certain dimensions and colour reproduction is largely known.** Screens have rapidly become a key medium for cartography, and we routinely now work with maps on screens in our offices, homes, and mobile devices. This means we no longer have a full map in front of us but a section. The map is multi-scale rather than single scale and insets are largely dispensed with. Content is responsive so as you interact with a map, the amount of detail changes depending on scale and other conditions. The rapid evolution of screen technology has meant that cartography has had to adapt at a similarly rapid pace from simply delivering slippy maps that are designed to be seen to fully interactive content delivery mechanisms, which are context and locationally aware as well as able to perform differently depending on input through touch, gestures, the spoken word, or augmented reality devices.

**How we organise and access information via a screen is fundamentally different from paper.** A single paper map presents information at once. A map consumed via a screen has the ability to support progressive or interactive reveal, pan, zoom, and multi-scale. The navigational structure of electronic cartography is different from its paper counterpart and demands that the design takes this into account. From simple pop-ups that reveal detail to complex interactions, there are many ways that cartographers can extend the potential of how maps are used.

**See also:** Anatomy of a map | Printing fundamentals | Scale and resolution

Electronic maps, web maps, can be navigated by going backward and forward, panning across or possibly scrolling up and down. These are some of the aspects that can hinder effective use of such products because we're effectively switching between screens or different views on the same screen. This can lead to change blindness as we're unable to hold onto the picture of what went before for comparative or aggregate cognitive processing. Print atlases get over this through consistent denotation. Similar principles should apply to web maps.

Different ways in which we interact with screens through gestures, pinch and zoom, and so on afford new opportunities, but as web technologies mature, they increasingly take on mechanisms that mimic our familiarity with printed maps and atlases. In developing maps for screen, we should at the least explore the characteristics that constrain map readers from getting to the information with as few clicks as possible. It's a delicate balance but navigating information in a time-efficient way, which is obvious to the reader and which supports the retrieval of information efficiently, is no different whether we're dealing with paper or screen.

Consider the classic 'you are here' map. In print it has a fixed position. You locate it. In digital, you are often the centre of the map itself and the map moves around you.

**Below:** The impact of user interaction on a static map.

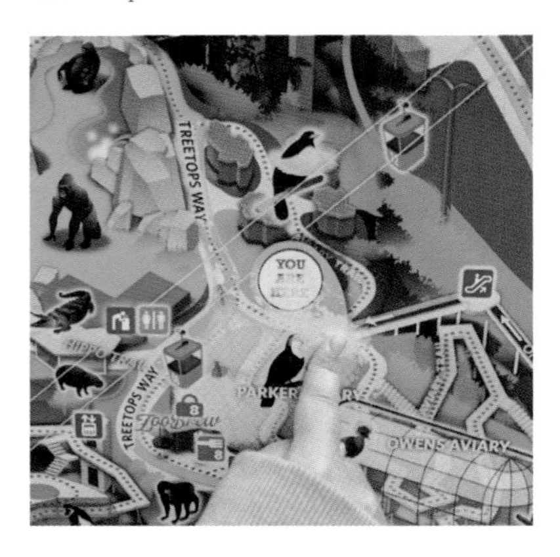

| | Paper | | Screen | |
|---|---|---|---|---|
| Search | Gazetteer | | Type / voice | ABC |
| Scale | Single and fixed but insets can expand the range | | Multi-scale and seamless through scales | |
| Zoom | Fixed | | Multiple | |
| Pan | Fixed, with heet edges | | Seamless and slippy | |
| Content | Fixed, determined by scale and map space | | Switchable, filtered, and personalised | |
| Routing | Distance markers and emphasised / drawn line | A B | Clickable locations and multiple options with live context (e.g. traffic) | A B |
| Context | Predetermined and fixed | | Personalised, clickable, and supports query | |
| Where am I? | Fixed icon | YOU ARE HERE | Pulsing pointer / coordinates and you're at the centre of the map | |

# Panoramic maps

## Highly illustrative landscape paintings, often overprinted with additional information and commonly used as a basis for winter ski trail maps.

**One of the main reasons why panoramas are useful is that they are less abstract than their two-dimensional counterparts.** As such, they might be easier as a mechanism to visualise a landscape by people with limited map reading skills. They present a landscape as if viewed from the perspective of looking at the landscape itself. They have become synonymous with mountain cartography and the creation of tourist maps for ski resorts or mountain bike trails.

**Despite their popularity as works of art and as a basis for a lot of tourist cartography, panoramas have been largely ignored by mainstream cartography.** They bear little resemblance to many of the familiar map types that are derived from ground survey and subject to usual cartographic design and control. Their accuracy is often relaxed in favour of artistic depiction; they do not make use of standard symbolisation; and their production is more often undertaken by artists or 'panoramists' than cartographers.

**Most panoramas are created through painting, illustration, or by computer rendering.** Key to the realism is the way in which blurring and haze effects are used in the distance to focus the reader's eye on the figural elements, which are exaggerated to emphasize the figure in its surroundings. Terrain is often distorted in relation to the point of view so that all elements of the mountain can be seen without recourse to insets.

**A progressive projection, with the foreground appearing much steeper than the background, usually characterises a panoramic map.** Each panorama is effectively projected into its own unique 3D space so different parts are contorted to face the point of view. For tourist mapping, this ensures that overprinted trails are all visible from a single point of view. The foreground becomes almost planimetric, the figural mountain dominant in the map view and with a background that rolls away from the viewer creating both a depth of field and a horizon. Panoramas also pay close attention to detail with individual trees and buildings being included and, in some examples, animals and other curiosities.

**See also:** Aspect views | Hand-drawn maps | Pseudo-natural maps

**Above:** *Mammoth Mountain* by James Niehues, 2008.

Heinrich Berann (1915–1999) is possibly the most highly regarded artist of the panoramic style whose work across Europe and also in North America for the National Park Service was exceptional. In North America, Hal Shelton (1916–2004), a cartographer and accomplished terrain artist, was commissioned by the State of Colorado to create a panorama of Colorado Ski Country in the 1960s. This was a superb illustration across the Colorado Rockies showing the location of the various ski resorts in the state.

By the turn of the century, James Niehues became increasingly responsible for making the winter and summer maps for many of North America's mountain resorts as well as a range of global ski resorts. Here, his Mammoth Mountain, CA, map is seen overprinted with the ski resort detail to create the final map. Each panoramist brings a style and technique much like any painter. You can spot a Berann from a Shelton from a Niehues because of their use of colour, stroke, and technique.

# Pattern fills

## Combining visual variables to apply an opaque material to areas and lines.

**Pattern fills are an extension of the use of visual variables to create symbology for areas and lines.** They are used in multiple ways to support identification, differentiation, quantifiable difference, emphasis, or simply to improve the aesthetic appearance of a map element. Fills can be developed from one or multiple visual variables, and in combination they form patterns, each of which can be created using any or a combination of different hues or values. Changes in value create different tonal patterns. Tonal patterns are created by changing the amount of a specific hue in a total area to create a percentage pattern fill. Each of the individual characteristics can vary considerably in their own right so the palette of possible pattern combinations is almost infinite.

**Combinations of different visual variables result in patterns that support different graphical functions.** Changing the shape of individual elements in a fill shows difference among features and is used to connote different feature types such as wavy lines (water) or tiles (karst landscape). Changing the orientation of individual elements extends the palette for any one fill. Lines can be positioned at different angles or superimposed to create cross-hatched patterns. Orientation can apply to smaller individual elements in an overall pattern. Texture can also be referred to as *coarseness* since individual elements in a fill are designed using heavier weights and the spacing between them varies. This can be used to show both qualitative and quantitative differences. Similarly, size is used to vary individual elements, but by keeping the spacing between elements constant, they provide patterns that are good for showing quantitative differences. By keeping weight of individual elements constant and varying spacing, a similar effect can be achieved. Using transparency in conjunction with a solid fill supports multiple gradient effects, such as bottom to top, side to side, diagonal, focal points, and radial transitions.

Pattern fills can equally be used for lines, borders, and frames on a map. Care should be taken when using patterns for lines on the map itself since the overall line weight and natural orientation will help determine how much of the pattern is distinguishable. Pattern fills for lines are usually used to differentiate between line type. Some can also be used for effects such as using transparency to create a gradient along a line that includes an arrowhead cap.

In general terms solid fills are best avoided on maps, particularly when using greyscale where solid black and white are at the extremities. They become dominant in perception so should be reserved for small areas where possible. They do tend to support the impression of totality (black) or emptiness (white). White can also be misconstrued as the same fill as the map background if care isn't taken to distinguish between the mapped area and its surround. Lighter fill hues, values, and finer patterns are generally used for larger areas.

Visual variation

Patterns are mostly seen as different

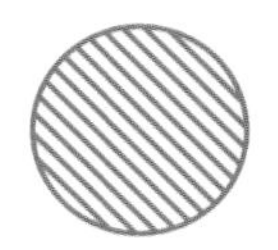
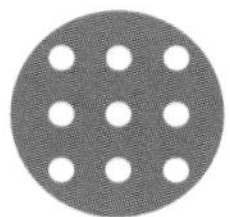

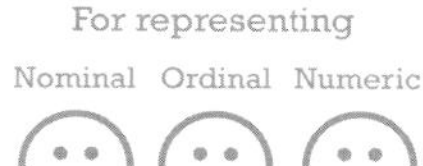

**See also:** Orientation | Shape | Texture | Typographic maps

**Opposite (bottom):** Pattern fills from *Elements of Topographic Drawing* by R. C. Sloane and J. M. Montz, 1930.

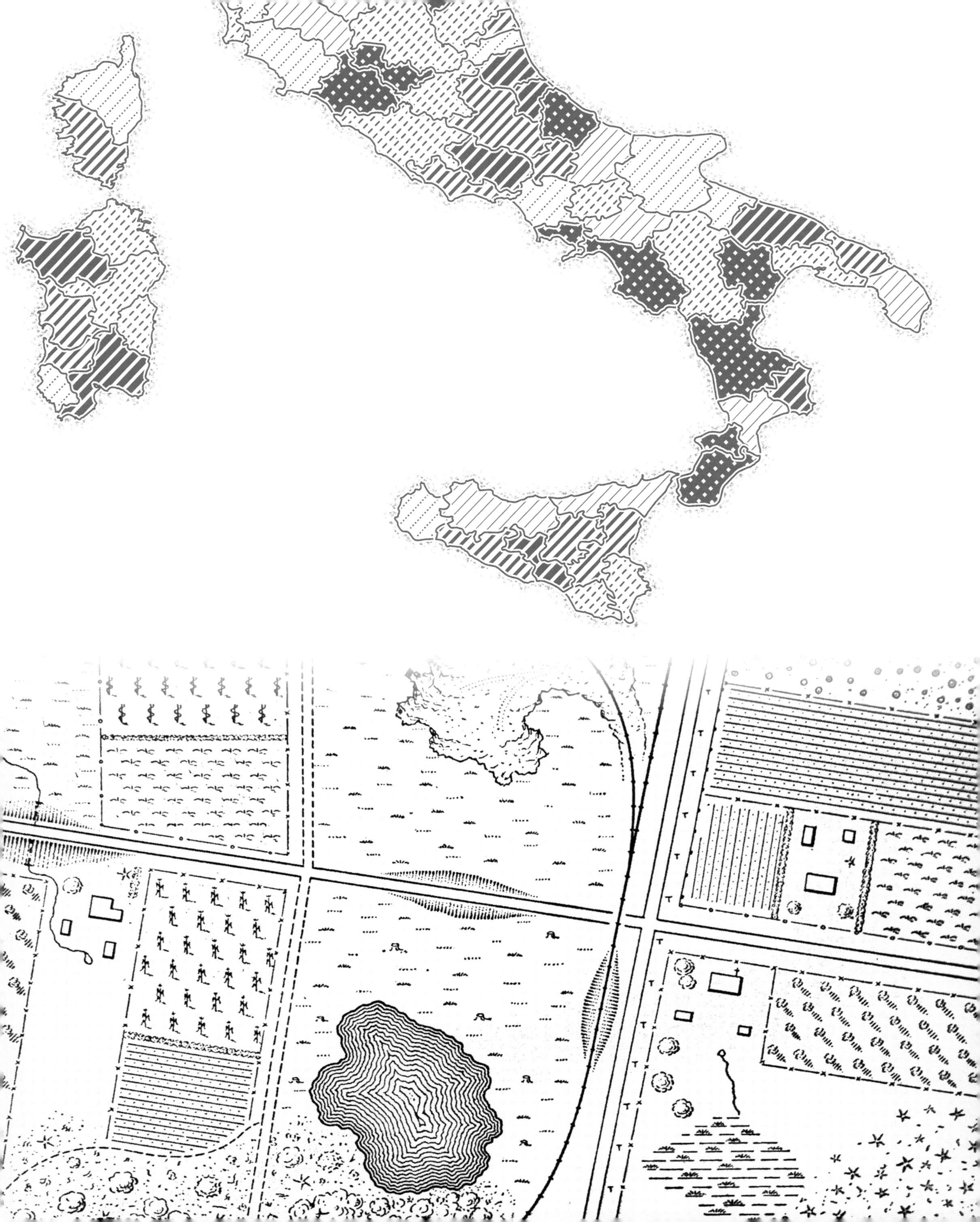

# Perceptual colour spaces

## Perceptual colour spaces provide a more accurate representation of non-linear colour.

**A wide variety of colour spaces and models help explain the nature of colour.** Most have three dimensions, including HSV (hue, saturation, value), HSL (hue, saturation, lightness), RGB (red, green, blue), and CMY (cyan, magenta, yellow). but treat colour in a linear way. Three-dimensional colour spaces can also represent the non-linear perceptual dimensions of colour. Ideally, you would use a perceptual colour space to select map colours. However, complete three-dimensional spaces are not computationally efficient and so the approximated HSV model is often used. It helps to understand the nature of colour to make better decisions even if you use HSV or, more directly, RGB or CMY colour models to select map colours.

**True perceptual colour spaces arrange hues in spectral order around an axis of lightness with saturation increasing outward from the axis.** The axes on perceptual models are non-linear and match our visual perception of colour to a specification. Equal distances in the colour space produce equal distances in human perception of colour. Two of the most widely used perceptually scaled colour spaces are the Munsell and CIELAB.

**The Munsell colour model was developed by the US artist Professor Albert H. Munsell.** He began work on his book of colour in 1898, and it was published in 1905 (Munsell, 1905) and is still in use today. His focus was to create a decimal system for colour specification to overcome the everyday descriptions of colour that were less than helpful when printers were attempting to colour-match prints to original painted artwork. The system allows you to specify colour using the variables hue, value, and chroma.

**The CIE model was proposed by the French Commission International de l'Eclairage (CIE) in 1931.** Red, green and blue primaries are correspondingly represented by x, y, and z. Each value is based on the sensitivity of the human eye to different wavelengths of light. Using equations, any colour can be assigned x, y, and z chromaticity values determined by its spectral power distribution.

**See also:** Colour cubes | HSV colour model | Mixing colours | Seeing | Seeing colour

In the Munsell colour space, colours are created from an arrangement of five basic hues, with intermediate hues being created through mixing proportions of the adjacent major hues. Variations of each hue are created by adding different proportions of white and black. The central axis in a Munsell colour space is a grey scale, which was designed so that it is perceptually equal from 0 (perfect black) to 10 (perfect white).

Munsell accounted for the non-linear response of the eye to reflected light in designing the grey scale into equal visual perceptual steps. The eye requires a larger difference of light energy stimulation as light intensity increases to produce the same perceptual difference that is perceived for darker intensities. Munsell also applied the same principles to his determination of perceptual steps for both hue and chroma. The solid form of the Munsell colour space is not symmetrical either because the eye is not uniformly sensitive to all hues.

The main advantage of the CIE model over other models is that colours can be plotted on a two-dimensional graph as opposed to a three-dimensional model where hue and saturation vary in two-dimensional space.

Study of the CIE model explains why cyan appears weak (low saturation) on a computer screen. Pure cyan is, in fact, midway along a line between the green and blue primaries, which falls close to the white point. This leads cyan to appear somewhat washed out Conversely, pure yellow falls midway on the line between red and green, and because it is some distance from the white point, it has a high chroma.

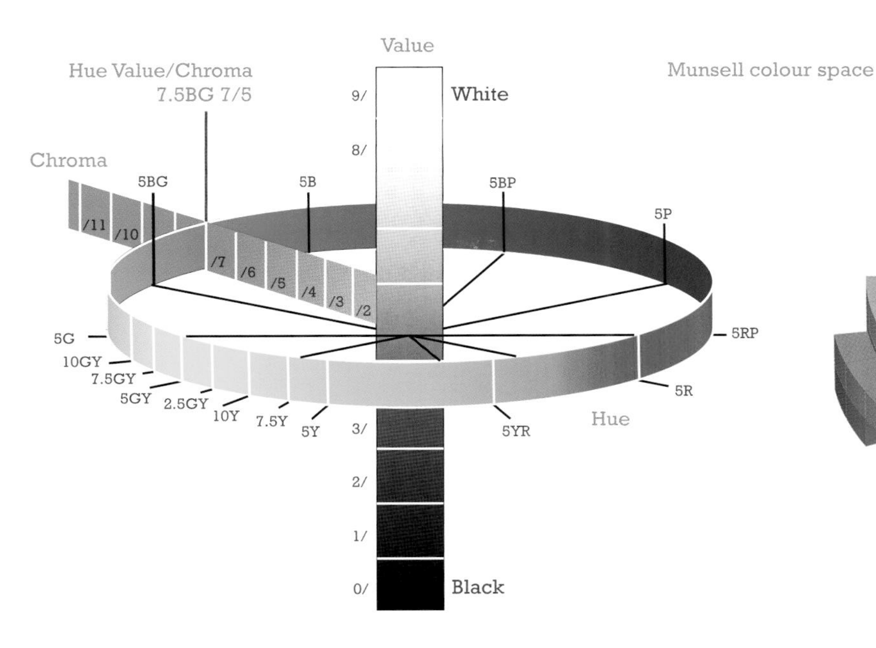


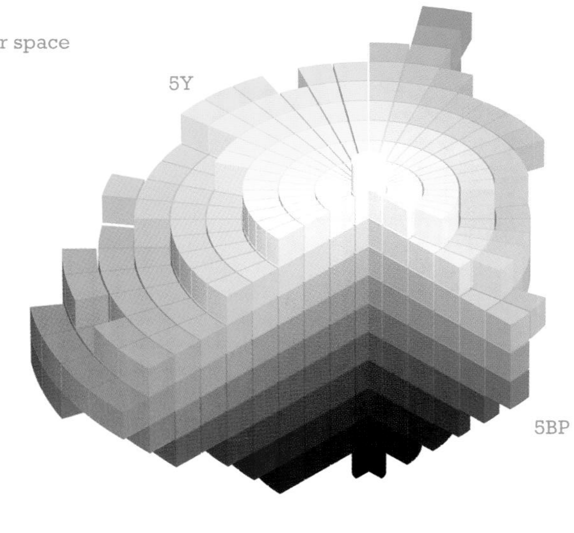


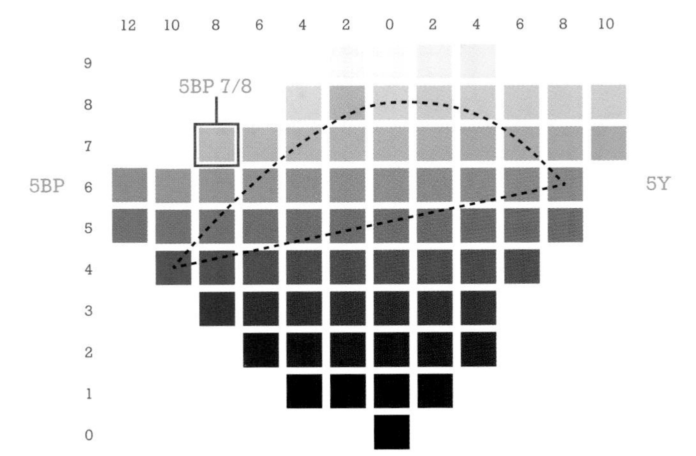


## Munsell colour slice

By taking a vertical slice through the Munsell colour space, you instantly see how chroma (saturation) increases away from the central axis of neutral greys.

The two opposing colour leafs (complimentary hues) also display different values of lightness. If you created a colour ramp that passes linearly through this space, you yield desaturated midtones.

Alternatively, by taking a systematic non-linear path through the colour slice, you better control the perceptual steps between adjacent colours.

By increasing the value in the midtones, you avoid overly desaturated colours. Because yellow is naturally lighter than blue/purple, you can then move to a more saturated (higher chroma) colour.

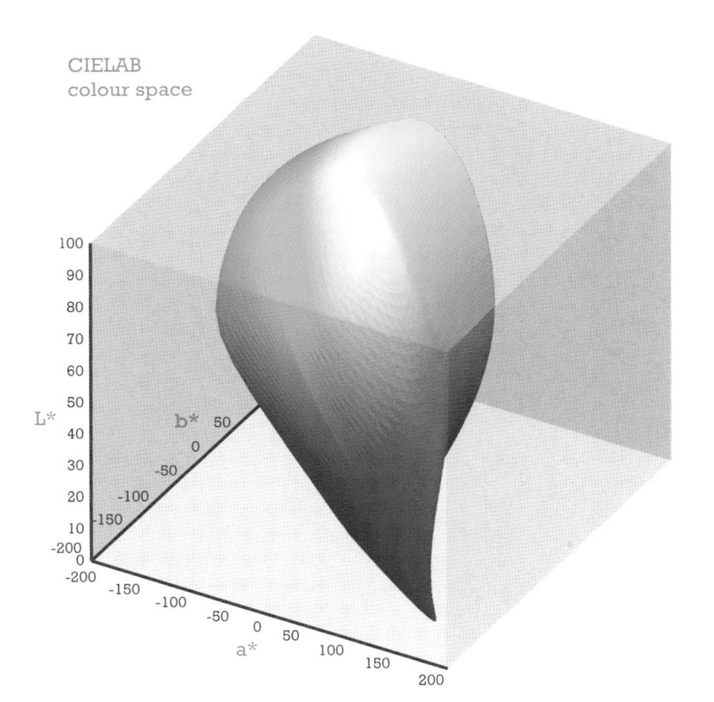


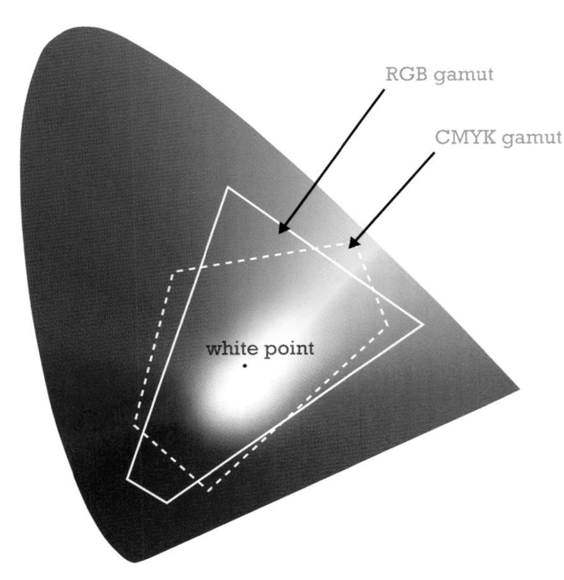

# Pictograms

## Encoding meaning in small pictorial graphics supports a wide variety of cartographic design needs.

**Lack of space is often a major cartographic dilemma as we try and fit detail on the map.** One of the main ways to overcome the lack of space is to use pictograms (or icons), which substitute for alternative depictions or obviate the need for lengthy typographic elements. Being able to summarize a definable feature or attribute in a single, perhaps black-and-white, symbol that may only be 5 × 5 mm, is, however, an art in itself. Designing one effective symbol is a challenge. Designing a set of related and graphically consistent symbols for multiple different features or for a map series is an even bigger challenge.

**Pictograms are not the unique preserve of maps, however, and we see them used in multiple ways in everyday life.** Pictograms are used routinely in transport, computing, and on our mobile devices. Pictograms on toilet doors are perhaps the clearest example of the success of encoding meaning. The human figure in different configurations is used almost universally to denote the location of toilet facilities. We know this through familiarity and convention. These two ideas are at the heart of pictogram design and how successful pictograms work.

**Standard symbol sets exist for signage, and similar sets exist for cartography.** Clear graphics with minimal geometry create a harmonious visual alphabet that is easily recognisable, and the meaning is most often understood without recourse to looking up the symbols in a legend. The key is simplicity and using the minimum amount of 'ink' in a small space to encode meaning. Different mapping agencies develop their own consistent sets, but for individuals seeking to make a unique pictogram or a new set, the challenge is to make them meaningful and different from other similar sets. A good example is seen in the study of pictogram sets from different Olympic Games over time. The events are the same but each organiser wants their own design. Some are more effective than others.

A pictogram that must be learned isn't particularly useful. Pictograms are supposed to help compensate for a lack of text and can be extremely useful for maps that are used in multiple language contexts.

Strong drawing is imperative to avoid ambiguity or misinformation. Put simply, although the graphic should be as simplified as possible, it must still look like the feature it represents. Oversimplification leads to ambiguity. Misinterpretation is more likely with pictograms, and special care should be taken to avoid problems arising from different cultural or religious beliefs. It's often useful to design at a larger size and then shrink the graphics down to the final size. Output medium should be borne in mind, and the finesse of the graphic will need to be modified depending on whether symbols will be reproduced as scalable vector graphics or rasterised for final production.

The design of pictograms broadly fits into the language of semiotics—the relationship between a sign, an object, and a meaning. Any pictogram will be recognised through a perceptual stage where the map reader becomes aware of the sign. They then interpret the sign with reference to objects they know and respond accordingly. Reducing multiple meanings is key to a good pictogram.

Beyond being able to command and inform, pictograms can also take on character. They can be seen as friendly, comforting, authoritative, aggressive, playful, or any number of other emotive states. The map's context will go a long way toward determining how to design a sympathetic set of symbols and will influence the way people react to the map.

**See also:** Literal comparisons | Semiotics | Shape | Simplicity vs. complexity | Symbolisation | Symbols

**Opposite:** Pictogram design for *Deaths in the Grand Canyon* by Kenneth Field, 2012.

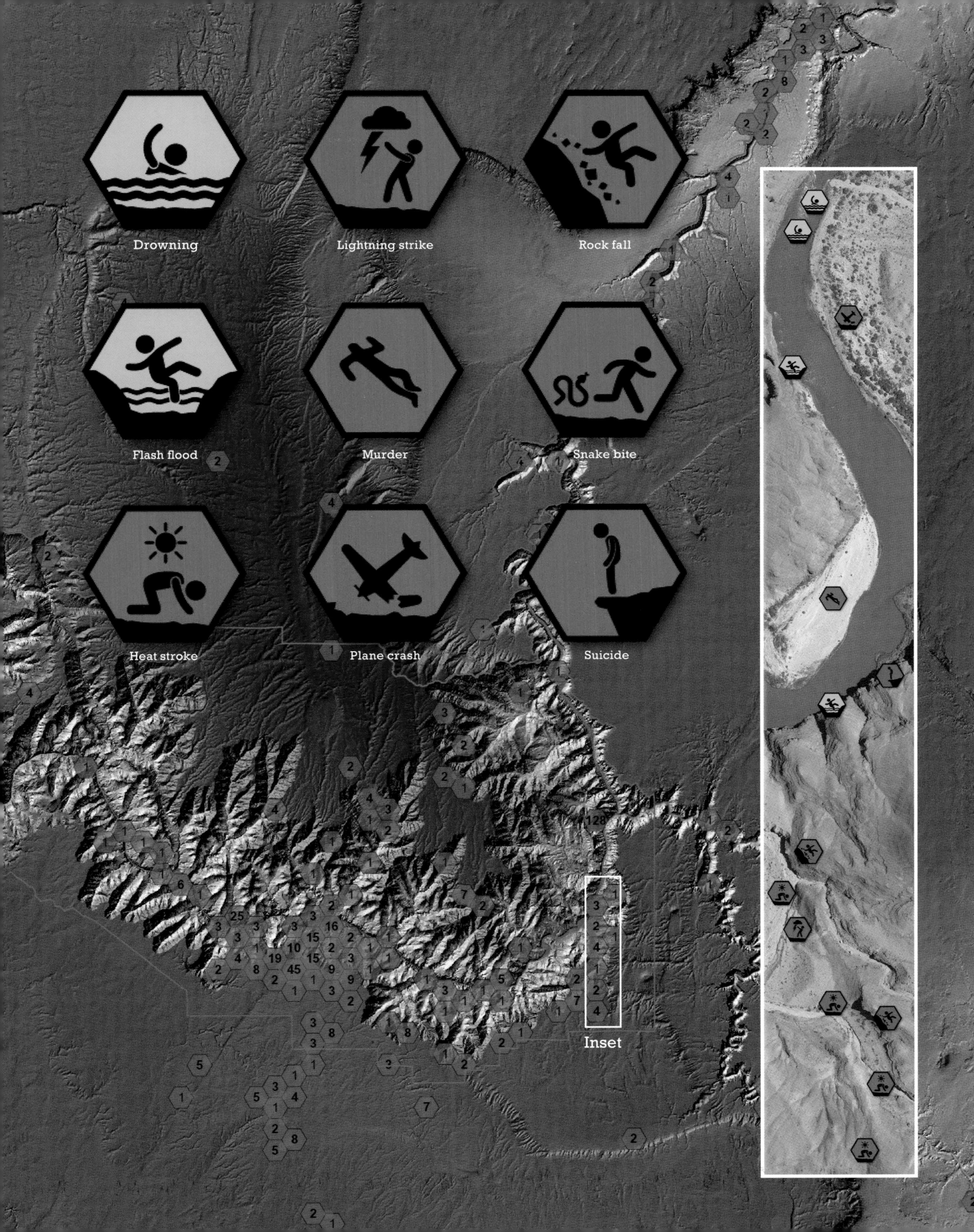
Drowning
Lightning strike
Rock fall
Flash flood
Murder
Snake bite
Heat stroke
Plane crash
Suicide
Inset

# Pie and coxcomb charts

## Pie charts and their variants can be used to show multivariate data.

**The much derided pie chart provides the basis for many different multivariate graphs that can be used for thematic mapping.** They are often criticised, usually through misuse rather than some fundamental structural problem. A pie chart is simply a circular graph that slices data into segments in proportion to data values. The arc length (and internal angle) of each slice is proportional to the quantity it represents. The key is to ensure that, conceptually, the sum of the parts equals the whole (i.e. 100%). On a map they can usefully be used to illustrate the distribution of data at a point or for enumeration areas. They can also be varied in size to encode an additional variable to create a proportional pie chart.

**A coxcomb chart is a variation of a pie chart in which each segment is constructed using equal values but where the radius of the segment differs.** This results in segments that differ in extent away from the centre of the graph. It is usually used to show temporal phenomena in line with the original use by Florence Nightingale in her classic 1858 diagram of the mortality of the armies of the east. Nightingale used 12 segments, each showing mortality data for a single month. The radius of the sector is calculated as a proportion of the square root of the data value (which can be modified using a constant to increase or decrease all segments in relation to one another). The entire graph becomes a collection of proportional symbols that encode magnitudes and also time. When nested they can also show additional categorical detail.

**Coxcomb charts, pie charts, and their derivatives can be useful to encode numerous variables into a single symbol.** It's worth bearing in mind that although this form of symbolisation has considerable benefit in many circumstances, as with any multivariate symbol there comes a point where symbols can easily become overloaded. There is also the persistent issue of overlapping symbols, and although some of the usual mechanisms for dealing with overlaps can be employed, there is a greater risk of elements of a chart being occluded and remaining unseen.

Doughnut charts, as a variant of the pie chart, usually clip out the centre of a pie chart, which can subsequently be used to show an additional piece of information or even simply the place upon which the pie chart is centred. It might also be used to embed a smaller pie chart, perhaps also proportional, showing the same variable but for a different time period. For instance, an inner pie chart might show the age distribution of the population for one decennial census and the outer one for a subsequent census. Thus this technique can provide a way of showing basic temporal information on the map.

Techniques such as exploding segments can be used to highlight certain characteristics. Ensure the separation distance between the segments and the main chart do not suggest altogether separate features.

A ring chart (sometimes called a 'sunburst chart' or 'multilevel pie chart') is a multilevel pie chart that is used to map hierarchical data using concentric circles. The central circle represents the first value with the hierarchy being illustrated moving outward from the centre. The key is to ensure that there is some meaningful hierarchical relationship between each concentric circle.

Where possible, avoid using pie charts in three-dimensional views. They are intended to be seen as circles so, cognitively, we see the proportions of different segments relative to one another. As soon as you tip the pie chart into 3D and, possibly, add depth, you're changing the surface area of the visual and that modifies how we see relative size.

**See also:** Descriptive maps | Graphs | Treemap | Waffle grid

**Opposite:** *Birdstrikes on Aircraft by Size of Animal and Month of Year*, 2014.

Bird-related aviation strike
December
January
November
February
October
March
April
September
May
June
August
July
Number of strikes
100
500
1000
Bird size
unknown
small
medium
large

# Placing type

**Placement of text goes a long way toward ensuring it is legible, visible, and presented in a way that makes it easily read.**

**Type should be oriented horizontally wherever possible.** This is normally horizontal to the map page and border. There are two main exceptions to the horizontal placement rule. Where the map contains a graticule with curved parallels (or curved meridians if oriented in the transverse aspect), lettering might be oriented along the curvature. Additionally, if the feature to be labelled is oriented diagonally (e.g. a road) or if it is curved linearly or areally, then the type should also be oriented in the same way.

**Where possible, avoid placing type over other graphical marks or map components.** This is termed *overprinting* and is most common when black type is placed over the top of black line work such that the type is difficult to read. Where it is not possible to place the type to avoid overprinting (i.e. where it would be too detached from the feature it relates to), techniques exist to mitigate the impact. Masks place a block of colour underneath the type. Halos can also be used to deal with problems of overprinting. They tend to obscure less map detail than masks since they are constructed by placing a larger outline of the same letter form around each letter in the word.

**Callouts combine a mask and a leader line.** These are rarely used in topographic maps since they appear quite dominant and are inevitably positioned away from the actual feature. They are more normally used in thematic maps or where a particular feature requires a certain emphasis (i.e. highlighting the position of a particular feature).

**Type should be clearly placed so that it relates explicitly to the feature it represents.** This often requires some reorganisation of type to work around other graphic elements. It is worth remembering that type also takes on levels of importance with larger type (representing more significant features) taking precedence over smaller type (representing less significant features).

Masks are normally the same colour as the background colour of the map to allow type to blend in but they can also selectively mask different layers. The danger of using masks is that they inevitably obscure some of the underlying map detail, which not only removes map detail but also promotes the type as appearing to be more important than the feature it represents. A careful balance should be the objective, and each instance of mask deployment should be carefully and uniquely considered to minimise its impact.

When using halos, instead of masks, map detail can be seen between letters as opposed to the use of a mask, which obscures anything under the typographic element. The use of alternative colours and transparency can also be effective ways to dissociate text from the background. For instance, using yellow halos can highlight a particular feature, or using a transparent halo allows the map detail behind to remain while giving just enough graphical contrast.

Placing larger type first, and smaller (or less important) type as a second pass, can assist in achieving well-balanced typography. This often leads to less repositioning for the more important textual components in comparison with those of lesser importance and can assist in effective graphical structuring of the overall map composition.

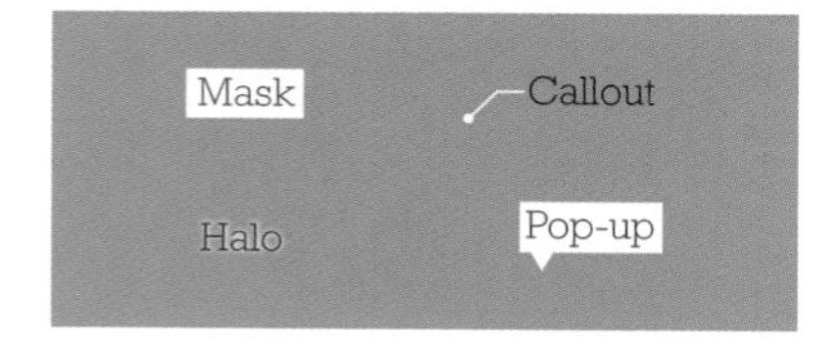


**See also:** Graphic and dynamic labelling | Guidelines for lettering | Placing type for areas | Placing type for lines | Placing type for points

**Opposite:** The complexity of a names 'plate' or layer isolated from the basemap.

# Placing type for areas

**In addition to general type placement guidelines, there are specific considerations for lettering area symbols.**

**For areas that have sufficient extent to fully contain the label, position it in the visual centre of the feature.** Aim for at least the equivalent of one and a half times the type size between the ends of the label and the feature's boundary. This guideline is designed to prevent the label from crowding the feature and dominating it visually. Where possible, orient the label along a horizontal baseline (or following the line of a curved graticule). Many features are, however, irregular in shape and have a dominant orientation, which should also be followed in the same manner as labelling a generalised linear path.

**Uppercase type is often used for areal features to help emphasise extent.** Emphasising areal extent can also be achieved using letter and word spacing effectively and is, again, usually applied to uppercase lettering. Applying letter spacing to lowercase type tends to leave it looking disjointed. Care should be taken when exaggerating letter or word spacing to avoid individual letters or words becoming isolated. This problem is usually a function of other underlying features that might cross a large area and, subsequently, cross the area's label. The visual relationship between letters and words should be retained despite the existence of other map detail across the same extent.

**Increasing the leading can also be used to emphasise areal extent.** The only caveat is to ensure that the relationship between words is maintained and that words on different lines do not appear isolated or take on a meaning of their own.

**Areal features that are too small to contain their label should be labelled in the same way as point symbols.** This might be through placement of the label in close proximity or by using thin leader lines. This approach is preferred over using much smaller type sizes, which might create different levels of visual hierarchy among the labels used for similar features.

Labelling areas internally can be horizontal, curved according to a graticule, or along the medial axis of the feature itself.

On some maps, area labels may be better shown on alternate sides of a boundary between areas. For instance, labelling areas on both sides of a political border, in a manner similar to an offset linear feature label, can help the map reader differentiate the areas and recognise the importance of the border itself.

In some circumstances, complex relationships between area features may exist such as land parcels containing buildings on large-scale maps or lakes containing islands (that themselves may contain lakes!). The key to these situations is to avoid conflicts and overlapping labels. It's often necessary to place labels outside of an area boundary in these situations or to begin a label in the area and have the label extend away. Using curved labels can help emphasise that the label relates to a small or contained polygon if it is otherwise not curved.

**See also:** Areas | Graphic and dynamic labelling | Placing type | Placing type for lines | Placing type for points

I N D O N E S I A
Democratic Republic of the Congo
Sweden
South Africa
Swaziland
Lesotho
Italy
Saudi Arabia
Bolivia
Puerto Rico
Russia

# Placing type for lines

## In addition to general type placement guidelines, there are specific considerations for lettering linear symbols.

**Labels that accompany linear features should be placed above the line wherever possible.** The line provides a visual base upon which the text appears to rest. Placing type below a line gives the impression it is hanging from the line and less visually harmonious. Further, the bottom edge of lowercase type is often less ragged than the upper edge so also causes less visual disharmony. The label should be close to but not touching the line. It is important to accommodate descenders of lowercase type (e.g. *y*, *g*, *p*) so that they do not touch the line symbol.

**Where a line curves, it is desirable to curve text along the same path.** However, some attention should be paid to the sinuosity of the line since a more sinuous line would be unsuitable as an anchor for curved text. In this case, following the general curvature of the line will result in a better visual appearance of the label. Rivers, in particular, often cause these sorts of placement issues, and following a smoother, generalised linear orientation works well.

**It is common practice to label lengthy linear features more than once across the map.** This is normally preferred over the use of exaggerated letter and word spacing since the sinuosity of lines might lead to dissociation between individual letters. The exception would be for street labels where the two words are obviously linked such as 'Burke Street'.

**Inherent orientation means that labels can easily be placed upside down or wrong-reading.** Correctly placed labels should be upright and also read from left to right. Some adjustment might be necessary considering the proximity of other map detail but these general principles should apply where possible. For vertical features, convention states that type should be read from the right side of the page (i.e. the label reads from the bottom to the top in normal orientation). Attention should also be paid to any part of a word that tilts upside down, and measures should be taken to avoid this situation.

Some linear features tend to have their own specific placement requirements, including street networks and contours.

Modern street datasets often have vertices at each intersection, so when labelling, multiple placements occur. By grouping features you'll be better able to label the street as a whole rather than individual segments. Placement for streets will also be determined by the scale of the map. On large-scale maps where streets are cased, there's often room to place the label within the casing, centred along the street itself. Labels can be straight or curved. Word and letter spacing can be modified to avoid placement near junctions. For short streets with long names, abbreviations, stacking, or overruns provide additional options for placement. Street maps also often show address ranges for properties on each side of the street. Placing street addresses along the street, relative to and adjacent to junctions, is optimum.

There are often far too many contour lines to individually label, and their length might ordinarily result in multiple labels per feature. Brevity is the best solution for contour label placement ... the fewer the better since contours themselves are somewhat artificial abstractions for representing relief and labels add visual clutter. Label a contour line as a whole, once and, where possible, align labels in ladders. This also allows you to avoid having to label every contour, perhaps only labelling index contours. Labels should be placed aligned to the page ensuring that they are never upside down (and curved to the contour line if desired) or, alternatively, aligned to the contour elevations so the top of the label is always facing uphill.

**See also:** Graphic and dynamic labelling | Lines | Placing type | Placing type for areas | Placing type for points

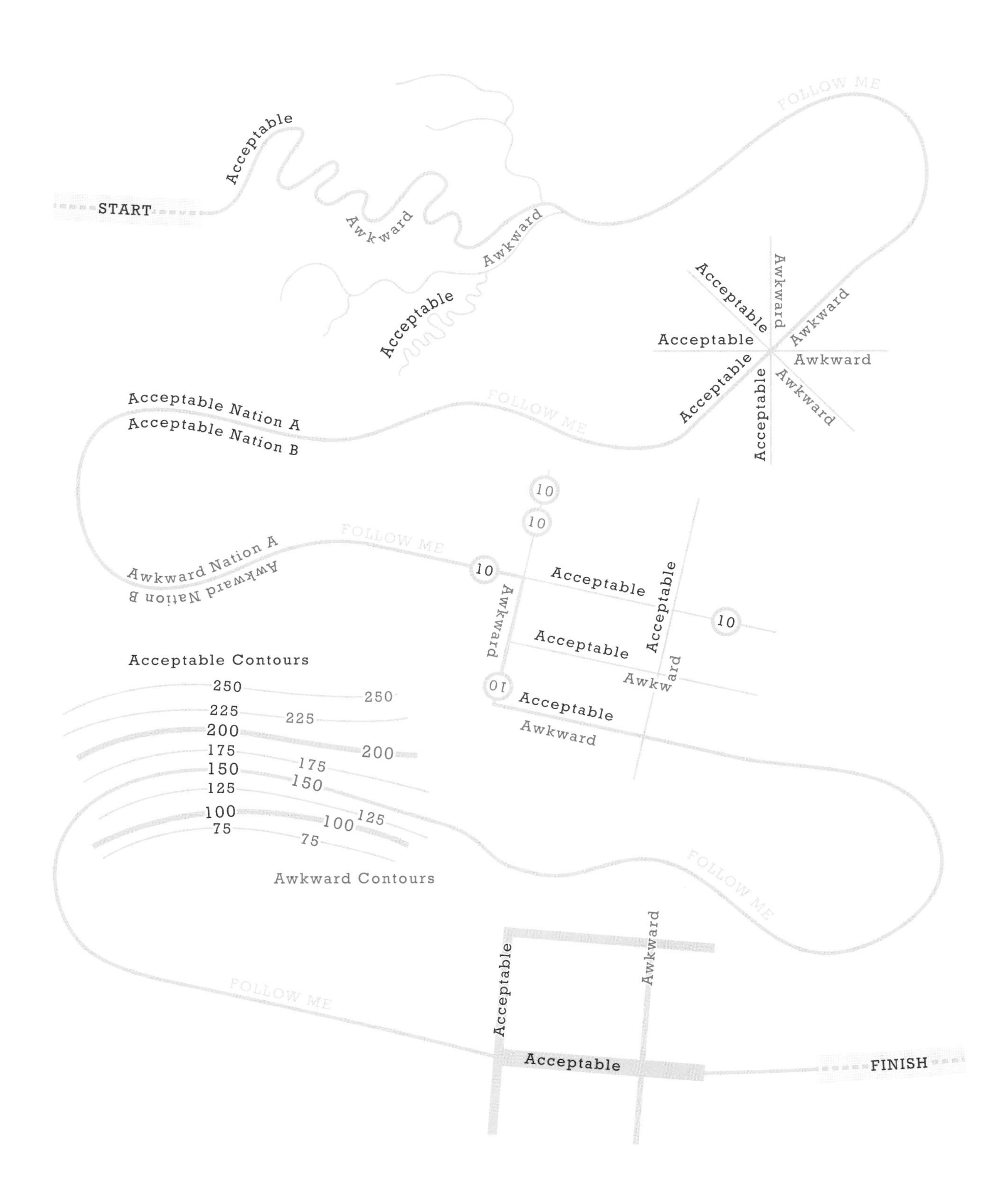
START
Acceptable
Awkward
Awkward
Acceptable
FOLLOW ME
Acceptable
Awkward
Acceptable
Awkward
Awkward
Acceptable
Acceptable
Awkward
Acceptable Nation A
Acceptable Nation B
FOLLOW ME
Awkward Nation A
Awkward Nation B
FOLLOW ME
10
10
10
Acceptable
Acceptable
10
Awkward
Acceptable
Awkward
10
Acceptable
Awkward
Acceptable Contours
250
250
225
225
200
200
175
175
150
150
125
125
100
100
75
75
Awkward Contours
FOLLOW ME
Awkward
Acceptable
FOLLOW ME
Acceptable
FINISH

# Placing type for points

## In addition to general type placement guidelines, there are specific considerations for lettering point symbols.

**Avoid placement positions that overlap with underlying map components.** The accepted order of placement suggests that the feature should be placed on the left and the label to the right with a suitable offset either above or below the baseline of the feature itself. Placement to the right or left on the same horizontal baseline as the feature should be avoided so that the map reader cannot mistake the symbol used to represent the feature with its label.

**Avoid the positioning of a label such that a map component is visible between it and the feature it represents.** Optimum location must often be compromised to avoid this problem and to ensure the label is optimally visually associated with its feature.

**In the event that there is no optimum position, a mask, halo, or callout may be the best solution to avoid conflicts between labels and map detail.** A further option might be the use of a simple leader line. When designing leader lines, they should be distinct from all other map detail, very thin (e.g. 0.25 point) or light grey, not include an arrowhead, and should 'lead' to the centre of the symbol without actually touching it.

**Where multiple labels are placed around a single feature, the same order of positioning should be followed.** Individual lines of text should be arranged so they are right or left of centre justified, which helps emphasise the association between the multiple labels and the feature.

**Exaggeration of word or letter spacing should be avoided for labelling point features.** Such an approach weakens the association between the label and feature. Exaggerated spaces are usually reserved for emphasising areal extent.

Labels that relate to point features along a coastline should be placed entirely on the land if possible to emphasise that the feature is on land rather than offshore. Failing this (given a large amount of land-based map detail), then placing the label entirely offshore is the next best solution. However, because of the relative lack of base detail in offshore areas, consideration should be given to the different levels of prominence that type placed offshore might be given. The principle in placing features that are on the coastline is to avoid overprinting the coastline with type.

Some subtle modifications to point placement can also significantly improve the overall composition. Although automated label placement will usually support a consistent gap between feature and label, adding a shift provides flexibility to nudge labels a little to accommodate other map detail. Small nudges are almost imperceptible but can greatly support clarity and legibility.

When using a curved graticule, it can sometimes make sense to align (and curve) the placement of point labels to the lines of the graticule.

Rotating labels  by a specified angle might allow you to associate a label with an attribute of the data you're mapping. For instance, if the feature itself has some sort of orientation, using a label that shares the orientation can emphasise the characteristic. For this, and other similar instances where type is not positioned horizontally, it's important to avoid flipping and upside-down text.

**See also:** Graphic and dynamic labelling | Placing type | Placing type for areas | Placing type for lines | Points

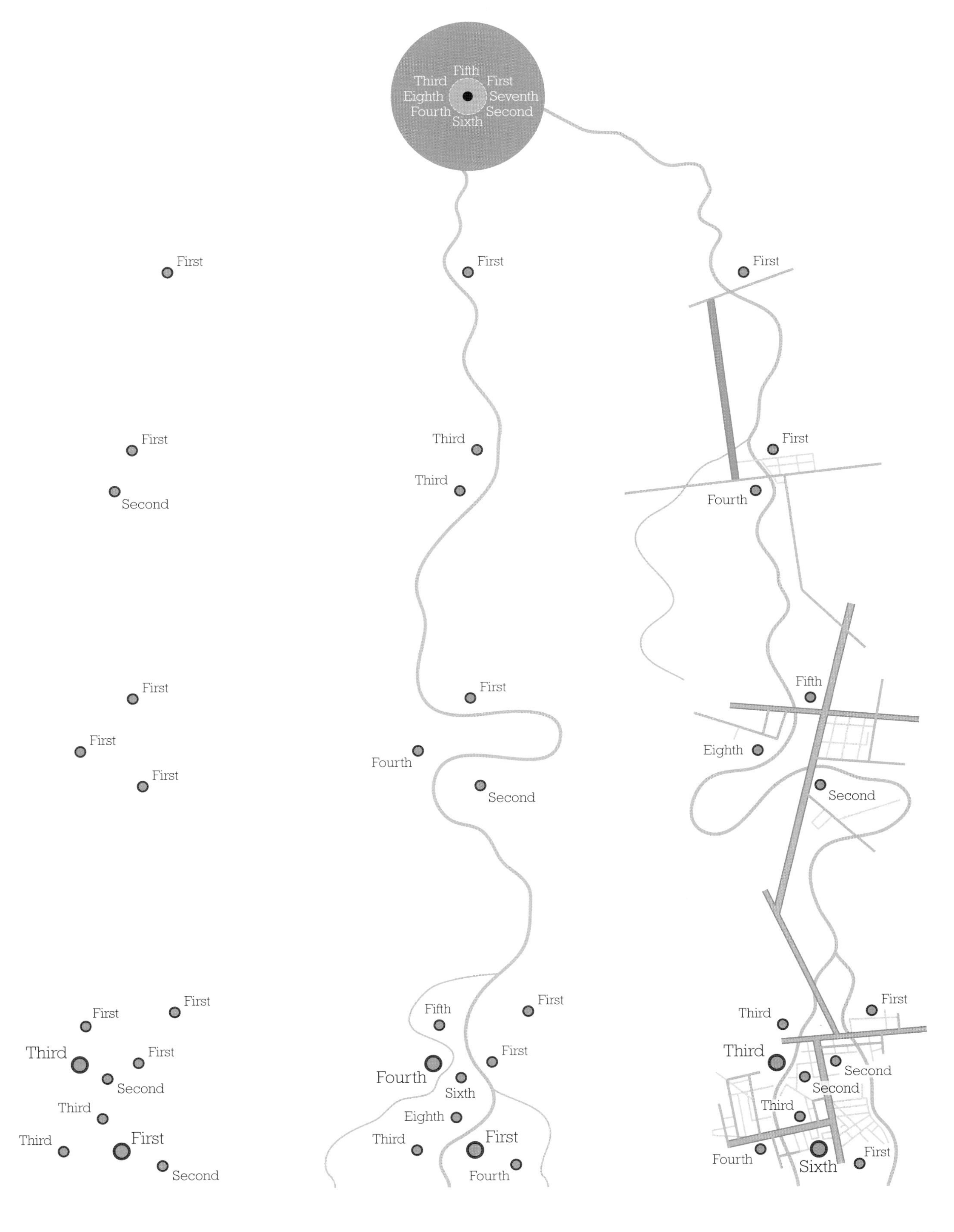

Fifth
Third
First
Eighth
Seventh
Fourth
Second
Sixth
First
First
First
First
Second
Third
Third
First
Fourth
First
First
Fifth
First
Fourth
Eighth
First
Second
Second
First
First
First
Third
First
Second
Third
Third
First
Second
Fifth
First
First
Fourth
Sixth
Eighth
Third
First
Fourth
Third
First
Third
Second
Second
Third
Fourth
Sixth
First

# Planetary cartography

## Mapping of extraterrestrial objects.

**Making maps is not limited to Earth or, even, to objects we can see with our own eyes.** Mapping of extraterrestrial objects has long been a discrete arm of cartographic practice. In Galileo's 1613 work *Istoria e dimostrazioni intorno alle maccie solari*, he combined words and images to show early maps of Saturn from astronomical observation. The simple diagram, oOo, clearly illustrated the main body of the planet and its rings. Planetary cartography is concerned with the totality of efforts among space agencies and researchers to calibrate, analyse, and present data about planetary surfaces. Before a new extraterrestrial object can be examined, it needs a reference coordinate system. The size, shape, and rotation of the object must be determined along with choices for latitude and longitude. Locations can then be determined, and that, in turn, supports comparisons with other observations and the development of theory. This is largely undertaken through remote sensing, which demands an understanding of the instrumentation, the sensor model, and the ability to interpret results. Issues of consistency, accuracy, uncertainty, and reliability are key. Maps provide the key mode of communication of, often, scientific results but also for determining landing site selection.

**Planetary maps tend to focus on physical characteristics of extraterrestrial surfaces.** For instance, topography can be relatively easily measured using remote-sensing techniques. Subsurface geology can be determined through the analysis of samples or from data transmitted back to Earth or even through analysis of the morphology of a surface and interpretation of past climate and conditions. Planetary imagers routinely provide data to enable us to better understand and map planets and other extraterrestrial objects.

**Space exploration and a desire to map other worlds has yielded a wide range of cartographic products.** Airbrush artistry has given way to photographic interpretation, analytical photogrammetry toward an analysis, and visualization of multiple digital data sources.

Maps of extraterrestrial objects perhaps exhibit some of the widest array of design approaches of any type of map. Topographic map sheets are a common form of representation of surface and subsurface features. Although primarily designed to support the scientific community, the maps are often highly artistic and colourful. Feature paucity is often replaced by a high density of typographic components through the scientific naming of features.

3D and animated visualizations give us a sense of what a planet might look like and whet the appetite for exploration and discovery. Many maps and visualizations are designed to excite the imagination and garner support for further missions. The map, therefore, becomes as much a marketing tool and an advert for a planet as it is a scientific tool.

Maps of other worlds have also been ripe for artistic interpretation, much like historical maps of Earth from the Age of Discovery. The analogy is clear. Many Victorian maps, for instance, illustrated Earth at a time when most of it was a mystery. Although we can see other planets, they, too, remain something of a mystery. The juxtaposition of detailed and accurate data about our nearest planets with a historical map aesthetic has provided a way to imagine alien worlds, which seem close by but remain distant and devoid of (actual) human contact. There's so much that is yet to be explored and understood that it lends itself to such stylistic treatment.

**See also:** Continuous surface maps | Digital elevation models | Shaded relief | Topographic maps

**Opposite:** *Here There Be Robots: A Medieval Map of Mars* by Eleanor Lutz, 2016.

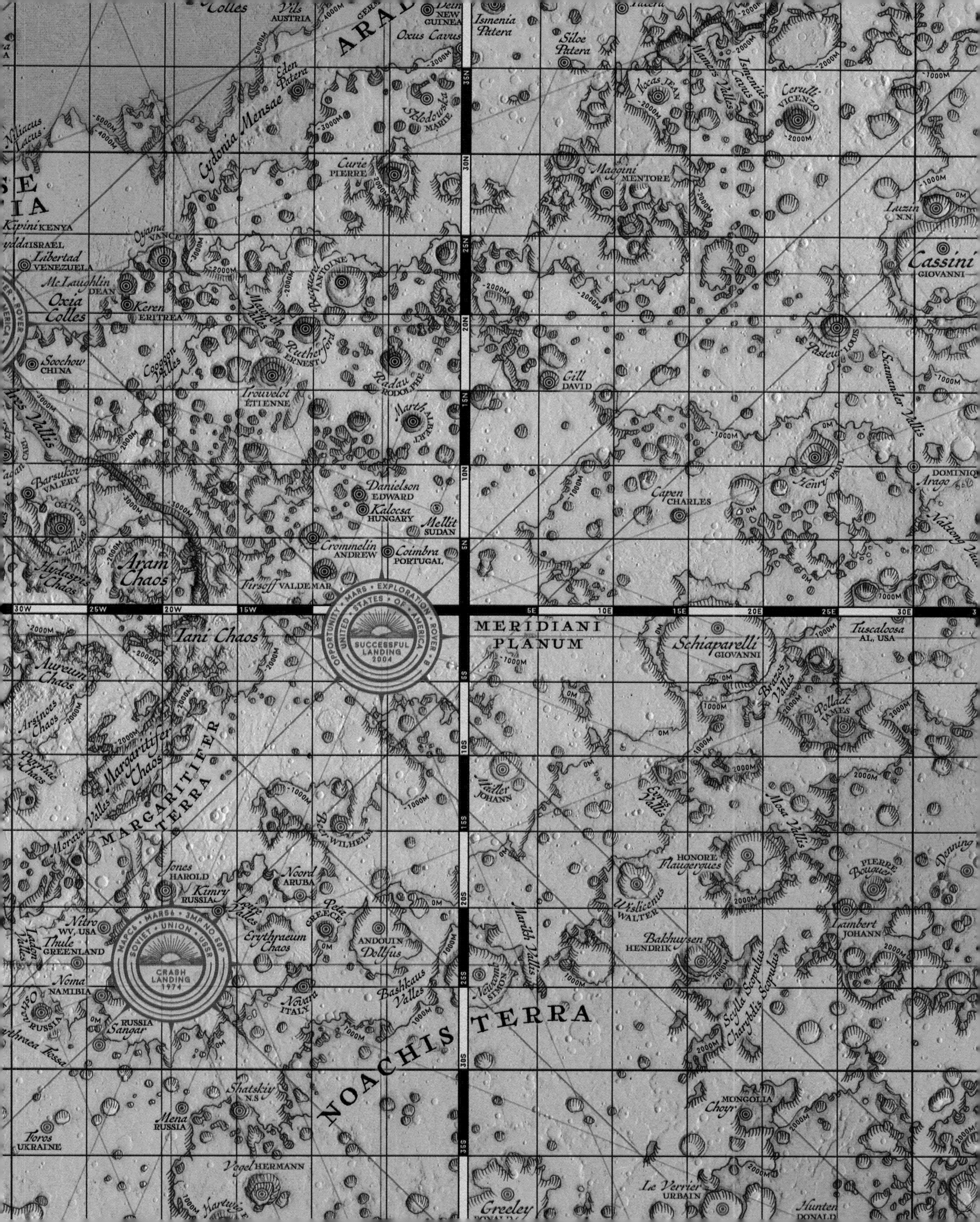

MERIDIANI
PLANUM
NOACHIS TERRA
MARGARITIFER
TERRA
OPPORTUNITY · MARS · EXPLORATION · ROVER · B
UNITED · STATES · OF · AMERICA
SUCCESSFUL
LANDING
2004
SOVIET · UNION · USSR
CRASH
LANDING
1974
Cydonia Mensae
Oxia
Colles
Aram
Chaos
Tani Chaos
Aureum
Chaos
Arsinoes
Chaos
Margaritifer
Chaos
Erythraeum
Chaos
Schiaparelli
GIOVANNI
Cassini
GIOVANNI
Curie
PIERRE
Maggini
MENTORE
Trouvelot
ÉTIENNE
Danielson
EDWARD
Kalocsa
HUNGARY
Mellit
SUDAN
Crommelin
ANDREW
Coimbra
PORTUGAL
Firsoff VALDEMAR
Gill
DAVID
Capen
CHARLES
Luzin
N.N.
Arago
Tuscaloosa
AL, USA
Pollack
JAMES
Mädler
JOHANN
Wislicenus
WALTER
Bakhuysen
HENDRIK
Lambert
JOHANN
Dollfus
ANDOUIN
Noord
ARUBA
Jones
HAROLD
Kimry
RUSSIA
Nitro
WV, USA
Thule
GREENLAND
Noma
NAMIBIA
Sangar
RUSSIA
Novara
ITALY
Shatskiy
N.S
Mena
RUSSIA
Toros
UKRAINE
Vogel HERMANN
Le Verrier
URBAIN
Greeley
Choyr
MONGOLIA
Barsukov
VALERY
Soochow
CHINA
Keren
ERITREA
McLaughlin
DEAN
Libertad
VENEZUELA
Kipini KENYA
Oyama
VANCE
Ismenia
Patera
Siloe
Patera
Oxus Cavus
Eden
Patera
Vils
AUSTRIA
Scamander Vallis
Mosa Vallis
Brazos
Valles
Bashkaus
Valles
Marikh Vallis
Scylla Scopulus
Charybdis Scopulus
30W
25W
20W
15W
5E
10E
15E
20E
25E
30E
35N
30N
25N
20N
15N
10N
5N
5S
10S
15S
20S
25S
30S
35S

# Point clouds

## Point cloud data, derived from lidar, creates highly detailed 3D renderings.

**Light detection and ranging (lidar) uses pulses of laser light to capture data. Lidar equipment is one form of remote-sensing device, and the data collected is in the form of points.** Each point is positioned in 3D space as an x, y, z coordinate and represents the surface from which emitted laser light is reflected back to the sensing device. Measurement can be calculated by determining the distance between the sensor and the reflecting surface based on an assessment of the speed of return of the laser light. The collection of all sensed points becomes a point cloud and often incorporates millions of measurements.

**Different returns can be captured simultaneously.** For example, a forest canopy can be recorded at the same time as lower-level vegetation and the forest floor. In urban areas, roofs of buildings can be captured at the same time as the general ground surface.

**Point clouds can be mapped in several ways.** Lidar obtained from airborne remote sensing can be used to create high-resolution digital surface models, often processed into polygonal meshes and symbolised using typical techniques such as hypsometric tinting to create planimetric representations of topography. These surfaces can be used to derive more generalised products or to support the creation of detailed 3D landscape models. The point clouds can be used directly in 3D applications with each point rendered using colours mapped from high-resolution digital photography captured at the same time as the scan. Together, the coloured points create a 3D picture of a scanned environment simply through the structure of the rendered coloured points. They can subsequently be interpolated to build 3D surfaces.

**See also:** Digital data | Digital elevation models | Points | Spatial dimensions of data

Lidar data tends to be large in terms of the number of datapoints and, subsequently, the amount of storage needed. It supports high-resolution mapping and can typically be used to create 1 m resolution digital surface models. These maps provide an incredible amount of detail, often showing very specific topographic features that would otherwise be missed from traditional surveying techniques.

Point clouds have become a form of mapping for the capture and representation of data that conventional survey would struggle to perform. Although large-data capture efforts have been undertaken to create high-resolution detail for mapping topography using airborne lidar, perhaps the mapping of relatively small areas at large and detailed scales using terrestrial lidar is a key use. For instance, the mapping of cave systems or archeological digs in incredible detail not only provides detailed scientific data (for instance, to accurately measure volumes and distance) but views of environments never previously possible. Being able to map an entire cave system, and then interact with the rendering on screen literally opens up hidden worlds. Traditional cave maps tended to be highly generalised. They would take planimetric and profile forms that, together, gave an impression of the three-dimensional characteristics of a system. Through the mapping of point clouds, the system can be seen and interacted with in 3D.

Mapping of point clouds is a good example of the impact of technological change on the art and science of cartography. The science of measurement is relatively known but applied using new equipment. The art of representation has changed to take advantage of new techniques of data capture. The medium of representation has also changed from 2D paper maps to 3D interactive screen-based cartography.

**Opposite:** *Digital Mapping of China's Miao Room Cave System*, by ixtract, 2017.

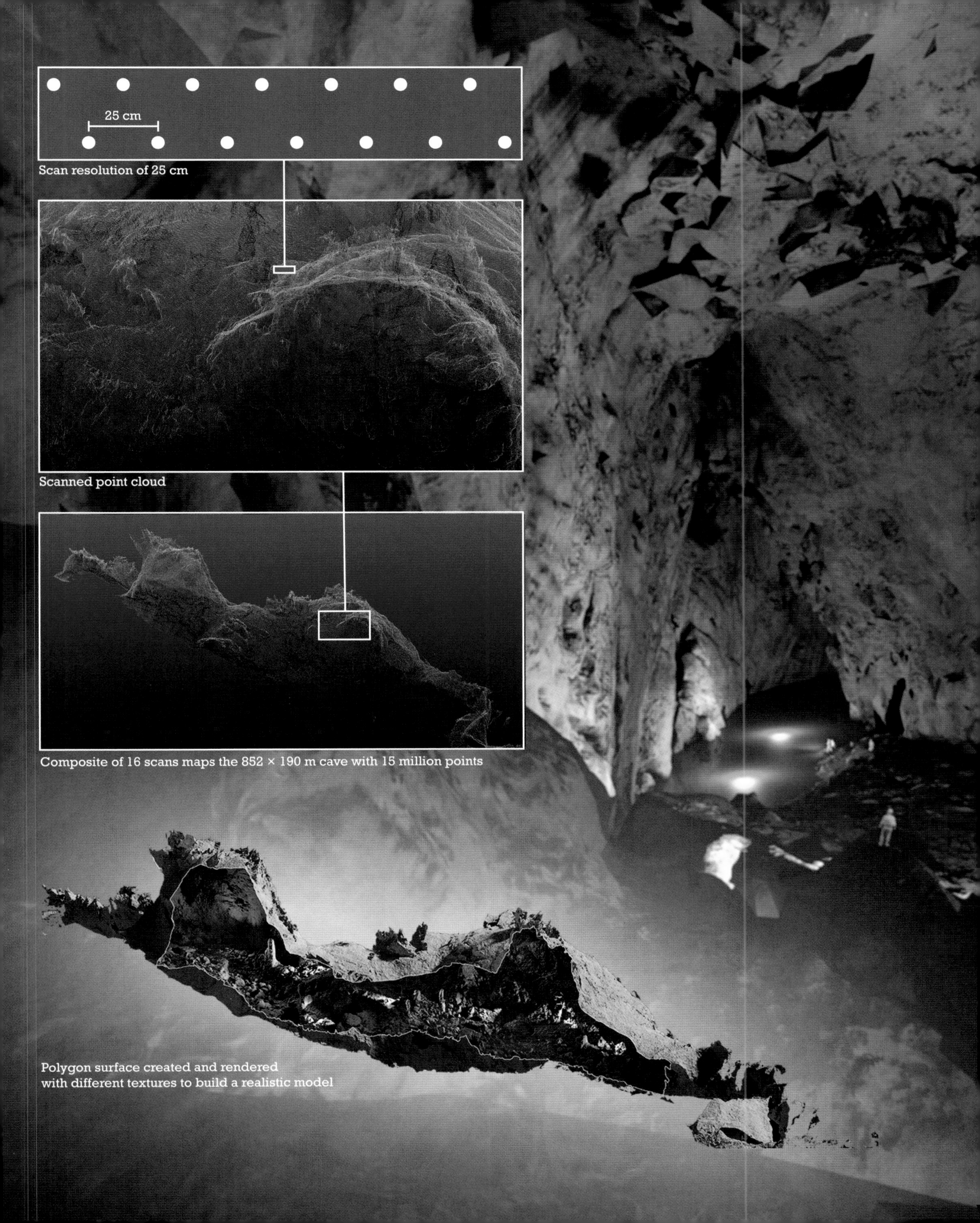


Scan resolution of 25 cm

Scanned point cloud

Composite of 16 scans maps the 852 × 190 m cave with 15 million points

Polygon surface created and rendered
with different textures to build a realistic model

# Pointillism

## The power of the dot can be used in many different ways.

**Pointillism refers specifically to a painting technique yet it has a value in cartography.** In art, the pointillism technique is defined by the application of small dots of paint. Each dot is of a different colour, and the overall pattern of the dots forms the overall image. It's a form of Impressionism developed by Georges Seurat and Paul Signac in the late 1800s. The technique relies on the limits of the human eye and brain, which are unable to resolve individual dots. The recognition of individual points of paint becomes blurred into an overall image and a fuller picture because it's impossible to see the fine detail of the original points.

**Cartography has always had a relationship with pointillism.** The four-colour CMYK offset printing process lays down small dots of the different inks. The dots are placed at different screen angles and at different sizes and spacing. Without magnifying the printed map, you would struggle to see the individual colours yet at normal reading distance they appear to merge to form a rich palette of colour. Many digital displays also lean on the principles of pointillism by emitting red, blue, and green colours as small pixels that we see as merged colours at a normal viewing distance.

**Pointillism can also be a useful symbology.** Maps often make use of circles as symbols, and we vary the size and colour to represent certain characteristics. Positioning dotted symbols of a uniform size in a particular array becomes dot density maps. By using the same approach in colouring dots and positioning them carefully, you can form a different hue in the mind of the map reader. For instance, making a dot density map of US political voting patterns from a distribution of blue and red dots can lead to the impression of purple in areas with similar voting numbers for the different candidates. This technique effectively encodes additional detail into the map than what the original symbology supports.

Pointillism can also be used creatively. One of the main criticisms of the widely used choropleth technique is that large areas always visually dominate the map. Even taking into account data that has been properly processed into a rate or some other relative and comparable metric, the fact remains that if the spatial pattern of the map contains areas of largely varying size, the fundamental problem persists. This can be acute since, often, the size of an area in ground cover is simply a convenient size to reflect a similar number of people who may live in that area. This is why we see administrative units that are very small in populated urban areas but larger in more sparsely populated rural areas. Using points can be a way to overcome the problem.

Choropleth maps use shaded fills as the symbol treatment for areas. The entire area receives the fill. An alternative approach would be to specify a size of area that acts as a threshold beyond which you have determined that the area's pure size is becoming too visually dominant. In all the areas above this threshold, a single point symbol is positioned in the centre of the area, symbolised by the same hue/value that you'd ordinarily apply to the area fill. The map becomes a pointillist choropleth hybrid. The dominant large areas now have a systematic symbology that reduces their visual weight across the map. The overall impression is now much more harmonious, and the focus shifts.

Pointillism can also be used to create shade on a map. The use of dotted pattern fills has a long history in the design of relief shading. By changing the size, amount, and positioning (usually random) of dots, you can modify the amount of shade and create an expressive, artistic look to relief shading. Dots have also been used to represent contour lines that tend to give a retro-appearance to maps.

**See also:** Choropleth maps | Dasymetric maps | Data density | Points

## Choropleth

Larger areas on a choropleth map will always visually dominate. This exaggerates their visual importance, often to the distraction of much smaller areas.

The irony is on most enumeration maps, smaller areas are those where more people live or where the population density is greater.

## Pointillist choropleth

Here, for areas over a specified size a dot is used instead of an area fill.

The dot colour matches the classification and symbolisation of the areas and is sized to be similar to the areas.

The result recalibrates the visual balance so that larger areas no longer dominate our perception.

# Points

## Possibly the most fundamental of marks on a map.

**Points serve many different functions on a map.** Graphically, they can be used to represent phenomena that are located at a point. This might extend to phenomena that are referenced to a point. Cartographically, point symbols can also be used purely for mapping purposes even if the phenomena they represent isn't itself a point feature.

**Point-based phenomena have zero dimensionality.** That is, they have no width or area, and the only characteristic is that they have an x,y coordinate. In fact, survey trig points are a good example of this phenomena since they are defined only by their coordinates. An example of a feature referenced to a point might be a statistical measure such as the average position of a set of point features. Although no real feature may exist, the average can be calculated and referenced. Two-dimensional features are often shown as points. Points of interest are a common map object on topographic maps. Point symbols may be geometric or mimetic, and they show the location of a phenomenon. Many phenomena are best shown as points to avoid the complexity of trying to draw their footprint. Indeed, many point features simply cannot be drawn to scale on many maps, and so a point symbol provides an excellent graphical solution.

**Larger areas such as towns and cities will increasingly be shown as points as map scale reduces**. There simply isn't space on the map, and there is rarely a need at medium or small scales to know anything other than position and name. Technically, this is one of the processes of generalisation to provide a suitable graphical way of encoding information.

Exploring these ideas using trees gives us a way of seeing the differences between how we represent points as well as their relationship with scale. Tree data might be captured by mapping the two-dimensional area of the crown of each tree. However, crowns vary in size as the tree grows. A better way to capture position is to map a point that represents the centre of the trunk.

At a medium scale, the trees might be shown as a geometric point symbol, perhaps a small circle coloured green to connote vegetation. As the map scale gets larger, more room emerges, and instead of a simple geometric point, the symbol may become proportional with a circle being used to represent the extent of the crown, which may be attached to the data as an attribute. The symbol may even get a green fill or, perhaps, a mimetic symbol is used to show difference in species. At the other end of the continuum of scale, it would be impossible to show individual trees as a collection of points. The map scale demands a different approach, and the individual points become amalgamated into an area symbol indicative of a stand of trees.

As scale changes and the depiction of trees is modified, the representation of a survey control point does not change. The point symbol remains the same across scales because it is simply an x,y coordinate with no characteristics that might be used to give it dimensionality for symbol purposes.

**See also:** Areas | Isotype | Lines | Pictograms | Shape | Size

**Opposite:** Points are shown isolated from the complete map. They sit atop the area and linear features.

# Position

## Position on Earth is measured using lines of latitude and longitude that form the graticule.

**The pattern of the lines that form latitude and longitude on Earth is called the *graticule*.** If we assume that a point is positioned 40°N and 60°E, then a latitude value of 40° is the angle formed between a line passing through the point on Earth's surface and the centre of Earth and a plane passing through the centre of Earth and the equator. Latitude is usually symbolised by the capital Greek letter phi (Φ). A longitude value of 60° is calculated by the angle formed between a plane passing through the point and the centre of Earth and a plane passing through the prime meridian at Greenwich. Longitude is usually symbolised by the lowercase Greek letter lambda (λ).

**A location on Earth's surface is usually stated as the value of latitude followed by the value of longitude.** So the location of the Statue of Liberty in New York Harbour is given as 40°41′20′′N and 74°2′42.4′′W. In decimal degrees this is specified as 40.6894, –74.0447.

**On a Cartesian plane, longitude is *x* and latitude is *y*.** Despite the usual form of specifying latitude and then longitude, we also often specify coordinates as *x* followed by *y* (x,y). Clearly this can lead to confusion.

**With increasing use of geodetic rather than geocentric datums, the position of the prime meridian has changed.** Modern satellites orbit around the centre of the mass of Earth. Their geodetic reference systems result in the modern prime meridian being 5′3′′ east of the astronomic Greenwich prime meridian. At the latitude of Greenwich, this difference is 102 m.

The problem of the order of specifying latitude and longitude ought to be covered by the fact that there is an international standard for representing geographical point locations—International Organisation for Standardization (ISO) 6709. The general rules state that first, the horizontal coordinate is specified (x, longitude) with negative values west of the prime meridian and positive values east of the prime meridian. Second, the vertical coordinate (y, latitude) is specified with negative numbers south of the equator and positive north of the equator. Third, the vertical coordinate of height or depth is specified (though optional), and finally, the identification of the coordinate reference system (CRS) is specified, again optionally.

Lines of latitude can be derived from the natural starting points of the poles and use the equator as zero. Longitude is slightly more complicated as it has no natural starting point or zero line. Up until 1884 maps of different nations used their own zero lines or prime meridians, but at the International Meridian Conference in Washington, DC, 22 countries voted to adopt the Greenwich meridian as the prime meridian of the world. The French abstained and continued using the Paris meridian until 1911.

**See also:** Datums | Earth coordinate geometry | Earth's shape | Globes | Latitude | Location | Longitude

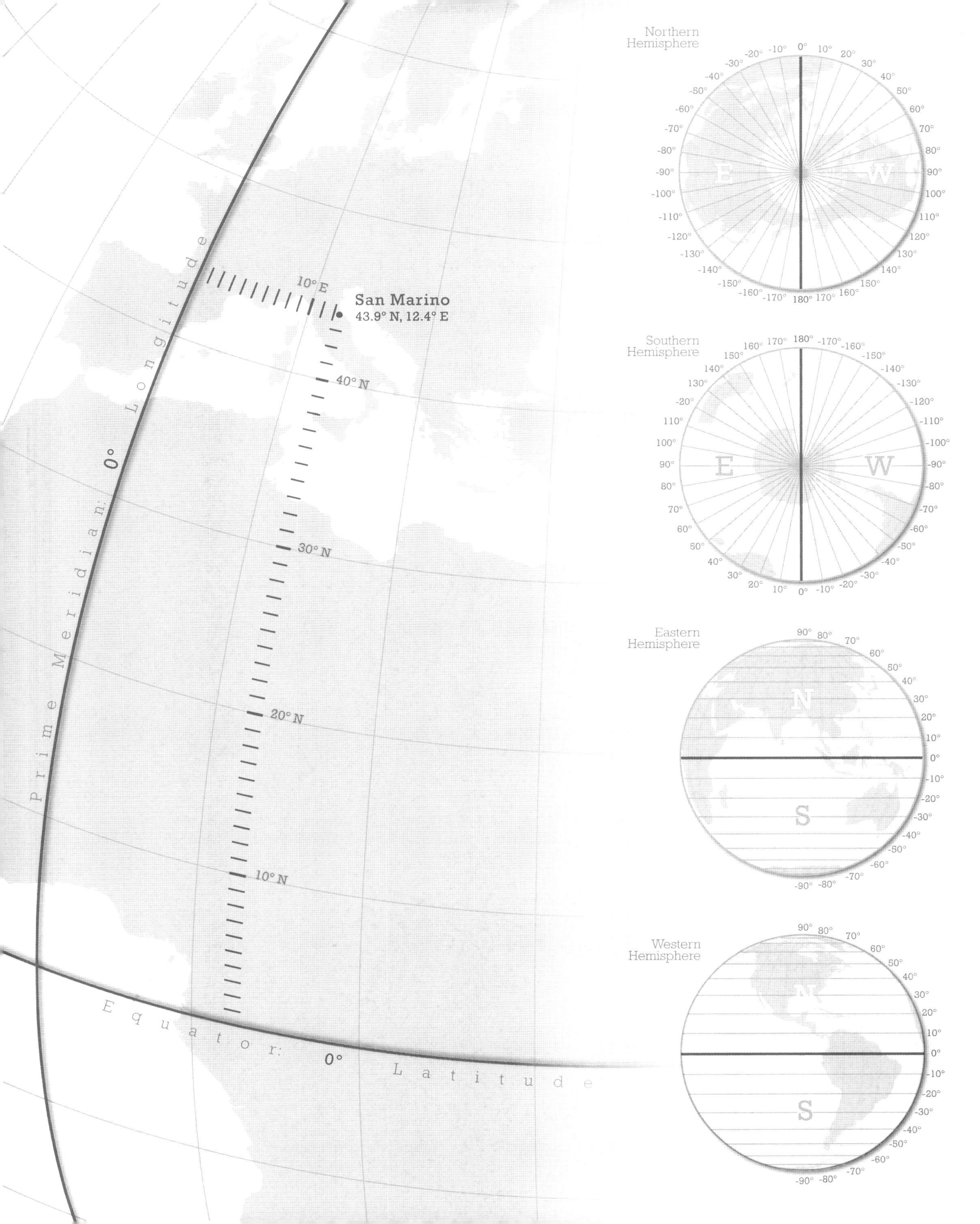

Prime Meridian: 0° Longitude
Equator: 0° Latitude
10° E
San Marino
43.9° N, 12.4° E
40° N
30° N
20° N
10° N
Northern Hemisphere
Southern Hemisphere
Eastern Hemisphere
Western Hemisphere
E
W
N
S

# Printing fundamentals

## Printed maps aren't dead. At least not yet.

**The design of your map is inextricably linked with how it will be produced.** At the outset of any mapping project, making the simple decision of whether you are making a map for print or for the web is key. Each medium presents specific opportunities and constraints. Defining your map's specifications requires a knowledge of the intended production process. Printing remains crucial for many so an understanding of how printing works is useful.

**Desktop printing is accessible for making a few copies but commercial production presses handle most printing.** Maps to be printed should always use the CMYK colour model. Printing uses each component of the colour model separately in the four-colour process. Conversions from RGB are rarely faithful when printed. Commercial printing of maps has closely followed the development of the letterpress (relief) through engraving (intaglio), lithography (planar), and digital. Lithography remains widely used. CMYK becomes separate layers on printing plates, and so colour registration is vital for the quality of the final product. If a map is to be commercially printed, registration marks and/or crop marks are added to the work outside the intended final print area. The prepress phase of making a commercial product demands that you spend considerable time editing and checking. Once you supply the files to your printer, you will lose control over the output. Obtain proofs, colour proofs, and check them. Mistakes on large print runs can be costly.

**Digital printing presses apply ink directly from the file, via the print head to the print material.** No plates are used. Dry ink is vaporised and sprayed onto electrically charged paper. The toner is attached to small electrically charged dots, which are then bonded to the paper by a heating mechanism. Digital presses produce short-turnaround, high-quality on-demand printing. It avoids much of the costly setup compared with lithography, although as an emerging technology, the quality can sometimes not be as detailed and the relative scarcity of large-format digital presses means costs are relatively high.

Desktop printing can be by inkjet, laser, or plotter. Laser uses four-colour toners to electrostatically adhere toner to the paper. Ink is applied as dots per inch (DPI), and higher dpi output gives sharper detail. Inks are laid down at different angles in the same way as lithographic technology to avoid overprinting. Inkjets use a print head that propels small drops of liquid ink onto the paper. Dots are applied in a random, non-continuous pattern with increasing colour achieved by laying down more dots. Some bleeding can occur, particularly with lower quality paper stock. Plotters are large-format inkjet printers for large-format output, conventionally up to 44 inches in width.

Maps to be printed commercially are usually supplied to the printer digitally as a file which supports embedding of vector objects, raster images, fonts, and symbols. Digital files are converted into four-colour separation plates, normally made of thin aluminium which has an emulsion layer that takes the image. Colour is applied by passing the paper through the press with the value of colour controlled by changing the size of the dots laid down. Dots for each separation layer are generated at different angles (15° C, 75° M, 90° Y, and 45° K). These screen angles allow for ink to be printed in the same area without overprinting, which, visually, gives the impression of a much wider palette of final colour. The offset lithographic process takes place when ink is applied to a plate attached to a drum. The ink sticks to the image on each plate. Water removes ink from parts of the plate that do not have an image. The drum rotates, transferring the inked image to a wrong-reading drum composed of material coated on fabric. This drum then transfers the image back to its right-reading form on the paper. Presses sometimes have more than four plates so they can apply additional spot colours.

**See also:** Additive and subtractive colour | Colour charts | Old is new again | Style, fashion, and trends

## Offset lithographic printing

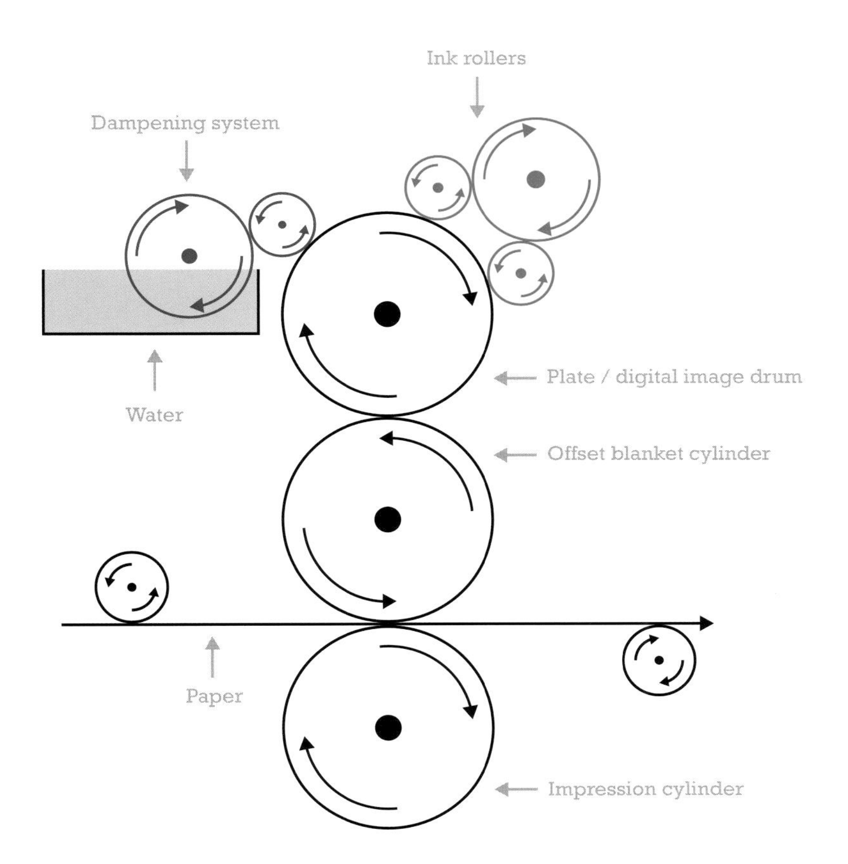


1. Moistening and application of colour to the printing plate

2. transfer of colour from the printing plate to offset blanket

3. Transfer of colour from the offset blanket to paper

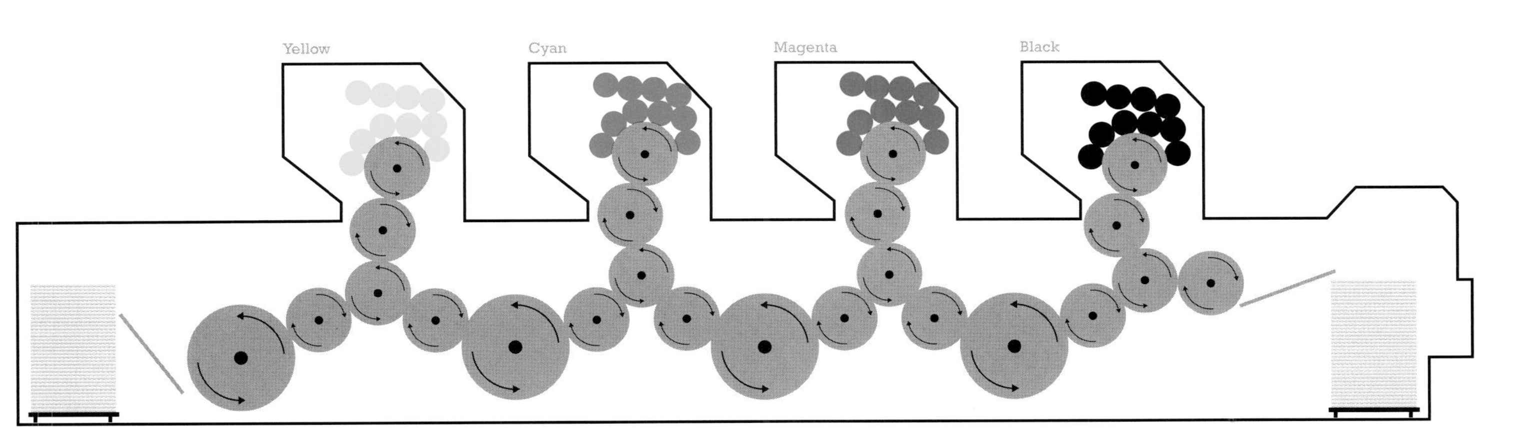

4. Process repeated for the remaining colours

# Prior (c)art(e)

## Learning from the past informs your work.

**Very little work is truly original.** One of the biggest disappointments in any creative process is finding that someone beat you to the idea. But instead of that becoming a problem, it can be used to your advantage. All creative work builds on what has come before. Immerse yourself in some of the past to explore ideas and generate a palette of interesting pieces of cartographic practice that you may find a use for. Your work can still demonstrate originality even if it riffs off other ideas, techniques, and motifs. Just because you decide to use a choropleth technique, it doesn't render the map invalid or any less useful than any of the millions of choropleth maps that have gone before.

**History teaches us what works and what doesn't.** The almost immeasurable depth and breadth of cartography through the ages has given us a richness of work to draw upon. From the truly artistic pieces to the technical maps, from painted topographies to meticulous thematics, all maps give us a learning opportunity. The art to making great new maps is to learn what works and to reapply it. Borrowing ideas from elsewhere to form your own new works isn't just useful, it's necessary. Becoming a collector of cartographic ideas and snippets of great work not only informs your own work but respects that work in the process. Furthermore, it means that most projects won't start off with a blank page because you may already have in mind a particular style or idea.

**Your work will be a mashup of your creative influences.** The sort of maps you like, the types of map you like to make, and your experience of prior cartographic work will shape your knowledge base. These influences will feed directly into your work. This is not a new idea either. Picasso once said, 'Good artists copy, great artists steal', and even this phrase has been reused through time, not least by Steve Jobs as he defined some of the creative process behind Apple Inc when he said, 'We have always been shameless about stealing great ideas.' Taking ideas, combining them, and transforming them is what generates new work which others can then use, and so the cycle continues.

**See also:** Critique | Ethics | Old is new again | Style, fashion, and trends | Who is cartography?

Much of this book isn't new. Trying to write a completely original book on cartography would be impossible. Artist Austin Kleon framed these ideas on creative originality as the phrase 'Steal like an artist'. I've stolen and modified his 10 principles to fit cartography:

1. Steal like an artist. This is not plagiarising. This is the study, credit, and remixing of cartographic work. Creativity stems from combining what went before.
2. Don't wait until you know who you are to start making maps. Start by doing. Immerse yourself in the work of others. Emulate, don't imitate.
3. Draw the map you want to see. Be invested in wanting to see the fruits of your labour as something you want to see.
4. Use your hands. Move away from the screen to doing something physical. Leaf through old atlases and map sheets. They are full of inspiration. Sketch and experiment.
5. Side projects are important. Hobbies keep you happy. They are also spaces to stretch yourself, perhaps beyond the sort of maps you may be paid to make.
6. Do good work and put it where people can see it. Share your work and thought processes. Get feedback and generate more ideas. Contribute to other people's ideas.
7. Geography is no longer our master. Travel and virtual escapism can widen your horizons and bring new creative insights.
8. Be nice. Show appreciation for what you see. Be critical but be constructive. The world of cartography is relatively small, and people are always willing to talk maps.
9. Be boring. Set a routine but include time for creativity and activities that expose you to new maps and new cartographies.
10. Creativity is subtraction. Focus. Avoid the daily detritus that fogs thinking. Make choices about your influences and what and who to learn from. The best way to be creative is to set constraints rather than be paralysed by limitless possibilities.

**Opposite:** An extract from *Steal Like an Artist* by Austin Kleon, 2012.

| GOOD THEFT | VS. | BAD THEFT |
|---|---|---|
| HONOR | | DEGRADE |
| STUDY | | SKIM |
| STEAL FROM MANY | | STEAL FROM ONE |
| CREDIT | | PLAGIARIZE |
| TRANSFORM | | IMITATE |
| REMIX | | RIP OFF |

# Prism maps

## A 3D representation that extrudes features by values.

**Prism maps represent numerical data by extruding the height of features to create a 3D statistical surface.** The most common form extrudes areas. They are sometimes referred to as *stepped* or *stepped relief maps* or *block diagrams*. The shape of the areas is maintained as the height of the prisms is extruded relative to the value being mapped. The higher the value, the taller the prism. The map is usually classless so each area is scaled to its specific values though class intervals can be used to create discrete heights.

**The stepped version of a prism map can be difficult to interpret.** Size of areas isn't taken into account so larger areas will dominate. The prisms do not equate to a spatial histogram because of the difference in size of areas. Perspective and occlusions can also cause problems for interpretation since they affect the ability to assess the absolute and relative height of prisms. Using an axonometric projection view can overcome this problem by equalising scale across the map. Interactive environments can also help in giving readers the ability to move around the map.

**Whereas point features can be extruded to create columns, point features are more normally interpolated to create a smoothed 3D map.** The result is a wireframe 3D representation where a grid of interpolated values is connected either using straight lines or ones that have been smoothed to fit a general surface model. Smoothed 3D maps are more normally used to represent data sampled at points whereas the stepped version is used where data is collected for areas. By interpolating in only one of the planar directions, a profile relief map can be constructed showing sliced cross-sections stacked side by side.

**Because of the difficulties in data recovery, prism maps are predominantly used for visual impact.** The technique does a good job of clearly illustrating peaks and troughs or outliers in comparison to a more traditional flat map. The unfamiliarity of prisms can create problems in understanding the geography of an area, particularly if a perspective projection is used.

**See also:** Globes | Height | Isometric views | Lettering in 3D

Scale and projection affect the form of a prism map. As scale decreases and the mapped area gets larger, areas begin to take the form of thin columns. Larger scale maps will give a blocky appearance, which can overemphasise often arbitrary boundaries between adjacent areas/prisms. A projected map will create a 3D version that sits on a plane with the vertical projection lines plotted orthogonally and parallel to the plane. Adjacencies are maintained vertically.

Tilt and rotation are considerations for the static form of map. Combined, they are instrumental in supporting readability. For small z-values, only a small amount of tilt might be needed. Higher peaks generally require increased tilt to bring them into view and lessen occlusion. A rotation of 45° and tilt of 25° are generally a good starting point. For a two-axis map where you're showing a consistent extruded depth, the amount of tilt and rotation can usually be greater to support the artistic requirement you're intending.

Labelling prism maps is a challenge, especially for print. At most, only a few labels might be added. For digital products, techniques that use billboarded labels, a progressive reveal, and which deal automatically with overlapping text offer improvements. Colour can show either categorical information or add a scheme that emphasises the data itself. Transparency should be minimised so occluded prisms do not bleed through other prisms to cause difficulties in legibility.

Prisms on a virtual globe might be constructed from area or point features but the main difference is that extrusions are perpendicular to the surface and are not orthogonal. The effect is a spiky ball. The inherent perspective and curvature causes gross distortions in perception, and so these maps should only be used as eye candy.

Eye candy

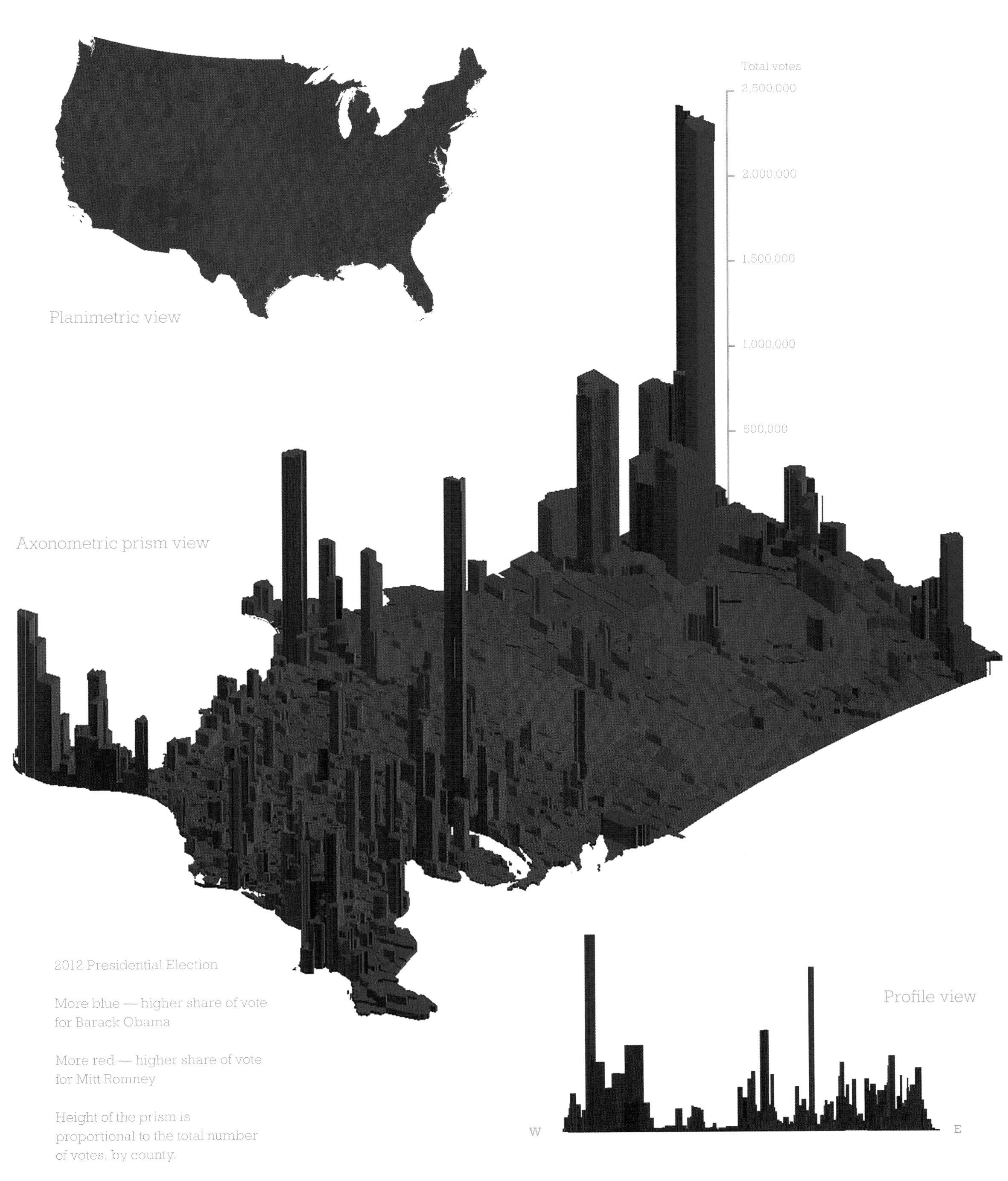

Planimetric view
Total votes
2,500,000
2,000,000
1,500,000
1,000,000
500,000
Axonometric prism view
2012 Presidential Election
More blue — higher share of vote for Barack Obama
More red — higher share of vote for Mitt Romney
Height of the prism is proportional to the total number of votes, by county.
Profile view
W
E

# Profiles and cross-sections

## Drawing a line across a map and representing features along it.

**Change in the character of a mapped variable can be plotted along a line drawn on a map.** Elevation profiles are a side-view of the changes in elevation (height or a statistical variable) along the line as if the terrain has been sliced open and can be seen from the side. For instance, straight-line profiles can be constructed from a contour map along any line drawn on top. The profile is constructed by plotting the position of contours along the line, and then interpolating a curve between the contour lines to represent the surface between these known points. Cross-sections add detail underneath the profile line to show additional variables.

**Vertical exaggeration is often applied to emphasise the profile.** This is the ratio between the map's vertical scale and horizontal scale. It is used to accentuate features and help the map reader see variations, some of which may be relatively slight. Overdoing vertical exaggeration can potentially lead to a misunderstanding because it increases the overall perception of height as well as steepness and length of parts of the profile so it's important to use with care.

**Although profiles are most often constructed along straight lines, irregular lines are also used.** This might be a useful way to show the elevation changes in a footpath for a hike or a bicycle ride, for instance. It is useful to include marks on the profile that show key changes in direction along the irregular line. These marks help readers understand the relationship of line segments on the map with those on the elevation profile.

**Profiles are often used as a framework to show changes in the character of the landscape.** They are typically used to create geological profiles by extending what we know of the surface geology to depth. These profiles result in cross-section diagrams, which combine an elevation profile with knowledge of the subsurface geology. Being able to accurately illustrate the subsurface features requires a knowledge of the different rock layers as well as the dip and strike angle.

**See also:** Aspect views | Digital elevation models | Geological maps | Graphs

Profile diagrams and cross-sections can be plotted by hand or computer generated using the following general construction guidelines.

Constructing a profile:

1. Create a straight or irregular profile line on the map.
2. Determine the highest and lowest contour lines on the map that the profile line crosses and use these lines to create the vertical scale for the profile.
3. Apply vertical exaggeration to the profile diagram if necessary.
4. Plot the intersection of each contour line on the map with the profile line onto the profile diagram at the correct vertical height.
5. Interpolate the plotted elevation points on the profile diagram to create a smoothed side-on view of the profile of the topography.

Constructing a cross-section follows the same general procedure as noted to create the surface line on the cross-section diagram. Because a cross-section is concerned with illustrating a theme above or below the line, the following steps relate to the theme itself—in this example, a geology cross-section:

1. Construct a profile diagram.
2. On the profile diagram, mark where the mapped theme changes (e.g. between rock types).
3. Use additional information (e.g. dip or strike angles) to plot lines that represent how the different mapped theme orients below the surface.
4. Symbolise the different mapped theme types using the same symbols that appear on the planimetric map (e.g. geological map colours).

**Opposite:** *Population Lines* by James Cheshire, 2013, showing population as a profile graph along lines of latitude.

Moscow
Paris
New Delhi
Tehran
Cairo
Dhaka
Lagos
Jakarta
Nairobi
Kinshasa

# Properties of a map projection

## Different map projections maintain properties that support use in particular contexts.

**The property of a map projection is perhaps the most useful characteristic to consider.** Properties relate to the projection's ability to maintain certain traits such as equivalence (equal area), conformality, (angular relationships), and equidistance (distance measurements).

**Equivalent map projections are referred to as *equal-area* and *preserve-area ratios*.** The trade-off is that distance distortion is often quite pronounced. It is also impossible for a single equivalent map projection to also maintain shapes so the shape of countries and continents is often sheared or skewed. The angles of intersection between parallels and meridians is not maintained at 90°. Equivalent projections are extremely important in mapping and might arguably be the most appropriate default projection. Because many general-interest maps are designed to convey information at a glance, the display of real area ratios supports the need.

**Conformal mapping (sometimes referred to as *orthomorphic*) employs projections that preserve angular relationships.** Because of this conformity, the shape of objects across small areas is also relatively well preserved (although technically distortions exist, they are minimal). Conformal projections maintain the angular relationship between parallels and meridians such that the 90° angle on a reference globe will also be mapped as a 90° angle on the plane surface map. Scale may change across the map but will be the same in all directions around a single point. Because scale can change substantially across a map, the shape and area of large objects can be severely distorted.

**Equidistance preserves great circle distances.** However, there are limitations to this projection in that distance is only accurate from one to all other points, or from a few points to others but not from all points to all other points on the map. Equidistance can never be applied consistently across the whole map. Scale is uniform along the lines of equidistance. Such projections are referred to as *equidistant projections*.

Although different properties can be preserved when choosing and applying a map projection, our perception of shapes, sizes, and angles is distorted before we even begin to view a map of the world.

Whenever we examine a globe and view a country or continent from directly above, we can only rest our eyes on a single point at any one time. Because of curvature, the image we see will fall away from the point we are viewing, and that curvature inevitably distorts our view of the shape of the object. If we view another point, again vertically, the view changes as does the shape of the object. This perspective is further complicated if we view the globe at an angle.

Thus, our impression of the shapes of countries and continents on a plane surface is hard to assess as we have a poor grasp of the real shape against which we are comparing the map.

This problem can be confounded by the maps we become more familiar with. Maps make powerful images and distort our views of reality. For example,Mercator's projection is extremely useful in the right context but is all too frequently used where conformality is unimportant and the distortion to area ratios misleading.

A compromise map projection attempts to balance the properties. It neither maintains any property nor is devoid of distortion. However, the fact that it maintains most properties well enough makes it a useful choice, particularly for world maps where distortions are spread throughout the map.

**See also:** Assessing distortion in map projections | Distortions in map projections | Families of map projection

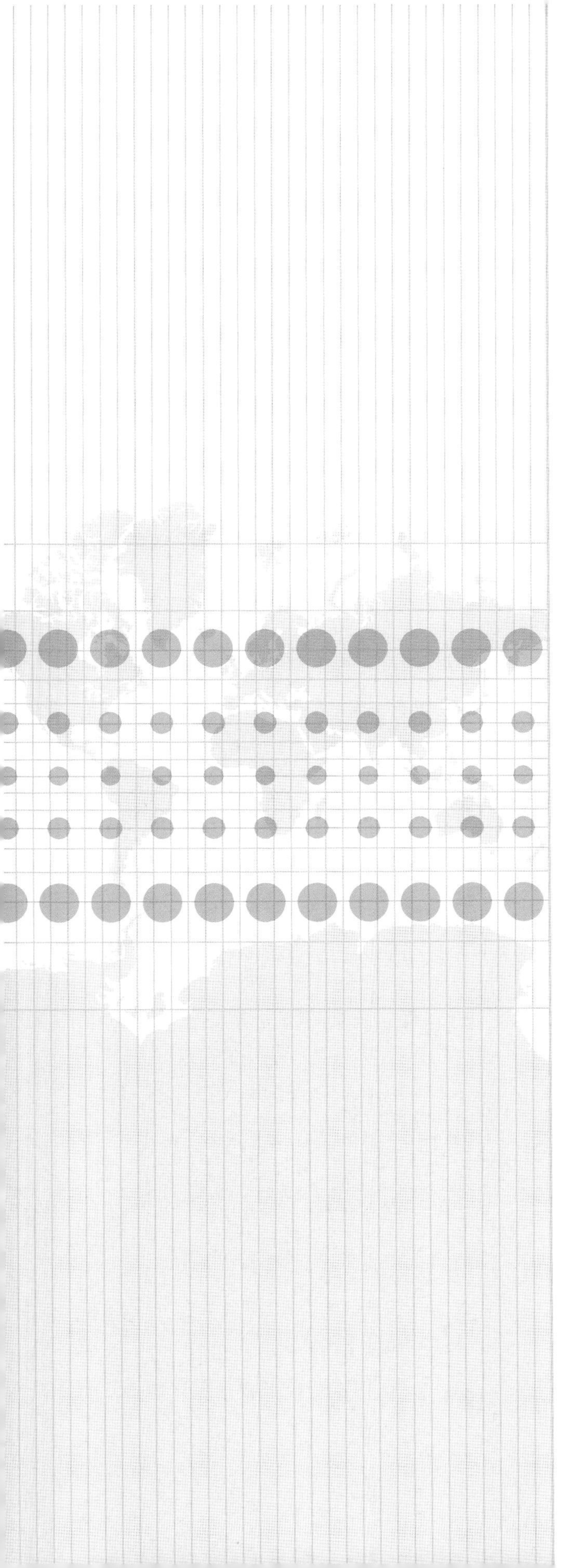

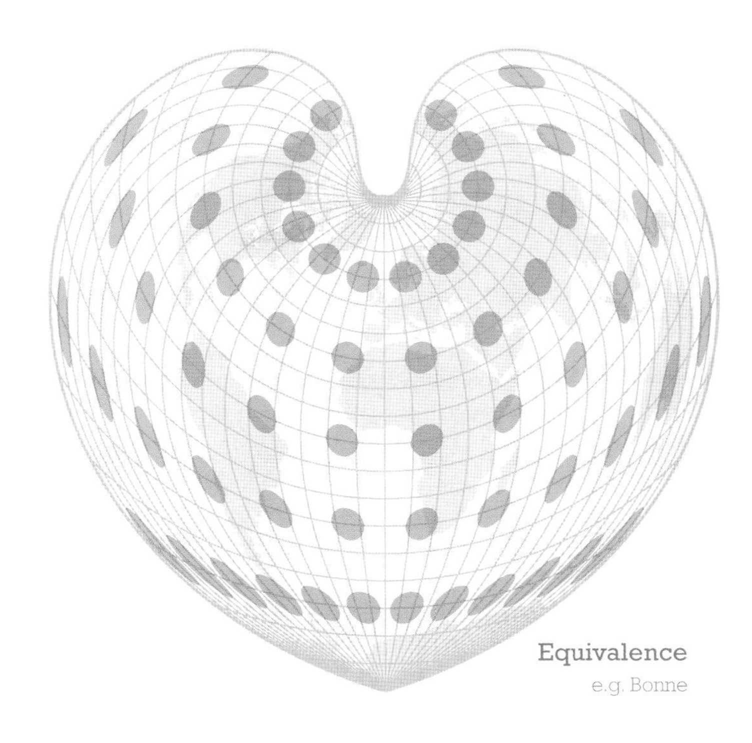

Equivalence

e.g. Bonne

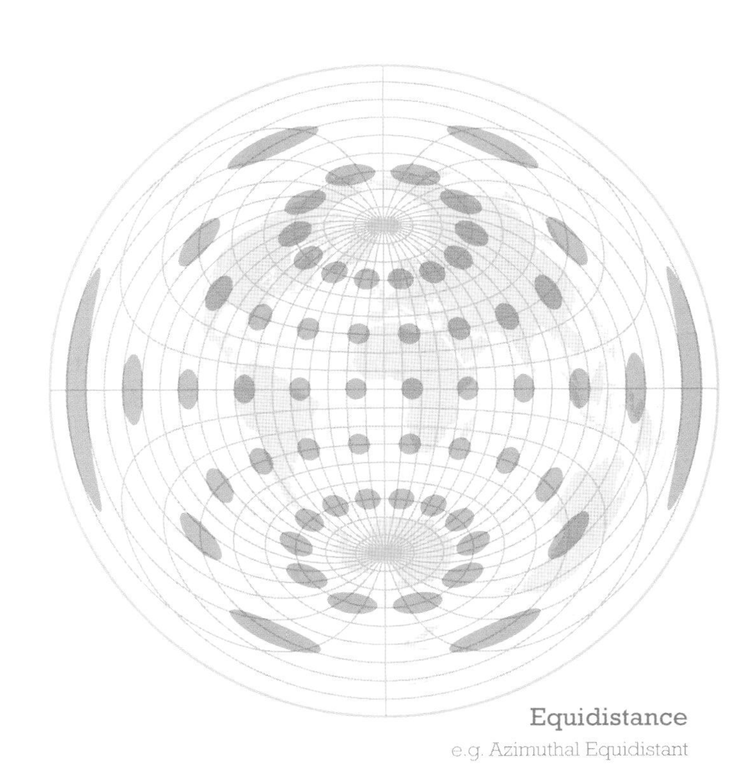

Equidistance

e.g. Azimuthal Equidistant

Conformality

e.g. Mercator

# Proportional symbol maps

## Proportional symbol maps show quantitative differences between mapped features.

**The purpose of a proportional symbol thematic map is to show how features differ in quantity for the theme being mapped.** The data should be numerical (quantitative) and represent differences between features on an interval or ratio scale of measurement. The map type requires data to be absolute (totals), which is crucial to the calculation of symbol size which scales the area of a symbol to the data. Although similar, on a proportional symbol map, each symbol on the map will be scaled according to its actual data value rather than classified and grouped on a graduated symbol map.

**Symbols should be designed so that different magnitudes of data can be easily distinguished through variation in the size of the symbol.** Symbols should be scaled so that the smallest are visible and the largest do not overly smother the map with the largest symbols representing the largest magnitudes. Some overlaps are inevitable but using techniques such as an outline or a cut-out allow different symbols to be seen.

**Generally, geometric symbol shapes work better for size estimation.** Circles are usually used because they perform best in enabling visual interpretation. Squares are an alternative, but the more complicated the shape, the harder it is for the reader to interpret the differences between symbols. Pictorial symbols can often be used to create a striking image but the downside is the difficulty in assessing quantities since the envelope of the shape will normally be used to set up the symbol scaling rather than the actual shape itself.

**The map reader should be able to efficiently estimate the different quantities mapped in different areas.** At the least, relative differences should be obvious and the reader should be able to determine a pattern across the map. When size is used as an ordering visual variable, we are ascribing more importance to the larger magnitudes of data. We visually interpret the symbols as differently sized so we perceive larger symbols as meaning 'more'.

The usual method of setting up symbol sizing makes use of the square root of the data value, possibly with an exponent that scales up or down to best fit the media size (or scale). This value then becomes part of the measurement of area such as the radius in $\pi r^2$ when calculating the area of a circle. If the data contains significant outliers, we can use a symbol of fixed size to designate values either all above or below a minimum or maximum threshold while intervening values are proportionally scaled.

The legend should include a representation of the symbols to enable size estimation. Because size estimation is generally a problem in human perception, a good legend design is vital for proportional symbol maps to create visual anchors that we can easily relate mapped symbols to.

Transparency is often used to deal with the problem of overlapping symbols yet overlaps create darker symbol colours. This colouring can be confused with areas that have a greater magnitude whereas it's more an artefact of the symbol overlap. For this reason, transparency is best avoided and other methods to deal with overlaps preferred. For multi-scale maps, symbols that are held at a constant size will naturally separate as the map scale becomes larger.

Proportional text maps are a variation of a proportional symbol map that uses type as a literal symbol. If the label is numeric, the data is encoded directly into the map symbol Text can also be used as proportionally sized labels and can create a striking alternative to more traditional maps. It's critical that overlaps do not compromise this type of map. Progressive reveal can be used for multi-scale versions. Because the data itself is encoded into the map, meaning can often be recovered without a legend.

**See also:** Dimensional perception | Flow maps | Graduated symbol maps | Points | Size | Threshold of perception

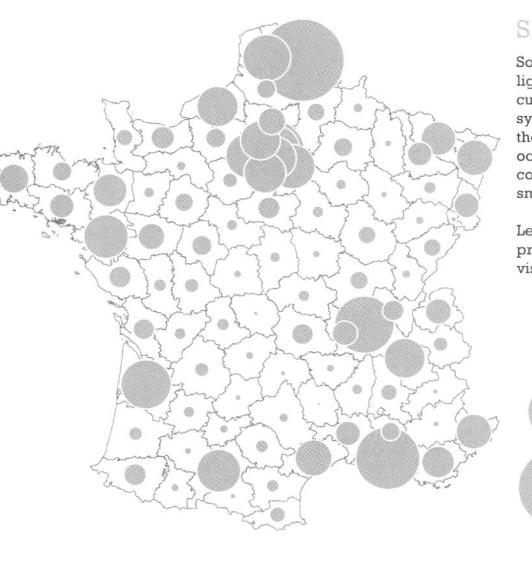

## Shaded

Solid fill symbol with light stroke to act as cut-outs. Smaller symbols sit on top though some occlusion occurs in congested areas at smaller scales.

Legend is strung out, providing several visual anchors.

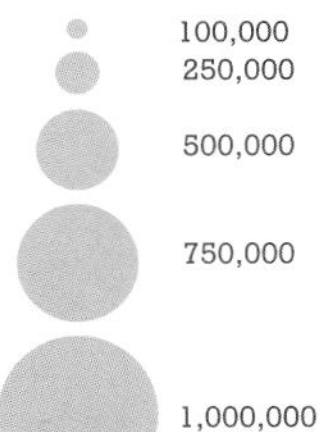


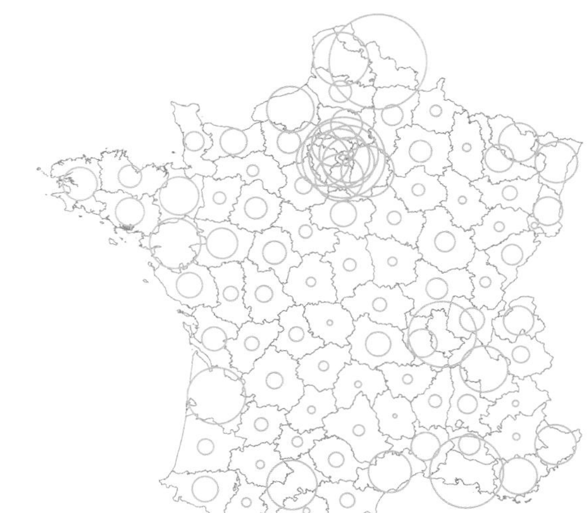

## Outlined

Symbols have an outline stroke but no fill. Allows individual symbols to be seen though can be congested in some areas.

Flannery compensation is applied to exaggerate larger symbols and mediate underestimation.

Legend is nested, providing several visual anchors while using less overall space.

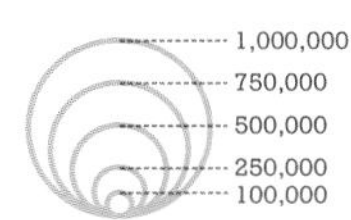


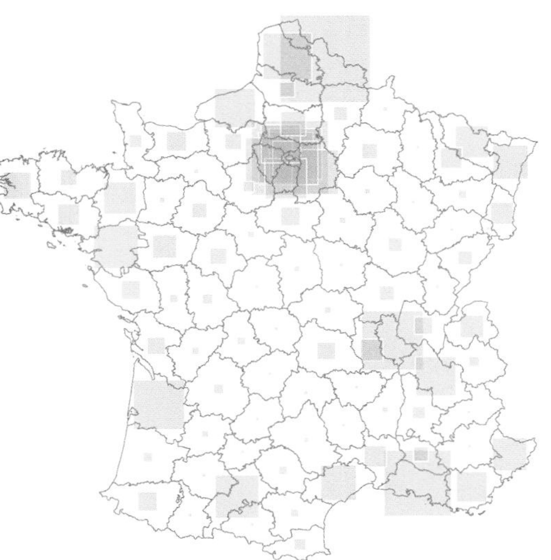

## Transparent

A square symbol replaces the more common circle.

Symbol fills are semi-transparent which can provide a way of seeing overlaps.

However, overlaps are an artefact of symbol size so the additive effect of several symbols creates a darker area which will be seen as 'more'.

Legend is nested, appropriate for the shape.

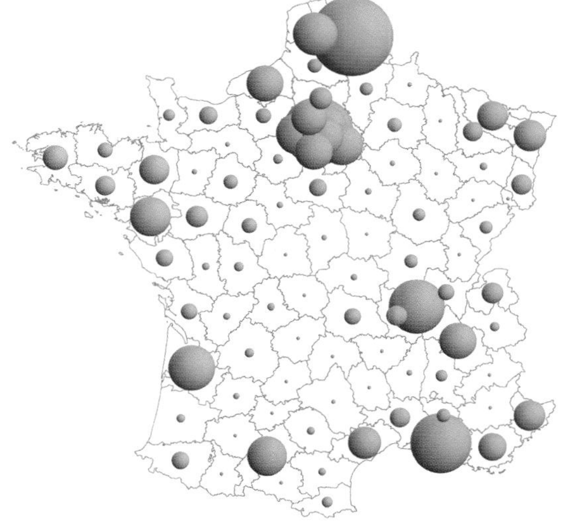

## Volumetric

Volumetric symbols are scaled to the data value, instead of area, making estimation of data values difficult.

Legend is strung out, providing several visual anchors.

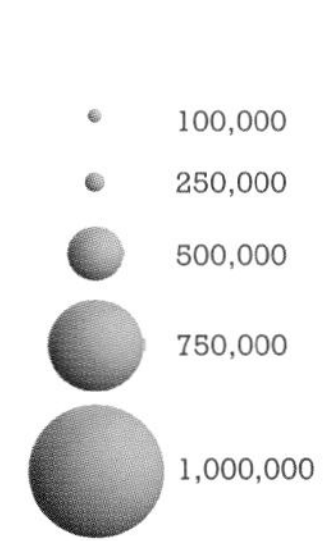


## Labels

Labels are sized proportionally giving the reader a sense of the pattern of the data distribution as well as the actual data value itself.

Some labels require leader lines for congested areas.

A legend is unecessary for this type of map since the symbol size is acting merely to emphasise the label's literal meaning.

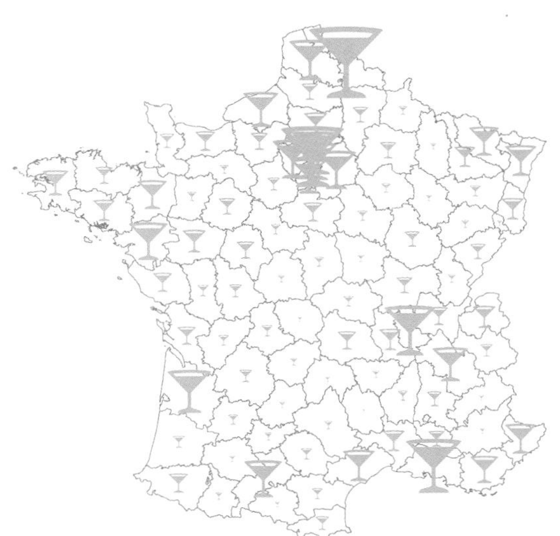

## Pictorial

Although pictorial symbols can provide an attractive link to the map's theme, the problems of overlap and value estimation are compunded.

The area of the symbol's bounding box is scaled. The symbol shape doesn't provide a good visual form from which to estimate values. Symbol height may be incorrectly assumed to be the scaling dimension.

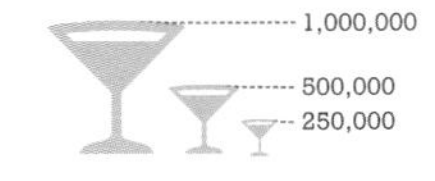

# Proximity in design

## Proximal grouping of mapped objects must be carefully handled to avoid optical illusions.

**At their basic level, maps are simply a list of different map elements—a spatial jigsaw puzzle that needs organising to give it meaning.** Lists are more easily understood when they have some sort of organising principle. Written lists tend to employ justified typesetting and consistent letterspacing, wordspacing, and linespacing to achieve control in the overall image that we see to help us decipher the content. The precise dimensions and type of justification do not particularly matter but the overall notion of controlling the proximity of different elements of the visual display is the important idea.

**Designing maps is an exercise in controlling the proximity of map elements as part of the overall display.** We use figure and ground to create layers of information such that like information is proximal on the same visual plane. We might use positive and negative design to give identity to like features, particularly in pictogram symbol design. We might also cluster features and map elements and leave that all-important white space.

**Properly organising map elements and the associated white space is key to making it easier for readers to see patterns and relationships and, therefore, to understand the map.** White space is often seen simply as a way to separate features that need to be seen together but it also gives the information space to breathe. Separation of grouped items is important but white space can also be used as part of the overall framework.

**Proximity should be used to group like features and elements where possible.** Pieces of information that are better understood by being associated with something else support the communication of the overall message. Tables of allied information, legends, and other marginalia are better organised together. Labels are better next to the feature rather than through the addition of leader lines. Alignment of related map elements also helps provide structure as the eye can move easily horizontally or vertically across features that may not be spatially proximal but which are related in other ways.

**See also:** Dispersal vs. layering | Foreground and background | Seeing | Seeing colour | Signal to noise

Proximity applies to our perception of grouped graphical objects. Objects that are near each other tend to be seen to be grouped together. Adjusting symbol position may seem to be in conflict with accurately locating mapped features but being aware of some of the pitfalls of proximity can help improve the map.

There are many illusions that result from the grouping of mapped objects. For instance, symbols of the same size that are regularly spaced can take on the appearance of being in columns or rows even if that structure is irrelevant. Lines of equal length can appear different depending on the style of the line's end cap. Perspective illusions can easily arise from the relationship between foreground and background.

The haze illusion, moiré effect, Hermann grid illusion, and Ebbinghaus illusion also show us that proximity between certain objects can have consequences. Some illusions cause us to see objects (or the gaps in between them) in peculiar ways, and depending on our gaze and focus these can differ as we scan an array of objects. There are many such illusions which we can easily overlook.

Proximity is, of course, modified by the design of symbols. For instance, we perceive values (e.g. on a grey scale) differently depending on their proximity to symbols that share the same value. The checker shadow illusion demonstrates this difficulty. Overcoming this effect comes down to reflecting on values across a map and ensuring they are uniquely identifiable.

## Which line is the longer?

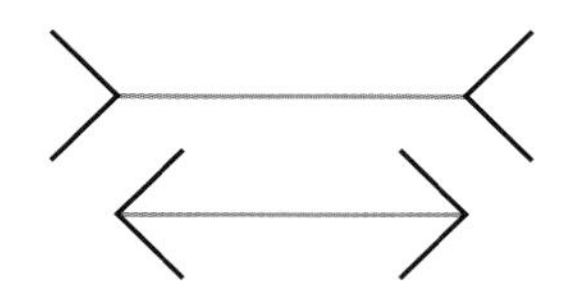

The Muller-Lyer illusion
The lines are the same length yet the segment with two arrow tails appears longer than the segment with two arrow heads.

## Which central circle is larger?

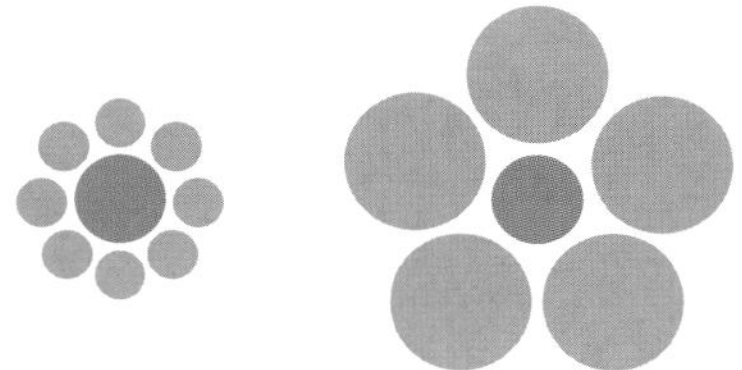

The Ebbinghaus illusion
The juxtaposition of nearby objects makes the central circle surrounded by larger circles appear smaller than the central circle surrounded by smaller circles.

## Can you make the grey disappear?

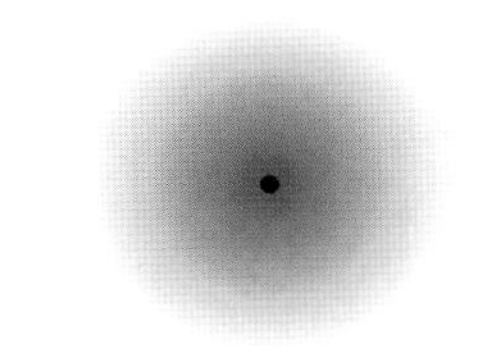

Focus
By staring at the focal point the surrounding grey haze begins to disappear. We're concentrating on the sharply defined object against the out of focus grey that recedes.

## How many black dots do you see?

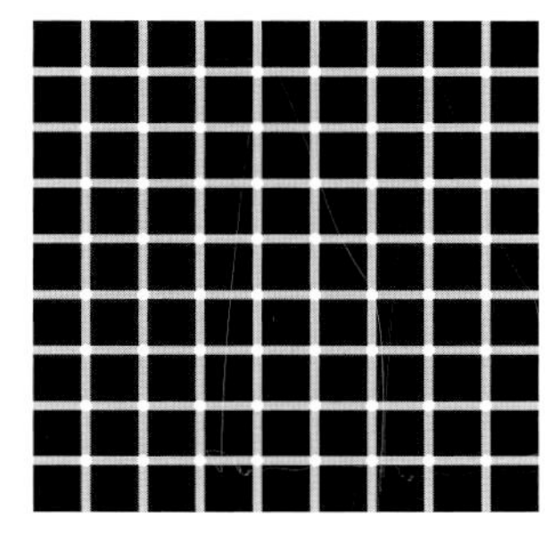

The Hermann grid illusion
Ghostlike grey appears at the intersection of a light grid on a black background. They disappear when you look at an intersection.

## Is the page moving?

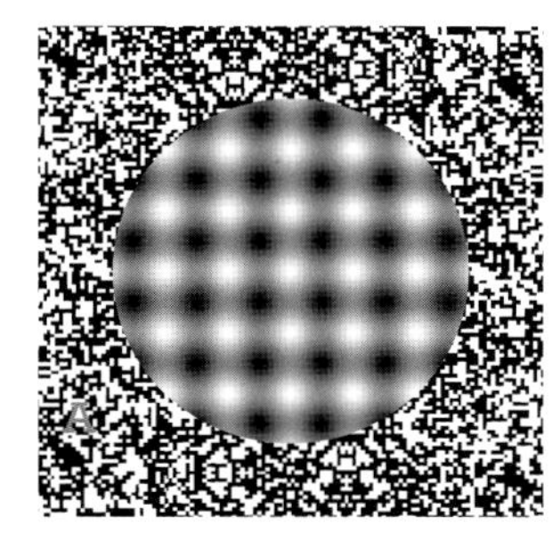

Movement
Certain patterns, juxtaposed with others, give the appearance of a fluctuating pattern. The pattern can appear to move.

## Which square is darker?

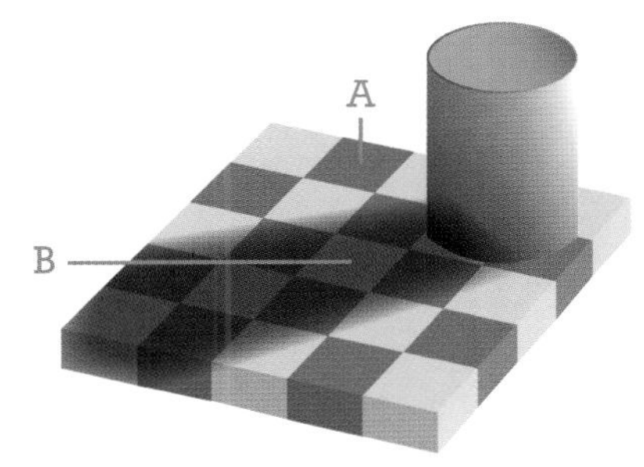


Checker shadow illusion
Despite the appearance of a consistent checker pattern of light and dark greys, the brain adjusts them with the presence of shadows. A and B are the same colour.

## Do the horizontal lines slope or not?

Checker shadow illusion
Despite the appearance of a consistent checker pattern of light and dark greys, the brain adjusts them with the presence of shadows. A and B are the same colour.

## How many drawn triangles?

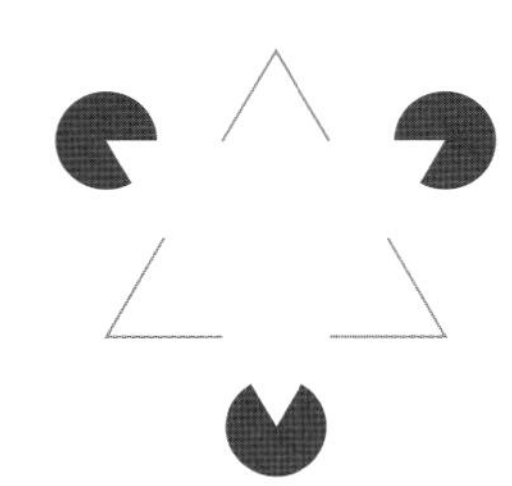

The Kanizsa triangle
A white equilateral triangle is perceived as white space even though non is drawn.

## Spots, dots, black and white

Look at the image below. Your brain is searching for a meaningful image. Once it begins to see familiar shapes it reorganises the dark and light to form something you know and can then see.

The image remains the same. The projection onto your retina doesn't alter. Your brain, however, is constantly and actively re-processing to discover meaning it can understand.

Got it? It's a picture of a Dalmation dog from a famous photograph by Ronald C. James, 1965.

Once you see it, you cannot unsee it!

## Which square is darker?

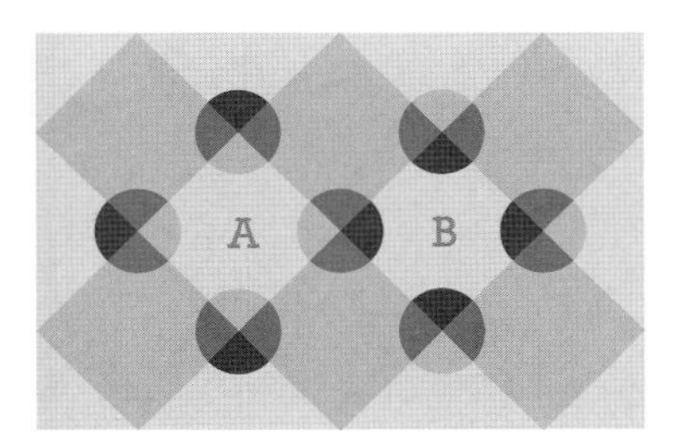


Illusion of haze
The two marked areas are the same shade of grey. The surrounding shades and arrangement of shapes makes A appear hazy and B appear sharp.

# Pseudo-natural maps

## Provides a sense of realism through texture and realistic symbolism.

**Most topographic maps are abstract representations of the world but pseudo-natural represents the features in an altogether more realistic way.** Geometries and symbols are normally designed to provide a simplified, one might call 'sterile', character to a topographic map. In pseudo-natural mapping, the rather flat approach to symbol design is replaced with more vivid colours, shades, and textures. For instance, on a standard topographic map sheet, a forested region would likely be represented by an area shaded in a uniform green. Similarly, every other forested feature would look the same so we have a clear expectation of what the feature is and where they occur across the map—and there's good reason for that consistency in a cognitive sense. On a pseudo-natural map, a forested area may incorporate many different shades of green, perhaps based on the typical colours of the canopy of certain species. They may differentiate between coniferous and deciduous trees. It may include surface roughness or texture to denote a rough canopy, and it may even map individual trees and include drop shadows and other techniques.

**The concept, then, is to use graphical techniques to achieve a sense of realism in the abstracted map while still applying the well-tried techniques of simplification on the overall content.** This is not an attempt to make a map look like an aerial image because the map's content has still been significantly modified for the scale of representation. Instead, it's the symbology that is heavily altered from what we might be used to. The intent is to give the map reader a more attention-grabbing, immersive experience which combines cartographic abstraction with a natural appearance.

**The technique has its roots in hand-painted panoramic landscapes.** These maps evoke a manual cartography that creates a realistic-looking visual summary of the view, often with a good degree of artistic licence that would shift features, exaggerate them, or show them in a graphically vivid way. Winter maps and natural-colour relief maps also tried to mimic the natural conditions of a landscape through symbology.

**See also:** Imagery as background | Shaded relief | Topographic maps

Contrast and texture can be applied to most area features such as forests, water, sand, and marshland. Additionally, features that you might not necessarily find on traditional topographic maps might be included such as snow-capped peaks in a mountainous area. Man-made features such as roads might still be rendered without texture to support the abstract quality of the map with man-made features sitting in what might otherwise look like a natural landscape.

In generating such maps, you can begin by using palettes of colour that are derived from photographs or other representations of the same landscape to inform your own colour choices. Texture generation can be created in most desktop applications to provide different types of surface. Shadows and pattern fills can also be used creatively to provide a pseudo-natural appearance. For instance, adding a wave-texture patterned fill to augment a watery landscape will make it look, well, more watery!

Using imagery can also be successfully used to introduce texture to a map. For instance, adding semi-transparent monochrome aerial imagery to a land cover map can introduce subtle variation. Individual field boundaries or crop types begin to be demarcated. Although the imagery itself may be too coarse in resolution to be used directly, its use to enhance maps is beneficial.

Pseudo-natural maps are not restricted to planimetric. Similar techniques can be applied to panoramic or plan-oblique representations, for instance, to add to the sense of realism the cartographer is trying to evoke.

**Opposite:** A range of textured layers build a pseudo-natural map. All elements are created digitally.

# Purpose of maps

## Maps share important characteristics that give them a specific purpose.

**Fundamentally, all maps are concerned with two basic elements of reality: locations and attributes.** Locations are the positions that an object occupies in two-dimensional space. Linear objects additionally carry orientation and can be viewed as the location of the same object at several, joined locations. Areal objects take on a shape and have an extent in geographical space; this is the same as mapping a bounding linear object where all points within the boundary are of the same type. Attributes are the characteristics or quantities that exist at locations and which we can map by varying the symbols used to represent the data at its different locations.

**Maps fulfil certain purposes on the basis of the organisation and combination of location and attributes.** You can establish relationships between locations when no attributes are involved (e.g. distance and direction); among different attributes at one location (e.g. the relationship between temperature, precipitation, and groundwater flow); among the locations of attributes of a particular distribution (e.g. the variation of rainfall from place to place); and among locations of derived or combined attributes of a distribution (e.g. the impact of varying levels of poverty on health status from place to place).

**The purpose of a map can be established by thinking about what spatial relationships are borne out of your data.** For instance, the following characteristics render a map not only a graphical storage device but a powerful tool for spatial analysis:

- distance
- direction
- adjacency
- enclosure
- clustering
- networks
- interactions

*'Maps have many functions and many faces, and each of us sees them with different eyes'.*
—R. A. Skelton, 1972

All maps, by their definition, are reductions of reality because they represent reality at a smaller size than it actually exists. A mathematical relationship between the map and reality is expressed as scale. Scale impacts the main function of the map because it defines the size of the mapped area in relation to the real-world area and, in so doing, determines the amount of space available for feature representation.

The scale of the map effectively sets a limit on the canvas size that you can work with, which, in turn, limits what information can be included. As scale is reduced, a larger Earth area is mapped but with less detail and with increasingly generalised symbols. Large-scale maps, conversely, depict small earth areas in detail and with less generalised symbols.

**Opposite:** Ernst Shackleton's hand-drawn map of the South Pole, atop a south polar chart and showing routes of Shackleton, Stackhouse, Bruce, and Koing. Shackleton's map perhaps reflects the purest of map documents as a first-person record of exploration, discovery, and reporting.

**See also:** Defining map design | Design and response | Emotional response | Information products | Informing | Map cube | Map traps | Types of maps

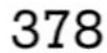

Oates Land
King George V Land
Queen Mary Land
Appearance of land
Luitpold Land
Coats Land
MAUD'S RANGE
ANTARCTIC Continent
S. AMERICA
S. AFRICA
Pole
E H Shackleton, March 7th 1914
Presented to me by my neighbour at the dinner, in order to explain his prospective plans. Clive Morrison-Bell
Committee.

# London A–Z

**Geographer's Map Co. Ltd.**
**1938–present**

Until the advent of the mobile phone and the development of small-screen interactive mapping at the start of the 21st century, there was only one way to get around London on foot: A to Z.

The Geographer's Map Company's A–Z map of London was developed in the 1930s and became dominant in city way-finding for the remainder of that century—every visitor to the city needed one.

Named after the alphabetic street index (that took up roughly half the pages of any A to Z city atlas), it was both map and gazetteer. The style is deliberately cluttered because the city is cluttered.

The maps were first produced in black and white, often printed on cheap paper: cluttered but clear. It succeeded because the type was varied in both size and style to highlight principal streets, stations, and key landmarks. A–Z maps didn't just show where to go, they showed how the city was structured.

The map and its style have endured, and although they are now printed in colour, they are unmistakable as descendants of the original developed some nearly 100 years ago.

—Danny Dorling

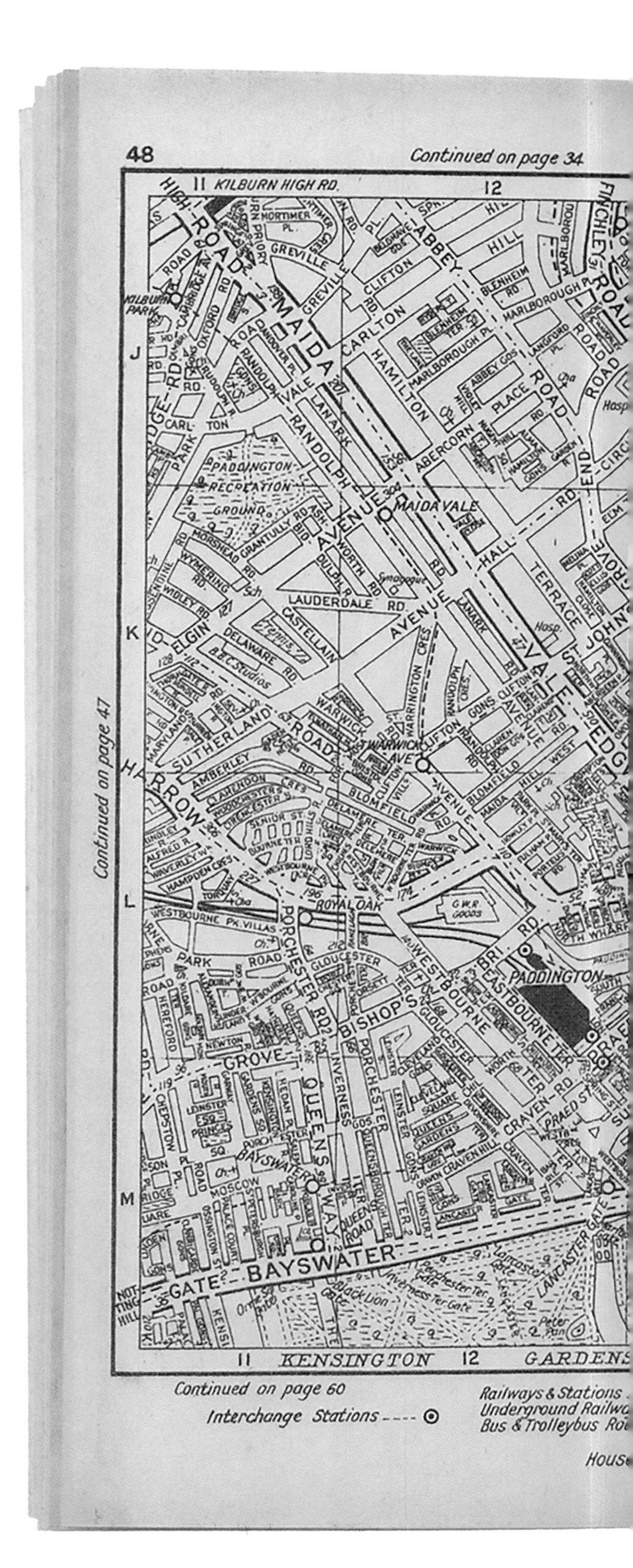

Continued on page 35
49
REGENTS PARK
ZOOLOGICAL GARDENS
Cricket Ground
QUEEN MARY'S GARDENS
Open Air Theatre
Bedford College
EUSTON
EUSTON ROAD
MARYLEBONE ROAD
BAKER STREET
OXFORD
WIGMORE ST
REGENT ST
HAMPSTEAD ROAD
TOTTENHAM COURT RD
GREAT PORTLAND ST
PICCADILLY CIRCUS
Continued on page 50
Continued on page 61
Scale
0
1/4
1/2 Mile
Divided into half-mile squares

# Quantitative statistical maps

## Depiction of quantitative data for features on a thematic map.

**Quantitative statistical maps show numerical data for points, lines, and areas on a thematic map.** Data may take the form of absolute values (totals) or derived values (rates, percentages, or ratios), and the choice of representation will in part be a function of data type. The data is usually depicted as a thematic layer over the top of a skeleton basemap. For some map types, such as a choropleth, the thematic layer becomes its own basemap because it exhausts space. Statistical data usually references the areas which define the collection framework (e.g. countries, states, counties), points (e.g. cities, stores, schools), or lines (the distances and lines of communication between places).

**For area maps, data can be encoded by area or shown using symbols.** Choropleth maps show quantitative difference by changing the value of a hue from light to dark. Alternatively, symbols can vary proportionally from one another to represent different magnitudes. Class intervals or an unclassed approach can be used for either technique although it's important to ensure data is normalized for representing on a choropleth. For point data, proportional symbology can also be used though problems of overlap may cause difficulties since point features are often more closely located to one another.

**Lines can be used to show connections or movement.** A distance map differs from a linear network map since it quantifies the distance in some way, perhaps by showing time by transport mode. Distance maps are often indicative and use curved or straight arrows of a consistent or tapering design. The connecting of points of equal value can create linear forms or bands, called *isograms*.

The main categories of statistical map can be extended by considering distorted versions or cartograms. In these examples, the map is not simply a container for the display of statistical data, the map itself is warped to reflect some character of the data. The value-by-area cartogram modifies the size and shape of an area to represent the data values. A Dorling cartogram uses abstract proportional circles and rearranges geography. Distorting the map allows one variable to be mapped. Additional variables can be mapped by modifying the symbols.

A different distorted map can be made by changing the distance between a central location and peripheral locations by warping the map to a fixed set of concentric distance bands. In this way, even if ground distance is the same between the origin and two locations, the road type and travel time might be significantly different and so the two remote locations would be plotted in different bands. This is, in effect, a linear cartogram where it is an attribute of distance that warps the map rather than an attribute of area.

Data type will likely lend itself to one type of map over another unless you are willing to convert the data to suit a different technique. The area itself may have specific geographical issues that hinder some map types. For instance, choropleths are often compromised when the map has large areas of little interest punctuated by a few small areas of more interest. The medium and user you are targeting may need a map type they are familiar with or, conversely, they may need their interest piqued by a more challenging design. The scope of choice for statistical maps is large but each has opportunities and constraints that must be carefully handled. The principle of being wary of statistics in general equally applies to statistical maps as depiction can easily be crafted to modify the message.

**See also:** Pie and coxcomb charts | Small multiples | Statistical literacy

## Deviation
Variations from a reference point

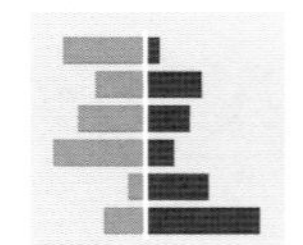
Diverging bar

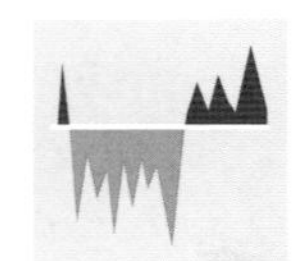
Filled line

Choropleth (diverging)

Literal symbol

## Association
Relationship between two or more variables

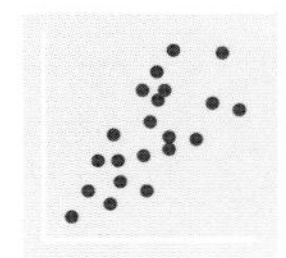
Scatterplot

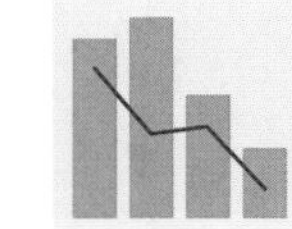
Line & Column

Value by alpha

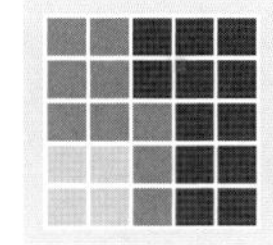
XY heat map

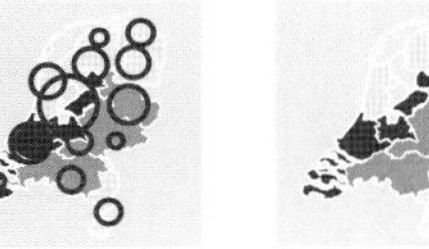
Multivariate map

Choropleth (bivariate)

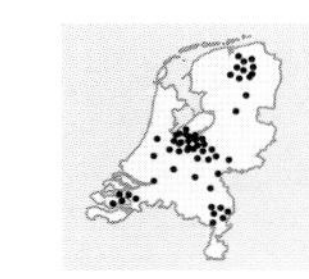
Dasymetric map

## Order
Highlighting position among variables

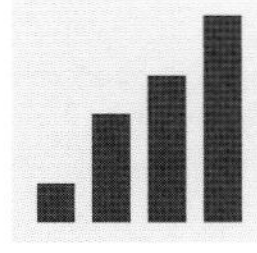
Bar chart

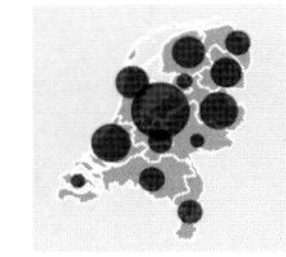
Proportional symbol

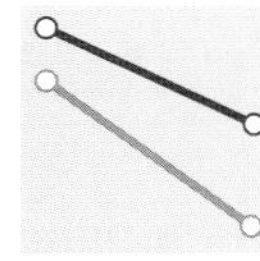
Slope graph

## Distribution
Values of data, its occurrence, and uniformity (or not)

Histogram

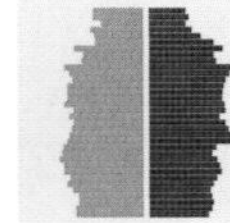

Population pyramid

Sequential choropleth

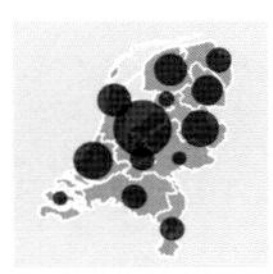
Proportional symbol

Isarithmic

Cartogram

Dot density

## Temporal
Trends over time

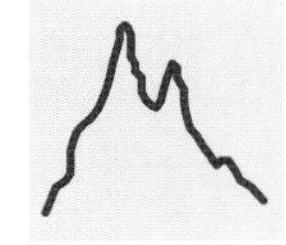
Line

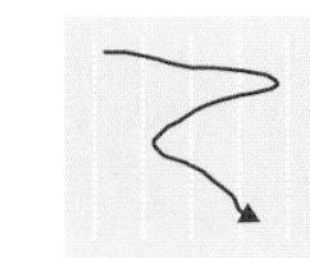
Timeline

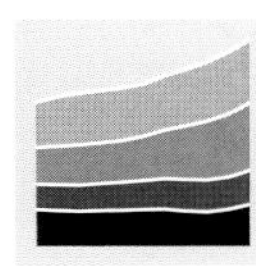
Area chart

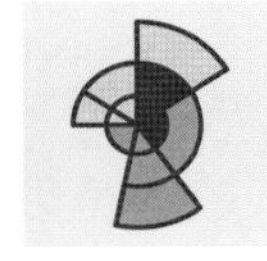
Coxcomb chart

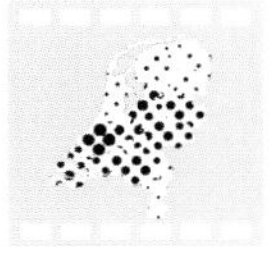
Animated

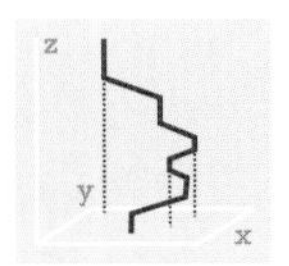

Space-time cube

## Magnitude
Comparisons of size (relative or absolute)

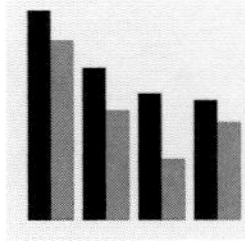
Column

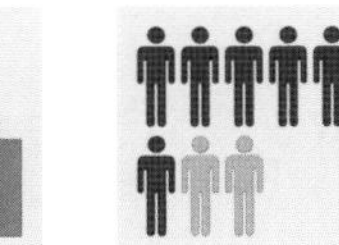
Isotype

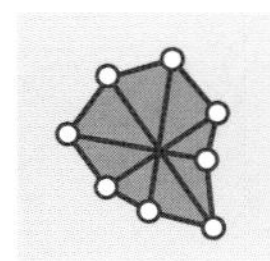
Radar chart

Prism map

## Part to whole
Breakdown of general to detailed

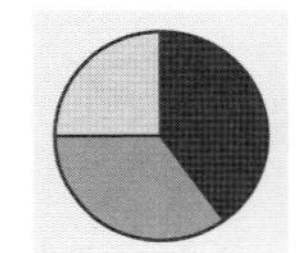
Pie

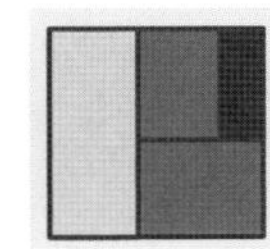
Treemap

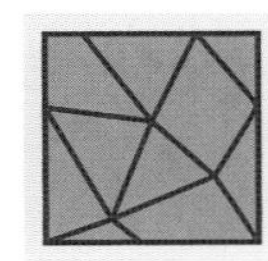
Voronoi

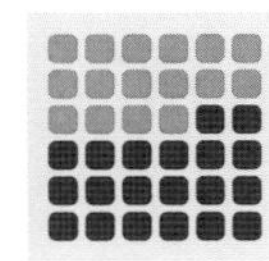
Waffle grid

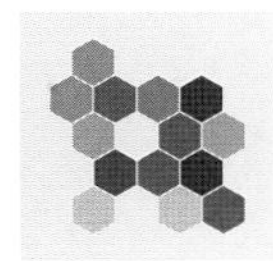
Binning

## Flow
Volume or intensity of movement

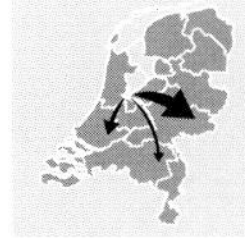
Flow map

Sankey

Chord diagram

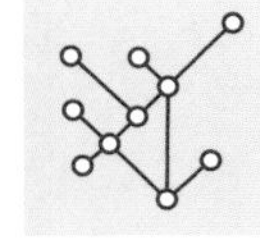
Network

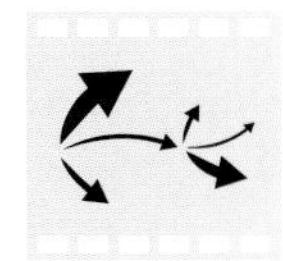
Animated

# Map of London's Underground Railways

**Harry Beck**

**1933**

A schematic diagram was Beck's solution for mapping the London underground for navigation. It's a simplified and heavily generalised map consisting of stations and colour-coded straight-line segments, which run vertically, horizontally, or at 45° diagonals.

Ordinary stations are differentiated from interchanges, the central area is exaggerated, and external areas contracted. The map shows no relationship to aboveground geography other than the River Thames. It's a perfect design solution for a specific mapping requirement. Although it was initially dismissed by the London Passenger Transport Board, rail users took to it, and it ranks as one of the most exquisite cartographic solutions perfectly marrying form and function.

The same approach is still in use today by Transport for London. Although the map has gone through countless revisions and design changes, the core characteristics remain. Beauty in simplicity and a model for many transport-related maps to this day.

For many, the map is their view of the city. It is London. Stations become the signposts of wayfinding and navigating. Moving between the same places above ground requires different spatial skills (and maps). It's my favourite city. It's also my favourite map.

—Kenneth Field

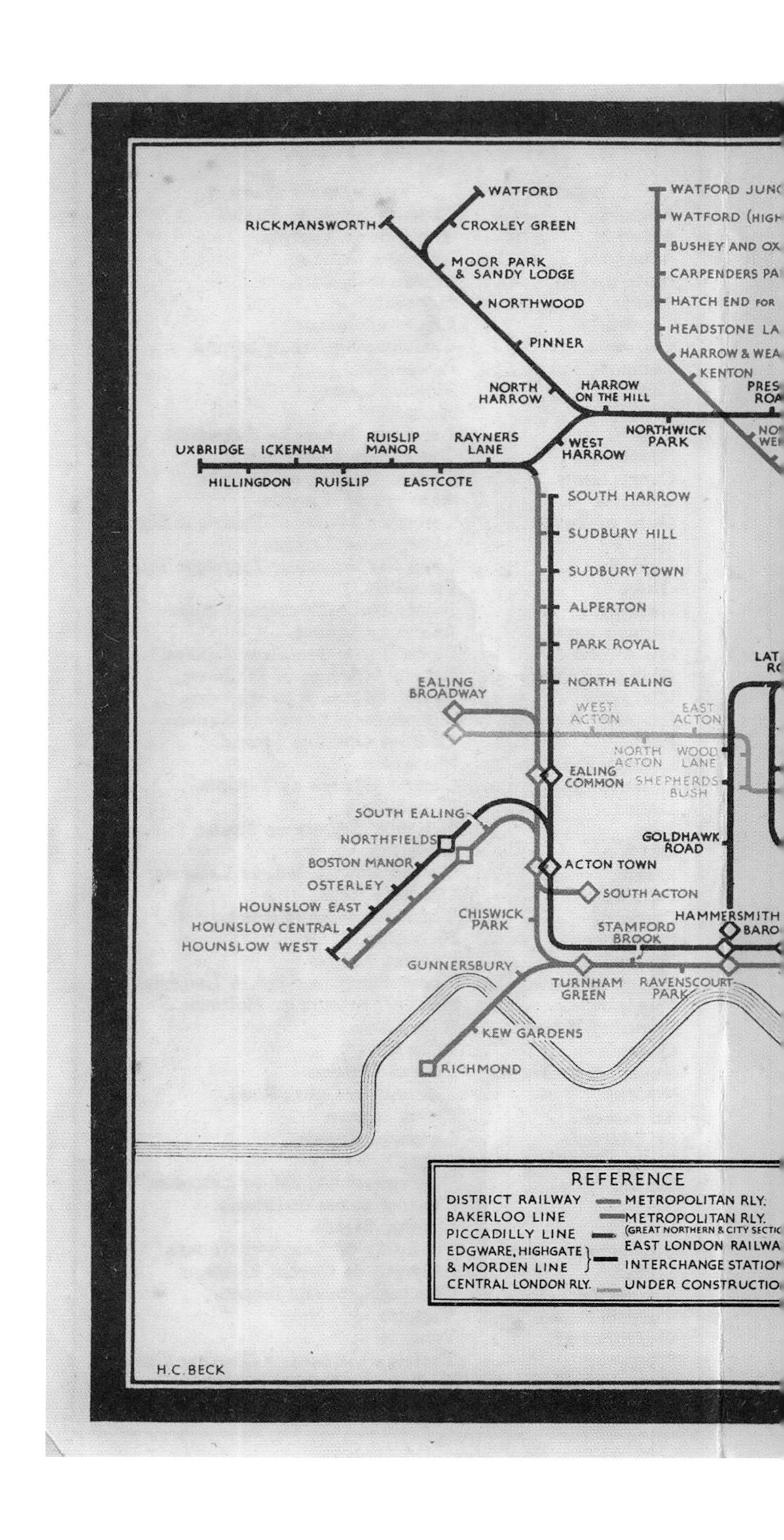

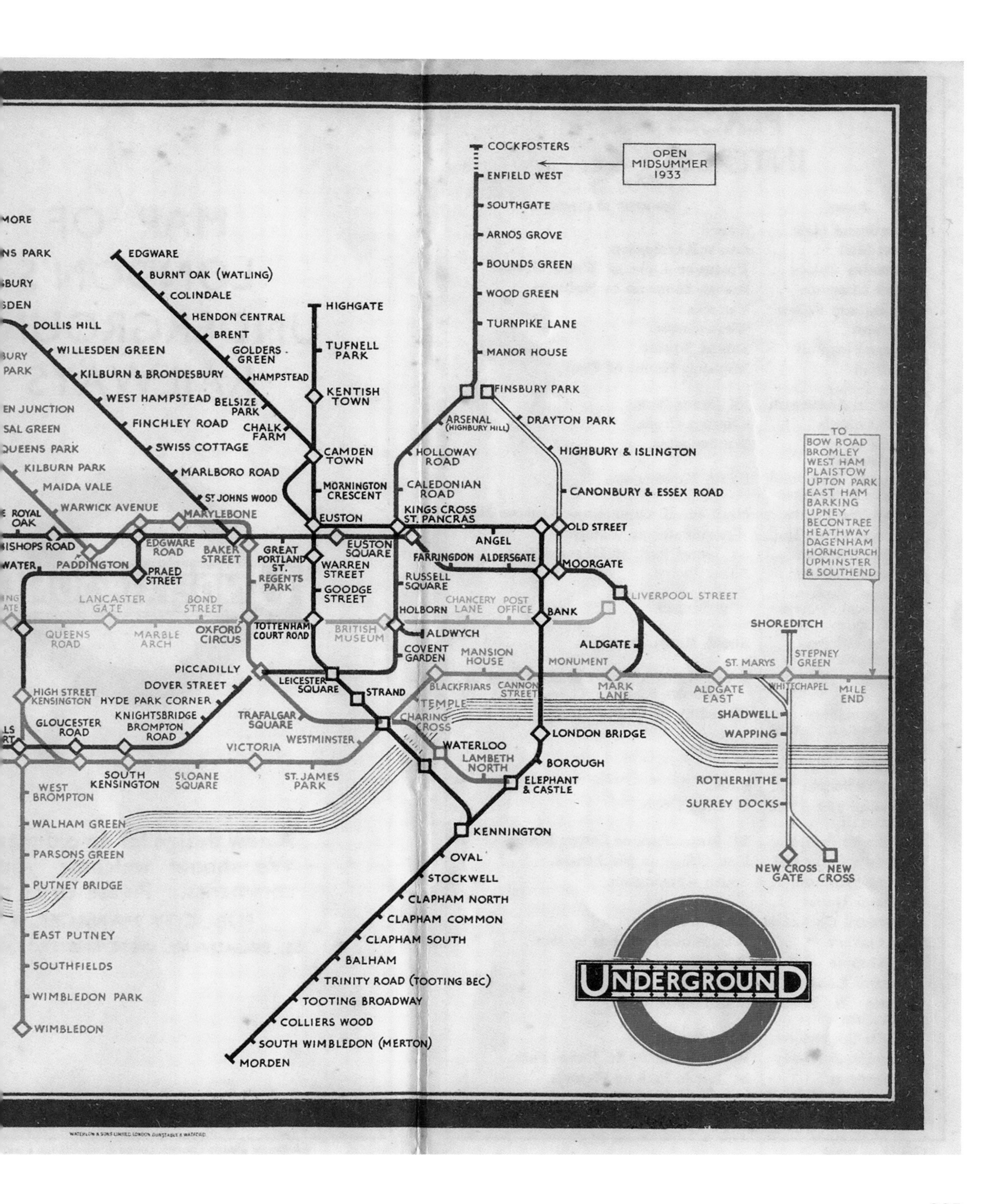

COCKFOSTERS
OPEN MIDSUMMER 1933
ENFIELD WEST
SOUTHGATE
ARNOS GROVE
BOUNDS GREEN
WOOD GREEN
TURNPIKE LANE
MANOR HOUSE
FINSBURY PARK
ARSENAL (HIGHBURY HILL)
DRAYTON PARK
HOLLOWAY ROAD
HIGHBURY & ISLINGTON
CALEDONIAN ROAD
CANONBURY & ESSEX ROAD
KINGS CROSS ST. PANCRAS
OLD STREET
ANGEL
EDGWARE
BURNT OAK (WATLING)
COLINDALE
HENDON CENTRAL
BRENT
GOLDERS GREEN
HAMPSTEAD
BELSIZE PARK
CHALK FARM
HIGHGATE
TUFNELL PARK
KENTISH TOWN
CAMDEN TOWN
MORNINGTON CRESCENT
EUSTON
DOLLIS HILL
WILLESDEN GREEN
KILBURN & BRONDESBURY
WEST HAMPSTEAD
FINCHLEY ROAD
SWISS COTTAGE
MARLBORO ROAD
ST. JOHNS WOOD
MARYLEBONE
KILBURN PARK
MAIDA VALE
WARWICK AVENUE
ROYAL OAK
BISHOPS ROAD
PADDINGTON
EDGWARE ROAD
BAKER STREET
GREAT PORTLAND ST.
EUSTON SQUARE
WARREN STREET
REGENTS PARK
PRAED STREET
FARRINGDON
ALDERSGATE
MOORGATE
RUSSELL SQUARE
GOODGE STREET
LIVERPOOL STREET
TO BOW ROAD BROMLEY WEST HAM PLAISTOW UPTON PARK EAST HAM BARKING UPNEY BECONTREE HEATHWAY DAGENHAM HORNCHURCH UPMINSTER & SOUTHEND
LANCASTER GATE
BOND STREET
CHANCERY LANE
POST OFFICE
HOLBORN
BANK
QUEENS ROAD
MARBLE ARCH
OXFORD CIRCUS
TOTTENHAM COURT ROAD
BRITISH MUSEUM
ALDWYCH
SHOREDITCH
ALDGATE
COVENT GARDEN
MANSION HOUSE
MONUMENT
ST. MARYS
STEPNEY GREEN
PICCADILLY
LEICESTER SQUARE
DOVER STREET
HIGH STREET KENSINGTON
HYDE PARK CORNER
STRAND
BLACKFRIARS
CANNON STREET
MARK LANE
ALDGATE EAST
WHITECHAPEL
MILE END
TEMPLE
KNIGHTSBRIDGE
BROMPTON ROAD
TRAFALGAR SQUARE
CHARING CROSS
GLOUCESTER ROAD
SHADWELL
WAPPING
LONDON BRIDGE
WESTMINSTER
VICTORIA
WATERLOO
LAMBETH NORTH
BOROUGH
SOUTH KENSINGTON
SLOANE SQUARE
ST. JAMES PARK
ELEPHANT & CASTLE
ROTHERHITHE
WEST BROMPTON
SURREY DOCKS
WALHAM GREEN
KENNINGTON
PARSONS GREEN
OVAL
STOCKWELL
PUTNEY BRIDGE
NEW CROSS GATE
NEW CROSS
CLAPHAM NORTH
CLAPHAM COMMON
EAST PUTNEY
CLAPHAM SOUTH
BALHAM
SOUTHFIELDS
UNDERGROUND
TRINITY ROAD (TOOTING BEC)
WIMBLEDON PARK
TOOTING BROADWAY
COLLIERS WOOD
WIMBLEDON
SOUTH WIMBLEDON (MERTON)
MORDEN

# Raised relief

## Models and visualizations that reflect Earth's topography.

**Planimetric maps use a variety of techniques to give the impression of relief.** These standard methods of representation support everyday needs on topographic maps. However, we are generally concerned with seeing high and low elevation or depth relative to where we are. We are rarely concerned with the absolute measurement unless the specific need (e.g. on an aeronautical chart) demands it. Instead, we think of terrain as hills, valleys, mountains, and plains. Viewing relative relief can be achieved using a variety of techniques that do not encode quantitative information such as hachures and hillshading.

**Three-dimensional views of relief can also be created to give the illusion when viewed in 2D.** For instance, plan oblique and perspective views give an impression of three-dimensional landscapes. Landscape drawings or panoramas are artistic representations that provide photorealistic terrain portrayals. Block diagrams are often created to show a cutaway of a part of Earth including the relief aboveground and the geological structure below ground. Wire-frame maps can also be created from either digital elevations models or statistical surfaces to create a three-dimensional profile.

**Many techniques can be replicated digitally to create interactive three-dimensional raised-relief models.** Plan oblique, perspective, axonometric, isometric, block diagrams, and wire-frame maps can all create compelling relief representations.

**Globes are perhaps the epitome of a model that can be used to show raised relief.** Globes often have textured surfaces that show an exaggerated form of relief. Virtual globes provide the same, albeit digital, experience and support highly detailed representations of the surface of terrestrial or celestial bodies. Raised-relief globes can also be used to isolate and show elevation or bathymetric information, often without any other topographic detail. These globes often have considerable vertical exaggeration to make peaks and troughs noticeable at scale.

**See also:** Hachures | Pseudo-natural maps | Shaded relief | Styling shaded relief

Plaster relief models can be constructed from layered plywood that conforms to contour lines. Layers are built up to create a stepped model and steps are filled using a pliable material. A model cast is made, and from this a plaster model is created. Variations of this construction technique ultimately lead to the creation of a negative cast of terrain so that plaster can be poured and the process repeated.

Raised-relief models can also be constructed from thermoplastic. Invented by the US Army in the 1940s, vacuum-formed relief models were produced in large quantities. The technique involves milling a solid, often plastic form from which a plaster mould is made. A reproduction mould is cast that is resistant to heat and pressure. Thermoplastic sheets are then moulded to the cast and become solid as they cool.

Subtractive machining starts with a block of material such as plaster, resin, or wood, and a milling tool creates relief by removing material on the block. Modified inkjet technologies can be used to paint the resulting models. Laser cutters can be used as milling machines to mill a single piece of material or to create layers of material of uniform thickness that can be combined to create a stepped contour surface. Additive techniques such as 3D printing build models from a flat base using layers of molten resin or plastic which dries to form individual layers.

**Opposite:** *Schiehallion*, by Karen Rann, 2016. The physical model uses two alternative three-dimensional forms to depict elevation of the mountain, illustrated in plan above.

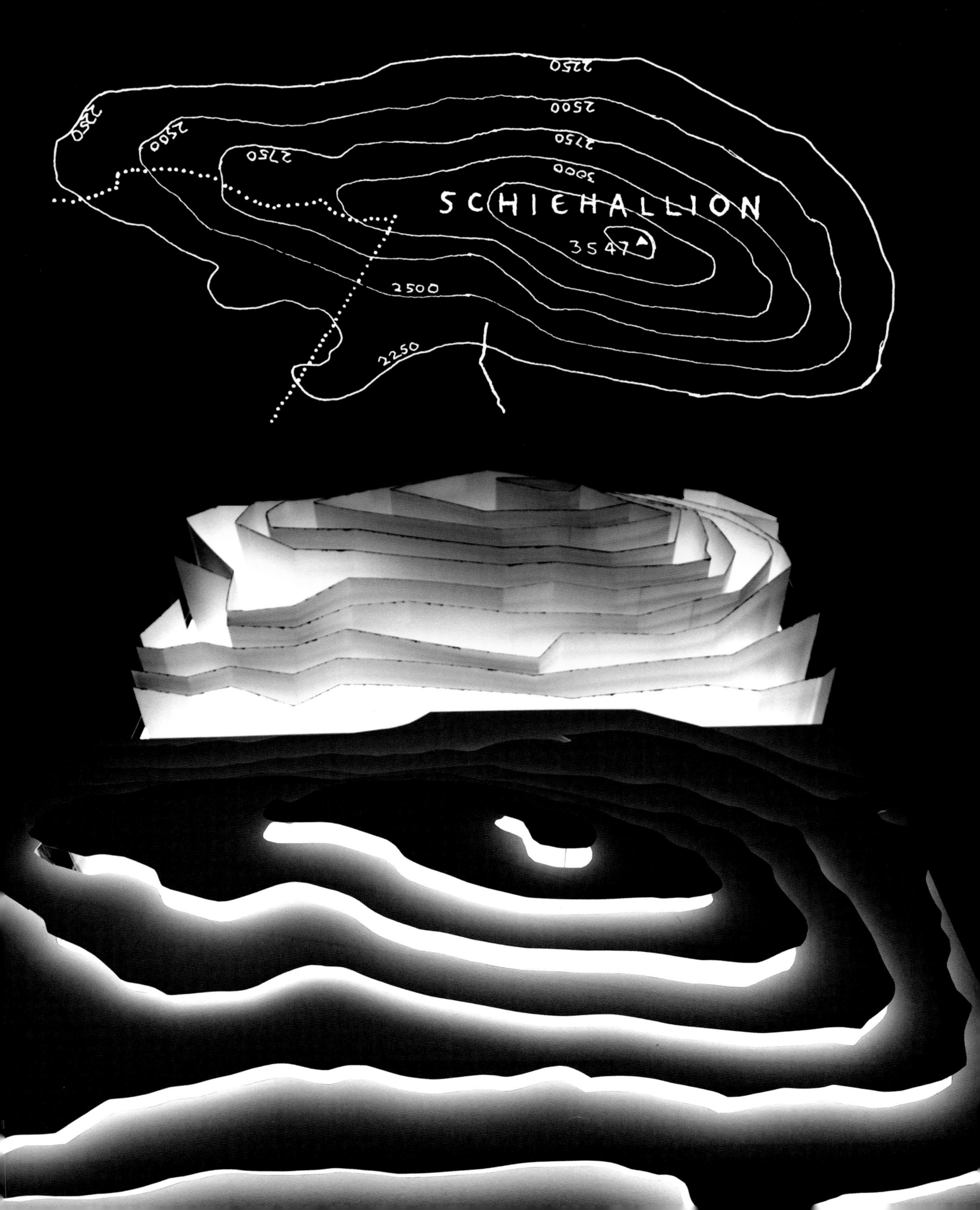

SCHIEHALLION
3547
3000
2750
2500
2250

# Ratio and interval data

## The mean and standard deviation are most commonly used to summarise ratio and interval data.

**Because ratio and interval data is numerically scaled, it has numerous statistical measures that can be mapped.** The mean is the most commonly used summary statistic for this purpose, and the standard deviation is used to describe dispersion of values around the mean. Arithmetic mean is calculated by

$$\overline{X} = \frac{\sum x}{N}$$

where $\overline{X}$ is the mean, $\sum$ is a summation sign, $x$ is a value in the array, and $N$ is the total number of values.

**You can also illustrate the location of the mean value along a number line to show the distribution of data values.** The mean is the data value around which other data values balance.

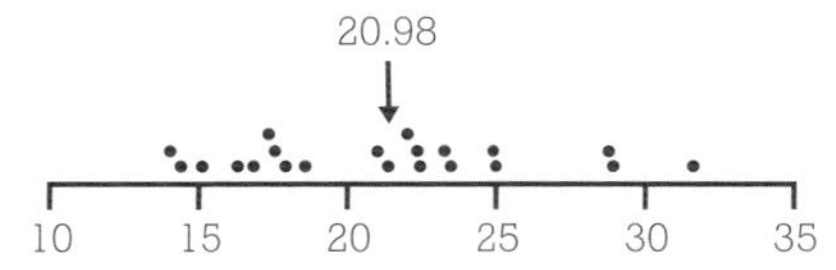


The mean does not illustrate how values are arranged around the mean itself. Two very different arrays may have exactly the same mean.

**The standard deviation is used to describe this characteristic of the data, calculated by:**

$$\sigma^2 = \frac{\sum\left(x - \overline{X}\right)^2}{N}$$

where $\sum^2$ is the variance, $\left(x - \overline{X}^2\right)$ is the computation of each value of $x$'s deviation from the mean value, and $N$ is the number of values.

**Standard deviation is useful when comparing two variables.** The variable with the smaller standard deviation indicates values that are clustered more closely around the mean. When examining standard deviation in a single dataset, certain relationships between $\overline{X}$ and $\sigma$ can be examined. Importantly, $\overline{X} \pm 1\sigma$ will always include just over two-thirds of the data values (68.27% of values).

**See also:** Levels of measurement | Making numbers meaningful | Nominal data | Ordinal data

The sum of 21 values in the table below is 440.5, so the mean is:

$$\overline{X} = \frac{440.5}{21}$$

$$\overline{X} = 20.98$$

Additionally, $\sigma = 4.71$. Thus, 14 of the data values are within 4.71 units (in our case % of value) of $\overline{X}$. The remaining seven values fall outside $1\sigma$ from the mean. The space between $\overline{X} \pm 2\sigma$ will contain at least 75% of data values, and the space between $\overline{X} \pm 3\sigma$ will contain at least 88% of data values.

| Value (%) | Rank | | |
|---|---|---|---|
| 30.8 | 1 | | |
| 28.4 | 2 | | |
| 28.3 | 3 | | |
| 24.9 | 4 | | |
| 24.8 | 5 | | |
| 23.4 | 6 | upper quartile | Interquartile range |
| 23.2 | 7 | | |
| 22.0 | 8 | | |
| 21.9 | 9 | | |
| 21.7 | 10 | | |
| 21.2 | 11 | median | |
| 20.8 | 12 | | |
| 18.6 | 13 | | |
| 18.1 | 14 | | |
| 17.6 | 15 | | |
| 17.4 | 16 | lower quartile | |
| 17.0 | 17 | | |
| 16.3 | 18 | | |
| 15.2 | 19 | | |
| 14.6 | 20 | | |
| 14.3 | 21 | | |

the Interval* & Ratio Outlook

| | Sparrow Sprint | Clover Loop | The Bear Crawl | Willow Run | Juneau Roll |
|---|---|---|---|---|---|
| Start Time* | 9:45am | 10:30am | 12:15pm | 2:00pm | 4:00pm |
| Race Day Temp (F)* | 66° | 72° | 68° | 71° | 74° |
| Distance | 0.1km | 100km | 42km | 16km | 120km |
| Spectators | 28,000 | 16,000 | 9,000 | 800 | 12,000 |
| Mean Grade | 0° | 0° | 12° | 17° | 4° |

# Ratios, proportions, and percentages

## Converting totals into a derived measure is often necessary for specific map types and to convey meaning in a more understandable way.

**Ratio is the relationship between data expressed as**

$$\frac{f_a}{f_b}$$

where $f_a$ is the number of items of data in one class and $f_b$ is the number of items in another class. For instance, the ratio of 24 acres of urban land compared to 12 acres of rural land is:

$$\frac{24}{12} = \frac{2}{1}$$

This gives a ratio of 2:1. Ratio is often used to calculate population density, defined as the number of people per unit area. If the population of a town is 5,500 and it is 10 sq km, then a density of 550 persons per sq km is derived by:

$$\frac{5500}{10} = \frac{x}{1};$$
$$10x = 5500;$$
$$x = 550$$
$$\therefore \frac{5500}{10} = \frac{550}{1}$$

**Proportion is defined as the number of items in one class of data in relation to the overall number of items expressed as**

$$\frac{f_a}{N}$$

where $f_a$ is the number of items of data in one class and $N$ is the total number of items. In a town of 5,500, of 3,000 males and 2,500 females, then the proportion of males would be

$$\frac{3000}{5500} = \frac{30}{55} = 0.55 \text{ or, as a percentage } 0.55 \times 100 = 55\%$$

**Percentages, or rates, are used a great deal in describing concentrations or how the character of one area differs from another.** Because areas differ dramatically in size and capacity, using ratios or percentages implicitly gives a common denominator. Visually, it means a map can be used to compare one place to another without visual bias.

**See also:** Frequency distributions and histograms | Variables, values, and arrays

In statistical mapping, matching the data type to the appropriate map type is not a trivial issue. Yet modern software does not have a filter to stop you from using a data type that is unsuitable for a particular map type.

If your data is in totals and you neither care, nor have a need to convert it to a rate or other derived measure, then your main choices for mapping are the proportional symbol, graduated symbol, or dot density map along with cartograms. None of these techniques demand that account be made of the underlying enumeration area.

If you convert your data to a rate or some meaningful derived measure, then the choropleth map becomes a possible map type to use.

In the examples shown here, the same data looks very different when mapped as totals compared with when it is mapped as a percentage. The map of totals takes no account of either the unit area or the base population, both of which can vary widely across the map. The map of percentages normalises the totals by dividing the values by the total number of votes cast per county. This normalization creates a consistent basis from which you can compare like-for-like quantities across the map.

Both maps are shown using a quantile classification scheme so each class shows the same number of counties, ranked. The totals map is misleading. The percentages map is not.

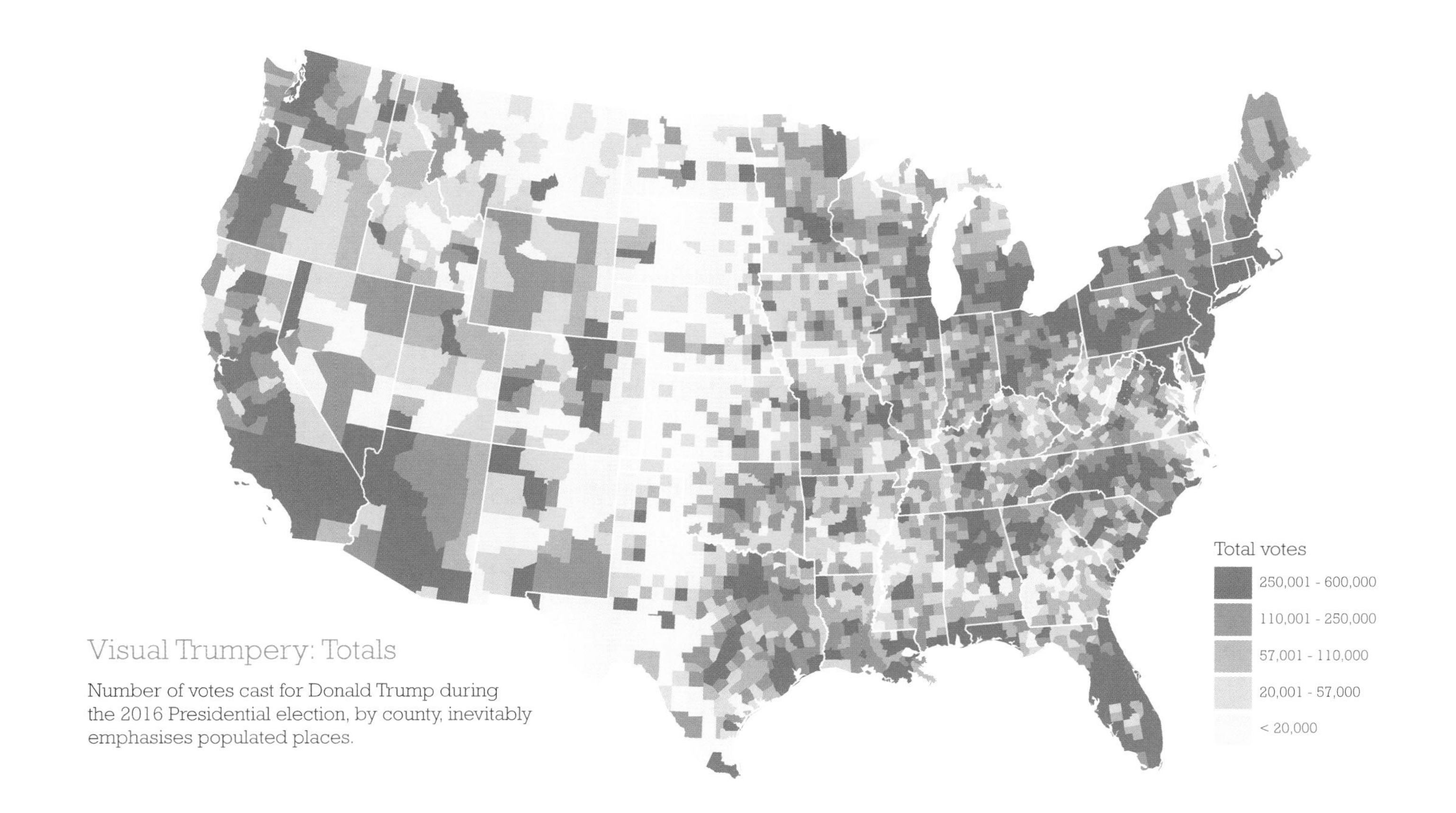


## Visual Trumpery: Totals

Number of votes cast for Donald Trump during the 2016 Presidential election, by county, inevitably emphasises populated places.

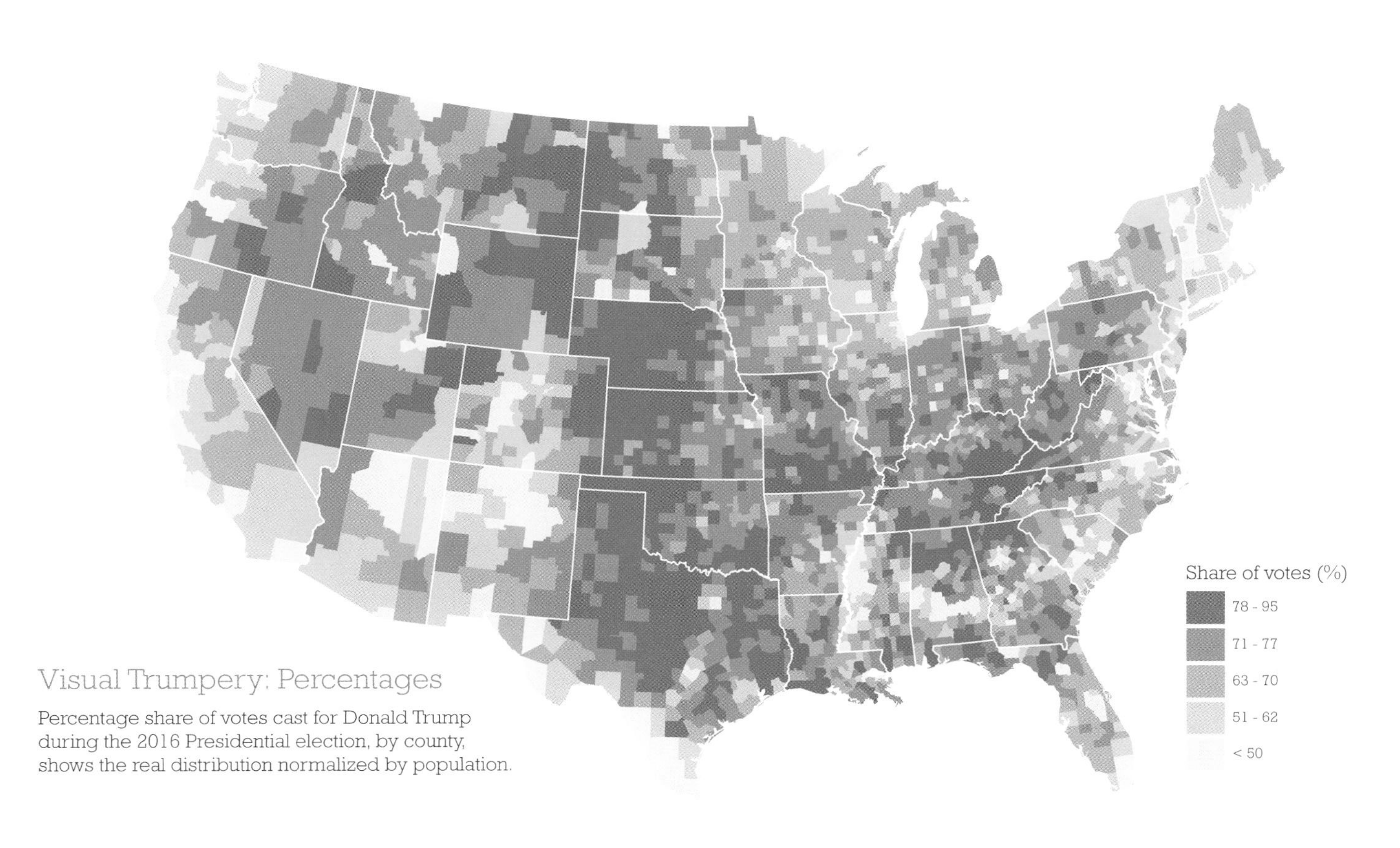


## Visual Trumpery: Percentages

Percentage share of votes cast for Donald Trump during the 2016 Presidential election, by county, shows the real distribution normalized by population.

* trumpery
Noun: trum·pery
1. attractive articles of little value or use
2. showy or worthless

# Reference maps

**Reference maps are often referred to as general-purpose maps because they support tasks such as finding locations and navigating.**

**A reference map (or general-purpose map) generally displays both natural and man-made objects that exist in the geographical environment.** The topographic map is an example of a reference map. It illustrates the location of physical features such as coastlines, rivers, forestry, and lakes along with man-made features such as urban areas and roads. Additionally, topographic maps also tend to show features that are non-tangible in the real world yet provide important contexts for describing the character of a region such as political or administrative boundaries (which may follow the line of a river or a road, for instance). Small-scale topographic maps (approximately > 1:500,000) illustrate whole countries or continents. Large-scale topographic maps (approximately < 1:50,000) might form part of a national map sheet series (or dataset) and provide greater detail. In a digital environment scales tend to work seamlessly from small to large.

**Large-scale reference maps (e.g. 1:1250) may be used for special purposes such as site location where positional accuracy is of paramount importance.** Such large-scale reference maps are collectively called *plans*: detailed maps showing buildings, roads, and boundary lines and are often the most detailed available for an area being the result of large-scale survey and data acquisition.

**A chart is a specific type of large-scale reference map used to aid navigation.** Charts assist navigation by aircraft (aeronautical charts) or by ships (nautical charts). Topographical features such as the location of mountains or submerged reefs and visible objects such as roads and small islands are often overprinted with additional navigation detail that allows the charts to be worked upon rather than simply viewed. Charts allow users to determine their position and to plot courses

Larger scale maps are usually derived from smaller scale maps and data acquired for the same area, leading to the term *derived mapping*; that is, a map whose detail has been derived from a map or dataset held at a larger scale and in greater detail. This process is quite often the normal method of preparing maps at different scales and avoids the cost of resurveying areas simply to create a map at an alternative scale.

Large-scale plans may form the basis of legal documents for determining a property boundary, for instance. A plan whose function is based on delimiting property and landownership zones is termed a *cadastral map*.

**See also:** Atlases | Basemaps | Descriptive maps | Shaded relief

**Opposite**
**Top:** Aeronautical chart.
**Middle:** Topographic map.
**Bottom:** Nautical chart.

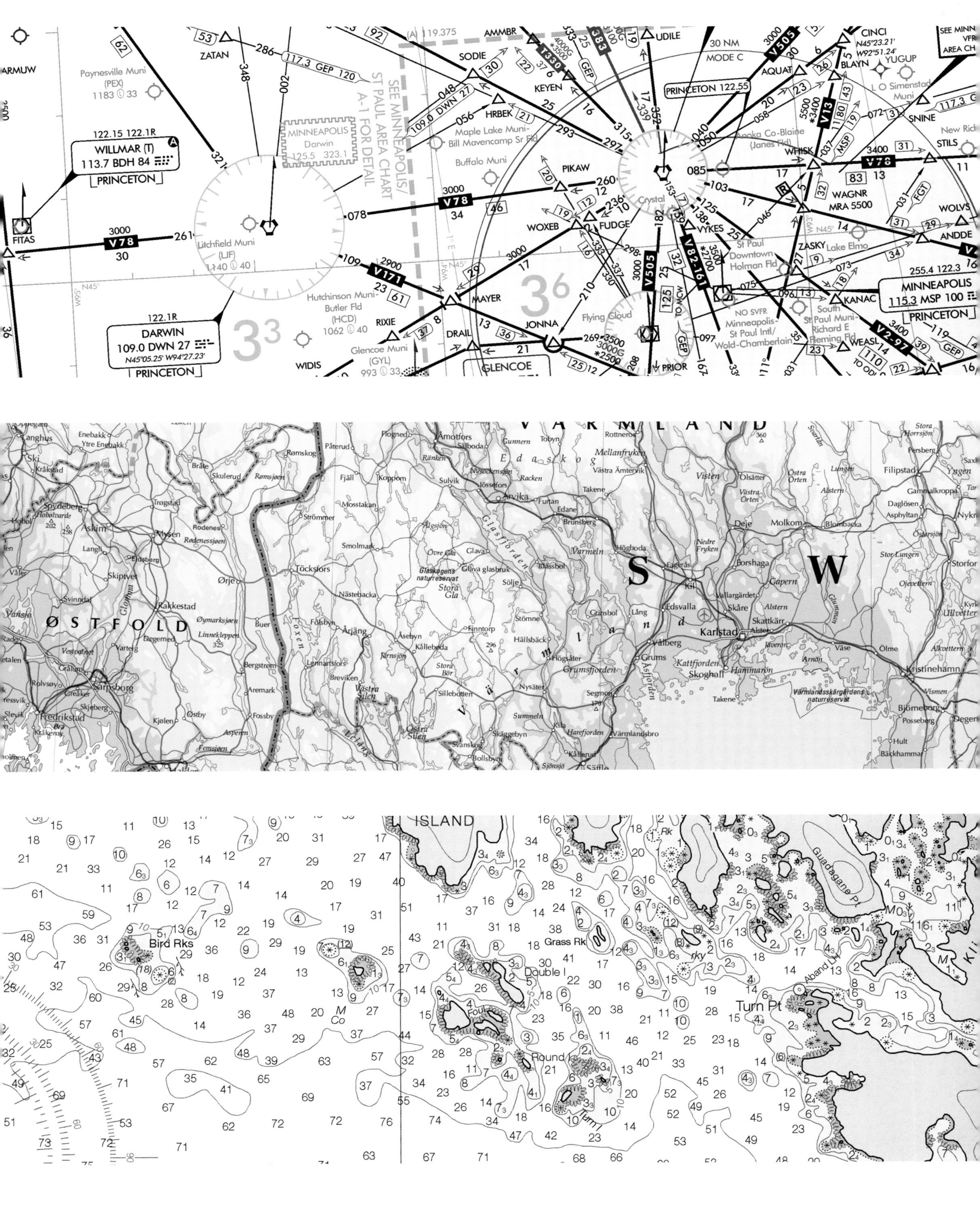

ZATAN
Paynesville Muni
(PEX)
1183 33
WILLMAR (T)
113.7 BDH 84
PRINCETON
122.15 122.1R
FITAS
3000
V78
30
261
286
117.3 GEP 120
SEE MINNEAPOLIS/
ST PAUL AREA CHART
A-1 FOR DETAIL
MINNEAPOLIS
Darwin
125.5 323.1
Litchfield Muni
(LJF)
1140 40
122.1R
DARWIN
109.0 DWN 27
N45°05.25' W94°27.23'
PRINCETON
Hutchinson Muni-
Butler Fld
(HCD)
1062 40
Glencoe Muni
(GYL)
993 33
WIDIS
RIXIE
DRAIL
MAYER
JONNA
GLENCOE
V171
2900
AMMBR
SODIE
HRBEK
KEYEN
109.0 DWN 27
Maple Lake Muni-
Bill Mavencamp Sr Fld
Buffalo Muni
PIKAW
WOXEB
FUDGE
Flying Cloud
UDILE
30 NM
MODE C
PRINCETON 122.55
Crystal
Anoka Co-Blaine
(Janes Fld)
WHISK
WAGNR
MRA 5500
AQUAT
BLAYN
CINCI
N45°23.21'
W92°51.24'
YUGUP
L O Simenstad
Muni
SNINE
117.3
STILS
New Rich
SEE MINN
VFR
AREA CH
V78
V505
V13
V82-161
V2-97
St Paul
Downtown
Holman Fld
ZASKY
Lake Elmo
WOLVS
ANDDE
255.4 122.3
MINNEAPOLIS
115.3 MSP 100
PRINCETON
KANAC
NO SVFR
Minneapolis-
St Paul Intl/
Wold-Chamberlain
South
St Paul Muni-
Richard E
Fleming Fld
WEASL
PRIOR
VYKES
V Ä R M L A N D
Enebakk
Ytre Enebakk
Ski
Kråkstad
Spydeberg
Askim
Mysen
Trøgstad
Skulerud
Rømskog
Rødenes
Langli
Eidsberg
Skiptvet
Ørje
Rakkestad
Svinndal
Glomma
Ø S T F O L D
Degernes
Varteig
Sarpsborg
Fredrikstad
Kråkerøy
Greåker
Skjeberg
Rolvsøy
Kjølen
Østby
Fossby
Aremark
Bergstrøm
Buer
Töcksfors
Foxen
Fjäll
Koppom
Mosstakan
Strömmer
Smolmark
Nästebacka
Årjäng
Fölsbyn
Lennartsfors
Breviken
Västra Silen
Östra Silen
Åsebyn
Kålleboda
Glava
Glaskogens naturreservat
Stora Gla
Glava glasbruk
Finntorp
Sölje
Glasfjorden
Arvika
Jössefors
Sulvik
Ämotfors
Edane
Brunsberg
Furtan
Värmeln
Klässbol
Stömne
Hällsbäck
Högsäter
Grumsfjorden
Nysäter
Sillebotten
Skåggebyn
Svanskog
Bollsbyn
Kila
Harefjorden
Värmlandsbro
Kållerud
Säffle
Segmon
Grums
Vålberg
Edsvalla
Karlstad
Skoghall
Hammarön
Kattfjorden
Värmlandsskärgårdens naturreservat
Takene
Kil
Fagerås
Högboda
Mellanfryken
Västra Ämtervik
Rottneros
Olsätter
Deje
Molkom
Forshaga
Gapern
Skåre
Skattkärr
Alstern
Väse
Ölme
Kristinehamn
Björneborg
Hult
Filipstad
S
W
ISLAND
Bird Rks
Grass Rk
Double I
Foul I
Round I
Turn I
Turn Pt
Guadagane Pt

# Refinement

**Often used to represent features using only a part of the original data, sometimes displaced or exaggerated to enhance their visibility.**

**Refinement weeds out some original features of the same type and proximity to display the general character more clearly.** For instance, a cluster of buildings may be adequately represented by the location of a single building at a much reduced scale. A stream network can be refined by making use of only first- or second-order streams. A road network might depict only motorways and major roads rather than the full network. Refinement, then, applies a form of classification to select features that typify the overall character.

**Exaggeration changes the geometry of a feature so that certain characteristics remain visible at smaller scales.** Amplification of topological dimensions is often desirable to avoid them being lost as line work and distances between features become progressively smaller. For instance, at smaller scales, the mouth of an inlet or bay might be lost so widening the bay so it retains its visibility is useful.

**Enhancement applies symbolisation to features to change their relative importance.** For instance, a road network depicted as a set of interconnected lines might not adequately depict intersections of overpass/underpass relationships. Cased lines can improve the depiction of roads with the main road being cased in a heavier line weight to give increased prominence. Additionally, a mask could be applied around the main road to ensure the minor road appears to pass underneath. The addition of some line work to portray a bridge or overpass might complete the enhancement.

**Displacement is similar to exaggeration but is used in specific circumstances when a reduction of scale results in a coalescing of features.** For instance, as scale is reduced, what might be a junction with offset roads may be seen as a crossroads. To ensure the junction cannot be confused with a crossroads, the roads would be moved slightly so that they intersect the main road at an increased offset distance from each other. This offset preserves the depiction of the original character of the junction.

Cartographic refinement offers a flexible set of approaches for modifying the appearance and position of features to give them clarity. Unlike other generalisation tools, they often add graphical embellishments or other symbols to give context to the feature which, in turn, makes it clearer.

Refinement also requires a sense of subjectivity in making decisions about how to represent features consistently. If bridge symbols are used for some parts of a road network, then for consistency they should also be applied elsewhere. Similarly, features that are exaggerated or displaced in one part of the map should also act as a guide for the same treatment elsewhere. This symbolisation not only results in a map where individual features are clarified but it also leads to a structural and visual balance for the whole map.

By taking a consistent approach across the whole map, you'll avoid a plethora of different generalisation and design treatments leading to a consistent denotation of features. By minimising the amount of different or obvious refinement techniques across your map, you help your map reader's expectations for what they are seeing and how it should be interpreted.

**See also:** Aggregation | Generalisation | Simplification | Smoothing

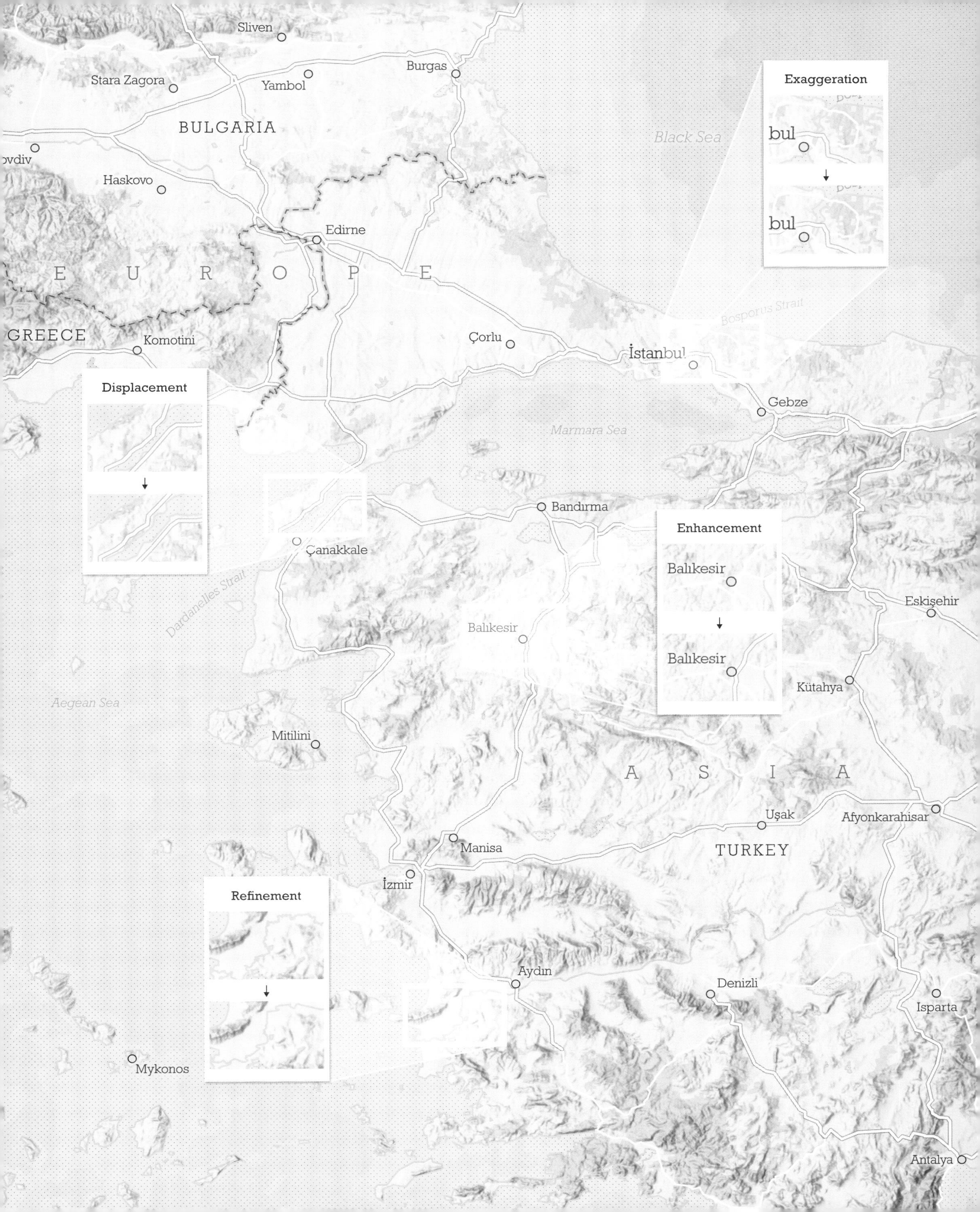

Sliven
Burgas
Stara Zagora
Yambol
BULGARIA
Black Sea
ovdiv
Haskovo
Edirne
E U R O P E
GREECE
Komotini
Çorlu
İstanbul
Bosporus Strait
Gebze
Marmara Sea
Bandırma
Çanakkale
Dardanelles Strait
Balıkesir
Eskişehir
Kütahya
Aegean Sea
Mitilini
A S I A
Uşak
Afyonkarahisar
Manisa
TURKEY
İzmir
Aydın
Denizli
Isparta
Mykonos
Antalya
Exaggeration
bul
bul
Displacement
Enhancement
Balıkesir
Balıkesir
Refinement

# Resolution

## The spatial precision of feature representation can be used as a visual variable.

**The resolution of features describes the spatial precision at which they are displayed.** This has as much to do with the choice and size of symbol as it does the coarseness of the data and the general way in which it is represented. Resolution is usually considered as a form of generalisation of map data. For raster data this would be the pixel size, and for vector data the number of nodes and edges used to represent a feature. Usually, as the map scale gets smaller, the resolution of data is generalised. As a visual variable, resolution can be used to express the character of the data in different ways in relation to other map detail.

**Higher resolution data will tend to be seen as figural.** Regardless of scale, data that is represented in more detail, that is at a higher spatial precision, will appear to be more important. This characteristic of the way in which maps are seen can be useful since different levels of generalisation might be applied to different layers on a map to give a sense of visual order. Rather than using the same level of abstraction to encode information across the map, different levels of precision enable you to create different visual levels.

**Larger map symbols can appear less important if their resolution is low.** For most maps, larger symbols are seen as more figural and, therefore, are interpreted as more important in the context of the surrounding information. However, somewhat counter-intuitively, large symbols might be used to reflect a lower level of spatial precision in comparison with smaller symbols appearing to represent more precise locations. In this case, particularly for point of interest symbols, the size of symbol will not necessarily be seen as more important but, instead, will be seen as indicating a general location.

**See also:** Data density | Dispersal vs. layering | Generalisation | Scale and resolution

The relationship between the scale at which map data was collected and how it is represented shouldn't be underestimated. National mapping agencies survey at large scale (e.g. 1:1,250) and create a detailed map from which all other smaller scale versions are derived. At each successive map scale, the resolution of data is reduced, making it a more generalised version of the original.

When data has been collected by less precise means, it is important not to imply a higher level of spatial precision than the data supports. For instance, data collected using a consumer-grade GPS receiver might contain positional errors of up to 15 m. For medium- to small-scale maps, this error will be imperceptible. The scale of the map itself masks the error. Making larger scale maps from data collected at a lower level of precision is, for this reason, inadvisable.

Symbols also have dimensionality. For instance, on a map at a scale of 1:50,000, a line that is 1 mm wide represents 50 m in map distance. This can assist in masking some level of error in map data because the symbol itself represents a larger map distance than the potential error. As map scale gets larger, the real-world distance represented by the same 1 mm line gets smaller so implied resolution increases. As a visual variable, increased symbol detail is enhanced at larger scales.

Visual variation

Increased resolution is seen as figural

For seeing: Distinct, Levels

For representing: Nominal, Ordinal, Numeric

**Opposite:** Resolution of mapped detail diminishes with distance in this classic front cover of the *New Yorker*, 1976.

Mar. 29, 1976

# THE NEW YORKER

Price 75 cents

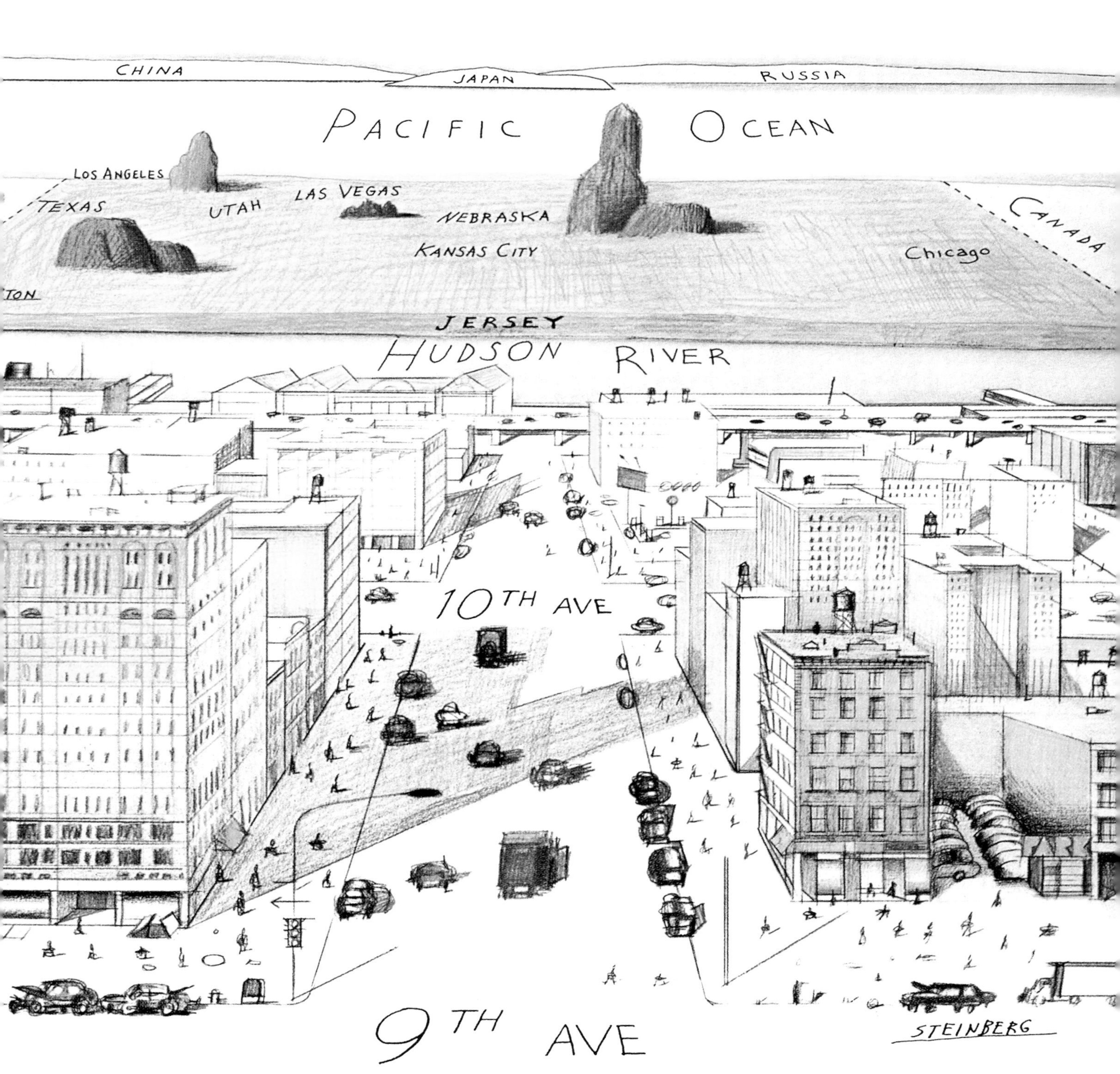

# Rock drawing

## Depiction of rugged, complex rocky areas on topographic maps.

**Rock drawing is used to depict rugged, often steep and complex terrain for which standard terrain rendering is insufficient.** For instance, contours often become indecipherable because the steepness and angular structure render them chaotic. Most techniques are artistic and designed to illustrate the landscape mimetically. Rock drawing might use interval contours coupled with skeletal or edge lines or rock hachures. With any rock drawing, the scale of the map and the particular nature of different rock formations mean that there is no universal technique that can be applied.

**Understanding rock forms helps to support effective rock drawing.** Geomorphology and landform characteristics help distinguish between types of formation. The type of rock, structure of surfaces, distinct patterns made by weathering and erosion on different rock types, and pattern of debris fields all lead to a variety of different techniques. Planimetrically, rock drawing must illustrate ravines, gullies, and crest lines as well as key erosional features.

**Rock contours on smaller scale maps generally downsize the techniques used for larger scale versions.** Contours become meaningless to depict rocky terrain at a small scale. The detail of rock hachures becomes too dark and heavy so is replaced by the skeletal outlines of major features with sharp contrasts in the direction of line work used to emphasise the effect. Black is usually replaced by dark brown or deep purple so the black isn't too dominant in the overall composition. Rock contours are constructed by eliminating parts of contour lines that run too close together. This type of drawing is the simplest form of suggesting a steep rocky outcrop but is not the most aesthetically pleasing and can be confusing as contours disappear and reappear.

Skeletal lines are constructed by drawing a framework that demarcates the natural edges of the surface of rocks, including ledges, cliffs, cracks, and ridges. In sufficient detail, the technique gives a good impression of form and structure. An element of rock shading is usually applied to accentuate the 3D nature of the drawing, and edge lines are often drawn slightly thicker to contrast between heavy and light line widths. Shading might be illuminated from the traditional northwest to place some areas in shadow, or shading can be applied using the principle 'the steeper, the darker'.

Shaded rock hachuring is considered to be the classic form of rock drawing and comprises contours, skeletal outlines, and rock shading. The result is a finely detailed 3D effect with tonal variation. Individual strokes follow terrain features on the slope itself though some geological types (e.g. limestone pavements) are better drawn with lines that represent both clints and grikes. Shading is usually denoted by changing line weights so lines in greatest shadow or steepness are thicker. Stroke density over the rock hachured area should be consistent. Distances between strokes vary only slightly. Sharp outlines of individual features should be avoided as they appear too harsh. Change appears through the subtlety of contrast from light to dark. Contrast between light and dark should also become sharper with increased elevation to create focus and avoid the impression of relief inversion.

Rock drawing is almost always black or dark grey. Colour can be used to delineate it from other surface types. Some Swiss maps use a light-pink tint behind rock drawing to clarify the difference between vegetation and rocky areas. Coloured contours (e.g. brown or orange) can be overprinted. For glaciated areas, rock drawing is depicted in blue to convert it to a representation of ice structures.

**See also:** Geological maps | Pseudo-natural maps | Small landform representation

## Planimetric Building Blocks

Steep Slope
Vertical Lines

Gentle Slope
Horizontal Lines

Width Creates Darkness

NW Light source

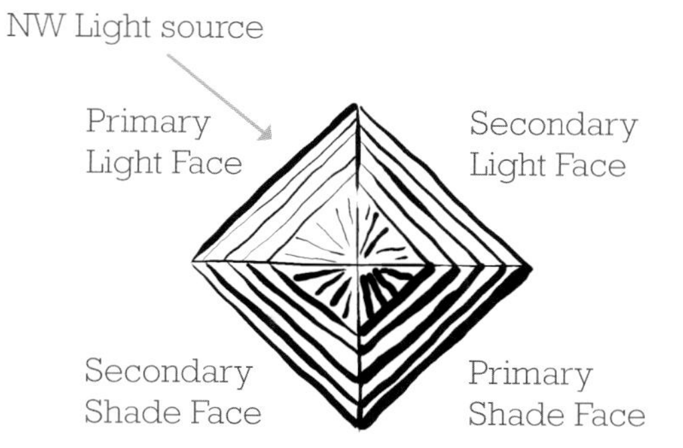


Final Rock Section

## Final Drawing

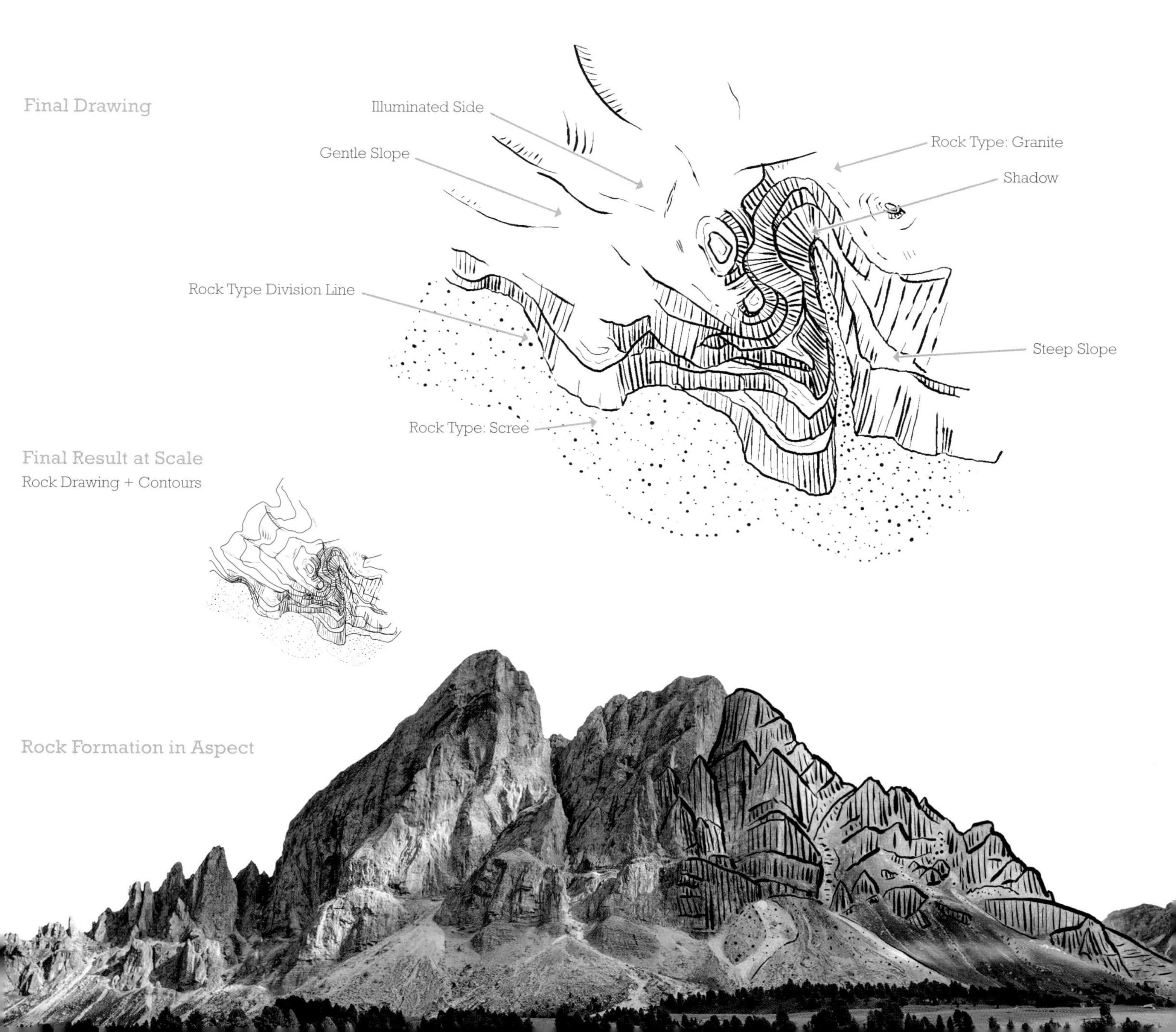


## Final Result at Scale

Rock Drawing + Contours

## Rock Formation in Aspect

## The Magnificent Bears of the Glorious Nation of Finland

Annukka Mäkijärvi

2014

Good maps are not boring, and even a simple univariate choropleth can be engaging under the guidance of a clever designer. Manual illustration gives this map individuality and character, which is the starting point for its appeal. Its subsequent popularity reminds us that audiences are hungry for uniqueness and personality in an era of mass digital production.

For centuries, maps have been appreciated for their aesthetics more than their content, and this one is no exception—most of the people who saw it were not seeking out data on bear populations. But this is not a complaint so much as an acknowledgement of reality, and the lesson here is simple: to engage an audience, cartographers would do well to avoid rote neutrality and boldly embrace a clear style. If no one wants to look at or share a map, it cannot fulfill its function of conveying information.

Would a choropleth with standard boundaries be as useful? Yes, possibly more so because the shapes would be more relatable to a familiar administrative structure. But would people have paused to explore, been captured by the novelty, or be willing to spend time enjoying the map? Unlikely. This is not a map for a statistical report. But it works magnificently as a one-off product.

—Daniel Huffman

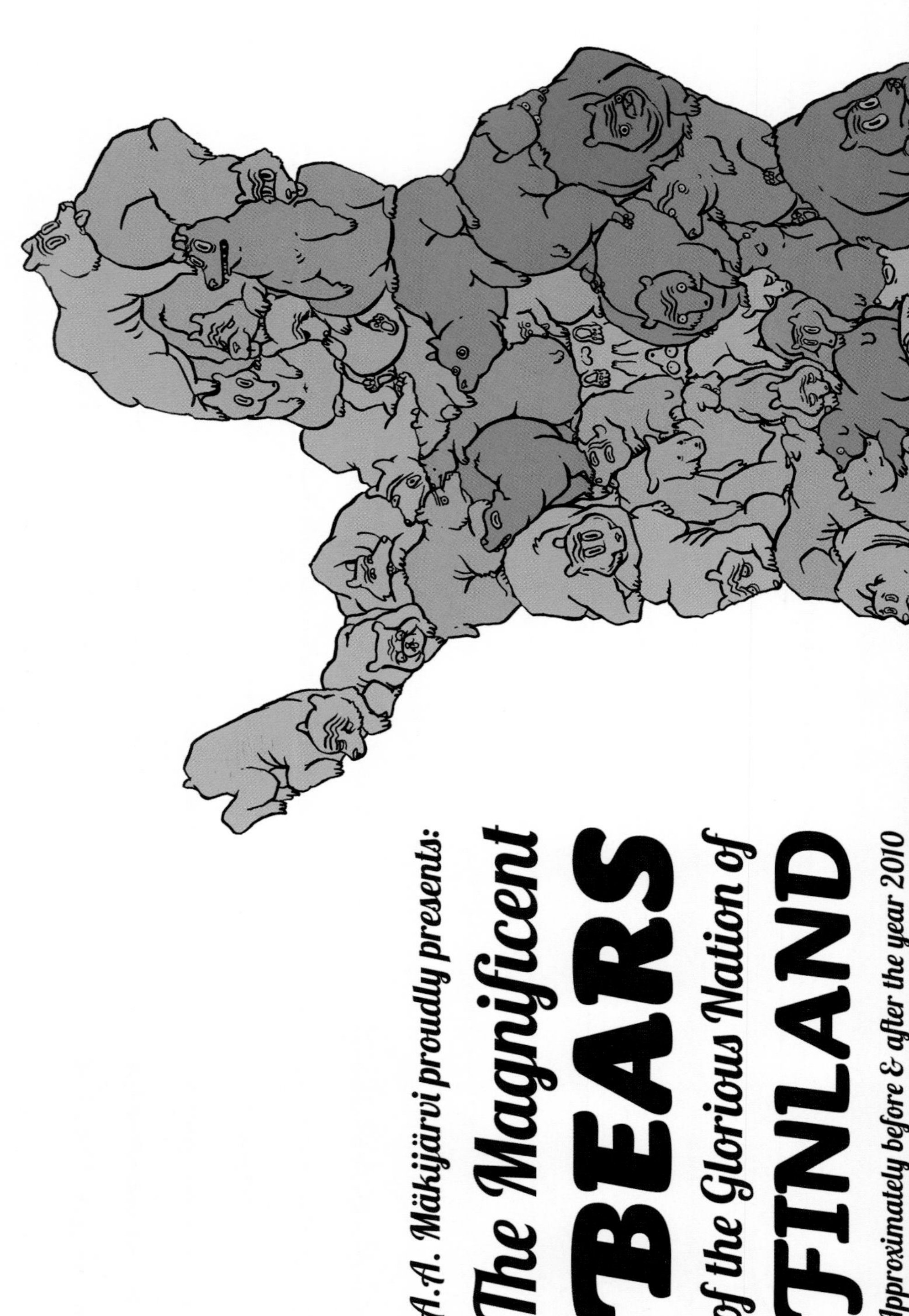

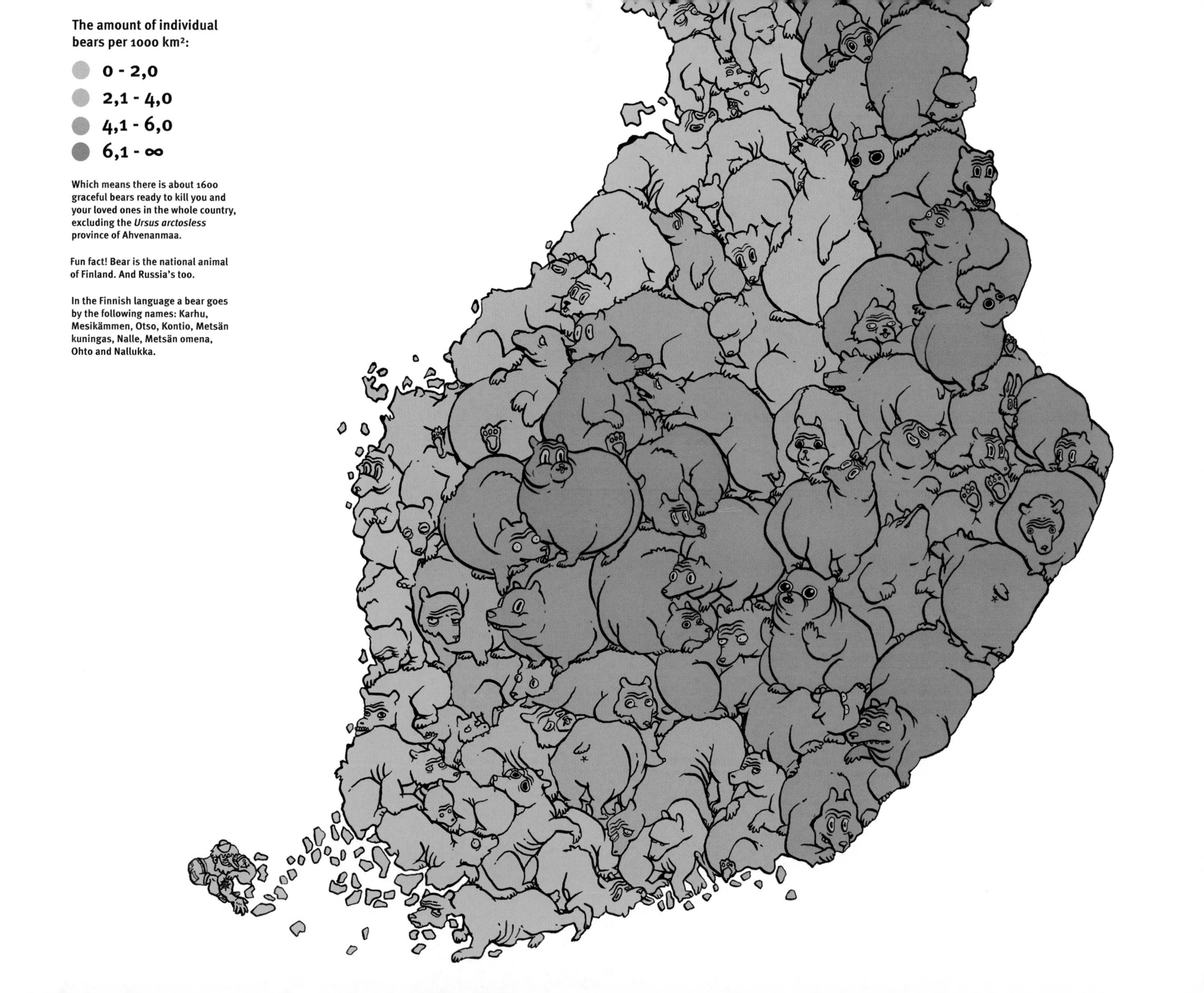

The amount of individual bears per 1000 km²:
0 - 2,0
2,1 - 4,0
4,1 - 6,0
6,1 - ∞
Which means there is about 1600 graceful bears ready to kill you and your loved ones in the whole country, excluding the *Ursus arctosless* province of Ahvenanmaa.
Fun fact! Bear is the national animal of Finland. And Russia's too.
In the Finnish language a bear goes by the following names: Karhu, Mesikämmen, Otso, Kontio, Metsän kuningas, Nalle, Metsän omena, Ohto and Nallukka.

# Saturation

## The relative peakedness of a hue used to create vivid or muted symbology.

**Saturation can be described as a mixture of grey and a pure hue of a dominant wavelength.** Saturation is, in fact, the intensity or peakedness of a hue as lightness is held constant. For instance, if we use a red hue and lightness is held constant, the colour's intensity goes from a desaturated greyish red to a fully saturated red. Bold or highly saturated hues reflect energy in a concentrated band of the visible electromagnetic spectrum whereas their counterparts, desaturated pastel hues, reflect energy evenly.

**Saturation can be used in a similar way to lightness for showing quantitative differences in data.** Lightness is often chosen ahead of saturation on the grounds that it is more aesthetically pleasing because bold, saturated hues tend to dominate heavily as figures in perception.

**To create a series of values across a single-hue progression, it's necessary to add white to get light values and black to get dark values.** Every hue has a pure version that has no white or black. This is the saturation level of a hue that is neither washed out (desaturated through the addition of white) nor soiled (darkened by the addition of black). This is often referred to as the *pure tone* of a hue. It is difficult to simply use saturation to represent quantitative differences in data because of the problem of creating sufficient perceptually different classes of the same hue. It's more usual to vary value at the same time to create a good progression.

**The saturated tone of each dominant wavelength in the visible spectrum does not hold the same value (or lightness).** This issue leads to some difficulties in the use of colour in cartography. If only pure tones are used, then colour variation also varies by hue and value. Perceptually, the eye sees value as dominant and so a natural order emerges but not in the spectral order. It's therefore necessary when using saturation as a visual variable to be cognisant that adjustments to the pure tone might need to be made to ensure that the eye does not see value as dominant across a range of different hues.

**See also:** All the colours | Hue | Mixing colours | Value

Saturation can be used to good effect to create a different aesthetic for a map. Traditionally, topographic mapping and atlas products have tended to use fairly desaturated colours that, when combined with low value, create muted colour schemes. The logic here was that the map should have subtle colours that print well. With increasing use of high-resolution electronic devices, the trend in cartography has been toward more saturated colours to allow features to stand out more. The use of saturated colours can also lend a dramatic appearance to the map and make it vivid. This can be useful when your map is competing with a plethora of other material that people see fleetingly across their social media feeds and which needs to grab attention.

Saturation is often labelled using a range of different terms such as *chroma*, *colourfulness*, and *intensity*. The use of different terms stems from different domains such as psychophysics. Alternatively, terminology such as *shades*, *tints*, and *tonal variation* stem from art. A tint is created by mixing white into a hue, a tone is created by mixing in grey, and a shade by mixing in black. Hues become desaturated the more they tend to neutral, and achromatic colours (white, black, and grey) have neither saturation nor hue.

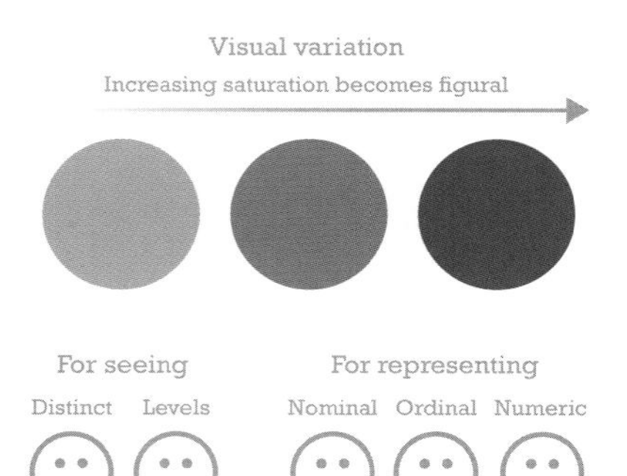


**Opposite:** *UFO Sightings* by John Nelson, 2015.

UFO
SIGHTINGS PER CAPITA
SIGHTINGS / 2014 US COUNTY POPULATION
CHELAN CO, WA
134 SIGHTINGS
74,588 POPULATION
YAVAPAI CO, AZ
300 SIGHTINGS
218,844 POPULATION
FAYETTE CO, TN
0 SIGHTINGS
39,011 POPULATION
HORRY CO, SC
387 SIGHTINGS
298,832 POPULATION
ST JOHN THE BAPTIST PAR, LA
0 SIGHTINGS
43,745 POPULATION
OKEECHOBEE CO, FL
0 SIGHTINGS
39,149 POPULATION
SHAPE TRENDING
AS A PROPORTION OF OTHER REPORTED SHAPES THAT YEAR.
TREND LINES BEGIN IN 1964.
"UNKNOWN" "CHANGING"
1986 20%
"LIGHT" "FIREBALL" "FLASH"
1998 34%
2013 38%
1983 09%
"DISK"
1970 32%
2014 03%
"CIRCLE" "SPHERE"
2014 23%
1987 10%
"OVAL" "EGG"
1965 15%
"TRIANGLE" "CONE"
1990 14%

# Scale and resolution

## The spatial extent of the mapped area and granularity of data are major constraints on map design.

**Scale refers to the spatial extent of the area on a map.** Large-scale refers to small areal extents, and small-scale refers to large areal extents. Expressing scale makes use of a representative fraction (RF). The RF represents the relationship between map and Earth distance. For example, an RF of 1:50,000 means one map unit is equivalent to 50,000 real Earth units. If we use real measures of distance, 1 cm on a map will equate to 50,000 cm of real-world distance, (or 500 m). Large-scale maps have a small RF, say 1:2,500. Conversely, a small-scale map will have a larger RF, say 1:250,000.

**Data resolution refers to the granularity of data used to create the map and arises from the various different sources and mechanisms used to collect geographical data.** For instance, census data can usually be mapped at a variety of resolutions from the smallest area to larger aggregated areas. Each level of resolution portrays the data in a different way with small areas referred to as high resolution and progressively larger areas having a lower spatial resolution. The principle of data resolution can also be applied to geographical data derived from remote-sensing sensors with coarse data from the 30 m Landsat Thematic Mapper providing a relatively low resolution and the 1 m Ikonos panchromatic sensor delivering relatively high resolution data.

**Increasingly, maps are constructed from databases that contain multi-scale and multi-resolution datasets, and digital maps themselves are multi-scale.** Each dataset will likely require different processing and cartographic treatment before it can be mapped alongside data held at a different scale or resolution. This is so that manipulation applied correctly to one dataset is not unsuitably applied to another dataset. Of course, the benefit of such databases is that it is often possible to construct multiple representations of the data, in map form, at a variety of scales. An important principle, though, is to ensure that data held at one scale or resolution is not purported to be of a higher resolution or larger scale than it was measured simply by virtue of including it on a map of more detailed data.

**See also:** Earth coordinate geometry | Earth's shape | Earth's vital measurements | Spatial dimensions of data

Once you know the map scale, you can use the information in two ways. Firstly, if the scale of the map is known to be 1:25,000, then the map distance which corresponds to a ground distance of 2 km is calculated by

$$\frac{\text{No. of m in 2 km} \times \text{No. of cm in 1m}}{\text{Scale denominator}} = \text{Map distance in cm}$$

$$\frac{2{,}000 \times 100}{25{,}000} = 8.0 \text{ cm}$$

If the scale of the map is 1:50,000, then the ground distance which corresponds to a map distance of 3.0 cm is calculated by:

$$\frac{\text{No. of cm on map} \times \text{Scale denominator}}{\text{No. of cm in 1 m} \times \text{No. of m in 1 km}} = \text{Map distance in km}$$

$$\frac{3.0 \times 50{,}000}{100 \times 1{,}000} = 1.5 \text{ km}$$

Multi-scale mapping accords to standard zoom scales used as a framework for web mapping in general: these scales are the stop points for cached map tiles. They are also used for vector tiles though vector tiles additionally support interpolated rendering between scales.

| Zoom Level | Scale 1: |
|---|---|
| 0 | 591,657,528 |
| 1 | 295,828,764 |
| 2 | 147,914,382 |
| 3 | 73,957,191 |
| 4 | 36,978,595 |
| 5 | 18,489,298 |
| 6 | 9,244,649 |
| 7 | 4,622,324 |
| 8 | 2,311,162 |
| 9 | 1,155,581 |
| 10 | 577,791 |
| 11 | 288,895 |
| 12 | 144,448 |
| 13 | 72,224 |
| 14 | 36,112 |
| 15 | 18,056 |
| 16 | 9,028 |
| 17 | 4,514 |
| 18 | 2,257 |
| 19 | 1,128 |

## WAYS TO EXPRESS MAP SCALE

Bar Scale (Graphic Scale or Linear Scale or Scale Bar)

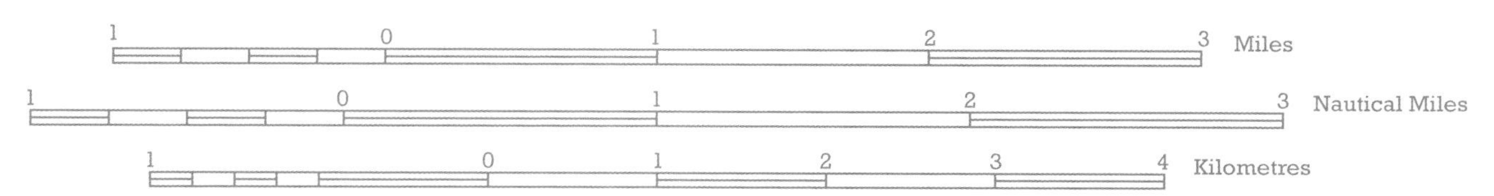


*Graphic representation of distance. Often, more than one bar scale is shown on a map so different units of measurement can be used.*

Verbal Scale (Statement of Equivalence)

1 cm on the map represents 50 km on the ground

*Verbal representation of distance. It is important that the unit be given with verbal scales.*

Representative Fraction (R.F.)

1:1,000,000

*Comparative representation of distance. One unit of length on the map represents 'x' units of the same unit on Earth.*

## SMALL SCALE AND LARGE SCALE

Small Scale
*Features look smaller*

pproximate Range 1:500,000 and beyond

Medium Scale

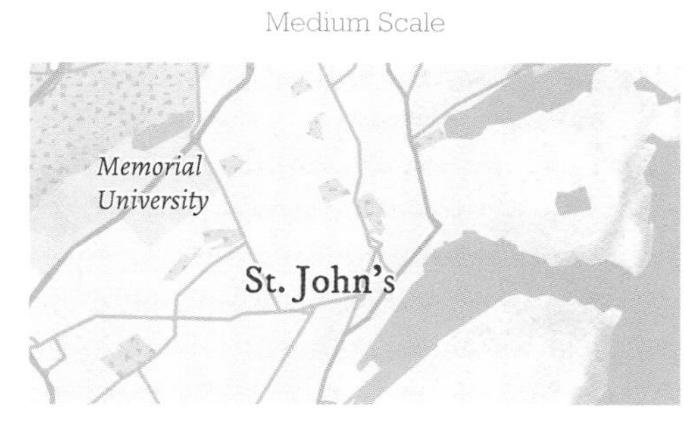


Approximate Range 1:50,000 to 1:500,000

Large Scale
*Features look larger*

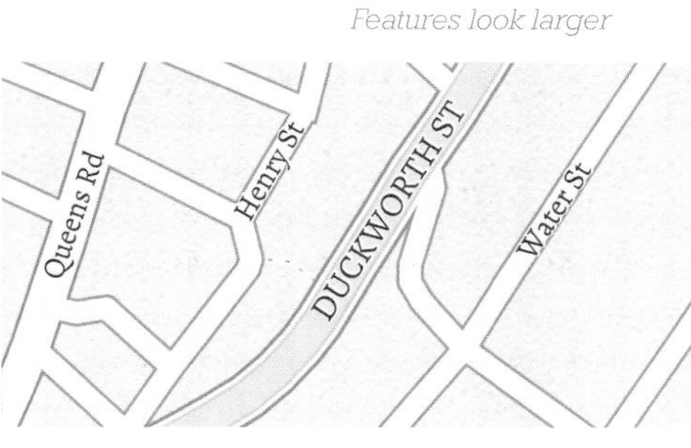


Approximate Range 1:1 to 1:50,000

**There is no official definition on when something is large or small scale.*

## IMAGERY RESOLUTION

| Map Scale | Detectable Size (m) | Raster Resolution (m) |
|---|---|---|
| 1:1,000 | 1 | 0.5 |
| 1:5,000 | 5 | 2.5 |
| 1:10,000 | 10 | 5 |
| 1:50,000 | 50 | 25 |
| 1:100,000 | 100 | 50 |
| 1:250,000 | 250 | 125 |
| 1:500,000 | 500 | 250 |
| 1:1,000,000 | 1,000 | 500 |

8 cm resolution at 1:8,000

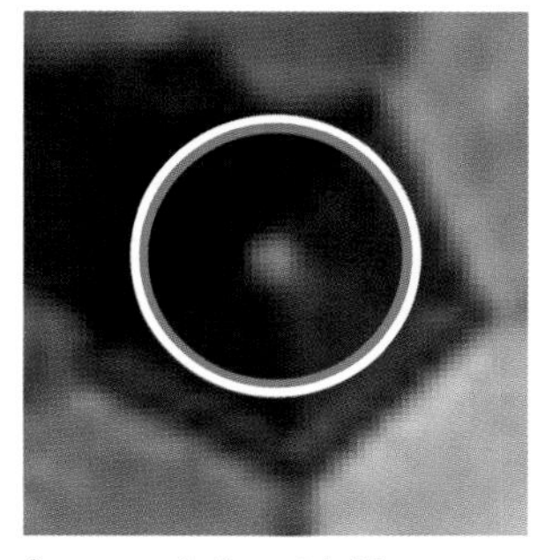

8 cm resolution at 1:75

# Schematic maps

## Simplifying geography to a simple set of lines and connectors supports the mapping of movement and interconnectedness.

**Geography is complex yet the way in which we need to map it to support a particular function has borne some remarkable cartographic products.** For people wishing to get from *A* to *B* via *C*, all they really need to know is the path they need to travel on and the places through which they connect. This would be a network diagram or a schematic map, which we can perhaps trace back to the Marshall Island stick charts showing the ocean currents as lines and islands as shells. A more modern version of the strip map was successfully designed by John Ogilby in his set of maps in the mid-1600s, which presented main roads in England as linear strips with only those places en route marked out. Schematic maps have subsequently almost universally been used to map transport networks and support navigation.

**While Ogilby's strip maps attempted some sense of accuracy in terms of distance, modern schematic maps eschew this measurement completely.** For instance, the *Map of London's Underground Railways* contracts the outer network and expands the central area to give more space to the increased network density. Although geographic maps have the advantage of being true to scale, and which support numerous functions, schematic maps have the advantage of being legible and easily remembered. They are free from clutter by definition. They open up congested map space and reduce empty space. They are extremely efficient maps for the purpose they support.

**Many schematic maps are abstract in appearance.** Many people refer to them as *diagrams* because of their regular and geometric appearance with clear rules often being followed such as lines only being horizontal, vertical, and at 45°. Increasingly, cartographers and designers have begun to think about how schematic maps might be reconstructed using different ideas. Curved lines and other geometries such as hexagons have provided interesting experiments as a grid upon which to hang a schematic network.

Are these maps or diagrams, and does it matter? Well, arguments have raged about the distinction but I take the view a *map* is something that describes something geographical. *Diagrams*, show how something works and can incorporate graphs and other visual devices. Since schematic maps deal specifically with geographical relationships, they are maps—though they are also clearly a type of diagram since both maps and diagrams intend to explain.

Perhaps the most commonly cited example of the schematic map is Henry (Harry) Beck's 1933 *Map of London's Underground Railways* (see page 384), which has, over the years, become regarded as a modern design classic. Geographical configurations, direction, and distance are irrelevant and do not correspond to the experiences of using the tube network to get from one station to another. Simply presenting the different lines and stations is a perfect design solution to support navigation. Most other metro maps around the world employ a similar approach. We do not think of units of distance when we are on a bus or a metro. We think in terms of stops, and so these maps place stops at their core.

Creation of schematic maps needs only consider the core elements, such as the stops, and the lines in between. They tend to place north at the top and some, like Beck's map with the River Thames, incorporate a key geographical feature to position the map—in his case it makes us understand we're looking at London even if a different city employs a similar style. Typography is crucial as we navigate by names of stations and similar core features. Colour is usually used to denote a qualitative difference between lines and services with clearly indicated interchange stations, distinct from single-line stations.

**See also:** Cartograms | Graphs | Simplification

**Opposite:** Extract of the schematic *Submarine Cable Map* by TeleGeography®, 2016.

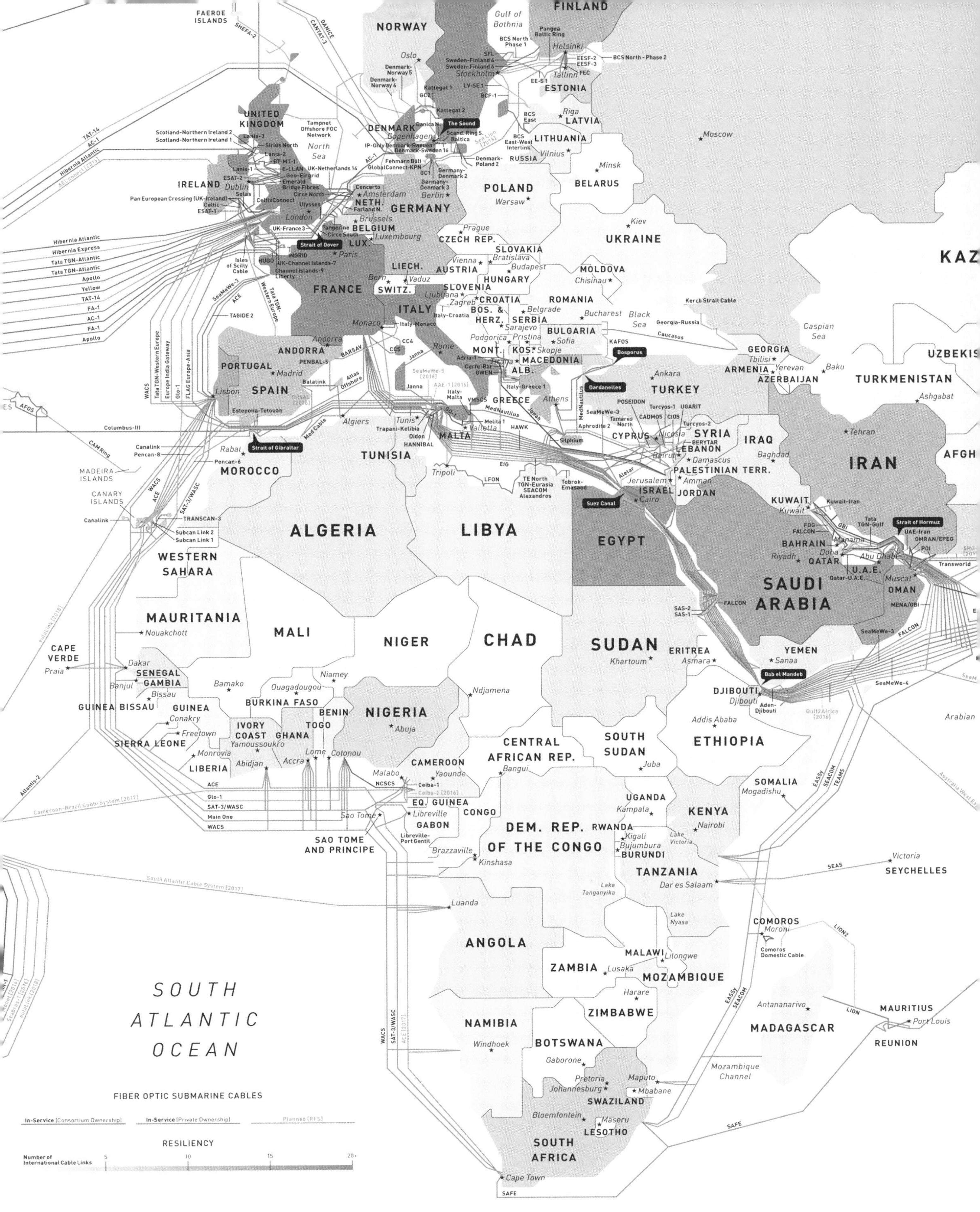

FAEROE ISLANDS
NORWAY
FINLAND
Gulf of Bothnia
Oslo
Helsinki
Stockholm
Tallinn
ESTONIA
Riga
LATVIA
LITHUANIA
Vilnius
RUSSIA
Moscow
Minsk
BELARUS
UNITED KINGDOM
North Sea
DENMARK
Copenhagen
The Sound
IRELAND
Dublin
London
NETH.
Amsterdam
Brussels
BELGIUM
LUX.
Luxembourg
GERMANY
Berlin
POLAND
Warsaw
Kiev
UKRAINE
Prague
CZECH REP.
SLOVAKIA
Bratislava
Vienna
AUSTRIA
Budapest
HUNGARY
MOLDOVA
Chisinau
Strait of Dover
Paris
FRANCE
LIECH.
Vaduz
Bern
SWITZ.
SLOVENIA
Ljubljana
CROATIA
Zagreb
ROMANIA
Bucharest
Black Sea
BOS. & HERZ.
Sarajevo
SERBIA
Belgrade
ITALY
Monaco
BULGARIA
Sofia
Podgorica
Pristina
MONT.
KOS.
Skopje
MACEDONIA
Tirana
ALB.
Rome
Andorra
ANDORRA
PORTUGAL
Madrid
SPAIN
Lisbon
Strait of Gibraltar
Rabat
MOROCCO
Algiers
Tunis
TUNISIA
MALTA
Valletta
Tripoli
GREECE
Athens
Bosporus
Dardanelles
TURKEY
Ankara
GEORGIA
Tbilisi
ARMENIA
Yerevan
AZERBAIJAN
Baku
Caspian Sea
KAZ
UZBEKIS
TURKMENISTAN
Ashgabat
CYPRUS
Nicosia
SYRIA
Damascus
LEBANON
Beirut
IRAQ
Baghdad
IRAN
Tehran
AFGH
PALESTINIAN TERR.
Jerusalem
Amman
ISRAEL
JORDAN
Cairo
Suez Canal
EGYPT
KUWAIT
Kuwait
Strait of Hormuz
BAHRAIN
Manama
Doha
QATAR
Riyadh
Abu Dhabi
U.A.E.
Muscat
OMAN
SAUDI ARABIA
MADEIRA ISLANDS
CANARY ISLANDS
ALGERIA
LIBYA
WESTERN SAHARA
MAURITANIA
Nouakchott
MALI
NIGER
CHAD
SUDAN
Khartoum
ERITREA
Asmara
YEMEN
Sanaa
Bab el Mandeb
DJIBOUTI
Djibouti
CAPE VERDE
Praia
Dakar
SENEGAL
GAMBIA
Banjul
Bissau
GUINEA BISSAU
GUINEA
Conakry
Bamako
Niamey
Ouagadougou
BURKINA FASO
Ndjamena
NIGERIA
Abuja
BENIN
TOGO
IVORY COAST
Yamoussoukro
GHANA
Freetown
SIERRA LEONE
Monrovia
LIBERIA
Abidjan
Accra
Lome
Cotonou
Addis Ababa
ETHIOPIA
SOUTH SUDAN
Juba
CENTRAL AFRICAN REP.
Bangui
CAMEROON
Yaounde
Malabo
EQ. GUINEA
Libreville
GABON
Sao Tome
SAO TOME AND PRINCIPE
CONGO
Brazzaville
Kinshasa
DEM. REP. OF THE CONGO
UGANDA
Kampala
KENYA
Nairobi
RWANDA
Kigali
Bujumbura
BURUNDI
Lake Victoria
SOMALIA
Mogadishu
Arabian
TANZANIA
Dar es Salaam
Victoria
SEYCHELLES
Lake Tanganyika
Lake Nyasa
Luanda
ANGOLA
COMOROS
Moroni
MALAWI
Lilongwe
ZAMBIA
Lusaka
MOZAMBIQUE
Harare
ZIMBABWE
NAMIBIA
Windhoek
BOTSWANA
Gaborone
Antananarivo
MADAGASCAR
MAURITIUS
Port Louis
REUNION
Mozambique Channel
Pretoria
Johannesburg
Maputo
Mbabane
SWAZILAND
Bloemfontein
Maseru
LESOTHO
SOUTH AFRICA
Cape Town
SOUTH ATLANTIC OCEAN
FIBER OPTIC SUBMARINE CABLES
In-Service (Consortium Ownership)
In-Service (Private Ownership)
Planned (RFS)
RESILIENCY
Number of International Cable Links
5
10
15
20+

# Seeing

## Reading maps requires a process mediated by the human eye and brain.

**When you read a map, the eye and the brain work in tandem to see and process the information received.** Light reflected off a map or emitted from a device is sensed by the eye and then reported to the brain, which has to disentangle the messages to form a picture which in turn supports interpretation and understanding. Knowing something of how the eye and brain work helps the map design process.

**Perception is the result of sensory detection of the map image.** Light enters the eye through the cornea, the amount is controlled by the iris muscle, focussed by the lens, and projected onto the retina at the back of the eye chamber. The retina is composed of rod cells, which determine brightness on a grey scale, and cone cells. Three types of cone cells are sensitive to either blue, red, or green portions of the electromagnetic spectrum. Light stimulates the rod and cone cells, which send electrical pulses via the optic nerve to the brain. The eye has natural limits of visual acuity. For instance, under low levels of illumination, the eye can only see differences in shade but not colour. Small objects are much harder to see than larger objects.

**Cognition is the process of comprehension performed by the brain on the basis of the signals it receives.** The brain converts the eye's signals into images. It also builds a picture of colour on the basis of how the cones are stimulated. Cognition is far more complex than just interpreting what the eyes see. It is based on thought and experience. It incorporates what you know, your memory, judgement, reasoning, problem-solving ability, and comprehension. It's both a conscious and subconscious process and might be concrete or abstract. Intuition also plays a part. In short—it's complex. For map reading, cognition is concerned with information processing and a person's ability to evaluate the map and reconcile the abstraction and symbology to rebuild an image that reflects the purpose of the map. Designing a map that minimises the amount of cognitive processing required will likely be more widely usable.

Visibility is normally measured by specifying the subtense, or angular measure, of the image as it strikes the retina. An understanding of visual acuity, or the capacity of the visual system to detect what it is seeing, helps establish minimum size thresholds for map objects. A second component of visibility is resolution acuity, which is the ability of the eye to recognise separation between objects. This helps establish the distance between map objects so they are individually recognisable.

Everyone is different, and as much as we can see differences between people such as hair colour and height, so people differ in their ability to see and process maps. The health and functioning of the eye can affect its ability to see. Refractive errors caused by different eyeball shapes, the shape of the cornea, and the flexibility of the lens cause myopia (short-sightedness) and hyperopia (far-sightedness). Other factors can affect the retina and the interpretive capabilities of the brain.

Once seen, making the map's information easy to extract limits the dependence on more complex cognitive processing. Of course, making a map for a particular user group—for instance, experts in a specific domain—means you can assume a different level of understanding of that domain and build more assumptions into your map for that readership.

**See also:** Abstraction and signage | Colour deficiency | Focussing attention | Navigating a map | Seeing colour | Vision | Threshold of perception

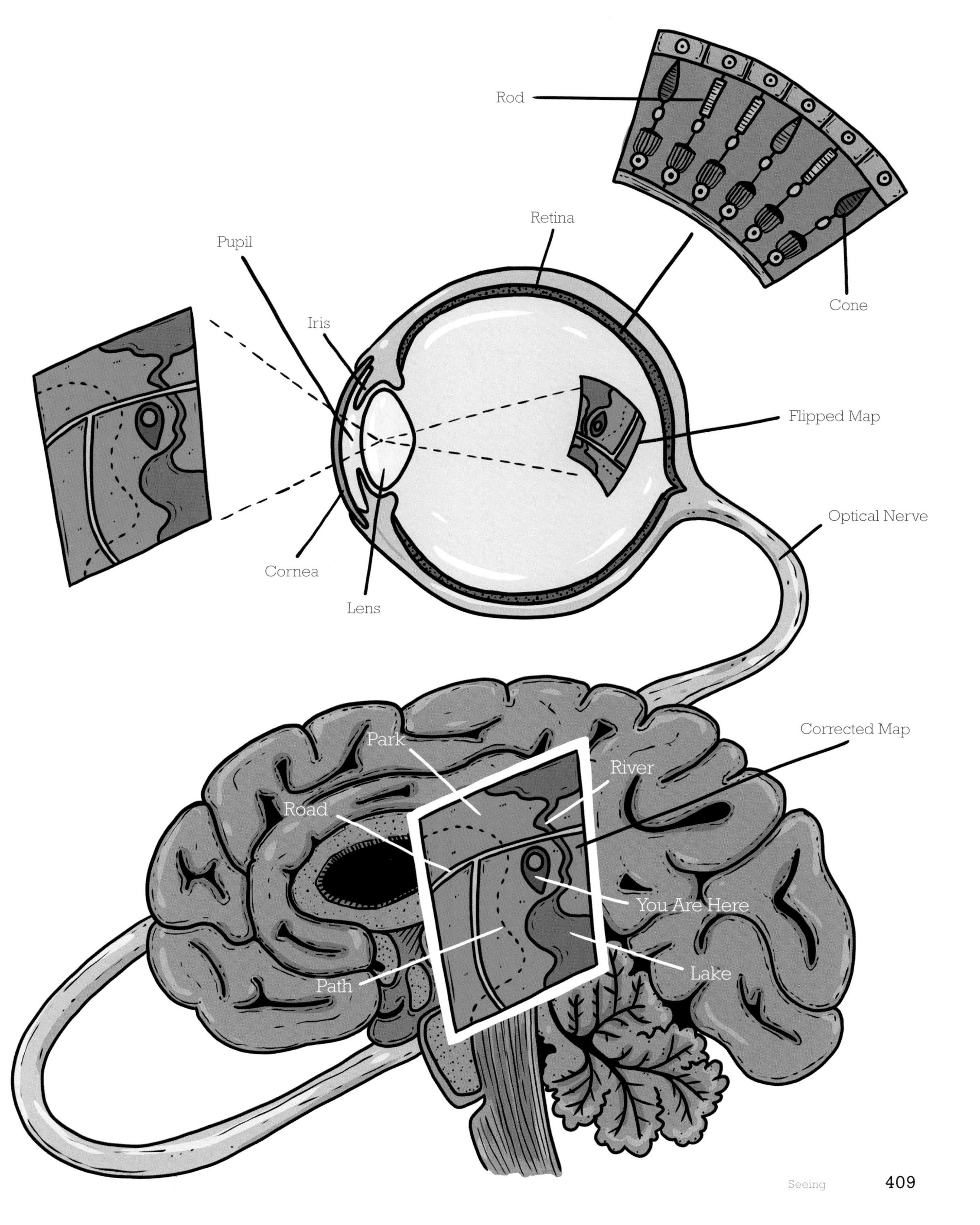

Rod
Retina
Pupil
Iris
Cone
Flipped Map
Optical Nerve
Cornea
Lens
Corrected Map
Park
River
Road
You Are Here
Lake
Path

# Seeing colour

## Surrounding colours make a difference.

**Colours are seen differently across a map depending on their surrounding colours.** This perceptual problem is often referred to as *simultaneous contrast*. When grey symbols of the exact same tonal value are surrounded by different hues they are perceived differently by the eye. If the grey symbol is surrounded by darker tones, tending toward black, it leads to a tonal shift in the central grey tone that is perceived as lighter. Conversely, a grey symbol appears darker if the surrounding lighter tone tends toward white. For different surrounding hues, the apparent hue of a grey shaded area will also shift toward the opponent colour of the surrounding colour. For instance, if the grey symbol is surrounded by green, then it appears slightly reddish, and if surrounded by blue, it appears yellowish.

**Understanding the relationship of colours as they are perceived allows you to deploy colour to minimise the simultaneous contrast.** The complex physiology of how the eye–brain system processes colour puts constraints on symbolisation and design alternatives. Being aware of the relationship of neighbouring or surrounding colours and how they interact with one another helps to inform better map design. Often it's not possible to do much about how map features are organised because location is the overriding visual variable, but selecting colours that don't cause too many complex visual distortions will help ensure that simultaneous contrast is minimised in your map.

**Simultaneous contrast affects both topographic and thematic maps.** Symbols on a topographic map will always be surrounded by an array of different symbols and colours. Although this is inevitable, trying to adjust a colour palette to ensure different colours can be recognised as distinct across the map is helpful. For thematic maps, such as a choropleth, sequential colour schemes will inevitably place colours at one end of the spectrum in proximity to those at the other end. Again, being cognisant of this schema and open to adjusting the colour scheme to minimise simultaneous contrast is helpful.

**See also:** Colour deficiency | Design and response | Elements of colour | Perceptual colour spaces | Seeing

One useful test you can apply to explore the impact of simultaneous contrast on your map is to check whether you can match different colours across the map with their respective legend item regardless of different backgrounds. This approach is particularly useful for choropleth maps to assess a map reader's ability to compare different colours between regions on the map and between maps in a series.

Of course, assessing how colours appear in relation to a legend where colours might be presented in order across a uniform background is only one test. There is often little you can do to change the distribution of data so checking the selection of colours in the final map pattern is vital. You should be able to identify each colour symbol and also see them distinctly across the map. Finding areas where one colour is surrounded by contrasting colours is the biggest test for your overall colour palette. Good contrast and perceptual steps between colour choices helps mitigate simultaneous contrast.

Colour processing is also different under different presentation and lighting conditions so care should be taken to ensure colours remain distinct for how the map is to be used. Your choice of publication medium has a big impact so whether the map is to be printed by inkjet, lithographic print, projected, or transmitted via digital display, be sure to test colour in these contexts. They will all look different, and adjusting for the intended media will avoid unintentional difficulties for your map reader.

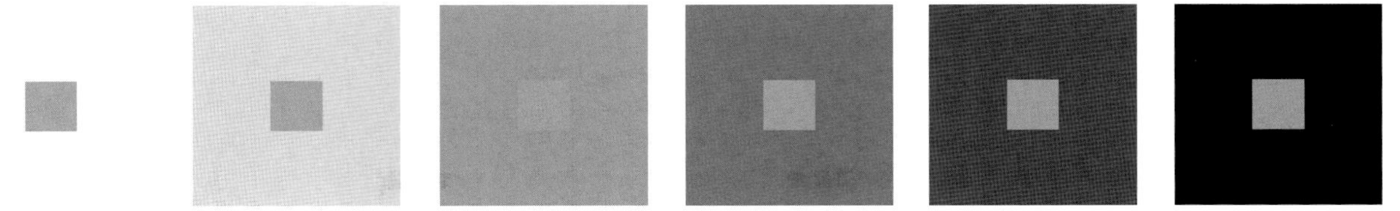

The middle grey square is the same shade. It appears darker on a lighter background and lighter on a darker background

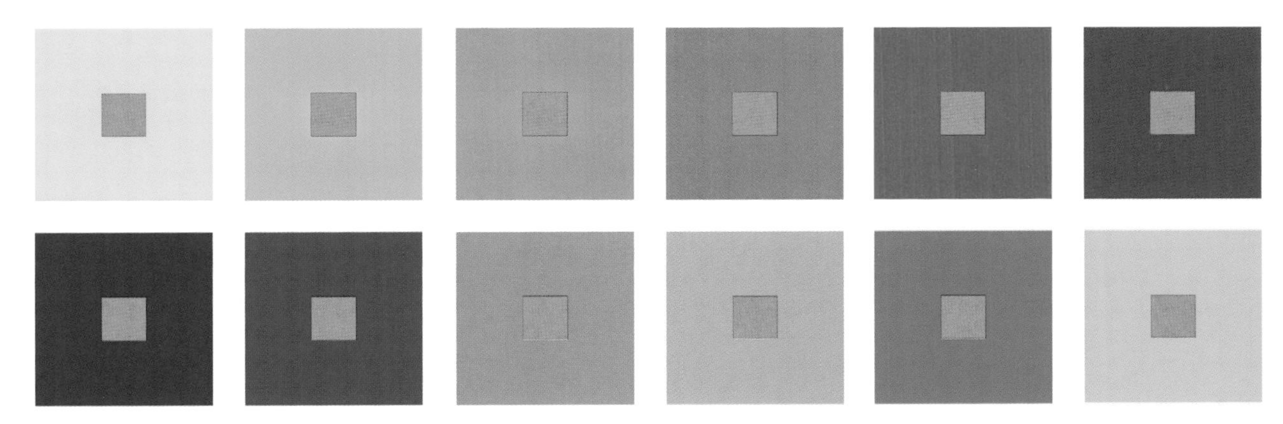

The middle grey square tends towards the opponent of its surrounding colour. For instance, for the red/green opponent colours, the grey square looks greenish on a red background and reddish on a green background.

Green or blue — what colour is Earth?

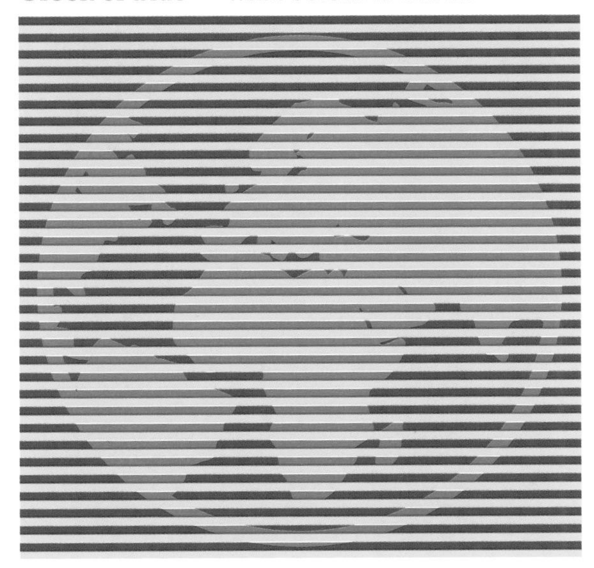

Yellow lines on top

Blue lines on top

# Semiotics

## Semiotics is the study of signs and symbols as communication devices.

**Cartographic semiotics, the study of signs and symbols and their meaning, has its roots in the work of French cartographer Jacques Bertin (1983).** Bertin's research explored how signs on a map were perceived and understood. From this, he developed a systematic way for cartographers to assign meaning to map elements through varying the properties of the visual marks used to represent them. The practical application of Bertin's work was the creation of a set of retinal variables, more commonly referred to as *visual variables*, which have provided the foundation for information visualization more generally. Different visual variables function in specific ways depending on the feature dimension and character. So, for instance, changing the size of a point symbol connotes a difference in magnitude of the mapped phenomena.

**Visual variables can be subdivided into those that differentiate and those that order.** Differentiating visual variables are used to show qualitative difference between features while ordering visual variables are used to show quantitative difference between features. More recently, the advent of animated and web-based mapping has reinvigorated the discourse of visual variables since the medium of communication is different in these forms. Additionally, abstract sound variables have also been proposed to communicate spatial information in an audible fashion (for instance, a larger sound might connote increased magnitude).

**Additional visual variables might be used in particular mapping circumstances.** For instance, if you are creating a 3D scene of a cityscape, you may also modify lighting by changing the angle of illumination, the brightness, the shadows according to the time of day, and the atmospheric conditions. These additional variables help us graphically connote a particular aesthetic appearance for our maps.

Bertin was not the first to explore the relationship between the form of a sign and its meaning. The pioneering work of William Playfair in the late 18th century sought to overcome the problems of the burgeoning growth of statistical and economic reporting by developing statistical graphs and charts. He converted numbers into meaningful visuals to aid reading and understanding. The power of his invention was to convey information and inform.

Playfair never described nor documented his process or explained the theory as a guide for others to implement. Bertin, nearly 200 years later, did that at the onset of the computerised revolution to graphically represent large amounts of data. In the face of the need to see patterns clearly and portray them graphically, Bertin (1983) defined the understanding of how to do it, inventing the grammar of graphics. Bertin's original seven retinal variables, listed below, expanded to form a grammar for cartography.

Location
Size
Shape
Texture
Orientation
Colour—hue
Colour—lightness (value)
Colour—same saturation (chroma)
Pattern
Crispness
Resolution
Transparency
Height

**See also:** Design and response | Dimensional perception | Isotype | Symbolisation

**Opposite:** An extract from *Carte de Déplacement des Baleines* by Matthew Maury, 1851.

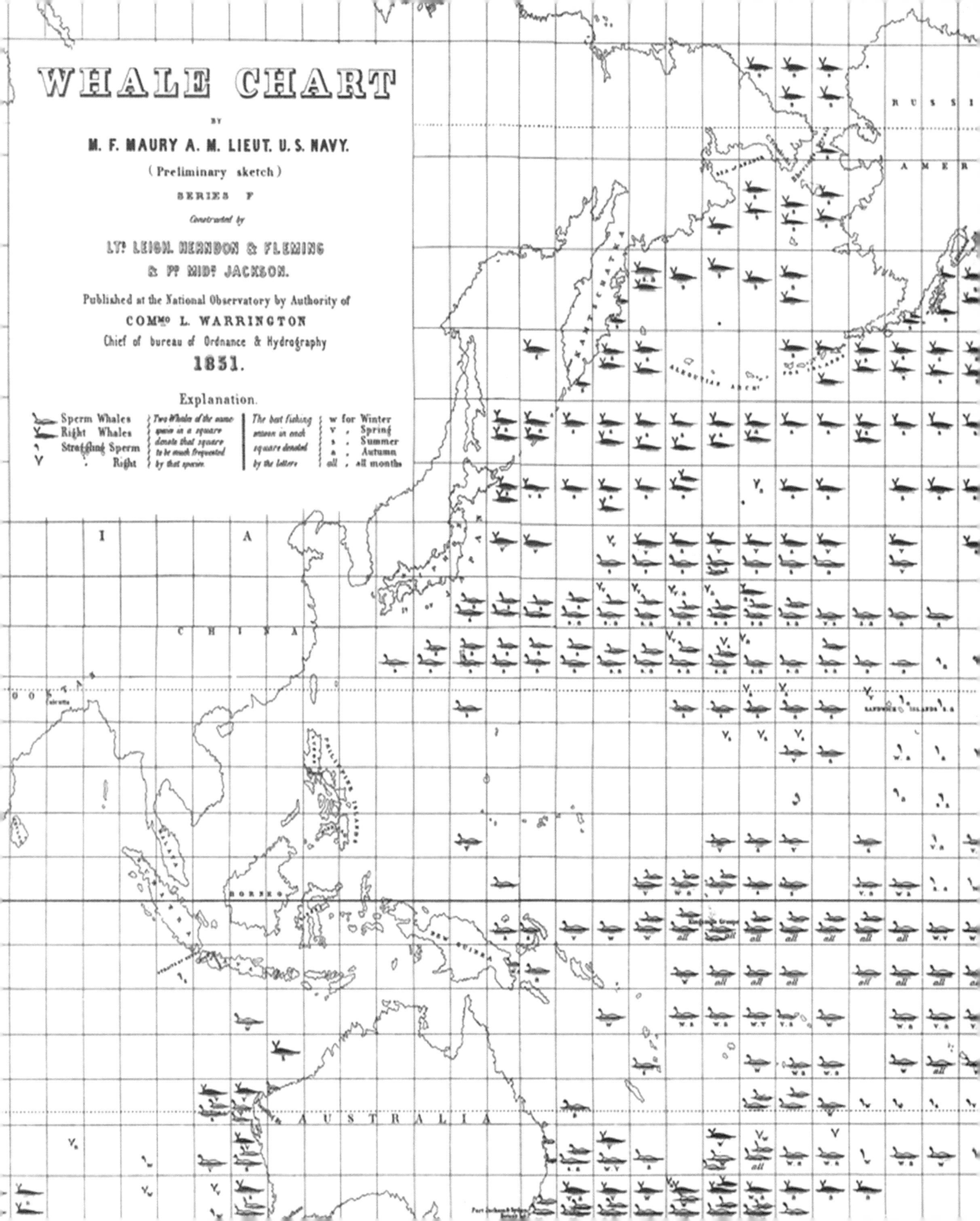
WHALE CHART
BY
M. F. MAURY A. M. LIEUT. U. S. NAVY.
(Preliminary sketch)
SERIES F
Constructed by
LT. LEIGH. HERNDON & FLEMING
& PT. MIDT. JACKSON.
Published at the National Observatory by Authority of
COMMO. L. WARRINGTON
Chief of bureau of Ordnance & Hydrography
1851.
Explanation.
Sperm Whales
Right Whales
Straggling Sperm
Right
Two Whales of the same species in a square denote that square to be much frequented by that species.
The best fishing season in each square denoted by the letters
w for Winter
v , Spring
s , Summer
a , Autumn
all , all months
RUSSI
AMER
I A
CHINA
KAMTSCHATKA
ALEOUTIAN ARCH.
BORNEO
NEW GUINEA
PHILIPPINE ISLANDS
SANDWICH ISLANDS
AUSTRALIA

# Sensory maps

## Going beyond sight to stimulate different senses.

**Maps are conventionally designed to be viewed but can take on different forms to support different senses.** Although we may think of the design of such maps as being for specific user groups such as the blind and visually impaired, these are not the only reasons to consider mapping for other senses. Maps can be made to be read using any of the five human senses.

**Tactile maps are produced by raising the surface of the product so the marks can be touched.** Tactile maps can be made using a variety of methods such as thermoform (moulded plastic), raised ink-jet, textured substrate, or 3D printed. Symbology tends to be highly generalised to account for touch, and typographic components may be printed in Braille. Audible maps are commonplace, particularly to support navigation. For instance, car devices routinely provide audible directions in addition to the visual map so that the driver can understand a route without recourse to the visual representation.

**Smell maps may work on several levels: some seek to visualise connections between emotion, memory, and place, some act as archives, while others spatialise qualitative olfactory perceptions focussing on the dynamic and ephemeral nature of the sense.** In practical terms, a map of a smellscape may be augmented with a physical scent either natural (derived from original source ingredients) or synthetic bespoke molecular combinations. Smell maps are based more in artistic realms of cartographic practice. Unrelated, the neurological condition synaesthesia can be used to create a taste map where words on a map trigger a sensory association. For instance, London Underground stations all have a distinct taste to James Wannerton, who made a taste map replacing the station names with the words that define the taste he perceives at that particular location. There are numerous flavours depicted on the map, including British favourites such as HP Sauce and wine gums, but the tastes are not restricted to food items. Tastes also include burnt rubber, pencil eraser, and dried blood.

**See also:** Continuous surface maps | Craft | Interaction | Map transformation process | Texture

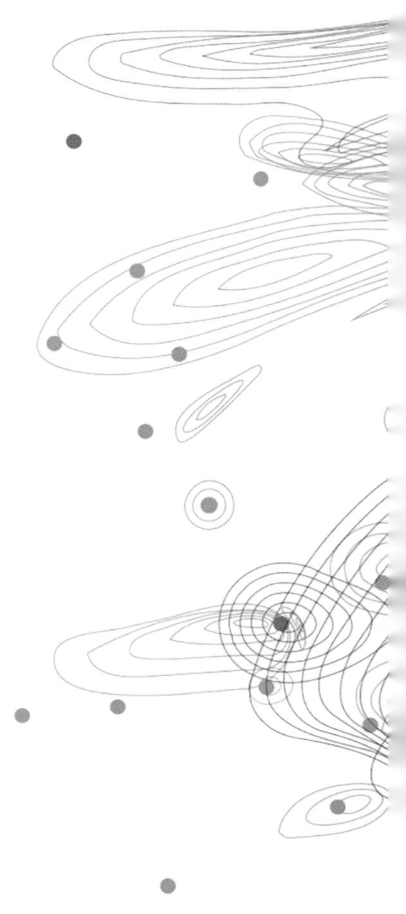

- *Sea, sand, beach*
- *Brewery malt fumes*
- *Vaults & underground streets*
- *Boys toilets in primary schools*
- *Fish & chip shops*
- *Penguins at the zoo*
- *Cherry blossom*
- *Newly-cut grass*
- *Coffee*

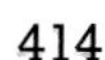

**Below:** *Smellmap of Edinburgh* by Kate McLean, 2011, which illustrates zones of different smells. The original map was accompanied by bottled scent. Kate McLean specialises in making sensory maps based on smell. Her focus is on creating alternative modes for individual and shared interpretation of place beyond the normal visual mode that data supports. She has created several smell maps of major cities as a way to capture the distinctive character of those places. Whereas her map of Glasgow reveals smells associated with reinvention, rebuilding, and regeneration, Amsterdam is quite different with warm, sugary powdery sweetness (from the predominance of waffles) along with Oriental spices from the many restaurants and the smell of old books from Amsterdam's house hotels.

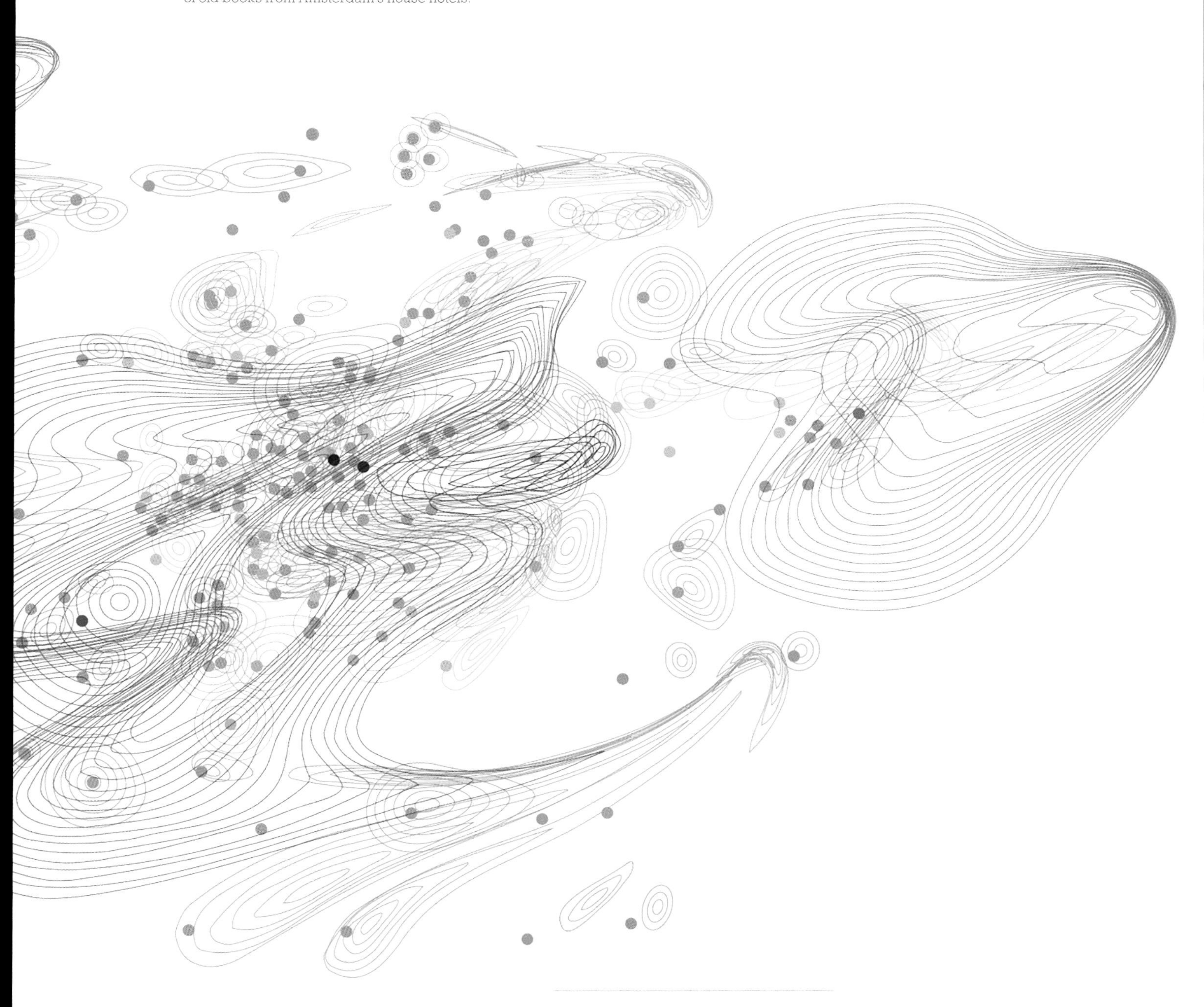

*Auld Reekie emits a plethora of scents and smells; some particular to Edinburgh, some ubiquitous city aromas. It is the smell combination, and how they are distributed by the prevailing south-westerly winds that make them city-specific. Coloured dots indicate the point of origin of the scents, the contour lines show where they blow to in the wind.*

# Shaded relief

## Casting light across terrain creates lighted versus shadowed slopes and an impression of relief shading.

**Planimetric mapping often incorporates some representation of relief shading.** It gives metrics to the topography but, frankly, it also looks great, and this is largely because of how we illuminate it on a map. Representing terrain can be achieved in many ways but all have a common need to show the third dimension in a way that the map reader can understand. One way is to mimic how light hits a landscape resulting in faces being lit or, conversely, which are in shadow. Our eye–brain system is well attuned to seeing the world in this way and assessing brightly lit surfaces from those in shadow. Illumination is therefore the amount of light that hits a surface and which can be used to determine which areas fall in shadow and by how much.

**Assessing illumination requires the calculation of azimuth and zenith.** Azimuth is the angular direction of the sun from any point on the terrain. This is usually measured from north in clockwise degrees from 0° to 360°. Zenith (or altitude) is the angle of the sun above the horizon measured from 0° to 90°.

**Once azimuth and altitude are defined, the position of the hypothetical sun in relation to the terrain is set and illumination can be calculated.** This is normally calculated on a digital elevation model (DEM) with an illumination value for each cell calculated relative to its neighbouring cells. A typical calculation of illumination would set the azimuth to 315° and a zenith of 45° giving an illumination from the northwest. This gives rise to the typical illumination of illuminated northwest slopes and shadowed southeast slopes.

**Illumination of terrain can be calculated to represent a specific time for a particular area by using solar ephemeris information.** This gives the actual position of the sun for places on Earth's surface at a given date and time. This information is often used to create realistic shadows in 3D models where the terrain might also include man-made objects such as buildings that also need illuminating.

Calculating illumination is the basis for creating hillshaded relief representations but illumination maps can easily lead to an optical illusion referred to as *relief inversion*. Hillshading produces an impression of a three-dimensional landscape that appears correct and is usually calculated using a light source in the upper left of the map. Care should be taken when relief shading to avoid the perceptual problem of relief inversion. By positioning the light source at different points, we see it cast in different ways. This can be useful and create a wide palette of possibilities but our eye-brain systems are attuned to seeing light cast from the upper left. If it is cast from the lower right, we can sometimes perceive hills as valleys and vice versa.

At its simplest, illumination is usually calculated using a single light source. Although this approach works well to assess how light hits an area perpendicular to the imaginary rays of light, it also presents problems for areas where the terrain is parallel to the light. Light tends to skip past features and they appear washed out. Relief shading often overcomes this problem by using multiple light sources to calculate illumination from different angles. Thus, illumination is the basis for more complex and realistic methods of relief shading.

Light and dark are a combination of angle of illumination, azimuth as well as terrain orientation, height, and steepness. Shadows should not be confused with relief shading as shadows are of a constant tonal value and extend beyond a feature, sometimes modifying the tone of an otherwise sunlit slope elsewhere.

**See also:** Greyscale | Hand-drawn shaded relief | Raised relief | Styling shaded relief | Value

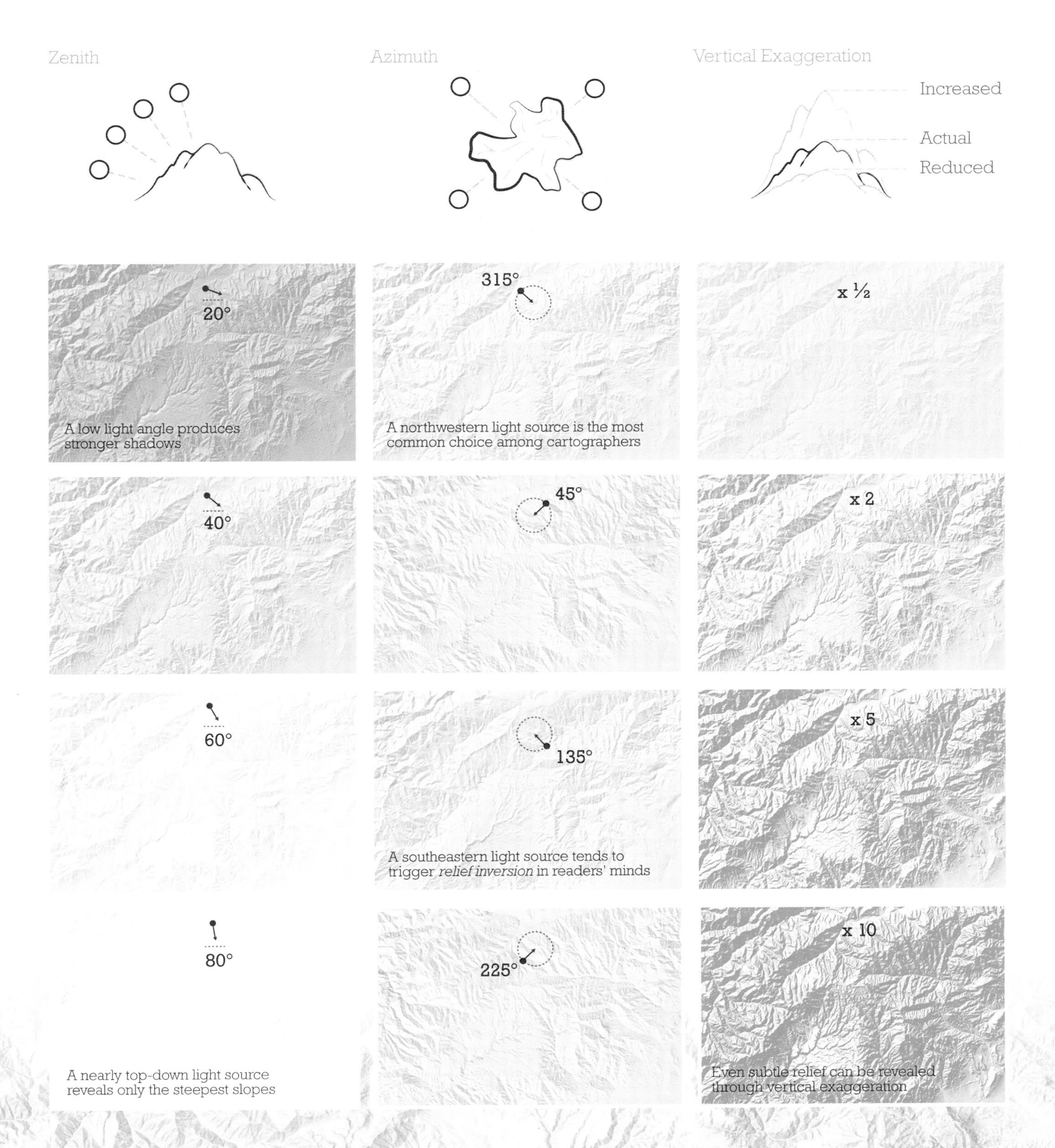


*Multidirectional* hillshading blends many light source angles together into a smoother, more topographically nuanced, surface.

# Shape

## One of the main ways in which symbols vary in appearance to give features different meaning.

**A graphical mark on a map can take on an almost infinite number of alternative shapes.** The ability to vary how a mark appears is fundamental to cartography since familiarity of shape is a key determinant for understanding. Shapes are broadly categorised as geometric or mimetic. They have simple geometric structure that we describe as circle, triangle, square, and so on, or they are drawn to take on the appearance of the object they represent. Well-designed mimetic symbols facilitate reading and understanding more than geometric shapes because of their lower threshold of recognition. Identification is easier, and the eye and brain find the job of seeing and interpreting much simpler. Similarity between the symbol and the real-world object is the key stimulus.

**Points, lines, and areas can all take on different shapes.** Representing points using geometric symbols only allows readers to recognize that two of the same type are similar. Numbers and letters are more recognizable. Mimetic shapes such as buildings, people, or vehicles evoke the same feature. Linear features can vary in shape through the use of different line styles—the combination of different linear segments, dots, dashes, and other marks that form an overall line. These too can be mimetic as, for instance, two parallel lines can mimic a dual carriageway or a series of dashes perpendicular to the main line can connote a railway line. Modifying the fill of an area can give rise to different textures when using geometric shapes to create, for instance, a stippling effect. Alternatively, by using repetitive mimetic shapes as fill patterns, you can generate patterns that denote different land use types such as forest, marsh, or sand.

**Different shapes are extremely effective in showing nominal differences between data.** They are not effective at illustrating quantitative difference. Topographic maps rely heavily on the use of shape as a differentiating visual variable for point, line, and area. Shapes are less often used in thematic mapping but they are useful. For instance, proportional symbols of mimetic shapes can be effective links to the overall map theme.

**See also:** Isotype | Orientation | Pattern fills | Pictograms

One of the key tasks in map reading is to determine what exists at a given place. Giving your map reader a sign that they instantly recognise makes the task of answering that question that much simpler. If you use a triangle, then the map reader almost certainly has to look at the legend. If you use a mimetic symbol, then they are more likely to resolve the problem.

That said, there's a temptation to think of a perfect map as not requiring a legend if symbols are designed to be mimetic. Unfortunately, map readers are not made equally. The message one person gets from seeing a footprint as a hiking trail might be different from the person that identifies it as the location of a podiatrist. Over time, people familiar with the symbol sets of particular map publishers will become familiar and have less recourse to the legend but you cannot assume your shapes will always be seen as you intend. Habit is the only way in which shapes become familiar over time as there is no universal shape signification.

Using shapes that are commonly used to signify a particular object is a good approach to map symbolisation by relying on people's general familiarity and understanding to reduce the potential for ambiguity and confusion.

Visual variation

Modify the shape's size to avoid showing levels

For seeing — Distinct, Levels

For representing — Nominal, Ordinal, Numeric

**Opposite:** Every snowflake has a different shape. How would you design a snowflake symbol? Which would you shoot down?

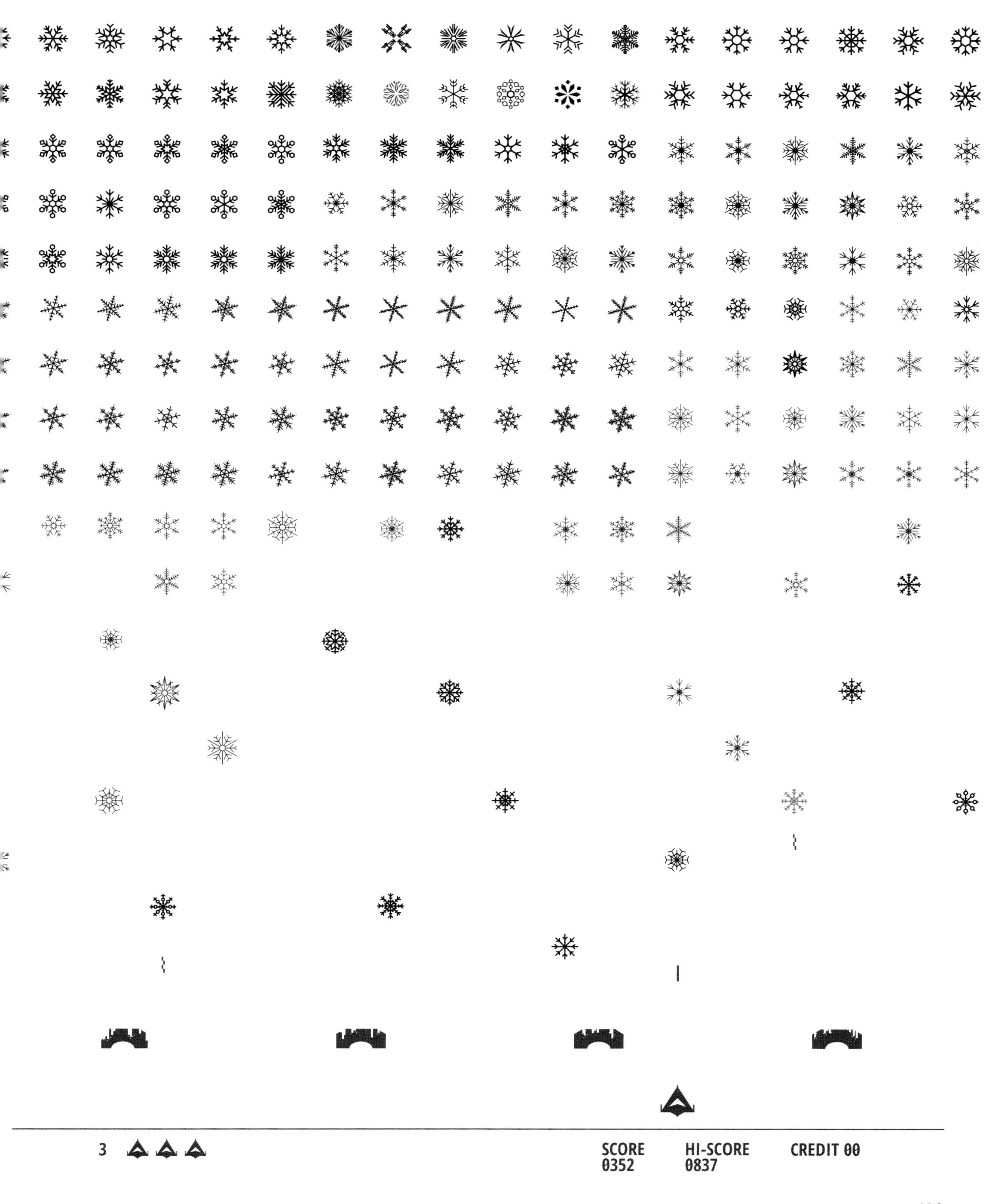
3
SCORE
0352
HI-SCORE
0837
CREDIT 00

# Signal to noise

## Giving clarity to the map's signal means reducing the amount of noise.

**In any form of communication the ideal situation is to optimise the signal while at the same time reducing the surrounding noise.** Clear-written or verbal prose is more easily understood than verbosity. The analogy of applying a Dolby® noise reduction to the map might help us understand what it is we're trying to cut out. During the era of magnetic tapes, background hissing was a perpetual problem. Dolby Laboratories invented several techniques to eliminate the hissing and allow the recorded audio space to be heard as intended. In cartographic terms, eliminating background visual noise will always help to give the map's key theme or message space to breathe and be seen.

**The signal-noise analogy (and Edward Tufte's related data-ink ratio) does not have a pure mathematical definition.** It's simply a ratio that defines the amount of detail that represents the data (the signal) relative to the amount of detail that represents nothing (the noise). Cartographic noise can come in many forms. Often, there's simply a case where the cartographer has the data and so there's a presumption that it should go on the map somewhere. It's also possible that the theme itself is struggling because of data paucity so extra content is added to fill in the gaps. Embellishment such as borders around legends or titles, underlined titles, ornate scale bars, and north arrows are all examples of visual noise. Simplifying them, making them smaller, or omitting them altogether is a quick way to reduce visual noise.

**Put simply, if the mark you are about to place on the map doesn't bring something to the message or the composition, then it's likely visual noise and can be omitted to improve the map's signal.** That said, the style of the map often contradicts this simple argument. Maps designed in an historical style or which rely on artistic flourishes will inevitably be built around such an aesthetic. Even so, the principle can be applied so that even if the map is artistic, it builds in white space and separation to bring focus to important map components.

Basemaps are often a key contributor to visual noise. They often come pre-built with all sorts of content atop of which you place your own content. Although you may not be able to modify the basemap yourself, you could make more appropriate choices. For instance, a neutral colour scheme (light grey, for instance) is preferable to a full-colour topographic map. It's certainly better than placing satellite imagery beneath the map, which introduces a 1:1 detail of reality that almost certainly skews your signal-noise ratio in the opposite direction.

Peer review is important to help you understand whether you have the signal-noise ratio well balanced. Ask people what they see from the map and gauge their first impressions. Do they describe key characteristics rapidly and easily or are they distracted by unimportant components? People's opinions on your work are important so use them to help improve your own sense of balance.

A useful way to critique your own map is to assess each component from the perspective of whether it can be removed from the map without the overall message losing content and context. If the message of the map can still be communicated without a component, there's a good argument for it to be removed as noise.

Removing all components that are considered noise will lead to minimalist cartography. Although this is deemed optimal for communication, bear in mind that your map reader may prefer a little context and adornment. People tend to like maps that contain a little embellishment as a totally minimalist approach can sometimes seem sterile or cold. Balance reducing noise with the emotional response you seek to have your map reader feel.

**See also:** Contrast | Focussing attention | Foreground and background | Resolution | Size

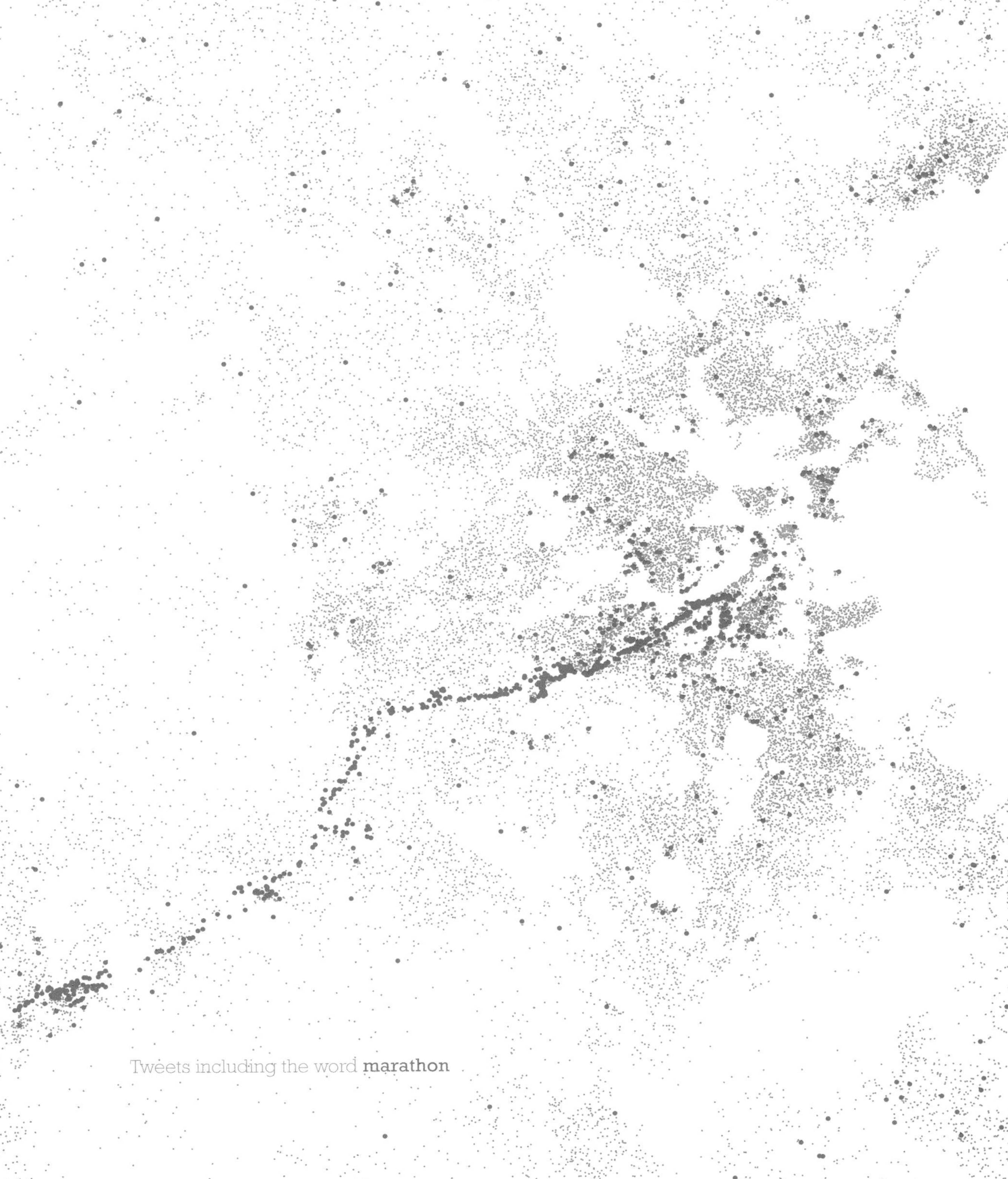

Tweets including the word **marathon**

# Simplicity vs. complexity

**Some information is simple. Some can be extremely complex. Either way, a map's purpose should reveal and make that information understandable.**

**A map is usually a window to a world, either geographic or thematic, that we cannot otherwise see.** We read maps of faraway places or of detailed datasets that, without the map, would remain a mystery. More than ever, the internet has given us unprecedented access to rich and often complex data, but revealing the key connections, highlights, and relationships is a challenge. Simplifying data into something meaningful has become more important to cartographic practice because there is ever more to sift through.

**There's a fine balance between simplifying and over-simplifying.** Being able to make a map that tackles a complex dataset doesn't, in and of itself, make the map work. Computing power supports complex maps and visualizations but it doesn't necessarily make them readable because our ability to cognitively process visual complexity is still a major constraint on our interpretive capabilities. Put simply, humans have evolved slowly. Computers continue to evolve rapidly. Being prudent with how we deploy computers is important so as not to just map complexity in ever more complex ways.

**Simplifying complex data is an important cartographic tenet but revealing complexity is possibly of more value.** Not interpolating or exaggerating a simple dataset beyond the realms of sensibility is the corollary. Maps that depict their subject matter in a simple, unambiguous way are often those that are most successful. Maps that express complexity through a complex design often don't work. However, that's not to say that you shouldn't reveal complexity by mapping a lot of data. Often, the complexity of a dataset is best revealed by not over-simplifying. The key is to ensure the final form is one that reveals the innate message in the map. This may be through casting aside a substantial portion of data to reveal the key element; conversely it might be through finding a neat visual solution to support the interpretation of complex dimensions. It's important to note that simplicity and complexity can both be very successful approaches as the illustrations on this spread show.

**See also:** Clutter | Data density | Space-time cubes

An example of how cartography has changed in the face of complex and large data has been the emergence of map-based art.

The rise of data-art maps is largely a reaction to the limitations of traditional cartography to provide a systematic solution for the visual representation of millions of pieces of information. Some of this may allude to the problems of representing 'big data' though in truth maps have always dealt with large datasets. What has changed is what we want to tease out of the data, and this relies more on analytics than cartography. Until cartography responds with new or innovative ways to handle big and complex data, we will likely see more data-art that simply dumps the data on the page.

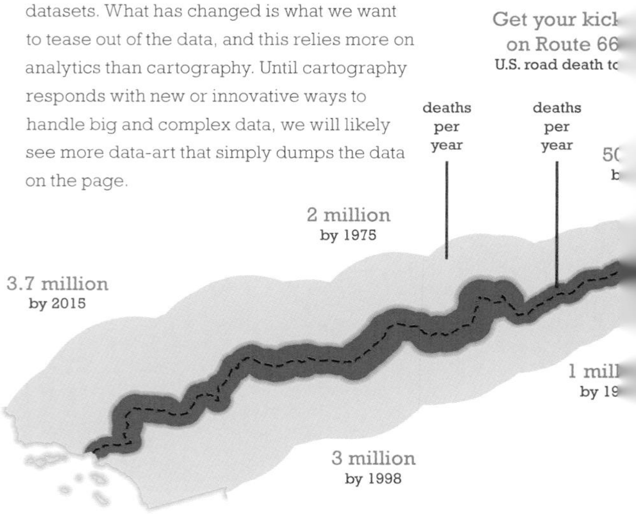

Although beautiful to look at, such maps do nothing to help us understand complexity. Yes, maps of data objects can be constructed from points, lines, or areas, but in some senses there's not much that's cartographic about these types of maps. They work on a visceral, artistic level but not necessarily as an information product. In some senses complexity is lost within the simplicity of representation. You end up with a cumulative artistic representation of a mass, not the nuance of outliers or key components.

**Above:** Simplicity: Cumulative road deaths in the USA over time represented as a single-line graph using Route 66.
**Opposite:** Complexity: *City of Anarchy* by Adolfo Arranz, 2013. 'This month' in the map description refers to March 2014.

# City of anarchy

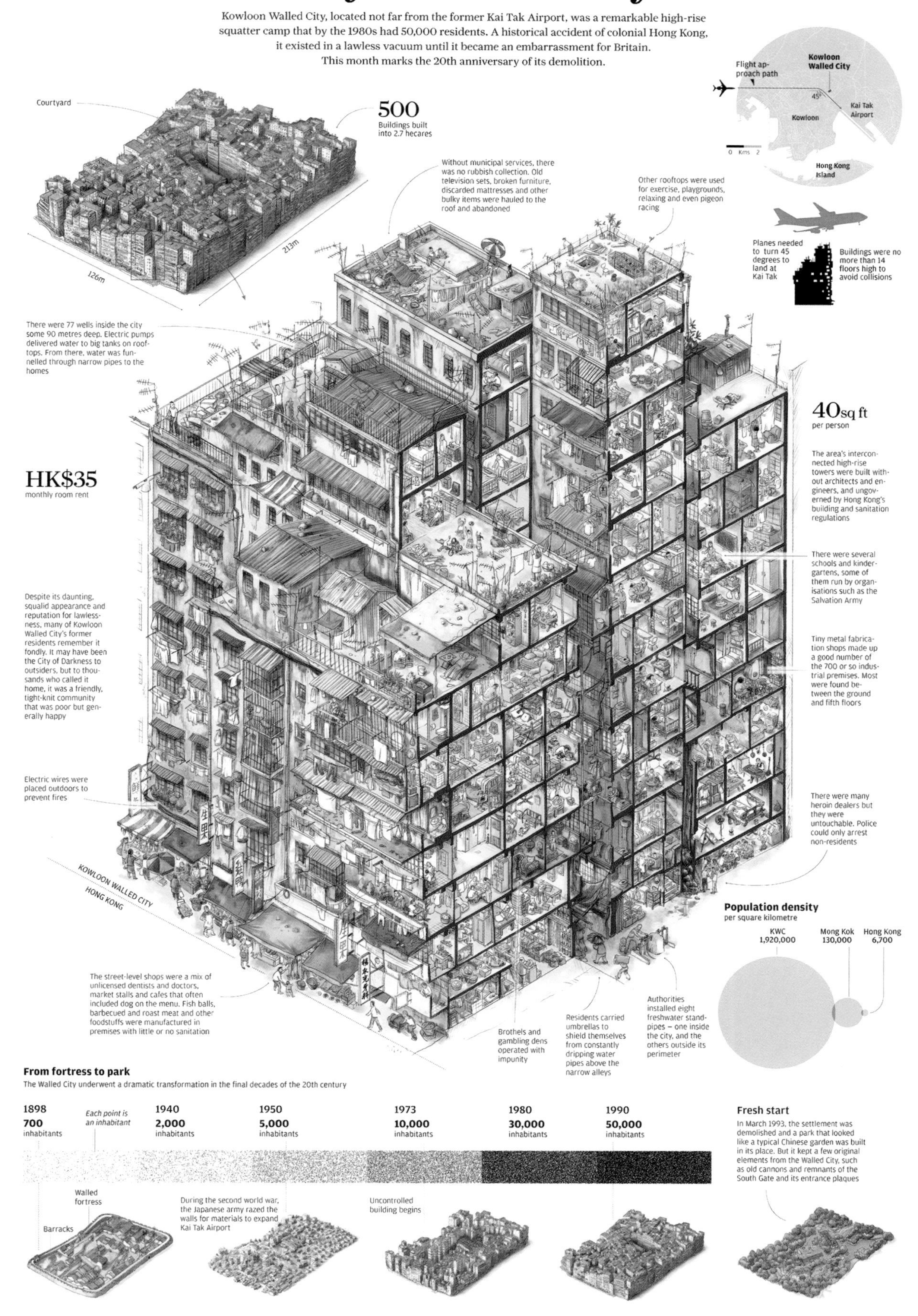

# Simplification

## Simplifying linear features to retain the essential geometric character with fewer coordinate points.

**Simplification is a key process to generalise linear features.** Conceptually, it involves weeding out data vertices with the goal to retain the essential geometrical characteristics of the feature with fewer coordinate points. Computationally, simplification is often performed by applying an algorithm that calculates complex distance and angular measurements along a line which are then used to select coordinate points critical to the line's orientation.

**One of the most commonly implemented simplification algorithms is referred to as the Douglas-Peuker algorithm.** The algorithm works by taking a global approach to simplification where it considers the entire line and iteratively selects critical points, weeding out those that are less important. It uses a linear distance tolerance measure to assess the proximity of the generalised line to the original. It creates a trend line between the first and last point of the line, selects the vertex on the original line that is farthest in perpendicular linear distance from the trend line, and then creates two new trend lines: one from the start point to the selected vertex and one from the selected vertex to the end point. The process of determining the vertex farthest from each of these trend lines is then repeated. The algorithm repeats this process until the specified linear distance tolerance is reached and any points that have not been selected thus far are deleted.

**The Visvalingam-Whyatt algorithm is a popular alternative for simplification.** It progressively removes points from the linear feature that have the least-perceptible change. It computes triangles formed by successive triplets of points along the linear feature, and then removes the smallest triangle. After a triangle is removed, the neighbouring triangles are recomputed and the process repeated.

**See also:** Clutter | Generalisation | Pointillism | Refinement | Smoothing | Symbols

SInce most maps are derived from larger scale surveys, it's inevitable that there will be many more survey points used to capture linear features than might be capable of being shown at smaller scales. If such data is left in its raw, detailed form, you'll find that even with the thinnest line weights possible, the small curves and bends will coalesce. The line becomes bulbous as a result. In almost all situations, simplifying linework to show a detailed dataset at a smaller scale will look so much better, graphically.

**Opposite:** These diagrams illustrate the effects of the Douglas-Peuker algorithm for two linear distance tolerances and the resulting 75% scale reductions. Note that many of the original points have been eliminated but, as with any automated approach, there is a lack of consistency. Some lines tend to be well approximated while others produce less satisfactory results. This effect tends to be particularly problematic where administrative boundary datasets are generalised and is especially true where certain angular relationships (such as 90° angles) exist which might not remain the same after generalisation. Angular relationships are not maintained when the automated simplification is applied. As with any automated procedure, manual checking for consistency is vital to ensure the resulting line work faithfully represents the original.

1:1,000,000
1:1,500,000
1:3,000,000
1:5,000,000
1:10,000,000

# Size

## Used to show differences in magnitude or relative importance.

**The size of a point or line object can be modified to show some quantitative difference between objects of the same type.** Varying the size of objects, while keeping other visual variables constant, provides a perceptual stimulus that is interpreted as relative importance. Changing the size of either the whole symbol or the marks that make up a symbol reflects quantifiable differences in the data on ordinal and numerical scales. The visual impression is one of a measurable difference in the magnitude of the underlying feature being mapped—larger graphical marks rise to become figural. As with spacing, changing size is not suited to depicting qualitative data on the nominal scale of measurement. Where possible, size is considered to be a more suitable alternative to spacing for representing quantifiable difference in data in map form.

**For point objects, the shape of the graphical mark can take many different forms and be graphical or mimetic.** The possibilities for varying the size are only limited by the graphical limits of the map page or screen though for practical purposes these maximum limits are rarely reached. Size is the key visual variable used in the construction of proportional symbol maps and can equally be applied to linear objects through varying line width.

**The size of the marks that are used to create the symbol can also be altered.** Although areas themselves cannot be changed in size (unless through being warped using a cartogram approach), the constituent graphical marks can. It's therefore possible to develop a range of textures that, perceptually, take on the same sort of stimulus as changing the size of a single symbol.

**Variations in symbol size can be compromised by optical illusions on the map.** For instance, the Ebbinghaus illusion shows that relative size perception is modified by other objects in the vicinity. Larger objects will reduce the perception of size compared with the same object surrounded by smaller symbols.

**See also:** Flow maps | Proportional symbol maps | Sizing type

The number of possible steps in a range of symbols of different sizes is technically unlimited. A drawback is that humans are not able to discern small differences between symbols, and they also find it difficult to see more than about 20 steps between two point symbols whose areas are in a ratio of 1 to 10.

As a modification to the use of differently sized symbols for underpinning ordered perception, a series of graduated sizes of symbols might instead be used on a classified version of the data. These graduated sizes generalise the data to an extent but provide the eye with an easier task of seeing difference between symbols.

Size variation is dissociative so any other visual variable that is combined will be dominated by it. For this reason, size is one of the main ways in which a cartographer can change the message of different phenomena on the map. Larger mapped objects become dominant in perception regardless of any other graphical treatment.

Visual variation

Increasing size becomes figural

For seeing

Distinct Levels

For representing

Nominal Ordinal Numeric

**Opposite:** Different-size countries representing different variables by Worldmapper.org.

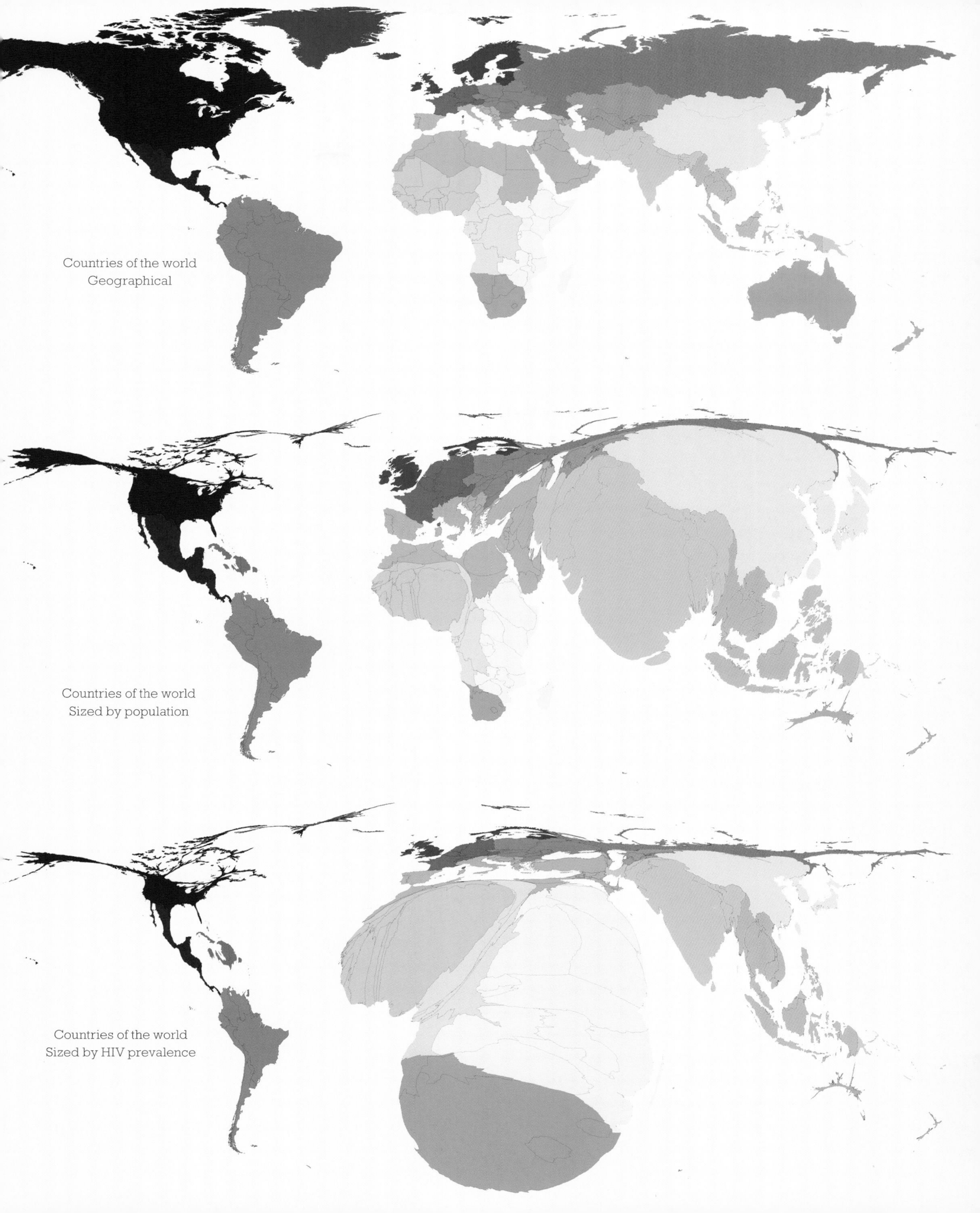
Countries of the world
Geographical
Countries of the world
Sized by population
Countries of the world
Sized by HIV prevalence

# Sizing type

## The sizing of type on a map goes a significant way toward shaping its overall look and feel.

**With reference to old foundry-type blocks, the height of the body of the type block specifies the size of a letter in points.** This height refers to the block size itself even though the letter may not extend to the full height (e.g. a capital letter does not have a descender so its block height will be larger than the individual letter). Type size is measured in point sizes even in modern digital environments on the basis of the legacy of its original specification.

**Determining the size of a typeface is the most common typographic decision facing the cartographer.** One guideline might be to base this decision on the size of the object being named or the space on the map to be filled but this might easily produce a map with type that dominates. A better principle to follow is that it must be large enough to be easily read. However, what constitutes 'large enough' can be a problem. Conventionally, maps are designed to be read at a normal distance of approximately 30–50 cm. However, when a map is to be displayed on a poster or projected for a mass audience, type size must be altered. The minimum threshold of visibility and legibility will be much higher than at normal reading distance. Consequently, the design of the map should take into account the final purpose and display environment.

**In digital environments, text size can benefit from responsive design.** In multi-scale environments, text size may increase in relation to feature size as scale decreases as well as selective omission weeding out unnecessary labels. User-specified type size can be used to allow people to set their own text size. Such an approach allows people with differing eyesight to set a map that suits their needs. Type size might need to differ depending on device. For instance, reading a computer display is often at a different distance than a mobile device so device-dependent sizing might be used to modify the size responsively. Higher resolution devices will commonly display type smaller than designed which might need to be accommodated.

Type size is not commonly specified in metric units and is based on the division of an imperial inch into 72 gradations referred to as *points*, one point being equal to 0.0138 in. (0.351 mm). Thus, 18-point type will have a distance between the ascender line to descender line of approximately 6 mm. Capital letters in 18-point type will be about 4 mm in height.

Numerous studies have found that 3-point type is the smallest that can be discerned by the human eye. However, this is particularly small and assumes 'normal' sight. It is much safer to assume 'average' sight, which would give a minimum lower limit for visible text as 5 point. This lower limit might be adapted if, for instance, the map user is aged or the map is to be used under low-light conditions.

Size is often used to show ordinal differences, but map readers are not sensitive to small differences in type size. Type is often spaced apart, which compounds the problem. Consequently, the following guidelines for differences between type size are helpful:

- less than 15% should be avoided;
- approximately 25% in height is optimal; and
- for the 5- to 15-point type size, differences of 2–2.5 points are appropriate.

The purpose of map lettering will go some way to defining its size. Setting up a visual hierarchy helps identify what a reader needs to know. A title should be one of the largest components as should major place-names or features.

If the map is interactive, it's possible that a label will act as a locator for a click event or rollover. Larger size type might differentiate so the map reader has a visual clue that they can interact with the label.

**See also:** Fonts and type families | Graphic and dynamic labelling | Printing fundamentals | Size

Face
Shoulder
Groove
Foot
Body
Nick
Point size (one em)

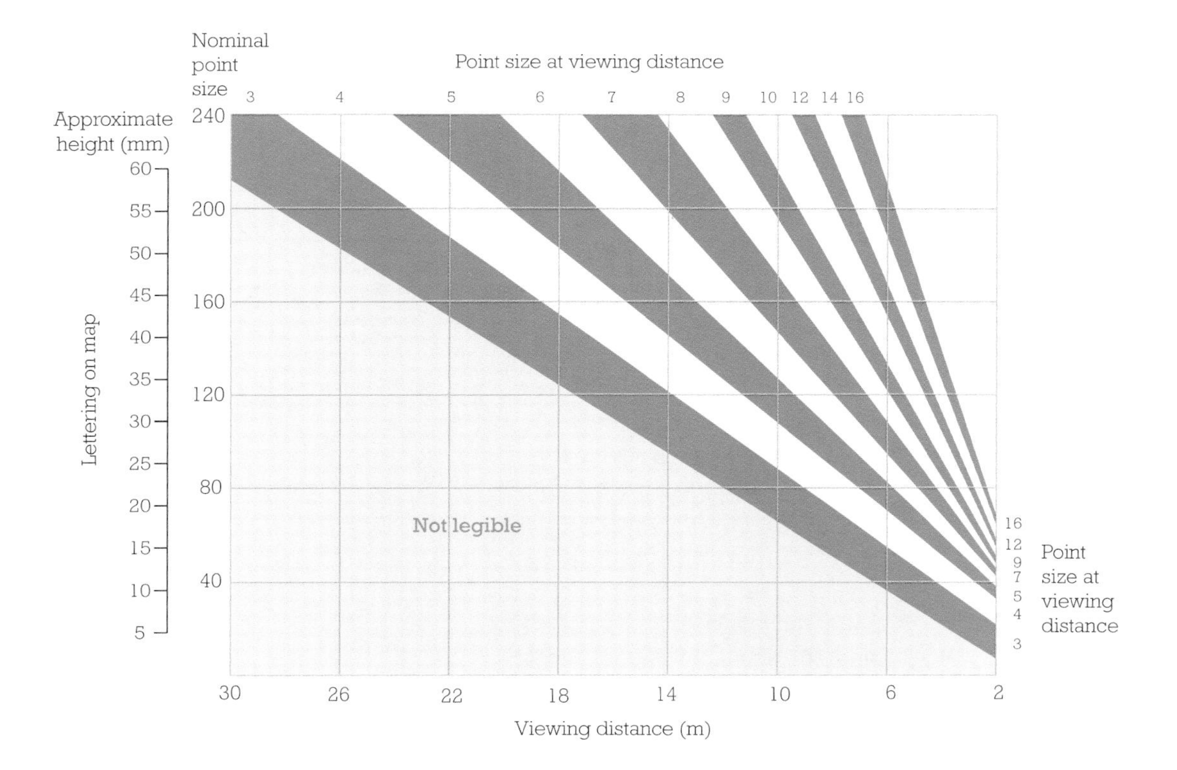


Point Scale
0 12 24 36 48 60 72 84 96 108 120 132 144

Centimetre (Metric) Scale
0 1 2 3 4 5 6 7 8 9 10

Inches (Imperial) Scale
1 2 3 4

# Slope, aspect, and gradient

## Processing a DEM gives several alternative terrain representations.

**Digital elevation models can be processed to extract alternative terrain representations such as slope, aspect, and gradient.** For each cell in a DEM, slope can be calculated as the maximum rate of change in value from that cell to its neighbours measured in degrees or percent. The maximum change in elevation over the distance between the cell and its eight neighbours can also be used to identify the steepest downhill descent from the cell. Aspect identifies the downslope direction of the maximum rate of change in value from each cell in DEM to its neighbours. It can be thought of as the slope direction that the landscape faces. Gradient is a property that helps describe a surface. It's a vector that measures magnitude and direction as both the maximum amount of vertical change between two points and the direction of that change. An assessment of gradient is often used as a way of finding the steepest path, the *gradient path*.

**Slope maps express lower values as flat terrain and vice versa.** Expressions may be ratio, percentage, or angle. The slope of a surface between two points can be expressed as a ratio between elevation difference and ground distance written as y/x or as a slope percentage if the ratio is multiplied by 100. A flat surface would be 0%, but as the surface becomes steeper, the slope percentage becomes larger.

**Aspect maps indicate the compass direction that the surface faces at that location.** It is measured clockwise in degrees from 0 (due north) to 360 (again due north), coming full circle. Flat areas having no downslope direction are given a value of −1.

**Calculating gradient requires the measurement of gradient magnitude and also gradient direction, usually as a true azimuth.** Using these values, gradient maps can be created either through the creation of a three-dimensional model, which builds the terrain using gradient magnitudes and direction, or the mapping of slope as a raster map or isolines and the addition of an overlay that illustrates direction using arrows.

**See also:** Curvature of terrain | Digital elevation models | Small landform representation

Slope maps are commonly used to support a variety of land use, landscape planning, and environmental assessment purposes. Slope maps can be simple rasters of the calculated slope values (rate of maximum change) for each pixel, or they can be developed into isoline maps showing classed slope zones separated by a meaningful contour interval.

Aspect maps usually use different hues to indicate different directions. They are usually categorised rather than allowing each pixel to have its own individual colour. For instance, categories that correspond to the main compass directions can be used to group pixels that share characteristics.

Aspect is quite often mapped in combination with detail that illustrates slope since the two surface characteristics are often related in reality. A spectral colour scheme would usually be used to map aspect as noted above, but saturation is additionally used to modify each pixel's hue to represent slope. Steeper slopes will be more saturated than flatter terrain which will appear desaturated, leading to a greyish flat terrain.

Gradient maps provide a suitable way of representing flow across a landscape. This may be water or, in fact, the movement of any phenomenon, including thematic data representing any movement of goods, services, or ideas.

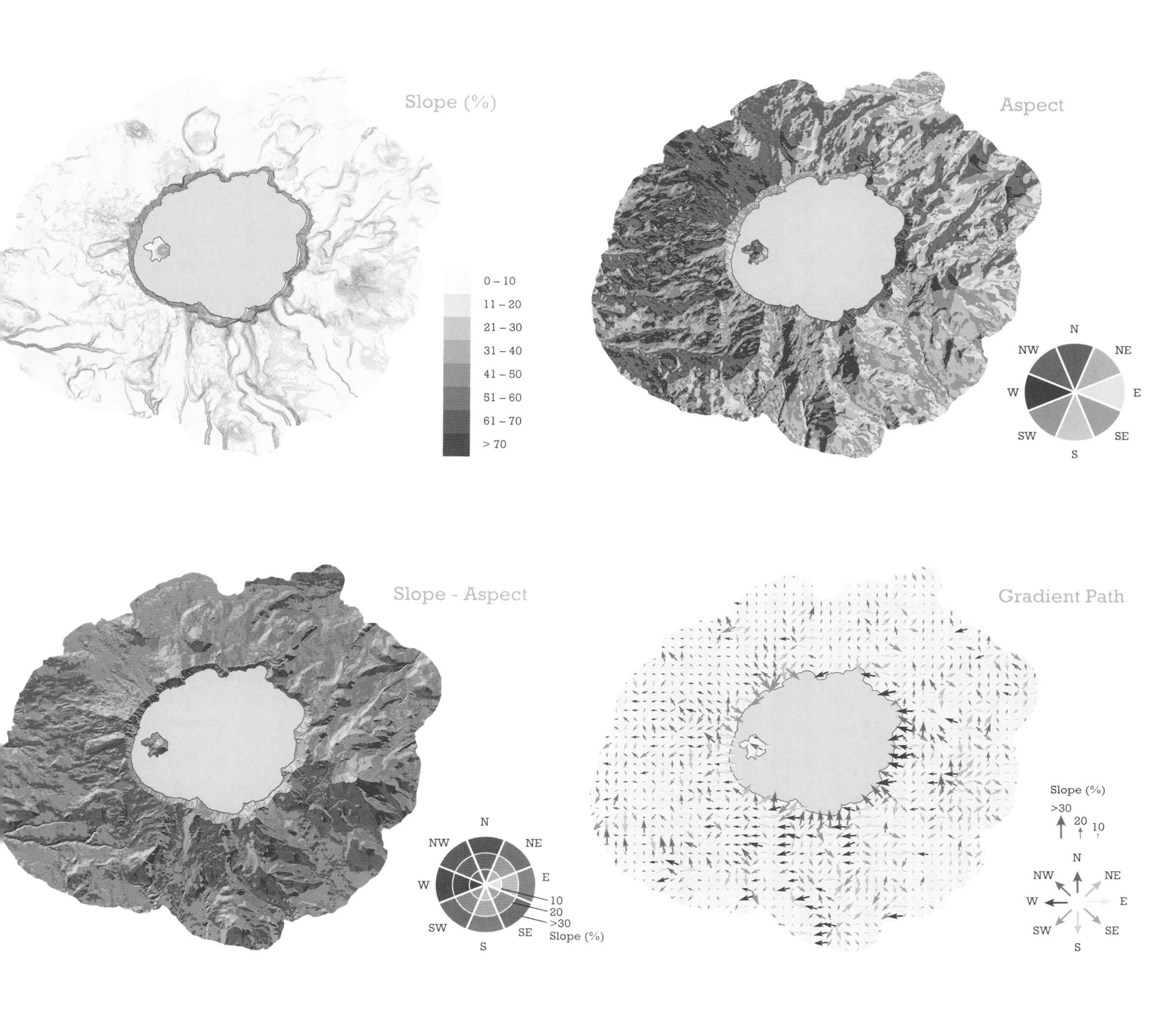

Slope (%)
0 – 10
11 – 20
21 – 30
31 – 40
41 – 50
51 – 60
61 – 70
> 70
Aspect
N
NE
E
SE
S
SW
W
NW
Slope - Aspect
N
NE
E
SE
S
SW
W
NW
10
20
>30
Slope (%)
Gradient Path
Slope (%)
>30
20
10
N
NE
E
SE
S
SW
W
NW

3D Gradient Path

# Small landform representation

## Depiction of small features on topographic maps.

**Some small landforms require specific depiction.** Natural and artificial features often exist that differ from their surroundings for which normal techniques, such as contours, are unsuitable. Features such as drumlins, scree, sand, gravel pits, and open cast mining often need to be mapped but can only be done so successfully through a symbology of their own. Pictorial symbols can be used but other techniques that extend the drawing of the landscape can also be applied.

**Graphically, hachures and dots are used to create an array of different representations that mimic the feature's salient character.** As far as possible the symbology should be as self-explanatory as possible. Tapering hachures are used for small, steep slopes. The horizontal hachure are more rounded and can be used to depict concave features. Dots provide a good symbology for sand, gravel, or scree fields. Density, dot size, and pattern can be varied to create a different appearance. Dots can also be used to create shadow effects and in conjunction with other symbology to give added depth. Shading is often used to depict small convex or concave features but only on maps that use hillshading more generally to avoid looking out of place.

**Scale plays a large part in how small landforms are represented.** As scale decreases, individual representations give way to area fills if the features are extensive, else they disappear altogether. Sometimes a single small landform feature will be used to represent a group of features, effectively aggregating the features for symbology purposes.

**Glacial forms are generally used only on large-scale maps where freeform slope hachures are often used to show crevasses and broken off ice forms.** Tapered hachures are used for gullies and moraines. Blue lines are often used to show glacial landforms and can be combined with black or brown symbology for showing scree, earth, or other rocky material that forms part of an older glacier. Volcanic structures tend to be haphazard in structure, and combinations of tapered hachures, shading, and dots are often used.

Tapered hachures differ from a general-slope hachure, which tends to depict whole stretches of uneven terrain. The tapered hachure is small and indicative of a uniform slope with a distinct sharp end-point on the downslope end. They are usually shown in brown if the material is earth or black if stone. They are evenly spaced and consistently drawn though may be longer to indicate a longer slope section within the feature. Hachures of any type are normally constrained to large-scale maps because of features being less legible via this technique as scale decreases. Ravines, gullies, and landslides can also be successfully shown using tapered hachures, often in conjunction with contour lines for larger features.

Pits and quarries are usually depicted using hollowed-out forms. These can be constructed by hachures or by rock drawing techniques. For quarries, the flat terraces may be indicated and the base of the pit shown using dots to indicate the material, or alternatively, sometimes with a small water feature. Sinkholes and other similarly small features are not well represented by tapered hachures because of their extremely small size. Instead, stylized symbols such as horizontal hachures, shaded depressions, or abstract symbology such as an outline with perpendicular inner ticks and a downslope arrow might be used.

Scree and debris fields are often made up of dot fills. Combined with contours, lines of scree are plotted in the downslope direction. Dot size varies so that finer material appears on upper slopes, and larger dots or, sometimes, small pieces of rock drawing to represent boulders appear downslope at the point of rest. Landslides can also be shown in this way as can sand dunes with a change in colour to brown or reddish brown and an emphasis on shading to show the overall structure of a dune field.

**See also:** Curvature of terrain | Geological maps | Hachures | Rock drawing

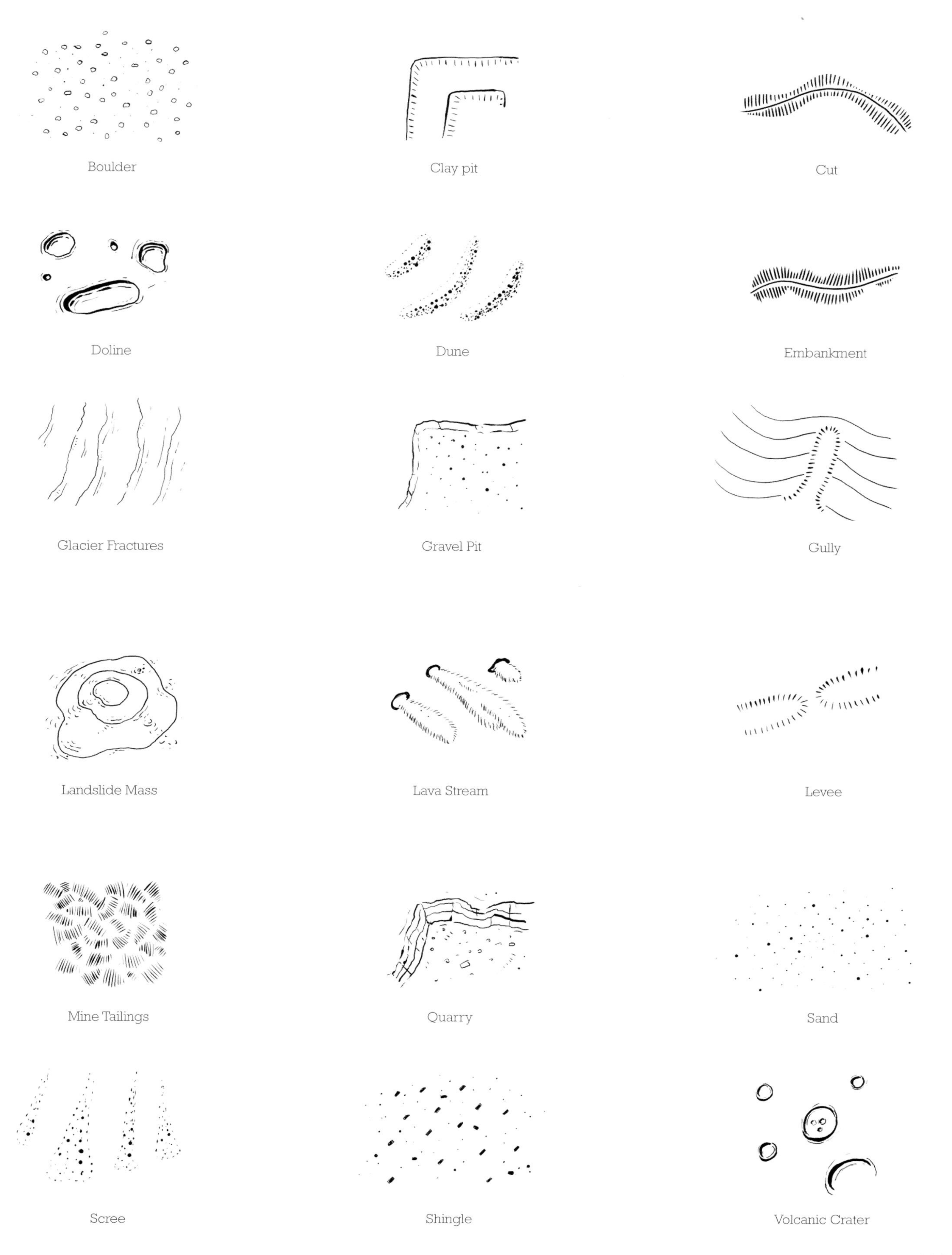
Boulder
Clay pit
Cut
Doline
Dune
Embankment
Glacier Fractures
Gravel Pit
Gully
Landslide Mass
Lava Stream
Levee
Mine Tailings
Quarry
Sand
Scree
Shingle
Volcanic Crater

# Small multiples

## A map or graphical composition that displays several small images to support the study of comparisons.

**One of the main ways to communicate a message visually is to provide comparisons.** Comparisons provide people with a baseline to assess other parts of the visual. It supports the determination of how something has changed or empirical assessments such as more or less. Comparisons on maps can be developed in different ways through multiple map pages, animation of change between map state, or through interactive layer selection. A drawback of all these approaches is that at any one time you can view only one map, one point in time, or one layer of data. Small multiples overcomes that limitation by presenting numerous small maps in a single layout. They are typically used when multivariate data is shown for the same area or where change over time is a key theme. Alternatively they can show the same dataset but for different areas to support comparisons of type.

**Small multiples tend to use highly generalised forms of the map and focus on the thematic detail.** Inevitably, you reduce the amount of space given over to any single map so simplification is important but this sort of map is designed to provide an overview across multiple maps. The major benefit is that all the information required to study the series is in view, and comparisons of different map states is immediate. The consistency of the design means people see changes in the mapped variable rather than changes in design.

**Small multiples of the same area are usually set out in a grid form.** They are often read from left to right and top to bottom, mimicking our normal (Western) reading mode. They contain minimal locational detail or typographic elements. They are uncluttered to enable the rapid movement of the eye across multiple images to gain an impression of the overall pattern. They are consistently designed—same size and internal consistency with only the variable of interest being altered. Usually only a single legend should be needed since it would apply to all maps in the series.

Some small multiples compositions can be successful with the absolute minimum of detail. In some respects, the geography of the underlying area is unimportant and the pattern of the overlaid data gives the maps their own structure. If necessary, a location map can be added to the composition for reference though the abstract character of minimal maps can provide a compelling aesthetic appeal.

White space becomes important in a small multiples composition to give spacing and regularity to the overall appearance. Alternatively, if you are presenting a dataset for different locations, arrangement of the small multiples diagrams can lead to a tessellated-cartogram appearance.

The use of small multiples is not restricted to maps. Many examples use multiple graphs in a sequence to show change. In fact, one of the earliest uses of small multiples can be found in the work of Eadweard Muybridge in the late 1880s. Muybridge is well known for his early studies of motion. His studies used stop-motion photography to capture different phases of, for instance, a galloping horse, which he then subsequently presented as a series of small photographs. In so doing, he demonstrated that during part of the gallop, all four hooves leave the ground. The use of small multiples was soon adopted in statistical atlases, particularly to show the change in population structure over time.

**See also:** Dispersal vs. layering | Graphs | Pie and coxcomb charts

**Opposite:** A mosaic of small multiple images of the maximum eclipse experienced across *African Solar Eclipse of 2001*, by Fred Bruenjes, 2002.

# Smoothing

## Smoothing modifies the position of linear feature vertices to improve a line's cartographic appearance.

**Smoothing shifts the position of a linear feature's vertices in x,y map space to improve its appearance by reducing angularity.** It differs from simplification in that the process generally doesn't remove vertices. Smoothing attempts to move points in a way that eliminates small linear changes but retains the essential trend in the line.

**There are many automated smoothing algorithms.** Weighted averaging techniques calculate an average value on the basis of the positions of existing vertices and neighbouring vertices. The end vertices remain unaltered. In general, the resulting line contains the same number of vertices as the original. Tolerances are used to create different smoothing conditions (i.e. include more or less neighbouring vertices and so on). Epsilon filtering uses geometrical relationships between points and a user-defined tolerance to smooth the line. Again, end vertices remain unaltered but in this technique the number of coordinate vertices in the resulting line may vary. Finally, mathematical approximation can be used to describe the geometrical character of a line. The number of vertices in the resulting line may vary and some of the original points may also be retained in their original position. As with the simplification algorithms, it is important to review the result of an automated smoothing algorithm since shape can be altered considerably.

**A common smoothing algorithm is the moving average.** This algorithm replaces an x,y vertex with an average x,y position calculated from those surrounding x,y vertices within a defined tolerance. Tolerance areas can take the form of different shapes (circles, rectangles, and triangles) and can also be weighted to give precedence to nearer vertices. The result tends to retain straight edges and convert angular change to curved corners. Alternatively, smoothing can be calculated by fitting Bézier curves to the input lines themselves, which results in a more sinuous result as Bézier curves will pass through the input vertices.

The idea of moving the x,y vertices of a linear feature might seem anathemic to some because it is moving geography itself. If the purpose of a map is to accurately show location, then modifying the very essence of that location seems at odds.

However, at small scales, when the thickness of a line itself might represent maybe 50 miles of map distance, the slight adjustment of vertices to render the character of a line can be a useful technique instead of simplifying the line to a straighter version.

Smoothing can also be successfully used to remove sharp angles in linear features to improve their cartographic appearance. For instance, contour lines which have been automatically derived from a raster-based digital elevation model will normally contain many angular discontinuities because the underlying elevation values in the raster conform to a square grid. Smoothing the resulting lines will create a much more pleasing appearance.

Smoothing is also a process that can be applied to raster datasets for cartographic purposes. For instance, performing focal statistics to calculate a cell value as an average of surrounding cells will result in a smoothed raster. This can then be used to generate smoother contour lines, which provides a way of preprocessing the raster for contour generation rather than post-processing the contour lines from an unmodified DEM.

**See also:** Generalisation | Refinement | Simplification

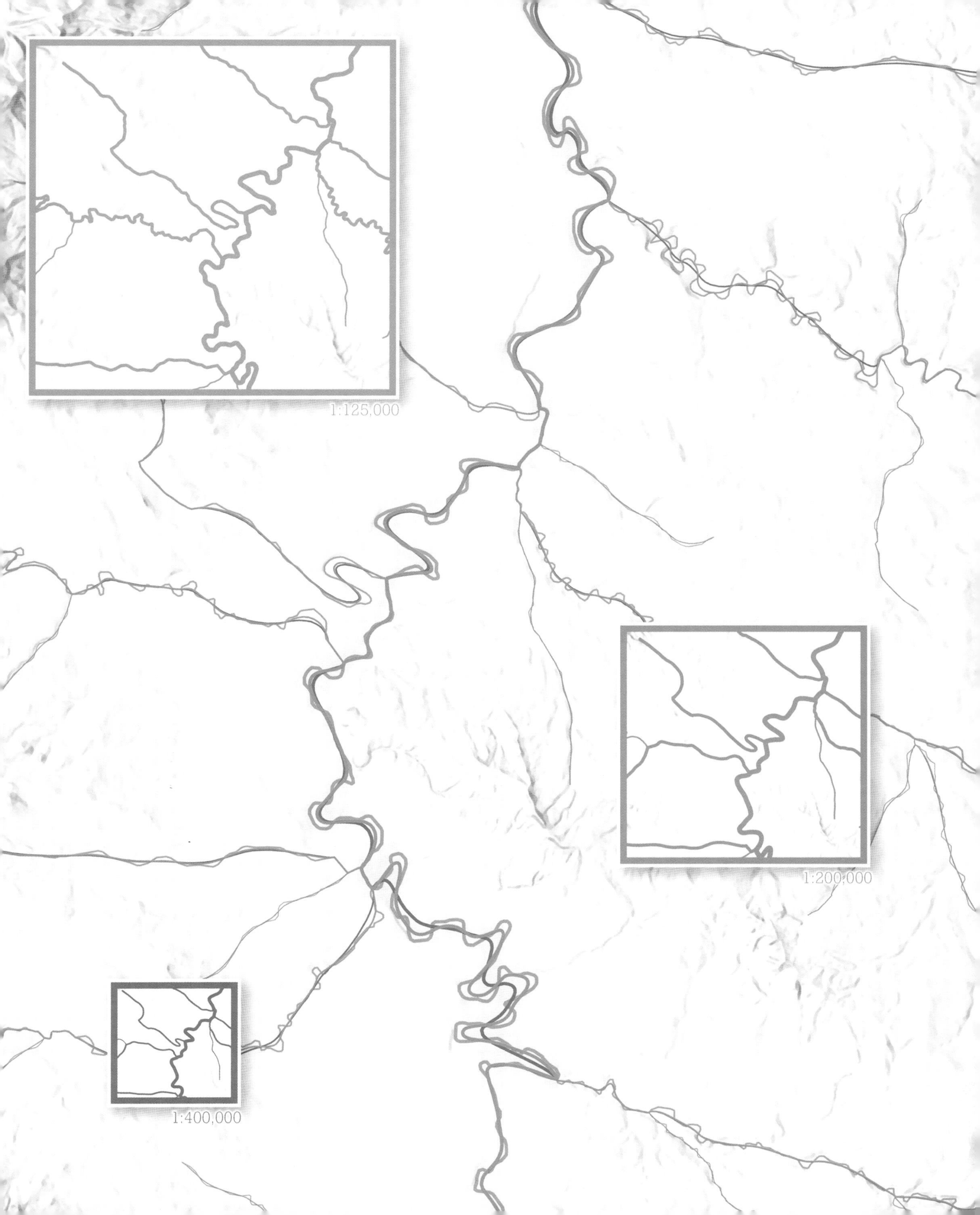
1:125,000
1:200,000
1:400,000

# Space-time cubes

## Space-time cubes can be used to map space and time together.

**The space-time cube is an effective way to model geographies in both space and time.** They were created by geographer Torsten Hagerstrand in 1970 as a framework to study the social interaction and movement of individuals. The space-time cube adds a z-axis denoting time to the familiar planar x,y coordinates of maps. For point data, each event can be plotted according to their x, y, z position, which allows you to explore clusters in both space and time. For linear events, individual segments join vertices with steeper lines indicating faster movement and vertical lines indicating stationary periods. Areas can be extruded and also moved in vertical space so that prisms depict both areal extent and temporal extent. Multiple events that exist for areas in different time periods would stack on top of one another.

**The space-time cube has seen sporadic use because of problems relating to its technical construction and usability.** For instance, when you add multiple pieces of data, it can suffer from visual clutter and occlusion, and estimating size and values are problematic if the cube is shown in a perspective view, particularly when multiple trajectories are displayed together. Using an isometric projection can alleviate problems of comparing position and value throughout the cube.

**Animated and interactive space-time cubes overcome many limitations.** They bring opportunities to develop ways of showing multiple related aspects of data, which give map readers a way of slicing and dicing the cube as an exploratory tool. In this way, they have been used to successfully illustrate flow and movement through time using linear features or points that build to show the cumulative result of multiple events. Additionally, transparency can be used to modify the outer faces of prisms but this can create complex illustrations as the amount of data you're trying to show increases. The effect of being able to see through some of the data into the heart of a prism-based space-time cube can create a compelling image but at the cost of easily being able to interpret the spatial patterns it represents.

**See also:** Animation | Graphs | Inquiry and insight | Interaction

Space-time cubes struggle with increasing data quantities, particularly when you try to show multiple polygonal layers together. Polygonal layers usually encode time by extruding a two-dimensional shape into a three-dimensional version. Different slices of extruded data can be stacked to build a cube of prisms. However, when you stack layers on top of each other, it's really only possible to see the outer faces of the prisms, and making sense of what's going on inside a column of stacked prisms is difficult. For this reason, space-time cubes of polygonal data should be avoided unless you're able to create an interactive version that allows you to strip away layers to reveal the inside.

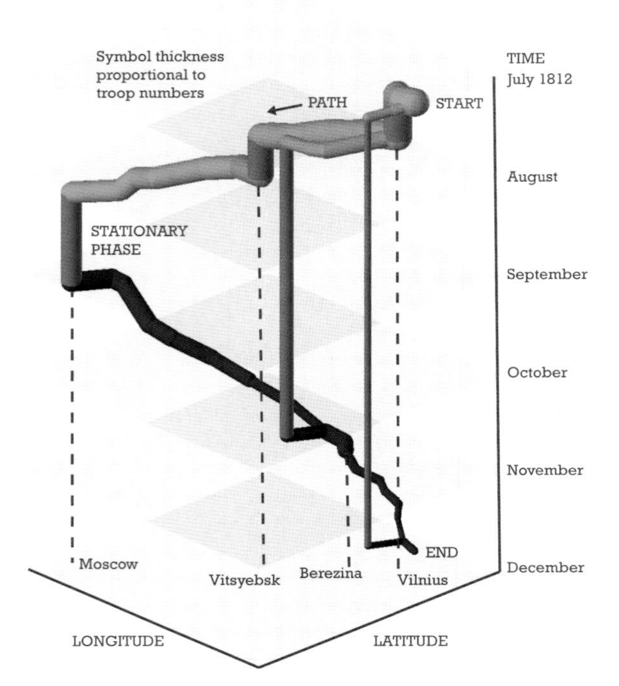


**Above:** Napoleon's march on Moscow as an isometric space-time cube.
**Opposite:** *Paths to the Future* by Andrew DeGraff, 2017, showing the main characters and plotlines of Robert Zemeckis's *Back to the Future* (1985). Hill Valley in 1985 in blue, Hill Valley in 1955 in pink.

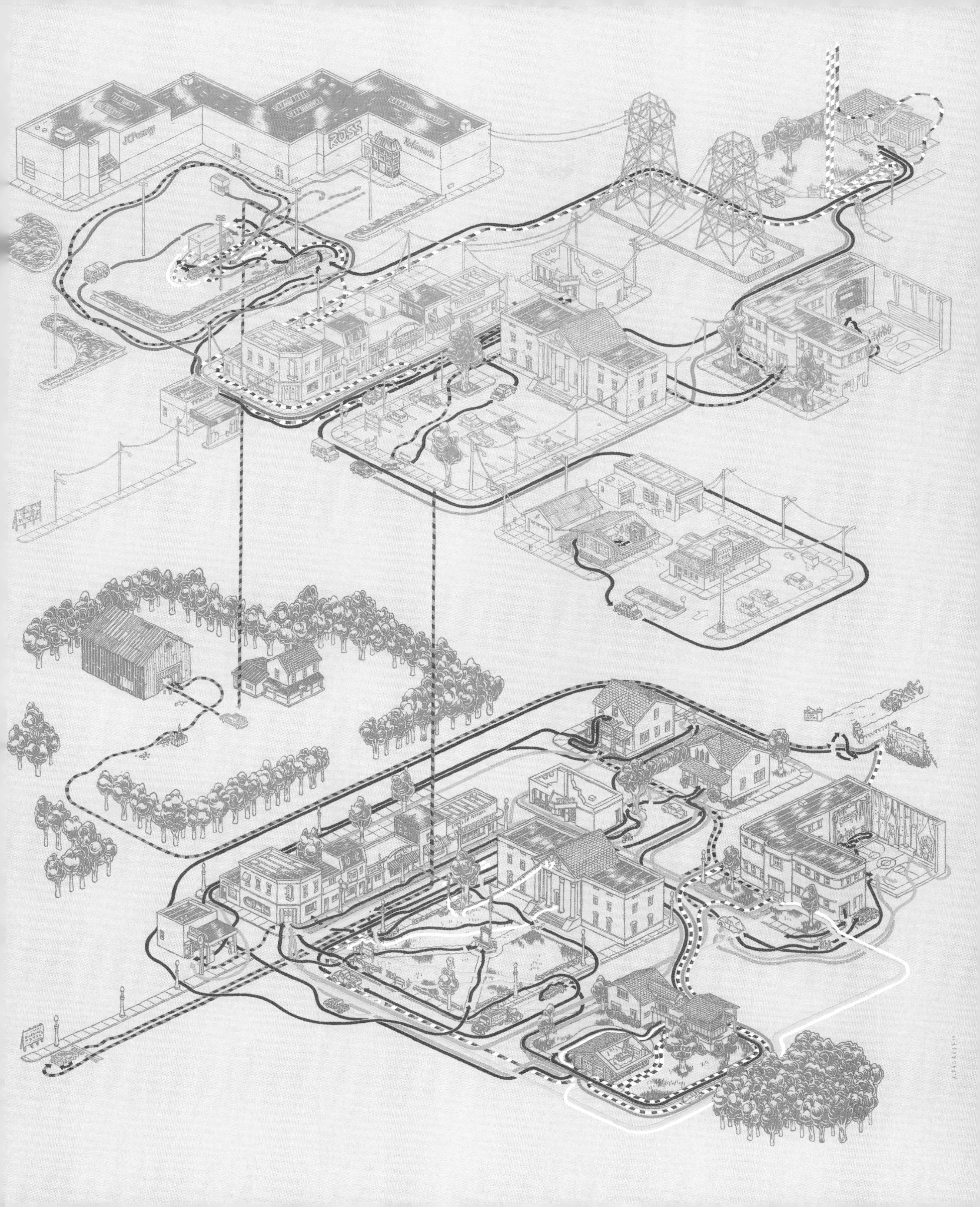

JCPenney
ROSS

# Spacing letters and words

## Spacing gives different effects and supports different uses of type on the map.

***Letter spacing* refers to the space between each letter in an individual word with *word spacing* referring to the space between individual words.** Compact type is achieved by minimising letter and word spacing, which can often make placement on a map easier considering the reduced amount of space needed. Increasing letter and word spacing creates a less cramped appearance, which, in turn, makes the text easier to read. Text set with increased spacing should be used where possible for this reason. Exaggerated spacing is often used for type displayed in uppercase to overcome the difficulty of reading words that are blocky in appearance. Exaggeration of text is also used in placement for areal features to support the emphasise of areal extent and avoid the potential of a label being seen as relating to a specific point on the map.

**Care needs to be applied when spacing words across a map to ensure that two related words are not viewed as two separate words.** This is particularly true when placed across varying underlying map content. Word spacing can cause problems when cartographers attempt to stretch words across a geographical area. In real terms, a balance that is visually stable will be achieved if each individual placement is carefully considered.

**The variation of space between two adjacent characters is referred to as *kerning*.** In digital environments, kerning is applied automatically once a typeface and size is determined. However, this automatic spacing does not necessarily result in spacing that is visually consistent.

**Leading is the vertical space between individual lines of type relevant to their baselines.** Alterations to the leading can be applied to bring lines closer together or further apart with the principle that they should be far enough apart to allow multiple lines to be read effectively. However, leading should always be balanced with the space constraints on the map.

Some letters cause particular problems in cartography. Uppercase forms such as *A, J, L, P, T, U, W,* and *O* contain few vertical strokes and occupy a large surface area when compared with other letters. Spacing of uppercase letters can be improved by following the neutralizing rule. Consider the word ROOM. The *O*s are too close in visual terms to the letters *R* and *M* such that a hole appears in the word. If increased letter spacing is applied for the part RO and OM such that it is equivalent to the inner part of the O, the hole disappears and the form of the word is much more visually pleasing. The neutralising rule states that a minimum spacing between capital letters should be equal to the optical value (visually equal space) for all capital letters.

As an example of kerning, consider the pairing of WA, which requires that space is removed from between letters to look consistent with the spacing between the letters MU. Digitally applied type includes preset kerning pairs, which sets kerning uniquely whenever pairs exist but it should always be checked and adjusted manually if necessary. Kerning is measured using ems. A single em is equivalent to the point size of the typeface being used.

Bold typefaces are often used to provide emphasis but care must be taken on how a bold form affects legibility. The character of a font can change dramatically when a bold form is used, which sometimes leaves an overly rounded appearance and, without sufficient letter spacing, a situation where one letter is too close to another, especially at small sizes.

Condensed forms are useful in cartography since they require less horizontal space but, conversely they often require increased letter spacing.

**See also:** Graphic and dynamic labelling | Placing type | Sizing type

LAND OF THE FRANKS
Bay of Biscay
ATLANTIC
Bordeaux
Garonne
GASCONY
Bilbao
Toulouse
Montpellier
Andorra
Gulf of Lion
Cantabrian Mtns
NAVARRA
Pyrenees
GALICIA
LEON
CASTILE
ARAGON
Rio Ebro
CATALONIA
Barcelona
IBERIA
Rio Duero
PORTUGAL
Porto
Madrid
Rio Tajo
VALENCIA
Valencia
Balearic Islands
Palma
Rio Guadiana
TOLEDO
Lisbon
ANDALUCIA
Baetic Mountains
GRANADA
Seville
Mediterranean
Gibraltar
Algiers
Straight of Gibraltar
(Pillars of Hercules)
Alboran Sea
Tangier
Tell Atlas
Rif
Oued Sebou
ALGERIA
Rabat
Casablanca
Oued Moulouya
Saharan Atlas Range
MOROCCO
Oued Draa
High Atlas Range
SAHARA
(The Great Desert)

# Spatial dimensions of data

## Geographical phenomena exist at different levels of dimensionality, which require translating before being represented in map form.

**Geography uses various observational techniques to identify phenomena and map their location and character.** Phenomena cannot themselves be analysed or mapped without first being translated into data. Geographical data is selected features that can be used to measure phenomena that have a spatial dimension. Measurement may be direct or indirect and numerical (quantitative) or non-numerical (qualitative). The data may be a subset, a summary, partial, or a sample of the phenomena and may have been subjected to a number of processes and choices made by others before the process of making a map begins.

**Geographical phenomena exist in reality at different levels of dimensionality.** Point phenomena have no spatial extent and are measured simply by their location. Linear phenomena are one-dimensional since they are characterised by length but do not have width (and, therefore, do not have an aerial extent). Linear phenomena are defined in space as a series of coordinates in two-dimensional space that do not close (i.e. do not form an area) and which share the same characteristics. Area phenomena are two-dimensional and are constructed as a series of related coordinates in two-dimensional space that close to form a region. Point, line, and area dimensionality are normally sufficient for many cartographic tasks, but when volumetric phenomena are considered, there becomes a further distinction: 2½ D phenomena can be characterised by a surface in which geographical location is defined by x,y coordinates at which a third value, height or depth around a zero point (z), is measured. This sort of volumetric phenomena differs from true 3D phenomena which might exhibit multiple values in the z-dimension. True 3D phenomena therefore have four parameters: x- and y-coordinates, a z-coordinate indicating height or depth, and the value of the phenomenon at that point.

Spatial dimensions can often be used to characterise the form of geographical data. For instance, a phenomenon that exists as a linear feature in reality (e.g. a road) may be equally represented by a linear object in a cartographic sense. However, spatial dimensions do not always map neatly in this way since scale and form are intricately linked. As scale and the level of inquiry are altered, the spatial dimension of a feature may also change. For instance, at a large scale, a city may be represented as an area with two dimensionality. However, at a small scale the same city might be better represented as a point feature with zero dimensionality. Map scale therefore plays an extremely important role in determining how the spatial dimensions of geographical phenomena are handled and will impact upon how they are cartographically represented.

A fourth dimension can be assigned to data when time is involved. The temporal dimension can apply to any of the lower orders of dimensionality. Points may move or change in character over time. Lines might do likewise, or signify flow or some change in character at different points along the length. Areas and volumes can also move over time or exhibit a change in character. Cartographically, the temporal dimension might be depicted using small multiples, animations, space-time cubes, or other mechanisms that support good illustration of the fourth dimension.

**See also:** Areas | Data distribution | Digital data | Lines | Points

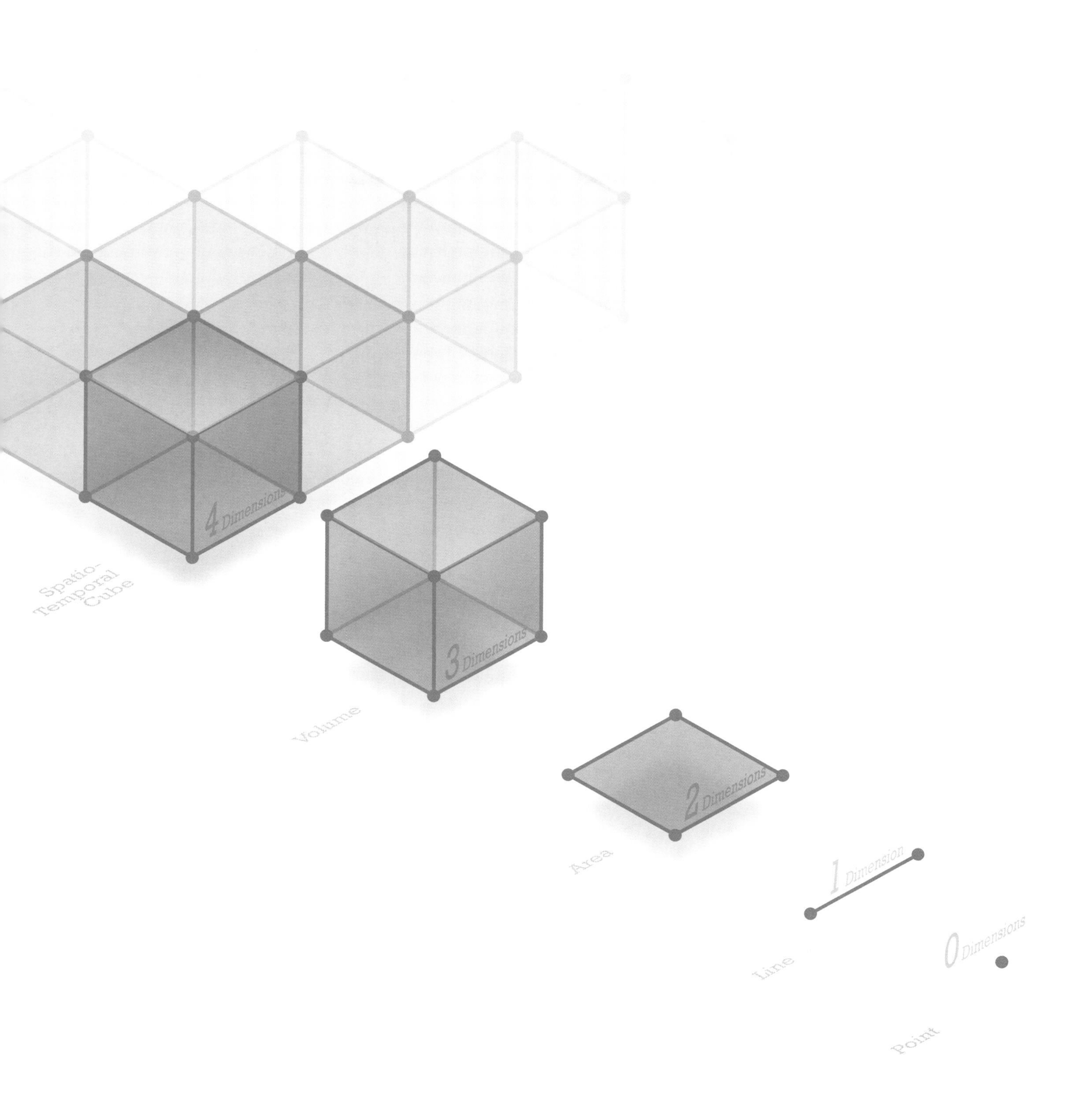
4 Dimensions
Spatio-
Temporal
Cube
3 Dimensions
Volume
2 Dimensions
Area
1 Dimension
Line
0 Dimensions
Point

# Statistical data classification

## Statistical classification schemes use an aspect of the data's summary statistics or distribution to form the class interval.

**Classifying data using statistical techniques for thematic mapping involves calculating the class interval with respect to the distribution of data values.** These techniques asses the distribution in statistical or mathematical terms and use the results to determine the class interval. They support an objective assessment of the data array though the use of statistical nomenclature can be difficult to interpret.

**The standard deviation classification method makes use of the data distribution as part of the classification.** Classes are formed by repeatedly adding or subtracting the standard deviation from the mean of the data. The standard deviation method works well with data that is normally distributed but can lead to problems such as empty classes if the data is skewed. One way to resolve this problem is to transform the data (for instance, by taking the logarithmic values), but, of course, if the intention of the map is to display raw data, then this transformation will in itself propagate to the map. A disadvantage of the standard deviation scheme is that interpretation of the resulting map will require some basic understanding of statistical methods. In making a map based on this classification method, you should be sure that the map user is likely to understand the concept of arithmetic mean and standard deviation.

**The maximum-breaks method involves examining individual data values and their relationship with neighbouring array values when in rank order.** Differences between each neighbouring pair of values are computed, and the largest of the differences provides the breaks between each class interval. As with the quantile method, it is normal to use the actual data values to define the class intervals, but it is also possible to use the break values themselves.

An advantage of the standard deviation method is that the mean provides a suitable dividing point for normally distributed data. The scheme provides a mechanism to easily identify values (and areas on the map) that are above or below the mean. This is best achieved if an even number of classes are selected so that, say, three classes depict data above the mean and three below.

The maximum-breaks method is simple to apply. It tends to group similar values well although in some cases it can seem anomalous. For instance, you might have several data values toward the upper end of an array but the gap between the highest value and the next two values forms one of the class breaks. It might seem more sensible to cluster these highest three values in a single class given their distance from the rest of the data.

**See also:** Arbitrary data classification | Data classification | Eyeball data classification

## Standard Deviation

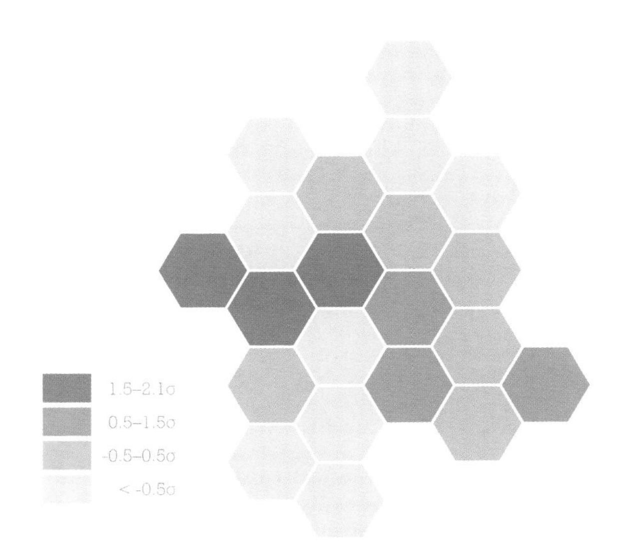


For this data, a Standard Deviation scheme shows that the majority of the data is more than -0.5σ away from the mean with fewer areas above the mean. It supports the interpretation of the dataset as being skewed.

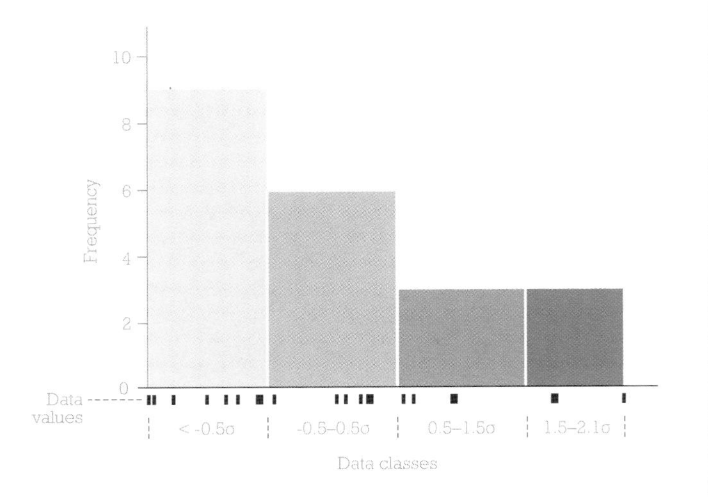


Regular class width
Irregular number of data items per class

## Maximum Breaks

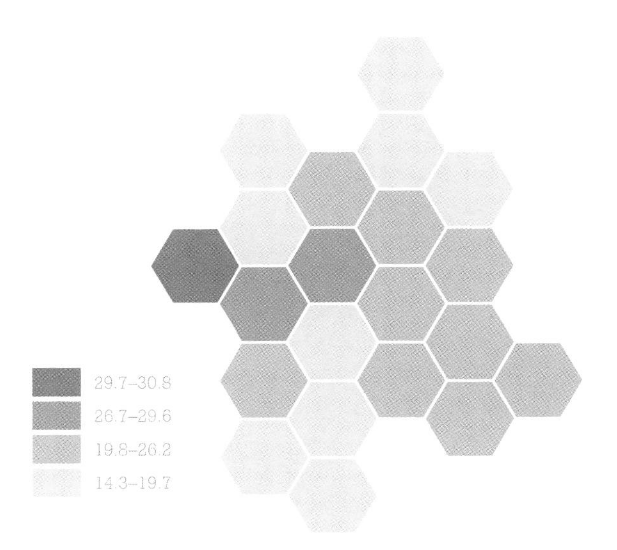


For this data, a Maximum Breaks scheme provides a good way of showing the similarity of the large number of low–mid values. At the same time it picks out the higher value outliers.

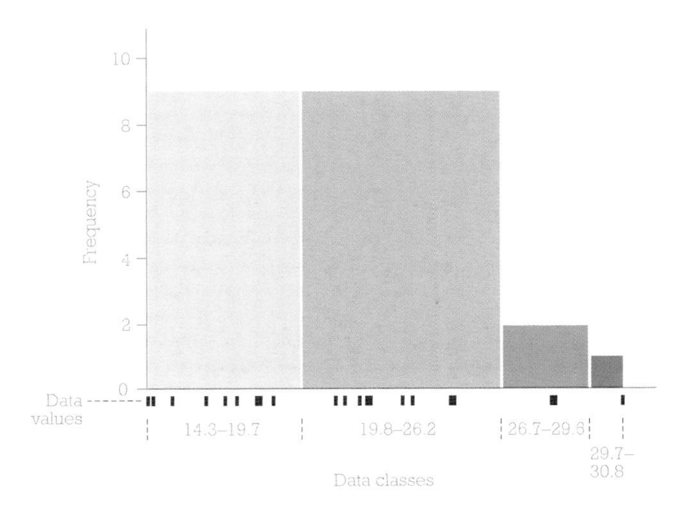


Irregular class width
Irregular number of data items per class

# Statistical literacy

## The needs of cartography demand a certain degree of numeracy.

**Maps are based on the manipulation of numbers.** Data is collected as numbers. Locations are referenced as numbers. Third-party data is supplied as lists of numbers. Attribute data also exists but, fundamentally, you'll likely need some numerical information to do anything with it. The uncomfortable truth for many is that you'll need an understanding of basic numeracy and, possibly, a more advanced understanding of particular techniques to make your map. This can cause trepidation in those who come to cartography from a design or artistic background. Most cartographers are not expert statisticians in the same way that most statisticians are not expert cartographers. Knowing enough to limit errors is the key, and asking for assistance where necessary is important.

**Cartography does not require an expansive knowledge of numeracy and statistics.** Basic mathematics allows you to handle data in a spreadsheet or database and manipulate several items of data to create new items. You're often interested in minimums, maximums, sums, and averages. You may need to go further and understand how to determine a suitable classification scheme for a thematic map. Ultimately, your needs for statistical knowledge will be largely based on univariate and multivariate statistical techniques. You are likely going to be interested in mapping quantitative values and exploring relationships between data. More advanced inferential statistics may be required if you want to do any data modelling or use techniques such as regression analysis as part of an analytical workflow to derive your mapped content. You may have a need to show relationships between variables or map correlations and causations. Getting into probability theory, testing hypotheses, forecasting, and modelling risk or uncertainty introduce further levels of complexity. It's not the purpose of this book to explore all the various possible numerical and statistical techniques at your disposal, but being aware of the need to become involved in some of this numeracy is important, if only to feel more comfortable about the role of numbers in cartography.

Statistical literacy is not just for cartographers. Map readers also need to come to a map with some basis of understanding of the techniques used so they can better interpret and understand the map. That said, you cannot assume any significant level of understanding unless your map is for an expert audience. People are inundated with references to statistics every day in advertising (8 out of 10 cats prefer a particular brand of cat food), news reports (approval ratings are down 15 percent), and general conversation (I'll be about 30 minutes based on average traffic). Numbers and statistical measures are often used to support a claim or counter-claim. They become powerful tools to back up assertions. It's important to be clear about how you have used or manipulated data shown on your map. Be forthright about units of measurement, classification schemes, and any relationships you set out. Being up front marshals your own thinking as well as helps readers and reduces the potential for misinterpretation.

'*Lies, damned lies, and statistics*', often attributed to British Prime Minister Benjamin Disraeli in the mid-1980s, is the well-worn phrase that warns of the ways in which statistics can be manipulated to tell different truths, or even worse, mask the truth. Being cognisant of the potential to tell mistruths guides cartographers to respect the science of numeracy and statistics in their own maps. Mark Monmonier's classic book *How to Lie with Maps* extends the cautionary tales of the use of numbers by exploring how maps themselves can twist and warp data, numbers, and statistics to represent geographies in any manner of truthful, persuasive, propagandist, or erroneous ways.

**See also:** Data classification | Earth's shape | Levels of measurement | Making numbers meaningful | Quantitative statistical maps

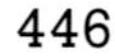

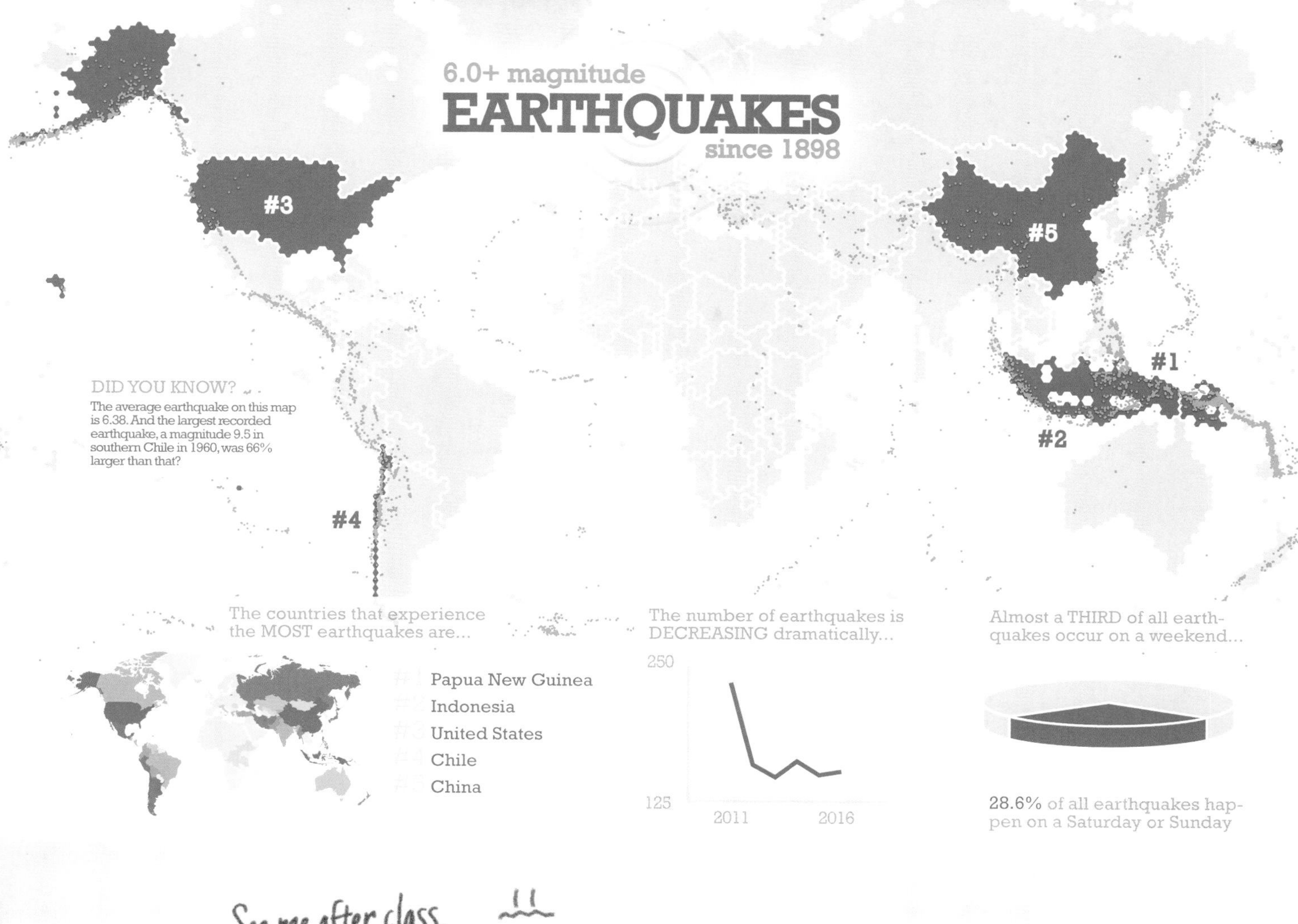


See me after class.

Dr. Field | Advanced Statistics

## CREATE AN INFOGRAPHIC

Create an "infographic" communicating the salient points of a data set of your choice. Feel free to use charts, graphs, maps, and summary statements, to reveal a phenomenon. Be careful to adhere to sound statistical statements of fact and effective design. Use what you have learned in this semester's exploration of *the dismal science*. This assignment will count towards 50% of your overall grade.

**NOTES:**

Earthquake magnitudes use a logarithmic scale. A magnitude 8 is 10 x stronger than a mag 7, and 100 x stronger than a mag 6. So you can't take a simple average (you might consider median). The earthquake in Peru was 3,162 x bigger than a mag 6.

Earthquakes don't care what day of the week it is, so all that section tells me is how frequently Saturdays and Sundays occur. While true, it's meaningless. This is a confusion of "statistical" vs "practical" significance. And 3D pie charts are horrible things.

The ranking of countries with the most earthquakes does not account for country size. The US and China are big, so it's an unfair comparison. Normalize this data. If you normalize for area, the top five countries become Vanuatu, Taiwan, the Solomon Islands, Papua New Guinea, and Costa Rica.

Fewer quakes? This statement, and chart, is called "cherry picking." You've isolated a small, downward trending, portion of the time range and only charted that. See my chart, below. What's more, the Y-axis is truncated, which visually biases the chart. Pretty misleading.

Pink? Anyway, where did this data come from? I checked... USGS. Always cite your sources.

Your chart!

1960

Now

# Stereoscopic views

## Techniques that encode three-dimensional information to create an illusion of depth.

**By showing a slightly different image of the same area to each eye, we can create the illusion of depth and a three-dimensional image.** A stereopair of aerial photographs or two maps, when viewed together, create the illusion of depth in the image because we're mimicking what our eyes do all the time. Our eyes are, effectively, looking at the same thing from two slightly different angles, and this allows us to see in stereo and see depth. Using stereopairs of imagery or maps is often used for photo interpretation and, historically, in photogrammetry to measure distances between mapped objects and determine contours of equal height.

**Anaglyphs are a special form of stereopair which uses two maps constructed from different vantage points, one of which is printed in red and the other in blue.** Rather than the two maps situated side by side, they are overprinted, though slightly offset from one another, and when we view the map through special anaglyph glasses with red and blue lenses, we see the map stereoscopically. This occurs because each lens filters out that particular wavelength of light, and so we see the blue map with one eye and the red with the other. Our brain sees two maps from different angles and builds the 3D image.

**Chromastereoscopy uses colour to carry the z information and gives a pseudo-3D holographic effect when you wear specialized glasses that contain prisms.** The benefit of this technique is that the map is still perfectly viewable in 2D without the glasses. Colour is used in a specific way. You assume a black ground and an RGB colour palette. The colour of an individual map element and its immediate surrounding colour determine its depth position. Red is viewed as foreground, dark blue as background, and green as middle ground. The specialised glasses contain prisms that bend light in different ways. Blue light is bent toward a sharper angle than red. Because the apex of the prism in front of each eye points toward the nose, the eyes turn more inward to see a red object than a blue object leading the brain to interpret, by means of parallax, that red is closer than blue.

Because of the way the image looks without the filtered glasses, anaglyphs have limited use in general-purpose cartography. They have been used to create pseudo-3D views through printing offset versions of aerial imagery using two colour channels. When printed in red/blue and using filters, the resulting image remains monochrome. Additional vector features such as contours can also be rendered using this technique.

Although most anaglyphs use the red/blue filtered lenses, images can be created using any combination of colour channels. For stereoscopic viewing, the two colours must be diametrically opposed. Impurities in the lens, printing, or in our eyes may result in double-imaging (ghosting). Red/blue combinations result in monochrome rendering but other combinations (e.g. red/cyan or green/magenta) result in colour rendering.

For chromastereoscopy, the most usual way of representing bathymetry, with dark blue meaning deeper water, is already well suited. It looks perfectly normal without the glasses but gives a pseudo-3D view when viewed using glasses that support chromastereoscopy. Above the water's edge you'd pass through the spectrum of colours reaching red for the highest elevations. Clearly this might give your map a peculiar appearance as bright-red mountain peaks are not a typical elevation tint.

For non-topographic uses, the same technique can be used to order layers of map detail with different depth cues. Red is reserved for the uppermost layer so administrative borders and labels might be represented in reds while background detail is in blues.

**See also:** Colour in cartography | Foreground and background | Seeing | Vision

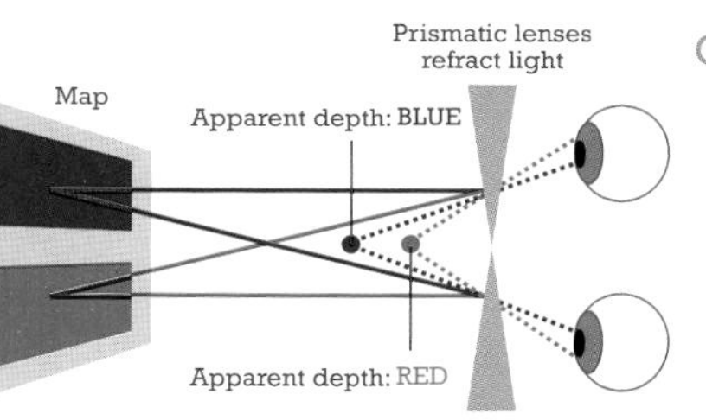


Chromastereoscopic
Best viewed with
ChromaDepth® glasses

Anaglyph
Best viewed with
red/blue glasses

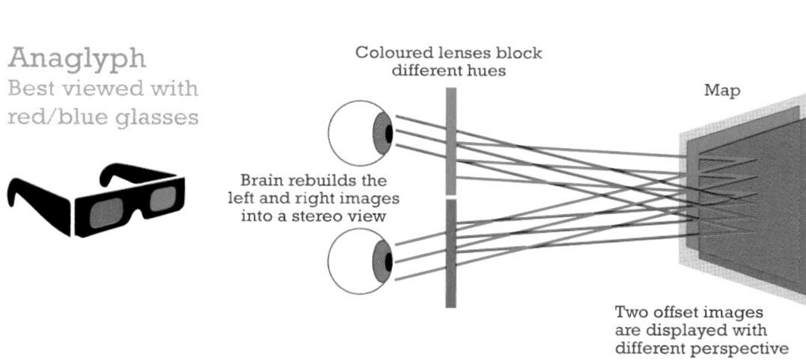


3D Anaglyph
Perspective maps can also
be viewed in anaglyph

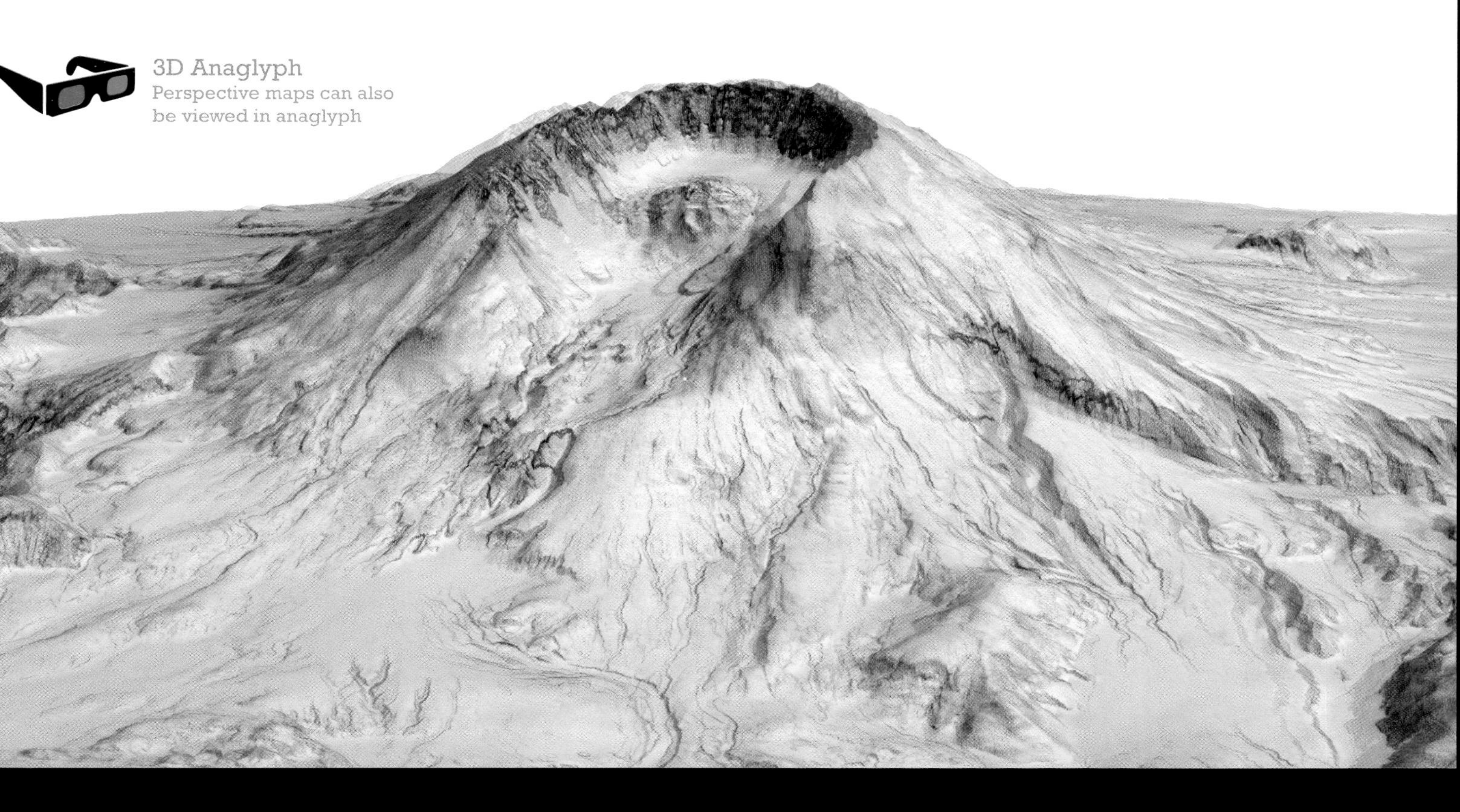

# Strip maps

## Representation of a linear feature, usually straightened and presented in strip form.

**Navigating using a standard topographic or reference map might be considered a fairly typical map reading task but how much of that map is superfluous to the task at hand?** There's often a lot of peripheral content and detail that is simply not needed to get from *A* to *B* on the map. Metro maps take a topological approach and dispense with the clutter to show us the network, but although they are often distorted to clarify detail in congested areas, they still have some general geographical qualities. Such maps generally still have a north, south, east, and west even though they are not necessary for using the metro itself.

**An even simpler, diagrammatic form of navigational map literally strips away peripheral detail to focus on a single feature.** Pioneered by John Ogilby in his 1675 road atlas, *Britannia*, he used a consistent scale of 1" = 1 mile but his two other innovations stand out. Firstly, he used a scroll technique, each strip of which illustrated a thin portion of the landscape through which was threaded a segment of the larger journey. Secondly, he oriented the strips in a heads-up fashion so each strip was read as a straight line from bottom to top with the compass orientation changing between strips.

**Ogilby's strip map was an ingenious solution for transport mapping and supportive of map reading for that task.** All people needed to do was follow the path of the road, and know which town was next, how many stops there were before their destination, and what the distance was. In many ways, Ogilby was the forerunner to modern schematic maps used for transport mapping and every other mapmaker who turns the transport network into a series of lines connecting nodes. Modern strip maps tend not to use the scroll technique but you'll often find strip maps in subway carriages that focus on a single line and present it as a straight line with the two ends of the map reflecting the two stations at each end of the line. As with most maps there are limitations, and Ogilby presented the maps in one heads-up orientation despite the fact that routes could be traversed in two directions.

**See also:** Cartograms | Flow maps | Lines | Reference maps

The linear strip maps you find in subway carriages are usually presented left to right and can be read in both ways as the two ends of the map naturally point to the two ends of the subway line.

Road atlases to this day often include modern versions of Ogilby's design to illustrate motorway networks and the important characteristics such as junction layouts.

In many ways, the modern sat-nav or mobile phone maps we use to navigate have developed an interface that feeds off Ogilby's ideas. Instead of orienting the network to heads-up, the map and its content is rotated so we face ahead at all times, thus creating a single straight line that rolls from top to bottom of our device as we move. In design terms, the map is no different from Ogilby with the route and important positions en route being the focus. The periphery falls off the edge of the cellphone screen in the same way as it did off Ogilby's strip.

Similar approaches have also been taken both for functional and artistic purposes. For instance, canals, bike paths, lake shores, coastlines, and rivers have been straightened out to show a line that supports the viewing of information in different ways. This is usually done to present a clear way of seeing the key nodes along a network (locks, sluices, bridges and so on) and the relative distance between them.

**Opposite (left):** Extract from *The Road from London to the City of Bristol* by John Ogilby, 1675, which presents the journey in parallel strips.
**Opposite (centre):** *The Manchester Ship Canal* strip map, 1923, which straightens the canal.
**Opposite (right):** *Lake Michigan Unfurled* by Daniel P. Huffman, 2015, which straightens out the shoreline of the lake.

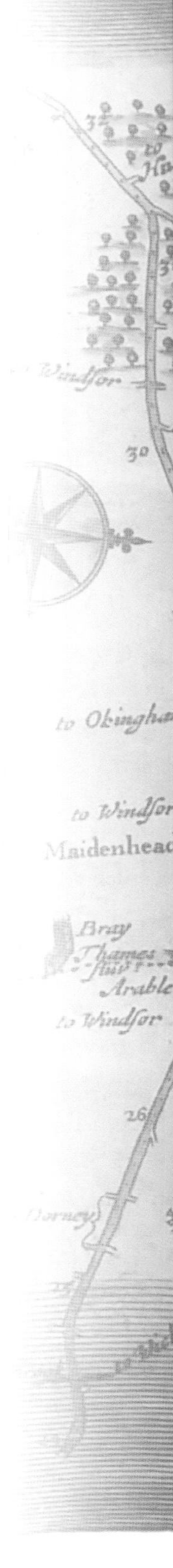

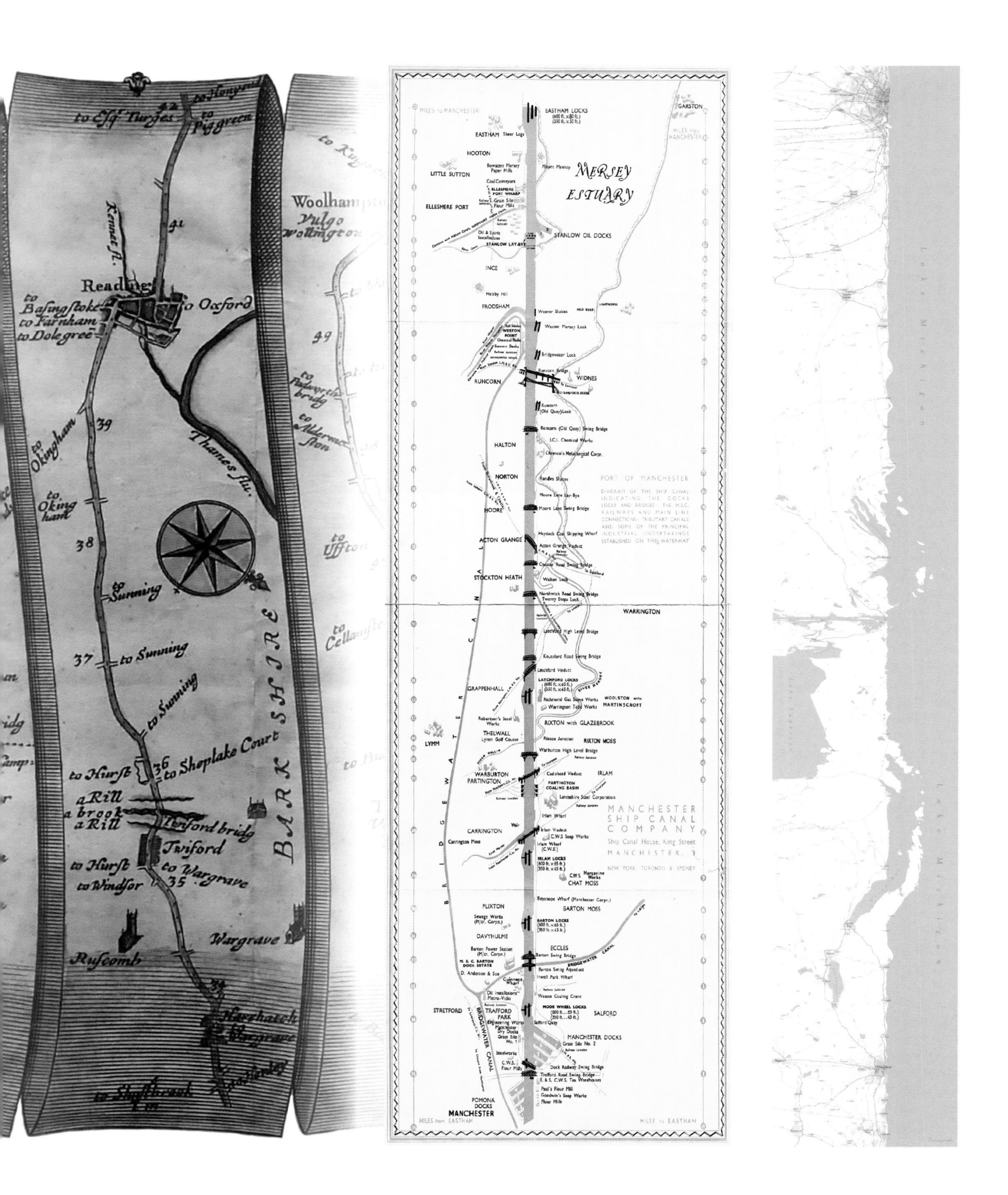

to Esq. Turses
to Piggreen
Kennet fl.
Reading
to Basingstoke
to Farnham
to Dolegree
to Oxford
Thames flu.
to Okingham
to Oking ham
to Sunning
37
to Sunning
to Sunning
to Shoplake Court
to Hurst
36
a Rill
a brook
a Rill
Twford bridg
Twiford
to Hurst
to Windsor
35
to Wargrave
Wargrave
Ruscomb
BARK SHIRE
38
39
41
42
Woolhampton
Vulgo Wollmgton
to Ufton
to Cellansfie
EASTHAM LOCKS
(600 ft. x 80 ft.)
(350 ft. x 50 ft.)
EASTHAM
HOOTON
LITTLE SUTTON
ELLESMERE PORT
MERSEY ESTUARY
STANLOW OIL DOCKS
STANLOW LAY-BYE
INCE
Helsby Hill
FRODSHAM
Weaver Sluices
Weston Mersey Lock
WESTON POINT
Bridgewater Lock
Runcorn Bridge
WIDNES
RUNCORN
Runcorn (Old Quay) Lock
Runcorn (Old Quay) Swing Bridge
I.C.I. Chemical Works
HALTON
NORTON
Randles Sluices
Moore Lane Lay-Bye
Moore Lane Swing Bridge
MOORE
ACTON GRANGE
Acton Grange Viaduct
Chester Road Swing Bridge
STOCKTON HEATH
Walton Lock
Northwich Road Swing Bridge
Twenty Steps Lock
WARRINGTON
Latchford High Level Bridge
Knutsford Road Swing Bridge
Latchford Viaduct
GRAPPENHALL
LATCHFORD LOCKS
(600 ft. x 65 ft.)
(350 ft. x 45 ft.)
WOOLSTON with MARTINSCROFT
Warrington Tube Works
RIXTON with GLAZEBROOK
THELWALL
Lymm Golf Course
Rixton Junction
RIXTON MOSS
LYMM
Warburton High Level Bridge
WARBURTON
PARTINGTON
Cadishead Viaduct
IRLAM
PARTINGTON COALING BASIN
Lancashire Steel Corporation
Irlam Wharf
CARRINGTON
Carrington Moss
Irlam Viaduct
C.W.S. Soap Works
Irlam Wharf (C.W.S.)
IRLAM LOCKS
(600 ft. x 65 ft.)
(350 ft. x 45 ft.)
C.W.S. Margarine Works
CHAT MOSS
PORT OF MANCHESTER
DIAGRAM OF THE SHIP CANAL INDICATING THE DOCKS LOCKS AND BRIDGES: THE M.S.C. RAILWAYS AND MAIN LINE CONNECTIONS: TRIBUTARY CANALS AND SOME OF THE PRINCIPAL INDUSTRIAL UNDERTAKINGS ESTABLISHED ON THE WATERWAY
MANCHESTER SHIP CANAL COMPANY
Ship Canal House, King Street
MANCHESTER, 2
NEW YORK, TORONTO & SYDNEY
Boysnope Wharf (Manchester Corpn.)
FLIXTON
BARTON MOSS
Sewage Works (M/cr. Corpn.)
BARTON LOCKS
(600 ft. x 65 ft.)
(350 ft. x 45 ft.)
DAVYHULME
Barton Power Station (M/cr. Corpn.)
ECCLES
Barton Swing Bridge
BRIDGEWATER CANAL
M.S.C. BARTON DOCK ESTATE
Barton Swing Aqueduct
Irwell Park Wharf
D. Anderson & Son
Guinness Wharf
Oil Installations Metro-Vicks
Weaste Coaling Crane
STRETFORD
TRAFFORD PARK
MODE WHEEL LOCKS
SALFORD
Salford Quay
BRIDGEWATER CANAL
MANCHESTER DOCKS
Grain Silo No. 2
Grain Silo No. 1
Steelworks
C.W.S. Flour Mills
Dock Railway Swing Bridge
Trafford Road Swing Bridge
Paul's Flour Mill
Goodwin's Soap Works
Flour Mills
POMONA DOCKS
MANCHESTER
MILES from EASTHAM
MILES to EASTHAM
MILES to MANCHESTER
GARSTON
MILES from MANCHESTER
BRIDGEWATER CANAL
Lake Michigan
Lake Michigan

# Style, fashion, and trends

## As an art form, maps evoke different styles, leading to fashions and trends.

**Maps are increasingly ephemera.** Whereupon once they took months or even years to survey, design, and produce, they are now made or updated rapidly, beyond the speed of thought in many cases. Many maps burn brightly for a short while yet also fade from the public consciousness rapidly. Although some of this might seem to be due to changes in map use, caused perhaps by dwindling attention spans and increased expectations, much of it also has to do with shifting cartographic fashions and styles: what's hot and what's not, much of which seems to work outside what we might consider cartographic norms.

**Maps can be stylish but they can also lack style.** Style, as in any art form, can define fashions which can change over time. Some styles become timeless and some are fleeting. Trends emerge and change over time. Style might be thought of as belonging to a particular look and feel, perhaps that of a national mapping agency or a news organisation which can go through updates from time to time to refresh their style. Increasingly it has more to do with an individual and the search for an approach that riffs off well-known and beloved imagery as a means of rapidly piquing interest. Style is designed to evoke a particular reaction.

**As Yves Saint Laurent famously said, 'Fashion fades, style is eternal'.** It's difficult chasing and keeping up with cartographic fashion because of the pace of change but style is impacted by other long-term factors. Fashion might see a technique become popularised, and then rapidly fade through overuse. Style is a character that embodies you as the mapmaker bringing a sense of style, borne out of your experiences, culture, politics, religion, and, not least, cartographic tastes.

You can exude cartographic style without necessarily being fashionable. Maps can exude style in their composure, clarity, harmony, and composition. They have a voice, authority, and a way of carrying their content and message. It's an overall impression that goes beyond the technical construction. You already have a personal style that is based on your tastes, developed over time. Your tastes are expressed through the music, films, clothes, and art you enjoy but they are also in part a function of where you work, your personal circumstances, and other impacts on your daily life.

Cartographic styles tend to be an expression of a particular aesthetic. This may come from expert knowledge and prior art or through experimentation and honing of a particular look and feel. In thinking about cartographic styles, it's worth exploring what has gone before that works. Evaluate how different elements hang together coherently.

Style goes beyond colouring in a basemap but, instead, builds coherence, balance, and impact. We can take design cues from art, furniture, architecture, film, and build these out to apply to maps. Asking yourself questions such as 'What do I like about this map and why?' helps establish your cartographic tastes. Think of how you mix colours and use fonts to create a mood, and determine whether that mood fits a particular theme you're mapping. Critically, it's sometimes important to take risks and experiment.

Styles can become associated with a particular person. As with changes in fashion, trends come and go, and instead of buying into a well-known style and following a trend, personalization and artisanal products have become an important market. Generating your own style certainly creates intrigue in your maps and defines your own brand of cartographic fashion.

**See also:** Branding | Critique | Elegance | Flourish | Knowledge and conviction | Map aesthetics | Prior (c)art(e) | Who is cartography?

**Opposite:** Evolution of the Swiss cartographic style by the Federal Office of Topography swisstopo.

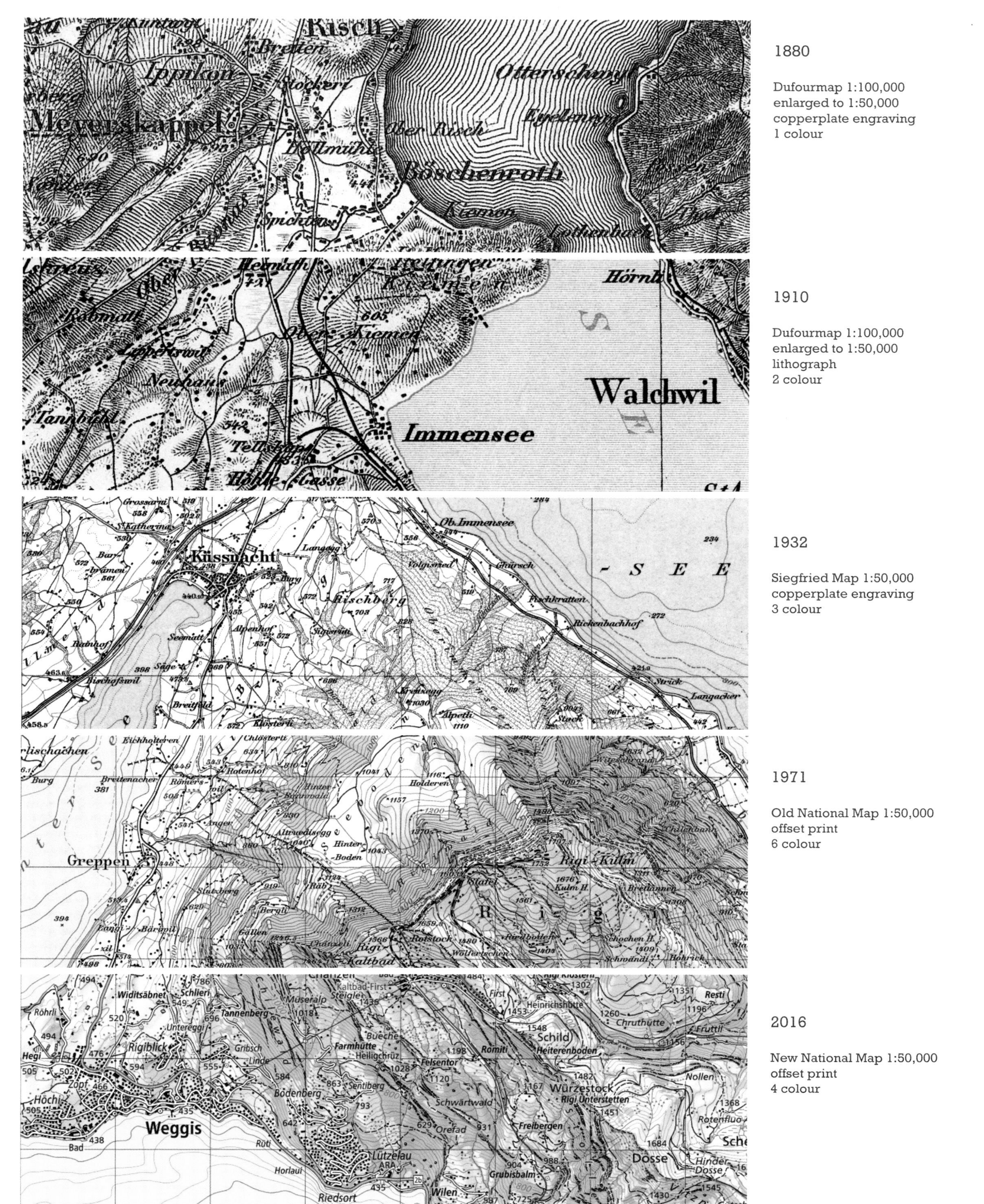

1880
Dufourmap 1:100,000
enlarged to 1:50,000
copperplate engraving
1 colour
1910
Dufourmap 1:100,000
enlarged to 1:50,000
lithograph
2 colour
1932
Siegfried Map 1:50,000
copperplate engraving
3 colour
1971
Old National Map 1:50,000
offset print
6 colour
2016
New National Map 1:50,000
offset print
4 colour
Meyerskappel
Böschenroth
Walchwil
Immensee
Küssnacht
Greppen
Rigi-Kulm
Weggis

# Styling shaded relief

## Relief shading doesn't have to be greyscale.

**The iconic Swiss cartographer Eduard Imhof felt that the base of almost all maps was the relief representation.** He further remarked that the 'proper rendering of relief is one of the primary tasks in cartography' (Imhof 2007). It's hard to argue with that statement as a striking relief representation is both impactful and useful.

**Relief shading usually creates tonal variation across a map using different shades of grey to illustrate the difference between sunlit and shadowed areas.** The grey can be replaced though. Imhof experimented with different stylistic treatment and developed a technique commonly referred to as the 'Swiss style'. Grey tonal relief shading can either be combined with, or replaced by, a bright-yellow tone on the sunlit slopes and blue-grey in the less illuminated areas. The blue simulates atmospheric haze and provides additional pale-blue shadows to add colour to the valley floors, flat land, and lower hilly regions. The yellow gives the impression of warmer colours on the sunlit slopes. The addition of different hues to the basic hillshade creates a connection to feelings of warmth and cold that strengthens how we perceive the relief.

**The Swiss style is not restricted to yellows and blues.** Hans Conrad Gyger, for instance, used yellow-green tones for sunlit slopes and dark green for shadowed slopes on his 1664 map of the Canton of Zurich. You can make some simple modifications to a basic hillshade to enhance it and create an effect that is similar to what you would see on a Swiss-style topographic map or to make different effects altogether. The basic principle remains, though, that echoes what Imhof set out to achieve, which is a representation that '...emphasises the major geographic features, minimises the minor features, smooths irregularities on the slopes, but maintains the rugged characteristics of ridge tops and canyon bottom. You can then simulate an aerial perspective that makes the higher elevations lighter and the lower elevations darker.'

**See also:** Hachures | Hand-drawn shaded relief | Pseudo-natural maps | Raised relief | Shaded relief

Many different colours can be useful in modulating the colour by illumination so it appears sharper at higher elevations or steeper slopes can also bring increased contrast to the map. By using colours as part of the representation of relief, we can also mimic different environmental conditions such as a sunrise or sunset. For this, we might use different pinks or orange hues on sunlit slopes, perhaps with a lower angle of illumination to cast a longer shadow. When combined with other effects such as a darker overall image and misty valley floors, you can re-create a style that mimics a particular time of day, say early morning or dusk.

The real value of a well-styled relief shading comes when it is combined with other techniques to create an overall sense of the terrain. Relief shading can be modulated by elevation or slope. For instance, higher altitudes or steeper slopes can be modified to accentuate their features, perhaps making them darker to emphasise their character. This is a form of generalisation used to exaggerate a feature so that it becomes more visible. Additionally, incorporating other techniques such as contour lines, rock and scree drawing, and picking out features such as glaciers all build an overall impression.

Cross-hatching is an artistic drawing method in which lines of variable thickness and orientation approximate tonal variations associated with relief shading and shadowing. Cross-hatched shadow maps apply cross-hatching to shadowed areas, with the length of these hatched lines based on the distance shadows are cast from point illumination sources at a number of discrete inclinations above the horizon. Thickness of lines increases within areas remaining shadowed at greater inclinations. Hatched lines can be added from multiple sources of illumination to create more diffuse results.

**Opposite**: Alternative styles for relief shading: detailed to generalised; realistic to artistic; and natural to abstract.

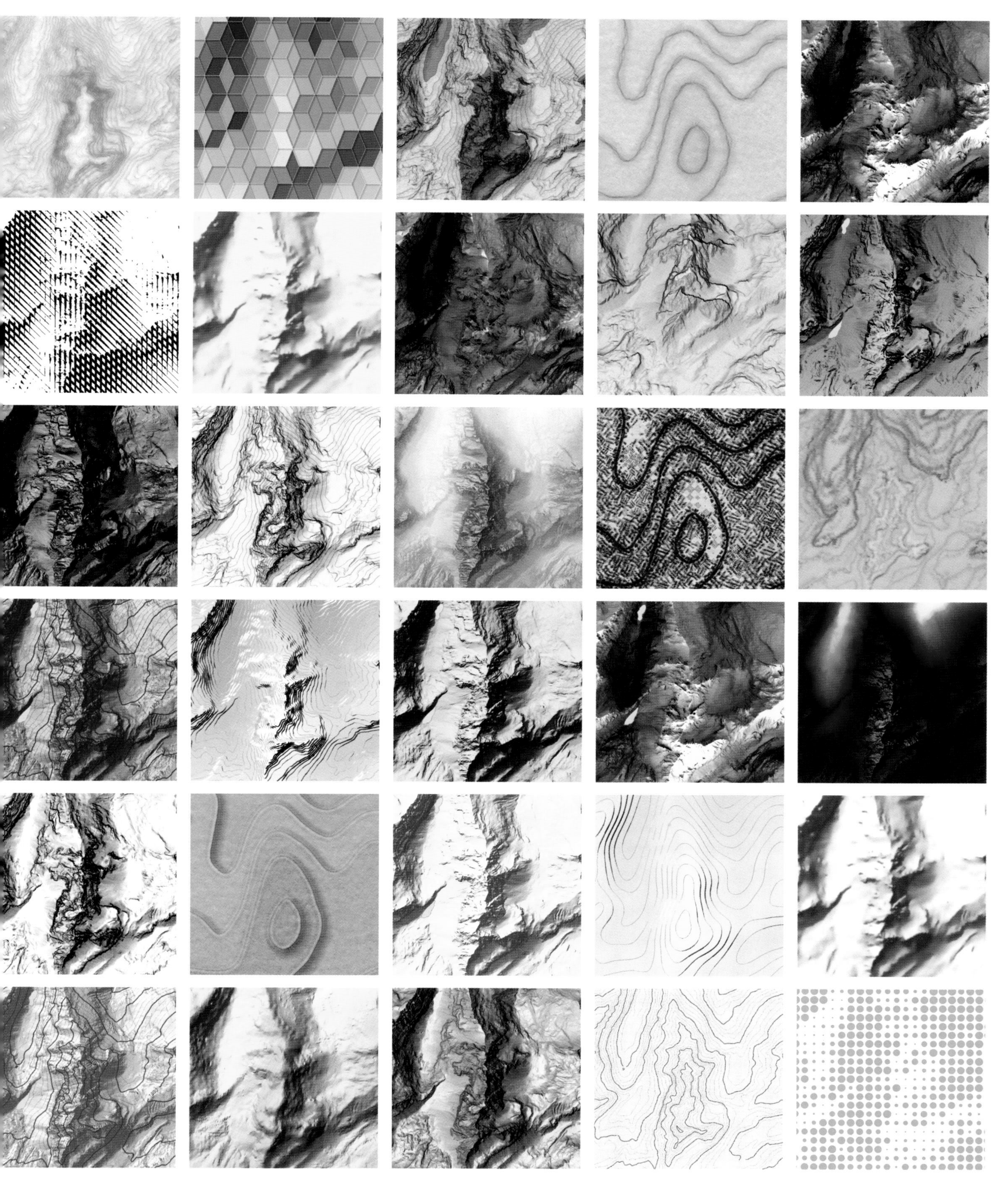

# Symbolisation

## Symbols are used in a variety of ways to imbue meaning in the map and convey a message.

**Symbolisation adds to the processes of generalisation through a range of graphical treatments for representing features on a map.** Symbols are used in various ways to imbue meaning into the many different marks on the map so that the desired message is conveyed to the map reader. It is not enough simply to know how to adapt a feature's geometry as a result of changes in scale—you must also attach meaning to the symbols so that the map user can discern difference in type or quantity of the data that the symbol represents. Choices are based on how map readers perceive, and then cognitively process the symbols to interpret them in a particular way. Symbolisation can be regarded as the coding of map signs so that generalisation is made visible.

**Symbolisation is predicated on an understanding of feature dimensionality and level of measurement.** Nominal data is often referred to as *qualitative* since it is used to measure difference between data items. The remaining levels of measurement are often referred to as *quantitative* since they give a numerical measure of magnitude associated with the item. Once you know how a feature is represented and measured, then you turn your attention to designing a symbol set appropriate for that particular combination.

**Symbol complexity can be both useful and problematic.** Symbols can be designed to portray many different characteristics at once, but with increasing complexity comes the risk of overloading the symbol and confusing your map reader. Symbols tend to have a natural point at which they become too complex to understand, and that is likely to vary depending on the mapping task at hand. The simple rule of thumb is to keep the symbols as simple as possible.

**Symbols are usually geometric but they can also be mimetic or literal.** Mimetic symbols take on a shape or character that looks like the phenomena it is mapping. For instance, an airplane is often used as a symbol for an airport. Type on a map is a literal symbol since the meaning is directly encoded in the words themselves.

**See also:** Consistent denotation | Information overload | Semiotics | Symbols | Varying symbols

Consider a set of circles of different sizes and reflect on what the symbols might convey considering that each circle represents a single feature at a unique location.

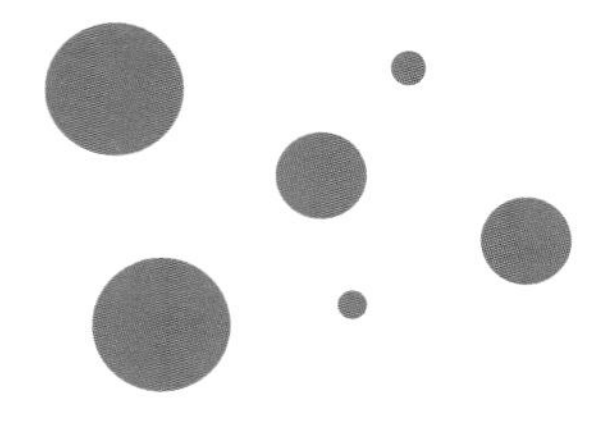

There are two characteristics that can be discerned—spatial arrangement and difference in character. The circles are different in size so you should have inferred from this that whatever exists at the locations differs in size or quantity in some way.

Size is a visual property that is recognised by the eye–brain system and processed in a particular way. Larger means more. Size is one of a number of properties of symbols that, collectively, are referred to as visual (or retinal) variables. By understanding how to apply visual variables effectively, we are better able to translate data into a meaningful map.

**Opposite:** The relationship between feature dimensionality and levels of measurement and the visual variables we might use to symbolise them.

| Feature dimension | Encoding | Qualitative | | Quantitative | |
|---|---|---|---|---|---|
| | | Nominal<br>(difference) | Ordinal<br>(ranked) | Interval<br>(quantity, value) | Ratio<br>(proportional, relative) |
| Literal<br>(words, numbers) | Single | airport, stadium, factory | small, medium, large | 5 ton<br>100 people<br>20 births | 5 ton per sq km<br>100 people per minute<br>1.2 births per family |
| | Multiple | City<br>*River*<br>Monument | **CITY**<br>Town<br>Village | **Nottingham** 100k—500k<br>Newark 10k—99k<br>Stapleford 0—9k | **Nottingham** 311,500 people<br>Newark 27,700 people<br>Stapleford 1,979 people |
| Point<br>(geometric, pictorial) | Single | **Shape**<br>airport, emergency phone, picnic site | **Crispness**<br>low, medium, high | **Value**<br>< 19, 20–29, 30+ | **Size**<br>50%, 75%, 100% |
| | Multiple | **Shape & Hue**<br>international airport, regional airport, airport (disused) | **Size & Hue**<br>small, medium, large<br>train, boat, plane | **Shape & Value**<br>170, 130, 65 | **Hue & Arrangement**<br>units sold, 2x, 4x<br>1st quarter, 2nd quarter |
| Line | Single | **Shape & Hue**<br>road, railway, river | **Shape, Size & Hue**<br>motorway, main road (dual), A road, B road, track | **Size**<br>10 vehicles, 10 vehicles, 10 vehicles | **Size & Transparency**<br>1 per hour, 4 per hour, 8 per hour |
| | Multiple | **Shape & Value**<br>railway, abandoned railway | **Orientation**<br>steep incline, moderate incline, no incline | **Shape, Orientation & Size**<br>45 miles W by air, 70 miles E by road | **Size & Orientation**<br>base speed, 2x speed, 4x speed |
| Area | Single | **Shape & Hue**<br>lake, building, forest | **Resolution**<br>smallest, medium, largest | **Saturation**<br>500, 1,000, 2,000 | **Size**<br>25%, 75%, 100% |
| | Multiple | **Hue, Shape & Arrangement**<br>industrial, greenbelt, urban | **Size**<br>class 1, class 2, class 3 | **Size & Transparency**<br>100, 500, 1,000 | **Value**<br>0–9%, 10–19%, 20–29% |
| Volume | | **Hue**<br>vote A, vote B, vote C | **Transparency**<br>third, second, first | **Height**<br>5, 50, 500 | **Size**<br>$10m^3$, $100m^3$, $1,000m^3$ |

# Symbols

## Symbols are small graphic representations and support varied uses.

**Symbols serve many functions on maps and graphs.** Symbols can encode multiple pieces of information including location, type, and quantity. Most symbols can, at their least, designate location simply through positioning on the map. Geometric or mimetic (pictorial) symbols are commonly used to locate data points on a map but also have multiple uses as other graphical elements of the map.

**Symbols often convey quantitative or qualitative information.** By changing the size, hue, shape, or value of a symbol, there is a large array of design opportunities to encode data. On both maps and graphs, varying the size of a symbol encodes magnitude. Repeating symbols in a systematic way achieves a similar result, proportional to the data values. Shape and hue is often varied to encode difference between feature type for points and lines. Mimetic symbols are also used to summarise or minimise textual elements or overcome language differences.

**Symbols have multiple uses on graphs.** For multiple data series, geometric symbols identify different series. On some charts (e.g. a flow chart), different-shaped symbols are used to differentiate by type, often as an enclosure for other details. Using symbols on charts in this way builds structure and clarifies overall patterns. Symbols are useful for highlighting detail. The most common use is the bullet point to differentiate between statements but arrows or other shapes are useful as a component of charts to emphasise certain aspects.

The form of a symbol can become the symbol itself. A simple example is in repeating shapes that collectively build up to a whole and a further piece of information. Isotype does this through repeating symbols, each one signifying one or more data values, the sum encoding the whole. Stacked cubes can communicate the same multiplicity.

Many pieces of information can also be encoded into the pictorial design of a symbol. For instance, sparklines provide a way of giving a quick representation of numerical or statistical information. Developed in the early 1980s by Edward Tufte, though not named 'sparkline' until 2006, the original sparkline takes the form of a small line graph designed to show change in time-series data. The function is to reduce a graph to a symbol or icon, a small, high-resolution graphic that can be embedded in a context among other words, numbers, or images, including maps. They contain none of the usual graphical furniture such as axis lines. The general shape is highly condensed and small enough to be used as a symbol within text or on a map. Several sparklines can be arranged in a small multiples array to support comparison.

Other chart forms such as column or bar charts or radar charts can also be formed into small versions and used in a similar way. They can be arranged to create a cartogram with each miniature graph representing one area.

Fonts can often be used to build mimetic symbology into a map. Numerous fonts exist where each of the normal alphanumeric characters is replaced by a graphic. Dingbat symbols (so named after characters used in typesetting known as *printer's ornament*) are a typical example but there are numerous others specific to different industries and which support a wide array of cartographic symbol needs.

**See also:** Information overload | Isotype | Pictograms | Semiotics | Symbolisation | Varying symbols

## Major functions of symbols

### Convey quantity

Sizes, shapes, and colours are varied relative to the magnitude of values the symbols represent. Quantity can also be shown by relating the number of symbols to the value.

### Differentiate

Symbols can vary in shape to show how they vary among other features of different types. Location is also encoded through position, and symbols often encode additional detail such as size.

### Indicate

Small (often geometric) point symbols are used to designate location or data points. Lines indicate linear features. Polygons designate boundaries of areas of the same type or as enclosures of a group of other feature types.

### Enclose

Symbols are often used to surround text or as components in graphical displays. They provide structure to an overall display and make distinct objects easier to see, read, and connect.

### Highlight

Symbols are often used to accentuate, or draw attention to a particular part of a map or graphic. They can provide a way of guiding the reader or to differentiate major aspects from subordinate aspects.

### Form meaning

The symbols themselves can often convey information literally. Comparison of icons can support the communication of more complex information, or symbols can be used to visually encode data.

## Gall-Peters projection

James Gall, 1885

Arno Peters, 1974

*'Wet, ragged long winter underwear hung out to dry on the Arctic Circle.'* So said no less a person than Arthur Robinson, the doyen of 20th-century cartography (Robinson 1985, 104) of the Peters map projection. And which was itself used as the basis for this wonderful cartoon by Wesley Jones.

In private, Robinson accused Peters of making a 'fetish' of the map and called Peters a 'crack pot' (J. Snyder Papers, Library of Congress). Indeed, the map does look awkward, even though of course it is perfectly correct. But who said that good design is only about aesthetics? Cannot 'good design' also include that which gets us thinking, or which challenges preconceived notions?

Part of design too is how much the map travels, and the Peters map cheekily made its way on to the popular US TV show *The West Wing*. Appropriate, considering that Peters designed a rectangular map for a screen-obsessed world.

Of course, the map's origin is disputed as James Gall had published his formulae and map in 1855 (shown as inset). The projection is often now regarded as the *Gall-Peters projection*.

—Jeremy Crampton

A.P. LAUNDRY SERVICE

# Temporal maps

## Showing change or time-series cartography.

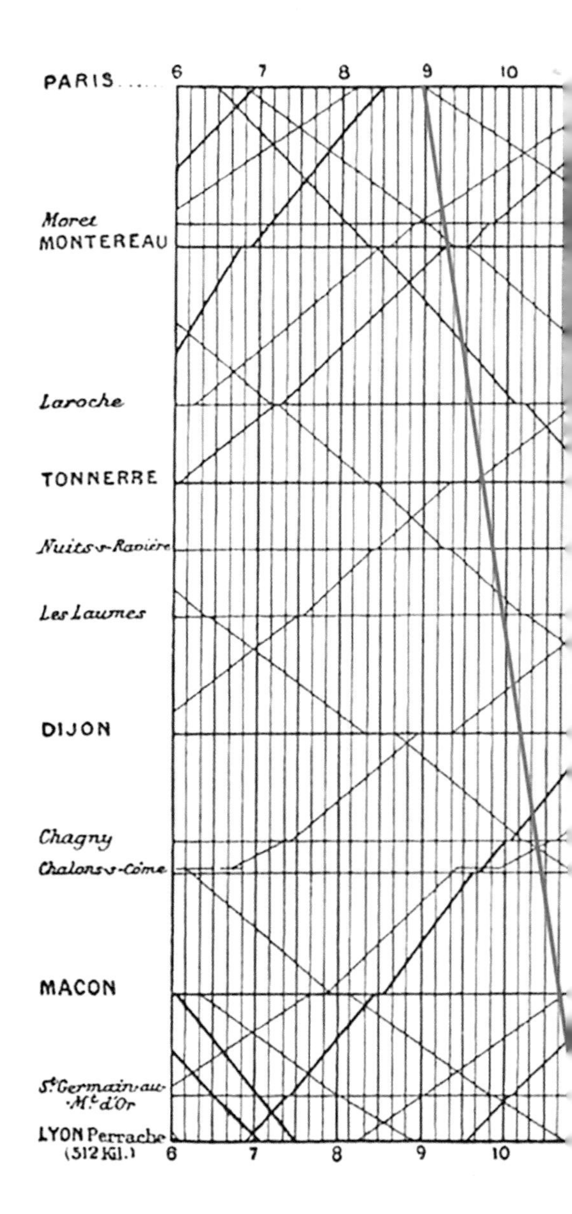


**Every map has a temporal component.** For many, it's a slice of time represented as a static image. It may be a map that combines multiple sources of data collected at different times or a predictive map of a future point in time. Temporal units can represent weeks, months, years, or any other delineation. In many web maps, the data might even represent real-time data feeds such as river flow or social media feeds. Other temporal maps might be used to try and understand how patterns or processes change such as migration, trade, weather, or geology, all across very different time-scales.

**Often, the cartographic challenge is to find a way to show change for time-series data.** One way is to use a single map to encode many variables, one of which is change. Building a temporal narrative in a single image is challenging. Alternatively, small multiples can show many slices of time in a sequence. Each map shows a particular map state using the same visual variables. It requires spatial deduction by the reader to interpolate the passage of time, and assessing velocity and rate of change between slices can be difficult. Animation shows a sequence of snapshots of time. A difficulty with animations is in memory recall and change blindness. Poor short-term memory means we find it difficult to retain different images from different periods in the animation. Animations should incorporate some way of monitoring progress such as a time line, a clock face, or a changing numeric counter. This helps accommodate change blindness and also navigation to specific points. Order, play rate, and duration can often be modified to move from slice to slice in equal time or changed to different periods.

**The type of change to be mapped creates different symbology needs.** Spatio-temporal data can display feature-, attribute-, and location-based change. Features can appear or disappear between time periods. The character of a consistent feature may change either qualitatively or quantitatively. A feature may move either singularly or as part of other movements.

**See also:** Animation | Interaction | Isochrone maps | Space-time cubes

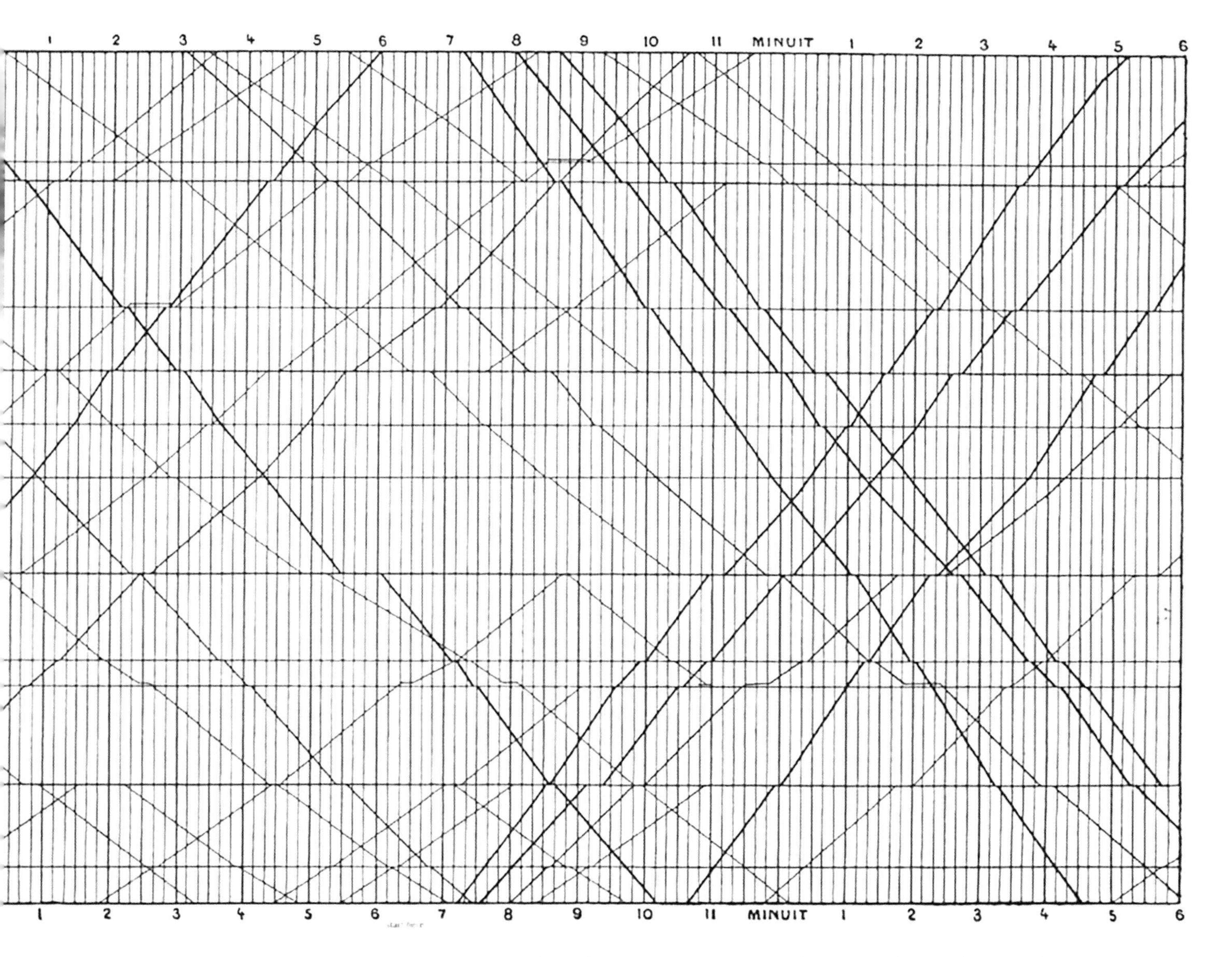


**Above:** The map illustrated here is E. J. Marey's *Schedule of Trains from Paris to Lyon* (designed by Ibry), 1880. Space and time are illustrated on the two axes. Arrival and departure times are linearly scaled on the x-axis. The locations of stations (in sequence) are indicated on the y-axis. These are scaled relative to their real distance apart. Space has been straightened yet the connectivity and relative distance between stations remains.

Each line drawn across the map illustrates a particular train. The line provides a means of seeing both space and time by examining the x- and y-axis values that the line intersects. The speed of the train can be assessed by comparing the steepness of different lines. A faster train will be a much steeper line as it traverses a longer length on the y-axis across a shorter length on the x-axis. Horizontal interruptions in the otherwise angled path of the train's plotted line show periods when the train is stopped—usually in a station. The intersection of two lines shows both the time and relative location of trains that are passing in opposite directions.

Marey's original has been overprinted with the current 9 a.m. TGV Paris to Lyon service, which has reduced the time of the journey to just under two hours. The example provides rich detail in a consistent but graphically efficient form.

# Texture

## The coarseness, arrangement, and spacing of a symbol's graphical marks.

**Textures of a symbol (usually areal) can be modified through changes to the coarseness of constituent graphical marks as well as their arrangement and spacing.** Historically, *texture* referred to the pattern of marks commonly used in photomechanical production such as a halftone screen tint of 50 points per inch. It was defined as the number of separate graphical marks contained in a unitary area, and different textures were used to mimic shading. A null texture would be where individual marks are imperceptible because they are so small. Coarse textures make use of large constituent marks though at this size limits may extend into ambiguity and confusion with single point symbol marks.

**Spacing between marks can be used to modify the relative density of marks, and arrangement can be both regular or irregular.** For instance, a line of uniform weight can be used within a circle to represent a point feature but the gap between each line can vary to create different textures. Spacing is a poor choice for depicting nominally measured data where qualitative differences are important. The impression given by increasing amounts of ink on the page (or pixels on a screen) much more effectively connotes quantifiable differences. Changing the spacing between marks modifies the texture and is more commonly used to represent ordinal or numerical data although it is less used in conventional cartography because alternative visual variables are considered more aesthetically pleasing.

**In its physical sense, texture can also be used to support sensory mapping.** That is, the marks on the map are texturised. Raised relief or 3D printing is often used to create maps for the partially sighted where the marks can be 'read' by touch, including the use of Braille rather than alpha-numeric typefaces. Content is normally sparse compared with a conventional map simply because distinguishing unique features is more challenging. Texture fills are more commonly used as a method of varying symbology since many of the other visual variables cannot be implemented, and texture fills are more easily distinguished through touch.

**See also:** Crispness | Orientation | Pattern fills | Sensory maps | Shape | Size

Historically, many publications demanded monochrome reproduction, so in order to display qualitative or quantitative difference among different areal symbols, different textures of symbol were required. With the shift from photomechanical to computerised production and the increase of full-colour printing, the use of texture has declined.

Texture might have been defined through the use of coarse screens (e.g. 50 ppi) to fine screens (e.g. 300 ppi) and a range in between to give a choropleth shading scheme with different perceptual steps using intermediate screens. Similarly different line widths, their arrangement, and spacing can be used well in monochrome designs to create a range of different textures.

In certain combinations, texture can create uncomfortable visual effects. This vibratory (or moiré) effect results from the way our eye responds to regular patterns. Resonance at a retinal level and your brain's difficulty in seeing figure and ground between the black and white components of a texture creates uncertainty. This vibratory effect can also be seen in line symbols—particularly dashed lines. For point symbols, vibratory effects can arise from internal complexity if the symbol comprises rings of different widths or if many points are organised close to one another.

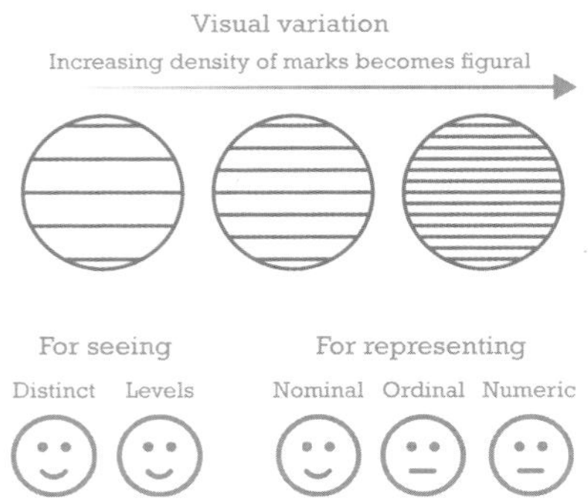

Visual variation
Increasing density of marks becomes figural

For seeing: Distinct, Levels
For representing: Nominal, Ordinal, Numeric

**Opposite:** An extract from the *Tactile Atlas of Switzerland* by Anna Vetter, 2016, which uses raised line work, symbology, and Braille.

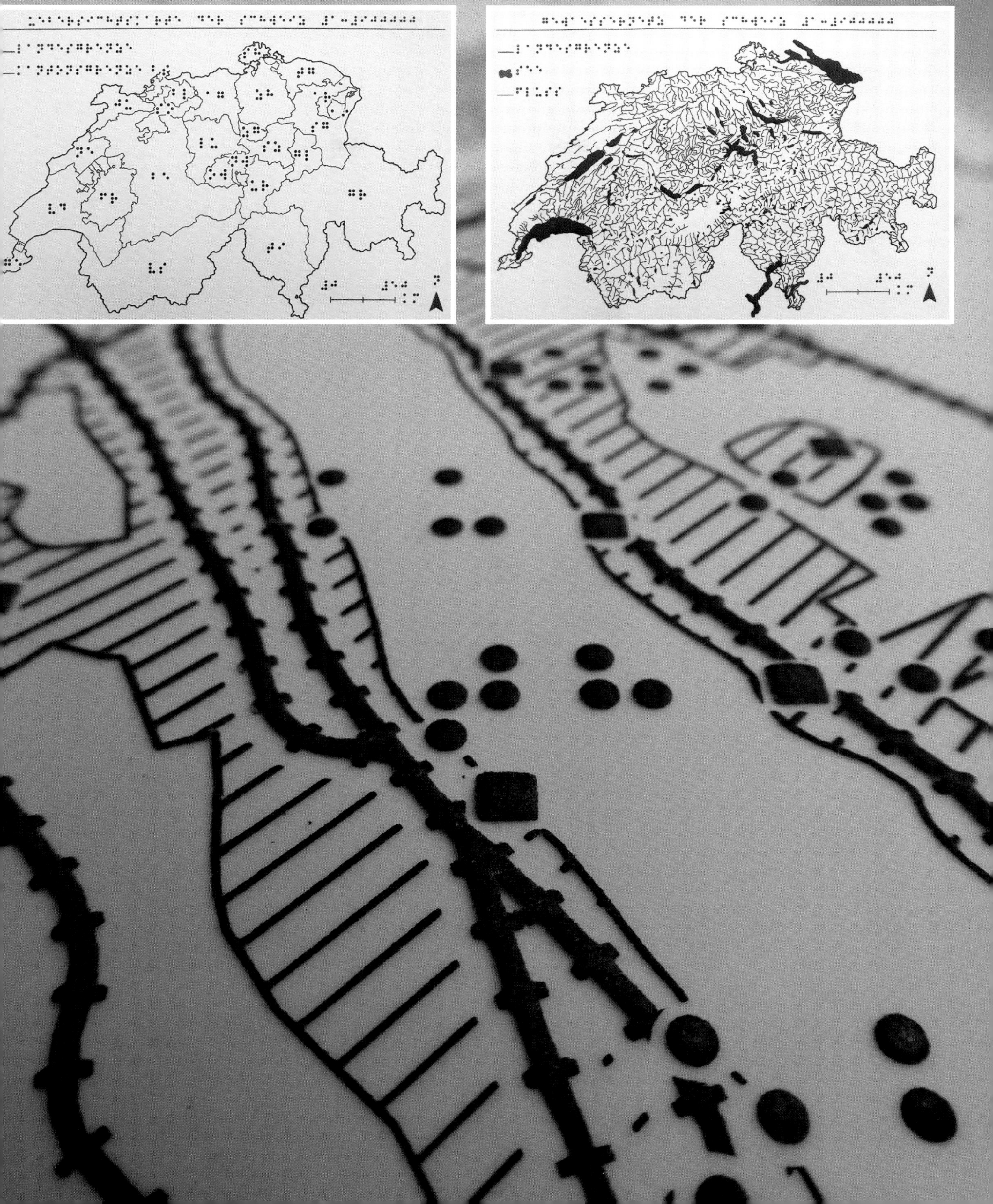

# Thematic maps

## Thematic maps tend to be special purpose and may be a one-off or part of a series focussed on a particular theme.

**Special-purpose maps, as opposed to general reference maps, are termed *thematic maps.*** These maps tend to display a single theme of information—hence the term 'thematic'—and describe a place rather than simply show where something is in space. This mapping may either be by indicating the spatial distribution of a single phenomenon, or the relationship among many phenomena. It involves mapping an abstraction of a physical or cultural phenomenon that might include the distance between objects and their directional relationship, patterns of location, or some magnitude of change, perhaps over time.

**Thematic map types vary considerably but examples include maps depicting precipitation, population distribution, hurricane incidence, and pollution.** Thematic maps tend to be small scale because the geographical distribution of a theme is often most appropriately shown, and accurately depicted, over a large area. Small-scale mapping allows the display of the overall structure of the distribution to become visible.

**Thematic maps may be further subdivided into two distinct types: qualitative and quantitative.** A qualitative thematic map illustrates the spatial distribution of data measured on a nominal scale; that is, data that is similar in kind. For instance, a map showing the location of atmospheric weather stations would be both thematic and qualitative and would illustrate both location and spatial distribution. At a relatively small scale (e.g. country-wide), this map would be designed to give an indication of the general pattern of distribution of weather stations rather than precise geographical position. Quantitative maps display spatial aspects of empirical data. Taking our weather station example, measurements of precipitation may be used to create a map showing the spatial distribution of the amount of rainfall in a 24-hour period. Other thematic maps based on measurements at the weather stations may include temperature and atmospheric pressure. The map displays variation in the theme from place to place, measured on either an ordinal or interval/ratio scale.

**See also:** Atlases | Choropleth maps | Descriptive maps | Flow maps | Mashups

Quantitative thematic mapping requires the transfer of empirical measurements into graphical signs that illustrate the spatial variation in difference between measured values. Essentially, aspatial, tabular data is converted into a spatial form, which adds further to the information by giving the map reader the distance, direction, shape, and location of the theme. This in itself is the purpose of a thematic map: that the transformation of data from tabular to map form aids spatial understanding. If spatial information is either lacking or not required, then there is no justifiable reason for presenting data in map form, and the original tabulated or graphed presentation will usually be sufficient. In transforming data into a map, a thematic map will normally involve the generalisation of the original data, meaning that data recovery is often not a key aim of thematic mapping.

Small-scale thematic maps often have little basemap detail since the map itself forms its own base. A choropleth map, constructed from administrative boundaries, rarely requires topographic detail although it's often useful to place some context above the map such as labels. At larger scales, detail becomes more contextually important.

Possibly the earliest thematic map was by Jodocus Hondius, who mapped dispersion of major religions in his 1607 *Atlas Minor.* Charles Dupin is credited with the first choropleth map in 1826. John Snow's cholera map of Soho, London, published in 1854, has become famous for the thematic mapping of disease distribution, providing the foundation for the study of epidemiology as well as anticipation the principles of geographical information systems and the overlay of mapped information.

**Opposite:** Extract from *Maps Descriptive of London Poverty* by Charles Booth, 1898-99, is an exquisite thematic map overlaying his surveyed data on life and labour in London atop a Stanford street map for context.

CHARING CROSS RAIL
FOOT BRIDGE
STAMFORD ST
CHRIST CHURCH
ST PETER
ST ANDREW
ST SAVIOUR
Southwark
ST JOHN THE EVANGELIST
ALL HALLOWS
NELSON SQUARE
ST MICHAEL
ST THOMAS
ST PAUL
ST ALPHEGE
ST GEORGE
THE MARTYR
Southwark
BOROUGH ROAD
HOLY TRINITY
ST THOMAS HOSPITAL
EMBANKMENT
HOLY TRINITY
ASYLUM FOR THE BLIND
BETHLEHEM LUNATIC ASYLUM
Ex. Par.
LAMBETH PALACE
ST JUDE
ST MARY
Lambeth
ST MATTHEW
EMMANUEL
ST PHILIP
Kennington
ST JOHN
Walworth
ST MARY THE LESS
ST MARY
Newington
ST MARK
East Str Walworth
ST PETER
Vauxhall
ST ANSELM
Kennington Cross
ST JAMES
Kennington
THE STREETS ARE COLOURED ACCORDING TO THE GENERAL CONDITION OF THE INHABITANTS, AS UNDER:—
Lowest class. Vicious, semi-criminal.
Very poor, casual. Chronic want.
Poor. 18s. to 21s. a week for a moderate family.
Mixed. Some comfortable, others poor.
Fairly comfortable. Good ordinary earnings.
Middle class. Well-to-do.
Upper-middle and Upper classes. Wealthy.
A combination of colours—as dark blue and black, or pink and red—indicates that the street contains a fair proportion of each of the classes represented by the respective colours.

# Threshold of perception

## Minimum size of graphics that can be seen.

**The human eye is only able to see and recognise objects of a minimum size under normal reading conditions.** Minimum sizes define a threshold of perception for individual objects as they might appear in black ink on a white background. The minimum size for the diameter of a point symbol is 0.1 mm and a linear object should be no smaller than 0.06 mm. However, most maps will not support optimal viewing conditions because of other map elements, information density, and colour. Minimum sizes should in practice be modified to:

0.2 mm—minimum diameter of a point symbol
0.1 mm—minimum thickness of a line symbol
0.4 mm—minimum length of the side of a solid symbol
0.6 mm—minimum length of the side of an open symbol

***Threshold of separation*** **is the minimum distance between two graphic elements.** Many map objects will be in close proximity to one another, a natural function of geography in many cases. In order to make each object graphically distinct, there must be a minimum threshold of separation between them of 0.2 mm. This threshold should be used to separate individual topographic features. For thematic maps, the same principle applies, and 0.2 mm should be considered the minimum between thematic symbology such as proportional symbols (e.g. 0.2 mm cut-out mask). Without a minimum threshold of separation, individual map elements will visually fuse together and be seen as a single entity.

***Threshold of differentiation*** **is the minimum difference between symbols of nearly the same type or size.** These guidelines ensure that different map elements remain visually unique. In many instances, maps can show subtle differences but if you make those differences too subtle they become imperceptible. Avoid shapes that are too similar. For instance, using different triangles or mixing circles with hexagons and octagons will all look similar. Avoid rotating symbols such as triangles since the rotation is insufficient to differentiate between types. Avoid symbol sizes that are too similar, and ensure sufficient differentiation between symbols.

Differentiation is important in topographic mapping to make different features individually distinct. It becomes vital in thematic mapping when one of the major cognitive tasks is the assessment of size compared with other symbols of a similar type. The scaling between one symbol and the next in sequence cannot easily be mathematically stated and is also dependent on factors related to the specific map being made but ensuring symbols are individually perceivable is the key.

There are many graphical techniques to assist in making symbols distinct. Line work should be capable of being perceived without ambiguity, and indecisive or poorly generalised line work has a tendency to simply look like bad drawing. Smoothing the geometry of line work helps sharpen an image though care should be taken not to change the course of a line's meaning. The threshold of perception of angular changes in a line is equivalent to two line widths. If angular change is below this threshold, it's likely you can smooth out the geometry. Area objects might also be left off the map entirely if they fall below a certain threshold. For instance, if an area is smaller than 0.4 mm, it's unlikely it can even be drawn with a line so that it still appears to enclose an area. In this case, collapse the line or remove the feature altogether.

When differentiating symbols, our eye actually tends to compare the top edge of the symbol first and use that for comparison. Quick and efficient differentiation is therefore achieved if you vary the top edge of a symbol more than if you vary the other geometries. This is also true when we read letterforms.

**See also:** Focussing attention | Information overload | Seeing

## Threshold of perception

| | | Minimum | 1mm |
|---|---|---|---|
| Point | 0.2 mm | · | • |
| Line | 0.1 mm | — | — |
| Full square | 0.4 mm | ▪ | ■ |
| Empty square | 0.6 mm | ▫ | □ |

## Threshold of separation

0.2 mm between objects

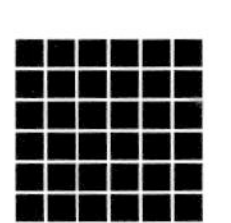

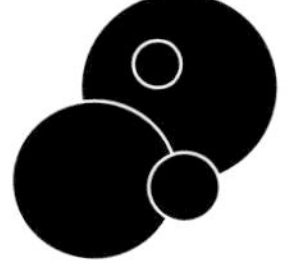

## Threshold of differentiation

Shape

Poor Good

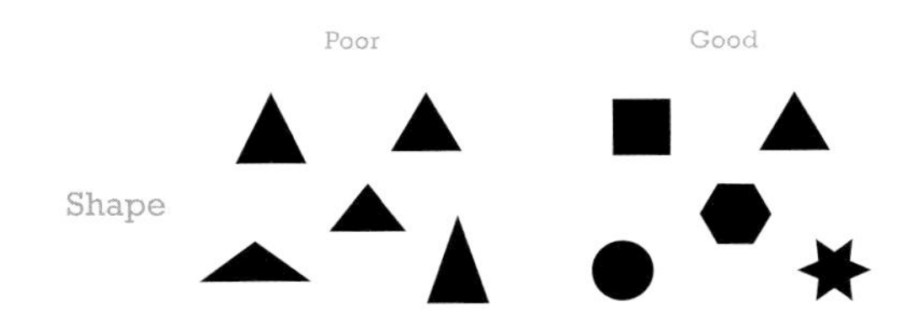

Value

Poor

Good

Size

Poor

Good

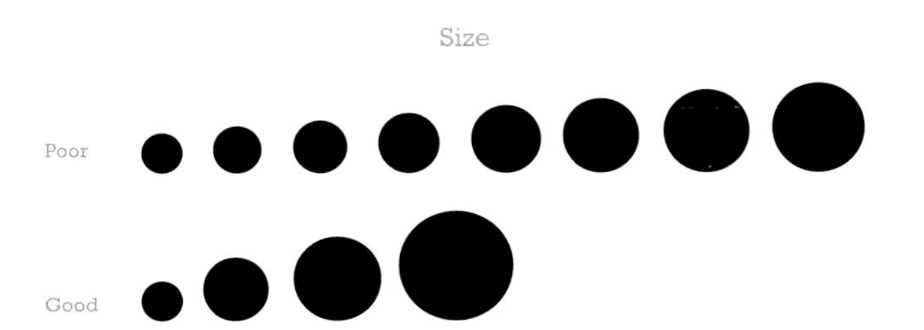

Boundaries

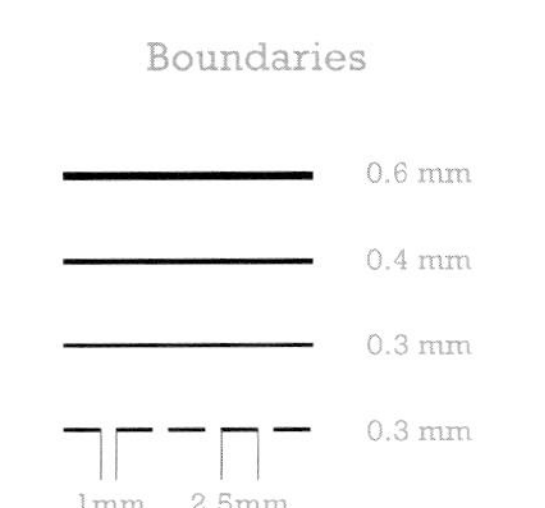


Type

## Threshold of recognition

Each map is designed in a 64 x 64 pixel grid for eventual 75% reduction

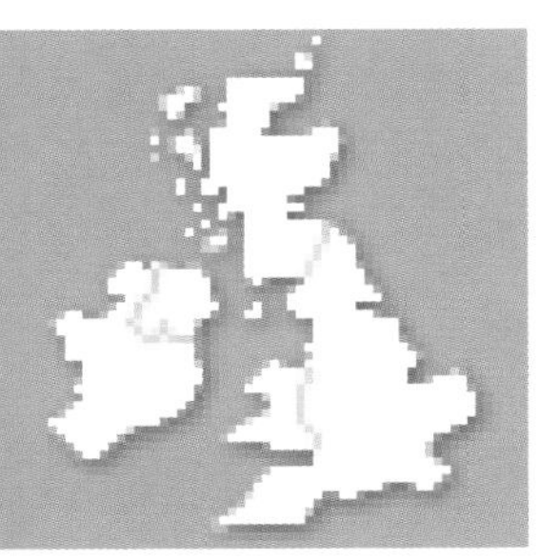

British Isles

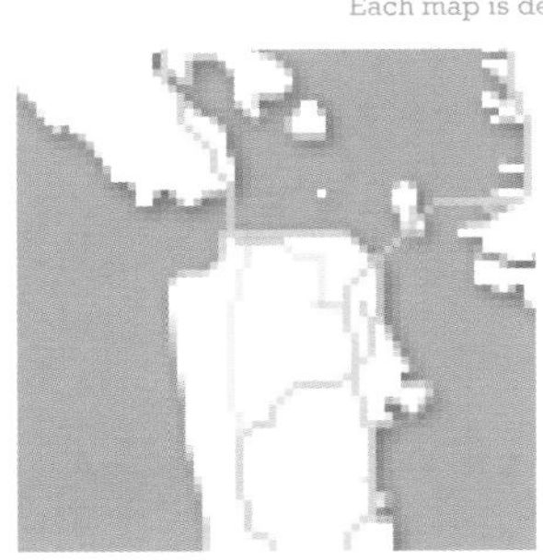

San Francisco

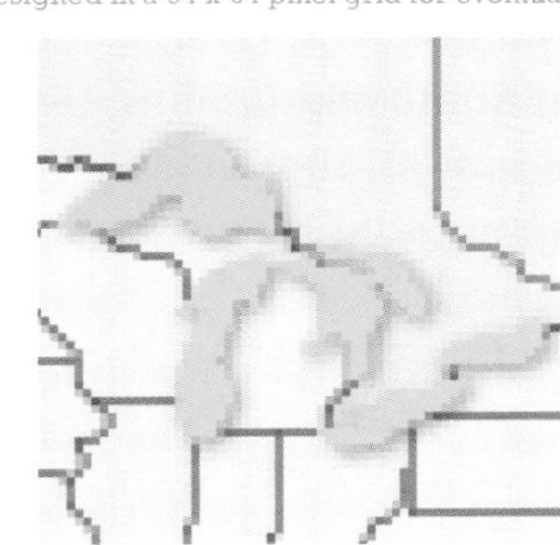

The Great Lakes

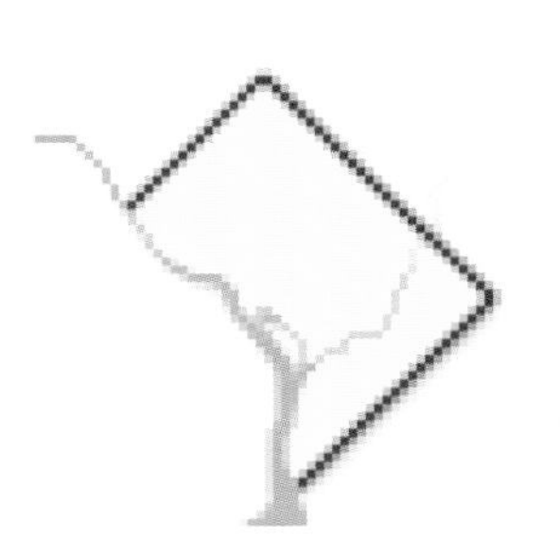

Washington, DC

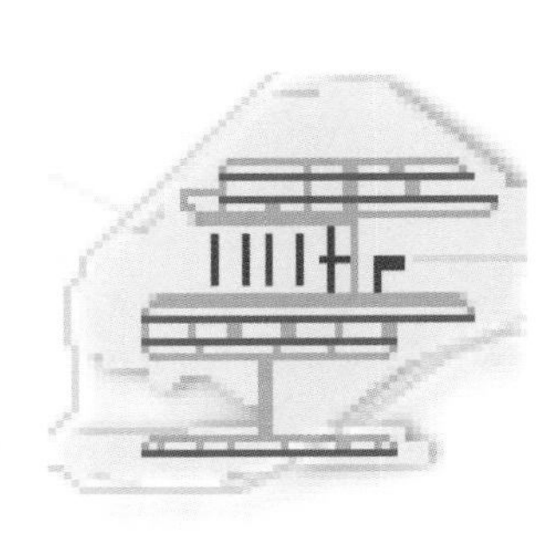

Hartsfield-Jackson International Airport

# Topographic maps

## A reference map that shows features on the surface of Earth.

**Topographic maps show physical and man-made features and their location on Earth.** They are generally described in relation to scale so a large-scale map would be up to 1:50,000, medium scale 1:50,000 to 1:500,000, and small scale 1:500,000 and smaller. Large-scale maps are usually constructed from data collected at a similarly large scale using horizontal datums for accurate measurement. As scale gets smaller, maps are derived, compiled, and generalised to ensure content fits the requirements of the intended scale.

**Topographic maps serve many basic purposes.** Principally they assist navigation and situation. They are an indispensable tool for administration, tourism, education, recreation, conservation, planning, and military purposes. Traditionally, they would be prepared in a map series of individual numbered sheets. There would be a relationship between sheets and also between scales for series made by the same national mapping agency. The series would cover a whole territory though the advent of the web map renders map sheet edges a paper-based anachronism. Topographic maps contain marginalia and legends vital to interpretation. Neatlines, graticules, north points, scale, source, date of publication, revision, and sheet index are all important.

**Relief portrayal is a key component of topographic mapping.** Relief might be achieved using a combination of methods such as contours, spot heights, hypsometric tinting, and hillshading. Additional pictorial representation for rocky outcrops and scree help represent geomorphological features. Elevation is measured against vertical datums so the map can be used to accurately locate vertically as well as horizontally.

**Geographical names distinguish topographic maps from many other map types.** Naming features is indispensable for orientation as well as description. Nomenclature comprises place-names of settlements and administrative areas, and the wide range of natural features, referred to as *toponymy.* Additionally, elevations, contour intervals, and coordinate reference systems are all labelled.

**See also:** Atlases | Basemaps | Descriptive maps | Hypsometric tinting | Reference maps | Shaded relief

Natural features include hydrography, vegetation, and natural land cover. Hydrography provides a skeleton for the map and a key focus for assessing the level of generalisation at each map scale. Coastlines, rivers, and lakes are usually shown in blue, but many different symbol treatments differentiate between features. Forest and woodland is generally shown in green fills at smaller scales and subdivided by type at larger scales, possibly with the inclusion of pictorial elements. Dunes, gravel, moraines, and salt pans are the topographic equivalent in desert regions.

The proportion of man-made features will be a direct function of the level of economic development of the mapped area. For many countries, only small-scale maps are available for areas with little or no economic development. Settlements, communication networks (road, rail, ports and so on), administrative boundaries, and a range of individual features are commonly shown. Depending on scale, urban areas may be an area fill or, at larger scales, individual buildings such as churches and factories shown. Linear features differentiate between type, such as road and rail and also class, such as road hierarchy. Administrative boundaries use a mix of styles to denote different levels, and pictorial symbols often show the location of specific features such as monuments, hotels, and campgrounds.

The importance of standardization of geographical names on maps has gained support from the United Nations, which established a Group of Experts on Geographical Names (UNGEGN) in 1960. The practical and scientific exchange of experience in toponymy has driven the establishment of geographical names authorities, which collect and organize names in gazetteers.

**Opposite:** Extract from *The Times Comprehensive Atlas of the World,* 14th edition.

PLATE 100
PLATE 99
PLATE 106
BRITISH COLUMBIA
ALASKA
WASHINGTON
CANADA
U.S.A.
PACIFIC OCEAN
COAST MOUNTAINS
ROCKY MOUNTAINS
CARIBOO MOUNTAINS
MONASHEE MOUNTAINS
Fraser Plateau
Chilcotin Ranges
VANCOUVER ISLAND
Queen Charlotte Sound
Queen Charlotte Strait
Juan de Fuca Strait
Strait of Georgia
Clear Hills
Prince George
Prince Rupert
Kitimat
Terrace
Smithers
Houston
Burns Lake
Vanderhoof
Fort St. James
Quesnel
Williams Lake
100 Mile House
Kamloops
Vernon
Kelowna
Penticton
Lillooet
Whistler
Squamish
Vancouver
North Vancouver
Burnaby
Coquitlam
Richmond
Surrey
Langley
Abbotsford
Chilliwack
Nanaimo
Victoria
Saanich
Port Alberni
Port Hardy
Powell River
Courtenay
Campbell River
Bellingham
Dawson Creek
Fort St. John
Chetwynd
Tumbler Ridge
Mackenzie
Grande Prairie
Jasper
Valemount
Revelstoke
Salmon Arm
Williston Lake
Wells Gray Provincial Park
Willmore Wilderness Provincial Park
Tweedsmuir Provincial Park
Conic Equidistant Projection
1:2 500 000
MILES 0 25 50 75 100

# Transparency

## Blending symbols with the background, by varying transparency, modifies how a symbol is seen.

**Transparency describes the graphical blending of a foreground symbol with either the background or with other symbols that occupy the same space.** Transparency has become a fundamental way for cartographers to extend the palette of available options for designing symbols. Its principal role is to modify a symbol to change its relative position in a visual hierarchy. By increasing the transparency of a symbol, the effect is to lower its position in a visual hierarchy. The converse also holds true, that fully opaque symbols will rise in visual hierarchy.

**Transparency can be used to blend symbols into the background.** Maps will usually be designed atop a background colour. The map background might be white, or a light colour, to distinguish the figural components (of the map) from the background or surrounding 'white space'. By increasing transparency, it's possible to subtly blend map symbols into the background to lower their relative importance in the overall design. This has the effect of muting the original symbol and might equally be based on data values rather than arbitrary application.

**Multivariate symbology can be constructed using transparency.** Bivariate and multivariate thematic maps comprising multiple layers can be blended if each layer is designed with constant transparency. For instance, a bivariate choropleth map comprising two layers, with each layer having a different shading scheme and set to 50% transparency, will blend the colours accordingly. Limiting the number of classes and providing clear legends are crucial for interpreting such maps. A similar process is used to create a value-by-alpha map, comprising a traditional choropleth map with a second layer placed above that varies a further dataset from opaque to transparent. This helps bring focus to some of the underlying choropleth while other areas tend toward increased opacity as they blend with the choropleth background.

When Jacques Bertin first proposed his original retinal variables in the 1960s, photomechanical production techniques didn't support the implementation of true transparency. Maps were generally limited to a few colours, and halftone patterns were used to create blended edges and modify the apparent transparency of a symbol's fill. A halftone pattern would comprise small dots of a pure colour spaced across the white background of the paper, which at reading distance gave the illusion of a lighter (alluding to a more transparent) colour. Modern design and production techniques now allow full-colour mixing as well as the modification of the alpha channel that controls transparency.

Transparency or opacity? In fact, these two descriptions are related. Transparency is the quality of being transparent, and relates to the amount of light passing through an object. 100% transparent is fully transparent. Opacity is the quality of a symbol being opaque, or not allowing light to pass through. So 100% opaque is a fill that allows no light to pass through. In cartography, the term *transparency* tends to be used, but in web design more generally, you'll likely hear the term *opacity*. Just remember that transparency is simply the inverse of opacity.

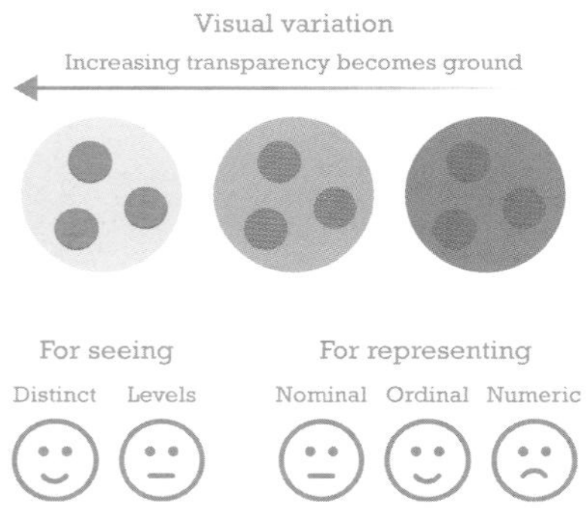

Visual variation

Increasing transparency becomes ground

For seeing: Distinct, Levels

For representing: Nominal, Ordinal, Numeric

**See also:** Crispness | Pseudo-natural maps | Value-by-alpha maps

**Opposite:** Extract from *Lights On Lights Off* by John Nelson, 2017.

Turkey
Georgia
Russian Federation
Cyprus
Armenia
Azerbaijan
Lebanon
Syria
Palestinian Territory
Kazakhstan
Israel
Jordan
Uzbekistan
Turkmenistan
Iraq
Kyrgyzstan
Tajikistan
Iran
Kuwait
Afghanistan
SaudiArabia
China
Qatar
Pakistan
UnitedArab Emirates
Nepal
Oman
Bhutan
Yemen
India
Bangladesh
Myanmar
A look at where nighttime lights have become Brighter or Dimmer between 2012 and 2016

# Treemap

## Technique to show hierarchical components of a whole dataset.

**A treemap is a form of tree-structured diagram.** It displays constituent parts of a whole dataset hierarchically and can be used on both spatial and aspatial datasets. A treemap layout is normally rectangular with the outer boundary of the diagram representing 100% of the theme being mapped. The internal area is segmented into different-size rectangles that each represent some part of the whole. The resulting appearance is of a tiled pattern. Hue can be used to show categorical differences between components or some secondary variable. A value-based sequential scheme can be used to show differences in a secondary empirical variable.

**The position and dimension of each component in a treemap is determined by a clustering algorithm.** The result is space-optimising but which also positions elements that are more closely connected near one another. In this way the treemap takes on a structure where more alike things are nearer in the diagram than those that are less alike. The result gives nested rectangles where each branch of a tree is assigned a branch, and sub-branches become smaller nested rectangles within the parent branch.

**A treemap provides a good visual representation of the constituent parts of a whole.** A quick scan of a treemap reveals high-level groupings of more alike components. Comparisons among those that share similar size characteristics can be easily read, and because size is ordered, it's easy to see ranked relationships between variables. Encoding of a second variable using colour will allow the reader to see if that variable follows the sizing pattern or not, establishing a visual association if evident. As with other diagrammatic representations of the whole (e.g. a pie chart) it's difficult to establish precise magnitudes though interactivity can support this through hover or click events.

**See also:** Binning | Cartograms | Graphs | Voronoi maps | Waffle grid

Treemaps tend to be more commonly used online where the reader can drill down into each tile to discover related information. This is particularly important for those extremely small tiles that are unable to hold labels easily. Labels themselves can often compromise the treemap since only larger tiles will be capable of showing a label and will, by definition, appear more prominent.

Treemaps are an extension of mosaic plots, which show the relationship among data for two or more qualitative variables. They are stacked column graphs although columns are different widths and each block in the graph is proportional to one another. Treemaps were developed by Ben Shneiderman in the early 1990s to include a recursive element which supported hierarchical data. The initial algorithm has been iterated to produce representations where each tile is as square as it can be.

Geographical space is often an inconvenient container for data because of the constraints of size and shape, and with the smallest areas containing the most diverse and important data. Spatial treemaps allow you to show relationships across the data and within the data at different scales.

**Opposite:** *Rectangular Hierarchical Cartograms for Socio-Economic Data* by Aidan Slingsby, Jason Dykes, and Jo Wood, 2010. Here, 1,526,404 postcode units in Great Britain are shown, sized by population and arranged so that geographical relationships and postcode geography hierarchy are maintained. The map is richly coloured according to a socio-economic classification comprising seven supergroups split into 52 subgroups that shows data concurrently. The map is beautifully arranged allowing patterns in the vast amount of information to become clear at local, regional, and national scales. In a single map, they have managed to effectively display detailed information about 60 million people recorded in 40 census variables in over 1 million places.

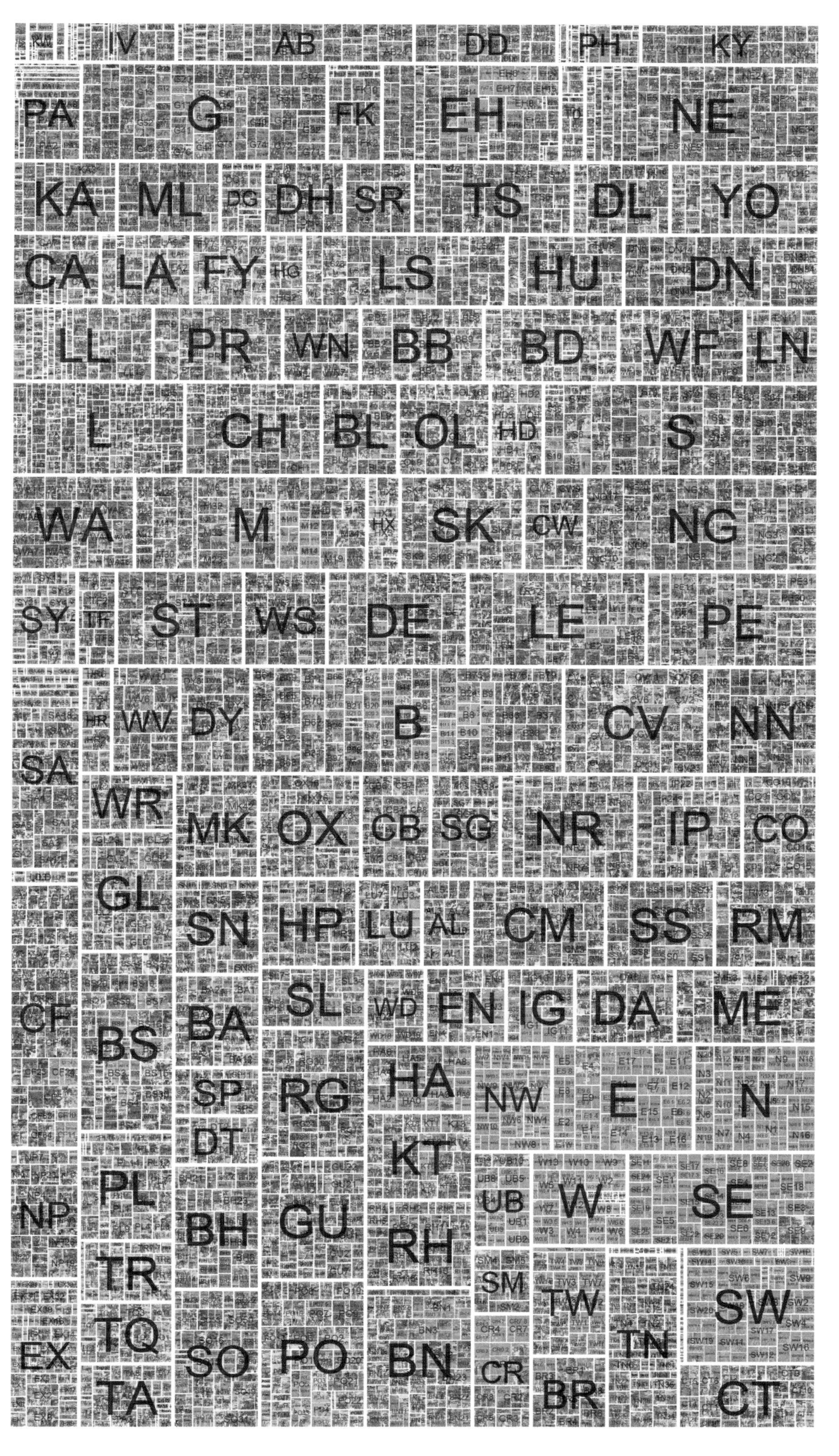


Hierarchical Rectangular Cartogram of Output Area Classification by Great Britain Postcode

The Output Area Classification (OAC) is shown below. Each area of the 3-level hierarchy is sized and ordered by population (i.e. the largest population proportion is classified as the "Prospering Suburbs"). We map these by unit postcode (1.52 million) on the left in a spatial hierarchy where each unit postcode element is sized by the number of residential postal delivery points used as a proxy for population (60,000 have no residential population so are not shown). Note the homogeneity of inner London postcodes (W, E, SE, SW) and the centres of large cities (B, G, NE).

Prospering Suburbs

- Semis
- Older Families
- Thriving Suburbs
- Younger Families

Typical Traits

- Least Divergent
- Settled Households
- Young Families (Terrace)
- Aspiring Households

Blue Collar

- Younger
- Older
- Terraced

Constrained by circumstances

- Older Workers
- Public Housing

Countryside

- Village
- Accessible
- Agriculture

Multicultural

- Asian
- Afro-Carribean

City Living

- Settled in City

# Type colour

## Type colour does not operate independently of the feature type and the background.

**Connotations and contrast are the key approaches to determining colour for typographic components.** The colour of type is often associated with the feature itself. Although black is the most legible colour for type, other colours such as blue for water labels, brown for contour labels, and green for forested areas or parkland are common colours used to emphasise a connotation with the feature. It's also common to see halos and masks of vibrant colours used to raise typographic components in the visual hierarchy of the map.

**The legibility of text on a map is a function of the background colour and the amount of visual contrast between the background and the type.** Most text on a map is dark (or black) on a light background. Black type on a white background is the most legible, but as the tonal value of the background approaches the tonal value of the lettering, so the type will be less distinguishable and less visible. This is a problem for map lettering since type is usually placed over large areas (for example, to show areal extent) or used to classify the same feature type (e.g. towns of a similar size). This results in a wide range of backgrounds over which type is placed, which will adjust the apparent hue and value of the lettering uniquely across the map.

**The intrinsic importance of a particular level of type can be modified simply because of the underlying map detail.** This is not something that can be easily overcome but it helps to appreciate the problem when designing and placing type effectively. The impact of this problem can be reduced by ensuring type is not placed over those areas or features that are most problematic. Label placement therefore becomes crucial in mitigating these issues and to achieving a consistent appearance of type across the map. It simply relies on good editorial practice and map checking to ensure that your typographic components are working in sympathy to other map detail and consistently across the map. Manual modifications to text placement are common to overcome local graphical problems.

**See also:** Colour in cartography | Elements of colour | Guidelines for lettering

Be aware of the various background colours in your map, and design your type colour accordingly. Four or five colours are typically more than enough across a map. Designing type colour for reference maps is mostly darker text on a lighter background but varying this approach can bring interesting results. For instance, using dark grey rather than pure black can lead to a more subtle effect and not promote the labels quite as much. Making the labels slightly transparent or giving them a drop shadow or a slight outer glow can also achieve different appearances.

Labelling features on imagery can be a challenge. For example, image quality may vary through and across different scales and resolutions of data. Seasonal and time of day variation in images (snow, shadows, and so on) and geographical differences such as deep-green vegetation, grey urban areas, and desert browns/golds also present difficulties for consistent colouring of labels. Labelling across imagery usually requires lighter labels to contrast across darker imagery. This might literally be through using lighter colours or by applying semi-transparent halos around labels to give them contrast. Be aware of the geography of your imagery and use it to your advantage. For instance, many urban areas have particular colours such as blue rooftops in Moscow or red rooftops in much of Spain. The colour of type is likely to vary more because of geography when labelling imagery because it is an abstract topographic map of the same area.

Using a colour in the legend in which the type is written can also be useful as it provides a direct link to the feature itself.

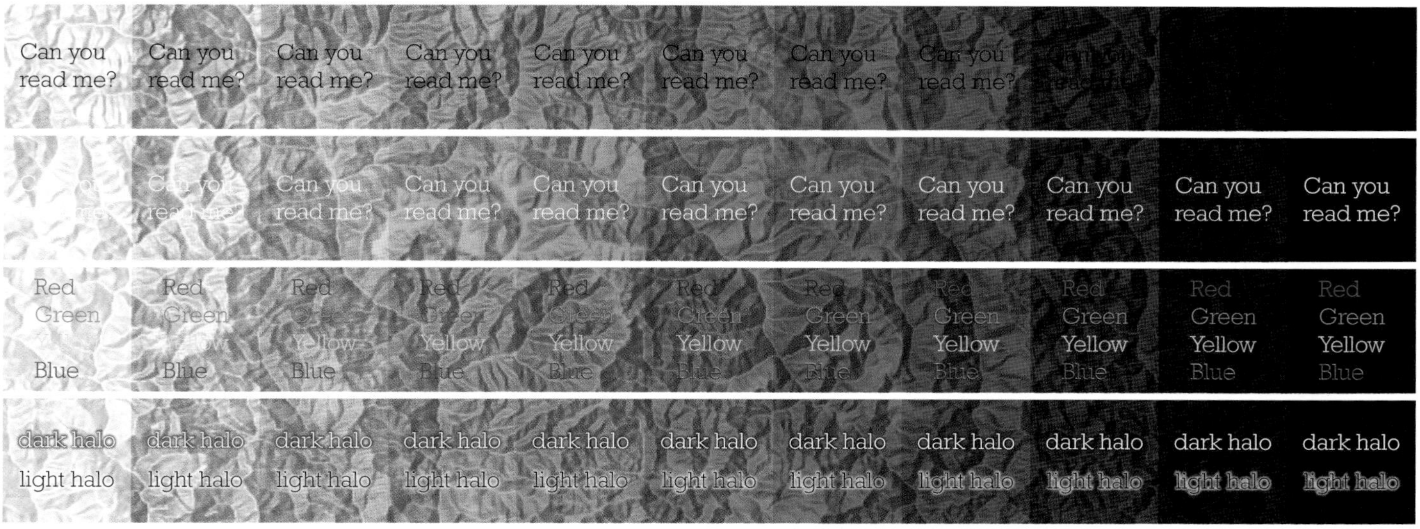

Can you read me?

Can you read me?

Red
Green
Yellow
Blue

dark halo
light halo

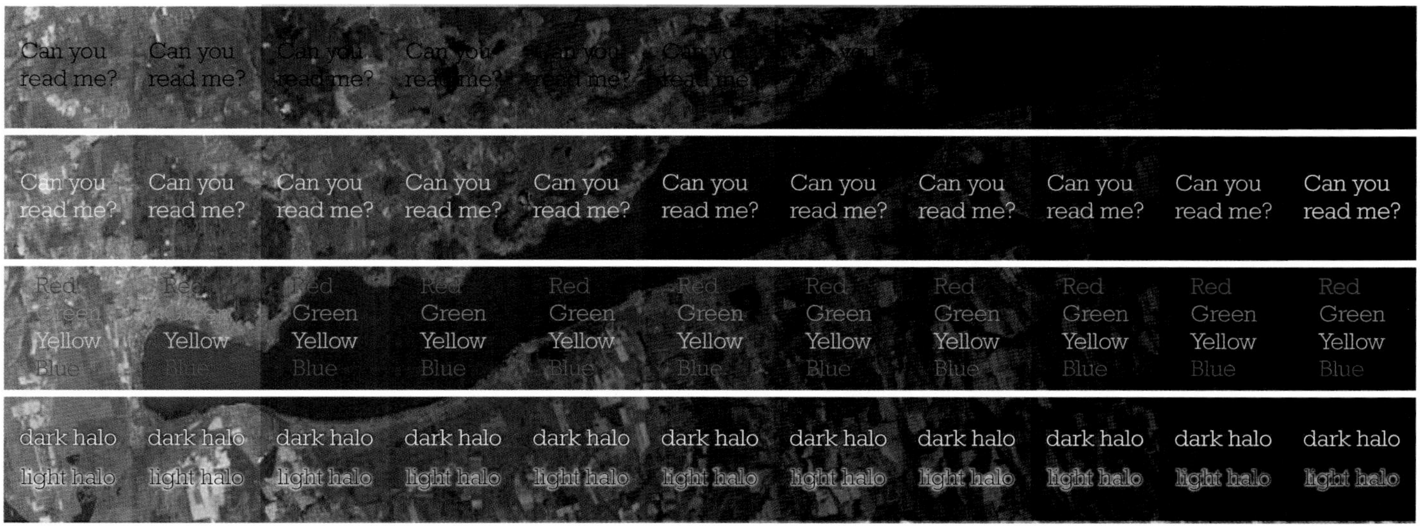

Can you read me?

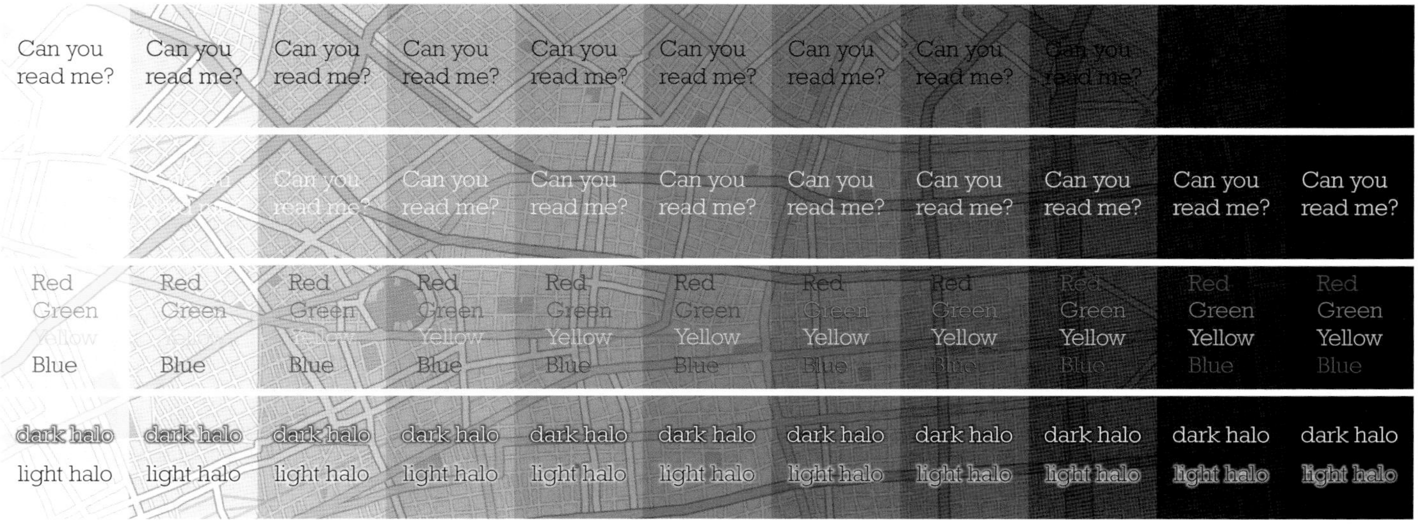

Can you read me?

Red
Green
Yellow
Blue

dark halo
light halo

Can you read me?

Can you read me?

Red
Green
Yellow
Blue

dark halo
light halo

# Types of maps

**Creating a neat taxonomy of map types is difficult since they vary considerably in form and function.**

**Cartographic possibilities are almost endless as the many ways that people refer to maps demonstrates.** However, trying to create a taxonomy helps you better understand the complexity associated with the specific requirements of each type and how they differ on the basis of scale and function. Although a general taxonomy is helpful, it is actually difficult to categorise maps neatly, and many fall into numerous categories.

**All maps might be classified as either general purpose (reference) or thematic but each of these may be further classified to better describe the rich variation.** Reference maps generally display the location of natural or man-made phenomena with an emphasis on location. These maps might include topographic maps, atlases, aeronautical charts, and cadastral plans. Thematic maps tend to be special purpose and single topic and usually describe a phenomenon rather than simply locate it. Examples might include population distribution, patterns of health outcome, or election results.

**A mental map is different from a paper map in that it is intangible and imagined in our brains.** These images are used daily to aid our navigation and recall where places exist in relation to others; they are not normally accurate in a geometric sense and tend to distort geographical distance and direction considerably. However, they give us our unique view of the world, and thus a mental map informs our spatial understanding and affects our spatial behaviour as well as our ability to read and understand tangible maps.

**Virtual maps are also intangible and might exist in a computing environment.** They can take many forms and at many different scales or even be multiscale. They are equally effective at illustrating the same range of phenomena as the tangible form but the medium of communication differs. The way map users interact with screens on computers and mobile devices differs from their interaction with paper, which has important implications for the design and production of digital map products.

**See also:** Mental maps | Thematic maps | Topographic maps

It's never been easy to create a taxonomy for maps, but historically you used to be able to define a map by general function (general purpose, reference, thematic ...), scale (small, medium, large ...) or type (choropleth, dot density ...) though clearly any single map can fall into multiple categories.

Digital mapping has brought about new terminology and there's also been a wave of vendor-specific terms as some try to claim a particular type of map through a name they ascribe (smart maps?). Perhaps more interesting has been the proliferation of new (or repurposed) terms for maps that so-called neo-cartographers have begun using. For instance, 'proportional symbol map' has a long and well-understood use in thematic cartography yet the emergence of the term 'bubble map' to describe the same map tends to be used by some mapmakers.

**Opposite:** 50 words for maps—more than an urban legend, Eskimos have many words for snow. Likely many more than 50. The Inuit dialect of Canada's Nunavik region includes *matsaaruti* (wet snow) and *pukak* (crystalline powder snow) as well as *aqilokoq* (softly falling snow) and *piegnartoq* (snow that is good for driving sledges). These different words describe different characteristics and uses. We might apply the same approach to how we define types of maps.

This list showcases a range of verbs used to describe a specific type of map. This vast range of different functions demonstrates the many alternative maps and uses they serve. It is by no means complete but demonstrates the rich variety of cartography!

They are reasonably self-descriptive but what is apparent is the overlap across many. It is for this reason that I've not attempted to corral these terms into a new taxonomy. A map is a map—whatever its type.

## A

Advertising
Aeronautical
Animated
Aspect views
Atlas (Book of)
Augmented reality

## B

Bad
Base
Binned

## C

Campus
Cartogram
Cartoon
Celestial
Chart
Chernoff
Children's
Choropleth
Contour
Community
Continuous surface
Crowd sourced

## D

Dasymetric
Descriptive
Digital
Dot density
Dynamic

## F

Fantasy
Flow

## G

General purpose
Geological
Globe
Good
Google
Graduated symbol

## H

Hand-drawn
Heat
Hot spot

## I

Info
Infographic
Information
Intelligent
Isarithmic
Isochrone
Isometric

## J

Jigsaw
Joke

## L

Land use
Large-scale

## M

Mashup
Medium scale
Mental
Mobile
Multivariate

## N

Navigational
Non-tangible

## O

Online
Open
OpenStreet

## P

Panoramic
Paper
Plan
Planetary
Pointillist
Print
Prism
Proportional symbol
Pseudo-natural

## Q

Qualitative
Quantitative

## R

Recreation
Reference
Risk

## S

Sandwich
Satirical
Schematic
Sensory
Single variable
Sketch
Small multiple
Small scale
Space-time
Statistical
Stereoscopic
Story
Strip
Subway

## T

Tangible
Temporal
Terrestrial
Thematic
Three-dimensional
Time-aware
Topographic
Transit
Transport
Treemap
Tube
Two-dimensional
Typographic

## U

Ubiquitous
Unclassed
Unique values

## V

Value-by-alpha
Viral
Virtual
Virtual reality
Visualization
Voronoi

## W

Waffle
Weather
Web
World

# Typographic maps

## Maps made purely of typography.

**Maps can be largely artistic endeavours.** Many cartographers and artists have experimented with the map as an art form, either using the map as a basis for an alternative expression or using the structure of common map elements artistically. Maps based entirely on typography are abstract representations of a landscape and have been used effectively as fills for land use and through repetition for linear networks. They are certainly artistic, and it's fascinating to see what used to be called the 'names plate' to show a geography that doesn't actually exist on the ground. Typographic elements take up a considerable amount of space on a map. Text adds meaning and allows us to interpret the landscape using labels and descriptions that we understand.

**Typography is more commonly used on the map as a literal symbol simply to add context to features that are symbolised using other graphics.** Typography is also used as a locative, quantitative, and literal symbol combined. Such maps might contain proportionally scaled labels that indicate a specific feature. Often there are no other graphics, and the maps have a certain aesthetic appeal. Typographic maps take a different approach by using typography to build the form of a planimetric map. The labels now stand out and work on their own to reference a feature. Alternatively, some typographic maps use type as a repeated literal symbol for area fills and to define the shape of lines. The repetition gives form to a larger feature. Linear features contain an array of repeated labels, and areas use labels as a pattern fill.

**Colour can also be used to accentuate the difference between features so we typically see labels that indicate water shown in blue and roads in black.** Green spaces might be in green and different tonal values help to create figure-ground relationships much as you'd see on a topographic map to make certain features stand out. The overall effect is appealing both at a distance where structure can be easily seen or close-up where the individual labels can be distinguished. They act as works of art.

**Opposite:** The two examples shown here juxtapose two different typographic maps of London.

*London's Kerning*, by NB Studio, was one of the first to gain wide attention and remains one of the most accomplished. Prepared for the London Design Festival in 2006 as a commentary on social space, the large-format poster went on to win the design week awards in 2007. The map shows only names of locations, streets, or places. Larger fonts reflect more important spaces with smaller fonts representing a less celebrated space. Smaller type is used as a replacement for roads, and viewed at a distance, the structure of the city emerges as the form, orientation, and positioning combine to create landmarks and shapes that can be easily identified.

The map is a great example of the power of typography in mapmaking and also illustrates how effective a single colour can be. Maps do not always need to be in colour to be visually stunning or effective. Indeed, this map shows you don't necessarily need points, lines, or areas either! The title is both clever and gives the work character to provide a rounded product.

The second map, *London*, by Axis Maps® takes a different approach to the genre to create coloured fills that define more of the structure of the city. At a distance, it appears more like a standard topographic map. The type used to represent water is particularly interesting as the random placement evokes movement and the idea that words are floating down river. Conversely, on the *London's Kerning* example, the river is a white space, and that negative appearance simply signifies a feature, sparsely labelled.

**See also:** Placing type | Spacing letters and words

PRIMROSE HILL
CAMDEN TOWN
ISLINGTON
DE BEAUVOIR TOWN
HAGGERSTON
REGENTS PARK
PENTONVILLE
HOXTON
SOMERS TOWN
FINSBURY
SHOREDITCH
BETHNAL GREEN
MARYLEBONE
EUSTON
CLERKENWELL
SPITALFIELDS
WHITECHAPEL
CITY
SOUTHWARK
BERMONDSEY
WAPPING
LAMBETH
NEWINGTON
WALWORTH
RIVER THAMES
BATTERSEA PARK
NINE ELMS
VAUXHALL
NEW COVENT GARDEN MARKET
SOUTH LAMBETH
BATTERSEA
CAMBERWELL
PECKHAM
STOCKWELL
CLAPHAM
CLAPHAM COMMON
BRIXTON
EAST DULWICH
PECKHAM RYE COMMON

## The Times Comprehensive Atlas of the World

HarperCollins
1895–present

*The Times Comprehensive Atlas of the World* has been considered by many—professional cartographers, designers, armchair users, politicians, educators, and travellers—to achieve optimum standards throughout its history, in terms of accuracy of data collection, production methods, cartographic aesthetics and presentation, and utility.

Its first edition (1895) incorporated maps specially produced, with English text and imperial units, by experienced publishers Velhagen & Klasing, of Leipzig, Germany. This presented comprehensive, systematic coverage of the globe, with some thematic maps.

Design and printing used individual flat hues (some boundary bands used line screen tints), relief representation shown by hachures, and hand-drawn typography. 20th-century innovations were driven by Edinburgh cartographic producers Bartholomews; in photolithography, including more flexible dot screens; in design, notably in hypsometric tinting; in varying and appropriate map projections; and in production, including use of GIS software-based workflows. This extract (from the 1922 2nd edition) shows the development of hypsometric tinting as a successful and beautiful approach to relief representation as well as the map cutting through its own border on the left edge.

Many derivatives of this atlas classic (historical, regional, multi-volume, concise) have enhanced its reputation as the self-styled 'Greatest Book on Earth'.

—David Fairbairn

JOHN BARTHOLOMEW & SON. LTD.

# UI/UX in map design

## Concepts for thinking about designing maps for an interactive environment.

**The screen-based medium is an established delivery mechanism for maps.** Rapid technological change in computing and, particularly, internet and web technologies means that maps are now routinely produced and consumed online or via mobile devices. They have become interactive mechanisms for information recovery. Critically, they now demand a consideration of UI (user interaction) and UX (user experience) in the map design process in a similar way to other digital information products.

**UI/UX has become a key aspect of web design generally.** Designing maps for web delivery means thinking about the interface as a tool to manipulate the map and discover information. Thinking about how a user interacts describes the process of map use. Your map readers use the interface of a screen, a browser, and the tools of navigation, and as part of that, they experience the interaction with the map itself, mediated by the mechanism of delivery. In design terms, a consideration of UI leads to the decisions that bring about an effective implementation, and UX describes how these changes lead to successful outcomes.

**Interaction is a cornerstone of successful UI/UX design for maps.** It supports exploration, insight, and visual thinking and provides flexibility in how maps might be represented. Importantly for cartography, it forms a practical component of the design of web maps because the map reader no longer sees the distinction between mapmaking and map use. Interaction has become a fundamental process of map use in digital environments. Interaction primitives can be used to formulate an overall strategy to support map reading and the definition of the overall experience. As with map design more generally, UI design is not to be considered as simply making the map look pretty. The form of a web map and its interface supports its function as with any map and goes a long way toward defining the tone, mood, emotional response, functional capability, and satisfaction of user experience.

**See also:** Balance | Focussing attention | Interaction | Legends | Mobile mapping | Page vs. screen | Web mapping | Wireframing and storyboarding

Interactions have different stages that require map readers to use perceptual, motor, and cognitive processes to view, manipulate, and interpret a map. Map interactions can be summarised into seven main phases (Roth 2017).

1. **Forming the goal:** establish what the map reader is trying to achieve via the interface.
2. **Forming the intention:** determining the specific map use objective that a map reader must achieve to meet their overall goal. This might be through identifying features, comparing features, ranking, or measuring.
3. **Specifying an action:** as a map reading task, the translation of an intention into a task by identifying how to operate the interface.
4. **Executing an action:** the use of devices (e.g. mouse, click, gesture, keyboard, speech) to command the device via the interface to return a particular response.
5. **Perceiving the system state:** viewing the response and feedback to clarify change in map state.
6. **Interpreting the system state:** making sense of the changed map state relative to the original intention.
7. **Evaluating the outcome:** comparing the insight gained from the changed map state to expectation to establish whether the goal has been reached.

Interface operators might include basic functions such as zoom, pan, and information retrieval but might extend to supporting interactive overlays of information, filtering, and graphical mechanisms to support comparison or suppression of zoom scales to only those that are relevant. Scope and freedom in UI design result in interface complexity, which is the number of ways in which a map reader can interact coupled with the possible changes in map state. The focus should be to achieve balance between flexibility and constraint to support a specific map use scenario.

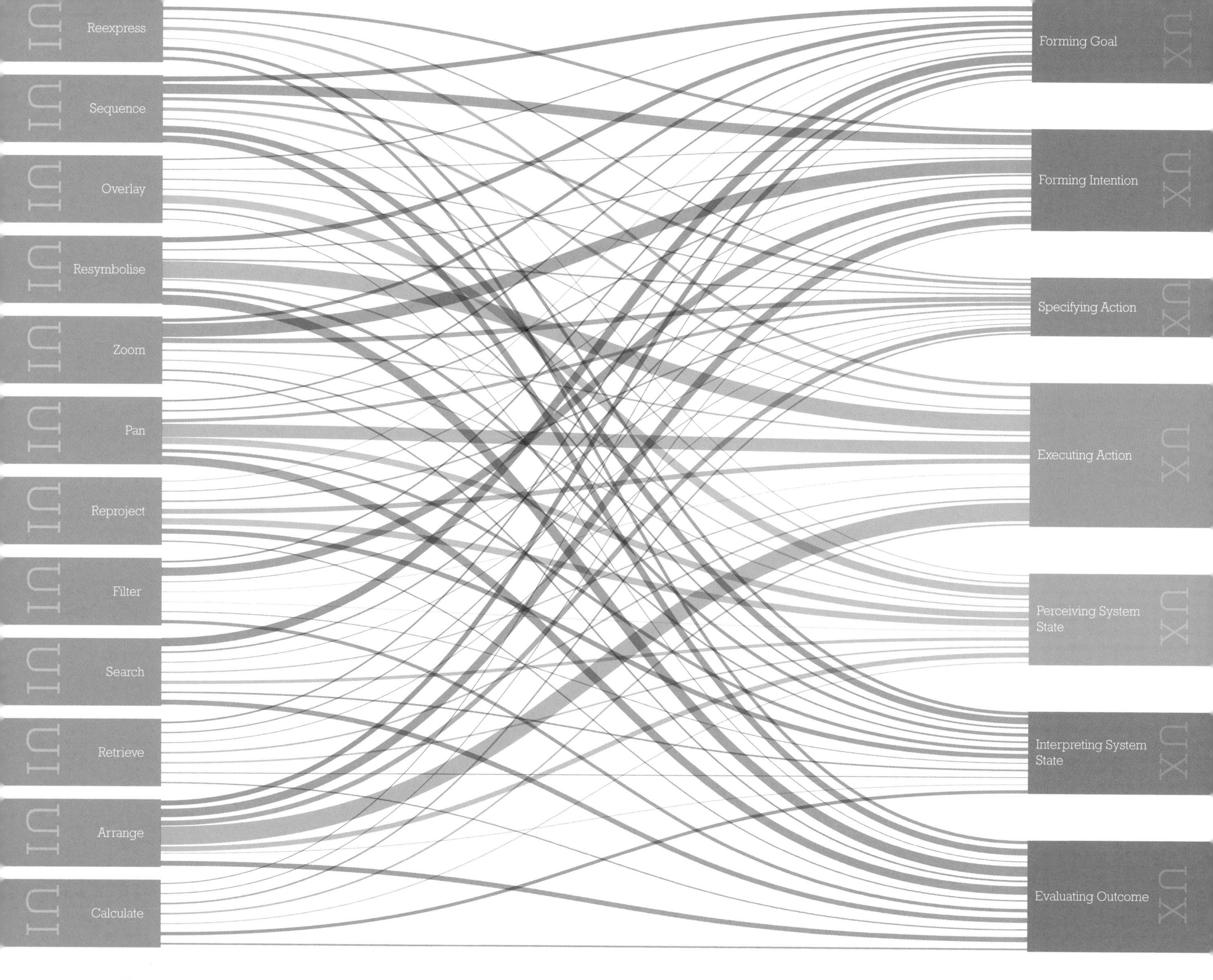

UI
Reexpress
UI
Sequence
UI
Overlay
UI
Resymbolise
UI
Zoom
UI
Pan
UI
Reproject
UI
Filter
UI
Search
UI
Retrieve
UI
Arrange
UI
Calculate
Forming Goal
UX
Forming Intention
UX
Specifying Action
UX
Executing Action
UX
Perceiving System State
UX
Interpreting System State
UX
Evaluating Outcome
UX

# Unclassed maps

## Unclassed maps use a unique symbol to show each value of data being mapped.

**Conventionally, thematic maps classify data to show areas that display similar characteristics.** The choropleth is a common technique for representing data for areas by variations in shading or pattern that represent classes of data. The classed choropleth is the norm because of the problems of the human eye being able to discriminate between too many symbol shades on the same map. Alternatively, a proportional symbol map uses unclassed symbols where each is scaled according to the value of the data it represents. The unclassed choropleth is not as widely used because of the problem of seeing subtle tonal variation in sequence but this drawback also has its own merits, principally that it visually maintains numerical relationships among data values by assigning a unique shade to each data value along a colour ramp.

**In effect, an unclassed choropleth is simply a unique values map.** Instead of being used to show a qualitative difference among nominal data, it is used to show a quantitative difference among data usually measured along an interval/ ratio scale. In symbol terms, a unique values map uses a scheme that signifies difference such as different hues. For an unclassed choropleth, the colour scheme would be sequential. The cognitive task of deciphering higher or lower is therefore supported, even though it's unlikely that readers will be able to accurately determine a particular value from the map.

**Proponents of the technique suggest that it obviates the problems of finding an often imperfect, and sometimes biased, classification scheme, which will inevitably be visually misleading.** Classifying data can often mask nuance or important details in a dataset, which may be some of the most important aspects. This can particularly happen at the extremes of a data range where outliers might be important to pick out. These values are sometimes incorporated in a class that has a wide interval. On an unclassed choropleth, they would be distinct.

The unclassed choropleth provides an unstructured and ungeneralised view of the data. One critique of this approach is that by its definition, it ignores cartography since teasing out meaning rather than presenting the data as a flat visual list is one of the key tasks in designing a map. The burden of interpreting patterns and finding areas that display similar characteristics falls to the map reader. Unclassed choropleths do not preserve the visual integrity of enumeration units, which is the purpose of a choropleth map. In that sense, unclassed choropleths have often been referred to as *tonal variation maps* since they bear little functional resemblance to a choropleth.

Unclassed maps attempt to let the data speak for itself yet most data is extremely noisy. It is this noise that classification attempts to deal with. Ultimately, the choice of whether to use a classed or unclassed map comes down to the extent to which you wish to pre-filter the data for the map reader. Sometimes it's vital. Sometimes it isn't at all important, and this is perhaps one of the perfect examples of knowing your map user, knowing your data, and then determining the cartographic approach to support the need.

**See also:** Choropleth maps | Data density | Web mapping

Unclassed
More
Less
2 class
6 class
10 class

# Unique values maps

## A unique values map shows the spatial distribution of a single theme of nominal data by showing variation of type.

**Unique values maps are perhaps the simplest form of thematic map.** The purpose is to map a single theme of nominal data across the map, perhaps showing variation by type. It is the predominant form of qualitative thematic map where the map is concerned simply in showing where things differ across the map and not by how much they differ. Most other thematic maps fall into the class of quantitative mapping because they are concerned with mapping the magnitude of a theme measured on an ordinal or interval/ratio scale.

**Typical unique values maps display the distribution of ecoregions, geology, soil type, or land use.** The map is subdivided into mutually exclusive areas that each record the predominant type of the general theme being mapped—for instance, different types of land use. There is no attempt on this type of map to show the quantity of a variable within each region though readers would be able to gauge the overall quantity, or proportion, of different area types as part of the whole map.

**Unique values maps can be point, line, or area based**. Area-based maps are by far the most common. For point or line versions, changing the shape of the symbol (e.g. square, triangle, or line style) while keeping the size constant can be used. Additional aspects might be incorporated to point-based unique values maps by changing the hue of each type of symbol. Of course, changing the value of a type of point symbol can also be used to introduce a quantitative component to the otherwise qualitative thematic map.

At its simplest, a unique values map might simply show the presence or absence of a variable resulting in a binary map. Symbology would be simple—two different colours or patterns to delineate presence or absence.

Geological maps are far more complicated and tend to include many tens of different categories of type, which requires many more unique symbols. Symbol design should take note of the fact that the map is supposed to show only difference in type, and no one symbol should stand out more than any other. For example, using a bright red as one colour among pastel shades of many others is going to send the wrong message to the reader. For area fills, changing the hue of a symbol connotes difference, and keeping value and saturation constant yields a good, visually consistent symbology scheme.

The addition of patterns across the top of coloured symbols is often used to stretch the palette further for area-based maps, particularly where there is a requirement for many symbols.

The legend on unique values maps becomes important since there is a likelihood that the symbology doesn't necessarily imply a specific class of the nominal data. Although some maps use standard symbology (e.g. geological colours are standardised), legends are important for context.

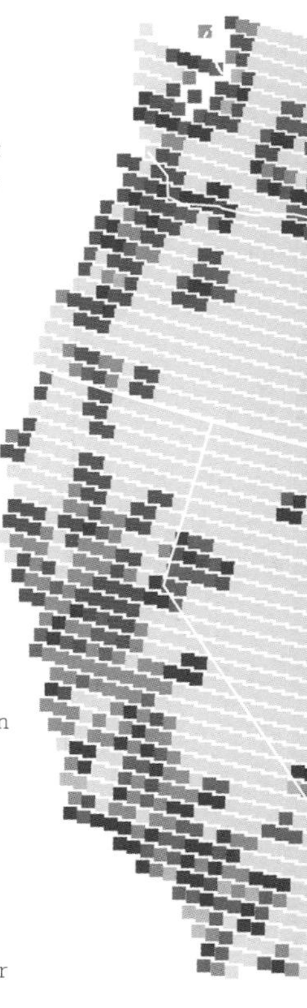

**See also:** Data (c)art(e) | Data density | Descriptive maps | Hue

**Opposite:** *Pizza Place Geography* by Nathan Yau, 2013.

## Nearest pizza chain

Pizza chains are everywhere, but some are more popular than others across the country. This map shows the nearest pizza chain among those listed.

# Using words

## Language and grammar are important aspects of a map.

**A picture paints a thousand words.** Well, this is true, and cartography is a real expression of the statement. The creation of a map has at its heart the distillation of spatial data into a meaningful picture to convey information. It is far harder to attempt to describe the spatial pattern of data or the characteristics of topography. Beyond this, the way in which you use words on a map is critical to its overall effectiveness.

**Words are used in many different ways on a map.** Lettering a map includes not only place-names but also marginalia, titles, explanatory text, statistical detail, and descriptions. Using language and grammar meaningfully is a skill in itself. There's no room for verbosity on a map but being concise and succinct is challenging. The correct use of spelling, punctuation, and grammar is important. There's nothing worse than a perfectly good map ruined by poor grammar or a blatant spelling mistake (I know—I've made some!).

**Typesetting has become part of the cartographic trade.** This mirrors many other discrete roles in the design and production process, which have melded together because of technological change. Typesetting demands an understanding of the differences between all manner of marks; between an apostrophe and the beginning of a single quote, for instance, or between a hyphen and a dash (or em). While the computer brings typesetting to the desktop for cartographers, it's still important to have an understanding of language and grammar to ensure your cartography sets expected standards.

**There are some simple ways to improve your map's language and grammar.** Many might seem obvious but are often overlooked. Keep textual components succinct. They should support the map and assist interpretation. Try not to write in an abbreviated form or overuse acronyms. This is particularly true for a general audience. Make sure you spell-check textual elements.

**See also:** Elements of type | Emotional response | Statistical literacy | Typographic maps

Most maps require a title. One of the only occasions when this isn't the case is if a map is presented within another document and has a caption. In other instances, a well-phrased title draws attention to the map. For a thematic map, the title should be succinct and clearly describe the theme of the map. For a topographic map, the title is usually the region of country being mapped.

A common mistake is making the title too wordy or using terms and abbreviations that might not easily be understood. For example, 'A map of the distribution of unemployment in the United Kingdom of Great Britain and Northern Ireland in 2017' is far too wordy. At the other extreme 'Unemp. UK 17' is almost unintelligible. Instead, 'Unemployment in United Kingdom, 2017' clearly expresses the mapped theme. Note also that the theme, geographical region, and date all appear in the title in that order. Of particular irritation to many cartographers is the error of using the word 'map' in the title of a map. Such use is tautological. It is obvious it is a map!

A subtitle can often be used to further expand on the main title where further information is required. For instance, to break down the previous example to the theme 'Unemployment' as the title and the region and date 'United Kingdom, 2017' as the subtitle. The region might be omitted if an area is easily recognisable. Italics and ornate styles are generally best avoided although the use of bold type can emphasise a title to create appropriate contrast or hierarchy. The title should be the largest piece of type on the map with the subtitle second largest. Where possible, position titles where map readers would expect to see them, often toward the top of the map.

Textual components in a legend should be clear and unambiguous and support interpretation of the symbology, including units of measurement where necessary.

Apocryphal Map Title Nomenclature and the Cartographer's Relative Employment of Language and Grammar to Inform and Describe *or* Possible Map Titles

Cartographer's Sense of Pith (Succinct — Flowery)

| | Then | | Now |
|---|---|---|---|
| Succinct | Occurrences of Cholera | Cholera Infections | Cholera |
| | A Charted Exposition Upon the Relative Occurrence of Severe Choleric Dysentery and the Loss of Vigors Thereby Rendered | Cholera Infections Both Known and Feared | Cholera: Confirmed & Suspected |
| Flowery | A Geographical Notice of the Constable Medical Examiner Hereby Depicting the Instances of Those Affected with the Most Terrible Bowel Complaints and Certain Exposure to Cholera, With the View of Affording Prompt and Gratuitous Assistance and Precaution of Vapours | A Mapping of Cases Known, or Thought, to be Bacterial Cholera Infection, Arranged and Printed by the National Medical Authorities for the Immediate Assessment of Public Risk and Requisite Prevention of Transmission | A digital spatio-visualization presenting the GPS-sourced, or rooftop-geocoded, locations of suspected and confirmed incidents of cholera within the current outbreak hot zone and contiguous districts, aiding an international response |

Vintage

## New York Map of Midtown Manhattan

Constantine Anderson
1985 (2nd ed.)

Some maps just 'work'. There is no need to learn how best to use them, or any confusion about interpreting their representations of geography. They are usable, and useful, immediately. For me, this is true of Anderson's isometric map of the major buildings and landmarks in New York. The projection allows buildings to be drawn at the same scale, angle, and relative height. Reading the map allows me to 'fuse together' an image of Manhattan, better than a 'normal' map would allow.

Vertical aerial photography was used to lay out the streets, oblique 45° photographs to capture the buildings, and street-level photography for the detail. Buildings were sketched from ground level and architects' plans used in the process of meticulously drawing the map block by block. Widening the streets means buildings are distinct from one another, which allows me to see more. Labels are well defined and unobtrusive. Good cartography normally omits detail, but here, Anderson includes such minute detail that it brings the map, and the city, to life. And seeing more affords me a greater understanding of the city.

The map may be criticised for being neither 'true' nor 'accurate' (but what is?). It conveys what, to me, are the essential elements of what makes New York, New York: landmark buildings claiming their part of the sky, broad avenues, and intersecting streets that convey citizens and goods—all crammed into that limited space between two rivers.

—William Cartwright

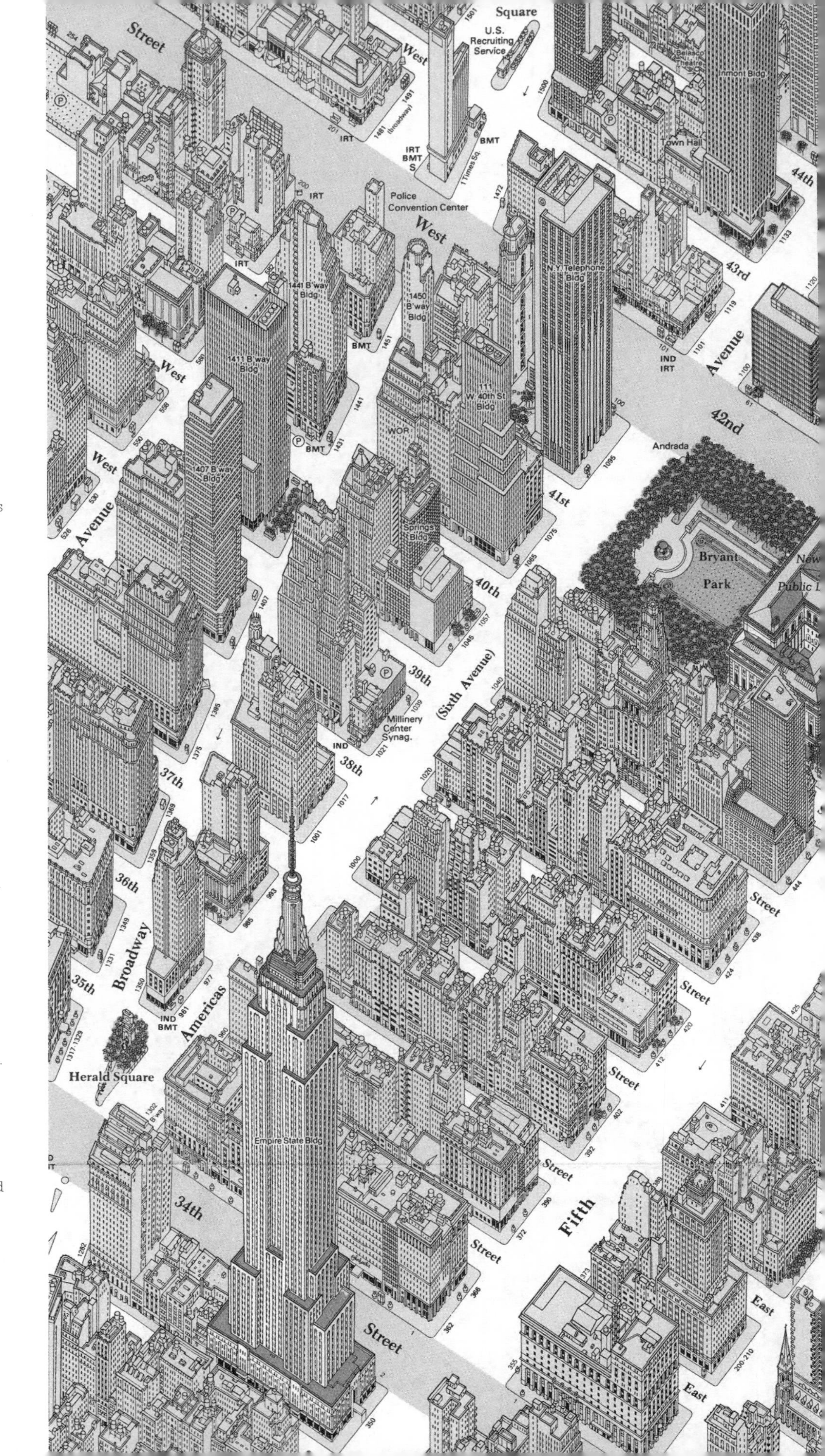

Grand Central Station
Pan-Am Bldg
Chrysler Bldg
Chanin Bldg
Graybar Bldg
Yale Club
Lincoln Bldg
Vanderbilt
Avenue
Madison
Park
Third
East
Street
42nd
41st
40th
39th
38th
37th
36th
43rd
44th
47th
48th
51st
American Brands Bldg
Bankers Trust Company
Chemical Bank Bldg
Westvaco Bldg
Colgate Bldg
French Bldg
535 5th Ave Bldg
529 5th Ave Bldg
Century Club
Yacht Club
U.S. Post Office
Children Aid Soc.
Teachers Ins. Bldg
St. Agnes Ch.
Chrysler Bldg East
Am. Home Bldg
150 E. 42nd St Bldg
Blue Cross & Blue Shield Bldg
Nat'l Distillers Bldg
Continental Group Bldg
Burroughs Bldg
Pfizer Bldg
WNEW
WPIX Plaza
Ch. of Our Saviour
Advertising Club
IRT
RAMP

# Value

## The perception of shading from dark to light to represent high to low.

**Value variation is the continuous progression from dark to light for a given hue.** Perception of shading results from the way in which different hue values reflect light so areas of light have high reflectance and areas of dark have low reflectance. Changes in value are often referred to as *lightness* and, on the whole, operate independently of the other two colour variables of hue and saturation. The exceptions are when you compare a hue relative to black, which will always appear darker.

**Value can most easily be described as greyscale though it applies to any hue.** It is equivalent to the surface ratio of the amount of paper (white) covered in black ink. So, 10% grey is an area of 90% white covered in 10% black to give a light-grey appearance at distance. A medium grey would have an equal amount of black and white to give a 50% mid-tone. Greyscale shading is usually achieved simply by specifying the amount of black in a CMYK printing process or by using equal values of red, blue, and green in a visual display.

**Value is most effective at representing ordinal data since it connotes quantitative differences between features.** Darker values are seen visually as representing greater quantities than features represented by lighter values. Keeping the number of classes low helps people see different unique shades. In some instances, it's inadvisable to use black, which might connote totality, or white, which might itself clash with the map background.

**Value is only marginally effective at representing numerical data**. It is only capable of showing order so you can determine which values are more than others. Even if you know the numerical magnitude represented by one value-shaded class, it's not possible to determine the numerical magnitude of any other class. The map reader is entirely dependent upon the legend unlike the size visual variable where ratios are perceptible and assessment of magnitudes possible. Because value is in itself ordered, it's impossible to reorder symbols in any meaningful way.

**See also:** All the colours | Greyscale | Hue | Mixing colours | Saturation

Dividing value equally does not create good visual perceptual shading.

It's relatively easy to discern class differences at the two ends of a shading scheme but not at the centre where mid-values appear similar. Good perceptual steps can be calculated using a logarithmic ratio of progression defined by:

$$r = \sqrt[n-1]{D/L}$$

Where:

L is the % value of lightest shade, $L = \frac{PL}{PD}$

D is the % value of darkest shade, $D = \frac{PD}{PL}$

r is the ratio of the progression, and
n is the number of classes.

Once a ratio of progression is calculated for known lightest and darkest shades and the number of classes, then the % shade for each of the intervening classes can be derived by:

$$P_i = \frac{100 \times (PL \times r^{i-1})}{PD + (PL \times r^{i-1})} \quad \text{for} \quad i = \{1,2,3,...n\}$$

So for a lightness of 5%, a darkness of 95%, and 10 classes, the array of % shades would be 5, 9, 16, 27, 42, 58, 73, 84, 91, and 95.

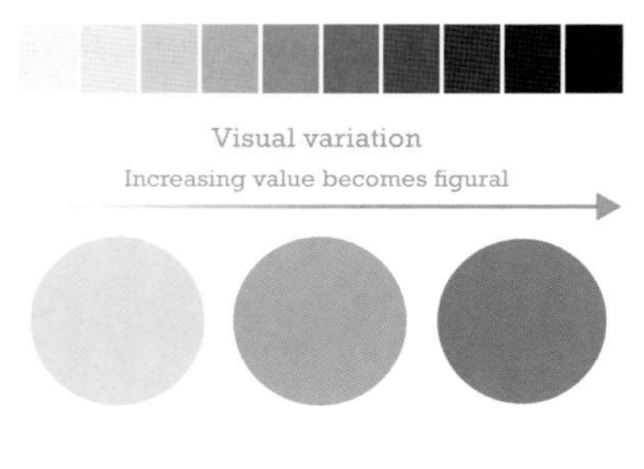


For seeing: Distinct, Levels
For representing: Nominal, Ordinal, Numeric

**Opposite:** *Tracking the Economic Disaster* by Daniel Mason (pictured), 2011, uses wood stain to create a laser-cut 3D choropleth map.

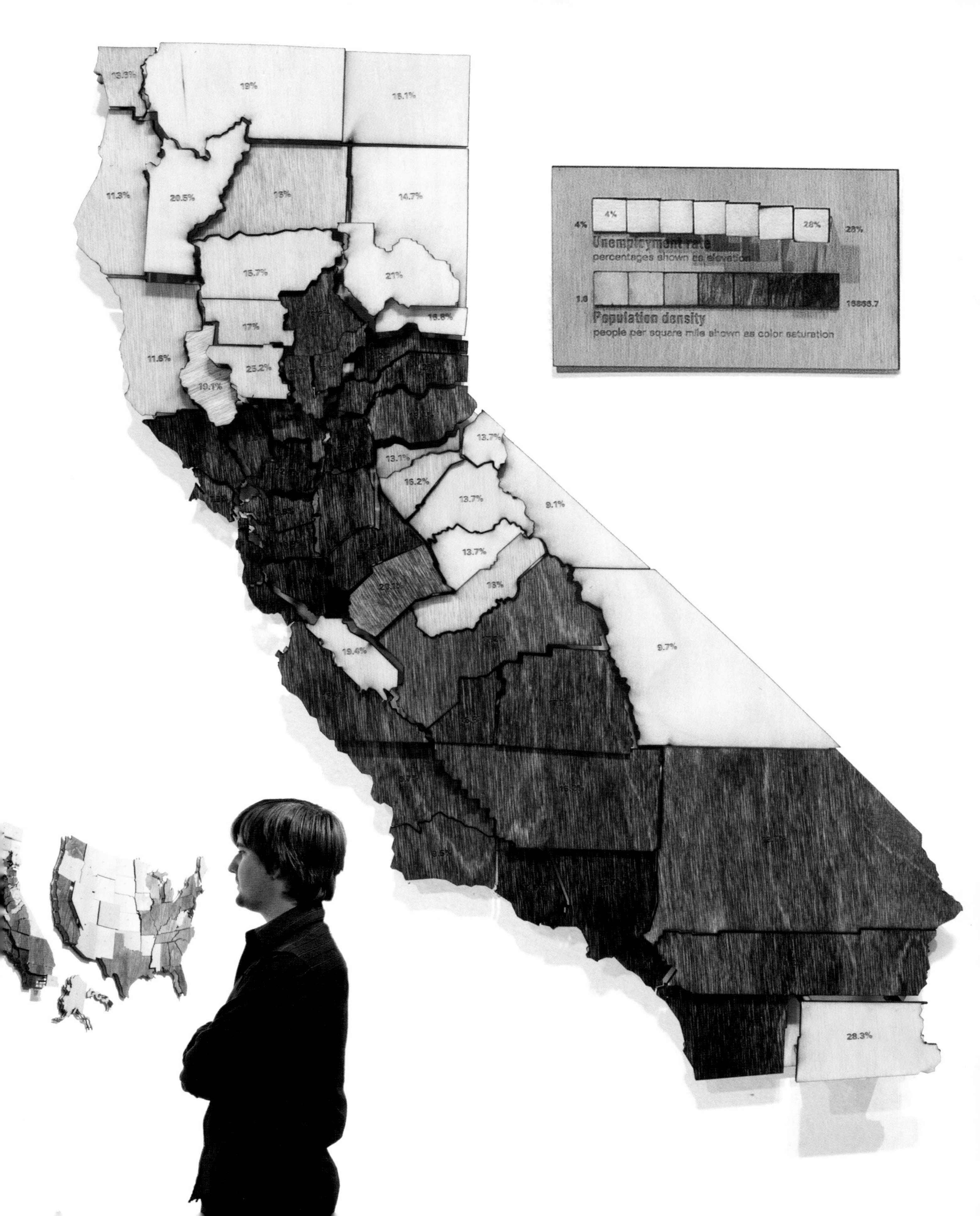
19%
16.1%
11.3%
20.5%
16%
14.7%
15.7%
21%
17%
11.6%
25.2%
19.1%
13.7%
13.1%
16.2%
13.7%
9.1%
13.7%
18%
9.7%
19.4%
28.3%
4%
4%
28%
28%
Unemployment rate
percentages shown as elevation
Population density
people per square mile shown as color saturation

# Value-by-alpha maps

## A variation of the bivariate choropleth map which modifies the appearance of one variable by changing the transparency of a second variable.

**Value by alpha is a way of adjusting for the problem of different base populations and different-size areas on a choropleth map.** Instead of modifying the shape of the geographical unit as a cartogram would, value by alpha effectively adds a second thematic layer over the top of the first and encodes a normalizing variable from opaque (low values) to transparent (high values). This has the graphical effect of bringing into focus areas that are more important on the map compared with areas that are less important. It adds a visual filter so that the reader is visually ushered to the key parts of the map. In essence, the map uses transparency rather than size to vary the visual impact of different enumeration units on the map. It equalises the inherent distortions that unequal-size areas and their unequal populations cause.

**Areas with low alpha values modify the underlying data layer so the net appearance is to fade into the background.** On population-based maps, the transparent overlay would encode population with low population densities given high opacity (low transparency) and high population densities given low opacity (high transparency). Therefore, the visual impact would be that the areas where population density is highest become more visible as the mapped variable on the base layer is exposed. The key for this equalising layer is that the data must be of consequence to the mapped variable so that together they make sense.

**The colours used to denote opacity in the value-by-alpha layer will depend on the background of the map.** Typically, if the background is white, then for lower-impact areas to fade to the back, the value-by-alpha layer would be symbolised from white to transparent. If the map background is black (for instance, on a screen), then the symbology would go from black to transparent. It's not recommended to use an alpha value of zero since that would leave areas completely invisible. Set the lower threshold in the 10%–20% range to allow some of the underlying data to show through. 100% transparent works well at the other end of the scale.

One of the drawbacks of the choropleth map is its inability to support mapping totals because of differences in the size and data values of areas. Mapping percentages or ratios solves this problem though the impact of the size of an area can still cause problems for map reading principally because larger areas will always visually dominate the map. There's no getting away from this simple cartographic dilemma without modifying the map type. Cartograms are one way of adjusting for the visual dominance of large areas, rendering the map as geometric shapes or warped geographies. Pointillist mapping is another alternative.

Value by alpha is a popular alternative since the geography of the map remains unaltered. This limits some of the potential confusion some people have with interpreting cartograms. However, as with any bivariate choropleth map, introducing a second layer and, effectively, mixing the colours of one layer with the second massively increases the number of colours on the map. This makes for a more visually complex map. For the underlying variable of interest being mapped, it's generally recommended that only a few classes are used. Keeping symbol classes to a minimum ultimately enables the reader to pick out patterns more easily.

The legend design for a value-by-alpha map would be similar to a bivariate choropleth in that it would show the variable of interest along one axis and the alpha channel variable along another.

Value by alpha can be used beyond equalising for population on a choropleth map. It might also be used as a way to encode a measure of statistical uncertainty across a map. This would bring into focus areas for which there is statistical certainty and mute areas where there is greater uncertainty in the data.

**See also:** Choropleth maps | Multivariate maps | Transparency

Value-by-alpha map

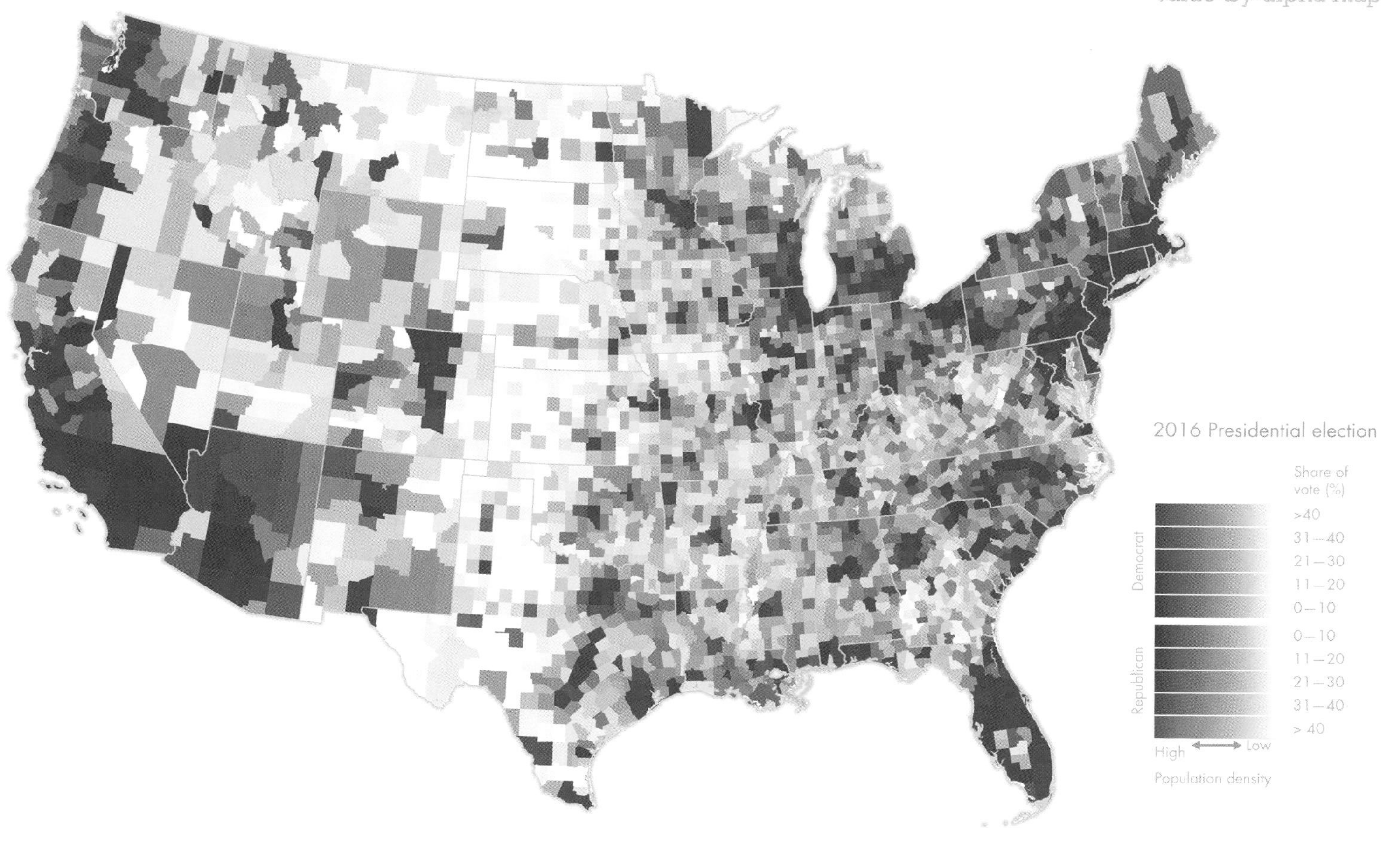
2016 Presidential election
Share of vote (%)
Democrat
>40
31–40
21–30
11–20
0–10
Republican
0–10
11–20
21–30
31–40
> 40
High
Low
Population density

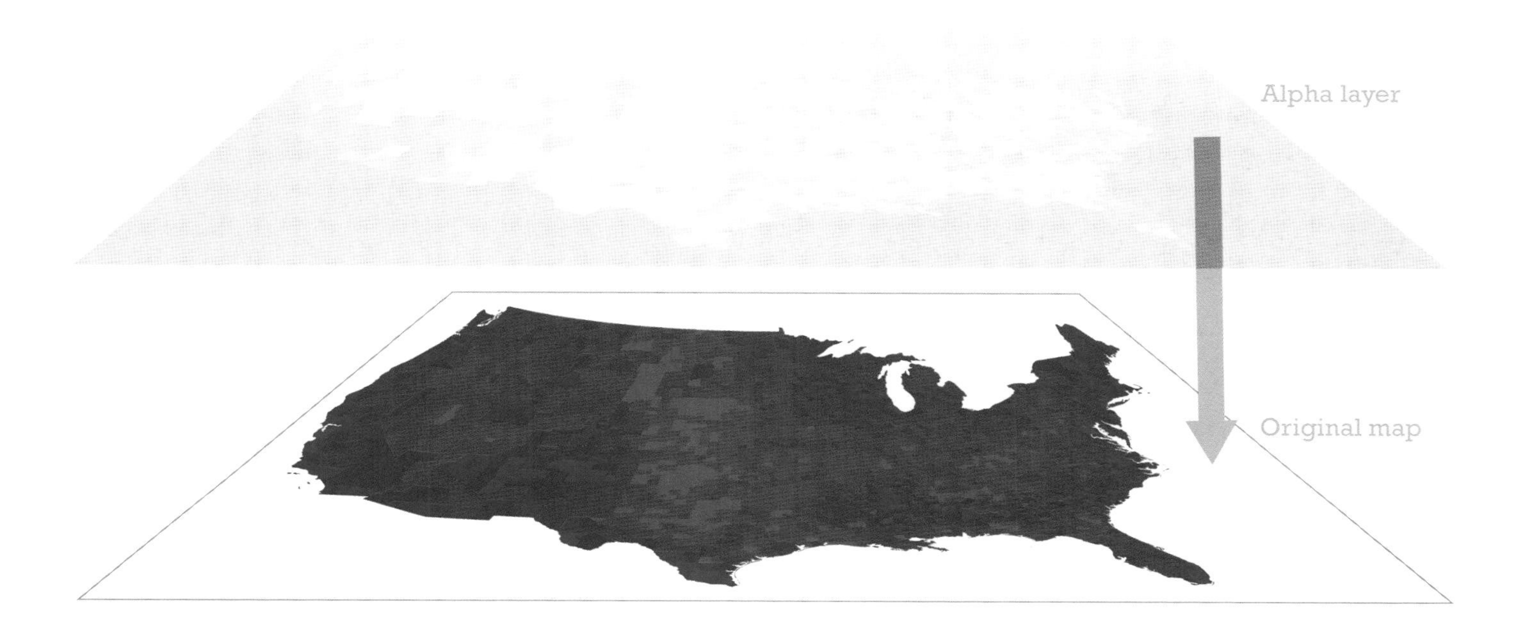
Alpha layer
Original map

# Variables, values, and arrays

## Exploring the patterns in numerical values can give us valuable insight into the messages mapped.

**Variables are the subjects used in statistical analysis.** They are most often the subject of thematic mapping (e.g. elevation, rainfall, employment). Individual observations of data in the variables are referred to as values (e.g. values of elevation could be 128, 163, 242, and so on). In common nomenclature, variables are designated with upper-case letters (e.g. $X$ or $Y$) and values with subscripts of the lower-case form of letters (e.g. $x_1$, $x_2$, $x_3 ... x_n$). The subscript is usually referred to as $i$ such that $x_i$ refers to an unspecified value of the variable $X$.

**Values of a variable can be arranged in order, which gives an array of the data.** An array is usually ordered in ascending or descending order of magnitude but it can also be used to determine an array of nominal data by grouping the values into similar groups. A nominal array might take the form:

oak oak oak rowan rowan birch birch birch ash ash ash ash

An array for ordinal data might take the form:

city city town town town town village village hamlet hamlet

An array for interval (or ratio) data might take the form:

1313 1245 1211 1005 987 765 656 631 543 213 198 153 124

**Organizing values into ranked order as an array can help establish and identify patterns worth mapping.** The values themselves might not be important but their relative position might be. For instance, if a particular value is in the top 10% of all values when ranked, this might be a useful aspect to map even if the value itself is unspectacular in the overall array.

As well as the patterns that numbers make, it's useful to have an appreciation for the numbers themselves because the way they are reported might lead to certain use cases or inferences.

An integer refers to a whole number. It can be positive or negative and does not include any fraction. For example, –50, –25, 0, 25, 50 is an array of integers.

Real numbers (also known as *floating numbers*) can be either whole numbers or hold a fractional component, represented using the decimal point to delineate the whole from the fraction. For example, –50.2, –25.8, 0, 24.6, 49.7 is an array of real numbers.

Strings are non-numerical and are used to store and reference words, feature types, or (often for mapping) labels. Strings consist of alphanumeric characters, and although they can include numerical symbols and other marks, they are treated as a string. For example, 'arable', 'pasture', 'fallow', and 'livestock' are examples of land use strings.

**See also:** Data distribution | Frequency distributions and histograms | Ratios, proportions, and percentages

**Opposite:** Seven days of fire coordinates in Central and South America detected by satellite.

# Varying symbols

## The design of symbols supports many different map functions.

**By varying different visual variables, symbols can be designed to fulfil many different functions.** Regardless of the basic form (shape) of a symbol, secondary visual variables develop and encode more complex meaning. Solid symbols stand out across map detail, usually designed in black to improve readability. In congested areas, overlaps can be handled by using cut-outs for symbols that need to remain solids. Alternatively, varying transparency can be used, and sometimes a fully transparent fill with only line work overcomes complex overlaps. The use of transparency tends only to work with simple geometric shapes rather than mimetic symbols. Colour, hue, value, and pattern can all be used to modify the general design character of a symbol.

**The view of the symbol can be oriented in different ways.** Symbols are often drawn planimetrically so the shape of a building footprint or a tree canopy might be represented as if you are looking down. These symbols tend to sit on a map with the appearance of the feature itself even if they are abstract representations. Mimetic symbols tend to be shown in aspect, sometimes using oblique or perspective for emphasis. Aspect is usually how we see features in real life such as a tree seen side-on so the symbol mimics what we expect to see.

**Two-dimensional symbols are most common.** They are often cleaner in design and contain the bare minimum of line work to support the need for clarity. Occasionally, three-dimensional symbols may be used, perhaps to add visual interest and more likely on a thematic or descriptive map than a topographic map.

**Descriptive information can be conveyed using different designs.** Symbols can denote a particular feature type or some attribute of the feature. Standard geometric shapes are good for differentiating but recourse to a legend is normally needed. Some symbols are specialised such as those used on a weather map. Pictorial or mimetic symbols are designed to resemble the phenomena they represent.

The humble symbol is of tremendous value to the cartographer. With almost infinite design possibilities there is potential to encode all types of data for quantitative and descriptive purposes. Most symbols are designed to represent a single piece of data and communicate a single phenomenon. Thus, the basic form of literal, point, line, or area can be combined with visual variables to support the encoding of all levels of measurement. The breadth of possible combinations forms a rich palette. Selecting the particular approach that matches your feature type and level of measurement goes a long way to ensuring that a map reader can decode your symbol design efficiently and in concert with its meaning.

Even within a single symbol, the variation that is possible is vast. For instance, taking a simple circle used as a small geometric shape, it can be used to show location just as a small black dot. Multiple locations build a picture of density with each dot encoded with a value. Qualitative characteristics can be shown by using a pattern fill inside the outline of the circle. Quantitative information can be encoded in many ways such as through changing the size, using light–dark fills, changing the shape of the outline, showing only a partial segment, or using repetition. Multiple variables can also be encoded. For instance, by changing size and value or pattern to encode two pieces of information or using a full pie chart with multiple segments or using different hues or values to show proportions of different types in a series of repeating symbols.

**See also:** Consistent denotation | Isotype | Symbolisation | Symbols

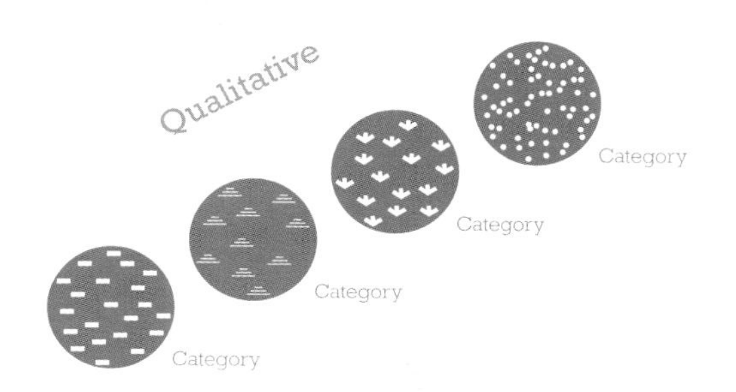

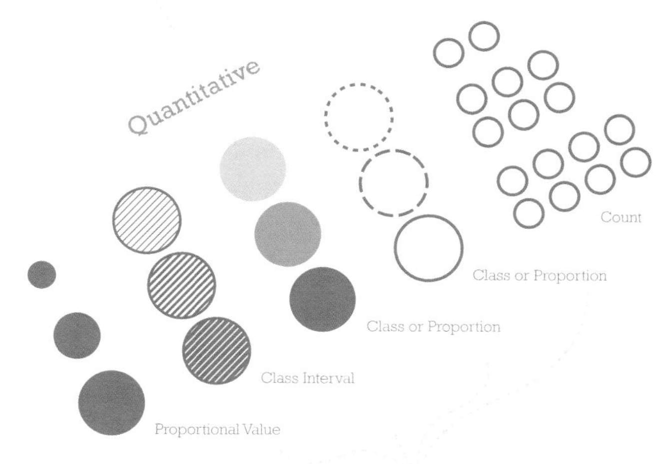


The depth of visual variability within a single symbol type is great. A symbol can encode spatial position or density, qualitative characteristics, quantitative information, and a host of combinations thereof. The humble circle is a popular and versatile shape used by cartographers for a multitude of representations, though other shapes can be used in a similarly expressive manner.

# Vignettes

**Vignettes are used to provide focus and contrast, often along coastal features but the same technique can be applied elsewhere.**

***Vignettes* is the general term for a graphical effect that provides additional contrast between features on a map.** They are used most often for coastal features and are particularly prominent on historical maps. They are often applied to give the map additional aesthetic appeal and an element of artistry that effectively adds a faded border to a coastline instead of a simple single line.

**Vignettes can be applied in many different ways, which largely depend on the mode of production or the specific look you are trying to create.** With most modern computer applications, the edge-glow effect is commonplace. Intensity of the colour fades from solid to transparent across a gradient perpendicular to the coastline. The technique can be applied in the inverse as well, particularly if the water bodies are dark compared with the land mass. It's simply an effect to bring extra focus to the object of the map.

**Alternatives exist such as the use of buffered lines.** The line weight and style of line changes, usually to become thinner or eventually dashed, as distance from the coastline increases. This approach has also been termed *waterlining*, particularly when it relates to inland water features such as lakes and rivers. Alternatively, horizontal lines drawn away from a coastline and which vary in length provide a coastal rake effect. A common technique is to use lines that become dashed or fragmented away from the coastline to mimic the effect of a wave. Ironically, of course, real waves tend to become more pronounced nearer to a coastline!

**The same sort of approach can also be used to create a mask that emphasises an inner detail.** For instance, a topographic sheet map might bleed to the edge of a sheet but all you are interested in showing is a specific area. Masking the area external to the area of interest with a whitewash effect and using an internal vignette to give a soft inner glow can be used to aid the differentiation between the features of interest and the surrounding area.

**See also:** Flourish | Foreground and background | Map aesthetics

Another way to think of vignetting is to explore its use in allied design fields. For instance, in photography a vignette results from the reduction of brightness or saturation toward the edge of the image. This creates the effect of a soft focus which works as a border to the central image. It brings the centre into focus by effectively blurring the edge or reducing its detail. The idea is to create a post-processed effect.

Vignettes can be created using various graphical blending techniques, which are created by combining one graphical layer in your map with another in some way. These techniques are created through the mathematical combination of one layer with another. For this reason, the layers will usually be rasterized to allow performing pixel-to-pixel calculations between layers. For instance, a buffered vignette around an island could be made partly translucent, which allows the layer beneath to show through by changing its transparency settings or blending it with the layer beneath.

Vignettes can also be used to accentuate the symbolisation of boundaries. On many historical maps a black-and-white print was often hand coloured. Boundaries often had watercolour applied, and the modern version creates a similar effect. This can be used not only to give a historical appearance but many mapping agencies also make use of the effect. National Geographic maps routinely have boundary vignettes to demarcate political or administrative areas.

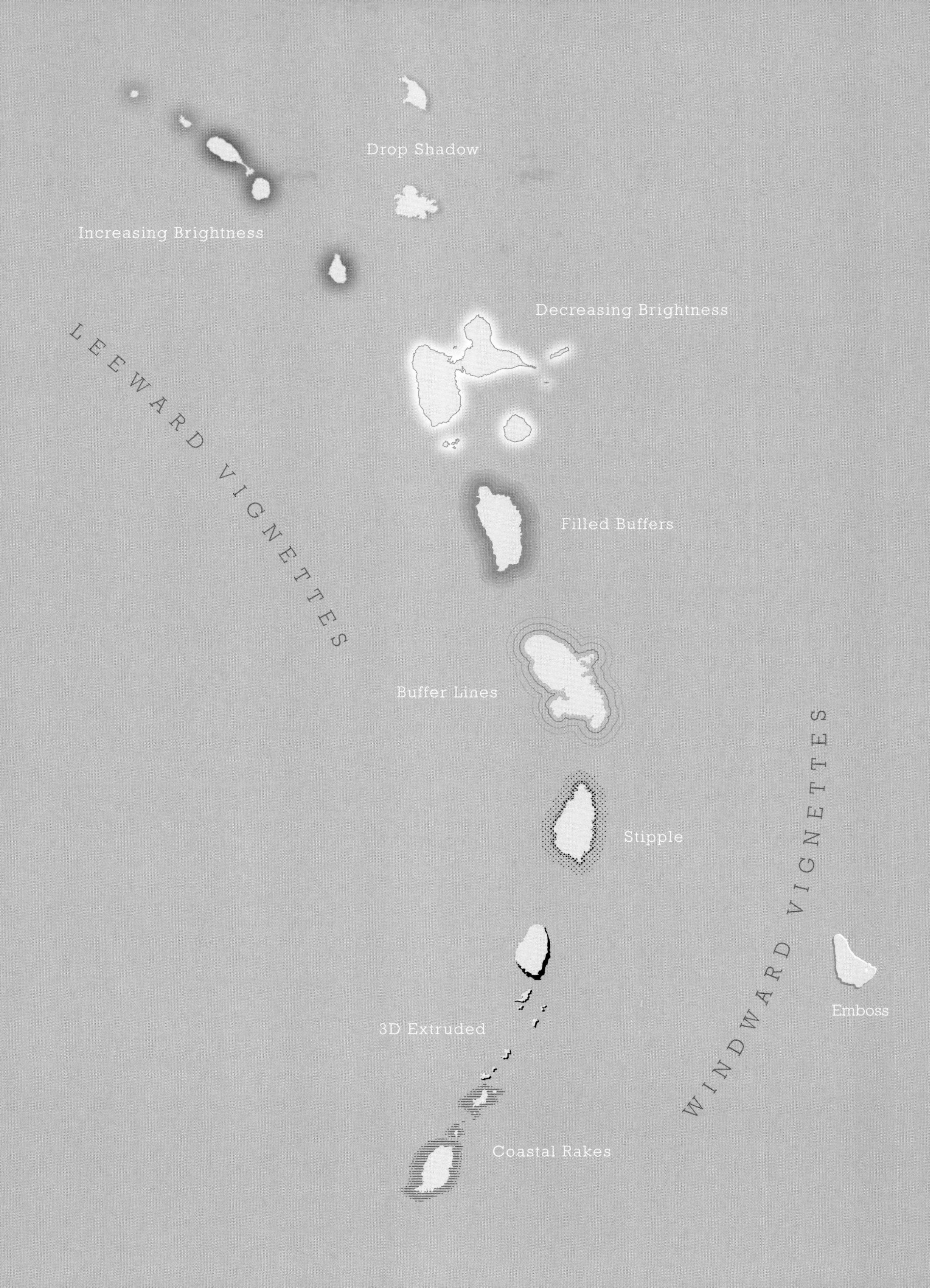

Drop Shadow
Increasing Brightness
Decreasing Brightness
LEEWARD VIGNETTES
Filled Buffers
Buffer Lines
Stipple
WINDWARD VIGNETTES
Emboss
3D Extruded
Coastal Rakes

# Viral cartography

## Anyone can make a map but how does it go viral?

**A lot of modern cartography is designed as a way to generate internet traffic, likes, and retweets.** This is a mixed blessing in terms of what might be called *professional cartography* since maps are usually made for a specific user. The trend of making a map for its own sake and simply posting it to the internet has formed a new mode of cartographic practice. Many may not evidence the best in cartographic design (an impetus for this book!) but they can sometimes generate enormous interest. They certainly pique interest from a consumer base with an insatiable appetite for maps of almost any theme imaginable. They illuminate and illustrate daily social media feeds and often shine brightly for a short time, and then slip out of the public consciousness just as rapidly.

**Viral maps tend to show a single theme simply.** They are almost always easy to understand, possibly overly simplistic yet unemcumbered by visual clutter. This supports the short-term attention span of the internet surfer but the downside is that the map often lacks nuance or the ability to support further exploration. Viral maps often use a familiar theme as a framework. For instance, the use of a metro map schematic onto which random themes are hung. The map type often makes no sense in the context of the rehashed theme but the familiar is the key to gaining immediate attention.

**The search for our place in the world is at the heart of many viral maps.** It's a natural fascination to see where we are in the world and what is close by so many maps support the ability to see what is closest to where you live. Stereotypes are often reinforced. The maps are often superficial as a result. The fact they may contain errors of a geographical or cartographic nature is almost inconsequential against the desire to simply see what exists where you live.

**Ugly maps are just as likely to go viral as well-designed maps.** The internet has no barometer of taste where viral maps are concerned. Style can often outweigh substance but a flashy map is no more likely to become a viral sensation than a plain, simple map.

**See also:** Fantasy maps | Form and function | Hand-drawn maps | Style, fashion, and trends

Traditional cartographic publishing was restricted to print—maps and atlases, generated over months or years by people who might be referred to as *professional cartographers*. Though it's true that not every map has, and has to be made by a professional cartographer, jump forward to the present day, and the internet has provided a publication mechanism for anyone with the savvy to jump online and post something for the world to see.

Internet bots scrape and republish maps from all corners of the internet. They have no filter for quality but they a generate huge amount of clickbait, which, in turn, places many eyes on the work at hand. The number of people that see the work of these new mapmakers is often way beyond the dreams of cartographers of yesteryear. Blogs and aggregators also collate and rehash maps to generate content that they know will bring people to their site. Clicks and likes are the new metric of success!

Reddit has provided a home for many aspiring viral cartographers. For instance, the subreddit r/mapporn is home to countless maps and a constantly updated feed of maps. This is perhaps the essence of what might be called the neo-geography revolution, which places the tools and the data of the geographer in the hands of the citizen scientist or, simply, someone who makes maps as a hobby.

Many of the maps that litter the internet generate their interest through humour. They take often partial pieces of information along a particular theme and make something interesting. If they are well designed and look good at the same time, then that is to their credit.

**Opposite:** *Vaguely Rude Places* by Gary Gale, 2013, and associated blog posts, from vicchi.org, explaining how his map mashup went viral.

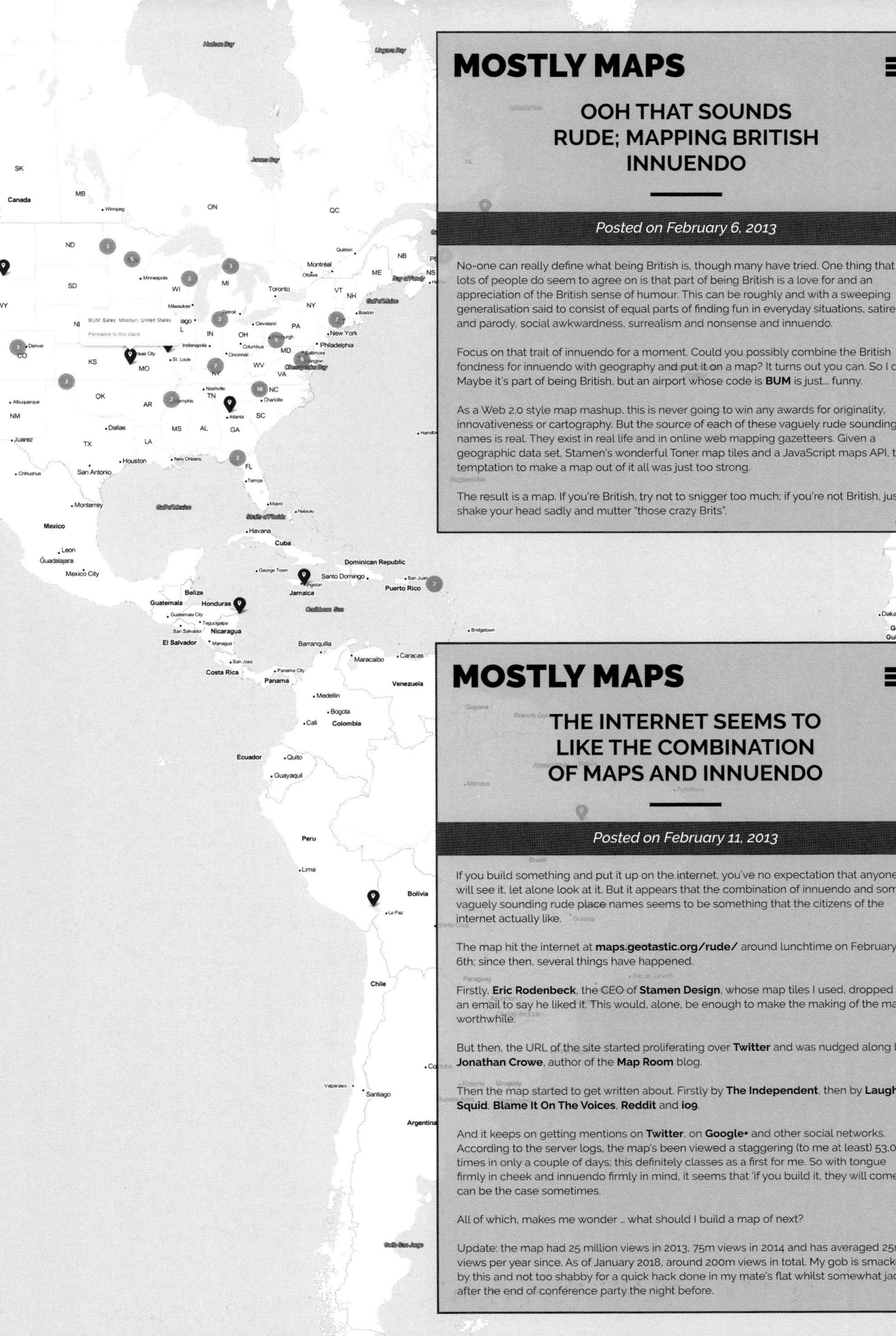

# MOSTLY MAPS

## OOH THAT SOUNDS RUDE; MAPPING BRITISH INNUENDO

*Posted on February 6, 2013*

No-one can really define what being British is, though many have tried. One thing that lots of people do seem to agree on is that part of being British is a love for and an appreciation of the British sense of humour. This can be roughly and with a sweeping generalisation said to consist of equal parts of finding fun in everyday situations, satire and parody, social awkwardness, surrealism and nonsense and innuendo.

Focus on that trait of innuendo for a moment. Could you possibly combine the British fondness for innuendo with geography and put it on a map? It turns out you can. So I did. Maybe it's part of being British, but an airport whose code is **BUM** is just... funny.

As a Web 2.0 style map mashup, this is never going to win any awards for originality, innovativeness or cartography. But the source of each of these vaguely rude sounding names is real. They exist in real life and in online web mapping gazetteers. Given a geographic data set, Stamen's wonderful Toner map tiles and a JavaScript maps API, the temptation to make a map out of it all was just too strong.

The result is a map. If you're British, try not to snigger too much; if you're not British, just shake your head sadly and mutter "those crazy Brits".

# MOSTLY MAPS

## THE INTERNET SEEMS TO LIKE THE COMBINATION OF MAPS AND INNUENDO

*Posted on February 11, 2013*

If you build something and put it up on the internet, you've no expectation that anyone will see it, let alone look at it. But it appears that the combination of innuendo and some vaguely sounding rude place names seems to be something that the citizens of the internet actually like.

The map hit the internet at **maps.geotastic.org/rude/** around lunchtime on February 6th; since then, several things have happened.

Firstly, **Eric Rodenbeck**, the CEO of **Stamen Design**, whose map tiles I used, dropped me an email to say he liked it. This would, alone, be enough to make the making of the map worthwhile.

But then, the URL of the site started proliferating over **Twitter** and was nudged along by **Jonathan Crowe**, author of the **Map Room** blog.

Then the map started to get written about. Firstly by **The Independent**, then by **Laughing Squid**, **Blame It On The Voices**, **Reddit** and **io9**.

And it keeps on getting mentions on **Twitter**, on **Google+** and other social networks. According to the server logs, the map's been viewed a staggering (to me at least) 53,000 times in only a couple of days; this definitely classes as a first for me. So with tongue firmly in cheek and innuendo firmly in mind, it seems that 'if you build it, they will come' can be the case sometimes.

All of which, makes me wonder ... what should I build a map of next?

Update: the map had 25 million views in 2013, 75m views in 2014 and has averaged 25m views per year since. As of January 2018, around 200m views in total. My gob is smacked by this and not too shabby for a quick hack done in my mate's flat whilst somewhat jaded after the end of conference party the night before.

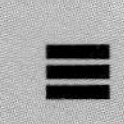

# Vision

## Seeing colour comes via different wavelengths of light, which people see differently.

**Seeing colour involves a stimulus (light), the reaction of the eye, and the interpretation by the brain.** Visible light (part of the spectrum of electromagnetic energy) takes on the form of a wave as it travels. A wavelength is measured in nanometres, nm, where 1 nm = 1 billionth of a metre, with visible wavelengths varying between 380 and 760 nm. Everyone sees and processes colour differently, which has consequences for the use of colour in map design.

**Visible light in fact contains a number of wavelengths rather than a single wavelength.** Different colours dominate specific wavelengths as Newton's classic prism experiment shows. The prism deflects individual wavelengths in visible light as they travel through the prism and project the different dominant hues. The human eye perceives wavelengths of light that are reflected, emitted, or transmitted.

**The human eye acts in much the same way as a camera.** It lets in visible light through the pupil (a diaphragm that changes to let in different amounts of light) and is focussed by a lens that then forms an image on the retina. The retina contains millions of light-sensitive rod and cone cells which, when stimulated, transmit an image to the brain via the optic nerve. Rods are sensitive to small amounts of light but do not process colour; whereas cones provide colour vision but require much more light to process the image. This explains why, in low-light conditions, humans struggle to see colours at all—a fact to bear in mind if your map is to be used in low light since you'll need brighter colours to counter the problem.

**Rods and cones are specialised nerve cells that contain light-sensitive chemicals that respond to light.** Cones themselves are of three types, each sensitive to a part of the visible spectrum. L-cones are sensitive to long wavelengths, M-cones to medium wavelengths, and S-cones to short wavelengths. Sensitivities overlap to some extent but the combination of stimulation that the cones receive because of the dominant wavelengths being received are passed to the brain that then 'creates' the image in colour.

**See also:** Focussing attention | Navigating a map | Seeing | Seeing colour

The trichromatic theory of colour perception suggests that long wavelengths of light will predominantly stimulate the L-cones, which the brain processes as the colour called red. At the other end of the spectrum, if light contains short wavelengths, then the S-cones will be those stimulated and you see blue. If the long and medium cones are stimulated to a certain extent, then you perceive yellow. If all three cones are stimulated equally to their maximum, then you perceive white. Conversely, if they all receive low stimulation, then you perceive black, or grey when stimulated equally.

Alternatively, opponent-process theory explains that colour perception is based on a lightness-darkness channel and two opponent colour channels: red-green and blue-yellow. Colours in each of the opposing colour channels work in opposition to each other so that you do not perceive mixtures of red-green or blue-yellow. Instead, you see mixtures of pairs from each channel: red-blue, red-yellow, green-blue, and green-yellow. Cones act to excite or inhibit cells that lead to particular colour perception.

The fact that there are two apparently conflicting ways of describing how the eye–brain system processes colour might seem contradictory. In real terms, it is a combination of the two theories that fully explains colour perception. The trichromatic theory is correct in that cones react differently and provide stimuli that produce the perception of colour. The way in which this information is processed is better explained by opponent-process theory. Of course, this works for what we might call 'normal' colour vision but many people do not perceive colour in this way, and that leads us to even more ways to think about how to use colour cartographically to support alternative colour perception.

**Opposite:** Light refracting through glass into separate wavelengths of colour which make up the visible part of the electromagnetic spectrum.

Ultraviolet
Infrared
Violet
Indigo
Blue
Green
Yellow
Orange
Red
wavelength in nanometers
400
500
600
700

# Visualization wheel

## Balancing design ideas is a constant tension between competing aims.

**Alberto Cairo's visualization wheel for infographics (Cairo 2013) provides a way to think about the competing aims of map design.** As you create a map you'll have an overall aim but the delivery of the product will require you to consider a range of tensions. The wheel has two hemispheres that each contain six features. The upper hemisphere reflects more deeply and more complex while the lower is more shallow and immediate.

**The wheel's features exist across a non-linear continuum.** Maps can take many shades along the abstraction-figuration axis. The more pictorial or mimetic, the more the map tends to figurative. A panorama will be more figurative and a cartogram more abstract but the same idea can be used for symbology throughout a map. Although maps can be both functional and aesthetically pleasing, the functionality–decoration axis refers to the choice of visual elements, which are not required directly to support understanding. Thinking of how much decoration your map contains helps to assess its balance with the functional content. The density–lightness axis forces you to consider the amount of material and the space it uses. Both dense and sparse can be good in design terms but assessing density of information helps reflect on the amount of space you take in delivering the message. The layers of detail in the map and the form of encoding is represented by the multidimensionality–unidimensionality axis. Some maps support multiple ways of seeing and interpreting. They are exploratory. Others are statements of a single idea, to be consumed as such. There's nothing wrong with using the familiar. Often it leads to clearer understanding. Originality is an expression of difference, and although new ways of mapping can be fascinating it's often for the sake of the technique over the map's message. This in itself is extremely valuable for cartography as a discipline but finding truly original forms is challenging. Finally, consider what you're trying to say and how you say it visually. Novelty is achieved by clearly expressing a single aspect once. Repeating the same idea differently leads to redundancy. The balance is often difficult to achieve. Novelty is important to gain interest. Some redundancy might be needed to make the map clear.

**See also:** Design and response | Form and function | Map cube | Types of maps

There's no optimal solution to the chart that helps describe the perfect map. It's more a conceptual tool you can use in evaluating your own work and whether your graphical solution actually matches your intentions. It can be used as a way to assist in map critique whether of your own work or in evaluating other maps.

The visualization wheel can be extended graphically into a radar chart that will reveal different shapes depending on the maps you're exploring. By evaluating each paired axis and determining where a map sits on the continuum, you'll end up with a series of data points that you can join to build the radar graph.

The use of any mechanism to support the conceptual thinking behind your map cannot be completed in isolation of other factors. The complexity of the map and its form and function are hugely modified by the intended audience. The more general your audience, the more shallow you'll likely need to make the map's form to support the function. A more specialised audience is likely to be able to handle more complex forms and deeper content.

Many map types and forms have become standard practice and support fundamental day-to-day cartography, but once they were new, original, and often contentious. The canon of cartography exists to provide a rich palette of opportunities to exercise the full spectrum of options on the visualization wheel.

Visualization wheel

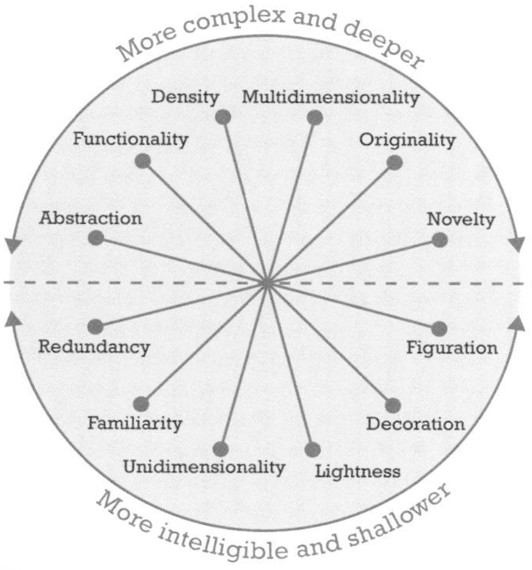

More complex and deeper

Density
Multidimensionality
Functionality
Originality
Abstraction
Novelty
Redundancy
Figuration
Familiarity
Decoration
Unidimensionality
Lightness

More intelligible and shallower

Visualization wheel favoured by scientists and engineers

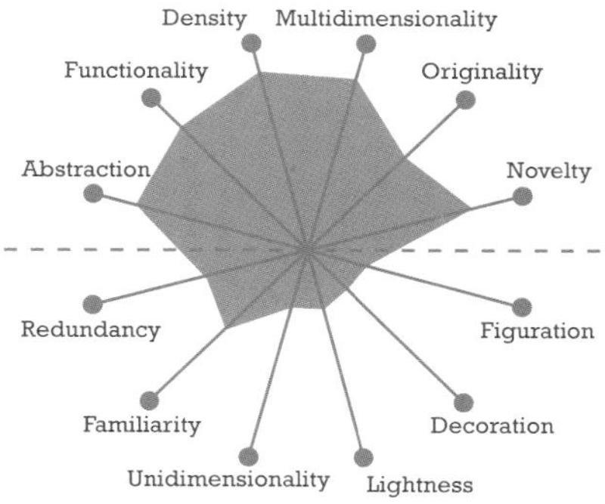

Density
Multidimensionality
Functionality
Originality
Abstraction
Novelty
Redundancy
Figuration
Familiarity
Decoration
Unidimensionality
Lightness

Visualization wheel favoured by artists, graphic designers, and journalists

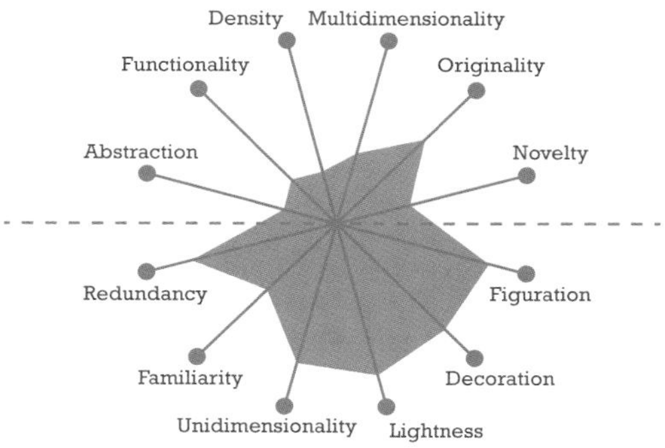

Density
Multidimensionality
Functionality
Originality
Abstraction
Novelty
Redundancy
Figuration
Familiarity
Decoration
Unidimensionality
Lightness

N IN THE U.S.A./NOT BORN IN THE U.S.A.

Visualization wheel for this album sleeve infographic

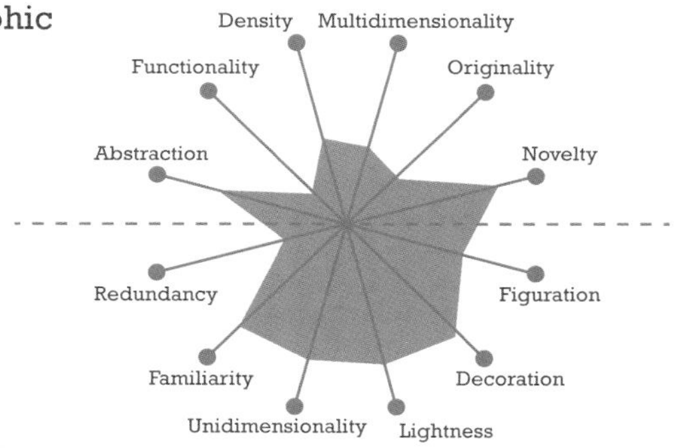

Density
Multidimensionality
Functionality
Originality
Abstraction
Novelty
Redundancy
Figuration
Familiarity
Decoration
Unidimensionality
Lightness

WHERE THE STREETS HAVE NO NAME

LIMITED EDITION
12" GIANT 45 RPM
X-14515
Kenneth Field
1. WHERE THE STREETS HAVE NO NAME
2. WHERE THE STREETS HAVE NO NAME

# Voronoi maps

## Sub-partitioning a map into regions on the basis of proximity to a subset of the theme.

**Voronoi diagrams have long been used in mathematics to partition a plane into regions on the basis of distance to points in a subset of the theme.** The result is a partitioning of a plane into regions where every point is contained in a subset because it is closer to the internal seed of that subset than any other subset.

**Voronoi maps are the spatial outcome of the process of calculating regions which tessellate.** They are also known as *Thiessen polygons* in which any location or value in one polygon is closer to its associated point than points in other polygons on the basis of Euclidean distance. Input data for making Voronoi maps can be point data or points that represent area data. For instance, taking a dataset of world airports, a Voronoi map would show the proximal zone where any location is included in the region that includes its closest airport.

**In some respects, Voronoi maps are merely abstract representations of a geography.** They turn point-based features into area-based features so it's vital to understand whether the area itself bears any relevance to the original data. For airports, yes, the resulting map shows proximity. If, however, data of the location of sports grounds was used to create a Voronoi map of the potential catchment area of support for that team, the resulting areas would likely be inaccurate. Support for a team is based on far more than proximity and so care must be taken to ensure the translation of the point data into an area makes sense. Voronoi maps are a great example of a technique that works well graphically but may not work quite so well when you think about the logic behind the resulting areas and shapes. It's therefore always useful to ask the simple question: Does the result make sense to represent the theme I'm trying to illustrate?

A development of the basic Voronoi map is the Voronoi treemap. This combines the basic Voronoi map design with additional secondary information that is incorporated in a spatially constrained treemap. One example, published by the *New York Times* (Aisch and Gebeloff 2014), presented a version of a Voronoi treemap which shows, for each state, where the people living in the state were born. Within each state, larger shapes represent a group making up a larger share of the population.

Using recognisable geographies is, like most cartography, a natural framework for making Voronoi maps but using abstract frameworks is also possible. For instance, using a circle or a square and subdividing based on some quantitative data can create a very different map.

John Snow's famous map of the outbreak of cholera in Soho, London, in 1865 is often lauded as a breakthrough in thematic cartography. In fact, Snow also used a Voronoi map to explore the relationship between the infected population and where they got their drinking water. Dotted lines were drawn on the map to denote equidistance between the infamous Broad Street pump and the nearest alternative pumps. In so doing, the Voronoi map illustrated that most of the people who died were closer to the Broad Street pump than an alternative and so were likely to have drawn their water there.

**See also:** Graphs | Quantitative statistical maps | Schematic maps | Treemap

**Opposite:** *The United States of Craigslist* by John Nelson, 2015, uses Voronoi polygons to demarcate zones.

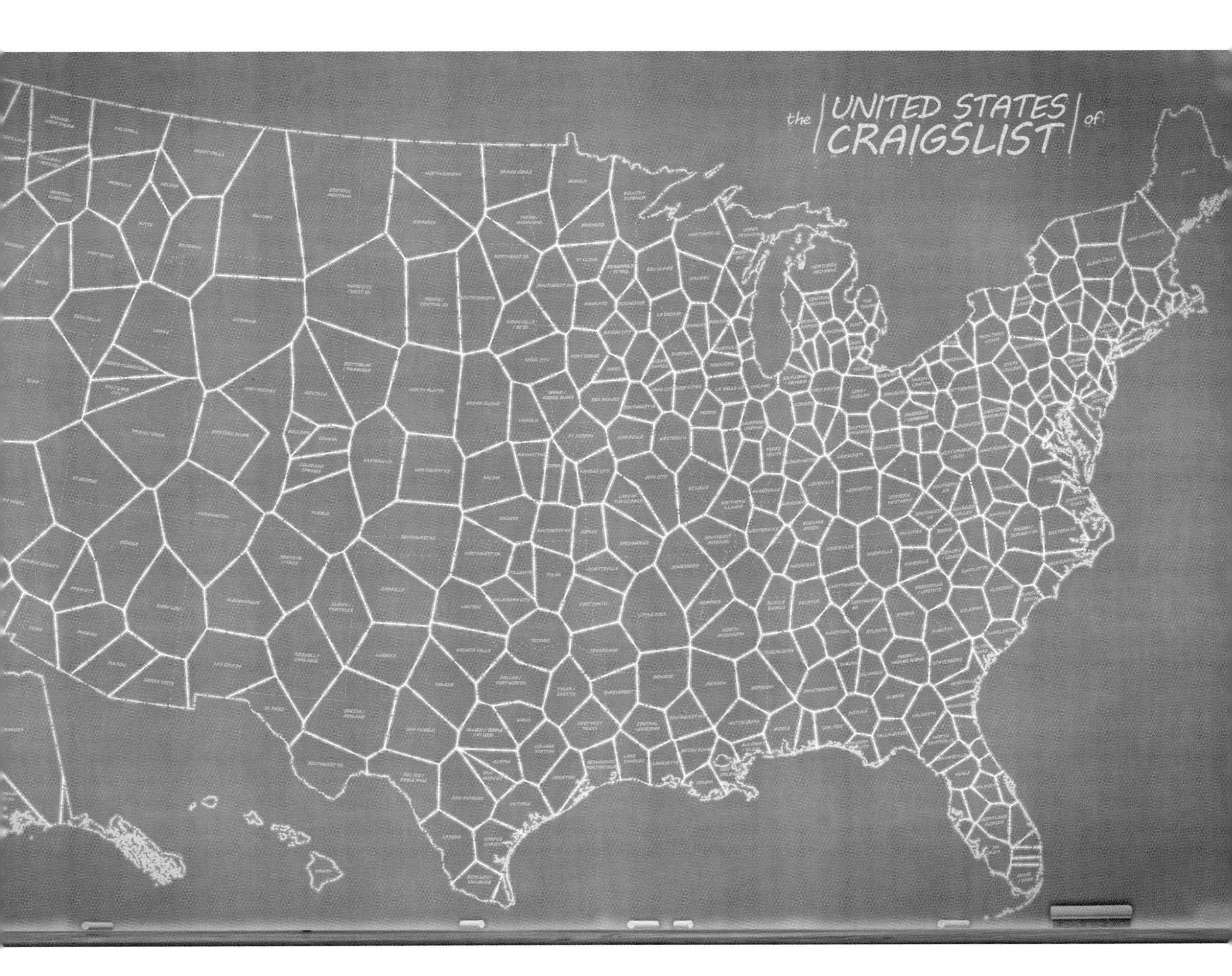
the UNITED STATES of CRAIGSLIST

# Where the Population of Europe Is Growing and Where It's Declining

**Berliner Morgenpost**

**2015**

There is something magical in being able to find yourself in a data graphic. Some visualizations tell a single story, but others are designed in ways that enable personal exploration and discovery. This is one of them. It illustrates how crucial the interplay between macro and micro views is when designing visualizations that aren't just informative, but whose depth is true to the complexity of the underlying data.

I was born in Galicia, in the northwestern corner of Spain, so the first time I saw this amazing map, my eyes quickly scanned it in its entirety, but then they immediately got directed to my birthplace. I zoomed in and marveled at the clarity of the patterns displayed: rural areas being depleted, urban regions booming; coastal cities receiving newcomers, towns in the inner plains fading and vanishing.

We've read stories about these migratory changes before, narratives by those who left their home to look for better futures for themselves and their descendants; now, we can see them all together in an aggregated portrait of sad blues and vibrant oranges.

—Alberto Cairo

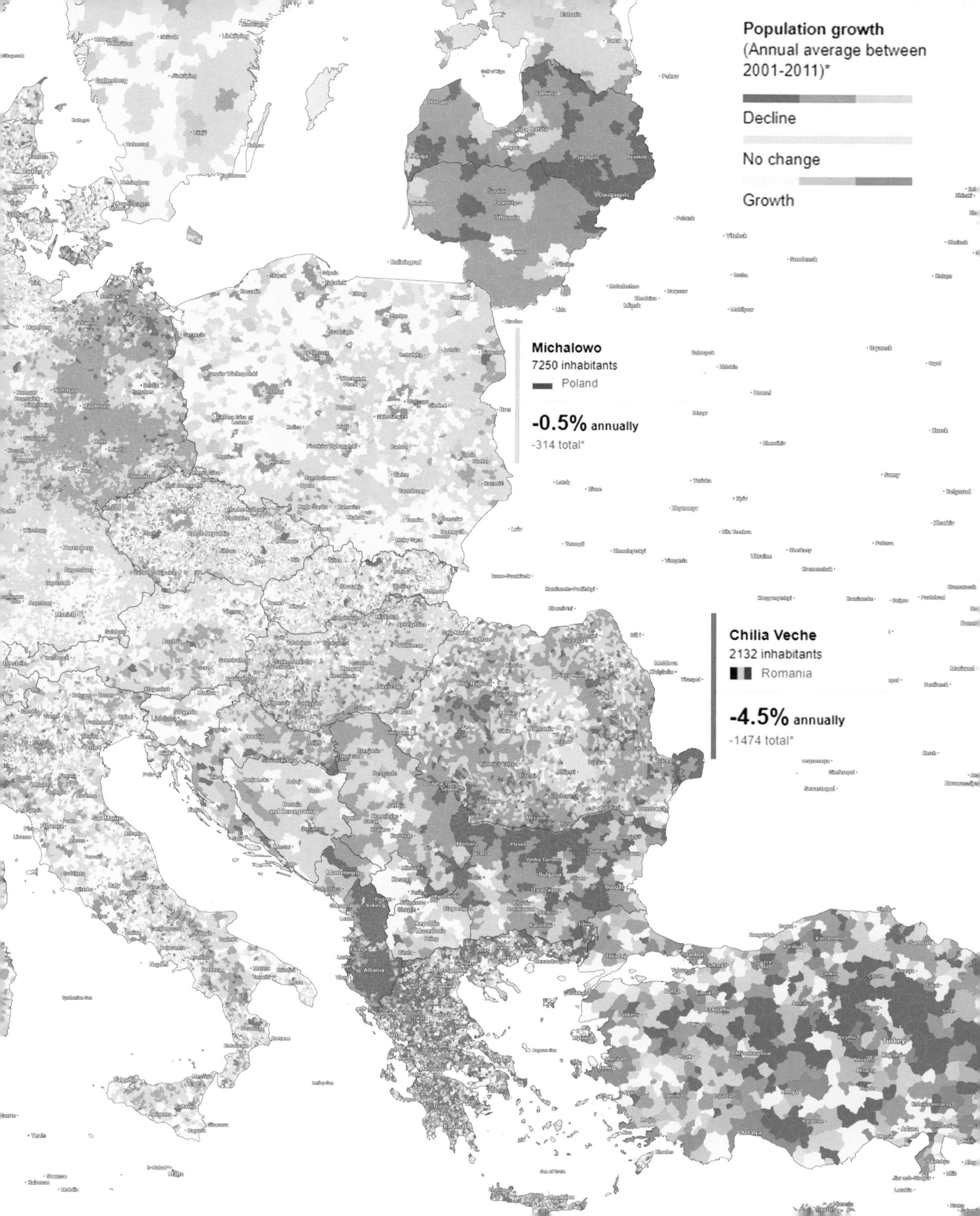
Population growth
(Annual average between 2001-2011)*
Decline
No change
Growth
Michalowo
7250 inhabitants
Poland
-0.5% annually
-314 total*
Chilia Veche
2132 inhabitants
Romania
-4.5% annually
-1474 total*

# Waffle grid

## Technique to show constituent components of a whole dataset.

**A waffle grid shows quantities of categories that form a whole.** It is similar to a pie chart but takes a square or gridded appearance. The grid is normally 10 × 10 and so contains 100 data cells. Each of the constituent categories is then displayed as a recurring symbol in the waffle grid. This symbolisation might be through using different hues to represent different categories. Alternatively, isotype might be used to create a pictorial representation of the category itself.

**The grid supports counting and data recovery.** Unlike a pie chart which relies on the reader being able to decipher the relationship between non-stated angles and proportions, a waffle grid supports direct measurement through counting the number of grid cells each category belongs to. A drawback is that the data recovery is limited to integers but the rapid recovery of information via this technique (in multiples of 10) is of considerable use.

**Small parts of the whole are much easier to see on a waffle grid than other charts.** By using a hue that contrasts or is more saturated, it is much easier to denote a single cell as a category that attempting to reconcile a very thin slice in a pie chart.

**Waffle grids can be used either on their own in support of a map or on the map themselves.** They lend themselves to being used in place of other graph types. By escaping from the 10 × 10 grid, they can also be used to show proportional differences between the same categories of a variable across a map. Repetition of symbology can still be counted as each individual icon or cell might represent a specific data value. Symbology is usually placed in a legend, especially when direct labelling on the grid is difficult to include.

**See also:** Cartograms | Graphs | Multivariate maps | Pie and coxcomb charts

Part of the success of the waffle grid is its basis on repetition. Perceptually, humans are good at recognising repetition among objects, and this familiarity helps in recovering meaning. These grids are therefore easy to read, are discoverable, and do not distort the data in ways that our cognitive system has difficulty disentangling. They provide a better solution than the pie chart for recovering data and are visually interesting.

The waffle grid provides one of the few examples of a chart that works relatively well in 3D. Given a consistent 10 × 10 grid, each layer of the waffle grid can be extruded to the same height as the basic dimensions of each cell. Additional layers of extruded cells sit on top of one another and so it's easy to count the number of layers and resolve these as multiples of 100. Ordinarily, occlusion would cause difficulties for data recovery, and perspective might distort the perception of magnitudes (e.g. a 3D pie chart) but the 3D waffle grid still functions and can be read rapidly. The only caveat is if different categories are included and symbol hue changes on a layer that is occluded other than the edge symbols.

Planimetric

Shear and drape

3D single layer

3D multiple layers

**Opposite:** *Applied Irrigation Water from the California Water Atlas* by William Bowen, 1979.

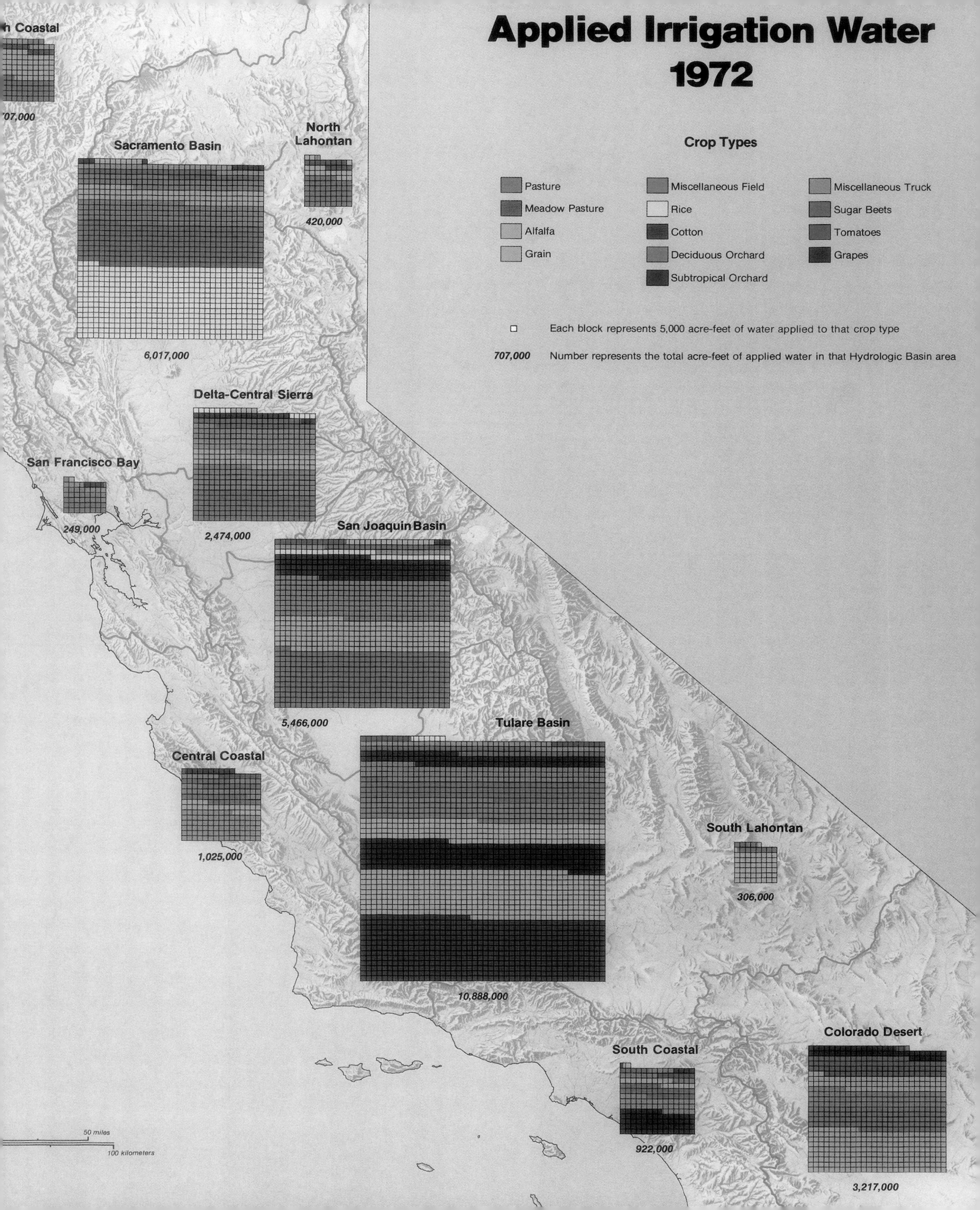

Applied Irrigation Water
1972
Crop Types
Pasture
Meadow Pasture
Alfalfa
Grain
Miscellaneous Field
Rice
Cotton
Deciduous Orchard
Subtropical Orchard
Miscellaneous Truck
Sugar Beets
Tomatoes
Grapes
Each block represents 5,000 acre-feet of water applied to that crop type
707,000 Number represents the total acre-feet of applied water in that Hydrologic Basin area
Coastal
07,000
Sacramento Basin
6,017,000
North Lahontan
420,000
Delta-Central Sierra
2,474,000
San Francisco Bay
249,000
San Joaquin Basin
5,466,000
Tulare Basin
10,888,000
Central Coastal
1,025,000
South Lahontan
306,000
Colorado Desert
3,217,000
South Coastal
922,000
50 miles
100 kilometers

# Weather maps

## A blend of different cartographic representations for a specific purpose.

**Weather maps are ubiquitous.** They are routinely broadcast on television, printed in daily newspapers, and available in realtime on mobile devices and weather apps. We use them every day to make a simple decision about what clothes to wear among many other reasons. Cartographically they are somewhat unique because they blend descriptive and statistical information with representations of temporal data such as flow and movement.

**Temperature is often depicted using coloured bands showing equal temperature ranges as a surface.** Although some favour a rainbow colour scheme to symbolise these isotherms, they would be difficult to interpret because different hues are not perceived as an empirical progression. Alternatively, a subtle hue progression from blues that show temperatures below a certain value (e.g. 0°C) through light greens to oranges and reds gives a good indication of relative temperature. Atmospheric pressure is usually symbolised by isobars—connected lines of equal value. Values are normally annotated on the isobars and the letters *H* and *L* used to show points of high and low pressure, respectively. Precipitation is sometimes shown using different pattern fills for areas of rain, snow, and ice. Warm and cold fronts show not only the type but the direction of the front using standard marker symbols along solid lines that are based on semicircles for warm fronts and triangles for cold fronts. Different symbol treatments and combinations give meaning to different types of front.

**Local weather can be detailed using meteorological symbols and numerical data.** Symbol sets are positioned at points across the map and show multiple pieces of information, including wind direction, wind speed, weather type, cloud cover type, and amount as symbols. Temperature, visibility, dew point, atmospheric pressure, atmospheric pressure change, and amount of precipitation are also shown using numerical detail.

Televised weather maps tend to be highly generalised and animated. Time-slice forecasts are stitched together and animated in a loop to give a sense of how weather has or is likely to change—the essence of the use of maps as a forecasting tool. Pulse-doppler radar techniques are also used to add detail such as the motion of precipitation. Radar is used to bounce a microwave signal off features and assess how the reflected signal has altered the frequency of the return signal. Other satellite-derived remotely sensed imagery can show snow cover, flooding, or other events.

The standard weather map is often augmented by a range of other maps that, combined, tell a comprehensive story of the environmental conditions. Indices give measures of solar radiation, often as a UV rating, air pollution, and pollen forecast. These maps are usually generalised rather than highly specific because of the difficulty in precisely predicting actual conditions.

Climate scientists and meteorologists have been notorious for their misuse of basic cartographic colour theory in developing the schemes they use for maps. The main culprit is the use of the rainbow (spectral) colour scheme. In addition to being illegible for many visual impairments, they also distort perceptions of data and alter meaning. They create false boundaries between values which lead to false impressions. Different hues bear no relationship to a sequence so there's no inherent meaning in the position of hues in relation to one another. Some hues (e.g. yellow) stand out far more than others despite not being at either end of the spectrum. They signify mid-values but they appear brighter. Detail is hard to see among colours. Far better would be the use of single colour changes where value or saturation are modified. These methods codify the data far better.

**See also:** Continuous surface maps | Contours | Isarithmic maps

## Weather map symbology graphic / realistic

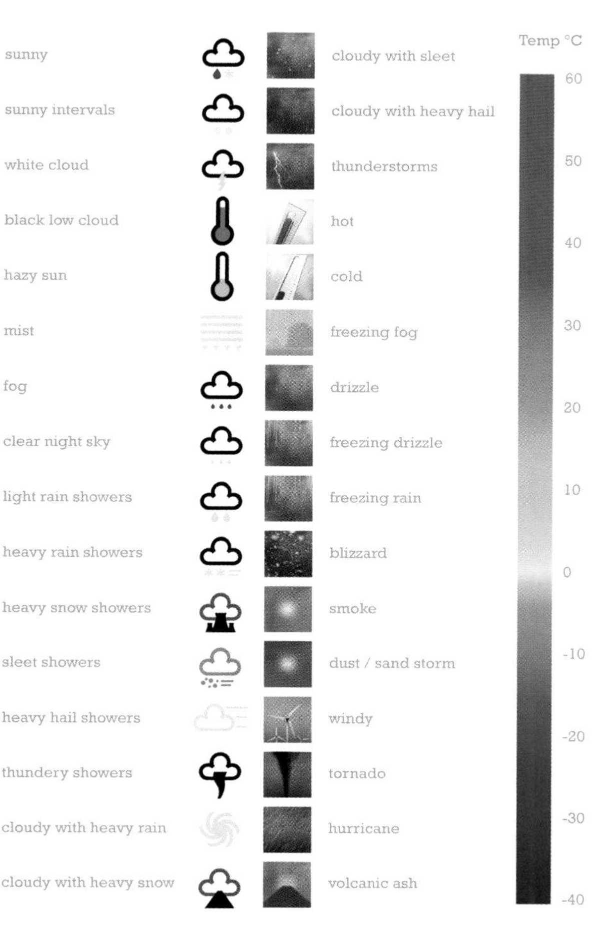


## The 1987 Great Storm - TV style weather map

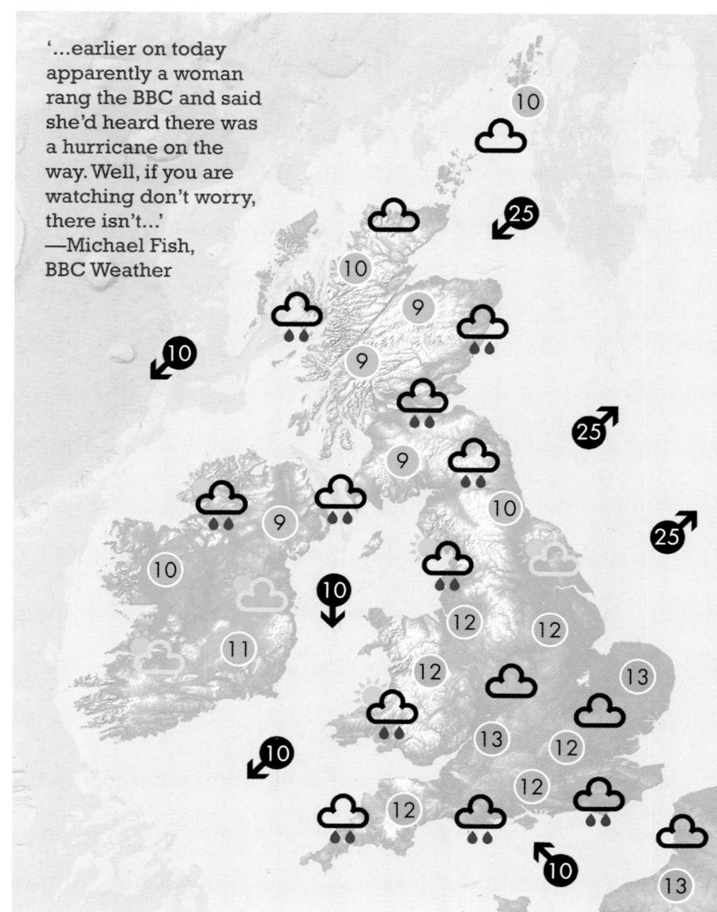


## Synoptic chart symbology

A wind direction
B wind speed
C extent of cloud cover
D barometric pressure
E air temperature

F weather condition
G visibility
H dewpoint
I pressure change
J cloud base

B A E D G F C I H J

### Common weather symbols

- intermittent drizzle
- continuous drizzle
- rain showers
- intermittent rain
- continuous rain
- heavy rain
- thunderstorm
- heavy thunderstorm
- haze
- squall
- tropical storm
- hurricane
- tornado
- sand/dust storm
- snow
- sleet
- hail shower
- freezing rain
- mist
- fog

### Cloud cover (eighths)

- clear sky
- 1/8th
- 2/8th
- 3/8th
- 4/8th
- 5/8th
- 6/8th
- 7/8th
- 8/8th
- sky obscured

### Wind speed (knots)

- calm
- 5
- 15
- 20
- 25
- 55

### Fronts

- cold
- warm
- occluded
- stationary
- trough
- dry line
- squall line

### Cloud types

- altocumulus
- altostratus
- cirrocumulus
- cirrostratus
- cirrus
- cumulonimbus
- cumulus
- nimbostratus
- stratocumulus
- stratus

## Synoptic chart - 15th October 1987 18:00

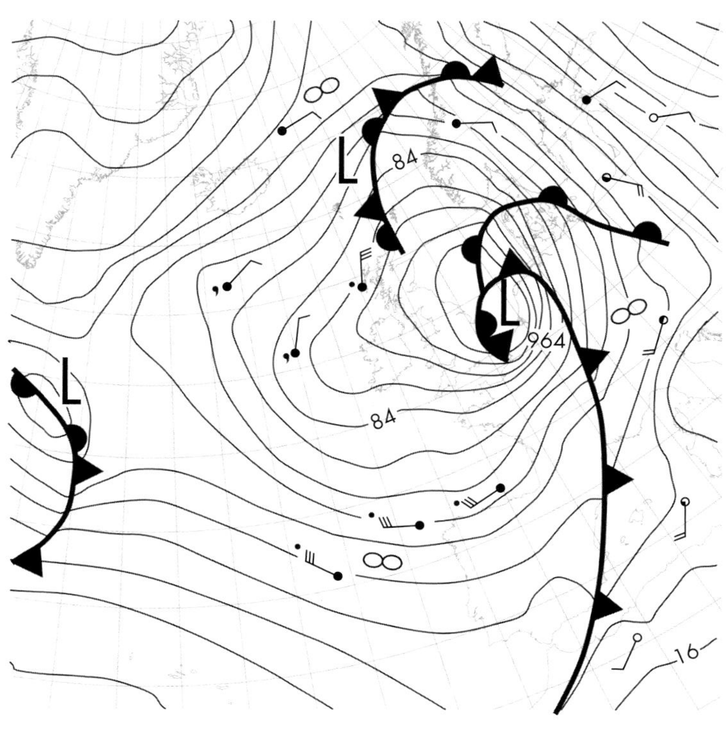

# Web mapping

**Print mapping is by no means dead but maps are increasingly published to the web. This transition has revolutionised map design and production.**

**Computing in general has revolutionised cartography.** It has underpinned the transition from a photo-mechanical to digital process, but more than that, it has led to a sea change in the way maps are designed. Maps are now consumed—and increasingly produced—for mobile devices, smartphones, and tablets. These web maps have replaced the somewhat tired paper street atlases in our glove boxes and to a large extent also the consumption of maps on our desktop computers. That all of this is happening at little or no direct cost to consumers is all the more remarkable. Maps have never been more accessible, nor cost less at the point of consumption.

**We carry the world in our pockets, and it's a detailed world whose cartography belies the simplicity of its operation.** The web in general has supported the rapid rise of mobile cartography with always-on, on-demand cloud-based mapping available to our always-on devices. Mobile devices are used in situ. They are hyper-localised to provide content relevant to who we are and where we are. The device becomes part of the cartography since the map is customised according to location. The interfaces to digital maps are now high resolution and support high-definition, fast, scalable cartography that has moved beyond the generic to become our own personal cartography.

**Maps can now be designed relatively easily using web-based tools that support numerous map types and requirements.** Additionally, many more traditional digital workflows, involving GIS particularly, support fast and efficient publication of the desktop map to a web map. Web maps might be referred to as the *construction zone* or *staging area* where modifications can be implemented. The final web map is more likely to be published to an app that provides a cleaner, more useful user interface and experience to the map user. Such apps can be configured and published using software designed for the purpose or through coded solutions.

Web mapping can trace its origins back to the birth of the worldwide web and the work of Sir Tim Berners-Lee at CERN in 1989, and then to the first map server: the Xerox® PARC Map Viewer in 1993. Only a decade later, the work of Lars and Jens Eilstrup Rasmussen at Where 2 Technologies was noticed by a fledgling search and online advertising company called Google. The company was acquired by Google, and the desktop program developed by the Rasmussens became a web application. Google Maps™ was launched on February 8, 2005 (coincidentally, my birthday).

After its initial launch, the map was subject to third-party reverse-engineering so that customized content could be introduced atop the map. The map mashup was born, and Google released the Google Maps API in June 2005 to support developers who wanted to integrate Google maps into their websites.

Many others to follow in almost infinite ways to now support web mapping, location-based services, and cloud-based GIS. The breadth of mapping activities extends far beyond mashups to complete online cartographic solutions and a wide array of geoenabled web platforms that integrate maps, imagery, and content.

The Google map pin has become synonymous with web mapping. Although many early map mashups suffered 'red dot fever' as content was simply dumped on a map, cartographic support for web delivery has now matured. The classic red push-pin is a metaphor for a tangible map pin and now strongly references the 'you are here' focus of our increasingly personalized maps.

**See also:** Animation | Basemaps | Imagery as background | Interaction | Mobile mapping | Page vs. screen | UI/UX in map design | Web Mercator

Google Maps on launch
8th February 2005

*'Today, Google's map includes the streets of every nation on earth, and Street View has so far collected imagery in a quarter of those countries. The total number of regular users: A billion people, or about half of the Internet-connected population worldwide. Google Maps underlies a million different websites, making its map A.P.I. among the most-used such interfaces on the Internet. At this point Google Maps is essentially what Tim O'Reilly predicted the map would become: part of the information infrastructure, a resource more complete and in many respects more accurate than what governments have....'*
*— 'Google's Road Map to Global Domination', The New York Times, December 12, 2013*

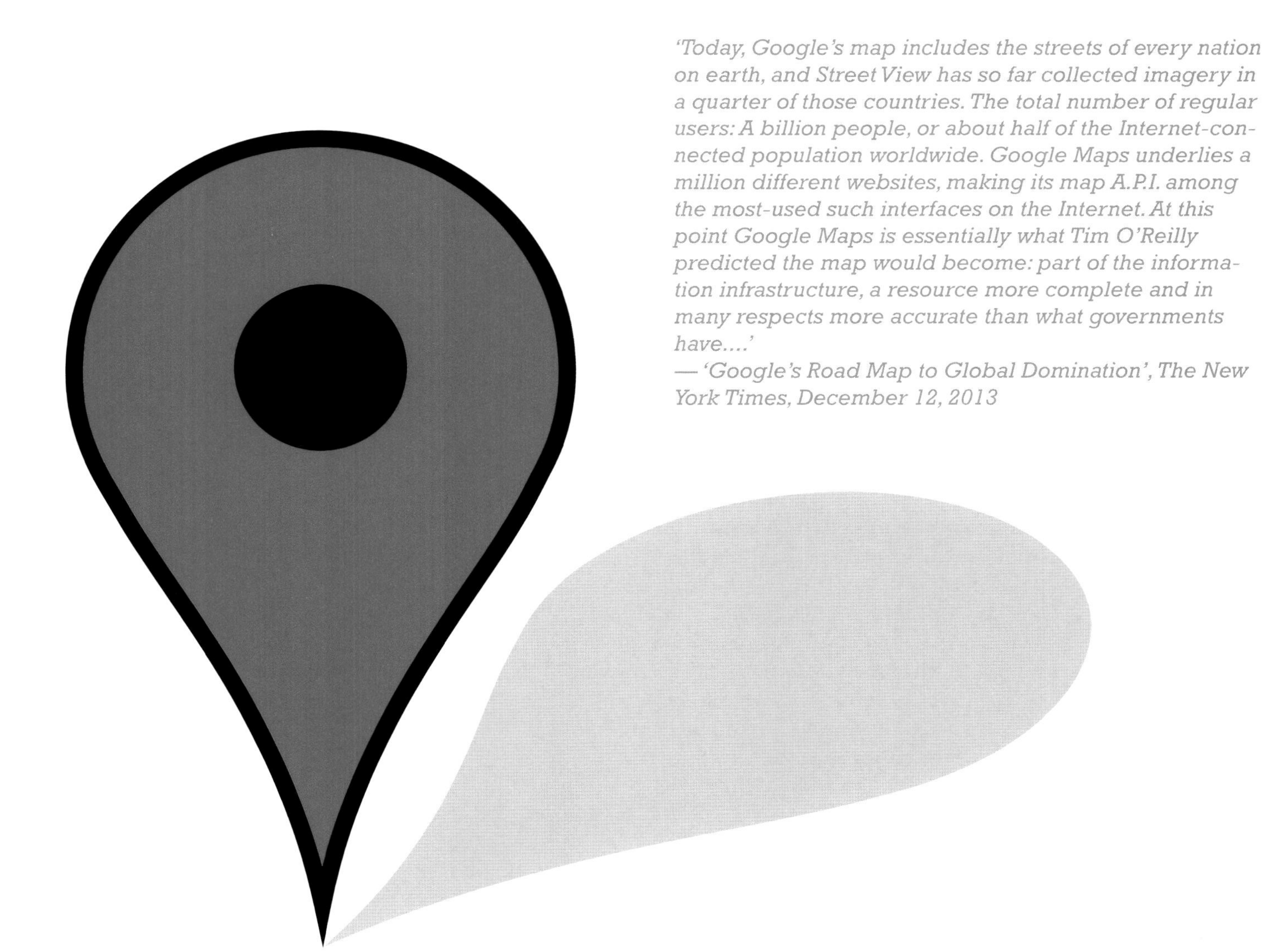

# Web Mercator

**Web Mercator has become a dominant, though much derided, web mapping projection.**

**Gerard de Gemor or de Cremer, better known by the latinized form of his name: Mercator (meaning 'merchant'), was born in Rupelmonde, Belgium, in 1512.** In 1569, he produced a new chart that contained parallel lines of longitude which for the first time meant that navigators could mark compass courses as straight lines on the map. His approach to plotting the world map went on to become possibly the most widely used and well-known map projection. The Mercator projection is a tangent case cylindrical conformal projection. Meridians are drawn equally spaced but parallels are unequally spaced; shape is preserved in small areas but with increasing distortions toward the poles, which are at infinity.

**The Mercator projection lives on in the most modern form of map production: the online web map.** Web Mercator is a spherical projection based on the WGS84 semimajor axis, which is used by most major online mapping services. The projection is reasonably well suited to the interactive form of an online map that can be zoomed seamlessly to large-scale local maps. Conformality means there is relatively little distortion at larger scales. A further benefit is that wherever you are on the map, up is due north, down is due south, and west and east are always left and right. This creates an attractive consistency and avoids some of the confusion that comes with other projections. Tiled, it works well for the entire globe, making the map seamless and ensuring imagery and other layers line up correctly.

**Web Mercator is routinely criticised because of problems associated with Mercator's projection in general.** The adoption of Web Mercator as a web mapping standard is routinely blamed on Google and its tile caching system but as one of the first to publish a seamless, zoomable web map in 2005 that's inevitable. The huge distortions in shape and area that are exposed in northern and southern latitudes present difficulties for comparisons. As such, it is a poor choice for thematic maps where equal-area projections are far more suited.

Gerhardus Mercator was primarily a craftsman of mathematical instruments and skilled engraver. He began working in the field of cartography at the age of 25 when he produced a map of Palestine, closely followed by his first map of the world in 1538. He produced several more maps, published his book on italic script, taught mathematics, and worked as a surveyor before being appointed Court Cosmographer to Wilhelm, Duke of Jülich-Cleves-Berg, in 1564. Other notable work by Mercator includes taking the word 'atlas' to describe a collection of maps, encouraging Abraham Ortelius to prepare the first modern world atlas (*Theatrum Orbis Terrarum* in 1570), producing his own atlas, and devising techniques to mass-produce both celestial and terrestrial globes.

The pre-rendering of Web Mercator map tiles has apparently odd scales that the different zoom levels represent. At its lowest level of detail (Level 0: smallest scale) the entire world map is represented by a square tile 256 × 256 pixels, centred on 0° latitude, 0° longitude (Null Island). For each integer increase in zoom level, there's a doubling of width and height, so at level 1 the map size is 512 × 512 pixels. There is also a doubling of the number of tiles in both the north–south and east–west direction. So, zoom level 1 covers the globe with a 2 × 2 grid of tiles and level 3 with a 4 × 4 grid and so on.

Because tile size (and pixel size) is fixed, each integer increase in zoom level reduces the size of the area on the ground represented by one pixel by a factor of 2. Ground resolution is the distance represented by each pixel but varies depending on latitude and level of detail. Map scale is the ratio between map distance and ground distance but is also a function of screen resolution.

**See also:** Map projections | Map projections: Decisions, decisions! | Web mapping

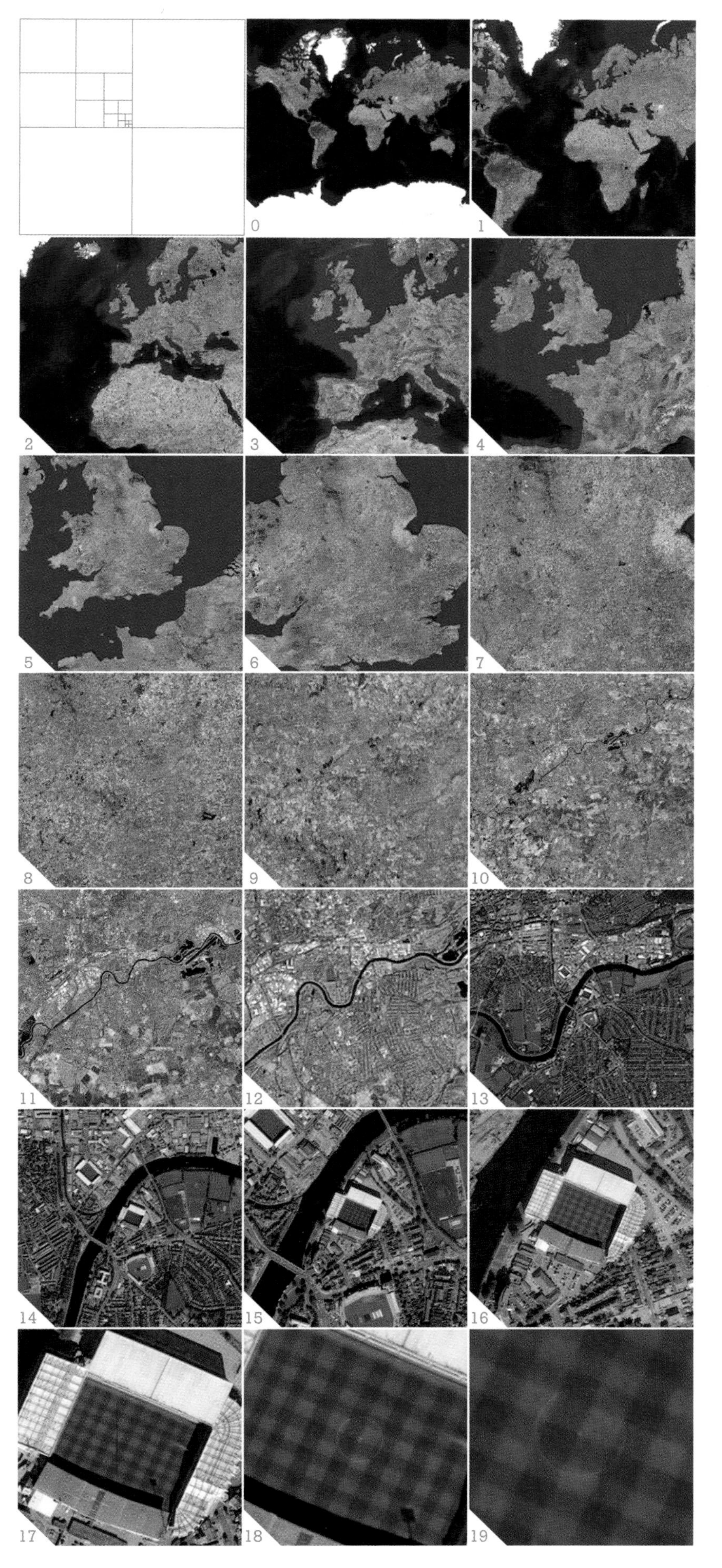


Level 16 | Reykjavík: 64.14° N

Level 16 | Singapore: 1.34° N

# Which way is up?

## Which way should north be? Is north always up, or can you change the orientation of your map?

**To many of us, our 'normal' view of the world on a map is with north positioned at the top.** That's the way most wall maps, printed maps, and atlases have been designed for decades. It's not always been the case. Many early T and O maps placed east at the top for religious reasons. There are even maps from Australian authors that invert the world to position the south at the top to make a satirical point.

**Many situational maps might position maps in different orientations.** Generally speaking, people prefer to see the way ahead as 'heads up', particularly when using a map for one of its prime purposes—navigation. This use has its roots in military mapping and references the need for pilots to have their heads up rather than looking down at instruments. In mapping this has developed into the standard of positioning the way the reader is facing at the top of the map. This position implicitly supports navigation because turns and directions are linked directly to the point of view and the map at hand.

**Most satellite navigation systems and maps made for mobile devices now automatically rotate to position the reader.** This rotation provides a heads-up display of the map which overcomes one of the perennial difficulties of navigating with a paper map, namely having to constantly rotate it and work out which way you are facing using a compass.

**The key to determining which way is up in your map is determined by its function.** If it's for navigation, then position with the view the reader is facing. Ensuring a digital map can rotate to meet a changing position is a standard expectation in modern digital mapping. If you are making a one-off map, it's fair to position it with north at the top of the page/screen, though with larger scale maps this becomes less important as the extent and shape of your content may dictate a particular orientation. If you change the orientation from what people might expect, it's particularly important to incorporate an indication such as a north arrow.

**See also:** Anatomy of a map | Earth's vital measurements | Globes | Measuring direction

Perhaps the most useful advice for determining which way is up is to work with people's expectations. You're trying to minimize the barriers to someone reading and understanding your map. If they do not understand which way up it should be relative to their understanding of the world, then that's a big initial hurdle.

There are psychological consequences of north being up. In general, north is equated to richer people and the south with poorer. This was emphasised with the use of the Peters projection and a stark red line demarcating north from south on the cover of the famous 1980 Brandt Report into international development issues.

For some maps the decision is a little more involved. For instance, maps made using azimuthal projections are commonly used to show the Northern or Southern Hemisphere, centred on the North or South Pole. For these maps, it's not a question of which way is up, but what rotation should be used. Convention places the Western Hemisphere to the left and the Eastern Hemisphere to the right but rotation could vary to ensure the visual centre of the map is coincident with the map theme.

Although there is no particular geographic, cartographic, or philosophical reason why north is represented as up, it has become convention. It's not a natural construction but a human one that has developed over the centuries and been promulgated by countless mapmakers of different nations, religions, and biases.

**Opposite:** *The Blue Marble*—Earth as photographed by Apollo 17 astronaut Harrison Schmitt in 1972 (top). North as up is such an entrenched view of our world that the image was rotated for publication to show Africa in its normal aspect (bottom).

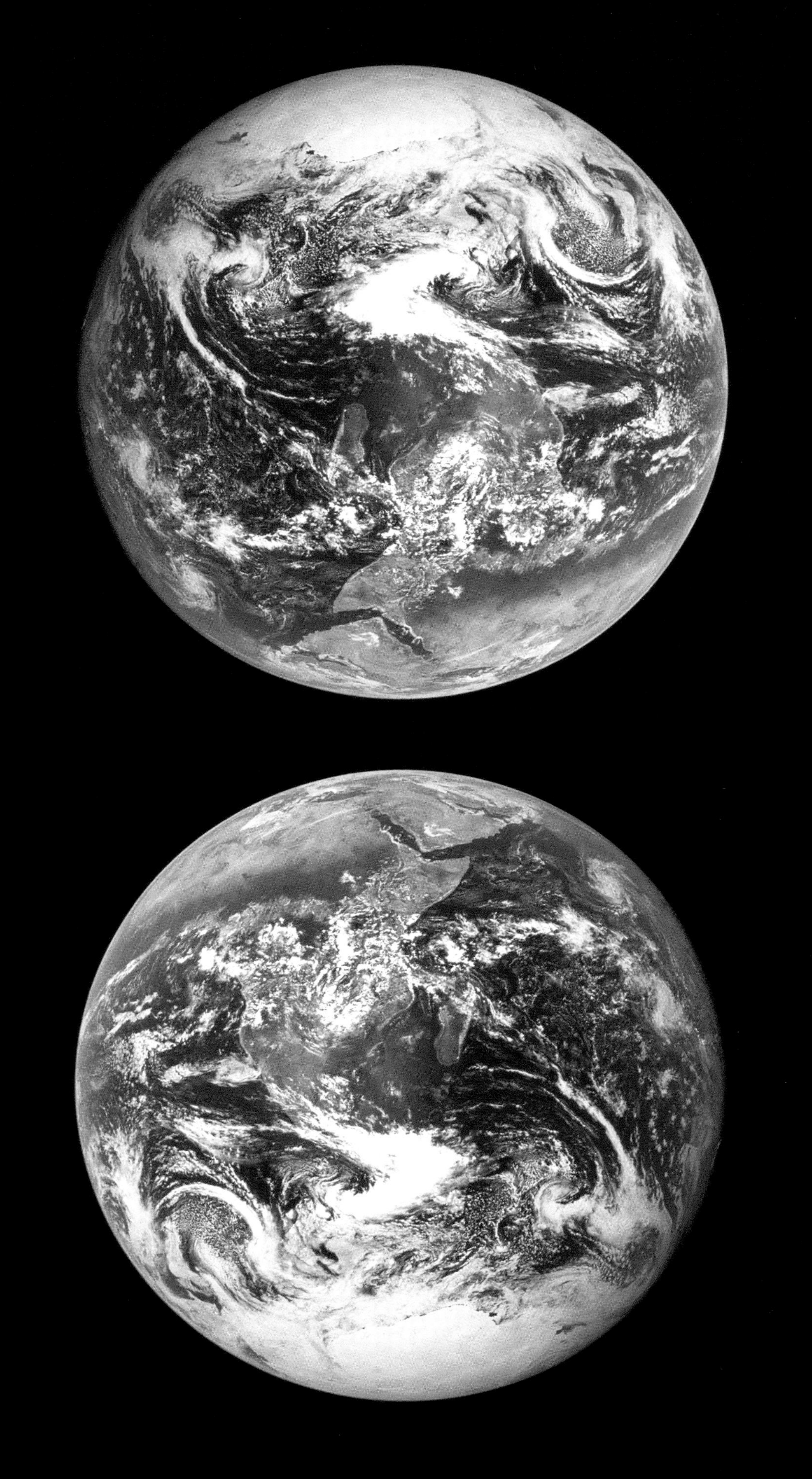

# Who is cartography?

## Cartography has changed from maps made by the few to maps made by the many.

**Maps have changed ... quite profoundly and quite irreversibly.** Maps are omnipresent, ubiquitous, transient, personalised even. Many are single-purpose, single-themed, and exist because of the availability of digital data (much of which is open). Far more people consume and author maps than they did even a few years ago. This upsurge in mapmaking has come about because of the web and has led to a profound refresh of cartographic process and practice.

**Geo-technologies are pervasive and support mapmaking by a wide range of people.** From professionals to the curious hacker; from government to the individual, making maps has become part of what you do with the web in the same way that you use it for search, navigation, restaurant reviews, finance, and dating. Marrying the on-demand nature of the web revolution with the portability of the mobile-device revolution has transformed people's use of maps. It's normal to have a modern, detailed, continuously updated, highly flexible, multi-scale map in your pocket. Smartphones and tablets provide a simple, effective, and accessible medium to consume these maps, which, in turn, has profoundly reshaped expectations and experiences of using maps. At the same time as mapping has become richer, faster, and more powerful, the barriers to entry have dropped so non-specialists are participating. These major shifts are both wonderful and disruptive and a cause for optimism about the future of mapping.

**Cartography as a profession and a discipline has always evolved as technology changes.** The printing press, computers more generally, and, now, digital data, mobile computing, and cloud-based mapping platforms have led to a new era for cartography. The web is now becoming the publishing mechanism of choice for many because the barrier to use has been reduced dramatically. Mapmakers are becoming empowered to create their own work and share their visions via maps, through the medium of the web. While in one sense disruptive, there's little doubt that cartography is experiencing a new golden age.

The type of person that makes maps has changed. The practice of mapmaking isn't the preserve of professionals working for national mapping agencies though, of course, their work is still vital. It now extends from the cartographic professional to data artists, journalists, and coders. The world is now awash with maps of all types and kinds which paint a fascinating picture and a tapestry of modern life.

With so many more maps flooding our social media feeds daily, it's sometimes difficult to sort the good from the bad but this should not detract from the fact that the number of really good, useful maps has also increased as a proportion.

Change is inevitable. Technological change has always followed patterns of general diffusion curves. Along the way, cartographers have always had to relearn and retool, and new practitioners join the practice. What we might call 'professional cartography' may have declined but there's no doubt that mapmaking is now practiced by more than ever before. Encouraging a desire for quality as part of that practice is perhaps the most useful contribution professionals and cartographic experts can now offer to the wider community

*'Quality is more important than quantity. One home run is much better than two doubles.'*
—Steve Jobs
(*Bloomberg Businessweek*, February 5, 2006)

**See also:** Craft | Ethics | Graphicacy | Hand-drawn maps | How maps are made | Maps for and by children | Mashups | Style, fashion, and trends

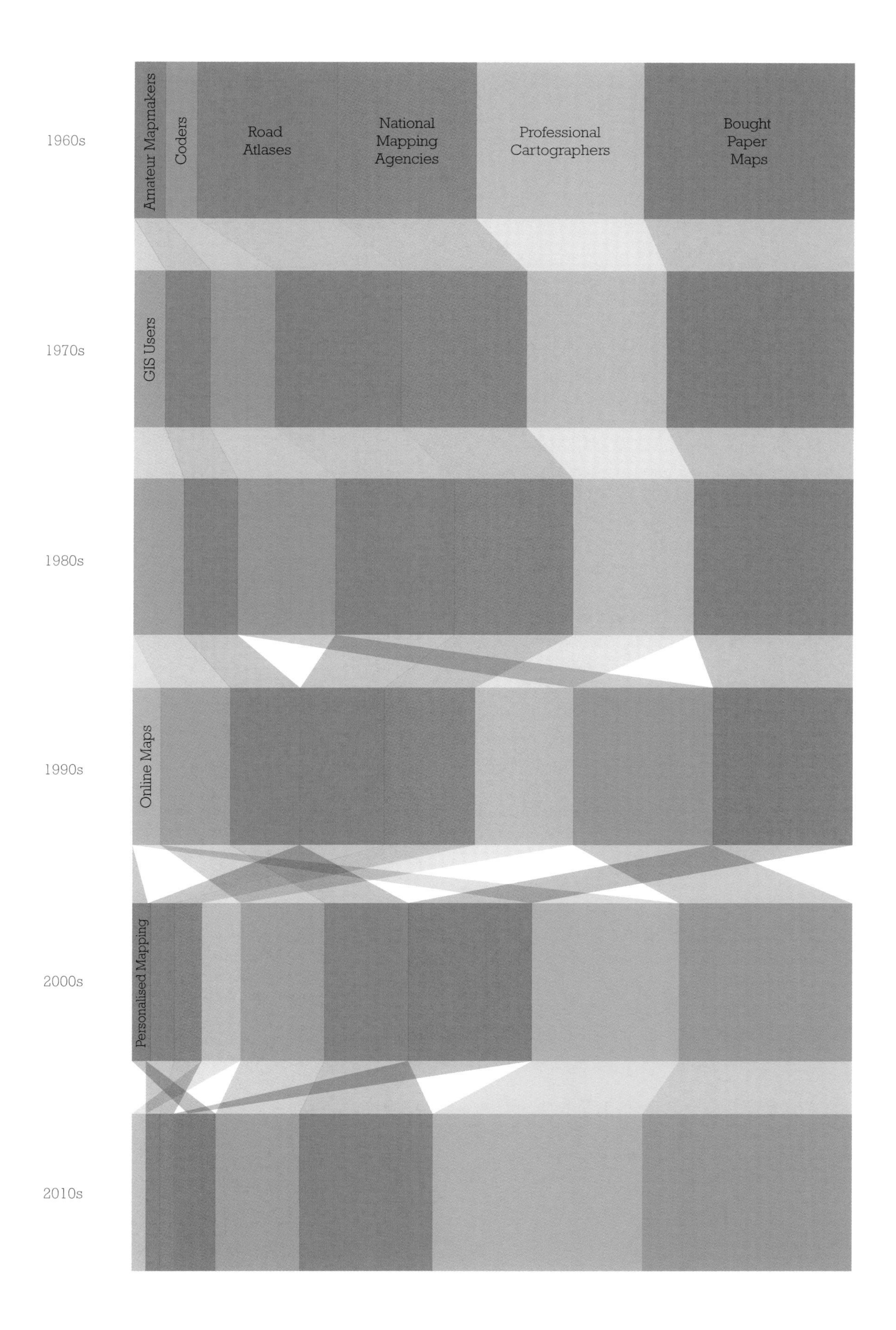
1960s
1970s
1980s
1990s
2000s
2010s
Amateur Mapmakers
Coders
Road Atlases
National Mapping Agencies
Professional Cartographers
Bought Paper Maps
GIS Users
Online Maps
Personalised Mapping

# Wireframing and storyboarding

## Developing the composition of your map.

**Wireframing and storyboarding provides the blueprint for your entire map composition.** It establishes the graphical framework and helps form the decision-making about all the map components, their interrelationships and visual properties, and the medium and interface design. The purpose is to build cohesive meaning to bring about a successful experience. Composing your final map can only begin when you've established your general design thinking and what you want the map to say. It's not about what to include but how to represent it.

**Composing your ideas sets out the visual content and establishes where everything goes, what the hierarchy is, and how you might sequence information delivery.** Thinking about the map as a narrative (a story) can help this process. Actually drawing it out helps even more as it's an iterative process, and seeing the framework appear helps reveal whether your thinking translates to an effective design. Established principles governing the location of titles usually slot them into place easily but it's the central message of the map that should be the focus and the design.

**Wireframing involves sketching different potential layouts for the map on a single page whereas storyboarding includes additional structure.** A one-off map or a single slippy map suits a wireframing approach, and any interactive controls would sit within the main map view if it's to be built for screen viewing. Generally speaking, web maps might involve more storyboarding to establish entry points, navigation, and sequencing. This is as important for linear storytelling (e.g. scrollified web delivery) as for non-linear where movement between different pages and map states needs careful handling.

Neither wireframing nor storyboarding should be considered to be detailed expositions of the final master plan. They should be concepts that are rapidly committed to paper to provide a record of your thinking. This may be as simple as using pen and paper or a whiteboard or even a graphics package. What you are not trying to achieve is a pixel-perfect mockup of your intended final design. The importance is in capturing the essence of your ideas and iterating until you've developed a solution that matches your overall goals for the map. As you progress you'll get an early indicator of any fundamental flaws you may not have considered. Through iteration, you will hone the design and the final solution.

The intent is to translate your thinking into an accessible design. To be accessible to your map reader, the design should at best clearly facilitate understanding and at worst not hinder it. Remember that map readers will come to your map fresh, so their brains will have to cope with trying to store and understand often detailed and complex information rapidly. You are dealing with the constraints of poor working memory so making clear the patterns and relationships in your map reduces the reliance on human fallibility.

Creating a harmonious structure helps people see things more clearly as they have less visual clutter to navigate. You are trying to make a design that is effortless to use (that's not the same as simply making things simple) rather than uncoordinated and staccato. If followed well, the effort you spend should result in an elegant design solution. Fine detail is what you resolve in the final design because the initial work you put in takes care of the bigger picture. When you are simply down to moving map elements by fractions or tweaking colours by percentage, you know your thoroughness has paid off.

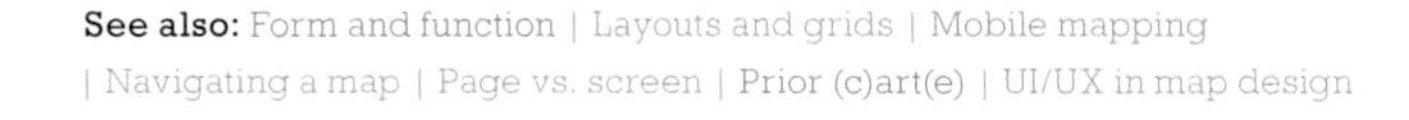
**See also:** Form and function | Layouts and grids | Mobile mapping | Navigating a map | Page vs. screen | Prior (c)art(e) | UI/UX in map design

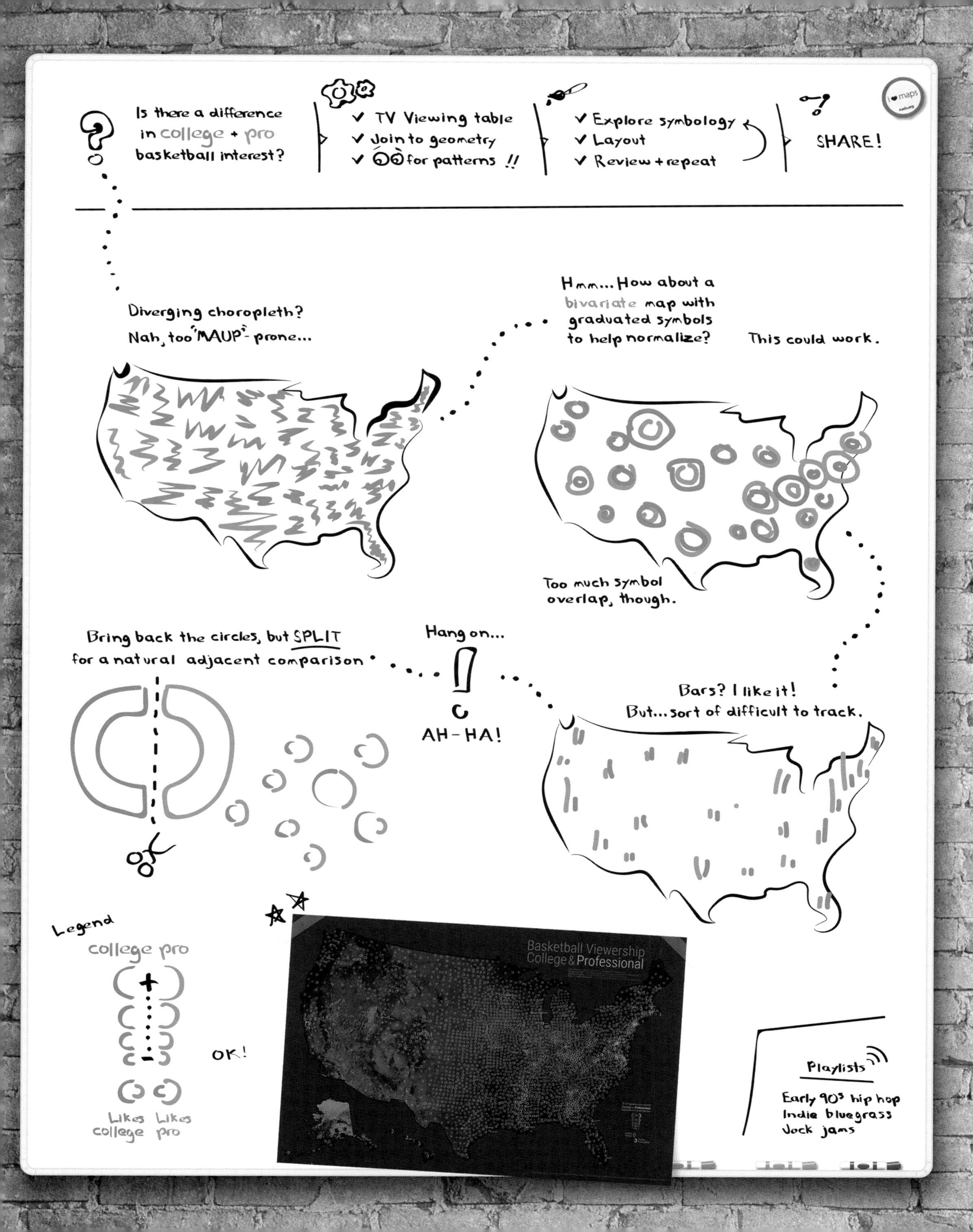
Is there a difference in college + pro basketball interest?
✓ TV Viewing table
✓ Join to geometry
✓ 👀 for patterns !!
✓ Explore symbology
✓ Layout
✓ Review + repeat
SHARE!
I ❤ maps
Diverging choropleth? Nah, too "MAUP"-prone...
Hmm... How about a bivariate map with graduated symbols to help normalize?
This could work.
Too much symbol overlap, though.
Bring back the circles, but SPLIT for a natural adjacent comparison
Hang on...
AH-HA!
Bars? I like it! But... sort of difficult to track.
Legend
college pro
+
-
Likes college
Likes pro
OK!
Basketball Viewership College & Professional
Playlists
Early 90s hip hop
Indie bluegrass
Jock jams

# World Geo-Graphic Atlas

**Herbert Bayer**

**1953**

In 1953, Herbert Bayer, a graphic designer, typographer, photographer, artist, interior designer, and architect, who studied and taught at the Bauhaus, produced this remarkable atlas. However, it's the informational graphics, and not the maps themselves, for which it's best known.

The atlas looks surprisingly modern today, with its sophisticated line-art style and careful use of colour. Five years in the making (on a huge budget), it was sponsored by the Container Corporation of America (CCA) to mark their 25th anniversary, and was never produced commercially, or reprinted. Some 30,000 copies were distributed to customers, schools, and universities.

Bayer researched the graphics himself, traveling extensively to assemble the information. He worked with a team of three designers to form a graphic language for the atlas, using the colour system that Egbert Jacobsen had developed for CCA. There are many influences in the pages, for example, the isotype system of pictograms, which Otto Neurath had developed in the 1930s and '40s.

Widely considered to be one of the finest examples of 20th-century information design, this atlas shows the staying power of precise, clear informational graphics.

—John Grimwade

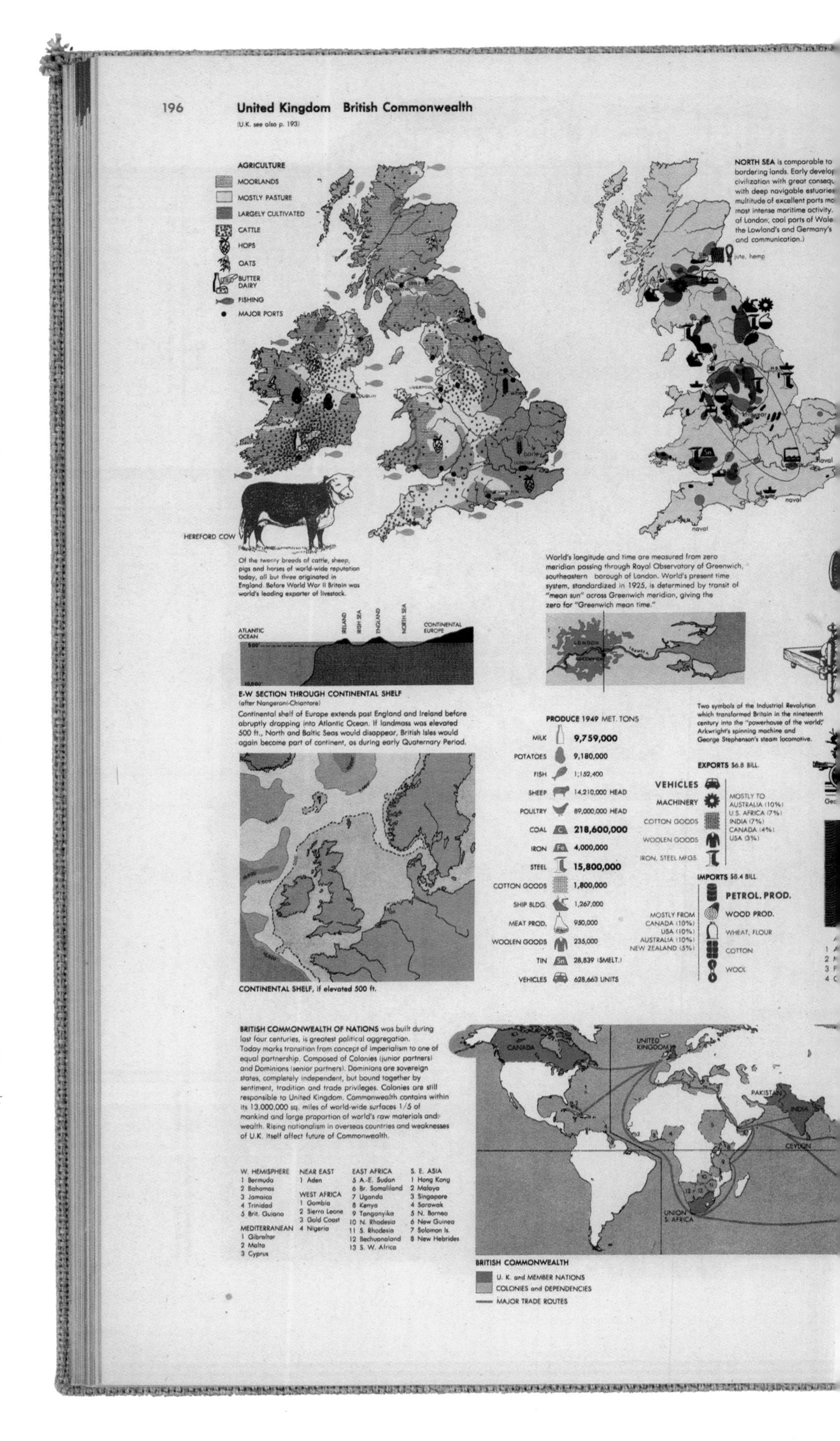

**SCANDINAVIA**

Scandinavian countries of Denmark, Finland, Norway and Sweden, grouped around almost closed sea, have regional unity from similarities of race, language, religion, severe northern climate, from rugged topography caused by glacial action of great ice age. Capable people of northlands (latitude similar to that of Alaska) also have in common a paucity of resources, harsh environmental conditions. Must use ingenuity to maintain high living standard. Limited agricultural dispositions allow primarily for grazing. Immense forests cover to 90% of North, supplying lumber for variety of wood products. Nature's gifts, mainly water, timber, fish, animal products plus manufacturing skills in selected fields are mainstays of Scandinavia's stable economy.
Historically known as explorers, conquerors, colonizers, Scandinavians have in recent years won international repute and respect for advancement of social ideas at home, where they live in peace and fight the elements.

(Norway and Finland see p. 199)

REINDEER
of northern Europe

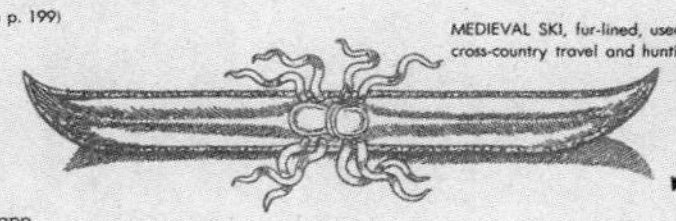

MEDIEVAL SKI, fur-lined, used for cross-country travel and hunting.

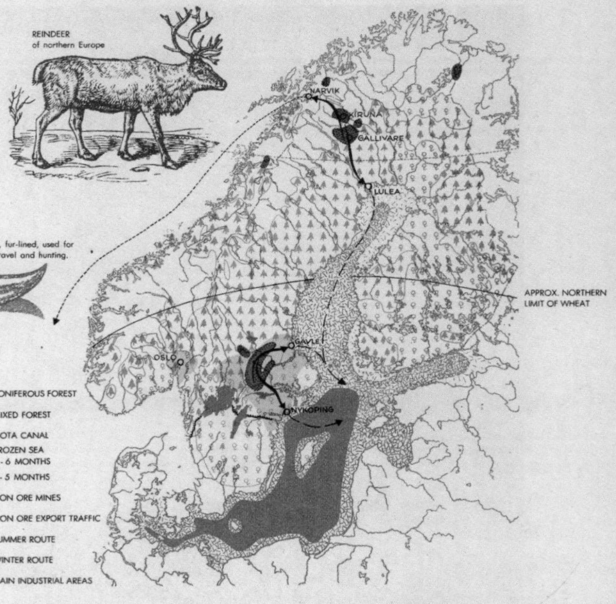

**SWEDEN**
**Sverige**
Kingdom
Area 173,423 Sq. M.
Population 7,017,000
Capital Stockholm
Language Swedish, Lapp

Largest and richest of Scandinavian countries, less limited by unfavorable environment. World's teacher of scientific timber cropping. Two major sources of economic distinction: high grade iron ore from Kiruna and Vastmanland districts shipped through specialized ports (formerly mainly to Germany); metallurgical industry, utilizing electrical power to manufacture arms and machinery sold mostly to U.S.S.R. Since domination of Baltic countries and Poland by U.S.S.R., Sweden is no longer strongest Baltic nation.

**PRODUCE 1949** MET. TONS

| | |
|---|---|
| POTATOES | **1,606,000** |
| OATS | 851,000 |
| FISH | 200,800 |
| IRON | **8,205,000** |
| WOOD PULP | **2,870,000** |
| CEMENT | 1,694,000 |
| STEEL | 1,371,000 |
| PAPER | 1,072,000 |
| SHIP BLDG. | 323,000 |

**EXPORTS** $1 BILL.

**WOOD PROD.**
MACHINERY
PAPER PROD.
IRON

MOSTLY TO UK NORWAY USA

**IMPORTS** $1 BILL.

MOSTLY FROM UK USA BELG., LUX.

**MACHINERY**
COAL, COKE
IRON, STEEL MFGS.

NARVIK
KIRUNA
GALLIVARE
LULEA
OSLO
GAVLE
NYKOPING

APPROX. NORTHERN LIMIT OF WHEAT

CONIFEROUS FOREST
MIXED FOREST
GOTA CANAL
FROZEN SEA
5 - 6 MONTHS
1 - 5 MONTHS
IRON ORE MINES
IRON ORE EXPORT TRAFFIC
SUMMER ROUTE
WINTER ROUTE
MAIN INDUSTRIAL AREAS

**DENMARK**
**Danmark**
Kingdom
Area 16,576 Sq. M.
Population 4,271,000
Capital Copenhagen
Language Danish

**PRODUCE 1949** MET. TONS

| | |
|---|---|
| DAIRY PROD. | **5,218,700** |
| POTATOES | 1,794,000 |
| BARLEY | 1,571,000 |
| CATTLE | 3,000,000 HEAD |
| POULTRY | 18,000,000 HEAD |

**EXPORTS** $672 MILL.
TO UK, GERMANY, NORWAY

**BUTTER**
MEATS
MACHINERY
FISH

**IMPORTS** $807 MILL.
FROM UK, USA, SWEDEN

**COAL, COKE**
IRON, STEEL MFGS.
OIL SEEDS

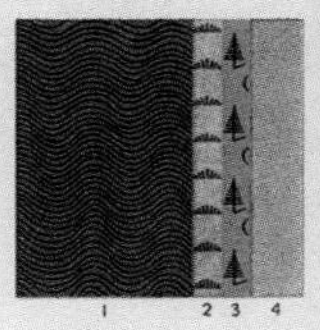

1:250 000
COPENHAGEN (KØBENHAVN)

8 9
STOCKHOLM
1:250 000

AREA 16,576 SQ. M.
1 Arable 62.5%
2 Meadow, Pasture 10 %
3 Forest, Woodland 9%
4 Other 18.5%

**DENMARK**

Flat, uniform terrain with mild climate. Specialist in dairying, sells standardized high value products (meat, butter, milk) for Europe's breakfast table, consumes lower valued ones at home. Densely populated country of educated businessmen-farmers, middlemen in cooperative trade. Dedicated to maritime commerce on account of peninsular position.

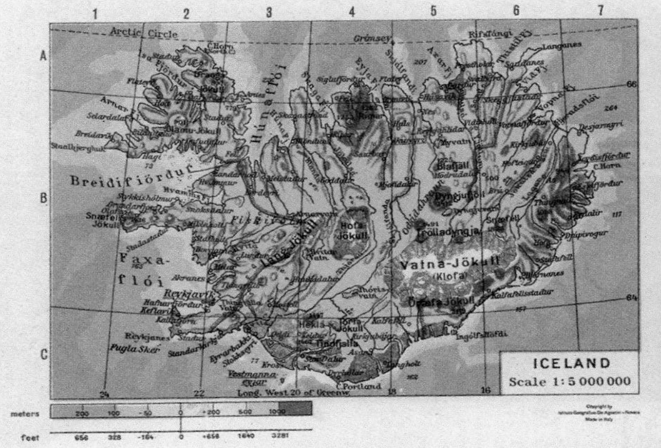

ICELAND
Scale 1: 5 000 000

meters 200 100 50 0 200 500 1000
feet 656 328 164 0 658 1640 3281

**ICELAND**
**Island**
Republic
Area 39,698 Sq. M.
Population 143,000
Capital Reykjavík
Language Icelandic

2/3 of Iceland is sand, bare mountains, weird lava fields, without vegetation, considered uninhabitable. In coastal valleys, farmers can raise little more than hay and potatoes. Tough hummocks constantly redevelop, making mechanized cultivation all but impossible. Fishing is main occupation.
Waterfalls supply large store of waterpower, now being developed. Numerous hot springs are exploited to heat homes, entire cities. In mid-13th century, Iceland joined Norway, passed (1380) to Denmark, endured political subjugation under both, became independent 1944. Most of population wiped out in early 1400s by plague. Later suffered great devastation from volcanic eruptions. Despite ill-favored topography and location remote from main currents of commerce, Iceland has in modern times developed into progressive nation.

**PRODUCE 1949** MET. TONS

| | |
|---|---|
| FISH | **394,000** |
| SHEEP | 454,000 HEAD |

VOLCANO HEKLA erupting 1947
from a postage stamp of Iceland issued 1948

# x and y

## The basic structure from which we define algebraic relationships and geometric interpretation of position.

**Earth is a complex shape that requires complex mathematics to provide accurate measurement of position but it has its basis in some very simple principles.** Cartesian (or plane) coordinate geometry is derived from the work of the French mathematician René Descartes in the 17th century. His system for supporting the geometric interpretation of algebraic relationships led to the branch of mathematics known as *geometry*. This system forms the basis for all subsequent determinations of position on Earth.

**Cartesian coordinate geometry is the graphical application of the mathematics that gives intersecting perpendicular lines on a plane.** The graph has two axes, the vertical one referred to as the y-axis and the horizontal axis referred to as the x-axis. Thus, any position on the graph can be described by the intersection of the x- and y-axes ($P_{xy}$) and plotted on the graph. Cartesian geometry uses a regularly spaced linear measurement for both axes. Each location therefore has a unique numerical combination of paired values. Each location thus has a unique position on the graph, and relative location can be shown by plotting several points.

**When used for mapmaking the x- and y-axes are often referred to as *eastings* and *northings*.** A measured point is described by its distance along its x-axis (easting) and y-axis (northing) from a point of origin. This origin is usually in the bottom left and supports the general map-reading advice of 'read right up' as a way of locating eastings and northings on the map. Although many cartographic coordinate systems use an origin to the south and west (bottom left) giving only positive easting and northings, Cartesian coordinate geometry allows negative values so an origin (0, 0) might be positioned in the centre of the map. In this configuration negative eastings and northings might be used, or a constant value added to all values to give what are often referred to as *false eastings* and *false northings*.

This basic system of measuring position works for all coordinate systems that exist at different scales and for all units of measurement (e.g. feet, metres). In fact, some coordinate systems are built upon map projections such as state plane (used throughout the United States) and the global Universal Transverse Mercator (UTM) coordinate system.

UTM is a conformal projection, which uses a Cartesian coordinate system to derive locations. It differs from most projections in that it divides Earth into 60 zones. Each zone is equivalent to 6° of longitude. Each of the zones then uses a secant transverse Mercator projection. The narrow zones and the fact that there are two standard lines resulting from the secant projection mean that each map is a region of large north–south extent with minimal distortion. Each UTM zone is split into 20 latitude bands, each 8° high and notated using a letter. The combination of the zone and latitude zone creates a grid zone—for instance, 30U–zone 30, and latitude band U. Because UTM uses a Cartesian coordinate system, a position is given by referring to the UTM zone number and the easting and northing planar coordinate pair for the zone. Each zone's point of origin is the intersection of the equator and the zone's central meridian.

Buckingham Palace, London, lies at:
Latitude: 51.501476
Longitude: 0.140634
DMS Latitude: 51°30'5.3136'' N
DMS Longitude: 0°8'26.2824'' W
UTM zone: 30U
UTM Easting: 698,451.53 m
UTM Northing: 5,709,470.23 m

**See also:** Aligning coordinate systems | Datums | Earth coordinate geometry | Spatial dimensions of data

1) Necessary information about geographic coordinates

Besides geographic coordinates of the point, one needs to have the information about a geographic coordinate system the point is in to correctly compute UTM Easting and Northing coordinates. Longitude $\lambda$ and latitude $\phi$ of Buckingham Palace in London are defined in ETRS 89 (European Terrestrial Reference System 1989) geographic coordinate system on GRS 80 (Geodetic Reference System 1980) reference ellipsoid. The necessary information of the reference ellipsoid is:

semi-major axis $\boldsymbol{a = 6,378,137.0\ m}$, and

inverse flattening $\frac{1}{f} = \mathbf{298.257\ 222\ 101}$.

From these two values an eccentricity $\boldsymbol{e}$ and second eccentricity $\boldsymbol{e'}$ are calculated using two equations below. Both values are needed in UTM equations.

$$e^2 = f \cdot (2 - f)$$

$$e'^2 = \frac{e^2}{1 - e^2}$$

2) Necessary information about the target Cartesian coordinate system

UTM uses the transverse Mercator projection to project geographic coordinates on a Cartesian coordinate system. UTM zone 30N (picture on the right) is defined with the following parameters:

false easting $\boldsymbol{fe = 500,000.0\ m}$,

false northing $\boldsymbol{fn = 0\ m}$,

central meridian $\boldsymbol{\lambda_0 = -3°}$,

latitude of origin $\boldsymbol{\phi_0 = 0°}$, and

scale factor $\boldsymbol{k_0 = 0.9996}$.

3) Computing necessary values for every point

Next step is computing, the distance $M$ along the meridian from the equator to the latitude $\phi$, the distance $M_0$ along the meridian from the equator to the latitude of origin $\phi_0$, and the radius of curvature $N$ of the prime vertical at latitude $\phi$.

$$M = a\Big[\Big(1 - \frac{1}{4}e^2 - \frac{3}{64}e^4 - \frac{5}{256}e^6 - \cdots\Big)\phi - \Big(\frac{3}{8}e^2 + \frac{3}{32}e^4 + \frac{45}{1024}e^6 + \cdots\Big)\sin 2\phi + \\ + \Big(\frac{15}{256}e^4 + \frac{45}{1024}e^6 + \cdots\Big)\sin 4\phi - \Big(\frac{35}{3072}e^6 + \cdots\Big)\sin 6\phi + \cdots\Big]$$

$$M_0 = a\Big[\Big(1 - \frac{1}{4}e^2 - \frac{3}{64}e^4 - \frac{5}{256}e^6 - \cdots\Big)\phi_0 - \Big(\frac{3}{8}e^2 + \frac{3}{32}e^4 + \frac{45}{1024}e^6 + \cdots\Big)\sin 2\phi_0 + \\ + \Big(\frac{15}{256}e^4 + \frac{45}{1024}e^6 + \cdots\Big)\sin 4\phi_0 - \Big(\frac{35}{3072}e^6 + \cdots\Big)\sin 6\phi_0 + \cdots\Big]$$

$$N = \frac{a}{\sqrt{1 - e^2 \sin^2 \phi}}$$

4) UTM Easting and Northing for every point

The final step is to compute Cartesian coordinates, easting $x$ and northing $y$, in the UTM coordinate systems with the following equation:

$$x = k_0 \cdot N\Big[A + (1 - T + C)\frac{A^3}{6} + (5 - 18T + T^2 + 72C - 58e'^2)\frac{A^5}{120}\Big] + fe$$

$$y = k_0\Big\{M - M_0 + N\tan\phi\Big[\frac{A^2}{2} + (5 - T + 9C + 4C^2)\frac{A^4}{24} + \\ + (61 - 58T + T^2 + 600C - 330e'^2)\frac{A^6}{720}\Big]\Big\} + fn$$

where $A$, $T$, and $C$ are pre-computed with the following expressions: $T = \tan^2\phi$, $C = e'^2\cos^2\phi$, and $A = (\lambda - \lambda_0)\cos\phi$.

Every point in geographic coordinates is transformed to easting $x$ and northing $y$ using values from (1) and (2) and equations in (3) and (4). The presented equations are after Snyder 1987 (p. 60–63). However, there are more sophisticated and mathematically complex equations used in today's software.

## A World of Lotus, a World of Harmony

Liao Zhi Yuan
2015

An amazing piece of cartographic design. This captivating map captures in an elegant way—and in a koi carp pond—what could be seen as an apocalyptic vision of the future of our planet, in which humans have disappeared from the surface of Earth to be replaced by beautiful, luxuriant lotus vegetation.

This might even be a post global warming vision since even Antarctica is covered by lotuses, and ocean levels have risen. Through this beautiful map, a post global warming world appears peaceful and harmonious.

Although it is not necessarily a constructive way to call for urgent action to stop climate change, this aesthetically pleasing map draws our attention and stimulates our imagination. This is also what good cartographic design should be about: to make us dream and think.

The map won an award in the International Cartographic Association Barbara Petchenik Children's Map competition in 2015 and was drawn by 15-year-old Liao Zhi Yuan.

—Sébastien Caquard

# Your map is wrong!

## Maps sometimes say one thing but demonstrate the opposite.

**The time it takes to make a map is affected by many factors.** You are fortunate if you have no deadlines and an open-ended amount of time to craft your map. That said, endless time can often become a time-sink as the danger to overthink and endlessly iterate can lead to the risk of never actually publishing the map. Feature-creep can also affect the map as you continually change the scope and modify the map to incorporate new elements. Then there are the vagaries of clients whose specifications are poorly formulated and which change once they've decided they don't like your first draft! The counter is that you may be rushed with unrealistic deadlines or demands that can impact quality of the finished product as you are almost required to take short cuts. Setting deadlines and meeting them helps you keep on track.

**The other component in the time spent on your map is that which the map reader expends.** Having, likely, spent a considerable amount of time making the map, you would hope that your map readers spend a reasonable amount of time exploring it. If the map reader has a definite need and value for your map, then this expectation might be realistic, but for general maps targeted at a general audience, how does time play out? The simple answer is 'not very long at all'. This is particularly true for web maps.

**Finding what you want via a map portal can be painful.** People don't search for maps or map layers, they search for specific details such as addresses or names. Making your map searchable can be important in getting it noticed in the first place. Making your map visible helps encourage more people to your map and to spend more time on your map. Provide what they want to see and make it easy to see. Give each map a unique URL to make it distinct. Most web map users are likely to spend less than a minute on your map and, at best, look at three different aspects. Don't hide features behind drop-downs and hidden legends. Put the data on the map and make it easy to navigate.

Metrics on map use for web pages are notoriously difficult to assess. Waiting for a map to load from a website can take too long and a user has gone elsewhere in the meantime. It's not untypical that the average time a person spends on a web map is around a minute. Map users have so little time to read the map that you have to design it to be extraordinarily clear. Only a fraction of what you put into the map will be retrieved by the reader. That said, using average page visit as a metric ignores the many people who will spend longer, and maybe it's those people whom you're designing for. As with any map-reading situation, people's behaviour varies dramatically.

Think of your web map as something that should stand the test of time. By that, it shouldn't go offline or suffer positive ageing where technology outpaces the map and it can no longer be supported. Try and build in negative ageing—building a map that has longevity and is unlikely to fail. This helps build trustworthiness in your work and encourages people to spend longer on the maps and site.

The first 10 seconds are crucial in convincing people to stay on your map. The probability of leaving is much higher during this phase. We all suffer poor web experiences so visiting any new map leads us to be initially sceptical of wasting time. Giving people a visual hook to get them past 10 seconds helps them stay longer. Only after they have stayed for 30 seconds or so does the rate of leaving flatten out to a much slower rate. People are assessing good versus bad in their own minds and creating impressions, rightly or wrongly. Think of meeting someone for the first time. You always form an instant impression. The same is true for your map. Quality will sell so make sure your map is well dressed and good to go.

**See also:** Consistent denotation | Data (c)art(e) | Different strokes | Signal to noise

# The Truest Size of Africa?

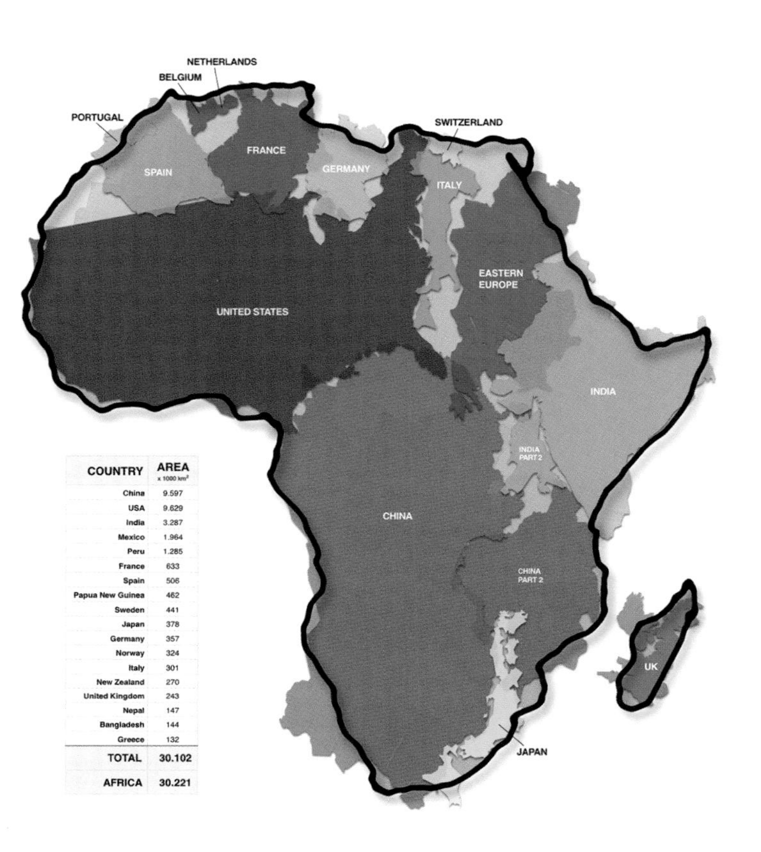


| COUNTRY | AREA x 1000 km² |
|---|---|
| China | 9.597 |
| USA | 9.629 |
| India | 3.287 |
| Mexico | 1.964 |
| Peru | 1.285 |
| France | 633 |
| Spain | 506 |
| Papua New Guinea | 462 |
| Sweden | 441 |
| Japan | 378 |
| Germany | 357 |
| Norway | 324 |
| Italy | 301 |
| New Zealand | 270 |
| United Kingdom | 243 |
| Nepal | 147 |
| Bangladesh | 144 |
| Greece | 132 |
| TOTAL | 30.102 |
| AFRICA | 30.221 |

The map on the left is by Kai Krause, a graphic designer whose *True Size of Africa* map became a viral sensation in 2011. It displays the true size of Africa in an eye-catching way using comparative geographies. At first glance it shows how large Africa actually is in comparison to other familiar shapes such as the outline of the United States of America. He wasn't the first to use this approach. He wasn't the last. But he received considerable criticism for his attempt as the internet cried foul.

There are some aspects of Krause's map that undermined its attempt to portray the relationship between the land area of various countries. The Republic of Ireland is included in the United Kingdom map; an odd collection of countries are labeled Eastern Europe; China and India are carved up to make them fit; and Alaska is omitted altogether which reduces the land area of the United States considerably. A different set of countries from those presented in the accompanying table is used. Approximately 45% of Africa is composed of the uninhabitable Sahara Desert, which also renders the comparison a little misleading and certainly points to a further issue: that land area isn't the best metric for comparison because we might be more concerned as a society with what constitutes habitable areas. Africa is certainly bigger than the combination of the countries Krause uses but it is a continent of over 50 countries and has a smaller overall population than either India or China. It's so much smaller that you can add the population of the world's third most populous country (USA) to Africa to approach the population total of China. The map shapes Krause used are also not equal area but, instead, are Web (or spherical) Mercator.

Although one can accept an element of artistic license in such work (which Krause also admits to openly), Krause suggests that the map is an attempt to bring attention to the poor levels of geographical knowledge displayed by many people. He cites the fact that a survey of American schoolchildren revealed the majority chose '1-2 billion' as the US population total and 'largest in the world' as the US land area. This may all be true. But why not make the map correctly in the first place? The intent of the graphic was laudable. The map is wrong.

The map on the right uses Goode's Homolosine equal-area projection and includes the countries as they appeared in the original table.

This (and other) corrected versions did not go viral. The timing was wrong, and Krause had already captured the public's interest and imagination. Whether maps are right or wrong, many become widely shared. They do not have to be correct—but they pique interest, speak to a theme of interest to a receptive readership. Crucially, if a map is first to engage a large audience they retain their popularity and infamy despite whatever design flaws may exist.

Next time you see a map on the internet, do some basic fact-checking and determine whether it is showing what it purports to show. If it's not, (politely) call it out and explain how the map is wrong. I did, and this very example is the reason my blog (cartonerd.com) was born.

## Yellowstone National Park

Heinrich Berann

1989

The late Austrian artist Heinrich Berann painted this panorama in 1989 at the end of his career spanning six decades. Published as a poster, it mimics the view from an airplane window looking from north to south over Yellowstone National Park.

In the background, the sharp peaks of the Teton Range jut into a sky decorated with wispy backlit clouds. Berann was a spiritual man who also painted baroque religious art. The arcing cloud patterns are an allegory for his belief in the "circle of life." He also thought that water was special. Lakes glimmer with sun glints, reflections, and wind riffles; waterfalls create mist; and geysers are revealed by soaring plumes.

In addition to these flourishes, Berann radically manipulated the underlying terrain, resizing, rotating, and moving mountains to show complex topography more clearly. The end result is a map that transports you to Yellowstone and invites you to dream.

The 2018 version shown here has been digitally remastered and restored to the colours Berann originally used. The National Park Service has added typographic components.

—Tom Patterson

PITCHSTONE PLATEAU
St. Anthony
Lewis Lake
SHOSHONE LAKE
MADISON PLATEAU
Grant Village
West Thumb
OLD FAITHFUL
WEST THUMB
Twin Buttes
Madison Junction
CENTRAL PLATEAU
HAYDEN VALLEY
Purple Mountain
UPPER FALLS
LOWER FALLS
Norris Geyser Basin
Mount Holmes
CANYON OF THE YELLOWSTONE
Canyon Village
Dome Mountain
Mount Washburn
Obsidian Cliff
Bannock Peak
Quadrant Mountain
Washburn Range
Gray Peak
GALLATIN RANGE
Tower Fall
Tower-Roosevelt
Bunsen Peak
Electric Peak
Sepulcher Mountain
Black Canyon of the Yellowstone
Lava Creek
Mammoth Hot Springs
Yellowstone River
Gardiner
BERANN

# Zeitgeist

## Cartography reflects the spirit of the time. Always has. Always will.

**Dominant ideals and beliefs that are characteristic of a period in time are often expressed as a creative Zeitgeist.** As I sit and write this last page of the book (and yes, it is the last page written—I needed something for Z), it's worth reflecting on the scope of the content and how it represents cartography. Clearly, much of the book is about methods and their technical implementation but more than anything it's about thinking. Cartography is a discipline that embodies change. Technology enforces it. New cartographers and cartographies emerge, and that shapes the contemporary Zeitgeist. The modern cartographic Zeitgeist is about channeling creative thought and navigating the technical challenges brought about by change. Many maps recycle old ideas. Many more challenge accepted convention. Some provide a point of reference that characterises the current mood. All the examples of maps included in this book do just that. They provide a reference point for a particular technique, the birth of an idea, or the demonstration of the very best of cartographic praxis.

**Cartography is always in a state of creative flux.** The beginning of the 21st century has brought perhaps a shift in the pace of change with technology so rapid that mapmaking has struggled to keep up. Maps are made rapidly and many die just as quickly. The time spent on creative thinking is less than it once was, and reliance on software defaults has become too common. Escaping from software constraints and re-engaging our creative minds will see cartography rise to the challenges it faces. New maps will emerge and become the points of reference for new cartographies. They will continue to be objects of art as well as tools of political power. They will remain a balance of objective argument and subjective design. They will continue to inform as much as stoke controversy. They will remain a pivotal medium for supporting our understanding of the world, and provide escapism as well as revealing new meaning. From national mapping to personalised cartography, the richness of the wide world of cartography will endure.

**See also:** 1st page | Style, fashion, and trends | Who is Cartography? | Your map is wrong!

Cartography doesn't necessarily drive a societal Zeitgeist. In fact, it tends to reflect the prevailing Zeitgeist of a society. For instance, many of the rapid changes we have witnessed in the fledgling years of the 21st century are reflected in cartography. Developments in the internet have led to profound changes in how maps are made, the medium in which they are published, and the way they are used. The development of 3D for film and gaming has seen similar developments in cartography. 3D geovisualizations have been motivated by the wider drive for 3D content and not specifically from cartography's need itself. Ultimately, developments in cartography become a product of developments of this wider Zeitgeist.

One of the consequences of Zeitgeist is the use of cartographic metaphors as a way of framing a theme. For instance, the established design of subway maps is well known. Using the design as a skeleton for other content has become a contemporary cartographic Zeitgeist of its own. Such maps tend to replace station names on familiar maps and assign different meaning to the coloured subway lines. These are often unconnected either topologically or thematically, and the transport map becomes a pointless framework. This kind of short-cut cartography bypasses thinking but reflects the current aesthetic Zeitgeist. People like subway maps as an object. The aesthetic is something relatable and familiar. It's easy to make a map like this with modern tools. They are immediately drawn to the map. It bypasses thinking. This will change as the Zeitgeist itself morphs into something new. Hopefully.

**Opposite:** *End of the Line: A Tube Map of Tube Maps Parody Map* by Kenneth Field, 2015, where each station is replaced by a link to a map that uses a tube map metaphor for their own cartographic efforts. I call it Becksploitation! esriurl.com/EndOfTheLine

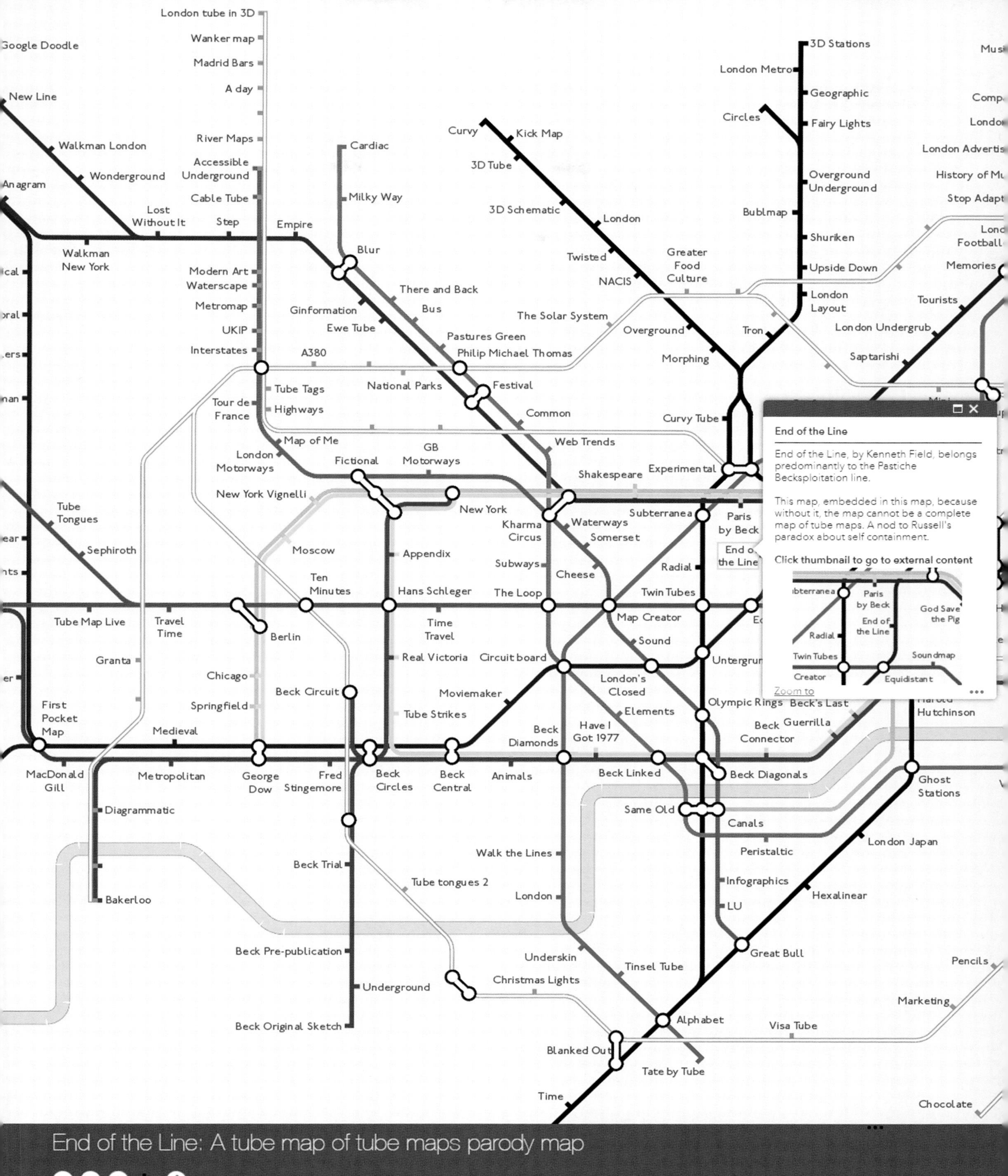

London tube in 3D
Wanker map
Madrid Bars
A day
River Maps
Accessible Underground
Cable Tube
New Line
Walkman London
Wonderground
Anagram
Lost Without It
Step
Empire
Walkman New York
Modern Art
Waterscape
Metromap
UKIP
Interstates
A380
Cardiac
Milky Way
Blur
There and Back
Bus
Ginformation
Ewe Tube
Pastures Green
Philip Michael Thomas
Curvy
Kick Map
3D Tube
3D Schematic
London
Twisted
NACIS
Greater Food Culture
The Solar System
Overground
Morphing
3D Stations
London Metro
Geographic
Circles
Fairy Lights
Overground Underground
Bublmap
Shuriken
Upside Down
London Layout
Tron
London Advertis
History of Mu
Stop Adapt
Football
Memories
Tourists
London Undergrub
Saptarishi
Tube Tags
Highways
Tour de France
National Parks
Festival
Common
Curvy Tube
Map of Me
London Motorways
Fictional
GB Motorways
Web Trends
Shakespeare
Experimental
New York Vignelli
New York
Kharma Circus
Waterways
Somerset
Subterranea
Paris by Beck
End of the Line
Radial
Tube Tongues
Sephiroth
Moscow
Appendix
Subways
Cheese
Ten Minutes
Hans Schleger
The Loop
Twin Tubes
Tube Map Live
Travel Time
Berlin
Time Travel
Map Creator
Sound
Untergrun
Granta
Real Victoria
Circuit board
Chicago
Springfield
Beck Circuit
Moviemaker
London's Closed
Olympic Rings
Beck's Last
Guerrilla
Beck Connector
Hutchinson
First Pocket Map
Medieval
Tube Strikes
Elements
Beck Diamonds
Have I Got 1977
MacDonald Gill
Metropolitan
George Dow
Fred Stingemore
Beck Circles
Beck Central
Animals
Beck Linked
Beck Diagonals
Ghost Stations
Diagrammatic
Same Old
Canals
Peristaltic
London Japan
Walk the Lines
Beck Trial
Tube tongues 2
Infographics
Hexalinear
London
LU
Bakerloo
Great Bull
Beck Pre-publication
Underskin
Tinsel Tube
Pencils
Christmas Lights
Underground
Marketing
Alphabet
Beck Original Sketch
Visa Tube
Blanked Out
Tate by Tube
Time
Chocolate
End of the Line
End of the Line, by Kenneth Field, belongs predominantly to the Pastiche Becksploitation line.
This map, embedded in this map, because without it, the map cannot be a complete map of tube maps. A nod to Russell's paradox about self containment.
Click thumbnail to go to external content
Paris by Beck
God Save the Pig
End of the Line
Radial
Twin Tubes
Soundmap
Creator
Equidistant
Zoom to
End of the Line: A tube map of tube maps parody map
Geographical

# Sources and resources


**REFERENCES AND SELECTED READING**

Aisch, G., and R. Gebeloff. 2014. 'Mapping Migration in the United States.' *New York Times*, August 15, 2014. Accessed 1 December 2016. https://www.nytimes.com/2014/08/16/upshot/mapping-migration-in-the-united-states-since-1900.html.

Anson, R., and F. J. Ormeling. 1993. *Basic Cartography for Students and Technicians*, 2nd ed., vol. 1. London: Elsevier Applied Science Publishers.

———. 1996. *Basic Cartography for Students and Technicians*, vol. 3. Oxford: Butterworth-Heinemann.

———. 2002. *Basic Cartography for Students and Technicians*, 2nd ed., vol. 2. Oxford: Butterworth-Heinemann.

Antoniou, A., and R. Klanten. 2015. *Mind the Map: Creative Mapmaking and Cartography*. Berlin: Gestalten.

Balchin, W. G. V., and A. M. Coleman. 1966. 'Graphicacy Should Be the Fourth Ace in the Pack.' *Cartographica: The International Journal for Geographic Information and Geovisualization*: 3 (1): 23–28.

Bertin, J. 1983. *Semiology of Graphics: Diagrams, Networks, Maps*. Madison, WI: University of Wisconsin Press.

———. 2010. *Semiology of Graphics: Diagrams, Networks, Maps*. Redlands, CA: Esri Press.

Brewer, C. 2015. *Designing Better Maps: A Guide for GIS Users*. Redlands, CA: Esri Press.

Bugayevskiy, L. M., and J. Snyder. 1995. *Map Projections: A Reference Manual*. London: Taylor & Francis.

Cairo, A. 2013. *The Functional Art*. Berkeley CA: New Riders.

———. 2016. *The Truthful Art*. Berkeley, CA: New Riders.

Chernoff, H. 1970. 'The Use of Faces to Represent Points in K-Dimensional Space Graphically.' *Journal of the American Statistical Association* 68 (342): 361–68.

Chilton, S., and A. J. Kent, eds. 2014. *Cartography: A Reader*. London: The Society of Cartographers.

Crampton, J. W. 2010. *Mapping: A Critical Introduction to Cartography and GIS*. Hoboken, NJ: Wiley-Blackwell.

Darkes, G., and M. Spence. 2017. *Cartography: An Introduction*. London: British Cartographic Society.

Dent, B., J. Torguson, and T. W. Hodler. 2008. *Cartography: Thematic Map Design*, 6th ed. Boston: McGraw-Hill Education.

Field, K. S., and A. J. Kent. 2014. *Landmarks in Mapping*. London: Maney Publishing.

Friendly, M., and D. J. Denis. 2001. 'Milestones in the History of Thematic Cartography, Statistical Graphics, and Data Visualization.' Accessed: 25 January 2018. http://www.datavis.ca/milestones/.

Gould, P., and R. White. 1986. *Mental Maps*. London: Allen and Unwin.

Harris, R. L. 2000. *Information Graphics: A Comprehensive Illustrated Reference*. Oxford: Oxford University Press.

Hessler, J., ed. 2015. *Map: Exploring the World*. London: Phaidon Press Ltd.

Holmes, N. 1991. *Pictorial Maps*. London: Herbert Press.

Imhof, E. 2007. *Cartographic Relief Presentation*. Redlands, CA: Esri Press.

International Cartographic Association. 2003. 'A Strategic Plan for the International Cartographic Association 2003-2011.' Accessed February 8, 2017. https://icaci.org/files/documents/reference_docs/ICA_Strategic_Plan_2003-2011.pdf.

Jenny, B. 2012. 'Adaptive Composite Map Projections.' *IEEE Transactions on Visualization and Computer Graphics* 18 (12): 2575–82.

Katz, J. 2012. *Designing Information: Human Factors and Common Sense in Information Design*. Hoboken, NJ: John Wiley & Sons.

Keates, J. 1982. *Understanding Maps*. London: Longman.

Kent, A. J., and P. Vujakovic. 2017. *The Routledge Handbook of Mapping and Cartography*. London: Routledge.

Kirk, A. 2016. *Data Visualisation: A Handbook for Data Driven Design*. London: SAGE Publications Ltd.

Klanten, R., and N. Bourquin., 2008. *Data Flow: Visualising Information in Graphic Design*. Berlin: Gestalten.

Klanten, R., and S. Ehmann. 2010. *Data Flow 2: Visualising Information in Graphic Design*. Berlin: Gestalten.

Kleon, A. 2012. *Steal like an Artist*. New York, NY: Workman Publishing.

Kraak, M-J. 2014. *Mapping Time*. Redlands, CA: Esri Press.

Kraak, M-J., and F. Ormeling. 2010. *Cartography: Visualization of Geospatial Data*, 3rd ed. New York, NY: Guilford Press.

Krygier, J., and D. Wood. 2016. *Making Maps: A Visual Guide to Map Design for GIS*, 3rd ed. New York, NY: Guilford Press.

MacEachren, A. M. 1995. *How Maps Work: Representation, Visualization, and Design*. New York, NY: Guilford Press.

MacEachren, A., and D. R. Fraser Taylor. 1994. *Visualization in Modern Cartography*. Oxford: Pergamon.

Meirelles, I. 2013. *Design for Information: An Introduction to the Histories, Theories, and Best Practices behind Effective Information Visualizations*. Beverly, MA: Rockport Publishers.

Monmonier, M. 1991. *How to Lie with Maps*. Chicago: University of Chicago Press.

Muehlenhaus, I. 2014. *Web Cartography: Map Design for Interactive and Mobile Devices*. London: CRC Press.

Munsell, A. H. 1905. *A Colour Notation*. Boston: Geo. H. Ellis Co.

Neurath, O. 1936. *International Picture Language: The First Rules of Isotype*. London: Kegan Paul, Trench Trubner & Co Ltd.

Openshaw, S. 1983. *The Modifiable Areal Unit Problem*. Norwich, UK: Geo Books.

Pye, D. 2015. *The Nature and Art of Workmanship*. New York, NY: Bloomsbury Academic.

Rendgen, S., and J. Wiedemann. 2012. *Information Graphics*. Cologne, Germany: Taschen.

Robinson, A. H. 1985. 'Arno Peters and His New Cartography.' *American Cartographer* 12: 103–11.

Robinson, A. H. 2010. *The Look of Maps*. Redlands, CA: Esri Press.

Rosenberg, D., and A. Graton. 2012. *Cartographies of Time: A History of the Timeline*. New York, NY: Princeton Architectural Press.

Roth, R. E. 2017. 'User Interface and User Experience (UI/UX) Design.' The Geographic Information Science & Technology Body of Knowledge (2nd Quarter 2017 edition). Edited by John

P. Wilson. Accessed 8 March 2018. http://gistbok.ucgis.org/bok-topics/user-interface-and-user-exerience-ui/us-design.

(The) Royal Society. 1955. *The Public Understanding of Science*. London: The Royal Society.

Sendpoints. 2016. *Cartographics: Designing the Modern Map*. London: Sendpoints Publishing Co. Ltd.

Skelton, R. A., 1972. *Maps: A Historical Survey of Their Study and Collecting*. Chicago: University of Chicago Press.

Sloane, R. C., and M. Montz. 1930. *Elements of Topographic Drawing*. New York, NY: McGraw Hill.

Slocum, T. A., R. B. McMaster, F. C. Kessler, and H. H. Howard. 2008. *Thematic Cartography and Geographic Visualisation*, 3rd ed. Upper Saddle River, NJ: Prentice Hall.

Snyder, J. P. 1987. 'Map Projections: A Working Manual.' Professional Paper 1395. Washington, DC: USGS, Department of the Interior.

Steele, J., and N. Iliinsky. 2010. *Beautiful Visualization: Looking at Data through the Eyes of Experts (Theory in Practice)*. Sebastopol, CA: O'Reilly Media.

Tanaka, K. 1950. 'The Relief Contour Method of Representing Topography on Maps.' *Geographical Review* 40 (3): 444–56.

Tobler, W. 1970. 'A Computer Movie Simulating Urban Growth in the Detroit Region.' *Economic Geography* 46 (Supplement): 234–40.

Transport for London (TfL). 2014. Design Style Guide. London: TfL. Accessed 8 February 2017. http://content.tfl.gov.uk/design-style-guide.pdf.

Tufte, E. 1990. *Envisioning Information*. Cheshire, CT: Graphics Press.

———. 2001. *The Visual Display of Quantitative Information*, 2nd ed. Cheshire, CT: Graphics Press.

Turner, E. 2004. Gene's Map Gallery. Accessed 9 January 2011. http://www.csun.edu/~hfgeg005/eturner/gallery/gallery.htm.

Vignell, M., and B. Noorda. 1979. *New York City Transit Authority: Graphic Standards Manual*. New York, NY: Unimark International.

Wood, D., and J., Fels. 2009. *The Natures of Maps: Cartographic Constructions of the Natural World*. Chicago: University of Chicago Press.

Yau, N. 2011. *Visualize This: The Flowing Data Guide to Design, Visualization, and Statistics*. Indianapolis, IN: Wiley Publishing.

———. 2013. *Data Points: Visualization That Means Something*. Indianapolis, IN: Wiley Publishing.

**SELECTED JOURNALS**

*Bulletin of the Society of Cartographers*: soc.org.uk/bulletin.

*Cartographica*: utpjournals.press/loi/cart.

*Cartographic Perspectives*: cartographicperspectives.org.

*Cartographic Journal*: tandfonline.com/loi/ycaj20.

*Cartography and Geographic Information Science*: tandfonline.com/loi/tcag20.

*International Journal of Cartography*: tandfonline.com/loi/tica20.

*Journal of Maps*: tandfonline.com/loi/tjom20.

**WEB RESOURCES**

Adaptive Composite Map Projections: cartography.oregonstate.edu/demos/adaptiveCompositeMapProjections.

Adobe Color Wheel: color.adobe.com.

Adobe Typekit: typekit.com.

ArcGIS Online: arcgis.com.

Color-Hex: color-hex.com.

ColorBrewer: colorbrewer2.org.

Color Oracle color-blindness simulator: colororacle.org.

DaFont: dafont.com.

David Rumsey Map Collection: www.davidrumsey.com.

Flex Projector: flexprojector.com.

Font Awesome: fontawesome.io.

Font Squirrel: fontsquirrel.com.

Maki: mapbox.com/maki-icons.

Map Advice Community: community.esri.com/groups/map-advice.

Natural Earth: naturalearthdata.com.

OpenStreetMap: openstreetmap.org.

Paletton: paletton.com.

Projection Wizard: projectionwizard.org.

Relief Shading: reliefshading.com.

Shaded Relief: shadedrelief.com.

Symbol Store: symbolstore.org.

The Noun Project: thenounproject.com.

Vischeck: vischeck.com.

**SOCIETIES AND ORGANISATIONS**

British Cartographic Society: cartography.org.uk.

Canadian Cartographic Association: cca-acc.org/.

Cartography and Geographic Information Society (CaGIS): cartogis.org.

ICA Commission on Map Design: mapdesign.icaci.org.

International Cartographic Association: icaci.org.

North American Cartographic Information Society (NACIS): nacis.org.

Society of Cartographers: soc.org.uk.

**BLOGS AND TUTORIALS**

Adventures in Mapping by John Nelson: adventuresinmapping.com.

All over the Map by Betsy Mason and Greg Miller: news.nationalgeographic.com/all-over-the-map.

ArcGIS Mapping by Esri: blogs.esri.com/esri/arcgis.

Cartonerd by Kenneth Field: cartonerd.com.

CartoTalk by NACIS: cartotalk.com.

Eagereyes by Robert Kosara: eagereyes.org.

FlowingData by Nathan Yau: flowingdata.com.

Jonah Adkins: jonahadkins.github.io.

Joshua Stevens: joshuastevens.net.

Making Maps by John Krygier and Denis Wood: makingmaps.net.

Mapping London by Ollie O'Brien and James Cheshire: mappinglondon.co.uk.

Maps Mania by Kier Clarke: googlemapsmania.blogspot.ca.

Map Smith by Stephen Smith: mapsmith.net.

Maps We Love by Esri: esri.com/products/maps-we-love.

Petrichor by Sarah Bell and Jacob Wasilkowski: petrichor.studio.

Tolomaps by Robin Tolochko: tolomaps.com.

SomethingAboutMaps by Daniel Huffman: cargocollective.com/somethingaboutmaps.

Spatial.ly by James Cheshire: spatial.ly.

Strange Maps by Frank Jacobs: bigthink.com/articles?blog=strange-maps.

Sxywu.com by Shirley Wu: sxywu.com.

The Functional Art by Alberto Cairo: thefunctionalart.com.

The Map Room by Jonathan Crowe: maproomblog.com.

The Work of Edward Tufte: edwardtufte.com.

## Sources and resources, *continued*

Transit Maps by Cameron Booth: transitmap.net.
Visual Communicator by Joshua Stevens: joshuastevens.net.
Visual Cinnamon by Nadieh Bremer: visualcinnamon.com.
Visualising Data by Andy Kirk: visualisingdata.com.

**SELECTED CARTOGRAPHERS AND 'GEOVIZ' EXPERTS ON SOCIAL MEDIA**

Adkins, Jonah @jonahadkins.
Bell, Sarah @sarahbellmaps.
Bostock, Mike @mbostock.
Bremer, Nadieh @NadiehBremer.
Brewer, Cindy @Colorbrewer.
Buckingham, Tanya @tammabuck.
Cairo, Alberto @albertocairo.
Carroll, Allen @AllenCarroll.
Cheshire, James @spatialanalysis.
Chilton, Steve @steev8.
Coe, Daniel @geo_coe.
Crampton, Jeremy @JeremyCrampton.
Dorling, Danny @dannydorling.
Feldman, Steven @StevenFeldman.
Field, Kenneth @kennethfield.
Fischer, Eric @enf.
Flanagan, Ben @benflan.
Foster, Mike @mjfoster83.
Fung, Kaiser @junkcharts.
Gale, Gary @vicchi.
Gamache, Martin @themappist.
Gamio, Lazaro @LazaroGamio.
Glynn, Charlie @charley_glynn.
Grimwade, John @johngrimwade.
Guidero, Elaine @elaineguidero.
Hall, Mike @thisismikehall.
Hennig, Benjamin @geoviews.
Holmes, Nigel @nigelblue.
Huffman, Daniel @pinakographos.
Jenny, Bernie @mappingbernie.
Jones, Wesley @wesleytjones.
Kelso, Nathaniel Vaugn @kelsosCorner.
Kennelly, Patrick @patkennelly.
Kirk, Andy @visualisingdata.
Kosara, Robert @eagereyes.
Lambrechts, Maarten @maartenzam.
Lutz, Eleanor @eleanor_lutz.
MacEachren, Alan @alanGeoVISTA.
McConchie, Alan @mappingmashups.
Moriarty, Dylan @DylanMoriarty.
Muehlenhaus, Ian @iMuehlenhaus.
Nelson, John @John_M_Nelson.
NYT Graphics @nytgraphics.
O'Brien, Ollie @oobr.
Ortiz, Santiago @moebio.
Parsons, Ed @edparsons.
Patterson, Tom @MtnMapper.
Preppernau, Charles @Geolographer.
Punt, Edie @epunt.
Robinson, Anthony @A_C_Robinson.
Rogers, Simon @smfrogers.
Rost, Lisa Charlotte @lisacrost.
Roth, Robert @RobertERoth.
Šavrič, Bojan @BojanSavric.
Shephard, Nathan @NathanCShephard.
Singleton, Alex @alexsingleton.
Smith, Stephen @TheMapSmith.
Stefaner, Moritz @moritz_stefaner.
Stevens, Joshua @jscarto.
Tait, Alex @taitmaps.
The Upshot @upshotNYT.
Thomas, Anton @AntonThomasMaps.
Tierney, Lauren @tierneyl.
Trainor, Clare @ClareMTrainor.
Transit Maps @transitmap.
Van der Maarel, Hans @redgeographics.
Velasco, Juan @juanvelasco.
Visualoop, @visualoop.
Wallace, Tim @wallacetim.
Watkins, David @DavidDWatkins.
Watkins, Derek @dwtkns.
Wesson, Christopher @ChrisWesson_UK.
Williams, Craig @williamscraigm.
Woodruff, Andy @awoodruff.
Yau, Nathan @flowingdata.

# Image and data credits

Cover: photographs © Angela Andorrer. Used with permission. andorrer.de/kartografen.

Foreword: Image courtesy of Kenneth Field (private collection).

How to use this book: original illustration by Kenneth Field and John Nelson, Esri.

Authors/cartographers: photographs courtesy of Eric Laycock, Esri.

Cover art: photograph © Kenneth Field. Used with permission (private collection).

Whose hands: photographs © Angela Andorrer. Used with permission. andorrer.de/kartografen.

Abstraction and signage: map extracts courtesy of Esri. Map data © OpenStreetMap contributors, CC-BY-SA, and Microsoft.

Additive and subtractive colour: original illustration by John Nelson, Esri.

Advertising maps: original illustration by John Nelson, Esri. Map data fabricated for illustrative purposes.

Aggregation: original illustration by Wesley Jones, Esri. Map data from Esri, Esri South Africa, Garmin, HERE, NASA, US National Geospatial-Intelligence Agency (NGA), USGS, Earthstar Geographics.

Aligning coordinate systems: original illustration by John Nelson, Esri. Imagery data from Microsoft. Additional data fabricated for illustrative purposes.

All the colours: image © David Naylor. Used with permission.

Anatomy of a map: original illustration by Wesley Jones, Esri. Map data from Esri, Garmin, HERE.

Animation: photograph (left page) © Kenneth Field. Used with permission (private collection).

Animation: original illustration (right page) by Wesley Jones. Made with Natural Earth, naturalearthdata.com.

Arbitrary data classification: original illustration by Kenneth Field, Esri. Data fabricated for illustrative purposes.

Areas: original illustration by Wesley Jones, Esri. Map data from Danish Geodata Agency, Esri, Garmin, HERE, Increment P.

Aspect of a map projection: Original illustration by John Nelson, Esri.

Aspect views: original illustration (top) by Kenneth Field, Esri. Made with Natural Earth, naturalearthdata.com.

Aspect views: *Guide for Visitors to Ise Shrine* (right page) © Jingu Administration Office. Used with permission.

Assessing distortion in map projections: original illustration (left page) by Kenneth Field, Esri. Made with Natural Earth, naturalearthdata.com.

Assessing distortion in map projections: original illustration (right page) by John Nelson, Esri. Made with Natural Earth, naturalearthdata.com.

Atlases: main photograph © Kenneth Field. Used with permission (private collection).

Atlases: photograph of *Earth Platinum* © Millennium House. Used with permission.

Atlases: *Atlas of the British Empire* image from GRANGER. Used with permission.

*100 Aker Wood* by E. H. Shepard: line illustration by E.H. Shepard © The Shepard Trust, 1926. Reproduced with permission from Curtis Brown Group Ltd, London, on behalf of The Shepard Trust.

Balance: original illustration by Wesley Jones. Made with Natural Earth, naturalearthdata.com.

Basemaps: original illustration by Wesley Jones. Map data from Esri, FAO, Garmin, HERE, National Oceanic and Atmospheric Administration (NOAA), USGS.

Binning: original illustration by John Nelson, Esri. Map data fabricated for illustrative purposes.

Branding: photograph © Kenneth Field. Used with permission (private collection).

*Airspace: The Invisible Infrastructure*: image courtesy of NATS, 2014. Used with permission.

Cartograms: original illustration by Kenneth Field, Esri. Map data fabricated for illustrative purposes.

Cartographic process: original illustration by Kenneth Field, Esri. *Nautilus Pompilius* image by Filip Fuxa/ Shutterstock.com.

Chernoff faces: original illustration by John Nelson, Esri. Map data from *ArcGIS® Living Atlas of the World*: Esri Country Boundaries; UN World Happiness Report 2016.

Choosing type: original illustration by Wesley Jones, Esri.

Choropleth maps: original illustration by Kenneth Field, Esri. Map data fabricated for illustrative purposes.

Clutter: *Diamonds Were a Girl's Best Friend* chart (left page) courtesy of Nigel Holmes. Used with permission.

Clutter: *Bomb Sight: Mapping the WWII Bomb Census* (right page) courtesy of The Bomb Sight Project (www.bombsight.org). Used with permission.

Cognitive biases: original illustration by Wesley Jones, Esri.

Colour charts: original illustration by Kenneth Field, Esri.

Colour cubes: excerpt from the *RGB Colorspace Atlas* by Tauba Auerbach, 2011, set of three books: digital offset print on paper, case bound book, airbrushed cloth cover and page edges, each book: 8 × 8 × 8 in. (20.3 × 20.3 × 20.3 cm), page count, each: 3,712 pages, Binding codesigned by Daniel E. Kelm and Tauba Auerbach. Bound by Daniel E. Kelm assisted by Leah Hughes-Purcell at the Wide Awake Garage, © Tauba Auerbach. Courtesy Paula Cooper Gallery, New York. Used with permission.

Colour deficiency: original illustration by Wesley Jones, Esri. Made with Natural Earth, naturalearthdata.com.

Colour in cartography: *PrettyMaps* by Aaron Straup Cope, Stamen Design. Data from Natural Earth, Flickr, and © OpenStreetMap contributors CC-BY-SA. Used with permission.

Colouring in: original illustrations by Kenneth Field and Wesley Jones. Map data from Esri, HERE, Increment P, NRCan, Parks Canada.

Colour schemes: original illustration by Kenneth Field, Esri.

Combining visual variables: original illustration by John Nelson, Esri. Map data fabricated for illustrative purposes.

Consistent denotation: *Those Who Did Not Cross 2005–2015* © Levi Westerveld, 2016. Used with permission.

Constraints on map colours: original illustration by John Nelson, Esri.

Continuous surface maps: original illustration by Kenneth Field, Esri. Map data fabricated for illustrative purposes.

Contours: original illustration by Wesley Jones, Esri. Map data from ASTERGDEMv2 and SRTM. ASTER GDEM is a product of NASA and METI. SRTM version 2.1 NASA/JPL/NGA.

# Image and data credits, *continued*

Contrast: original illustration by Richard Coyles and Kenneth Field. Used with permission.

Copyright: original illustration by Wesley Jones and Pete Schreiber, Esri.

Craft: extract from *The North American Continent* © Anton Thomas, 2017. Used with permission, all rights reserved. antonthomasart.com.

Crispness: original illustration (left page) by Kenneth Field, Esri.

Crispness: *Health Hangars* (right page) © Lateral Office (Lola Sheppard, Mason White), 2011. Used with permission.

Critique: *Map Evaluation Checklist* by Kenneth Field, Esri.

Curvature of terrain: original illustration by Wesley Jones, Esri. Imagery data from NED 1/9 arc second, USGS.

*Carte Figurative des Pertes Successives en Hommes de l'Armée Française dans la Campagne de Russie 1812–1813*: map by Charles Joseph Minard, 1869. Work is in the public domain.

Dasymetric maps: original illustration by John Nelson, Esri. Map data fabricated for illustrative purposes.

Data (c)art(e): *Visualising Friendships* © Paul Butler, 2010. Used with permission.

Data accuracy and precision: original illustration by Wesley Jones. Made with Natural Earth, naturalearthdata.com.

Data arrangement: original illustration by Kenneth Field, Esri. Map data fabricated for illustrative purposes.

Data classification: original illustration by Kenneth Field, Esri. Map data fabricated for illustrative purposes.

Data density: original illustration by Kenneth Field, Esri. Map data fabricated for illustrative purposes.

Data distribution: original illustration by John Nelson, Esri. Map data fabricated for illustrative purposes.

Data processing: original illustration by John Nelson, Esri. Map data from USGS.

Datums: original illustration (left page) by Kenneth Field, Esri.

Datums: photograph (right page) © Kenneth Field. Used with permission (private collection).

Defining map design: original illustration by John Nelson and Kenneth Field, Esri.

Defining maps and cartography: original illustration by Wesley Jones, Esri. Quotes by Menno-Jan Kraak, Waldo Tobler, and Ed Parsons used with permission.

Descriptive maps: *Cloud Cover and the 2017 American Eclipse* by Joshua Stevens, NASA Earth Observatory. Used with permission.

Design and response: original illustration (right page, far right) by John Nelson, Esri.

Design and response: *Japan, the Target: A Pictorial Jap-Map* (right page) by Ernest Dudley Chase, courtesy of the David Rumsey Map Collection, davidrumsey.com. Used with permission.

Different strokes: original illustration by Wesley Jones, Esri. Map data from Esri, HERE, Garmin, NASA, NGA, NRCan, Parks Canada, USGS, Earthstar Geographics.

Digital data: original illustration by Kenneth Field, Esri. Map data © OpenStreetMap contributors CC-BY-SA.

Digital elevation models: original illustration by Kenneth Field, Esri. Map data from NASA Elevation data, provided by the MOLA instrument on Mars Global Surveyor. ~463 m/px, 2017.

Dimensional perception: original illustration by John Nelson, Esri. Map data from *ArcGIS Living Atlas of the World.*

Dispersal vs. layering: *trees-cabs-crime (in San Francisco)* by Shawn Allen, Stamen Design. Data from Friends of the Urban Forest (fuf.net), Cabspotting.org, and the San Francisco Police Department. Used with permission.

Distortions in map projections: original illustration by John Nelson, Esri. Map data from *ArcGIS Living Atlas of the World.*

Dot density maps: original illustration by Kenneth Field, Esri. Map data from *ArcGIS Living Atlas of the World.*

Dynamic visual variables: original illustration by Kenneth Field, Esri. Made with Natural Earth, naturalearthdata.com.

Dysfunctional cartography: *Taxonomy of Ideas* illustration by David McCandless © David McCandless/ InformationIsBeautiful.net. Used with permission.

*Detail of Area around the Broad Street Pump*: map by John Snow, 1854. Work is in the public domain.

Earth coordinate geometry: original illustration by John Nelson, Esri. Map data from *ArcGIS Living Atlas of the World.*

Earth's framework: original illustration by John Nelson, Esri. Map data from *ArcGIS Living Atlas of the World.*

Earth's shape: original illustration by John Nelson and Bojan Šavrič, Esri. Map data from Esri and NGA EGM Development Team.

Earth's vital measurements: original illustration by John Nelson, Esri. Map data from *ArcGIS Living Atlas of the World.*

Elegance: *Pitch Perfect* by Kenneth Field, Esri. Imagery data from DigitalGlobe and Microsoft used to create derived product.

Elements of colour: *Election Pollocks* by Kenneth Field, Esri. General election results data from data.parliament.uk. Contains Parliamentary information licensed under the Open Parliament Licence v3.0. Map data contains OS data © Crown copyright and database right 2017 under a UK Open Government Licence (OGL).

Elements of type: original illustration by Wesley Jones, Esri.

Emotional response: photograph (left page) © Kenneth Field. Used with permission (private collection).

Emotional response: *Geo-Genealogy of Irish Surnames* (right page) by Kenneth Field and Linda Beale, Esri. Surnames data from the 1890 census. Map data from *ArcGIS Living Atlas of the World.*

Error and bias: *Locals and Tourists (London)* by Eric Fischer 2010. Attribution-ShareAlike 2.0 Generic (CC BY-SA 2.0). Basemap © OpenStreetMap, CC-BY-SA.

Ethics: original illustration by Kenneth Field, Esri. Photograph © Trey Yingst, 2017. Used with permission.

Eyeball data classification: original illustration by Kenneth Field, Esri. Map data fabricated for illustrative purposes.

*Diagram of the Causes of Mortality in the Army in the East*: map by Florence Nightingale, 1858. Work is in the public domain.

Families of map projection: original illustration by John Nelson, Esri.

Fantasy maps: *Treasure Island* by Robert Louis Stevenson, 1883. Copy from the 1885 edition originally sourced from Beidecke Library, Yale University. Work is in the public domain.

Flourish: original illustration (left page) by Kenneth Field, Esri.

Flourish: original illustration (right page) by John Nelson, Esri. Data from data.gov.
Flow maps: original illustration (right page insets) by Kenneth Field, Esri.
Flow maps: *Flight Paths over Stage II of the Tour de France* (right page) by Craig Taylor, 2017, using Ito Motion, Ito World Ltd. Flight data derived from flightradar24.com. Terrain data: CCGIAR-CSI SRTM 90m database. Basemap: NAVTEQ StreetMap Premium. Used with permission.
Focussing attention: original illustration by Wesley Jones, Esri.
Fonts and type families: original illustration by Wesley Jones, Esri.
Foreground and background: original illustration by Wesley Jones, Esri. Data from Esri, GEBCO, HERE, NGA, NASA, USGS, and Natural Earth, naturalearthdata.com.
Form and function: *Breweries of the World* by Kenneth Field, Esri. Data from beerme.com. Used with permission.
Frequency distributions and histograms: original illustration by John Nelson, Esri. Map data from *ArcGIS Living Atlas of the World*: World Happiness Report 2016.
Functional cartography: *Hurricanes since 1851* by John Nelson, Esri. Map data from NOAA.
*The Distribution of Voting, Housing, Employment, and Industrial Compositions in the 1983 General Election*: map by Danny Dorling, 1991. Print 151 from Danny Dorling's PhD thesis (1991), 'The Visualization of Spatial Social Structure.' University of Newcastle upon Tyne. Used with permission.
Generalisation: original illustration by Wesley Jones, Esri.
Geological maps: *Strata of England and Wales* by William Smith, 1815. Work is in the public domain.
Globes: original illustration (left page) by Kenneth Field, Esri.
Globes: photograph (right page) © Kenneth Field. Used with permission (private collection).
Graduated symbol maps: original illustration by Kenneth Field, Esri. Map data from GitHub.
Graphic and dynamic labelling: original illustration by Wesley Jones, Esri. Map data from Esri, Esri Netherlands, Garmin, HERE, Kadaster, Land NRW, NASA, NGA, Rijkswaterstaat, USGS.
Graphicacy: *Anson Island* by Kenneth Field, Esri. Based on an original idea by Roger Anson.
Graphs: original illustration (left page) by Kenneth Field, Esri.
Graphs: *London's Tube DNA* (right page) by Kenneth Field, Esri. Data from TfL Open Data, RODS dataset, 2015 under a UK Open Government Licence (OGL).
Graticules, grids, and neatlines: original illustration by Wesley Jones, Esri.
Greyscale: original illustration by Wesley Jones, Esri. Data from Esri, FAO, Garmin, GEBCO, HERE, NOAA, USGS.
Guidelines for lettering: original illustration by Wesley Jones, Esri.
Dymaxion projection: *Non-stop* by Alberto Lucas López/*South China Morning Post*, 2015. Data from openflights.org, Google Flights, Hong Kong International Airport. Used with permission.
Hachures: original illustration (left page) by Kenneth Field, Esri.
Hachures: original illustration (right page) by Wesley Jones, Esri.
Hand-drawn maps: original illustration by Wesley Jones, Esri. Inspired by Henry Holliday's *empty space* map in Lewis Carroll's *The Hunting of the Snark*.
Hand-drawn shaded relief: image by E. Mitchell for Rand McNally maps, 1947. Original image courtesy of Sean Breyer (private collection). Photograph © Kenneth Field. Used with permission.
Heat maps: original illustration by Kenneth Field, Esri. Map data compiled from original broadcast footage.
Height: original illustration (left page) by Kenneth Field, Esri.
Height: *Shan Shui in the World* (right page) by Weili Shi, 2016. Used with permission.
Hierarchies: original illustration by Wesley Jones, Esri. Made with Natural Earth, naturalearthdata.com.
How maps are made: original illustration by John Nelson, Esri. Made with Natural Earth, naturalearthdata.com.
HSV colour model: original illustration by Kenneth Field, Esri.
Hue: original illustration (left page) by Kenneth Field, Esri.
Hue: extract (right page) from *The Alluvial Valley of the Lower Mississippi River* by Harold Fisk, 1944, courtesy of USGS.
Hypsometric tinting: original illustration by Kenneth Field, Esri. Map data from SRTM 1 arc second, NASA.
*Earth Wind Map*: map by Cameron Beccario, 2017. *Hurricanes Jose and Maria* from earth.nullschool.net. Used with permission.
Illuminated contours: original illustrations by Kenneth Field, Esri. Data from USGS.
Imagery as background: original illustration by John Nelson, Esri. Map data from Earthstar Geographics.
Information overload: *Television, radio & cinema 1977–1987* by Kenneth Field, 1991 (private collection). Used with permission.
Information products: *Stick chart of the Marshall Islands* of undetermined origin. Work is in the public domain. Photograph © National Library of Australia. Used with permission.
Informing: New York's annual carbon dioxide emissions by Carbon Visuals, 2012. Licensed under a Creative Commons Attribution 2.0 Generic (CC BY 2.0).
Inquiry and insight: extract from the *Environment and Health Atlas for England and Wales* by Imperial College London, 2014. Licensed under a Creative Commons Attribution-NoDerivatives 4.0 International License.
Integrity: original illustration by Kenneth Field, Esri.
Interaction: original illustration by Wesley Jones, Esri. Map data (Japan) from Esri, Garmin, USGS. Map data (hologram) from Natural Earth, naturalearthdata.com. Photograph by Green Chameleon (Unsplash.com).
Isarithmic maps: original illustration (top) by Kenneth Field, Esri.
Isarithmic maps: *Hurricane Sandy* map extract (bottom) by James Eynard, 2017. Used with permission.
Isochrone maps: original illustration by Kenneth Field, Esri. Map data fabricated for illustrative purposes.
Isometric views: original illustration (left page) by Kenneth Field, Esri.
Isometric views: *Visitor map for the Goodwood Festival of Speed* (right page) by Mike Hall, 2015. Used with permission.

# Image and data credits, *continued*

Isotype: *Nuclear Nations* by Kenneth Field, 2017. Map data from Wikipedia.

*Geologic Map of the Central Far Side of the Moon*: map by Desiree E. Stuart-Alexander, 1978, USGS.

Jokes and satire: *San Serriffe*, 1977 © Guardian News & Media Ltd 2018.

Jokes and satire: *Comic San Serriffe* by Craig Williams, 2014. Used with permission.

*Atlas of Global Geography*: map by Erwin Raisz, 1944, maps from the *Atlas of Global Geography* courtesy of the David Rumsey Map Collection, davidrumsey.com. Used with permission.

Knowledge and conviction: *Interpretation* by Wayne Peterkin, 2016 (WANE, COD). Used with permission.

*Google Maps*, extract from Google Maps by Google. Map data © 2017 Google. Used with permission.

Latitude: original illustration by John Nelson, Esri. Map data from *ArcGIS Living Atlas of the World*.

Layouts and grids: original illustration by Kenneth Field, Esri.

Legends: original illustration by John Nelson, Esri.

Lettering: original illustration by Wesley Jones, Esri. Map data from Esri, HERE.

Lettering in 3D: original illustration by Wesley Jones, Esri. Map data from Esri, Esri Netherlands, City of Rotterdam.

Levels of measurement: original illustration by Wesley Jones, Esri.

Lines: original illustration by Wesley Jones, Esri. Map data from Danish Geodata Agency, Esri, Garmin, HERE, Increment P.

Literal comparisons: *Pro-Life/Pro-Choice* by Nigel Holmes (unpublished, private collection). Used with permission.

Location: original illustration (left page) by Kenneth Field, Esri.

Location: *Cartography Corridor* (right page) by Wesley Jones, Esri.

Longitude: Marine timekeeper H1 (left page) © National Maritime Museum, Greenwich, London. Used with permission.

Longitude: Original illustration (right page) by John Nelson, Esri. Map data from *ArcGIS Living Atlas of the World*.

*The Heart of the Grand Canyon*: map by the National Geographic Society, 1978 © NgMaps/National Geographic Creative. Used with permission.

Making numbers meaningful: *Exports and Imports to and from Denmark & Norway from 1700 to 1780* by William Playfair, 1786. Work is in the public domain.

Map aesthetics: original illustration by Wesley Jones, Esri. Map data from Esri, HERE, USGS, and Natural Earth, naturalearthdata.com.

Map cube: original illustration by Wesley Jones based on *The Map Cube* by Alan MacEachren and D. R. Fraser Taylor, 1994.

Map projections: original illustration by John Nelson, Esri.

Map projections: Decisions, decisions!: chart (top) *Choices for Map Projections*, after Lev Bugayevskiy and John Snyder, 1995.

Map projections: Decisions, decisions!: chart (bottom) *Adaptive Composite Projection Chart*, after Bernhard Jenny, 2012.

Map transformation process: original illustration by Kenneth Field, Esri. Photograph © Kenneth Field. Used with permission (private collection).

Map traps: *Fake Settlement of Agloe* © Langenscheidt KG (liquidated).

Map traps: London A–Z map extracts, reproduced by permission of Geographers' A–Z Map Co. Ltd.

Map traps: Switzerland map extracts, courtesy of swisstopo, reproduced by permission of swisstopo (BA18005).

Map traps: *Fake Town of Argleton* map extract, map data © 2017 Google. Used with permission.

Map traps: Gold Coast Survey Department 1923 map is in the public domain.

Map traps: background image, designelements/ Shutterstock.com.

Maps for and by children: photograph (left page) © Kenneth Field. Used with permission (private collection).

Maps for and by children: *Antarctica* (right page) from the *Collins Children's Picture Atlas*, 2015 © HarperCollins Publishers. Used with permission.

Maps kill: original illustration by John Nelson, Esri. Background image by Stella Caraman from unsplash.com/photos/T4wso6sVAaA. Aircraft image is in the public domain. Map data from ArcGIS Living Atlas of the World.

Mashups: *Housing Maps* by Paul Rademacher, 2005. Map data © 2005 Google. Used with permission.

Measuring direction: original illustration by John Nelson, Esri. Map data from *ArcGIS Living Atlas of the World*. Photograph by Suhyeon Choi (Unsplash.com).

Mental maps: *Ye Newe Map of Britain* by the Doncaster and District Development Council ca. 1975.

Mixing colours: original illustration by Kenneth Field, Esri.

Mobile mapping: original illustration by Wesley Jones, Esri. Map data from Esri, HERE. Photograph by Daniel Tseng (Unsplash.com).

Multivariate maps: *Five Years of Drought* by John Nelson, 2016. Data from the National Drought Mitigation Center (NDMC), US Department of Agriculture (USDA), NOAA.

*Islandia*: map by Abraham Ortelius, 1603. Work is in the public domain.

Navigating a map: original illustration by Wesley Jones, Esri.

Nominal data: original illustration by John Nelson, Esri. Data from NASA Visible Earth.

*Jack-o-lanterns*: from *Everything Sings: Maps for a Narrative Atlas*, 2nd edition, by Denis Wood. Siglio, 2010. Used with permission.

Old is new again: *The United States, Her Natural & Industrial Resources* (top), courtesy of Stephen Smith (mapsmith.net), 2014. Used with permission.

Old is new again: *Great Britain, Her Natural & Industrial Resources* (bottom), by British Information Services, 1944. Digital Commonwealth. Accessed April 17, 2018. https://ark.digitalcommonwealth.org/ark:/50959/0r96fn18w.

OpenStreetMap: original illustration by Kenneth Field, Esri. Map data © OpenStreetMap contributors, CC-BY-SA.

Ordinal data: original illustration by John Nelson, Esri. Data from NASA Visible Earth.

Orientation: original illustration (left page) by Kenneth Field, Esri.

Orientation: *Beyond the Sea* (right page), courtesy of Andy Woodruff, 2017/Axis Maps. Used with permission.

*Karte der Gegend um den Walensee*: map by Eduard Imhof, 1938. Courtesy of Esri Press and the estate of Eduard Imhof.

Page vs. screen: photograph (left page) © Kenneth Field. Used with permission (private collection).

Page vs. screen: original illustration (right page) by Wesley Jones and Kenneth Field, Esri.

Panoramic maps: *Mammoth Mountain* by James Niehues, courtesy of Mammoth Mountain Ski Area, LLC, 2009 and 2017. Used with permission.

Pattern fills: original illustrations (left page and top, right page) by Kenneth Field, Esri. Map data fabricated for illustrative purposes.

Pattern fills: extract (right page) from *Elements of Topographic Drawing* by R. C. Sloane and J. M. Montz, 1930. Work is in the public domain. Used with permission.

Perceptual colour spaces: original illustration by Kenneth Field, Esri.

Pictograms: extract from *Deaths in the Grand Canyon* by Kenneth Field, Esri, 2012. Map data from Esri, DigitalGlobe, NASA, and inspired by *Over the Edge: Death in Grand Canyon* by Michael Ghiglieri and Thomas Myers (Puma Press, 2001). Used with permission.

Pie and coxcomb charts: map *Birdstrikes on Aircraft* by Kenneth Field, Esri, 2014. Map data from Federal Aviation Authority.

Placing type: original illustration by Wesley Jones, Esri. Map data from Esri, Garmin, NGA, NASA, USGS.

Placing type for areas: original illustration by Wesley Jones, Esri. Made with Natural Earth, naturalearthdata.com.

Placing type for lines: original illustration by Wesley Jones, Esri.

Placing type for points: original illustration by Wesley Jones, Esri.

Planetary cartography: *Here There Be Robots: A Medieval Map of Mars*, © Eleanor Lutz, 2016. Used with permission.

Point clouds: *Digital Mapping of China's Miao Room Cave System*, courtesy of ixtract, 2017. Used with permission.

Pointillism: original illustration by Kenneth Field, Esri. Map data fabricated for illustrative purposes.

Points: original illustration by Wesley Jones, Esri. Map data from Danish Geodata Agency, Esri, Garmin, HERE, Increment P.

Position: original illustration by John Nelson, Esri. Map data from *ArcGIS Living Atlas of the World*.

Printing fundamentals: original illustration by Kenneth Field, Esri. Photograph of Wisley © Kenneth Field. Used with permission (private collection).

Prior (c)art(e): excerpted from *Steal Like an Artist*. Copyright © 2012 by Austin Kleon. Used by permission of Workman Publishing Co., Inc., New York. All rights reserved.

Prism maps: original illustration (right page) by Kenneth Field, Esri. Map data from data.gov.

Prism maps: original illustration (left page) by Kenneth Field, Esri. Map data fabricated for illustrative purposes.

Profiles and cross-sections: *Population Lines*, courtesy of James Cheshire, 2013, University College London. Used with permission.

Properties of a map projection: original illustration by John Nelson, Esri. Map data from *ArcGIS Living Atlas of the World*.

Proportional symbol maps: original illustration by Kenneth Field, Esri. Map data fabricated for illustrative purposes.

Proximity in design: original illustrations by Kenneth Field, Esri.

Proximity in design: Dalmatian dog photograph by Ronald C. James, 1965. Courtesy Special Collections, University Library, University of California Santa Cruz.

Pseudo-natural maps: original illustration by Kenneth Field, Esri. Map data from Esri, swisstopo, HERE, Garmin, Increment P, USGS, METI/NASA, NGA.

Purpose of maps: © The Royal Geographical Society. Used with permission.

*London A–Z*: map reproduced by permission of Geographers' A–Z Map Co. Ltd., 1938.

Quantitative statistical maps: original illustration by Kenneth Field, Esri.

*Map of London's Underground Railways*: map by Harry Beck, 1933, courtesy of London Transport Museum. Used with permission.

Raised relief: drawing (top), courtesy of Karen Rann, the Great Lines Project, 2016. Drawing from OS map of Schiehallion, 1873–90, from series How to Draw a Mountain. Photo credit: Karen Rann. Used with permission.

Raised relief: models (lower), courtesy of Karen Rann, the Great Lines Project, 2016. Models of Schiehallion (bottom) according to current OS contours and (middle) according to Charles Hutton's 1778 figures. Photo credit: Tim Bird. Used with permission.

Ratio and interval data: original illustration by John Nelson, Esri. Data from NASA Visible Earth.

Ratios, proportions, and percentages: original illustration by Kenneth Field, Esri. Map data from data.gov.

Reference maps: extract (top) of aeronautical chart by FAA.

Reference maps: extract (middle) of *Oslo and Stockholm* from the *Times Comprehensive Atlas of the World*, 14th edition. Map © Collins Bartholomew Ltd 2014, HarperCollins Publishers Ltd. Used with permission.

Reference maps: extract (bottom) of nautical chart by NOAA.

Refinement: original illustration by Wesley Jones, Esri. Map data from Esri, Esri South Africa, Garmin, HERE, NASA, NGA, USGS, Earthstar Geographics.

Resolution: original illustration (left page) by Kenneth Field, Esri.

Resolution: main image (right page) by Saul Steinberg, View of the World from 9th Avenue cover of *The New Yorker*, March 29, 1976 © The Saul Steinberg Foundation/Artists Rights Society (ARS), New York cover reprinted with permission of *The New Yorker* magazine. All rights reserved.

Rock drawing: original illustration by Wesley Jones, Esri. Photograph by Luca Bravo (Unsplash.com).

*The Magnificent Bears of the Glorious Nation of Finland*: map courtesy of Annukka Mäkijärvi, 2014. Used with permission.

Saturation: original illustration (left page) by Kenneth Field, Esri.

Saturation: *UFO Sightings* (right page) by John Nelson. Image © 2015 IDV Solutions LLC, an Everbridge company. Used with permission. All rights reserved.

Scale and resolution: original illustration by Wesley Jones. Map data from Esri, Garmin, GEBCO, HERE, NRCan, Parks Canada, USGS.

Schematic maps: extract of the *Submarine Cable Map* courtesy of Telegeography, 2016. Used with permission.

Seeing: original illustration by Wesley Jones, Esri.

Seeing colour: original illustration by Kenneth Field, Esri.

# Image and data credits, *continued*

Semiotics: extract from *Carte de Déplacement des Baleines* by Matthew Maury, 1851. Work is in the public domain.

Sensory maps: *Smellmap of Edinburgh* courtesy of Kate McLean, 2011. Used with permission.

Shaded relief: original illustration by John Nelson, Esri. Map data from USGS.

Shape: original illustration (left page) by Kenneth Field, Esri.

Shape: original illustration (right page) by Kenneth Field and John Nelson, Esri. Snowflake symbols from the Noun Project (Lluisa Iborra, AomAm, Vasily Gedsun, LSE Designs) licensed under a Creative Commons Attribution 3.0 Unported (CC BY 3.0).

Signal to noise: original illustration by John Nelson, Esri. Map data fabricated for illustrative purposes.

Simplicity vs. complexity: original illustration (left page) by Kenneth Field, Esri. Map data from Wikipedia.

Simplicity vs. complexity: *City of Anarchy* (right page), courtesy of Adolfo Arranz/ *South China Morning Post*, 2013. Used with permission.

Simplification: original illustration by Wesley Jones, Esri. Made with Natural Earth, naturalearthdata.com.

Size: original illustration (left page) by Kenneth Field, Esri.

Size: map extracts (right page) courtesy of Benjamin Hennig/worldmapper.org. Used with permission.

Sizing type: original illustration by Wesley Jones, Esri.

Slope, aspect, and gradient: original illustration by Kenneth Field, Esri. Map data from USGS.

Small landform representation: original illustration by Wesley Jones, Esri.

Small multiples: *African Solar Eclipse of 2001*, courtesy of Fred Bruenjes, 2002. Used with permission.

Smoothing: original illustration by Wesley Jones, Esri. Map data from USGS, Esri, EPA, NGA, NASA.

Space-time cubes: original illustration (left page) by Kenneth Field, Esri, based on *Carte Figurative des Pertes Successives en Hommes de l'Armée Française dans la Campagne de Russie 1812–1813* by Charles Joseph Minard, 1869.

Space-time cubes: *Paths to the Future* (right page), courtesy of Andrew DeGraff, 2017. The map first appeared in *Cinemaps: An Atlas of 35 Great Movies* by Andrew DeGraff and A. D. Jameson, (Quirk Books, 2017). Used with permission.

Spacing letters and words: original illustration by Wesley Jones, Esri. Map data from Esri, Garmin, HERE.

Spatial dimensions of data: original illustration by John Nelson, Esri.

Statistical data classification: original illustration by Kenneth Field, Esri. Map data fabricated for illustrative purposes.

Statistical literacy: original illustration by John Nelson, Esri. Map data from USGS.

Stereoscopic views: original illustration by Kenneth Field, Esri. Map data from USGS.

Strip maps: map extract (left) of *The Road from London to the City of Bristol* by John Ogilby, 1875. Work is in the public domain.

Strip maps: map extract (centre) of *The Manchester Ship Canal*, 1923. Work is in the public domain. Map courtesy of Martin Dodge (private collection, used with permission).

Strip maps: map extract (right) of *Lake Michigan Unfurled*, courtesy of Daniel P. Huffman, 2015. Used with permission.

Style, fashion, and trends: swisstopo map extracts courtesy of swisstopo. Reproduced by permission of swisstopo (BA18005).

Styling shaded relief: original illustration by Kenneth Field and John Nelson, Esri. Map data from SRTM 1 arc second, NASA.

Symbolisation: original illustrations by Kenneth Field, Esri.

Symbols: original illustrations by Kenneth Field, Esri. Made with Natural Earth, naturalearthdata.com.

Gall-Peters projection: original illustration by Wesley Jones, Esri.

Gall-Peters projection: extract of *Gall's Orthographic Projection* from the *Scottish Geographical Magazine*, Vol. 1, 1855. Work is in the public domain.

Temporal maps: *Schedule of trains from Paris to Lyon* by E. J. Marey, 1880. Work is in the public domain.

Texture: original illustration (left page) by Kenneth Field, Esri.

Texture: extract (right page) of *Tactile Atlas of Switzerland*, courtesy of Anna Vetter, Esri Switzerland, 2016. Used with permission.

Thematic maps: extract from *Maps Descriptive of London Poverty* by Charles Booth. Work is in the public domain.

Threshold of perception: original illustration (top) by Kenneth Field, Esri.

Threshold of perception: extracts of *Tiny Maps* (bottom), courtesy of Stephen Smith (mapsmith.net), 2017. Used with permission.

Topographic maps: map extract of British Columbia from *The Times Comprehensive Atlas of the World*, 14th edition. Map © Collins Bartholomew Ltd 2014, HarperCollins Publishers Ltd. Used with permission.

Transparency: original illustration (left page) by Kenneth Field, Esri.

Transparency: extract (right page) from *Lights On Lights Off* by John Nelson, Esri, 2017. Map data from NASA.

Treemap: extract from *Rectangular Hierarchical Cartograms for Socio-Economic Data* by Aidan Slingsby, Jason Dykes, and Jo Wood, 2010. Census Output Area Classification (OAC) and postcode locations provided by Office for National Statistics. © giCentre, City, University of London. Used with permission.

Type colour: original illustration by Wesley Jones, Esri. Map data from Esri, CGIAR, Earthstar Geographics, Garmin, HERE, NASA, NGA, USGS.

Types of maps: original illustration by Kenneth Field, Esri.

Typographic maps: extract of *London's Kerning* (main), designed by NB Studio, London, 2007. Courtesy of NB. Used with permission.

Typographic maps: extract (inset) of *London*, courtesy of Axis Maps, 2011. Used with permission.

*The Times Comprehensive Atlas of the World*: map extract of the *Malay Archipelago* from the first edition, by John Bartholomew and Son, 1922. Work is in the public domain. Image courtesy of the David Rumsey Map Collection, davidrumsey.com. Used with permission.

UI/UX in map design: original illustration by Wesley Jones, Esri.

Unclassed maps: original illustration by Kenneth Field, Esri. Map data fabricated for illustrative purposes.

Unique values maps: *Pizza Place Geography*, courtesy of Nathan Yau, 2013, FlowingData, flowingdata.com/2013/10/14/pizza-place-geography. Used with permission.

Using words: original illustration by John Nelson, Esri.

*New York Map of Midtown Manhattan*: © Manhattan Map Co. Ltd / Yale Robbins Inc. Used with permission.

Value: original illustration (left page) by Kenneth Field, Esri.

Value: *Tracking the Economic Disaster* (right page), courtesy of Daniel Mason, 2011. Photograph © Daniel Mason (private collection). Used with permission.

Value-by-alpha maps: original illustration by Kenneth Field, Esri. Map data from data.gov.

Variables, values, and arrays: original illustration by John Nelson, Esri. Map data from NASA.

Varying symbols: original illustration by John Nelson, Esri.

Vignettes: original illustration by Wesley Jones, Esri. Made with Natural Earth, naturalearthdata.com.

Viral cartography: courtesy of Gary Gale from vicchi.org CC-BY-SA. Used with permission.

Vision: original illustration by John Nelson, Esri.

Visualization wheel: original illustration by Kenneth Field, Esri. Map data from Esri, HERE.

Voronoi maps: *The United States of Craigslist* by John Nelson, 2015. Made with Natural Earth, naturalearthdata.com.

*Where the Population of Europe Is Growing and Where It's Declining*: map by *Berliner Morgenpost*, 2016. Used with permission.

Waffle grid: original illustration (left page) by Kenneth Field, Esri.

Waffle grid: *Applied Irrigation Water* (right page) by William Bowen, 1979, from *the California Water Atlas*, courtesy of California Department of Water Resources. Used with permission.

Weather maps: original illustration by Kenneth Field, Esri. Symbols from Mike Afford Media (mikeafford.com).

Web mapping: extract of Google Maps, 2005. Map data © 2005 Google. Google Maps red pin element © 2018 Google. Used with permission.

Web Mercator: original illustration by John Nelson, Esri. Imagery data from Earthstar Geographics, Microsoft, DigitalGlobe.

Which way is up?: photograph *The Blue Marble* by Harrison Schmitt from Apollo 17 in 1972. Image courtesy NASA Johnson Space Center.

Who is cartography?: original illustration by Wesley Jones and Kenneth Field, Esri.

Wireframing and storyboarding: original illustration by John Nelson, Esri. Background photograph by Shoot N' Design (Unsplash.com).

*World Geo-Graphic Atlas*: extract from the *World Geo-Graphic Atlas* by Herbert Bayer, 1953. Image courtesy of WestRock, Inc., westrock.com, and the David Rumsey Map Collection, davidrumsey.com. Used with permission.

x and y: original illustration by Wesley Jones and Bojan Šavrič, Esri. Made with Natural Earth, naturalearthdata.com. Equations from J. P. Snyder (1987). *Map Projections: A Working Manual*, US Geological Survey, Washington, DC.

*A World of Lotus, a World of Harmony*: map by Liao Zhi Yuan, 2015. © International Cartographic Association (ICA), 2015. Used with permission.

Your map is wrong!: map (left) *True Size of Africa* by Kai Krause, 2011. Licensed under Creative Commons CC0.

Your map is wrong!: map (right) *The Truer Size of Africa* by Kenneth Field, 2011. Made with Natural Earth, naturalearthdata.com.

*Yellowstone National Park*: map by Heinrich Berann, 1989. Map courtesy of US National Park Service, Harpers Ferry Center. Used with permission.

Zeitgeist: *End of the Line* by Kenneth Field, 2015. Map data fabricated for illustrative purposes.

Wizards, end matter: original illustration by Wesley Jones, Esri.

Happy Mapping!

W Jones